ADVANCES IN CHEMICAL PHYSICS

VOLUME XCI

EDITORIAL BOARD

Advances in CHEMICAL PHYSICS

Edited by

I. PRIGOGINE

University of Brussels
Brussels, Belgium
and
University of Texas
Austin, Texas

and

STUART A. RICE

Department of Chemistry
and
The James Franck Institute
The University of Chicago
Chicago, Illinois

VOLUME XCI

AN INTERSCIENCE® PUBLICATION
JOHN WILEY & SONS, INC.
NEW YORK • CHICHESTER • BRISBANE • TORONTO • SINGAPORE

This text is printed on acid-free paper.

An Interscience® Publication

Library of Congress Catalog Number: 58-9935

ISBN 0-471-12002-2

10 9 8 7 6 5 4 3 2 1

CONTRIBUTORS TO VOLUME XCI

ANDRZEJ R. ALTENBERGER, Department of Chemical Engineering and Materials Science, University of Minnesota, Minneapolis, Minnesota

JAMES B. ANDERSON, Department of Chemistry, The Pennsylvania State University, University Park, Pennsylvania

JOHN S. DAHLER, Department of Chemical Engineering and Materials Science, University of Minnesota, Minneapolis, Minnesota

G. DEBIAIS, Groupe de Physique Théorique, Université de Perpignan, Perpignan, France

J. L. DEJARDIN, Groupe de Physique Théorique, Université de Perpignan, Perpignan, France

J. F. DOUGLAS, Polymers Division, National Institute of Standards and Technology, Gaithersburg, Maryland

E. J. GARBOCZI, Building Materials Division, National Institute of Standards and Technology, Gaithersburg, Maryland

M. CYNTHIA GOH, Department of Chemistry, University of Toronto, Toronto, Ontario, Canada

NIELS ENGHOLM HENRIKSEN, Chemistry Department B, Technical University of Denmark, Lyngby, Denmark

C. NICOLIS, Institut Royal Météorologique de Belgique, Brussels, Belgium

G. NICOLIS, Center for Nonlinear Phenomena and Complex Systems, Université Libre de Bruxelles, Brussels, Belgium

ELIGIUSZ WAJNRYB, Department of Chemical Engineering and Materials Science, University of Minnesota, Minneapolis, Minnesota

INTRODUCTION

Few of us can any longer keep up with the flood of scientific literature, even in specialized subfields. Any attempt to do more and be broadly educated with respect to a large domain of science has the appearance of tilting at windmills. Yet the synthesis of ideas drawn from different subjects into new, powerful, general concepts is as valuable as ever, and the desire to remain educated persists in all scientists. This series, *Advances in Chemical Physics*, is devoted to helping the reader obtain general information about a wide variety of topics in chemical physics, a field which we interpret very broadly. Our intent is to have experts present comprehensive analyses of subjects of interest and to encourage the expression of individual points of view. We hope that this approach to the presentation of an overview of a subject will both stimulate new research and serve as a personalized learning text for beginners in a field.

I. PRIGOGINE
STUART A. RICE

CONTENTS

ATOMIC FORCE MICROSCOPY OF POLYMER FILMS

M. CYNTHIA GOH

Department of Chemistry, University of Toronto, Toronto, Ontario M5S 1A1, Canada

CONTENTS

Advances in Chemical Physics, Volume XCI, Edited by I. Prigogine and Stuart A. Rice.
ISBN 0-471-12002-2

I. INTRODUCTION

Microscopy, as the word is traditionally used, refers to the visualization of a small object, typically by way of focusing an appropriate probe on it, and a subsequent image reconstruction. The scanning tunneling microscope (STM), invented in 1982 by Binnig and Rohrer [1, 2], is the first of the new class of instruments generally referred to as "near-field" or "scanning probe" microscopes (SPM). In traditional microscopes, the resolution is determined by the wave nature of the probe: Since the capability for image reconstruction is restricted by diffraction, the best resolution that can be attained is about one half the probe wavelength. In near-field microscopy, this diffraction limit is circumvented because the measurement involves highly local interactions between a probe tip and the sample, which are separated by at most a few nanometers (see, e.g., [3]). The image is obtained by moving the probe across the surface, and the resolution is determined by the size of the probe, the nature of the interaction, and the separation distance between probe and sample. The STM, which relies on the very short-range nature of the tunneling process, is now used almost routinely at resolution down to the atomic level. The atomic force microscope (AFM) [4], the use of which in polymer studies is the subject of this chapter, has similarly been utilized to obtain images at subnanometer resolution.

A. Scope

With the availability of commercial instruments in the past 5 years, and the current trend for publication of conference proceedings,[1] the number of reported studies using SPMs, especially the AFM, is rapidly growing such that any review will be out of date as soon as it is published. Hence, rather than attempting a comprehensive review, this chapter will center on a discussion of the imaging process, and the general applicability of force microscopy to studies of polymer films as exemplified by reports in the literature. As is typical of a new technique, one encounters growing pains, such that some earlier reports may contain artifacts, or are supplanted by newer ones, as researchers learn to use the technique better. Unfortunately, the nature of the experiment does not lend itself to a simple scheme for printed corrections, unlike the case of equations or numerical data, for which revisions are published. Hopefully, by presenting a discussion on various aspects of the imaging process in Section II, this chapter can provide the reader with enough information to aid in perusal of the literature, as well as serve as a useful guide for somebody contemplating to use the technique.

Since the major use of the AFM is for visualization of structures, much of this chapter will be devoted to such. If we can see atoms, molecules, and their spatial relation to each other, then we have come a long way towards understanding properties of a substance. Hence, visualization of atomic and molecular detail is the holy grail of microscopy, and scanning probe techniques are very seductive because they offer the possibility of reaching this goal. A discussion of high-resolution imaging of polymers, and examples of reported images with submolecular resolution are presented in Section III. A major issue in high-resolution imaging is the interpretation of the pictures that are obtained, especially with the abundance of image enhancement procedures, which may serve to either aid or mislead the researcher. These issues are addressed in Section III.

While features at the submolecular level are fascinating, they are not necessarily the most important aspect of structural organization in a substance. For polymers in particular, morphological features at intermediate length scales abound, such as lamellar patterns and spherulitic growth, which are interesting both from a fundamental and a technologi-

[1] Proceedings from conferences predominantly dealing with AFM are reported in *Ultramicroscopy*, 44–44 (1992); *J. Vac. Sci. Technol. B*, **12** (1994); *Colloids Surfaces A: Physicochem. Eng. Aspects*, **87** (1994).

cal perspective. Herein lies the best unequivocal use of the AFM: Studies of structures at mesoscopic length scales are presented in Sections III–V.

With respect to biomolecules, the reader is referred to recent excellent reviews that specialize on the AFM of biological systems [5, 6]. The brief discussion presented in Section VI will be restricted to studies of biopolymers.

While visualization is the most popular and appealing use of the AFM, it should be brought to mind that the image one sees is a reflection of a series of measurements that are quantifiable. Thus, in addition to visualization of the surface topography, the AFM can be used in other ways that provide information about the surface. These will be discussed in Section VII, with emphasis on applications to polymer samples.

Finally, a look towards the future is given in Section VIII.

B. Applicability of AFM to Problems in Polymer Physics

The understanding of the evolution of macroscopic properties from the molecular structure is a prevailing theme in condensed matter studies. Nowhere is this goal more important than in the case of polymers. At the fundamental level there is the challenge of understanding why and how certain structures form, and how to control them. This information has major implications in technology since properties of polymeric materials depend not only on their atomic constitution but also on their morphology. Because of the molecular complexity involved, one encounters important morphological patterns at different length scales. First, there is the arrangement of the monomers within the polymer. While new monomers of different compositions are being investigated, the development of novel materials rely more on the arrangement of monomers, forming linear or various branched polymers, possibly with blocks of different composition [7]. The chains can assume different conformations in solutions, and depending on the nature of the solvent and the way it is removed, the resulting material can form crystalline, amorphous, or a mixture of structures, which give rise to lamellar arrangements (see, e.g., [8]). In block copolymers, there can be microphase separation and aggregation producing spherical, cylindrical, and other patterns [9]. At the micron scale, larger domains can be found due to phase separation and the formation of cracks, defects, as well as fibrils and crystallites. Empirically, it is known that the morphology is influenced by external conditions, such as the way the material is made, the solvent that is used for casting and the interactions with the surface in which a film is grown. All these hierarchy of structures influence the final property of the material, including its tensile strength, translucence, and hardness.

Because of the importance of morphology to the properties of polymers, new techniques that come up are immediately employed for its study. Surface or interface morphology has been of particular interest because of its role on various technologies, which rely on factors such as the formation of protective coatings and adhesive characteristics [10]. A comparison of the various methods that are used for investigations of polymer interfaces is given by Stamm [11]; the AFM serves as one of these. In its traditional role of visualization, there are three aspects that the AFM can address: It is capable of (1) seeing structures at various length scales with ease and minimal sample preparation, from subnanometer to hundreds of microns, (2) seeing the surface, with very high (<1 Å) vertical resolution, (3) imaging samples in a relatively nondestructive manner, and (4) monitoring changes in morphology as conditions are varied, even in real time. In addition, while most studies utilizing the AFM to date have centered on these advantages, slight modifications in usage that have been recently introduced enables the measurement of material properties, such as elasticity, at a local level, and the manipulation of surface molecules.

II. THE ATOMIC FORCE MICROSCOPE

The idea of tracking the surface by a probe tip as a measure of topography predates the AFM: A record player works on a similar principle, and the stylus profilometer, which is a common instrument in studies of surface hardness, has been used for examination of surface roughness as well as topography [12, 13]. The profilometer, however, has found limited use, since the probe mounted on a heavy lever arm that rides on the surface has a tendency to scratch the sample it was meant to study, and its most useful application is in determining ease of scratching. The advantage offered by the AFM is twofold: a much higher resolution and an enormous reduction in the perturbation on the sample due to the reduction of size and forces involved. A comparison of the AFM with its predecessors is given by Burham et al. [14].

The basic principles of operation for all the different SPMs are similar to that of the STM in terms of the control and data acquisition systems; they differ simply in the interaction that is being monitored, and hence in the way such interaction is measured, during the scan. The rapid rise in use of the AFM over other variants is due to the ubiquity of forces. The AFM thus offers an advantage in general utility, even if other SPMs may be more appropriate for specific situations. In particular, unlike the STM, which requires a conductive surface for tunneling, the AFM can be applied to any material. (While STM has been applied to studies of nonconducting samples laid upon a substrate, the mechanism of tunneling in these cases is not clear, raising numerous questions about image interpretation [15]). However, there are two trade-offs: First, the tunneling current monitored by STM decays exponentially with distance between the probe and sample, while surface forces generally have a much longer range [16]. Hence, the interaction probed by the STM is more local and would translate to its better resolving power. Second, since the STM measurement involves the local density of states of the sample, a change in the bias voltage that drives the tunneling makes it a spectroscopically sensitive probe [17]. This mode of operation, sometimes referred to as scanning tunneling spectroscopy, enables differentiation between some species at a surface. Atomic force microscopy has no equivalent spectroscopic capabilities so far, although several methods have been employed to distinguish certain species, as mentioned later.

A. Instrumentation

1. *General Considerations*

An excellent description of the instrumental details for an AFM can be found in the monograph by Sarid [18], to which the reader is referred. Only a short discussion will be provided here.

The essential components of an AFM are illustrated schematically in Fig. 1.1. The probe tip, which interacts with the sample, is mounted onto a cantilever, whose role is to translate the forces acting upon the tip into a measureable quantity. Either the sample or the cantilever assembly is mounted on an XYZ peizoelectric scanner, which is controlled from a computer that enables raster scanning in the X and Y direction, where Y is taken as the direction of the slow scan. Typically, to simplify the detection scheme, it is the sample that is moved, although it is possible to scan the tip instead [19].

As a scan is made, the probe tip, and thus the cantilever, is deflected by its interaction with the surface. The motion of the probe tip can be measured in several ways. In the original design [4], the deflection of the cantilever in the Z direction is monitored using an STM: A sharp metallic wire is held above the cantilever, and the tunneling current between the wire and the cantilever is measured. Since the STM is capable of detecting changes in vertical distances of much less than an angstrom, this is the most sensitive way of measuring cantilever deflection, theoretically setting the AFM resolution at sub-angstrom level. However, the force that is exerted upon the sample by the combined probe tip–STM tip assembly can be quite large in this arrangement. In addition, the $1/f$ noise in tunneling [20] sets a limit to the sensitivity of the STM. Various detection schemes based on optical interferometery [21–25] have been

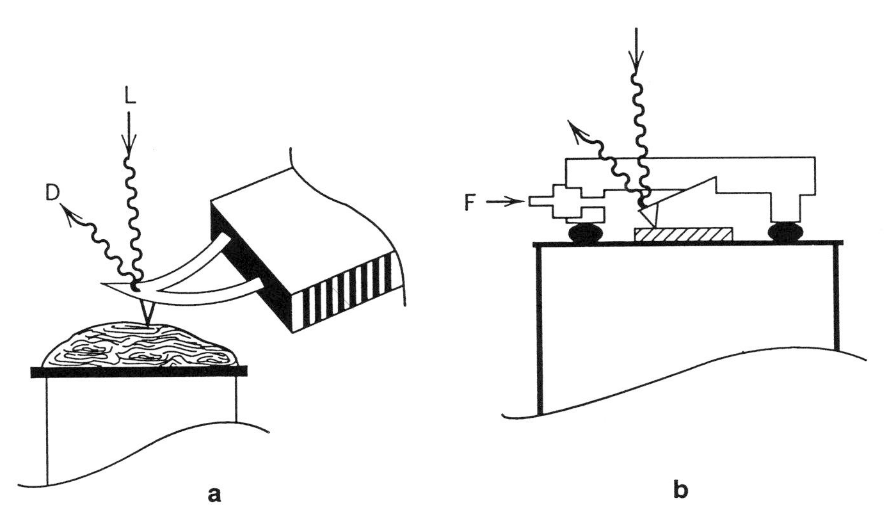

Figure 1.1. (a) Parts of an atomic force microscope using optical lever detection: L = diode laser source, D = bicell or quadrant photodetector. The sample is mounted on an XYZ piezoelectric scanner. (b) Example of a fluid cell for imaging under solution, which can be introduced through F.

employed, but the most common technique, which is simple and reliable although not as sensitive, is by a measurement of beam deflection, generally known as the "optical lever" method [26, 27]. In this latter case, light from a diode laser is reflected from the back of the cantilever, translating the cantilever deflection into a change in the reflectance angle, which can then be monitored by the use of a bicell photodetector.

The AFM imaging process is shown schematically in Fig. 1.2. Data of the deflection of the cantilever in the Z direction is collected as a raster scan along the surface is performed. The coordinates of points on the surface are collected in this manner. These coordinates can then be stored in a computer and processed to produce a space-filling three-dimensional image. While the scan is performed in both the X and −X direction, only one set is usually collected, except in the case of lateral force measurement, which is discussed later and this measurement can be accomplished by utilizing data from bidirectional scanning [28].

2. *Modes of Imaging*

The force between the probe tip and sample can be monitored in different ways, and there are several modes of imaging that are currently employed. It is important to bear in mind that the pictures resulting from the imaging process are simply a collection of data points from these

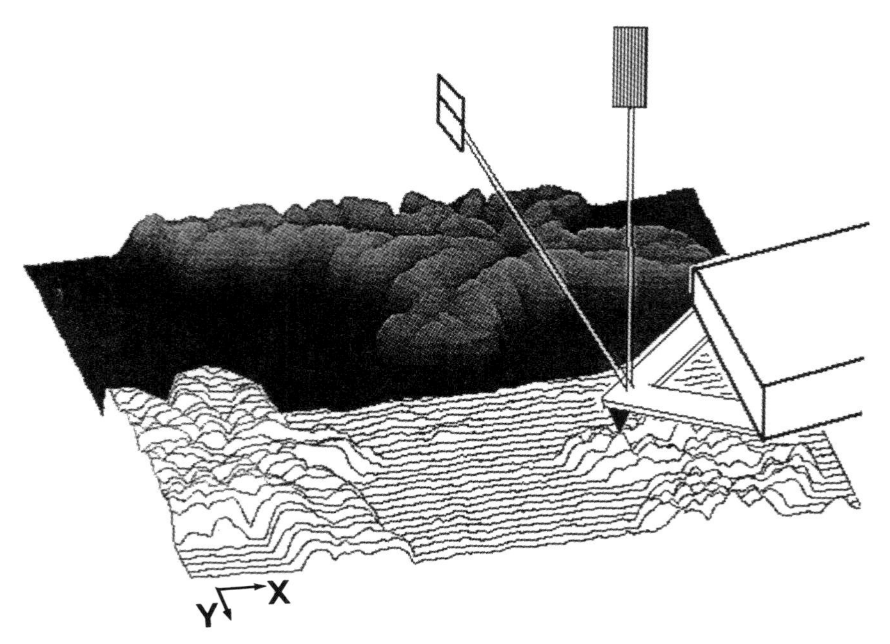

Figure 1.2. Illustration of the AFM imaging process. The fast (X) and slow (Y) directions of the raster scan are indicated.

measurements; hence, different modes may highlight different aspects of a sample.

a. Force Mode versus Height Mode. In the force (or "constant height" mode) the cantilever assembly and the sample mount are kept at a fixed Z position in the lab frame. The sample is then scanned in the X and Y direction, and the force, as measured by the cantilever deflection, is recorded at each point. This mode is quite sensitive to small changes in height, enabling examination of very small features, including those at the molecular scale. It is, however, unsuitable for examining rough surfaces and steep gradients in the Z direction because in these cases, the deflections during a scan can be large enough such that the tip is unable to track the surface. Artifacts may also arise in the presence of very steep gradients since the rapid changes in features may be beyond the flexibility of the lever. In the worst case, the tip could crash into the sample producing deformations, or it could pick up some material that will obscure the resolution of subsequent imaging.

Alternatively, the probe tip can be made to track the contour of constant force along the surface in the height (or constant force) mode. This is done by maintaining a constant deflection of the cantilever, and using a feedback loop to adjust the height of the sample with respect to the probe through adjustments of the Z component of the piezoelectric scanner. One advantage of imaging in the height mode is the capability of providing accurate information about height differences on the surface, provided the Z response of the scanner is properly calibrated. This calibration can be done with the use of standards of known height on a flat substrate, as discussed later. Note, however, that during the process of imaging in the height mode, there is a finite deflection of the cantilever due to the force between it and the sample. This deflection can change the measured value by as much as several nanometers, and thus should be taken into account for accurate height measurements [29].

An alternative imaging mode that is not commonly used is the error-signal or deflection mode [30]. In this case, the machine records the changes in the cantilever deflection that are used to signal the feedback loop to make the height adjustments, thus accentuating the location of edges in surface features.

b. Contact, Noncontact, and Tapping Modes. In images discussed in this chapter, unless specified, the tip tracks the surface by pressing against it, in the "contact" mode, since this is the most common practice currently employed for visualization. The main problem with this procedure is that the tip exerts both a vertical and a lateral force on the sample, which can cause deformation or drag. Alternatively, the AFM can be operated in a noncontact mode [21], which means that the tip is held a few nanometers away from the surface, and set to oscillate at a frequency of about 100 kHz. The presence of surface features changes the amplitude of oscillation during the raster scan due to the long-range attractive forces between the tip and surface, which can be detected even in liquids [31]. While offering the advantage of minimal sample perturbation, the resolution is also reduced, and their use so far has been limited.

A similar approach, which holds promise for minimizing sample perturbation, is the "tapping" mode [32]. In this mode, the cantilever is also made to oscillate at high frequency during the scan, but contact between the tip and the sample occurs at the extremes in the modulation cycle. The lateral force on a sample is thus reduced in this mode, making it effective for imaging materials laid upon a substrate without dragging them, as exemplified by studies of DNA on mica [33]. It has recently been shown that the tapping mode can be implemented for imaging under liquids [34, 35].

3. *Imaging Under Solution*

Surfaces under a layer of solution can be examined by the AFM, by simply submerging the whole cantilever assembly, as illustrated in Fig. 1.1. This capability is particularly useful for biological specimen, allowing structural studies to be performed in their native hydrated state. Studies of processes at the solid–solution interface may also be conducted, which enables applications in electrochemistry [36]. In addition, the presence of the solution can reduce imaging forces, or can result in pinning molecules at the surface, both of which translate to obtaining better resolution: For example, it has been shown that DNA imaging improves considerably under propanol [37]. Furthermore, the solution may be changed *in situ*, or additional components may be introduced, facilitating observations of dynamical changes in the morphology. A number of biological studies have utilized this aspect [38–40].

4. *Calibration*

Accurate measurements of distances parallel and perpendicular to the surface rely on being able to calibrate the instrument. Along the surface, a grating serves as a good calibration standard for large scans, while the molecular dimensions of known crystalline materials serve to calibrate high-resolution images. Typically, the interatomic distances in graphite or mica are used. It should also be noted that when the sample is mounted on an XYZ scanner, the elevation of the sample above the scanner has to be taken into account in the calibration [41].

While gratings and grids may also be employed for calibration in the Z direction, problems arise in the case of a wide tip, which might not be able to reach the bottom, hence producing an erroneous measurement of depth, as is evident from Fig. 1.5. The use of latex spheres [42] and colloidal gold [29, 43–45] cast onto a flat substrate, such as mica, as standards for calibration has been reported. Figure 1.3 shows a section through a monolayer of latex spheres and the corresponding cross-section for height measurement, which can be compared with results from other techniques, such as electron microscopy (EM) and light scattering. Both latex and colloidal gold offer the advantage of commercial availability as reasonably monodispersed suspensions. One problem with the use of such standards is that they tend to be dragged by the tip during imaging. However, if spread under conditions such that close packing of several spheres are attained, such drag can be reduced [42]; the faint horizontal lines in Fig. 1.3 are evidence of a slight drag. In addition, polystyrene

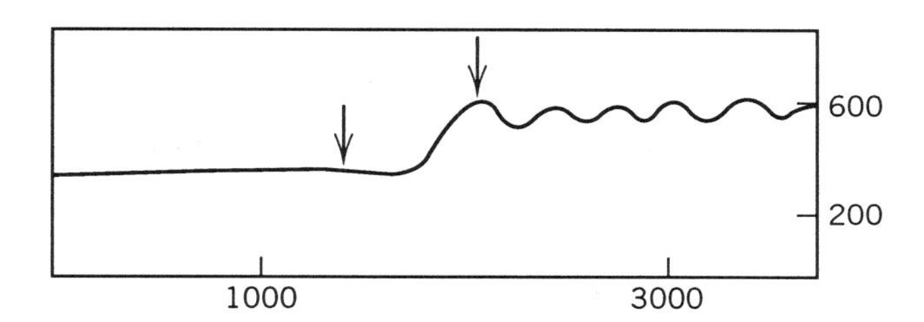

Figure 1.3. Latex spheres at submonolayer coverage on mica substrate. A cross-section taken along the indicated line enables measurement of height above the substrate.

latex spheres are available with different terminal groups, providing either negative or positive charges, which allows for a control of their interaction with the surface, facilitating a reduction in drag. The tobacco mosaic virus, which is a commonly used standard for calibrating electron micrographs, has also been utilized for the AFM [43]. However, when using biological materials or other soft matter for calibration, it is important to bear in mind that there is the possibility of sample deformation, and hence an error, due to the pressure of the tip on the soft surface.

Cantilevers that are commercially available come with nominal force constants, which are calculated using their (expected) dimensions and mass [46]. Due to inconsistencies in the manufacturing process, or to debris pick-up (see Fig. 1.4), such values could be off by as much as 50%. Thus, whenever an accurate measurement of forces is desired, the spring constant of the cantilever has to be calibrated separately. Schemes for such calibration are reported, based on the measurement of the resonant frequency [47, 48] or the mass of the assembly, as well as more sophisticated theoretical models [49, 50].

B. The Probe Tip

The imaging process relies on an interaction between the sample and the probe tip; hence, the probe tip is the crucial part of an AFM. It is also the main weakness of the technique: Most problems concerning irreproducibility in imaging may be attributed to uncertainties about the tip geometry, its composition especially when it picks up materials from the surface, and the nature of its interaction with the sample. A number of artifacts due to tip effects are reported, a summary of which is given in [51]. Hence, the characterization of existing probe tips as well as the design of new ones has been a primary consideration among users of the AFM.

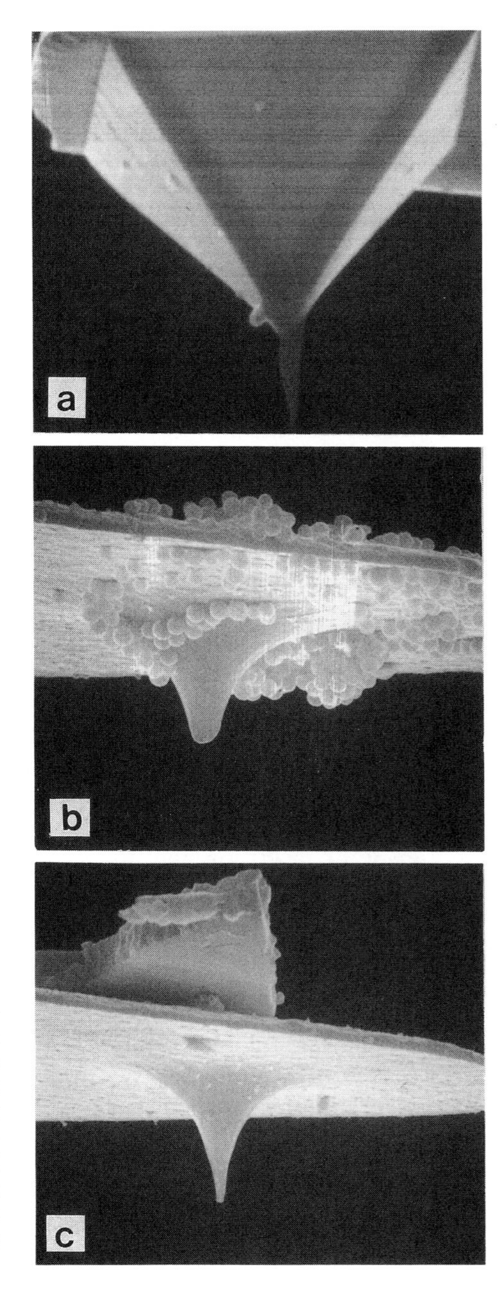

Figure 1.4. Scanning electron micrographs of samples of probe tips. (a) A square pyramidal silicon nitride tip, with sharp microtip grown by electron beam deposition. (b) An Ultralever™ that accumulated latex spheres during imaging. The apex may have been blunted during use. (c) An Ultralever™ that is sharper than (b), but with debris on the cantilever, possibly obtained during its fabrication.

1. Characteristics

The measureable quantity is the deflection of the cantilever, which is a reflection of the forces acting upon it. These forces are due to the mass of the cantilever itself, the mass of the probe tip, the spring constant of the lever, and the interaction between the probe tip and the sample. The force sensor is thus the whole cantilever-plus-stylus assembly. To maximize the sensitivity of the instrument, as well as to minimize the force exerted upon the sample, it is important to keep the masses small [4]. The stylus should be sharp, and the cantilever beam should have a small force constant and a high resonance frequency [46].

In principle, any material can be used as the force sensing device: A bent piece of metal, such as nickel [52] and diamond dust mounted on a metal foil [4], have been used in earlier studies. Rapid expansion in the use of the AFM is partly due to the introduction of microfabrication techniques for ease of production of tips with reasonably reproducible geometry. Hence, most work done today utilizes commercially available cantilevers with integrated tips of silicon nitride, silicon oxide, or silicon. These are made by using Si(100) wafers, and conventional batch microfabrication techniques [46]. In addition, these probe tips can be used as substrates for deposition of other materials, enabling changes in the interaction between tip and sample.

There are two aspects to consider with regards to tip geometry: the size at the apex, which is the actual contact with the surface (assuming no deformation), and the aspect ratio. The former will determine the area, and hence, the force of interaction between tip and sample, while the latter is an important consideration in examining steep gradients and rough surfaces.

A scanning electron micrograph (SEM) of a typical silicon nitride tip, which is square pyramidal in shape, is shown in Figs. 1.4a and 1.10a. Note that while ideal depictions of the AFM is for a single atom tip, in reality the probe has a radius of about 20–50 nm at the apex. The square pyramidal tip opens to an angle of about 70°, the shape of which is fixed by the crystal facet of silicon nitride [46]. This is advantageous in that it makes the stylus geometry reproducible; however, the wide opening of the pyramid prevents it from penetrating crevices, as illustrated in Fig. 1.5, such that details at the bottom of a deep well will be missed, and depth calculations will be erroneous.

To address this problem, which is particularly important in looking at integrated circuits, tips of high aspect ratios have been made. A relatively simple method for growing very sharp tips is to start with the standard pyramidal tip and expose it to the electron beam of a SEM, which

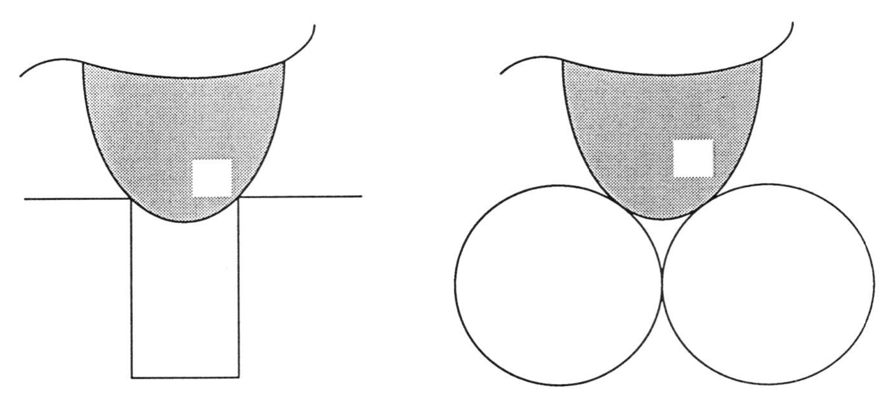

Figure 1.5. The size and geometry of the probe prevents it from accessing crevices.

deposits a thin carbon microtip on top of the pyramid [53–55], an example of which is shown in Fig. 1.4a. The mechanism involved in this process is not clear, but the whisker obtained in this manner is capable of imaging steep structures [54]. A combination of wet and dry etching techniques on single crystals of silicon has been used for the manufacture of high aspect ratio Si tips [56] shown in Figs. 1.4b and c, which are now commercially available. While this type of stylus enables a more accurate depth measurement, the radius at the apex is still nevertheless about 5–10 nm. Ion beam milling is a promising approach for the production of sharp tips [53, 57, 58], but so far, reproducibility of the tip geometry needs to be improved. Tips have also been fabricated for specific purposes, such as polished diamond for studies of hardness [59] and ion-milled tungsten [53] and iridium [57] with vertical walls for imaging integrated circuits and steep posts. A comparison of the characteristics of the most common commercially available tips is given in [51].

While it might seem desirable to have an infinitesimal cylinder tapering to a single atom to get atomic resolution, in reality such a stylus would be unstable. However, even in the traditional square pyramidal tip, high-resolution images of flat surfaces, such as mica, can be obtained routinely. These are presumably due to atomic scale corrugations at the tip apex that could result in effectively probing the surface with a single atom; however, questions have been raised about the mechanism for generation of atomic resolution images by the AFM [14, 60], which will be discussed in Section II.D.

2. *Artifacts at Mesoscopic Scales Due to the Probe Tip*

Since the probe tip is not of atomic dimensions, its geometry plays an important role in the imaging process, especially in the case of structures that are of comparable size, such as those of samples deposited on a flat substrate. As a result, images that do not accurately reflect the sample topography can be generated [51, 61].

Artifacts at mesoscopic length scales can be classified as follows: First, as mentioned previously, depth resolution may be lacking due to the inability of the tip to probe crevices. This lack could also lead to an erroneous estimate of surface roughness and particle size, as well as the smearing out of features. Second, the tip–sample interaction can produce sample deformation by the tip, such as indentation or an erroneous height determination. It can also cause movement of the sample, dragging it around, or accumulation of debris, which causes degradation of the quality of subsequent images. This deformation can be alleviated by minimizing the force of operation and immobilization of the sample by various means; alternatively, it can be utilized for manipulation of surface structures. Both aspects are discussed further in other sections. Finally, there is a tip–sample convolution, which is inherent in the imaging process even if the tip–sample interaction is minimal, which will be addressed at this point.

The image reconstruction in the AFM imaging process relies on the assumption that the point of contact of the sample with the tip is at the apex. This may be true in the case of a flat surface, but when the tip encounters a protrusion on the surface, the first point of contact between it and the sample is no longer at the apex. A single line scan depiction of this is shown in Fig. 1.6a and b, for a tip encountering an object with a circular and a rectangular cross-section. Two things are clear from these illustrations: First, while the object will appear wider, the height measurement will be accurate provided the object is not deformed by the tip.

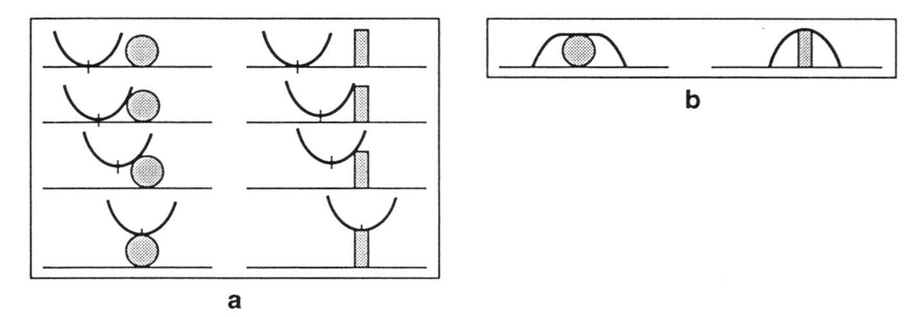

Figure 1.6. When a tip encounters an object of comparable size, it is prematurely raised (a). Since the AFM takes data assuming the point of contact is at the apex, the resulting image (b) appears wider, but the height is accurate. The image of a steep object is more a reflection of the tip than the sample geometry.

Second, the image obtained contains features of the tip, which is exacerbated in the case of a very steep slope, as in Fig. 1.6b.

Since the tip apex is about 10 nm in radius, the distortion in width is immediately apparent for samples a few nanometers in size. This is especially evident in reports of images of biological samples [51, 62], in which the samples were found to be much wider than expected. These width distortions can be explained by using a simple geometric argument (Fig. 1.7). When imaging samples with a circular cross-section, such as DNA, the actual sample height d is obtained. Assuming a stylus that is spherical at the apex with radius R, the measured width w is related to both the tip and the sample size by

$$w = (2Rd)^{1/2} \tag{1.1}$$

Images of a truncated tip are quite apparent in numerous reported examples with large sample corrugations (~100 nm), including those of polymer surfaces. The picture of the tip is truncated depending on the height of the surface feature. Because the typical probe tip is square pyramidal in shape, the appearance of pyramids and/or triangles in images is a sign of the presence of this artifact. Moreover, a distorted tip will generate less obvious shapes, but the presence of a repeated geometric pattern, such as elongations, squares, and multiple images (Fig. 1.8a), should alert the user to a possible tip artifact.

Using tips with a high aspect ratio can alleviate the appearance of these obvious geometric patterns, but the image will always contain tip as well as sample features. This makes the tip geometry of major consideration, especially when examining a sample of unknown features (i.e., where no external checks are easy): A deformed tip can easily mislead the researcher about the sample features.

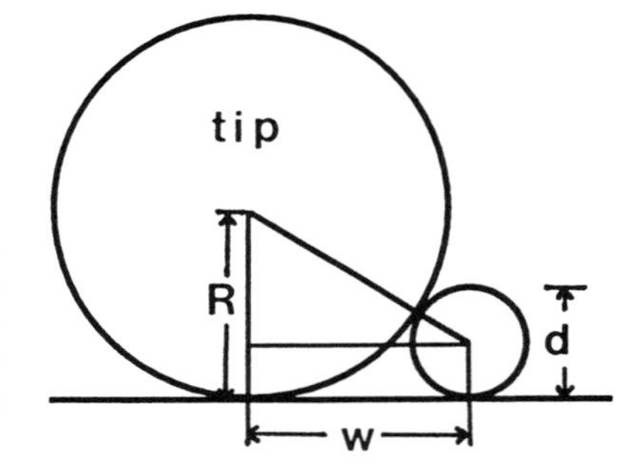

Figure 1.7. Schematic of a large tip, approximated by a sphere at the apex, imaging an object with spherical cross-section. The enlarged width (w) measured by the AFM can be accounted for by geometrical considerations [Eq. (1.1)].

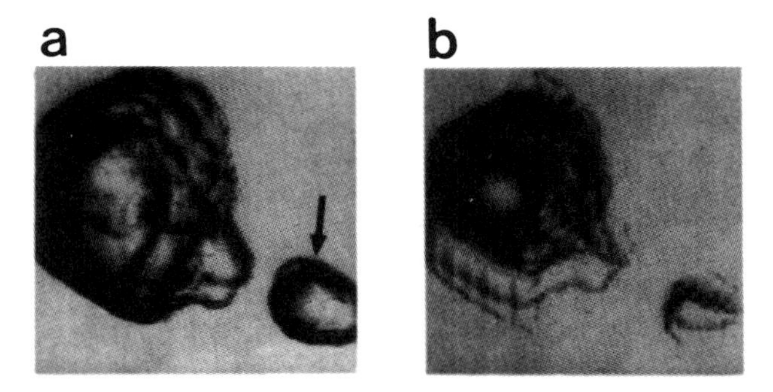

Figure 1.8. Images of a latex sphere using the deformed tip shown in Fig. 1.10b. (a) As obtained. Note the presence of a ghost particle, indicated by the arrow, due to the "double tip." (b) After numerically removing the tip shape.

Tip distortions can arise due to inconsistencies in the manufacturing process (Figs. 1.4b and 1.10b), or simply because the tip has picked up surface debris during a scan (Fig. 1.4c), which is always a possibility when the surfaces being studies are not ultraclean, or under vacuum. A specific example is a "double tip effect"; Figure 1.8a shows an image of a polystyrene sphere obtained using the tip in Fig. 1.10b. Not only does the sphere appear distorted, but there is a ghost image present due to the "second" tip.

How does one get rid of artifacts? There is a push for better tip-making efforts, both in terms of improving geometry, changing the chemical composition, and consistency in the manufacturing. In practice, an experienced user recognizes the presence of obvious artifacts, and can thus change the tip and the conditions of imaging if necessary. However, this approach causes irreproducibility and unreliability: Does the sample really look this way, is the tip bad, is the sample being deformed, does the sample preparation need improvement, and so on?

To make imaging more systematic, it is useful to obtain a picture of the tip. Scanning electron micrographs of the tip can be taken, examples of which are in Fig. 1.4, but the tip is altered during the process by the coating needed for SEM. Furthermore, it is not a practical procedure for routine imaging studies, especially if the deformation is caused by debris pick-up, as is usually the case for soft samples and films: It is only useful in examining whether or not a batch of tips has a systematic manufacturing problem.

Attempts have been made to examine the probe tip by using the AFM itself [43, 45, 57, 63–67]. With the use of Eq. (1.1), an idea of the picture of the probe tip has been derived by examination of a series of different sizes of biomolecules [43] and gold colloids [45, 67]. The AFM can also be used for direct imaging of the stylus, which is laid upon a substrate (Fig. 1.9) [64, 65].

The three-dimensional picture of the stylus may be visualized *in situ* with the use of the numerical deconvolution procedure due to Markiewicz and Goh [66], which is discussed in Section II.B.3. This procedure relies on imaging a sample of known geometry, such as polystyrene spheres [66] or calibration grids [68], which can be deconvoluted out of the image, leaving a picture of the tip. Samples of the tip shape obtained thus are shown in Fig. 1.10, together with the SEM pictures taken subsequently. The tip appears asymmetric, due to the way it is mounted in the AFM, which amounts to a tilt angle of about 16°. This procedure makes use of the imaging artifact mentioned previously, that imaging steep samples provide more information about the tip than about the sample. Quantitatively, the accuracy of the tip geometry obtained in this manner is

Figure 1.9. An image of a square pyramidal tip mounted on a substrate, obtained with the AFM.

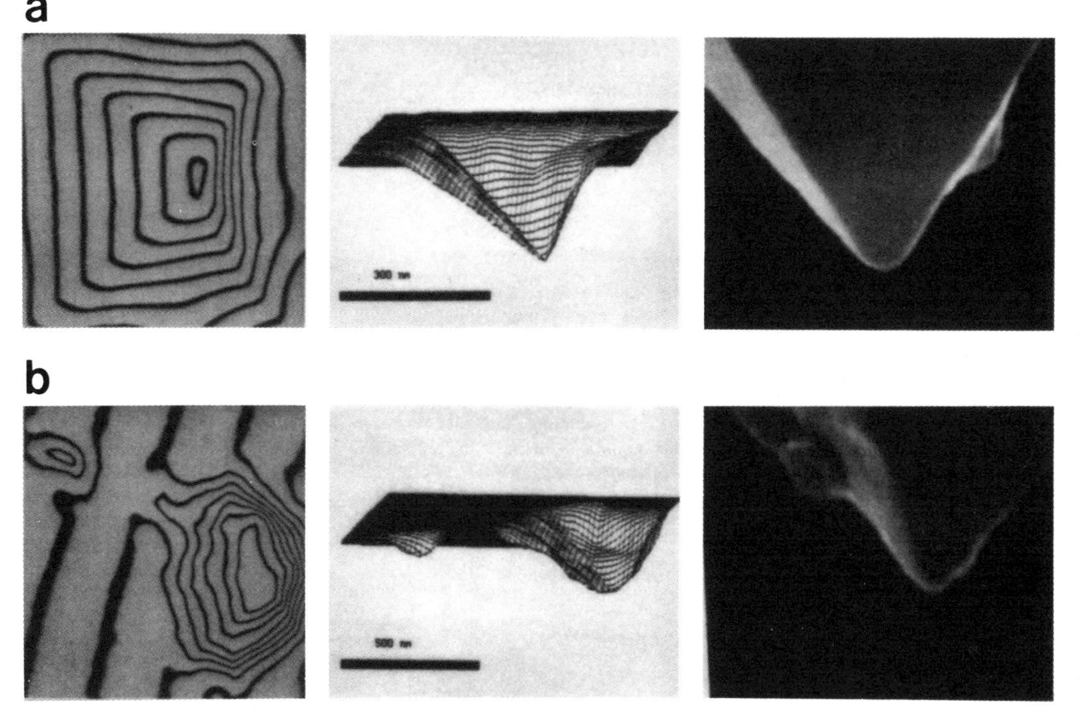

Figure 1.10. Tip information obtained numerically using polystyrene spheres with nominal radius of 260 nm. (a) A square pyramidal silicon nitride tip of good quality. (b) A deformed tip. The first column shows contours (33-nm line spacing). The middle column is the same data viewed from an oblique angle for comparison with the SEMs in the third column. The middle column contains an approximate 16° rotation due to the way the cantilever is mounted on the AFM (scale bar = 500 nm). [Reprinted with permission from P. Markiewicz and M. C. Goh, *Langmuir* **10**, 5 (1994). Copyright © 1994 American Chemical Society.]

dependent on having a well-characterized calibration standard: For example, uncertainty in the latex size translates to an erroneous estimate of the tip size [68]. Nevertheless, having an image of the tip provides a systematic criterion for its rejection or acceptance for use.

3. *Image Reconstruction by Deconvolution*

Since it has been recognized that the pictures obtained with the AFM are distorted by the tip, efforts have been made to reconstruct the actual sample topography from the image.

A theoretical approach to image reconstruction with the use of Legendre transforms of the distorted image and the tip has been used for both STM [69] and AFM [70]. Analytic solutions are available in the simple cases of a parabolic sample and parabolic tip; otherwise, a numerical procedure is feasible provided certain conditions are satisfied. First, the exact tip geometry must be known. This condition is not that restrictive, since only the geometry at the scale of the sample is important, and in the case of small corrugations ($<$10 nm), a sphere or a parabola is a reasonable approximation for the tip apex. However, the procedure fails when the tip and sample have two simultaneous points of contact, (Fig. 1.6), which is a common occurrence. Furthermore, since the transformation relies on taking derivatives, the height data have to be smooth: Since AFM images usually contain high-frequency noise, these as well as other sporadic fluctuations have to be removed, in which process some sample features may also be lost.

A numerical algorithm that effectively removes the tip geometry from the AFM image based on the imaging process illustrated in Fig. 1.6 has been reported [66]. This scheme is feasible provided the tip and the sample are in contact, the interaction between them is assumed to be only by excluded volumes, and the tip does not deform the sample. Basically, the tip and the sample cannot occupy the same space, and therefore if the tip geometry is known, it can be removed from the picture point by point, taking advantage of the fact that the AFM image is simply a matrix containing height information at every point. The deconvolution scheme is illustrated in Fig. 1.11 for a single line scan, but the actual process is done by moving a two-dimensional (2D) matrix containing information of the tip geometry, across the 2D matrix containing the AFM image, and comparing the heights at each point.

The tip geometry can be obtained by the same deconvolution procedure using standards of known geometry, or in the case of small samples, it can be assumed spherical at the apex, the radius of which can be estimated with Eq. (1.1).

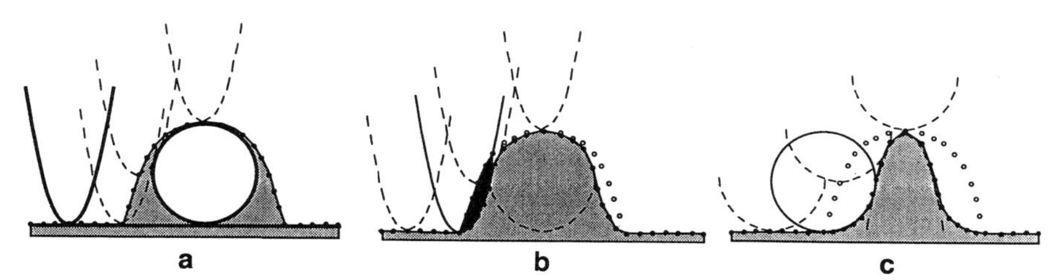

Figure 1.11. A single line scan depiction of the numerical deconvolution procedure [66]. (a) The shaded area corresponds to the original AFM image, which contains both the tip (parabola) and sample (sphere) geometry (b). As the tip is rolled along the sample, values of the sample height are adjusted to allow for the tip geometry, by placing a facsimile of the tip at each point and removing the dark region. (c) Conversely, if the sample geometry is known, it can be removed from each original data point, leaving a picture of the tip.

This numerical procedure allows for the examination of tips in a nondestructive way, which is fast and convenient. It is recommended that prior to examining an unknown sample, images of standards should first be obtained, from which the tip information can be derived. This information can then be used in two ways: the tip shape should be deconvoluted out of the subsequent images, and, if the tip is substantially deformed, it should be replaced. The utility of this procedure is demonstrated in Fig. 1.8, where even the "double tip" effect can be taken into account. An application of the deconvolution procedure to the study of biological paired helical filaments showed a dramatic result of a twisted ribbon structure that was obscured by the tip [71], as shown in Fig. 1.12.

C. Tip–Surface Interaction

In the simplistic view of imaging with an AFM, the tip tracks the surface and gives a measure of the sample topography without any interaction that would complicate the picture. In reality, complex intermolecular and surface forces are involved in the tip–surface interaction, which has been the subject of considerable study since the early days of the AFM [14, 18, 60, 72, 73]. In particular, questions are raised with regard to what really is being imaged, and whether or not single molecule resolution is indeed attained by the AFM.

1. A Consideration of Forces

Intermolecular forces arise from a combination of Coulomb interactions, polarization or dispersion forces, and quantum mechanical effects due to

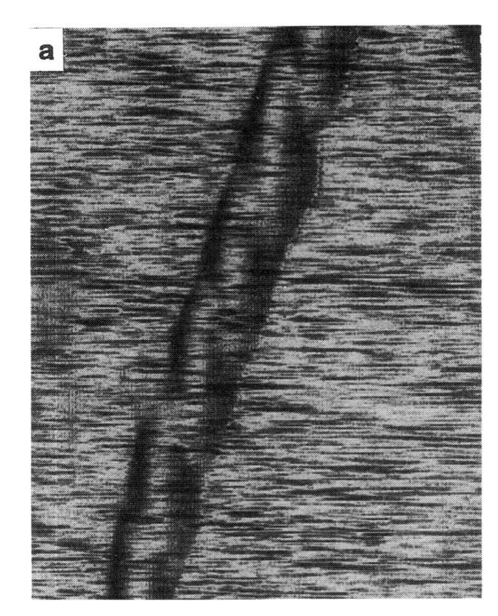

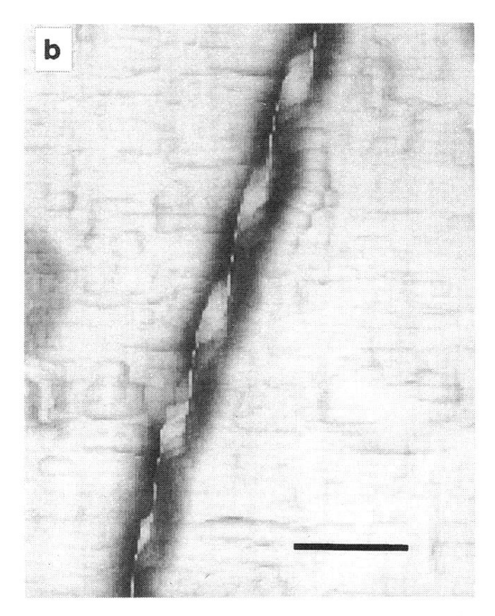

Figure 1.12. Portion of paired helical filaments (PHF) from Alzheimer's diseased brains: (a) raw data, and (b) after processing to remove the tip geometry, exposing a twisted ribbon structure (scale bar = 50 nm).

the exclusion principle, a comprehensive discussion of which is given by Israelachvili [16]. These interactions are thus atom specific, and cannot be described by simple force laws, making detailed theoretical considerations of molecular resolution imaging difficult. Nevertheless, to endow the AFM with spectroscopic capabilities, it is these atom-specific interactions that have to be exploited.

In general, phenomenological interaction potentials that contain both a long-range attractive and a short-range repulsive part explain a wide range of phenomena. The most commonly used example is the Lennard–Jones potential,

$$w(r) = -a/r^6 + b/r^{12} \tag{1.2}$$

which explicitly contains the correct form of the attractive part due to van der Waals forces.

The idealization of an AFM working in the contact mode depicts a single atom at the apex of the probe tip interacting with an atom at the surface, or with the planar surface, by a repulsive force due to the overlap of their electron clouds. Hence, the imaging is supposed to reflect molecular dimensions. Since the repulsive part of the interaction decays very rapidly, a hard sphere potential can be used to describe it. Alternatively, it can be modeled by a power law,

$$w(r) = (s/r)^n$$

where $n = 12$ for the Lennard–Jones model, or by an exponential decay,

$$w(r) = k\exp(-r/s)$$

In any of these cases, as long as only short-range forces are in effect, the tip–surface interaction will be quite local, and interpretation of AFM images will be straightforward. The measured forces would then reflect interactions at the atomic level.

At distances large compared with atomic dimensions, only average effects from all the atoms are involved, and the interactions can be considered as that between two macroscopic bodies. For a van der Waasls attractive pair potential of the form $w(r) = -C/r^6$, surface forces can be calculated either by summing the atomic interactions in a pairwise manner, or by considerations of polarization in continuum media (Lifshitz theory). The nonretarded potential as a function of the distance of separation $W(D)$, for a sphere of radius R near a flat surface obtained from either approach has the same form, and differ only in the expression for the Hamaker constant A:

$$W(D) = -AR/6D \tag{1.3}$$

An exact expression for the Hamaker constant is given by Lifshitz theory, and involves the frequency-dependent dielectric responses of the media. An approximate form [16] for two macroscopic phases 1 and 2 interacting across a medium 3, all of which have the same absorption frequencies, suffices for this description,

$$A = \frac{3}{4}kT\left(\frac{\varepsilon_1 - \varepsilon_3}{\varepsilon_1 + \varepsilon_3}\right)\left(\frac{\varepsilon_2 - \varepsilon_3}{\varepsilon_2 + \varepsilon_3}\right) + \frac{3h\upsilon_e}{8\sqrt{2}}\frac{(n_1^2 - n_3^2)(n_2^2 - n_3^2)}{(n_1^2 + n_3^2)^{1/2}(n_2^2 + n_3^2)^{1/2}\{(n_1^2 + n_3^2)^{1/2} + (n_2^2 + n_3^2)^{1/2}\}} \tag{1.4}$$

where ε refers to the static dielectric constant, and n is the index of refraction. It can be seen from Eq. (1.4) that the Hamaker constant is always positive in vacuum, whereas in the presence of an intervening medium, the dielectric properties of the medium should be considered, and A can take either positive or negative values.

Therefore, note that in the geometry considered in Eq. (1.3), the attractive part of the intermolecular potential leads to long-range surface forces, which increases with the size of the sphere. For a sphere of molecular dimensions interacting with a planar surface, the attractive interaction in vacuum gives rise to a force of the order of 10^{-9}–10^{-11} N.

2. *The Force Curve*

The force between the AFM probe tip and the sample can be obtained by multiplying the spring constant with the measured cantilever deflection induced by moving the sample towards the probe. A typical force curve under ambient conditions, which is a plot of the cantilever deflection as a function of distance of separation, is shown in Fig. 1.13. At large separations, the cantilever, located at height H_0, is unaffected by the approach of the sample. As the tip and samples get closer to each other, there is a slight bending of the tip towards the sample followed by a

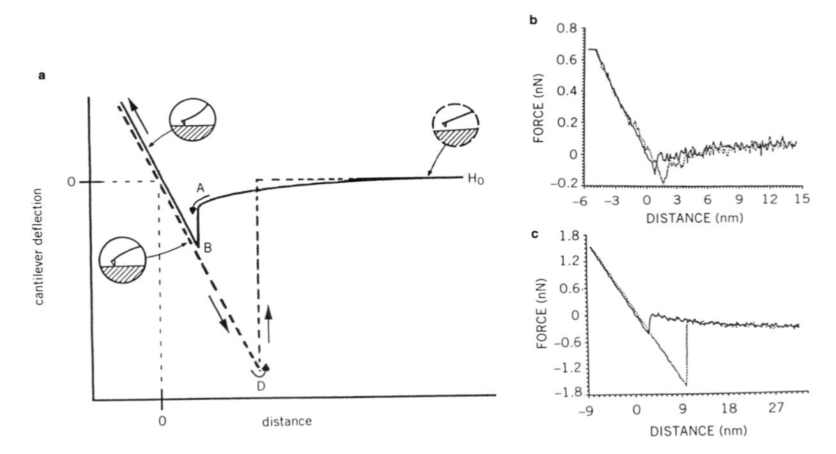

Figure 1.13. Force versus distance curves are obtained by monitoring the cantilever deflection as the sample is moved towards and away from the tip. The dashed line corresponds to the approach, and the solid line to the retraction. (a) A schematic of a force curve under ambient conditions where H_0 corresponds to the nontouching regime when the tip and sample are independent. As the separation decreases during the approach, the cantilever is deflected towards or away from the sample, depending on whether the van der Waals forces are attractive or repulsive. At A, the tip jumps into contact with the sample, and tracks its motion. Upon pullout of the sample, the tip moves with it until it breaks free at D, which is the point of minimal force. (b) and (c) Experimentally obtained force curves for mica in ethanol (b) and water (c) using a silicon nitride cantilever. (Data from [74].)

discontinuous jump. At point *B*, the tip and the sample are in contact. Upon further approach, the cantilever is deflected upwards, moving together with the sample. Retraction of the sample produces a hysteresis in the force curve: The cantilever tracks the sample and dips below its equilibrium, noncontact position H_0, usually by a substantial amount, before it finally breaks free of the sample at point *D*. The "dip" corresponds to an extra pull-out force that must be overcome to separate the tip from the sample. The amount of hysteresis is dependent on the spring constant of the cantilever, decreasing with stiffness.

The shape of the force curve changes when imaging under solution [74], as shown in Fig. 1.13 b and c for ethanol and water, respectively. In general, the hysteresis during the retraction is decreased, sometimes to approximately zero. Furthermore, in the initial approach the tip may bend either towards or away from the sample, depending on the nature of the solution.

The behavior of the force curve can be understood in terms of surface forces between the macroscopic tip, and the sample [18, 75], and have been addressed by simulations [76, 77]. During the initial approach the tip bends due to the long-range van der Waals interactions. These interactions are always attractive in vacuum but may become repulsive in the presence of an intervening liquid, which explains the direction of bending of the cantilever prior to contact. If the cantilever spring constant is small enough, as is desirable for maximum sensitivity of the instrument, the gradient of the surface force can exceed it. This gives rise to an instability resulting in a jump-to-contact (*A* to *B* in Fig. 1.13). Assuming the probe tip to be a sphere of radius 50 nm, and using typical values for the cantilever spring constant of 0.5 $\mathrm{N\,m^{-1}}$ and a Hamaker constant under vacuum of 10^{-19} J, the van der Waals attractive term in Eq. (1.3) indicates that the tip will jump into contact when the distance of separation is about 2–3 nm.

If the separation is reduced beyond the point of contact, the sample pushes against the tip and the cantilever. A load is thus applied, which produces a deformation of either sample or tip, or both, the effect of which is to create a larger area of contact between the two. During the pull-out process the tip is held onto the sample by a force of adhesion, which can be greater than the initial force of attraction due to the interaction within the contact area. By explicitly including both the attractive and repulsive surface forces, a hysteresis in the force curve can be produced [18, 75].

Measurements of the force curve between a tungsten tip and different surfaces (mica, graphite, sapphire, polytetrafluorethylene, and others) under ambient conditions show that the adhesive forces during retraction are always larger than the attractive forces in the approach, and both are of the order of 10^{-9}–10^{-7} N [78].

3. *Implications in Imaging*

From the point of view of imaging with minimal sample perturbation, it is important that the imaging force is minimized, the appropriate value being dependent on the specific sample [73, 79, 80]. The imaging force of the instrument is the difference between the point of minimal force and the operating point. This force can be lowered by reducing the operating point and the hysteresis in the force curve. Note that the force between the topmost atoms of the tip and sample is always repulsive at short distances, and this short-range repulsion is the key to attaining high resolution in the AFM contact mode. However, due to the attractive surface forces, a large load is applied to these topmost atoms, resulting in large local pressures, enough to cause irreversible deformations and to dislodge the atoms [80].

Operationally, lowering the imaging force involves retracting the sample in order to produce a net negative load on the tip, such that there is a partial balance between the interatomic repulsion and the attractive van der Waals forces. However, there is a regime of separation distance that is unattainable because the cantilever jumps into contact with the sample at point A. This regime could be minimized by increasing the cantilever stiffness. Alternatively, a negative load can be applied to the tip from an external source to fix the separation distance [81].

Another approach, evident in the form of the surface forces in Eqs. (1.3) and (1.4), is to manipulate the Hamaker constant by immersing the tip-sample assembly under a liquid with appropriate dielectric properties, so as to minimize the net load [74].

D. Limitations

1. *Theoretical Limits: Is Single-Molecule Resolution Possible?*

The limits imposed by the instrumentation itself are on the atomic scale: The lateral resolution of the machine, dictated by the piezoelectric scanner, is about 0.1 nm, while a vertical resolution of 0.01 nm is feasible with the available detection schemes.

One important limit to the resolving power is thus the ability to distinguish between features being observed, which will vary from sample to sample. Furthermore, the characteristics of the tip, and the way it interacts with the sample, play a major role. The presence of drifts, due to thermal gradients, vibrations, and molecular motions, further reduces resolution in practice, although in some cases these can be minimized with choice of experimental conditions.

Whether the instrument is indeed able to obtain single-molecule resolution is still an important and contentious issue. Experimentally, there are numerous reported images of ordered Langmuir–Blodgett (LB) films and other crystalline surfaces with periodic patterns of symmetry and distances that correspond to the expected molecular arrangements. Furthermore, substrates such as mica and graphite can be imaged routinely at subnanometer resolution, and the distances measured between the atomic features are in fact used to calibrate the AFM. How is it possible to obtain atomic resolution with a large tip? The STM is known to be capable of imaging atomic details, the basis for which lies in the short-range nature of the tunneling interaction such that probing of the sample is localized to the atom on the tip that is closest to it. Hence, a typical argument for the feasibility of atomic resolution AFM imaging is that there is roughness on the tip apex, such that a minitip of one or two atoms is doing the actual imaging, much like that of the STM. However, one should realize that not only is the probe tip relatively large at the apex, but that the long-range van der Waals forces, which are manifested in the force curves, can dominate the interactions as well as induce deformation that serves to increase the area of contact between tip and sample. The analysis of Weihs et al. [73] indicate that even at zero applied force, the presence of adhesive interactions creates a finite area of contact, and thus sets a limit to the imaging resolution in the contact mode of operation. The mechanism of AFM imaging has thus been a subject of considerable thought since its early days [14, 72, 73, 79, 82].

Several authors have raised the question as to whether or not the AFM can function as a true microscope at the atomic scale, which is capable of single-atom resolution [60, 73, 80, 83]. High-resolution images inevitably show perfect order, and it has been argued that the atomic patterns obtained are a superposition of several images that are the result of interactions of numerous minitips with the surface [60, 83]. These patterns will still reflect the molecular structure at the surface, and can thus manifest different lattice structures [84] and polymorphism [85]. However, it implies that high-resolution imaging of nonperiodic samples is not possible.

The preceding arguments suggest that true molecular resolution can only be obtained either by reducing the size of the tip to atomic dimensions [86], or by reducing the adhesive interactions. The latter was first employed by Marti et al. [87] by imaging under liquids. The former has its limitations, in that increasing the tip sharpness can result in inelastic deformation [73].

One indication of true molecular resolution is the observation of defects, examples of which are few. Garnaes et al. [88] reported the detection of grain boundaries in LB films at molecular resolution. By making a comparison between successive upward and downward scans, Radmacher et al. [83] were able to identify defects in Cd arachidate films. Ohnesorge and Binnig [80] were able to image defects in calcite by lowering the adhesive forces and minimizing lateral drifts. These studies indicate that in order to obtain true single-molecule resolution, forces have to be reduced to 10^{-11} N even for hard samples; otherwise the force between single atomic sites and the front atom of the tip is enough to obscure any defect in the sample, or to dislodge the front atom.

Nevertheless, these results demonstrate that true molecular resolution with the AFM is indeed feasible under certain conditions. However, under routine imaging conditions, which involve relatively large tip–sample interactions ($\sim 10^{-7}$ N), features that look like atomic details are obtained; caution is advised when interpreting such images.

2. *Time-Dependent Studies*

Studies of changes in the surface morphology with time can be followed with the AFM. The limit to the time scale that can be probed is determined by the rate of scan that is accessible with the tip still tracking the surface. Theoretically, the scan rate can be increased until close to the resonance frequency of the cantilever, or the low pass filter, whichever is lower. However, in order to track the surface, the response to the changing surface features of the cantilever deflection (in the force mode), or the Z component of the piezo scanner (in the height mode) determines the time resolution. For high-resolution imaging, Butt et al. [89] derives upper limits for the scan speeds of 0.1 μ s^{-1} for hard samples and 2.2 μ s^{-1} for deformable materials.

A limited number of examples of time-resolved studies are in existence, most of them biologically related and dealing with time scales of the order of minutes to hours. In the first example, Drake et al. [38] followed the formation of aggregates of fibrin from fibrinogen upon introduction of the clotting agent thrombin. The aggregation was attributed to the polymerization of fibrinogen into fibrin, which is the main reaction in the clotting of blood. However, Lea et al. [90] obtained a similar aggregation pattern on fibrinogen without addition of the clotting agent.

The binding of a protein to a lipid bilayer [91], and changes in processes at the surface of virus-infected cells in real time have also been reported [40].

Picosecond time resolution has been obtained with the STM [92], but no comparable attempts have been made with the AFM. Radmacher et al. [93] were able to follow the reaction of the enzyme lysozyme with a substrate with subsecond resolution. This was accomplished indirectly, by monitoring fluctuations in the amplitude of oscillation of the cantilever (in the tapping mode), which correlated with the reaction. This type of experiment, which needs to be tested further, could open enormous possibilities for studies of enzyme kinetics at the molecular level.

3. *Inherent Limitations*

An inherent limitation of any scanning surface technique, such as AFM, that cannot be circumvented is the inaccessibility of information about the underside of a feature. This means that a sphere on a substrate will look exactly the same as a hillock of similar dimensions; this fact should be remembered when interpreting images. Furthermore, the actual amount of information obtained is limited by the tip shape (discussed in Section II.B). As seen in Fig. 1.6, the only part of the image pertinent to the sample feature is that from the first to the last point of contact, even after the tip is deconvoluted from the image.

It should also be noted that the data for AFM images is taken at discrete intervals. Typically, 400×400 pixels are used for an image. Hence, the actual grain size that can be resolved is limited to the scan size divided by the number of pixels, which should be kept in mind when zooming in digitally on an image originally taken at large scans.

4. *Comparison with Electron Microscopy*

The strongest point of the AFM is that it enables the examination of a sample in air, or under a liquid, with minimal preparation, and resolution at the nanometer scale is easily obtained. While transmission electron microscopy (TEM) can be capable of similar lateral resolution, the vertical capabilities of the AFM is superior. Furthermore, the use of stains in TEM can obscure some features, distort images, or produce an error in the measured sizes of objects because of the thickness of a deposit. With the AFM, it is the sample itself that is observed, and not the location of stains. However, the use of stains in EM offers an advantage in that it helps identify the structure of interest, at the same time as making other features transparent. This finding is particularly noticeable in biological studies: The AFM image of a preparation that contains a specific feature that is clearly identifiable in EM may look like a complete mess because the feature of interest can be obscured by the presence of materials such as salts and other deposits.

Surface morphology is studied with the AFM, and in comparison with SEM, which is another surface technique, the AFM is less intrusive to the sample, although both can cause damage. An operating force of nanonewtons or less, which is the main perturbation on the sample, is routinely accessible in the AFM. In contrast, the typical SEM operates at electron energies of about 25 keV. In studies of polymer surfaces, SEM requires the deposition of a conductive coating, which is a disadvantage in several ways: First, the extra layer of coating could make features appear wider; second, some features may be obscured, resulting in a lower vertical resolution for SEM; and third, the sample prepared in this manner cannot be used for subsequent studies. This third point is illustrated in the AFM study of the annealing latex [94] discussed in Section V.B.

Studies showing a direct comparison of images obtained with EM and AFM provide a good idea of the comparative merits of the two techniques [95, 100], and it is useful to employ both as complementary techniques when possible.

E. Sample Preparation

One of the main advantages of the AFM over other microscopic techniques is the ease of sample preparation. In the first few years of use, images have been obtained from materials as is, under ambient conditions. Lately, it has become apparent that the imaging could be made more effective by appropriate sample preparation and control of imaging conditions, especially with the recognition that the perturbation on the sample could be nonnegligible. The goal is to prepare the sample in such a way as to minimize perturbations upon it and to enable the highest desired resolution. In general, imaging conditions and sample preparation procedures are derived empirically.

Studies at atomic length scales can only be conducted on samples that are hard enough such that tip-induced deformation is minimal. In general, one can image single crystals as is, although to diminish surface contamination, it is important to expose a new face just prior to study. Moreover, knowledge of the orientation of the crystal axes can provide assistance for image interpretation [101–104]. In certain cases, imaging is simpler under a layer of solution, which serves to reduce capillary forces. High humidity can have adverse effects on the quality of images since a thin film of water may form at the surface, which serves to increase the adhesive forces [104, 105]; hence, imaging under a nitrogen atmosphere is worthwhile in some cases.

High-resolution images have been obtained for close-packed structures deposited by LB techniques, examples of which are found in [84, 88, 106–108]. Self-assembled monolayers of long-chain alkanethiols on gold have also been observed at molecular resolution [109–111]. In these cases, close packing of the film is essential, and it helps if the substrate is molecularly flat. For LB films, mica and silicon oxide are commonly used, while for the formation of alkanethiol self-assembled monolayers, a flat gold surface, typically made by chemical vapor deposition on mica, is utilized.

There are two simple ways of producing polymer films from solution: solvent casting and spin coating. In the first case, a solution of the desired material is applied to a substrate, and the solvent is allowed to evaporate. This rate may be controlled by regulating the vapor pressure of the solvent above the cast film. Spin coating allows a fast evaporation rate by mounting the sample on a rotating disk, thus thinning the droplet uniformly and expelling the solvent outward. It is generally believed that spin-coated films have lower surface roughness than solvent cast films. While this is presumably true at macroscopic dimensions since the curvature of the droplet is eliminated by spin coating, AFM studies at the nano-to-micro level indicate otherwise. Polystyrene films that are cast from benzene and other solvents are smooth to within 1 or 2 nm. Furthermore, studies performed on latex films of poly(butyl methacrylate) indicate that the solvent cast film has a higher degree of order than the spin coated one (Fig. 1.14) [112].

When examining individual particles, it is customary to simply deposit them onto a substrate that is flat in comparison with the particle. The typical choices are mica and graphite, since they are flat at the molecular scale, but glass, gold, and silicon with its native oxide layer are also commonly used. The particle is held in place simply by its interaction with the substrate. However, since the scanning procedure imposes a lateral force on the sample, there is a tendency for the tip to drag the particle around. While the substrate may be coated to control the substrate–sample interaction and diminish the lateral motion of the desired particle, little has been done in this direction. Polytetrafluoroethylene (PTFE) deposited on glass by rubbing has been shown to be capable of acting as a substrate for orienting samples [113] due to the high degree of order at the molecular level [113, 114, Section III.C], and has been used for immobilizing biomolecules [115].

One technique that has been employed for immobilizing silica colloids is to embed the particles in a matrix [116]. While good for immobilization, this technique presents a problem with regard to measurement of particle size, which varies depending on the depth of embedding. However, the latex held thus can serve as the surface upon which adhesion studies may be conducted. Embedding in epoxy resin has also been used to hold polymer fibers in place; the resin can be cleaved subsequently with a microtome in order to expose the desired cleavage plane [102–104].

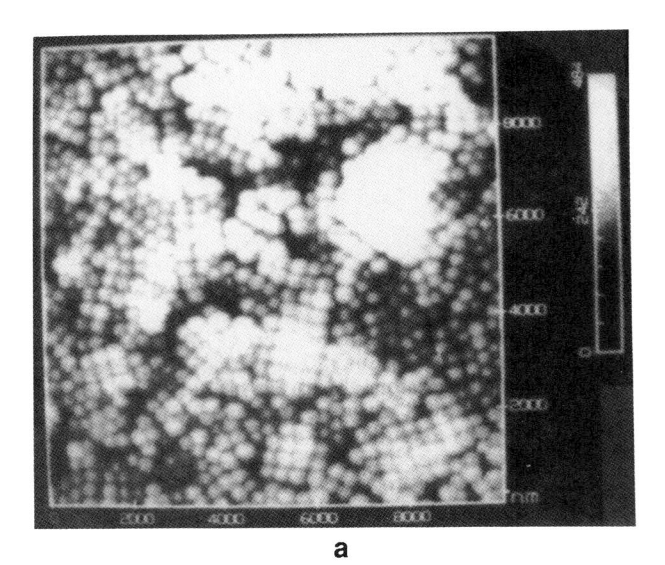

a

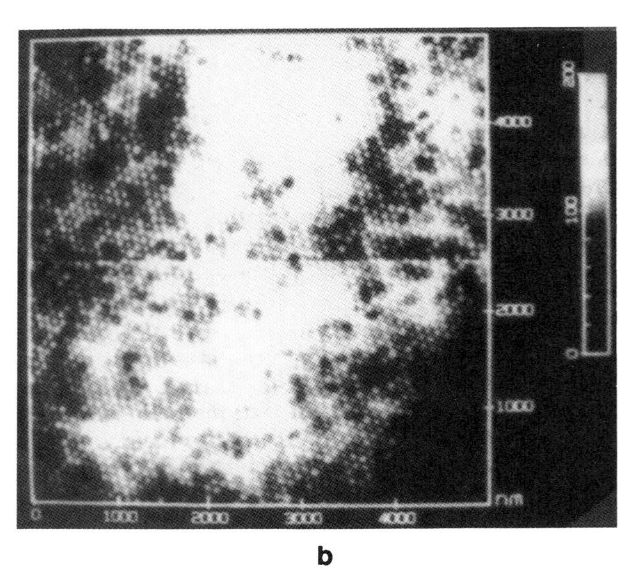

b

Figure 1.14. A film of PMBA latex spheres prepared by (a) spin coating and (b) solvent casting. [Reprinted with permission from Y. Wang, D. Juhue, M. A. Winnik, O. M. Leung, and M. C. Goh, *Langmuir* **8**, 760 (1992). Copyright © 1992 American Chemical Society.]

Studies on biological systems produced the most attempts at detailed sample preparation. There are two main aspects to be considered in these preparations: first, the immobilization of the sample upon the substrate so as not to be swept away by the tip during the scan, and second, the preservation of the native structure that one wants to see.

While the simplest way of immobilizing is by drying onto an appropriate substrate, it might not be the most suitable for soft biological samples, which could deform upon removal of the aqueous environment. Furthermore, there is a potential problem with the possibility of substrate-induced distortions on the sample: While a large interaction between substrate and sample is desired to minimize motion of the sample, too large an interaction may result in a deformation. Example of this is probably present in studies of collagen on mica, where the width of the collagen fibril is unusually wide (beyond what is accounted for by the tip size), presumably due to interaction with the mica [29].

Imaging conditions and sample preparation have to be investigated for specific samples. Numerous schemes have been employed for improving images of DNA [33] and proteins [117]. Various chemical attachment procedures have also been employed for other samples, such as coating the substrate to act as a ligand for the sample [118], and covalent cross-linking of the sample with the substrate [119]. The STM studies show that controlling the electrode potential of the gold substrate can immobilize DNA [120]; while not yet employed in AFM studies, this approach may be useful provided the electrochemical forces do not prove deleterious.

A technique that has not yet been applied for immobilization in AFM studies, but may potentially be useful, is by optical tweezers [121], which has already been shown capable of holding and moving biomolecules such as DNA [122] and biological molecular motors [123]. Another approach that is worthwhile in providing information albeit indirectly, is the use of markers, in a manner similar to EM. Immunogold labels have been utilized for locating structures at the cell surface that are beyond easy resolution of the AFM because of the fluidity of the cell [124].

While the AFM can also be used for replica imaging in a manner similar to EM, no advantage is gained by such; however, studies on replica images could provide a direct comparison between EM and AFM, which may be useful in examining certain sample preparations. Furthermore, samples that have been coated for SEM studies can also be imaged with the AFM [98, 125]. Not only can this serve as a cross-check between the two techniques, but in certain cases the coated sample may be more robust and thus less deformable by the AFM.

III. POLYCRYSTALLINE POLYMERS

A polycrystalline polymer solid is very fascinating from the materials perspective: The average molecular weight is an important parameter for polymers, but in addition, the organizational arrangement between molecules plays an important role in determining the final bulk properties (see, e.g., [8]). Whereas a simple molecular solid may be brittle, ductile, tough, or elastic, a polycrystalline polymer fiber may exhibit all of these properties due to the presence of crystalline and noncrystalline domains, the relative proportions of these domains, and their arrangement with respect to each other, all of which may vary depending on the way the fiber is made. After the fiber is formed, it can be stretched uniaxially to produce a highly oriented sample that, in comparison with metals, has a higher mechanical performance relative to its weight. Another aspect is the existence of polymorphism within the crystalline domains of the solid sample, in which the coexisting structures may possess very different properties. Thus, structural considerations from the submolecular level to the arrangement of domains, and finally to the fiber themselves, are important aspects in the design of materials.

A. Imaging at Submolecular Resolution

The AFM has been utilized for studies of polycrystalline polymers at high resolution. Numerous reports of submolecular resolution exist for several linear polymers, such as isotactic polypropylene (iPP) [102, 104, 126], poly(*p*-phenyleneterephthalamide) (PPTA or Kevlar) [85], PTFE [95, 114], polycarbonate [95], polyethylene [127–130], polyoxymethylene [131], and various polymerized LB films [132–134].

There are two main considerations in high-resolution imaging: sample preparation and image interpretation. For the former, methods abound in the polymer literature, some of which are outlined as follows, and the latter will be discussed more fully.

Since the AFM is able to see only surface structures, the fiber has to be prepared in such a way as to expose the features of interest. That is, in order to examine submolecular features, the chain has to lie along the plane of the substrate. In addition, only in cases where there is a high degree of order can high-resolution imaging be possible. Thus, preparation involves production of crystals, fibers, ordered layers using LB techniques, and highly oriented samples by mechanical means. In general, a polycrystalline or fibrous sample is fractured and/or cleaved with an ultramicrotome along an appropriate direction. Embedding in epoxy resin prior to cleaving serves to keep the sample in place. Depending on the polymer, other approaches may be utilized. The polymer samples can be crystallized directly on a flat substrate such as mica, or epitaxially crystallized on a substrate that is later removed [126]. Solvent casting is one way of depositing an easily crystallizable polymer, such as polyethylene oxide or polycarbonate. This approach, while convenient for studies of mesoscopic features, is unlikely to be of much use in high-resolution studies because the long axes of the chains are typically not along the plane of the surface. Moreover, the resulting crystalline surface may be quite rough, as well as randomly oriented. Highly oriented samples may be formed by mechanical stretching, pulling during crystallization, or with the use of an elongational flow [135].

1. *Image Enhancement for Submolecular Resolution Studies*

In high-resolution imaging, image interpretation is not straightforward. In most cases, one relies upon comparison with structures that are expected based on either bulk data or molecular models. Further complication exists due to the nature of the imaging process itself, as well as its technical limitations. As discussed in Section II.C, the forces that operate between the tip and sample during the imaging can be quite complex, and questions have been raised as to whether the AFM is in fact capable of resolving single molecules. Certainly, subnanometer scale corrugations are observed at the surface, which presumably are related to the molecular arrangement. Noise from sources such as external vibrations, thermal drifts, acoustic mechanisms, as well as the motion of the sample and complex tip–surface interaction are superimposed upon these corrugations. In a majority of cases, image enhancement procedures are employed to emphasize the desired phenomenon.

Amplification of signal over noise as well as curve-fitting and smoothing procedures for data interpretation are important aspects of any measurement. The AFM imaging is no exception to this. The machine itself can incorporate real-time low-pass and high-pass filters, and gain settings that control the response of the feedback loop, which could enable the operator to increase the signal quality while the data is being collected. While artifacts may arise from inappropriate choices of settings, an experienced user can detect these.

After an image is captured, it can then be subjected to a variety of digital processing to highlight the desired features. The simplest one involves the choice of lighting, view angle, contrast, and false color for presentation: This is done for all images reported. The only danger here is that beautiful pictures are seductive, and one may sometimes lose track of what makes up the actual measurement. Various filtering procedures may subsequently be employed to get rid of noise and to enhance the image. The simple ones involve digital high-pass and low-pass filters, where the low- and high-frequency ends of the spectrum are eliminated. Their removal can be justified since under typical conditions these frequencies correspond to most of the external noise, including vibrations and drifts, as well as the sampling frequency. Plane fitting is also sometimes employed, to correct for tilts or bows of regions within the sample. A flattening routine may be utilized to reduce spurious noise.

The less straightforward filters involve examination of the image in frequency space, which can be informative, and the removal of certain frequencies that appear in the spectrum. When the Fourier transform (FT) in the X and Y directions are taken separately, it is clear that there are high-frequency components in the Y direction, which is the slow direction of the raster scan, that are not present in the X direction. These components indicate the slight drift from one line scan to the next, and may justifiably be filtered out.

A fast Fourier transform (2DFFT) procedure can be used to obtain the 2DFT of an image. For samples of high symmetry, measurements of distances can be facilitated in the Fourier image. Furthermore, minor changes in symmetry are sometimes quite apparent in this representation, as exemplified by studies of annealing in LB films [108]. When one wants to highlight features in these high-symmetry cases, filtering is simple since the important frequency components are unequivocal in the Fourier image. One then simply puts a passbox around the high-intensity spots, and rejects all other frequencies. The enhanced image is obtained in a subsequent reconstruction by inverse transformation. Note, however, that the filter window is imposing a very sharp cut-off on the allowed frequencies, and variations in the size and shape of the passbox may result in slightly different images upon reconstruction. This is particularly problematic if there is a substantial amount of background noise present, as well as when dealing with a sample complex enough such that there are numerous frequency components present in the Fourier image. In these cases, the choice of filter windows becomes equivocal and the judgment of the operator can have an enormous effect in the final reconstructed picture: symmetries and periodicities that are not present in the original image may be imposed unknowingly. While no new information should have been added in the image enhancement process, pictures are so seductive that one can be led to believe that the reconstructed image is showing something not evident in the original. A further problem with a filter that relies on frequency selection is that any defect that is originally present will be obscured or smeared out in the enhanced image.

The autocorrelation pattern (AP) provides a measure of the correlation of intensity peaks in the 2D Fourier image of the sample. It is calculated by taking the product of the 2DFT of the image and its complex conjugate, then performing the inverse transformation to produce a pattern in real space. The uncorrelated (random) signals, which are presumably due to noise, are obliterated in the process. This operation is well defined in that it does not involve the judgment of the user, unlike the 2DFFT filter. However, while the autocorrelation is useful for measurement of distances and symmetries, and can provide an idea of the degree of order, one should be careful to note that the pattern obtained after image reconstruction may have very little to do with the real image. The nature of the operation smears out individual events and imposes a symmetry on the reconstructed picture.

In general, image enhancement techniques when used properly can aid in elucidating features in an AFM image; however, they must be used with care, keeping in mind that no new information can be added by the enhancement process. Reports of results at high resolution that have been processed should also include the original images, and the image in Fourier space with indications of the filter boxes used (as in Fig. 1.19), which should all be examined together. Image enhancement by digital processing is, after all, akin to curve fitting of numerical data, albeit at a more complex level; therefore, the reader should have a way of evaluating the "goodness of fit," for which there is no simple measure.

For studies at mesoscopic scales (features >1 nm), processing is unnecessary.

2. *Image Interpretation*

A further challenge for high-resolution imaging of polymers is the interpretation of the obtained subnanometer periodic corrugations in terms of atomic details. Unfortunately, the AFM has no spectroscopic capabilities that would enable one to distinguish a particular atom. In general, distances between features are measured, and are compared with available crystallographic data, as well as with molecular models. Since these comparisons rely on absolute values for distances, calibration of the piezo scanner in both the lateral and vertical dimensions is crucial, a point that is sometimes overlooked in some reports, where error bars are given in terms of reproducibility and not of accuracy.

Relative measurements of lateral and vertical distances may serve to aid in image interpretation with regards to orientations of the crystal plane, and assignment of submolecular features. For example, Snetivy and Vancso [131] interpreted locations of carbon and oxygen atoms in polyoxymethylene (POM) based on height profile along a longitudinal section (Fig. 1.15), and the known packing of the chains.

Model systems can provide an important aid to image interpretation. In the past, linear alkanes have been used as models for polyethylene in studies of its crystal structure. Stocker et al. [126] has similarly done so

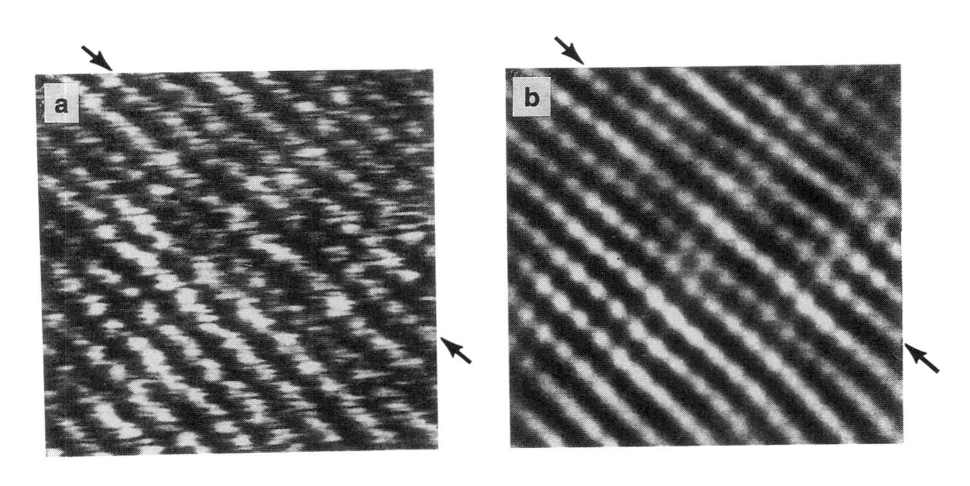

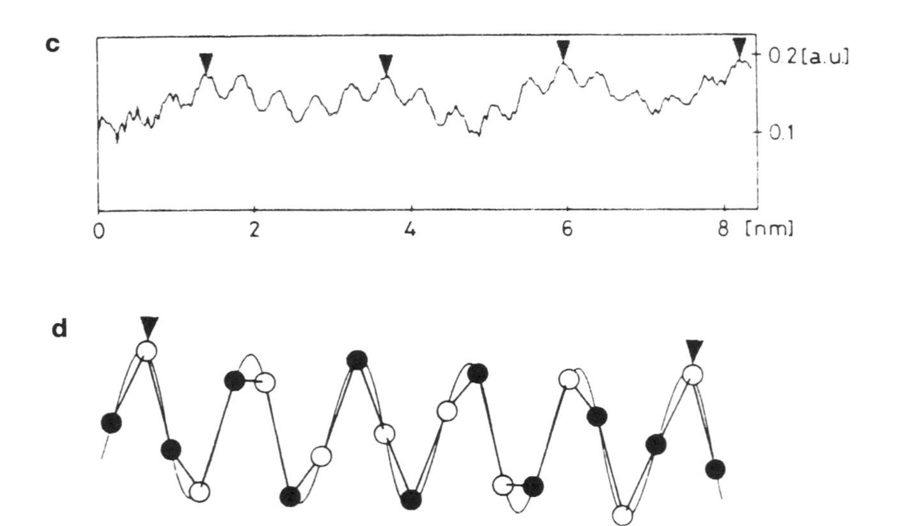

Figure 1.15. High-resolution image (7×7 nm) of POM. (a) Raw image. (b) Image after filtering in Fourier space. (c) Height profile along a section in (b) indicated by the arrows. (d) Schematic of the POM chain showing carbon (filled circles) and oxygen (open circles) atoms. [Reprinted with permission from D. Snetivy and G. J. Vancso, *Macromolecules* **25**, 3320 (1992). Copyright © 1992 American Chemical Society.]

with the AFM. In studies of linear and cyclic alkanes, they ascribed the corrugations in the AFM images to the methylene groups by comparison with known alkane crystallographic data. On this they based subsequent interpretation of images of polyethylene and iPP.

Molecular models have been employed as a guide in distinguishing between competing structures [85, 126], as well as in identifying the exposed crystal plane. While generally used in other AFM studies, this is particularly relevant to polymers wherein polymorphism is a common occurrence. Furthermore, in contrast with scattering techniques, the localized nature of the AFM measurement allows for the identification of coexisting structural forms, as well as the observation of forms that may be present only in small amounts [85].

Note that these interpretations of AFM images by comparison with other results assume that the surface structure is very similar to, if not exactly the same as, that of the bulk. To date, unlike the case of metals and semiconductors, no surface reconstruction has been reported for polymers. Whether this is a fundamental result, or simply due to the way the images are interpreted, is not clear. Furthermore, it should be reiterated that questions still arise with regards to the mechanism of imaging, which has impact on the presence of the periodic features usually identified as submolecular detail. For example, Annis et al. [128] reported observing subnanometer corrugations in polyethylene that were unidentifiable upon comparison with models. Whether this discrepancy corresponds to the existence of a surface structure that is drastically different from expected based on bulk results, or it is simply a product of complications in the imaging mechanism, is not easily determined.

B. Crystallization Patterns

The long-chain nature of polymers makes its crystallization different from that of smaller molecules. Solution-grown crystals are typically wide in two dimensions, but thin in the third, leading to the picture of molecules crystallizing in a folded-chain lamella with the backbone along the thin dimension, producing coexisting crystalline and amorphous domains (Fig. 1.16). Because a nucleation process is typically involved, the domains are organized into spherulitic structures (Fig. 1.18b), which could grow as large as a few centimeters.

The morphology of polymer crystallites have been studied extensively by optical and electron microscopy. The AFM offers the combined advantage of high resolution, especially in the vertical direction, which is difficult with EM, and ease of sample preparation. At the nanometer-to-microns length scale, the imaging is simple, minimal image processing is needed, and the features are unequivocal, except for the previously

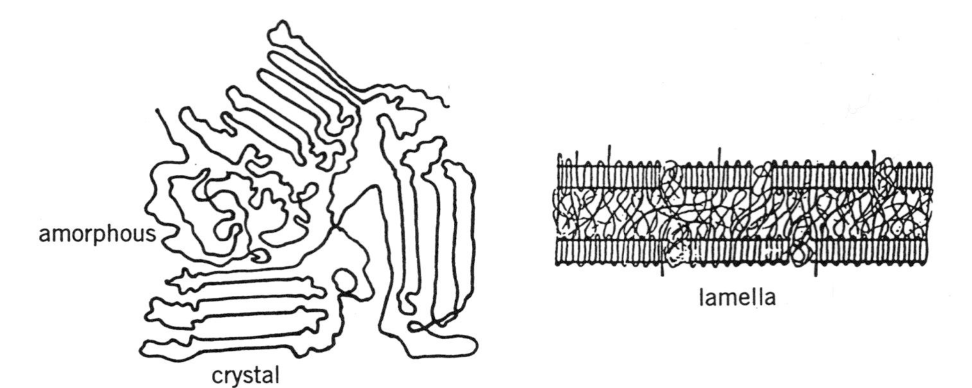

Figure 1.16. Polycrystalline polymers contain coexisting amorphous and crystalline regions. The chains fold into lamellar arrangements.

discussed tip convolution effects. At these length scales, the AFM could make an enormous impact.

The utility of the AFM was immediately evident in the first measurement reported on dendritic crystals of polyethylene. Patil et al. [136] found a spiral terrace near an apex of a basal lamella (Fig. 1.17), which upon looking at the cross-section, indicated a three-layer thickness, at a

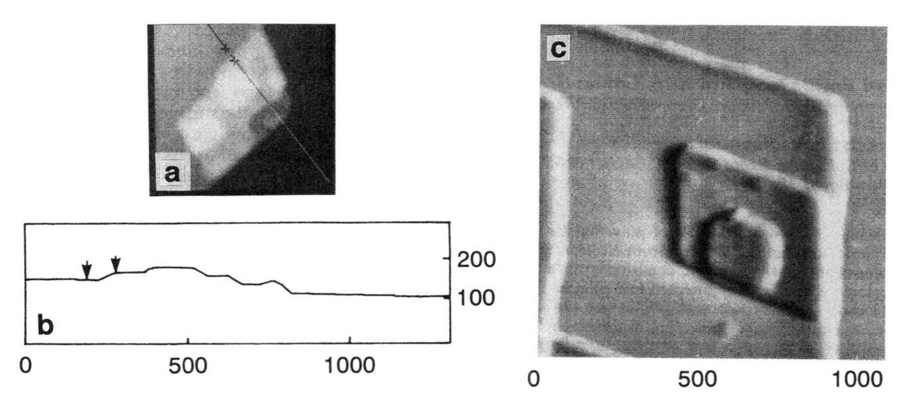

Figure 1.17. A polyethylene crystal showing the area near an apex of a basal lamella that contains a three-layer thick spiral terrace. Height profile along the indicated direction in (a) is shown in (b). (From [136], by permission of the publishers, Butterworth Heinemann Ltd. ©.)

resolution that is difficult to ascertain by EM. From this they were able to obtain a lamellar thickness of 15 nm.

Lamellar features, spherulites and dislocations produced by crystal growth, have since been reported on other systems [95, 131, 137, 138], some examples of which are shown in Fig. 1.18.

C. Specific Examples

Several polycrystalline materials have been imaged with the AFM at varying degrees of resolution, a number of which have been analyzed in detail. Some examples are presented here.

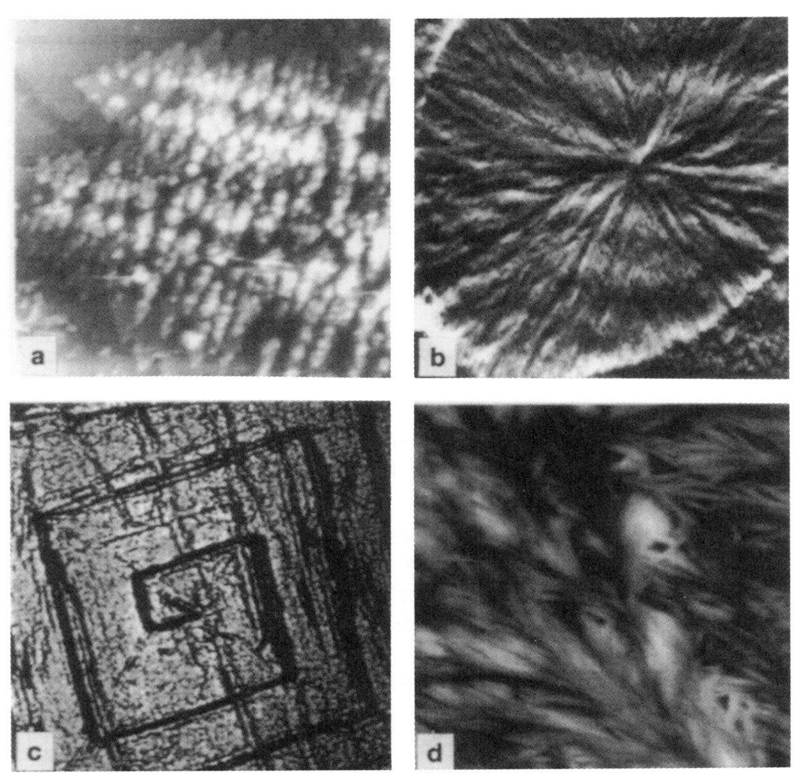

Figure 1.18. Morphology exhibited by various polymers. (a) Dendritic pattern in polyethylene (size: $7 \times 7\ \mu$). (b) A spherulite in isopropylene (size: $15 \times 15\ \mu$). (c) A screw dislocation in a polyethylene oxide (PEO) lamella (size $4 \times 4\ \mu$). (d) Solvent cast PEO (size: $3 \times 3\ \mu$).

1. *Isotactic Polypropylene*

Isotactic polypropylene is a commonly used synthetic polymer. The term "isotactic" refers to an arrangement wherein the methyl groups are all located on the same side of the carbon-chain backbone in the crystalline domain. The backbone traces a helix in 2D, which makes a full turn after every three monomer units. Two different crystal forms are known, labeled α and γ, with the former, a monoclinic crystal, being favored for high molecular weight polymers [126]. The known details about the iPP structure serve to make it a good test case for the AFM. In turn, the AFM can provide information regarding the surface layer, specifically with regards to the chirality of the helical arrangement of the chains at the interface.

The STM has been used to look at iPP after the deposition of a platinum–carbon coating to make it conductive [139]. These images showed surface roughness and features that correspond to lamellae, but not molecular detail. High-resolution images have been obtained by AFM for two different sample preparations, oriented [102, 104], and epitaxially crystallized [126].

The oriented samples were made by stretching the iPP in a tensile test instrument, and then cut with an ultramicrotome in planes containing the orientation direction. Epitaxial crystallization was done on a benzoic acid or nicotinic acid substrate, which was subsequently dissolved to expose the iPP contact face. These samples were then studied by the AFM, and by EM after further preparation (either shadowing with Pt/C, or Au decoration followed by backing with a carbon film, floating on water, and mounting on copper grids).

Lower resolution images were first obtained, which aided the interpretation at high resolution, since these provided information about the chain orientation direction. By performing EM, AFM, and electron diffraction on the iPP face originally in contact with the acid substrate during epitaxial crystallization, Stocker et al. [126] concluded that the surface corresponds to the (010) crystal plane and consists of left-handed helical chains, although the analysis of the data is not obvious. Snetivy and Vancso, on the other hand, utilized computer visualization to ascertain the crystal plane as (110) for the oriented sample. Examination of the images by autocorrelation (Fig. 1.19c) and Fourier filtering (Figs. 1.19b and f), in comparison with molecular models (Figs. 1.19d and e) led them to conclude that right- and left-handed helices are both present at the free-surface of oriented iPP.

Activation of the iPP surface by treatment with a corona discharge improves its adhesion characteristics and is thus widely used in technical

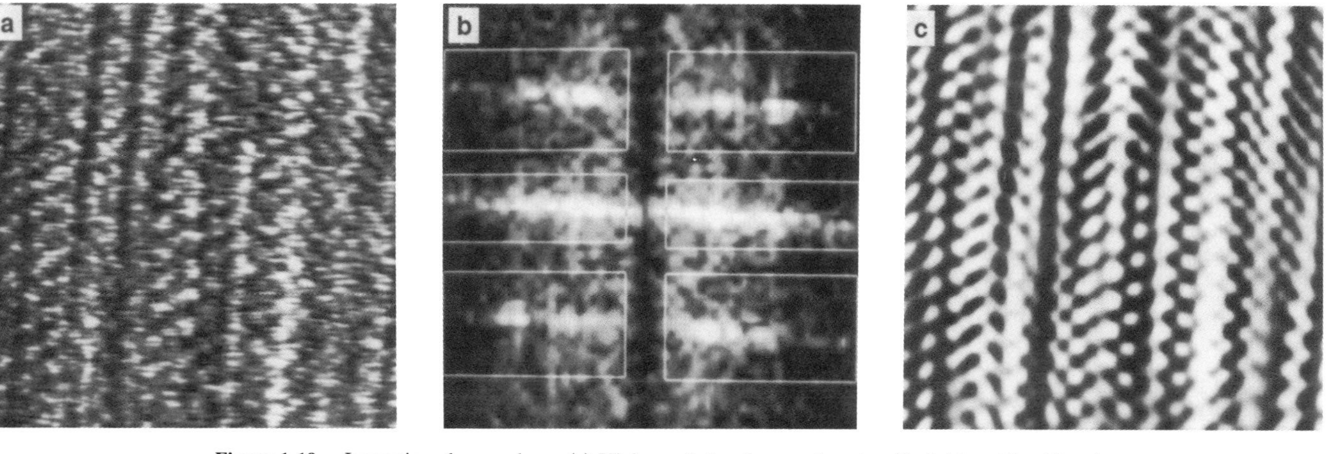

Figure 1.19. Isotactic polypropylene. (a) High-resolution image of a microfibril (size: 10×10 nm). (b) The 2D of (a). The filter boxes used to obtain (c) are indicated. (d, e) Mirror images obtained from computer visualization of the (110) crystal facet. (f) A 2.7×2.7-nm enlargement of an area in (c), showing contrast that could be attributed to left- and right-handed helices. (From [104], by permission of the publishers, Butterworth Heinemann Ltd. ©.)

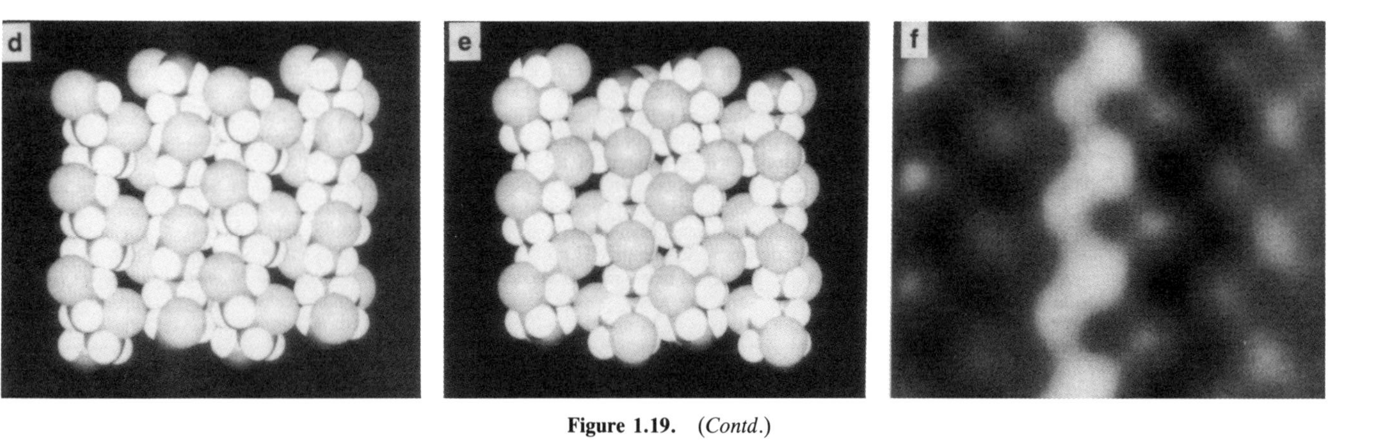

Figure 1.19. *(Contd.)*

applications. Overney et al. [140] used the AFM to monitor changes in the iPP surface morphology as a function of energy dosage. In uniaxially oriented samples, droplets were formed at the surface beyond a threshold value, the sizes of which increased with dosage. These results were correlated with measurements of the force for peel-off of self-adhered films, as well as vibrational spectroscopy for characterization of chemical composition.

2. *Poly(p-phenylene terephthalamide)*

Poly(p-phenylene terephthalamide) (PPTA) fibers, known commercially as Kevlar, possess ultra-high modulus and tensile strength, and thus has found use in a variety of demanding applications. The structure of PPTA associated with different forming conditions has been studied extensively [141], driven in part by the desire to improve its properties through understanding the source of its high-mechanical strength.

Studies using wide-angle X-ray diffraction (WAXD) have shown that PPTA fibers consist of extended, highly oriented chains existing in two different modifications of the crystal structure [142]. High-resolution images by AFM [85] verified this polymorphism. Images obtained of annealed and "as is" spun fibers manifested different submolecular arrangement, especially with respect to the configuration of the phenylene rings, the assignments for which were made by comparison with results from WAXD and computer simulations (Fig. 1.20). It appears that a third modification of the crystal structure, which was predicted by simulations but not previously found experimentally, exists in the as-spun fibers.

The supramolecular organization of the fibers play an important role in the unique properties of PPTA, and various models exist based on electron and optical micrographs, a summary of which is given in [143]. Li et al. [144] examined the fibrillar arrangements in PPTA, for samples that were microtomed along the fiber axis after epoxy embedding, and at 45° with respect to the axis (Fig. 1.21). The latter samples, which were expected to lead to fractured fibers, showed a banded pattern of about 200–300-nm period in the core region (Figs. 1.21b and c). These were interpreted as corresponding to molecular crystallite edges, which may be the weak point that is vulnerable to mechanical damage. The ease of imaging at this length scale should enable the correlation between submicroscopic details and the properties of the fibers; this is potentially useful, but has not yet been carried out.

3. *Polytetrafluoroethylene*

Polytetrafluoroethylene (PTFE, Teflon) is a highly crystallizable polymer, which has a tendency to form extended chain crystals. It has been shown

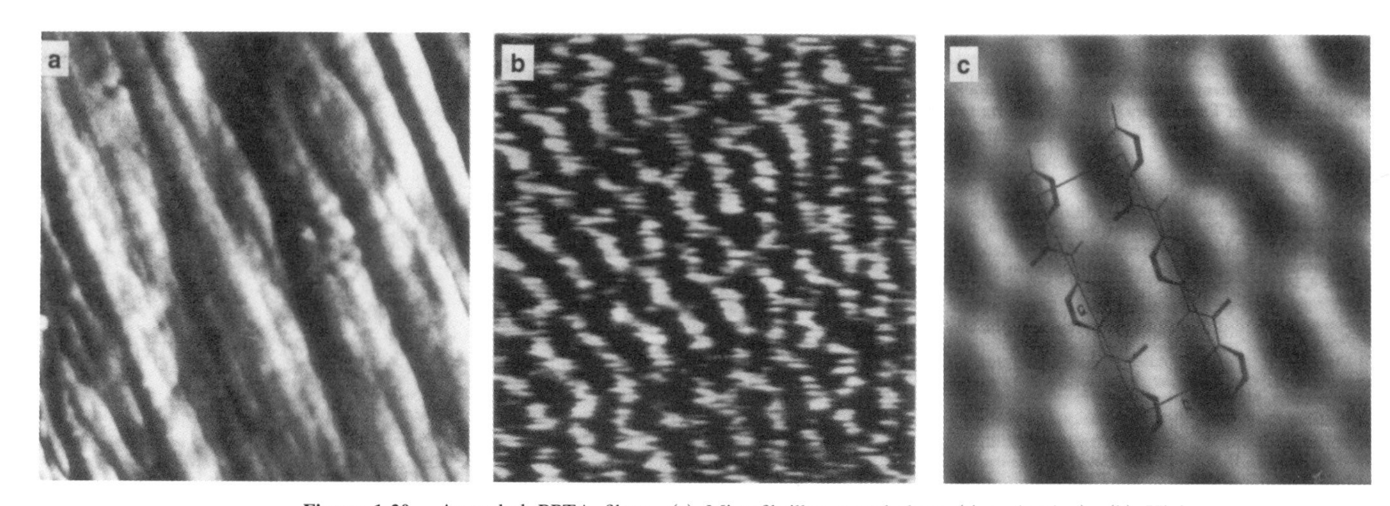

Figure 1.20. Annealed PPTA fibers. (a) Microfibrillar morphology (size: $1 \times 1\ \mu$). (b) High-resolution image (5×5 nm). The expected chain direction is indicated by the arrows. (c) Image after application of filters to the 2DFT; overlay shows the corresponding facet of the unit cell. (From [85], by permission of the publishers, Butterworth Heinemann Ltd. ©.)

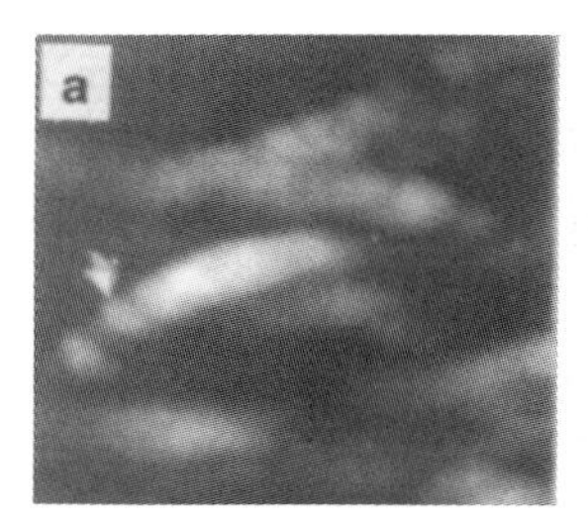

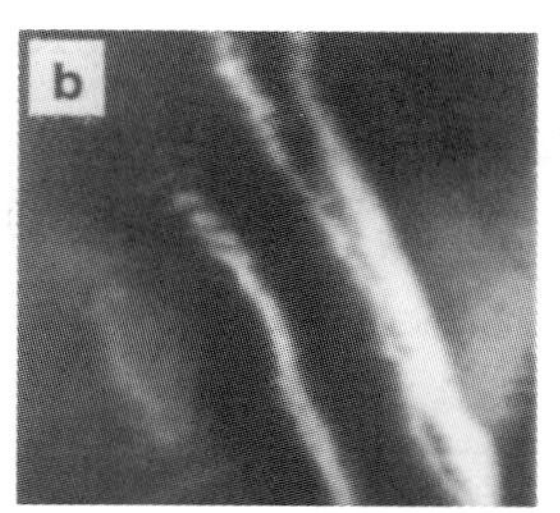

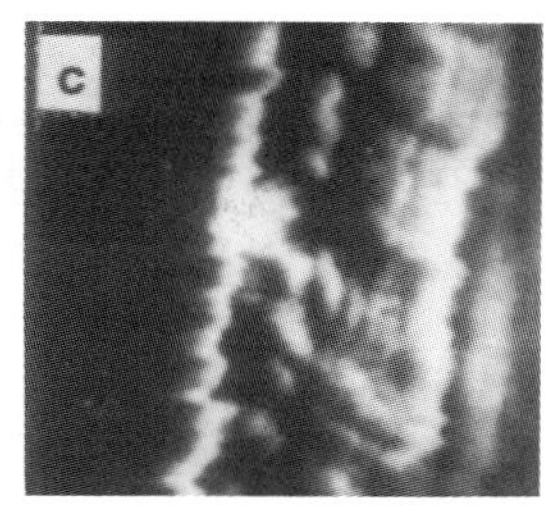

Figure 1.21. Supramolecular structure in PPTA fibrils. (a) Image of several fibrils near the core of the fiber, exhibiting a pseudoperiodicity of 30–50 nm (size: $1.05 \times 1.05\ \mu$). (b) A 45° section of a fiber showing the skin-core structure (size: $14.5 \times 14.5\ \mu$). (c) The core fiber. Bands of segmented fibrils of about 200–300 nm length and about 20–30-nm diameter are observed, which may be due to fracture of the fibrils at specific planes along the axis (size: $0.85 \times 0.85\ \mu$). (From [144], by permission of the publishers, Butterworth Heinemann Ltd. ©.)

that a PTFE surface can serve as a substrate for oriented growth of layers deposited upon it [113]. The PTFE surface is prepared as a mechanically deposited thin oriented layer, which is formed by dragging a piece of material on a smooth glass surface kept at about 200 °C at a controlled speed and pressure. Polytetrafluoroethylene oriented in this manner has been studied with the AFM [95, 114], from subnanometer to micron scales, to determine the source of its orienting capabilities. At low resolution, ridges parallel to the direction of rubbing were found. The spacing between ridges varied from less than 25 nm to more than 1 μ. At higher resolution, periodic features with distances corresponding to intermolecular spacings in the PTFE crystal lattice were observed. These results were in agreement with electron diffraction studies [113], and provided a clue about the orienting properties of the PTFE surface. One advantage that the AFM could provide with regards to this issue, is the capability of probing local structure, possibly relating it to the growth of the subsequent oriented layer. Such a systematic study has not been pursued.

IV. BLOCK COPOLYMERS

A diblock copolymer consists of two polymer sequences joined by a covalent bond. Since the two blocks may only be partially miscible, their coupling can result in microphase separation. The bulk morphology of microphase separated diblock copolymers have been studied intensively, both experimentally and theoretically, a recent review of which is given by [9].

Optical and transmission electron microscopes are the typical tools, and until recently, the surface morphology had been difficult to examine since the conditions for operation of traditional scanning electron microscopes (SEM) have been unable to provide fine enough contrast. This is in part due to the harsh conditions for SEM, such as deposition of a coating on the sample, and the use of an approximately 20-keV beam, which could result in sample damage or surface melting in some cases. The presence of a surface, be it a solid substrate or the free surface, during the microphase separation process can play a role in the resulting morphology by a preferential interaction with one of the blocks, leading, for example, to the formation of lamellar microdomains parallel to the interface.

Diblock copolymers where the two parts are noncrystalline, such as polyisoprene–polybutadiene, have been examined extensively using TEM, the results for which inspired the theoretical description of structures found during microphase separation, which is described in [9]. Depending on the relative length of the blocks, domains of spheres, cylinders, lamellae, and ordered-bicontinuous double diamonds are formed. These bulk results indicate interesting morphological patterns at the 1–10-nm scale, ideal for AFM studies. Nevertheless, only a few examples have been reported so far [96, 98, 145–147].

Studies of the polystyrene(PS)–polybutadiene(PB) diblock system [96, 145] illustrate the use of complementary microscopic techniques. The TEM relies on contrast enhancement by deposition of stains, OsO_4 in this case, which identifies the location of the PB block by oxidation of the double bond. This raises a question regarding the effect of staining on the morphology. Schwark et al. [96] addressed this issue by comparing stained and unstained samples using the AFM: The results indicated that staining has no obvious effect on the morphology. Conversely, TEM aided in explaining why AFM images of some samples manifested surface structures that correspond to lamellae, while others did not. Analysis of the vertical section of the same samples by TEM indicated that in the former, the lamellae were oriented perpendicular to the surface, while in the latter case, they were parallel (Fig. 1.22).

Diblock copolymer films can show complex changes in morphology as the chain–chain and chain–solvent interactions are varied. This is particularly true in the case where one block is crystallizable, while the other is not, an example of which is the polyethylene oxide (PEO)–polystyrene (PS) system, which is commonly used for coatings because of its amphiphilic character. The PEO block is crystallizable and hydrophilic, whereas the PS block is amorphous and hydrophobic. There are thus more possibilities during the crystallization process, and varying the

a

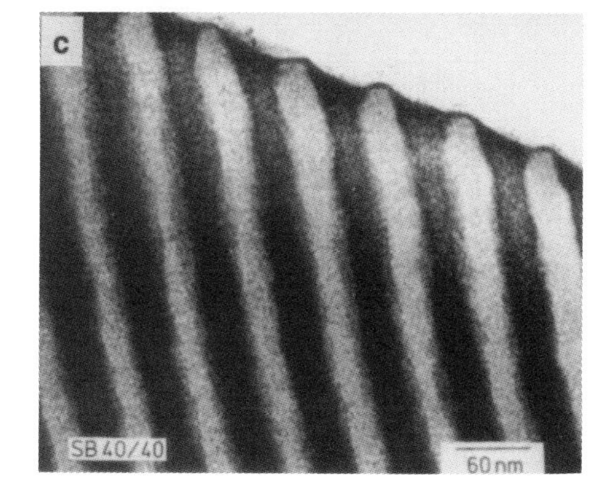

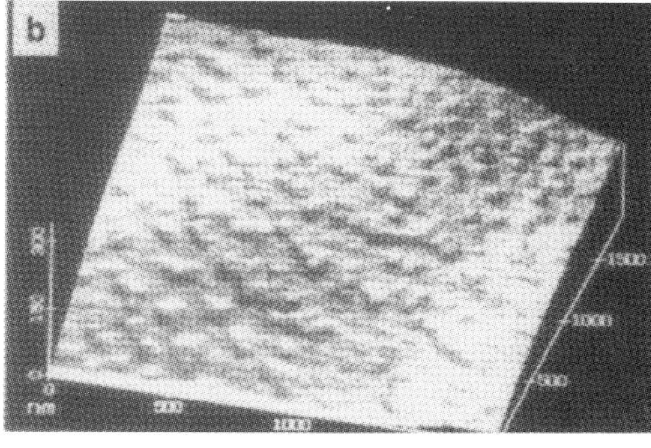

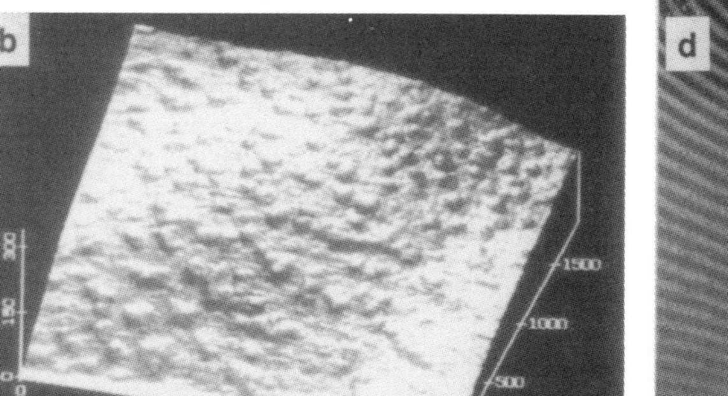

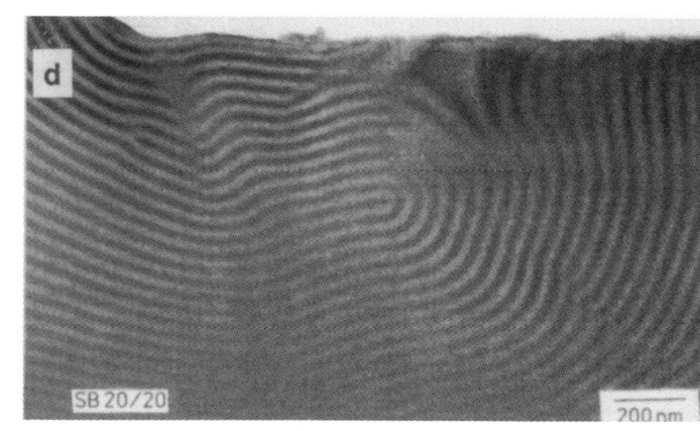

Figure 1.22. Diblock copolymers of styrene and butadiene, showing a comparison of AFM (a, b) and EM results (c, d). Both samples are about 55 vol% polybutadiene and show bulk lamellar structure. The polystyrene block has a molecular weight of 42,000 for the sample in (a) and (c) and 20,000 for that in (b) and (d).

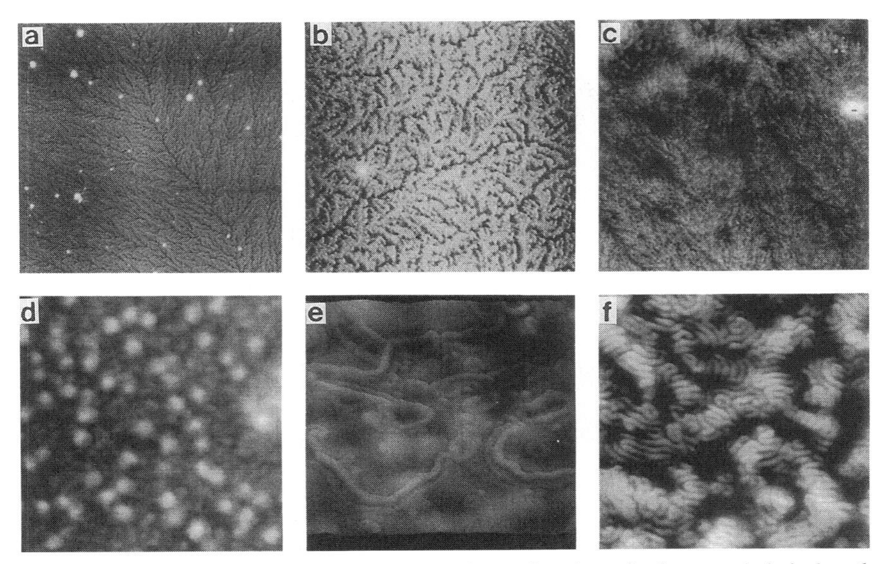

Figure 1.23. Various morphologies obtained as a function of solvent and chain length for PS–PEO diblock copolymers. (a) 30 wt% PS cast from nitrometane ($15 \times 15\ \mu$). (b) A 30 wt% PS case from nitromethane casting ($6 \times 6\ \mu$). (c) A 30 wt% PS cast from chloroform ($5 \times 5\ \mu$). (d, e, f) A 50 wt% PS, case from chloroform.

relative chain lengths of the blocks, as well as the solvent from which the film is cast can have profound effects on the morphology. Examples are shown in Fig. 1.23. The structures obtained for this copolymer include a featureless film, spherulites, dendrites, fibrillar patterns, micelles, and an intricate pattern of tangled tubule-like structures [148]. These latter structures, which were found when a nonpreferential solvent for the two blocks is used, may correspond to either distorted lamellae or tubular micelles.

The blocks can also be chosen so as to enable the spreading of a copolymer as an insoluble monolayer at the air–water interface. Compression of the monolayer may then result in the formation of 2D aggregates called "surface micelles" [149]. The micelles exhibit polymorphism, creating spherical, rodlike or planar structures, depending on the length of the blocks. Li et al. [98] visualized spherical and tubular domains in the polystyrenealkylatedpoly(vinyl-pyridine) diblock system, which were prepared first as a spread monolayer, then transferred to form an LB film (Fig. 1.24).

The AFM has also been used to study the behavior of a cast thin film of a diblock copolymer of polybutylmethacrylate–polystyrene (PBMA–

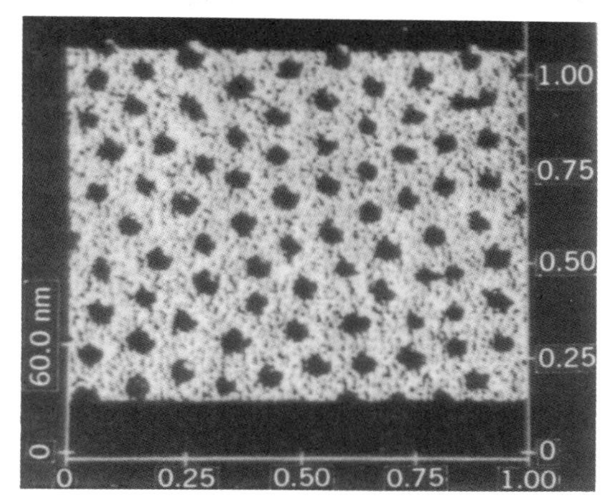

Figure 1.24. Surface micelles formed by a spread monolayer of a diblock copolymer, deposited as a LB film. [Reprinted with permission from S. Li et al., *Langmuir*, **9**, 2243 (1993). Copyright © 1993 American Chemical Society.]

PS) [146] upon annealing. This system is expected to eventually form an ordered lamellar arrangement upon heating past T_g. Since the lamellae would be oriented parallel to the surface, they will not be visible by AFM; however, Collin et al. [146] argued that the root-mean-square (rms) roughness from the AFM images provided a measure of the ordering, and will reach a plateau once lamellar order is established. With the use of the rms roughness, as well as the FT of the image, they were able to follow the evolution of the film morphology.

V. FILM FORMATION

Polymer films are widely used to provide coatings with desirable mechanical, chemical, or optical properties. The conditions under which it is formed governs the film morphology, which in turn is strongly related to the properties of the resulting film. While other microscopy techniques have been used in investigation of polymer films [11], the AFM offers an advantage of higher resolution and, due to its relatively nondestructive character, the capability of repeated imaging during the course of film formation.

A. Spin Coating

One of the most widely used processes for the preparation of smooth films and coatings is spin coating. This process is accomplished by the application of the solution of the desired coating material onto the chosen substrate, which is then spun at about 2000 rpm. Various parameters, such as the rate and time of spinning, the solvent used, and the concentration of the initial solution, can affect the properties of the resulting film. These parameters have been addressed empirically, although the mechanism of the spin coating process is still not understood.

Strange et al. [150] used the AFM and STM to examine the formation of a film of polystyrene spin coated on a silicon substrate, as a function of the molecular weight and the initial concentration (Fig. 1.25). At very low surface coverages, structures were seen of a size that could correspond to individual polymer molecules. At slightly higher concentrations, a network pattern that looks like a 2D foam structure was observed. An analysis of the edges and angles of the polygons in this network indicated that they can be represented by Voronoi tasellation patterns, which has been used to describe structures found in microemulsions and in insoluble monolayers at the air–water interface [151]. Voronoi patterns can be generated by starting with randomly located points on the surface, which are then allowed to expand simultaneously as circular regions. Contact

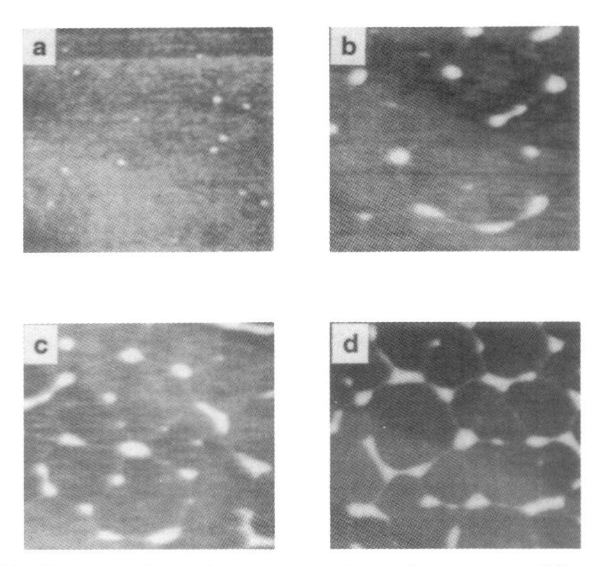

Figure 1.25. Patterns obtained by spin coating polystyrene at different initial concentrations. (a) 0.005 wt%. (b) 0.01 wt%. (c) 0.02 wt%. (d) 0.025 wt%, showing a tasellation pattern. A continuous film is formed at concentrations higher than 0.2 wt%. [Reprinted with permission from T. G. Strange, R. Mathew, D. F. Evans, W. A. Hendrickson, *Langmuir* **8**, 920 (1992). Copyright © 1992 American Chemical Society.)

between regions result in deformations into polygons. With these considerations, Strange et al. [150] concluded that there is a minimum thickness for the formation of a continuous film by spin coating. Below this thickness, rupture occurs simultaneously across large areas, producing the tasellation pattern.

A further increase in polymer concentration produced a thick layer, and eventually a defect-free film. A measurement of contact angles, which is commonly used to indicate the quality of a coating, was found to approach a constant value for the contact angle at a lower concentration than what the AFM sees as corresponding to a defect-free film. This indicates that contact angle measurements are insensitive to small defects in a film, and the utility of the AFM for examination of such.

B. Wetting

The formation of a film starts with its wetting the substrate, which is dependent on spreading parameters. For simple liquids, spreading is determined by a comparison of the tension of the three interfaces (solid–liquid, liquid–vapor, and solid–vapor) using the Young–Laplace equation. The wetting–dewetting behavior of polymers, which is an important consideration for the formation of homogenous films, can be quite different than that of simple liquids, in that the spreading of the film is deemed to occur by reptation through a precursor region that is roughly the thickness of the polymer radius of gyration [152]. This result has been tested experimentally by Silberzan and Leger [153] by observing optical interference on polymer droplets.

The tasellation pattern produced by the spin coating described previously disappears upon heating the sample above its bulk glass transition temperature; instead, droplets were observed. While not presented as such by Stang et al., [150] these patterns can be viewed as corresponding to dewetting of the solid substrate by the film. In a more detailed study, Zhao et al. [154] used the AFM in conjunction with X-ray scattering to study liquid films of polyethylene polypropylene on the surface of oxide-covered silicon. A transition from wetting (homogenous film) to dewetting (formation of islands) was found when the initial thickness of the prepared film was less than the polymer radius of gyration.

In a similar manner, Liu [147] investigated the effect of the nature of the substrate on the spreading of PS homopolymer. The substrates, which consist of diblock and triblock copolymers of polyvinylpyridine–polystyrene (PVP–PS) of varying block lengths, were prepared in such a manner as to form lamellae that exposed the PS block to the spreading PS homopolymer film. A change from wetting to dewetting was observed in the AFM images as the homopolymer length was increased (Fig. 1.26). In

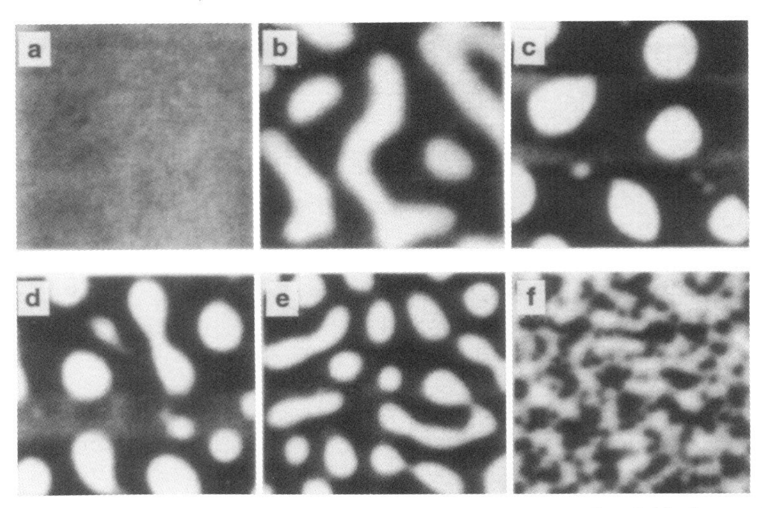

Figure 1.26. Wetting of polystyrene homopolymer on an ordered block copolymer substrate. Films are about 100-Å thick and are annealed for 24 h at 185 °C (size: $15 \times 15\ \mu$). The polymerization indexes for the homopolymer (NH) and the blocks of PVP–PS or PVP–PS–PVP are as follows: (a) NH = 650; 200–200, (b) NH = 650; 91–950–91, (c) NH = 2200; 200–200, (d) NH = 6700; 200–200, (e) NH = 6700; 510–540, (f) NH = 6700; 800–870. (From [147].)

conjunction with neutron scattering and secondary ion mass spectrometry data, these results were interpreted in terms of the interpenetration of the homopolymer and the copolymer surface.

C. Latex

Latex is a general name given to a colloidal dispersion of polymeric spheres. There are generally two kinds, "hard" and "soft" latex, which differ in the constituent material, leading to different film-forming characteristics. Colloidal polystyrene, whose use as a calibration standard for AFM was discussed previously, is an example of hard latex. When deposited on a substrate at high coverage, aggregates of the particles may be formed; however, the integrity of each particle is retained. With soft latex, solvent evaporation and contact of particles with each other produce deformations and coalescence, eventually forming a continuous film. In this section the discussion is confined to the film formation of soft latex.

Soft latex forms the base for most paints, and the process of film formation is of enormous industrial significance in the production of coatings. While numerous theoretical and experimental work exist on the subject, the coalescence of latex particles to produce the resulting film still provides interesting unsolved questions, and involves a variety of considerations [155]. At the molecular level, diffusion of polymers across the interface between particles is an important aspect of coalescence, and has been addressed by elegant experiments involving fluorescence techniques [156]. From the macroscopic level, particle deformation and wetting ideas are important. Electron microscopy [157] and light scattering [155] have been employed in these studies. From the technological point of view, there are desirable film characteristics, such as transparency, strength and resistance to fracture, and adhesive properties, that are dependent on the film formation process.

The formation of the film starts with the evaporation of the solvent. Under certain conditions (low salt, no ionic surfactant, and monodispersed particles), it was shown that spontaneous ordering into a face-centered cubic (fcc) arrangement may take place prior to evaporation [155]. As the water evaporates, the particle–particle distance diminishes, until they come into contact. Further evaporation drives particle deformation to produce a space-filling film with a honeycomb structure (rhombic dodecahedra), remnants of the original FCC packing. The presence of salt and surfactant destroys the order, and leads to a random packing instead. These results have recently been visualized by freeze fracture electron microscopy (FFTEM) for a poly(butyl methacrylate) (PBMA) film, with and without surfactant [157].

The AFM has proven to be a very useful tool for the study of latex film-formation because (1) the size scale of latex particles (tens to hundreds of nanometer) are well within easy reach of the AFM, and images obtained requires no further processing; (2) information about the surface, which is complementary to studies of the bulk film using FFTEM, is provided; (3) the process of annealing, which is an important part of film formation, can be monitored since the AFM is relatively nondestructive in that no complex sample preparation is required (unlike FFTEM and SEM); and (4) quantitative measurements of the surface roughness and particle sizes as a function of film-forming parameters can be obtained.

Solvent casting has been the preferred procedure for depositon since it produces a much more ordered film than spin coating Wang et al. [112], as shown in Fig. 1.14. The cast films were allowed to dry slowly above the known minimum film-forming temperature (MFT): When formed below this temperature, the resulting films are not translucent, and are less robust to mechanical perturbations. The AFM images provide an explanation for this, in terms of the presence of microscopic cracks visible in

Fig. 1.27. Note the difference between this and the smooth film shown in Fig. 1.14b.

The arrangement of particles in both Fig. 1.14a and b are consistent with an FCC packing of the spheres. One unexpected result of the AFM studies is the presence of a well-ordered surface for the film in the presence of surfactant [112], in direct contrast with the bulk results from FFTEM studies [157]. Annealing studies on the PBMA film [94] were carried out by heating a sample to 70 °C for a time t, after which it was imaged again. This process was repeated several times, and the corrugations of the sample as a function of annealing time was measured. This study was possible because of the existence of large ordered regions in the film: otherwise, quantification of the corrugations would not be simple. The results, which are summarized in Fig. 1.28, showed a decrease in surface roughness with annealing time, as the particle–particle boundary became more diffused. (The correction that would be introduced by the

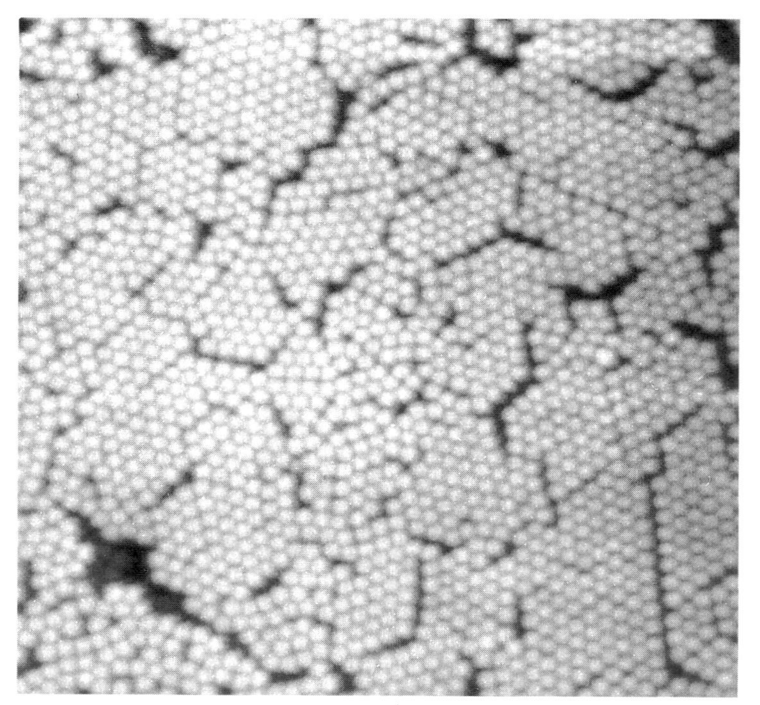

Figure 1.27. A film of PBMA latex allowed to dry below the minimum film-forming temperature has ordered domains separated by micro-cracks (size: $12 \times 12\ \mu$).

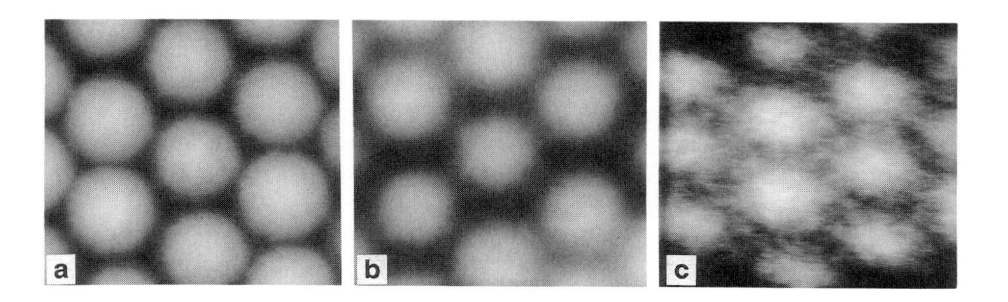

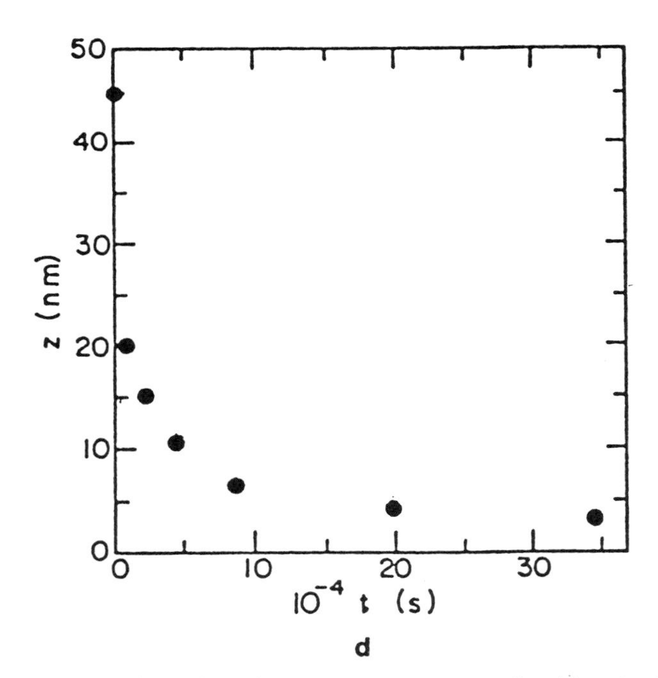

Figure 1.28. Evolution a film of PBMA latex upon annealing (size: $1 \times 1\ \mu$). (a) At $t = 0$, (b) $t = 12$ h, and (c) $t = 70$ h. The full scale of the false color is 400 nm in (a), 100 nm in (b), and 21 nm in (c). (d) As the film is annealed, the boundaries smear by polymer diffusion across interfaces. The decrease of the corrugations with time can be measured with the AFM. [Reprinted with permission from M. C. Goh, D. Juhue, G. Wang, O. M. Leung, and M. A. Winnik, *Langmuir* **9**, 1319 (1993). Copyright © 1993 American Chemical Society.]

tip size is not expected to be significant since the corrugations have very low amplitudes in comparison with the particle size). However, the size of the circular edges that correspond to the particle boundaries do not change with time. A primitive estimate of the polymer diffusion across boundaries for the surface layer was obtained, by using the data in Fig. 1.28d, which gave a value that is 4 orders of magnitude faster than found in the bulk by fluorescence measurements on the same system [156].

The role of additives on film formation is of major commercial interest, and has been studied by AFM [158, 159]. Juhue and Lang [158, 159] examined the role of the surfactant sodium nonylphenolpoly(glycol ether)sulfate postadded to the surfactant-free PBMA dispersion (Fig. 1.29). By looking at the degree of order and packing, it was concluded that an optimal surfactant concentration corresponding to a fully covered latex surface produces the best film surfaces. The quantitative comparison of images at different concentration is not straightforward because of the high degree of disorder in the films.

The mechanism of action of Texanol™ (2,4,4-trimethylpentan-1,3-diol monoisobutyrate, TPM), a paint additive known empirically to aid in the production of smooth films, has also been studied using the AFM [159]. Images obtained indicated that one role this additive plays in the formation of a film from a surfactant-stabilized precursor is to facilitate the exudation of the surfactant to the surface, which can be washed away with water.

VI. BIOPOLYMERS

Discussions of atomic force microscopy as applied to biological systems and reviews of recent literature are given in [5, 6, 33], and will not be repeated here. Instead, a few key issues will be presented briefly, and specific instances involving biopolymers other than DNA will be examined.

Since biomaterials are typically soft in comparison with crystals and metals, the imaging force could easily produce distortion or damage. From the early days of the AFM, the issue has been raised whether or not it is possible to image biological structures such as DNA [79]. Substantial progress has been made in this direction [33], and high-resolution images have been obtained for crystalline biomaterials.

The main strength of using the AFM is its capability for imaging under solution, enabling the preservation of the native hydrated structure of most biological systems. Nevertheless, appropriate conditions and sample preparation still have to be worked out for each individual case addressed. Since lateral forces during the scan tend to drag the sample and

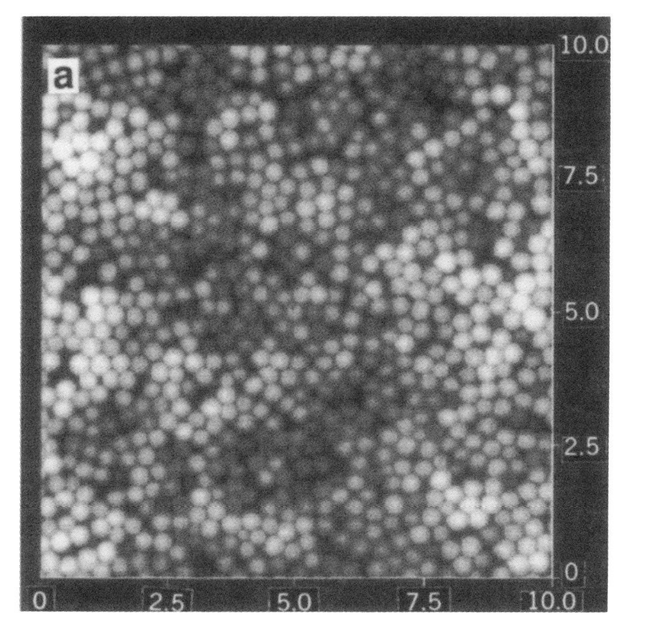

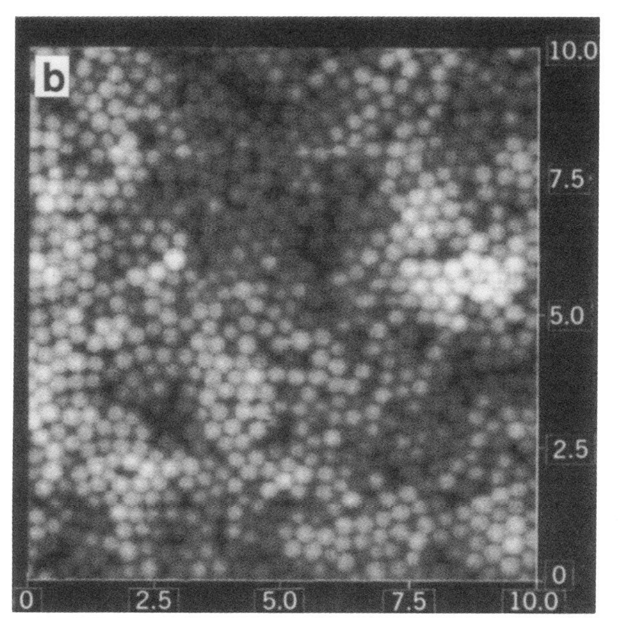

Figure 1.29. The PBMA latex film with surfactant [sodium nonylphenolpoly(glycol ether) sulfate] added to the dispersion prior to film formation, at concentrations of (a) 0.9 wt% and (b) 1.8 wt%. [Reprinted with permission from D. Juhue and J. Lang, *Langmuir* **2**, 792 (1993). Copyright © 1993 American Chemical Society.]

thus reduce the resolution, various means of anchoring the sample onto the substrate have been, and should be, investigated. Further progress in the technology, such as the newly introduced capability of performing studies in liquids with the tapping mode [34, 35], will aid in future studies.

The typical length scales involved in biological studies are such that one should always consider the effect of the finite tip width, which should be deconvoluted out of the image in order to extract appropriate dimensions, as well as to obtain a more accurate picture of the sample. This is particularly relevant because the identification of isolated structures is sometimes not simple since, unlike EM, one cannot rely on the location of stains to highlight the desired object and hide the rest.

The AFM has been particularly useful in studies concerning biopolymers such as actin [39], collagen [29, 160, 161], neurofilaments [100], intermediate filaments [162], and paired helical filaments [71, 163] for several reasons: (1) These materials are typically robust enough to withstand the imaging forces and retain their structure in air, (2) in general, they are not easily dragged by the tip, (3) their small dimension is larger than 1 nm, and (4) they are easily distinguished from other deposits by their known periodicity.

Collagen is a protein found in most connective tissues. Collagen fibrils extracted from various sources [160, 161] and assembled from the acid soluble monomer [29, 161] have been studied with the AFM. The characteristic banded pattern of the fibril that appears in EM with a period of about 70 nm are easily observed by the AFM. Additional periodicity has also been reported [161]. The evolution of the fibril from the collagen monomer has been studied using the AFM, in conjunction with traditional turbidimetry measurement [29]. Figure 1.30 illustrates stages in this assembly with an image of the mature fibril with fully developed periodicity, surrounded by smaller filamentous precursors. The AFM was found particularly useful in elucidating the early stages of assembly, before any significant change in turbidity was observed.

Actin filaments, which are major constituents of the cytoskeleton, have been studied by the AFM. Henderson et al. [39] reported the observation of actin in a living cell. It appears that the AFM tip was able to sense the more rigid actin network through a layer of the more pliable plasma membrane.

The structure of actin is well studied: The combined use of X-ray crystallography and cryo-EM provided information down to the near-atomic scale. The resolution of the AFM images of actin is nowhere comparable, although by using tapping mode under a liquid, Radmacher et al. [93] were able to clearly see the 36-nm period of the filaments.

A notable success for the AFM as a structural tool for biopolymers is

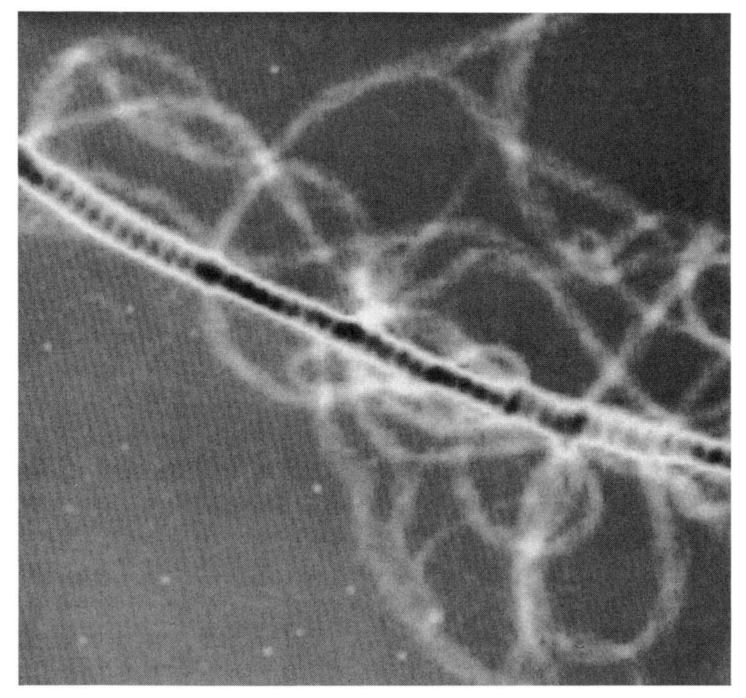

Figure 1.30. A collagen fibril showing the banded pattern with an approximate 70-nm period. Protofibrils that are precursor to the assembly are also visible.

the reported images of the paired helical filaments (PHF) of Alzheimer's disease [71, 163], previously shown in Fig. 1.12. These are polymers of the protein τ_{PHF} assembled into 12–20-nm wide filaments, showing about 80-nm periodicity. The name PHF describes the generally accepted structure, which is derived from 30 years of studies by EM. The AFM images after tip deconvolution instead revealed a twisted ribbon configuration.

VII. POLYMER STUDIES OTHER THAN IMAGING

Since the imaging process is a result of the interaction between the probe and the surface, the AFM has been employed for uses other than visualization. One aspect of this is the manipulation of materials at the surface by utilizing the finite imaging force. Another aspect is utilization of the AFM itself for measurement of properties relevant to the surface:

With minor modifications of the AFM, studies of surface forces, friction effects and elastic properties at very local levels can be performed.

A. Nanomanipulation of Polymers

Creation and control of structures at the nanometer length scale has been a major area of research, in large part driven by technology and the desire for microminiaturization. It is well known that the STM can move atoms one at a time. While the AFM has not been able to carry out structure creation at that scale, various ways of manipulating the surface have been reported. In most ways, one simply takes advantage of the weakness of the technique when it comes to imaging, by operating it at conditions under which the tip-sample interaction is large enough to cause deformation.

1. Nanoindentation and Nanodissection

Since the early days of the AFM, attempts have been made to understand the interaction between the tip and sample by examination of the behavior of the force curve. A decrease in the tip-sample separation distance beyond the point of contact results in a positive load, which is born by the sample in the typical case of a hard silicon nitride tip. The ideal response of the material under loading and unloading conditions can be characterized as either elastic or plastic, examples of which are seen in the AFM measurements in Fig. 1.35. In an ideally elastic material, the deformation as a function of load tracks the same curve whether in the loading or unloading direction. The response of an ideally plastic material, on the other hand, is characterized by a hysteresis, with the deformation persisting upon unloading. For most materials, a mixture of plastic and elastic behavior are found.

Examination of the details of the AFM force curve described in Section II.C allows extraction of materials parameters, such as hardness and adhesive characteristics. Burnham and Colton [52] described the use of an AFM as a nanoindenter, which is capable of a 1-nN force resolution and a 0.02-nm depth resolution, much higher than other types of indenters. This they tested on graphite and gold surfaces, with the use of an etched tungsten wire with a tapered cone geometry as a tip.

At sufficiently high loading, the response of the material is plastic, and the deformation produced by the loading persists after the tip is withdrawn. For soft samples, including most polymer films, forces close to the AFM operating conditions (hundreds of nanoNewton) are sufficient, and indentation is easy, and sometimes unavoidable, since a tip crashing into a soft surface will result in such.

The size of the indentation depends on the loading force on the tip, and the material used. Working at less than 10^{-7} N, Jung et al. [164] made pits 50 nm in diameter and 2.5 nm in depth on a film of polycarbonate. The depressions that can be created and distinguished must be larger than the roughness of the film, hence imposing a limit on what is possible with amorphous polymer films. On an inorganic, atomically flat thin film of MoO_3, Kim and Leiber [165] have written clear features about 10 nm wide and about 5 nm deep by controlling the applied load. By using a diamond tip, Bushan and Koinkar [59] have been able to measure indentation hardness in silicon, making pits 1 nm in depth.

If the goal is to use the AFM for data storage, nanoindentation using the above procedure suffers from several limitations, such as the piling up of debris, the amount of load possible, and the writing speed accessible. A scheme for high-speed writing on a film of polymethyl methacrylate (PMMA) has been implemented with the use of an AFM and a modulated laser beam [166, 167]. The laser beam is focused onto the tip, resulting in local heating above the PMMA softening point. When the tip is in contact with the PMMA film, an indentation is made, the size of which is determined by the tip contact area, the energy deposited, and the load on the tip. Typical features made are 150 nm wide and 15 nm deep, as seen by the AFM. (Taking into account tip convolution effects, the width is probably >150 nm. By mounting the sample on a rotating spindle, features at a repetition rate of 100 kHz were written (Fig. 1.31). With improvements in the tip geometry and the characteristics of the film, it should be possible to achieve writing speeds at megahertz frequencies with this approach.

One aim in the use of near-field techniques for nanomanipulation is the possibility of control of features at the molecular length scale. Bond breaking through the application of fields that are localized has been demonstrated with the STM. While incapable of such localized effects, the AFM has been utilized in a similar manner for "nanodissection." Scission of a DNA chain laid upon a substrate [37] and of a live cell [39] was effected by scanning the probe tip at high loading forces. Controlled dissection of a liver gap junction membrane was performed by the gradual removal of the top bilayer of the double bilayer structure through repeated scanning [168].

Nanoindentation relies on the actual contact area between tip and sample, thus surface features that are produced in this manner will be limited by the tip geometry. Since the commonly used square pyramidal Si_3N_4 tip has an apex of about 20–50-nm radius and a 2–1 aspect ratio, the minimum size of indentations will mirror such geometry, assuming no tip deformation occurs upon contact. While tips with higher aspect ratios

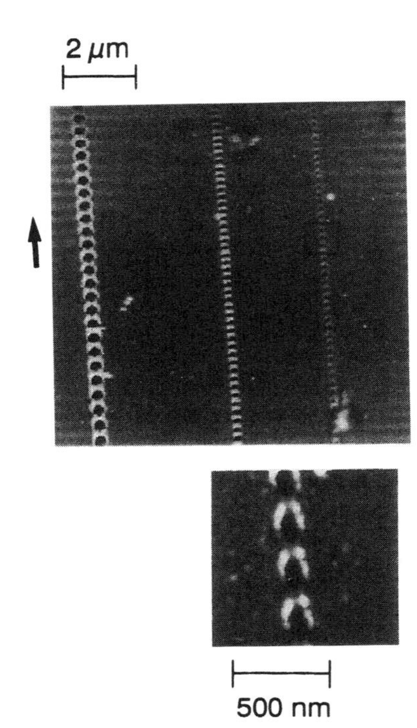

Figure 1.31. Tracks written on a PMMA film at 50–100 kHz by the AFM tip with an infrared laser focused on it. The inset shows details of the indentations, which are about 150 nm wide and about 15 nm deep. (From [166].)

are available, the apex is still a few nanometers in radius and in most cases, they are too fragile to carry a large load. Smaller features will thus have to await further developments in tip making.

An approach opposite to indentation for nanostructure creation is suggested by the molecular dynamics simulations of Landman et al. [76, 77], which show how the interaction between a Ni tip and an Au surface creates an atomic scale instability as their distance of separation is decreased prior to contact. These simulations indicate that the jump-to-contact described in Section II.C is a result of an inelastic deformation, pulling the Au atoms towards the Ni tip. Advancing the tip beyond the point of contact creates a plastic deformation (indentation), and the withdrawal produces a connective neck of atomic dimensions, which subsequently fractures. Should the structure persist after fracture, this connective neck would be of atomic dimensions. However, presumably due to the large surface energy difference between Au and Ni, or because the fracture is much too sudden, the Au atoms end up wetting the Ni tip. While these simulation results are in agreement with the measured force curves, no nanostructure has been created in this manner on the Au surface. However, this idea has been utilized for creating lumps on a polystyrene surface [169], the argument being that, due to entanglements, the polymer chains are less likely to jump into the tip in the retraction. While structures were created in this manner, it was noted that as the height of such increased, so did the width, probably mirroring the tip geometry.

2. *Use of Lateral Forces*

While the previous section relies on forces in the Z direction, the process of scanning involves forces in the lateral direction that can also be utilized to create surface structures. The terms nano-bulldozing and nano-ploughing have been used to describe this effect. Basically, one starts with a soft sample, usually made of polymers or biopolymers cast on a substrate as a thick film, and utilizes the AFM scanning action to sweep material, piling them up at the edges before the scan changes direction, where the lateral force is zero. In this manner, Jung et al. [164] were able to write lines that are 70 nm wide on a polycarbonate surface. Brumfield and co-workers [170, 171] described the use of nanoploughing as a possible tool for microelectrode fabrication, by first creating a hole in an electrochemically deposited film of poly(phenylene oxide), which was subsequently filled with either a dielectric or a conductive polymer.

The process of scanning, however, can result in structures more complicated than a simple piling up at the edges. Repeated scans on the same area of the surface of polystyrene films resulted in the production of a periodic pattern oriented perpendicular to the scan direction [169], as shown in Fig. 1.32. This bundling of polymer molecules becomes better delineated with increasing scan time. Further scanning produced aggregation of the bundles, and only much later in time are the materials piled at the edges of the scan window. The mechanism of formation of these patterns is not fully understood, but they have been observed in a variety of other systems [90, 170, 172]. Meyers et al. [172] speculate that these patterns are related to the macroscopic periodic patterns that occur in stretched rubber, known as Schallamach waves. While the molecular weight dependence of the periodicity is unclear, the pattern is not formed below the entanglement molecular weight in the case of polystyrene [148, 172]. The need for some mobility at the molecular level appears obvious, and was verified by cross-linking experiments using polymethylstyrene, where it was found that the pattern could no longer be generated after prolonged exposure to UV light (Fig. 1.33). Nevertheless, no dependence on the bulk glass transition temperatures of the polymers was observed: Similar patterns were created on polystyrene, polymethyl styrene, polymethyl methacrylate, and polyacenaphthalene [148].

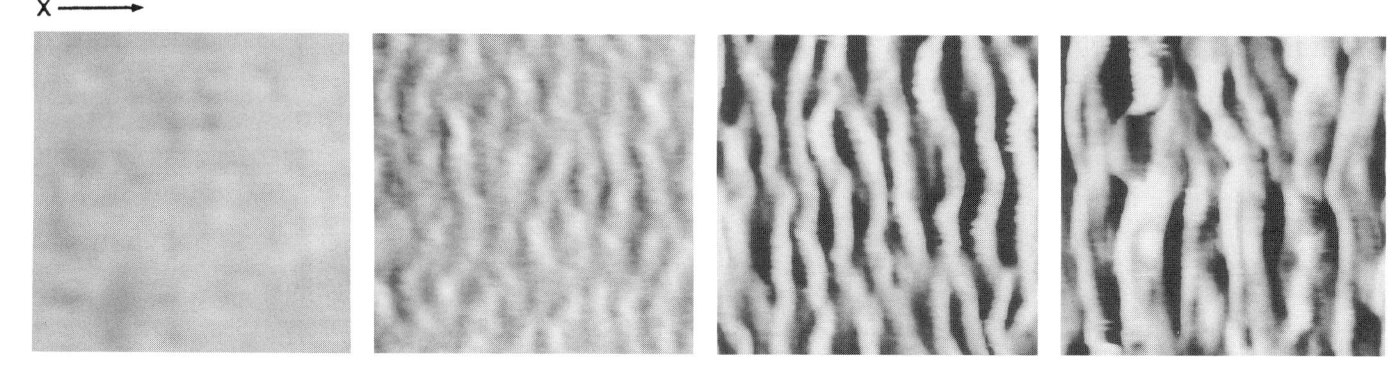

Figure 1.32. Patterns formed by scanning the tip across a solvent-cast polystyrene film. The direction of the slow scan, indicated with an arrow, is perpendicular to the periodic structures. The patterns get more distinct with increasing time of scanning (shown are $t = 0$, 1 min, 5 min, and 2 h). (Data taken from [169].)

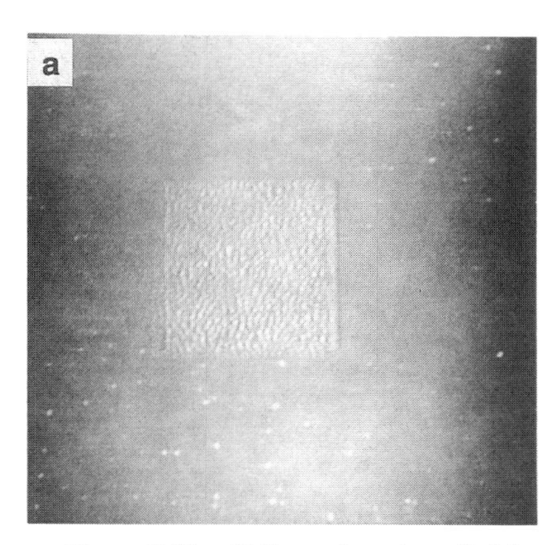

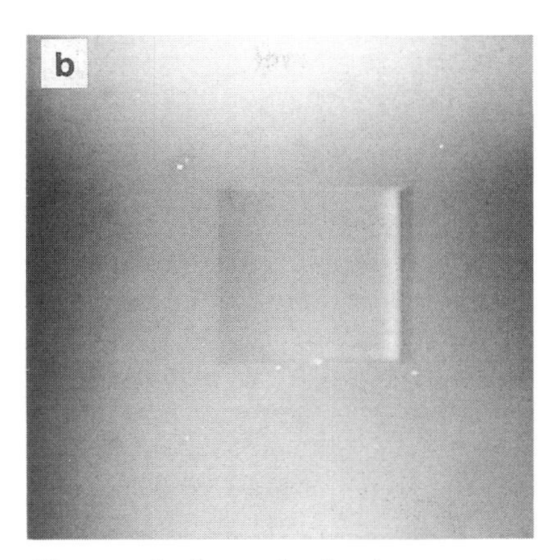

Figure 1.33. Patterns in polymethylstyrene. The area in the center has been scanned for (a) a cast film (5 min) and (b) a UV-irradiated film (30 min). The periodic patterns do not form in the UV-irradiated film.

A weakness in the use of the AFM for surface manipulation is the absence of an external parameter that can be controlled easily, both in magnitude, duration, and location. Unlike STM, where the field at the tip serves as such, forces cannot be turned on and off at will. Furthermore, the long-range nature of forces is such that the tip geometry inevitably plays a major role in the structures that are created.

B. Measurement of Surface Forces

The force between two objects depends on the distance of separation in a manner that is dictated by the geometry of the objects, and the nature of the intervening medium, an excellent discussion of which is given by Israelachvili [16]. For a few simple cases, analytical expressions for the surface forces, which include the dielectric properties of the objects and the medium, can be obtained. Surface forces play a particularly important role in colloidal systems, the stability of which depends on the interaction between particles in the presence of an electrolyte solution. A theoretical framework for colloidal interaction is provided by DLVO theory (see [16]), which describes colloidal stability in terms of a balance between an attractive van der Waals force, and a repulsive double-layer force. Comparison of experimental results with theory, however, is feasible only within a well-defined geometry. The surface force apparatus (SFA) [173] is a major advance in this direction, providing direct measurement of surface forces in solution, but only in limited cases that can be configured in a crossed cylinder geometry.

Since force is the basis of the AFM, it is a natural step to quantify the force measurement [52, 74, 78, 174]. To do so, the points of zero force and of zero separation distance have to be defined. The former is set to the flat region in the force curve (Fig. 1.13). The zero of distance is usually defined as the point where the motion of the tip and sample becomes coupled, that is, the displacement of the lever as monitored by the photodiode becomes a linear function of the sample displacement. For an accurate measurement of force, it is important that the cantilever spring constant and the piezo scanner be calibrated properly.

The AFM can be used for measurement of forces between any two particles that can be mounted in its sample compartment and on the probe tip. Moreover, effects of changes in conditions both in the surrounding medium and in the particles themselves can be examined, and a number of studies have been reported. Early studies of force measurement were done under ambient conditions [21, 52], which were subsequently extended to samples immersed in liquids [174, 175]. Working in aqueous solutions, Tsao et al. [176] found evidence for a long-range attractive force between hydrophobic surfaces, which is consistent with results from the SFA. The surfaces they examined were formed by coating the square pyramidal Si_3N_4 tip, the mica substrate, or both, with cationic detergents.

One problem with performing measurements of surface forces with the AFM is the ill-defined geometry of the probe tip. This has been circumvented with the use of a silica microsphere (typically a few microns in radius) glued onto the cantilever as the probe (Fig. 1.34), instead of the usual square pyramidal tip [177, 178]. In this configuration, quantitative analysis of force measurements and comparison with DLVO theory have been reported [177–183]. Furthermore, interaction between different surfaces may be studied by derivatization of the silica, or by using a

Figure 1.34. A silica microsphere glued onto a cantilever, for use in measurement of surface forces.

different material, such as a polystyrene sphere [182]. Rabinovich and Yoon [181] report on forces between hydrophobic surfaces made by deposition of silanes on the silica.

Force measurements between polymer-coated mica surfaces have been conducted using the SFA. Future use of the AFM for such is expected.

C. Mechanical Properties of the Sample

Small modifications on the original design of the AFM allows measurement of properties of the sample. These would be useful for polymer studies, but are not yet routinely performed because they require some changes in the instrument. In the past year, however, newer commercial instruments have incorporated these features, and more investigations along these lines have been performed.

1. Elasticity

One problem with imaging soft samples is the possible deformation due to the tip, which results in an erroneous height measurement. This is undesirable during imaging of the sample topography, and utilized in surface manipulation by nanoindentation. In addition, the indentation can be quantified for measurement of elasticity parameters of the sample [52]. Assuming that the elastic properties of the system are homogeneous within the sampling area, they can be characterized by force versus depth measurements, which can be done by monitoring the tip response as a function of the tip-sample separation distance. Figure 1.35 illustrates the different responses obtained for elastic and plastic materials for loadings high enough to cause indentation ($>10^{-6}$ N).

In a different approach, the elastic properties of a sample can also be examined nondestructively at low forces. When the stiffness of the sample and the cantilever are comparable, the measured height also reflects the elasticity of the sample, and hence can be used to assess such at a very local level [83, 184–186]. While the tip geometry is needed for an accurate assessment, it can be approximated by a sphere provided it is smooth on the scale of the sample deformation. The depth of indentation δ, can be related to the Young's modulus E, which is an intrinsic property of the sample using the Hertz model, provided that the tip does not deform and that deformation of the sample is due to an elastic indentation instead of adhesion [83, 185]. To a first-order approximation, E can be obtained by using a spherical tip with radius r with an applied load F,

$$\delta^3 = (\tfrac{9}{16})(F^2/Er)$$

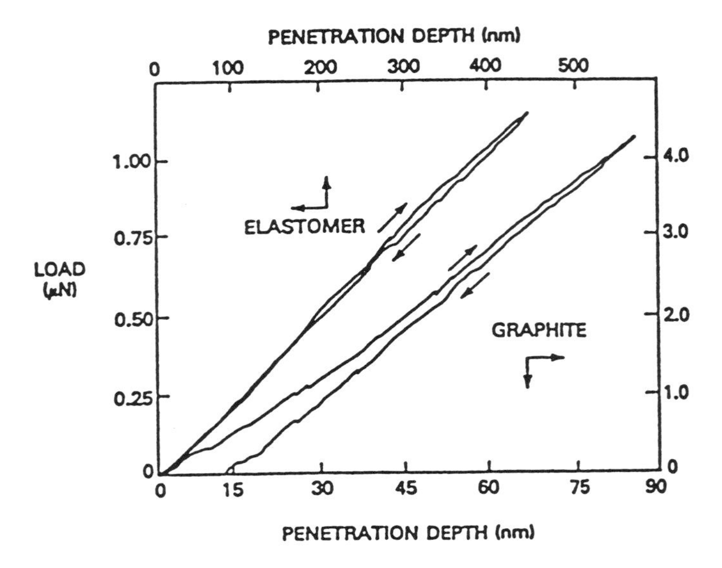

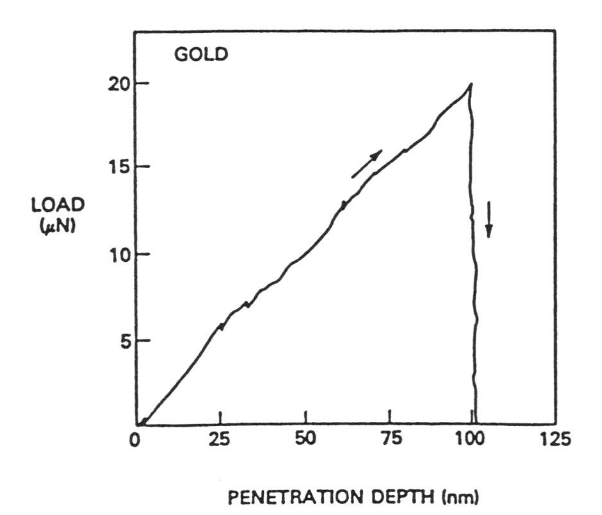

Figure 1.35. The behavior of different materials as a function of loading force, as measured by the AFM. (Data from [52].)

The expression for the load F can be expanded about the equilibrium load F_0, as $F = F_0 + (dF/d\delta)\delta$, from which an effective stiffness for the system can be obtained as,

$$K_{\text{eff}} = dF/d\delta = \{(6/\pi^2)(ErF)\}^{1/3}$$

While measurement of the absolute value for Young's modulus may not be accurate given the uncertainties in the contact area, nevertheless variation in compliance across a sample may be ascertained by a measurement of changes in the modulation amplitude of a vibrating cantilever, which is described in [83, 185]. Moreover, by performing this measurement at various parts of the sample, that is, as a function of the raster scan, an image of the mechanical topography of the sample can be obtained [83, 184–186], an example of which can be seen in Fig. 1.37c.

2. *Friction*

Even in these early days of AFM technology it is clear that the scanning process involves lateral forces, one effect of which is to tilt the cantilever with respect to the surface under study [187]. Various instrumental designs for monitoring lateral forces based on measuring the torsion or the horizontal bend in the cantilever have been reported. Mate et al. [187] used optical interferometry to measure the tilt of a tungsten wire cantilever during a scan of graphite. The image thus obtained (Fig. 1.36) during the raster scan is a picture of the variation of the lateral force between tip and sample, which is ascribed to friction. It is interesting that the regions of high and low friction correlate with atomic positions, showing stick-slip conditions at the atomic scale.

In the standard design of an AFM, lateral forces may be monitored by recording two images scanned in opposite directions (+X and −X), and taking their difference [83]. This facilitates measurement, but can be problematic if the tip is not symmetric with respect to a reversal of scan direction, or if the drift from one measurement to the next is nonnegligible. Newer commercial instruments enable lateral force measurement based on the fact that in the optical lever method for detection, the deflection of the cantilever is due to a combination of normal and lateral forces. Thus, lateral force measurement can be performed concurrent with topographic imaging, by the use of a quadrant photodiode [188].

Lateral force measurement is useful from several perspectives. At the very least, it provides an additional mode of imaging in the AFM that may aid in distinguishing different domains at the surface, such as the case of phase separated LB films [185, 186, 189]. Studies of tribology can be conducted from molecular to macroscopic length scales. Atomic scale

Figure 1.36. Atomic scale friction at the surface of graphite. (From [187].)

friction on highly ordered pyrolitic graphite [187], mica [190], and a self-assembled surfactant monolayer on mica [191] have been reported. Lubrication properties of materials, especially polymers, may presumably be evaluated.

There have been some reported images of polymers with the lateral force mode. Images of friction on PMMA and PC surfaces were found to be the same before and after surface scratching by the AFM tip [192]. The block copolymer PS/PEO grafted on mica has been examined in air, and under several liquids [193]. The air and liquid results differ significantly in the behavior of the friction as a function of the applied load on the tip, which could be partly due to the presence of water as a contaminant in the air sample. The lateral force map of a single hexagonal crystal of polyoxymethylene shows variation in friction within the crystal between domains defined by its diagonals and sides. This result was interpreted in terms of the existence of molecular loops arising from the chain folds at the surface [194].

For qualitative evaluation of topography, it has been shown that the lateral force mode provides additional information about the surface, and a number of reports of images in the friction mode has recently increased due to the availability of commercial instruments for such studies. However, quantitative interpretation of the measurement requires further reflection. As with the normal force measurement, one can question what is really being measured, especially since frictional forces are even harder to quantify than surface forces, involving not just the area of contact but the elastic properties of the materials as well. One aspect of the problem is the uncertainty in the actual relevant contact between tip and surface: The macroscopic tip may be deformed or may have several microscopic points of contact with the sample. Thus, results of the measurements may even be dependent on the length scale considered simply because of local tip structure. However, in the case of a hard AFM probe tip interacting with a soft material at submicrometer scales, it is reasonable to assume a single rigid asperity at contact [186, 195]. For grafted polymer surfaces, O'Shea et al. [193] showed that their data are consistent with such. As for friction at the atomic scale, in view of the questions raised regarding single-atom imaging (Section III.A), it is unclear what information is provided by lateral force measurement: If the force between the tip and the sample is substantial enough to create deformation and/or dislodge atoms, then the difference between normal and lateral force measurement becomes ambiguous. Theoretical considerations and simulations concerning friction measurements are lacking at this point.

The utility of the AFM as a "multimode" machine, able to provide a deeper insight into the mechanical properties of materials at a local level, is beautifully illustrated in a series of studies on LB films prepared from phase separated domains of fluorinated and hydrogenated samples [185, 186]. Figure 1.37 shows the same area where simultaneous measurements of topography, friction, and elasticity have been performed. Interestingly enough, friction on the fluorocarbon domains are higher than the

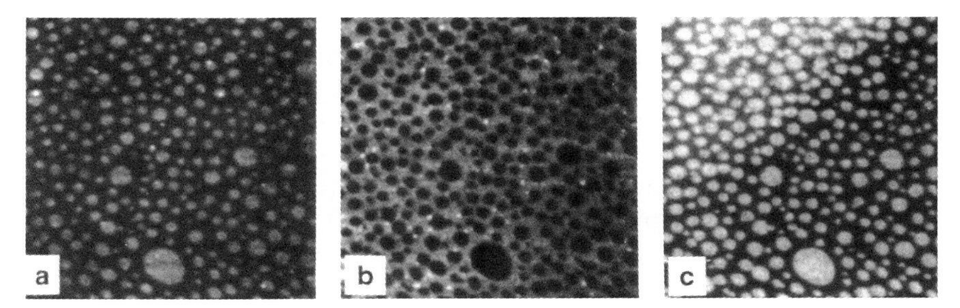

Figure 1.37. Simultaneous measurements of topography, friction, and elasticity for a phase separated LB film (size: $3 \times 3\ \mu$). Islands correspond to hydrocarbon domains in a sea of fluorocarbon film. (a) Topography: dark = low, bright = high. (b) Friction force map shows lower friction (dark) for hydrocarbon islands and three times higher friction for fluorocarbon sea. (c) Elasticity map shows higher Young's modulus (bright) on the hydrocarbon domains. [Reprinted with permission from R.M. Overney et al., *Langmuir*, **10**, 1281 (1994). Copyright © 1994 American Chemical Society.]

hydrocarbon domains. This was explained by the elasticity results, which showed the fluorocarbons to be softer, and hence more susceptible to deformation, which thus increases friction.

VIII. FUTURE DIRECTIONS

A. Instrumentation

Progress in the design of the AFM instrument itself has been rapid in the past 5 years, and reports abound in the literature. Moreover, new models of commercial instruments have improved both in terms of ease of use and in hardware. In addition, other scanning near-field microscopes, which could provide complementary information about surfaces by measuring various parameters, have been devised. Microscopes based on measurements of thermal [196, 197], mechanical [198], optical [199–217, 202], acoustic [203, 204], electrical [205, 206], electrochemical [207–209], nuclear magnetic resonances (NMR) [210–212] properties, and various combinations thereof [213–215] have been reported. The near-field scanning optical microscope [199] is particularly promising because, with the use of light as the probe, it offers the advantage of possible spectroscopic capabilities and nonsecond-to-picosecond time resolution. This microscope can be capable of obtaining signal from a single molecule [216].

To diminish sample deformation by the probe, SPMs operating in the noncontact mode are used, where the probe is set at distances of tens to hundreds of nanometers away from the sample. Magnetic and electrostatic forces have been monitored in this manner under ambient conditions [217], and under liquids [31]. Recently, the sensing of magnetic forces has been utilized for NMR studies with spatial resolution as high as about 2 μ [212]. This is still in its early stages, and it holds exceptional promise after further developments in the design.

While it is convenient that the AFM is normally operated under ambient conditions, being able to change the temperature would be a useful feature. Operation at low temperature is deemed desirable for reducing molecular motion, and several versions of a low-temperature AFM have been reported [218]. Access to higher temperatures might also be useful for polymer studies, enabling examination of morphological changes in situ.

B. Assessment

Eight years after it was invented, the AFM has made the submicroscopic world easily accessible. While there is still drive to get molecular and submolecular resolution, we found along the way that there are numerous interesting problems in the nanometer to micron length scale. The AFM is quite suited for these studies, and it possesses many advantages over other techniques. This is particularly relevant to polymers, for which numerous morphological considerations occur at these length scales. In addition, polymer studies will benefit from new variants in the AFM that allow measurement of mechanical properties.

The tip is still the most uncertain part of the instrument. At the worst, deformed tips lead to artifacts that may mislead the researcher; at the least, the search for a suitable tip wastes many a researcher's time. A numerical deconvolution procedure, developed to visualize the tip as well as remove its geometry from AFM images is recommended for use as part of the routine of measurement. However, this is still of limited utility especially with high-resolution imaging, since there may exist small protrusions on the tip that would affect its characteristics. Hence, original approaches to tip making are still needed.

So far the AFMs utility has been limited in regard to submolecular resolution studies of polymers: While numerous reports exist, the AFM provided information otherwise unobtainable only in a few cases. Furthermore, questions still remain in regard to the actual imaging process, and whether or not images of subnanometer corrugation correspond to true molecular resolution. While it is generally believed that imaging forces have to be reduced to about 10^{-11} N for single-molecule resolution, observations of "molecular features" at high forces continue to abound. To what then do they correspond?

The lack of spectroscopic capabilities of the AFM is a weakness with regards to identification of structure. Unlike the STM, there is no control parameter that will allow for the identification of the atom being imaged. In principle, operating in the attractive regime of the interaction potential should enable this. Unfortunately, the long-range nature of the attractive part would lower the resolution, and high-resolution imaging may not be feasible.

The mesoscopic scale is easily accessed with the AFM. For these studies, image enhancement is unnecessary, and the features obtained are unequivocally present. Thus, the AFM can be used as a quantitative technique. If proper care is taken, the surface features are accurate. Nevertheless, image interpretation still has to be considered. For biological systems, markers may be used to aid in highlighting the desired structures. In studies that rely on pattern formation, one notes that similar images may be obtained for different systems, and for different reasons. In some cases, spectroscopic information would be helpful. Thus, there is a need to use different techniques, and some of them, like fluorescence microscopy [219], have been physically integrated with the AFM. It has been shown that measurements of elasticity and lateral forces, together with the topography, can lead to a deeper understanding of lubricating systems, although some additional theoretical framework, including simulations, for these types of measurements would aid future studies. Others reported the use of the AFM in conjunction with electron microscopy, X-ray and neutron scattering, fluorescence microscopy, surface forces apparatus, among others, to address real issues in surface science. In fact, this is where the instrument should be heading. When its capability for visualization of topography is combined with other types of measurement, the AFM can be a very powerful tool.

Pictures are appealing, and the rapid progress in the development of microscopy of different types in recent years (various SPMs, confocal, fluorescence, video, etc.) and the accompanying video technology, has made visualization a big part of scientific research. Unfortunately, it is so simple to collect numerous beautiful pictures that one can easily lose sight of the science, or the fact that the pictures are data and should be analyzed as such. To date, there are excellent, but relatively few, results using the AFM as a quantitative tool; it is to be hoped that this proportion will increase.

At this point, the AFM is mature enough that it can be used as an almost routine tool for surface examination, together with other surface techniques. I look forward to the time when every lab has the capability of inspecting their surfaces with an AFM, in a diagnostic manner the way contact angle and spectroscopic measurements are employed. Thus we will be able to focus on addressing real issues and making relevant measurements. We should no longer be satisfied with the novelty of just having a picture.

ACKNOWLEDGMENTS

Support from NSERC Canada and the Ontario Laser and Lightwave Research Centre is gratefully acknowledged. I would like to thank Peter Markiewicz for assistance in preparation of the figures and careful reading of the manuscript.

REFERENCES

1. G. Binnig and H. Rohrer, *Helv. Phys. Acta*, **55**, 726 (1982).
2. G. Binnig, H. Rohrer, C. Gerber, and E. Weibel, *Phys. Rev. Lett.*, **40**, 178 (1982).
3. H. K. Wickramasinghe, *J. Vac. Sci. Technol. A*, **8**, 363 (1990).
4. G. Binnig, C. F. Quate, and C. Gerber, *Phys. Rev. Lett.*, **56**, 930 (1986).

5. R. Lal and S. A. John, *Am. J. Physiol.* **265**, C1 (1994).
6. H. G. Hansma and J. Hoh, *Ann. Rev. Biophys. Biomol. Struct.* (1994).
7. A. Halperin, M. Tirrell, and T. P. Lodge, in *Advances in Polymer Science*, Springer-Verlag, Berlin, 1992, Vol. 100, pp. 31–72.
8. R. J. Samuels, *Structured Polymer Properties*, Wiley, New York, 1974.
9. F. S. Bates and G. H. Fredrickson, in *Annual Review of Physical Chemistry*, H. Strauss and G. Babcock, Eds., Annual Reviews, Inc., Palo Alto, CA, 1990, Vol. 41, pp. 525–557.
10. W. J. Feast and H. S. Munro, Eds., *Polymer surfaces and interfaces*, Wiley, Chichester, 1987.
11. M. Stamm, in *Advances in Polymer Science*, Springer-Verlag, Berlin, 1992, Vol. 100, pp. 357–400.
12. R. Young, J. Ward, and F. Scire, *Rev. Sci. Instrum.*, **43**, 999 (1972).
13. E. C. Teague, F. E. Scire, S. M. Bacher, and S. W. Jensen, *Wear*, 83, 1 (1982).
14. N. A. Burnham, R. J. Colton, and H. M. Pollock, *J. Vac. Sci. Technol. A*, **9**, 2548 (1991).
15. R. J. Wilson, K. E. Johnson, D. D. Chambliss, and B. Melior, *Langmuir*, **8**, 3478 (1993).
16. J. Israelachvili, *Intermolecular and Surface Forces*, Academic, San Diego, CA, 1992.
17. R. J. Hamers, in *Annual Reviews of Physical Chemistry*, H. L. Strauss and K. Babcock, Eds., Annual Reviews, Inc., Palo Alto, CA, 1990, Vol. 40, pp. 331–359.
18. D. Sarid, *Scanning force microscopy with applications to electric, magnetic and atomic forces*. Oxford series in optical and imaging sciences, Oxford University Press, New York, 1991.
19. D. R. Baselt and J. D. Baldeschweiler, *Rev. Sci. Instrum.*, **64**, 908 (1993).
20. S. Park and C. Quate, *Appl. Phys. Lett.* **48**, 112 (1986).
21. Y. Martin, C. C. Williams and H. K. Wickramasinghe, *J. Appl. Phys.*, **61**, 4723 (1987).
22. R. Erlandsson, G. M. McClelland, and S. Chiang, *J. Vac. Sci. Technol. A*, **6**, 266 (1988).
23. D. Rugar, H. J. Mamin, R. Erlandsson, J. E. Stern, and B. D. Terris, *Rev. Sci. Instrum.*, **59**, 2337 (1988).
24. D. Sarid, *Opt. Lett.*, **13**, 1057 (1988).
25. M. Nonnenmacher, M. Vaez-Iravani, and H. Wickramasinghe, *J. Vac. Sci. Technol. A*, **11**, 758 (1993).
26. G. Meyer and N. M. Amer, *Appl. Phys. Lett.*, **53**, 1045 (1988).
27. S. Alexander et al., *J. Appl. Phys.*, **65**, 164 (1989).
28. D. Brodbeck, L. Howald, R. Luthi, E. Meyer, and R. Overney, *Ultramicroscopy*, **42–44**, 1580 (1992).
29. M. Gale, P. Markiewicz, M. S. Pollanen, and M. C. Goh, *Biophys. J.*, **68**, 2124 (1995).
30. C. A. Putman, K. O. V. d. Werf, B. D. Grooth, N. V. Hulst, and J. Greve, *SPIE*, **1639**, 198 (1992).
31. R. Giles et al., *Appl. Phys. Lett.*, **63**, 617 (1993).

32. Nanoscope III Digital Instruments Inc., Santa Barbara CA.
33. H. G. Hansma et al., *Scanning*, **15**, 296 (1993).
34. C. A. Putman, K. O. V. d. Werf, B. G. D. Grooth, N. F. V. Hulst, and J. Greve, *Appl. Phys. Lett.*, **64**, 2454 (1994).
35. P. K. Hansma et al., *Appl. Phys. Lett.*, **64**, 1738 (1994).
36. S. Manne, P. K. Hansma, J. Massie, V. B. Ekings, and A. A. Gewirth, *Science*, **251**, 183 (1991).
37. H. G. Hansma et al., *Science*, **256**, 1180 (1992).
38. B. Drake et al., *Science*, **243,** 1586 (1989).
39. E. Henderson, P. G. Haydon, and D. S. Sakaguchi, *Science*, **257**, 1944 (1992).
40. J. K. H. Horber et al., *Scanning Microsc.*, **6**, 919 (1992).
41. D. Snetivy and G. J. Vancso, *Langmuir*, **9**, 2253 (1993).
42. Y. Li and S. M. Lindsay, *Rev. Sci. Instrum.*, **62**, 2630 (1991).
43. T. Thundat et al., *Scanning Microsc.*,**6**, 903 (1992).
44. J. Vesenka, S. Manne, R. Giberson, T. Marsh, and E. Henderson, *Biophys. J.*, **65**, 992 (1993).
45. S. Xu and M. F. Arnsdorf, *J. Microsc.*, **173**, 199 (1994).
46. T. R. Albrecht, S. Akamine, T. E. Carver, and C. F. Quate, *J. Vac. Sci. Technol. A*, **8**, 3386 (1990).
47. J. P. Cleveland, S. Manne, D. Bocek, and P. K. Hansma, *Rev. Sci. Instrum.*, **64**, 403 (1993).
48. J. L. Hutter and J. Bechhoefer, *Rev. Sci. Instrum.*, **64**, 1868 (1993).
49. D. Sarid and V. Elings, *J. Vac. Sci. Technol. B*, **9**, 431 (1991).
50. R. Neumeister and W. A. Ducker, *Rev. Sci. Instrum.*, **65**, 2527 (1994).
51. U. D. Schwarz, H. Haefke, P. Reimann, and H. Gunthrodt, *J. Microscopy*, **173**, 183 (1993).
52. N.A. Burnham and R. J. Colton, *J. Vac. Sci. Technol. A*, **7**, 2906 (1989).
53. D. Keller, D. Deputy, A. Alduino, and K. Luo, *Ultramicroscopy*, **42–44**, 1481 (1992).
54. D. J. Keller and C. Chih-Chung, *Surf. Sci.*, **268**, 333 (1992).
55. M. Yamaki, T. Miwa, H. Yoshimura, and K. Nagayama, *J. Vac. Sci. Technol. B*, **10**, 2447 (1992).
56. O. Wolter, T. Bayer, and J. Greschner, *J. Vac. Sci. Technol. B*, **9**, 1353 (1991).
57. D. A. Grigg, P. E. Russell, J. E. Griffith, and E. A. Fitzgerald, *Ultramicroscopy*, **42–44**, 1616 (1992).
58. M. J. Vasile, C. Biddick, and H. Huggins, *Appl. Phys. Lett.*, **64**, 575 (1994).
59. B. Bushan and V. N. Koinkar, *Appl. Phys. Lett.*, **64**, 1653 (1994).
60. G. Binnig, *Ultramicroscopy* **42–44**, 7 (1992).
61. P. Grutter, W. Zimmermann-Eding, and D. Brodbeck, *Appl. Phys. Lett.*, **60**, 2741 (1992).
62. M. J. Allen et al., *Ultramicroscopy*, **42–44**, 1095 (1992).
63. J. E. Griffith, D. A. Grigg, M. J. Vaslie, P. E. Russell, and E. A. Fitzgerald, *J. Vac. Sci. Technol. B*, **9**, 3586 (1991).
64. L. Hellemans et al., *J. Vac. Sci. Technol. B*, **9**, 1309 (1991).
65. F. Jensen, *Rev. Sci. Instrum.*, **64**, 2595 (1993).
66. P. Markiewicz and M. C. Goh, *Langmuir*, **10**, 5 (1994).

67. J. Vesenka, R. Miller, and E. Henderson, *Rev. Sci. Instrum.*, **65**, 2249 (1994).
68. P. Markiewicz, M. C. Goh, *J. Vac. Sci. Technol. B*, in press.
69. R. Chicon, M. Ortuno, and J. Abellan, *Surf. Sci.*, **181**, 107 (1987).
70. D. Keller, *Surf. Sci.*, **253**, 353 (1991).
71. M. S. Pollanen, P. Markiewicz, C. Bergeron, and M. C. Goh, *Am. J. Pathol.*, **144**, 869 (1994).
72. J. B. Pethica, W. C. Oliver, *Phys. Scri.*, **T19**, 61 (1987).
73. T. P. Weihs, Z. Nawaz, S. P. Jarvis, and J. B. Pethica, *Appl. Phys. Lett.*, **59**, 3536 (1991).
74. A. L. Weisenhorn, P. Maivald, H.-J. Butt, and P. K. Hansma, *Phys. Rev. B*, **45**, 226 (1992).
75. E. Meyer et al., *Thin Solid Films*, **181**, 527 (1989).
76. U. Landman, W. D. Luedtke, and A. Nitzan, *Surf. Sci.*, **210**, L177 (1989).
77. U. Landman, W. D. Luedtke, N. A. Burnham, and R. J. Colton, *Science*, **248**, 454 (1990); W.D. Luedtke and U. Landman, *Computational Mater. Sci.*, **1**, 1 (1992).
78. *N. A. Burnham, D. D. Dominguez, R. L. Mowery, and R. J. Colton*, Phys. Rev. Lett., **64**, 1931 (1990).
79. B. N. J. Persson, *Chem. Phys. Lett.*, **141**, 366 (1987).
80. F. Ohnesorge and G. Binnig, *Science*, **260**, 1451 (1993).
81. S. A. Joyce and J. E. Houston, *Rev. Sci. Instrum.*, **62,** 710 (1991).
82. F.F. Abraham, I. P. Bhatra, and S. Ciraci, *Phys. Rev. Lett.*, **60**, 1314 (1988).
83. M. Radmacher, R. W. Tillman, M. Fritz, and H. E. Gaub, *Science*, **257**, 1900 (1992).
84. D. K. Schwartz, R. Viswanathan, and J. A. N. Zasadzinski, *Langmuir*, **9**, 1384 (1993).
85. D. Snetivy, G. J. Vancso, and G. C. Rutledge, *Macromolecules*, **25**, 7037 (1992).
86. F. Franke and D. Keller, *SPIE*, **1891**, 78 (1993).
87. O. Marti, B. Drake, and P. K. Hansma, *Appl. Phys. Lett.*, **51**, 484 (1987).
88. G. Garnaes, D. K. Schwartz, R. Viswanathan, and J. A. N. Zasadzinski, *Nature (London)*, **257**, 508 (1992).
89. H. J. Butt et al., *J. Microsc. Oxf.*, **169**, 75 (1993).
90. A. S. Lea et al., *Langmuir*, **8**, 68 (1992).
91. A. L. Weisenhorn, F.-J. Schnitt, W. Knoll, and P. K. Hansma, *Ultramicroscopy*, **42–44**, 1125 (1992).
92. G. Nunes and M. R. Freeman, *Science*, **262**, 1029 (1993).
93. M. Radmacher, M. Fritz, H. G. Hansma, and P. K. Hansma, *Science*, **265**, 1577 (1994).
94. M. C. Goh, D. Juhue, Y. Wang, O. M. Leung, and M. A. Winnik, *Langmuir*, **9**, 1319 (1993).
95. S. N. Magonov, S. Kempf, M. Kimmig, and H. Cantow, *Polym. Bull.*, **26**, 715 (1991).
96. D. W. Schwark, D. L. Vezie, J. R. Reffner, E. L. Thomas, and B. K. Annis, *J. Mater. Sci. Lett.*, **11**, 352 (1992).
97. B. K. Annis and D. F. Pedraza, *J. Vac. Sci. Technol. B*, **11**, 1759 (1993).
98. S. Li et al., *Langmuir*, **9**, 2243 (1993).
99. F. Boury, A. Gulik, J. C. Dedieu, and J. E. Proust, *Langmuir*, 1654 (1994).
100. S. Karrasch, S. Heins, U. Aebi, and A. Engel, *J. Vac. Sci. Technol. B*, **12**, 1474 (1994).

101. A. L. Rachlin, G. S. Henderson, and M. C. Goh, *Am. Mineral.*, **77**, 904 (1992).
102. D. Snetivy, J. E. Guillet, and G. J. Vancso, *Polymer* (1992).
103. D. Snetivy, G. Rutledge, and G. J. Vancso, *Polymer Prep.*, **33**, 786 (1992).
104. D. Snetivy and G. J. Vancso, *Polymer*, **35**, 461 (1994).
105. T. Thundat et al., *J. Vac. Sci. Technol. A*, **10**, 630 (1992).
106. R. M. Overney et al., *Langmuir*, **9**, 341 (1993).
107. A. Schaper, L. Wolthaus, D. Mobius, and T. M. Jovin, *Langmuir*, **9**, 2178 (1993).
108. L. Bourdieu, O. Ronsin, and D. Chatenay, *Science*, **259**, 798 (1993).
109. C. A. Widrig, C. A. Alves, and M. D. Porter, *J. Am. Chem. Soc.*, **113**, 2805 (1991).
110. C. A. Alves and M. D. Porter, *Langmuir*, **9**, 3507 (1993).
111. J. Pan, N. Tao, and S. M. Lindsay, *Langmuir*, **9**, 1556 (1993).
112. Y. Wang, D. Juhue, M. A. Winnik, O. M. Leung, and M. C. Goh, *Langmuir*, **8**, 760 (1992).
113. J. C. Wittmann and P. Smith, *Nature (London)*, **352**, 414 (1991).
114. H. Hansma, F. Motamedi, P. Smith, P. Hansma, and J. C. Wittman, *Polymer*, **33**, 647 (1992).
115. J. R. Rasmusson et al., submitted for publication, 1993.
116. K. M. Shakesheff et al., *Surf. Sci. Lett.*, **304**, L393 (1994).
117. G. J. Leggett et al., *Langmuir*, **9**, 2356 (1993).
118. C. R. Ill et al., *Biophys. J.*, **64**, 919 (1993).
119. J. Yang, L. K. Tamm, T. W. Tillack, and Z. Shao, *J. Mol. Biol.*, **229**, 286 (1993).
120. S. M. Lindsay et al., *Biophys. J.*, **61**, 1570 (1992).
121. A. Ashkin, J. M. Dziedzic, J. E. Bjorkholm, and S. Chu, *Opt. Lett.*, **11**, 288 (1986).
122. S. Chu, *Science*, **253**, 861 (1991).
123. A. Ashkin, K. Schultze, J. M. Dziedzic, U. Euteneuer, and M. Scliwa, *Nature (London)*, **348**, 346 (1990).
124. C. A. J. Putman, B. G. D. Grooth, P. K. Hansma, N. F. V. Hulst, and J. Greve, *Ultramicroscopy*, **48**, 177 (1993).
125. G. J. Leggett et al., *Langmuir*, **9**, 1115 (1993).
126. W. Stocker, S. N. Magonov, H. Cantow, J. C. Wittmann, and B. Lotz, *Macromolecules*, **26**, 5915 (1993).
127. S. N. Magonov, S. S. Sheiko, R. A. C. Deblieck, and M. Moller, *Macromolecules*, **26**, 1380 (1993).
128. B. K. Annis, J. R. Reffner, and B. Wunderlich, *J. Polym. Sci. B. Polym. Phys.*, **31**, 93 (1993).
129. D. Snetivy, H. Yang, and G. J. Vancso, *J. Mat. Chem.*, **2**, 891 (1992).
130. R. Patil and D. H. Reneker, *Polymer*, **35**, 1909 (1994).
131. D. Snetivy and G. J. Vancso, *Polymer*, **22**, 422 (1992).
132. S. Arisawa, T. Fujii, T. Okane, and R. Yamamoto, *Appl. Surf. Sci.*, **60–61**, 321 (1992).
133. N. Auduc, A. Ringenbach, I. Stevenson, Y. Jugnet, and T. M. Duc, *Langmuir*, **9**, 3567 (1993).

134. V. V. Tsukruk, M. D. Foster, D. H. Reneker, A. Schmidt, and W. Knoll, *Langmuir*, **9**, 3538 (1993).
135. Y.-M. Zhu, Z.-H. Lu, X.-M. Yang, Y. Wei, and F. Qian, *Phys. Lett. A*, **181**, 183 (1993).
136. R. Patil, S.-J. Kim, E. Smith, D. H. Reneker, and A. L. Weisenhorn, *Polym. Commun.*, **31**, 455 (1990).
137. W. Stocker et al., *Ultramicroscopy* **42–44**, 1141 (1992).
138. H. Schoenherr, D. Snetivy, and G. J. Vancso, *Polym. Bull.*, **30**, 567 (1993).
139. G. W. Zajac, M. Q. Patterson, P. M. Burell, and C. Metaxas, *Ultramicroscopy*, **42–44**, 998 (1992).
140. R. M. Overney, R. Luthi, J. Frommer, E. Meyer, and J. Guntherodt, *Appl. Surf. Sci.*, **64**, 197 (1993).
141. A. K. Dhingra, Ed., *Encyclopedia of Polymer Science and Engineering*, Vol. 6, Wiley, New York, 1987.
142. K. Haraguchi, T. Kajiyama, and M. Takayanagi, *J. Appl. Polym. Sci.*, **23**, 915 (1979).
143. E. J. Roche, M. S. Wolfe, A. Suna, and P. J. Avakian, *Macromol. Sci.-Phys. B*, **24**, 141 (1985).
144. S. F. Y. Li, A. J. McGhie, and S. L. Tang, *Polymer*, **34**, 4573 (1993).
145. B. K. Annis, D. W. Schwark, J. R. Reffner, E. L. Thomas, and B. Wunderlich, *Makromol. Chem.*, **193**, 2589 (1992).
146. B. Collin, D. Chatenay, G. Coulon, D. Ausserre, Y. Gallot, *Macromolecules*, **25**, 1621 (1992).
147. Y. Liu et al., *Phys. Rev. Lett.*, **73**, 440 (1994).
148. O. M. Leung and M. C. Goh, unpublished results.
149. J. Zhu, A. Eisenberg, and R. B. Lennox, *J. Am. Chem. Soc.*, **113**, 5583 (1991).
150. T. G. Stange, R. Mathew, D. F. Evans, and W. A. Hendrickson, *Langmuir*, **8**, 920 (1992).
151. K. J. Stine, C. M. Knobler, and R. C. Desai, *Phys. Rev. Lett.*, **65**, 1004 (1990).
152. R. Bruinsma, *Macromolecules*, **23**, 276 (1990).
153. P. Silberzan and L. Leger, *Macromolecules*, **24**, 1267 (1992).
154. W. Zhao et al., *Phys. Rev. Lett.*, **70**, 1453 (1993).
155. M. Joanicot et al., *Prog. Colloid Polym. Sci.*, **81**, 175 (1990).
156. Y. Wang, C. L. Zhao, and M. A. Winnik, *J. Chem. Phys.*, **95**, 2143 (1991).
157. Y. Wang et al., *Langmuir*, **8**, 1435 (1992).
158. D. Juhue and J. Lang, *Langmuir*, **9**, 792 (1993).
159. D. Juhue et al., *J. Polym. Sci. B. Polym. Phys.*, **33**, 1123 (1995).
160. *E. A. G. Chernoff and D. A. Chernoff, J. Vac. Sci. Technol. A*, **10**, 596 (1992).
161. D. R. Baselt, J.-P. Revel, and J. D. Baldeschweiler, *Biophys. J.*, **65**, 2644 (1994).
162. M. S. Pollanen, P. Markiewicz, L. Weyer, M. C. Goh, and C. Bergeron, *Am. J. Pathol.*, **145**, 1140 (1995).
163. M. S. Pollanen, P. Markiewicz, C. Bergeron, and M. C. Goh, *Colloids Surfaces A: Physicochem. Eng. Aspects* , **87**, 213 (1994).
164. T. A. Jung et al., *Ultramicroscopy*, **42–44**, 1446 (1992).

165. Y. Kim and C. M. Lieber, *Science*, **257**, 375 (1992).
166. H. J. Mamin and D. Rugar, *Appl. Phys. Lett.*, **61**, 1003 (1992).
167. S. Hoen, H. J. Mamin, and D. Rugar, *Appl. Phys. Lett.*, **64**, 267 (1994).
168. J. H. Hoh, R. Lal, S. A. John, J. P. Revel, and M. P. Arnsdorf, *Science*, **253**, 1405 (1991).
169. O. M. Leung and M. C. Goh, *Science*, **255**, 64 (1992).
170. J. C. Brumfield, C. A. Goss, E. A. Irene, and R. W. Murray, *Langmuir*, **8**, 2810 (1992).
171. C. A. Goss, J. C. Brumfield, E. A. Irene, and R. W. Murray, *Langmuir*, **8**, 1459 (1992).
172. G. F. Meyers, B. M. DeKoven, and J. T. Setiz, *Langmuir*, **8**, 2330 (1992).
173. J.N. Israelachvili and G. E. Adams, *JCS Faraday Trans. I*, **74**, 975 (1978).
174. A. L. Weisenhorn, P. K. Hansma, T. R. Albrecht, and C. F. Quate, *Appl. Phys. Lett.*, **54**, 2651 (1989).
175. W. A. Ducker and R. F. Cook, *Appl. Phys. Lett.*, **56**, 2408 (1990).
176. Y.-h. Tsao, D. F. Evans, and H. Wennerstrom, *Science*, **262**, 547 (1993).
177. H.-J. Butt, *Biophys. J.*, **60**, 1438 (1991).
178. W. A. Ducker, T. J. Senden, and R. M. Pashley, *Nature* (*London*), **353**, 239 (1991).
179. T. J. Senden, C. J. Drummond, and P. Kekicheff, *Langmuir*, **10**, 358 (1994).
180. M. W. Rutland and T. J. Senden, *Langmuir*, **9**, 412 (1993).
181. Y. I. Rabinovich and R. Yuen, *Langmuir*, **10**, 1903 (1994).
182. Y. Q. Li, N. J. Tao, J. Pan, A. A. Garcia, and S. M. Lindsay, *Langmuir*, **9**, 637 (1993).
183. S. J. O'Shea, M. E. Welland, and T. Rayment, *Appl. Phys. Lett.*, **60**, 2356 (1992).
184. P. Maivald et al., *Nanotechnology*, **2**, 103 (1991).
185. R. M. Overney et al., *Langmuir*, **10**, 1281 (1994).
186. R. M. Overney et al., *J. Vac. Sci. Technol. B*, **12**, 1973 (1994).
187. C. M. Mate, G. M. McClelland, R. Erlandsson, and S. Chiang, *Phys. Rev. Lett.*, **59**, 1942 (1987).
188. G. Meyer and N. M. Amer, *Appl. Phys. Lett.*, **57**, 2089 (1990).
189. R. M. Overney et al., *Nature* (*London*), **359**, 133 (1992).
190. R. Erlandsson, G. Hadziioannou, C. M. Mate, G. M. McClelland, and S. Chiang, *J. Chem. Phys.*, **89**, 5190 (1988).
191. Y. Liu, T. Wu and D. F. Evans, *Langmuir*, **10**, 2241 (1994).
192. E. Hamada and R. Kaneko, *Ultramicroscopy*, **42–44**, 184 (1992).
193. S. J. O'Shea, M. E. Welland, and T. Rayment, *Langmuir*, **9**, 1826 (1993).
194. R. Nisman, P. Smith, and G. J. Vancso, *Langmuir*, **10**, 1667 (1994).
195. R. M. Overney et al., *Nature* (*London*), **359**, 133 (1992).
196. M. Nonnenmacher and H. K. Wickramasinghe, *Appl. Phys. Lett.*, **61**, 168 (1992).
197. A. Majumdar, J. P. Carrejo, and J. Lai, *Appl. Phys. Lett.*, **62**, 2501 (1993).
198. B. Cretin and F. Stahl, *Appl. Phys. Lett.*, **62**, 829 (1993).

199. E. Betzig, J. Trautman, T. D. Harris, J. S. Weiner, and R. L. Kostelak, *Science*, **251** (1991).
200. M. Vaez-Iravani and R. Toledo-Crow, *Appl. Phys. Lett.*, **62**, 1044 (1993).
201. E. Betzig, S. G. Grubo, R. J. Chichester, D. J. DiGiovanni, and J. S. Weiner, *Appl. Phys. Lett.*, **63**, 3550 (1993).
202. J. M. Guerra, M. Srinavasarao, and R. B. Stein, *Science*, **262**, 1395 (1993).
203. U. Rabe and W. Arnold, *Appl. Phys. Lett.*, **64**, 1493 (1994).
204. K. Yamanaka, H. Ogiso, and O. Kolosov, *Appl. Phys. Lett.*, **64**, 178 (1994).
205. F. Saurenbach, D. Wollmann, R. D. Terris, and A. F. Diaz, *Langmuir*, **8**, 1199 (1992).
206. R. A. Said, G. E. Bridges, and D. J. Thomson, *Appl. Phys. Lett.*, **64**, 1442 (1994).
207. P. K. Hansma, B. Drake, O. Marti, S. A. C. Gould, and C. B. Prater, *Science*, **243**, 641 (1989).
208. A. J. Bard, F. R. F. Fan, J. Kwak, and O. Lev, *Anal. Chem.*, **61**, 132 (1989).
209. C. B. Prater, P. K. Hansma, M. Tortonese, and C. F. Quate, *Rev. Sci. Instrum.*, **62**, 2634 (1991).
210. D. Rugar, C. S. Yannoni, and J. A. Sidles, *Nature* (*London*), **360**, 563 (1992).
211. O. Zuger and D. Rugar, *Appl. Phys. Lett.*, **63**, 2496 (1993).
212. D. Rugar et al., *Science*, **264**, 1560 (1994).
213. E. Betzig, P. L. Finn, and J. S. Weiner, *Appl. Phys. Lett.*, **60**, 2484 (1992).
214. K. Lieberman and A. Lewis, *Appl. Phys. Lett.*, **62**, 1335 (1993).
215. L. M. Eng, K. D. Jandt, and D. Descouts, *Rev. Sci. Instrum.*, **65**, 390 (1994).
216. E. Betzig and R. Chichester, *Science*, **262**, 1422 (1993).
217. Y. Martin and H. K. Wickramasinghe, *Appl. Phys. Lett.*, **50**, 1455 (1987).
218. C. B. Prater et al., *J. Vac. Sci. Technol. B*, **9**, 989 (1991).
219. C. A. J. Putman, H. G. Hansma, H. E. Gaub, and P. K. Hansma, *Langmuir*, **8**, 3014 (1992).
220. D. Snetivy and G. J. Vancso, *Macromolecules*, **25**, 3320 (1992).

INTRINSIC VISCOSITY AND THE POLARIZABILITY OF PARTICLES HAVING A WIDE RANGE OF SHAPES

J. F. DOUGLAS* AND E. J. GARBOCZI†

Polymers and Building Materials Divisions†*
National Institute of Standards and Technology, Gaithersburg, MD 20899

CONTENTS

The intrinsic viscosity $[\eta]$ and the electric $\boldsymbol{\alpha}_e$ and magnetic $\boldsymbol{\alpha}_m$ polarizabilities of objects having general shape are required in the calculation of some of the most basic properties of solid–solid composites and fluid–solid mixtures. Specifically, the leading order virial coefficients of diverse properties (viscosity, refractive index, dielectric constant, magnetic permeability, thermal and electrical conductivity, and others) can often be expressed in terms of these functionals of object shape. These virial coefficients also provide basic input into effective medium theories describing higher concentration mixtures. The electric and

Advances in Chemical Physics, Volume XCI, Edited by I. Prigogine and Stuart A. Rice.
ISBN 0-471-12002-2

magnetic polarizability tensors have independent interest in applications involving the scattering of electromagnetic and pressure waves from objects of general shape. We present an argument that the ratio of $[\eta]$ and $\langle \alpha_e \rangle$ (the average electric polarizability tensor trace) is an *invariant* to a good approximation. Many analytical and numerical finite element results for a variety of shapes are presented to support the conjectured relation. Our *approximate* relation between $[\eta]$ and $\langle \alpha_e \rangle$ complements the *exact* relation between the hydrodynamic virtual mass $\mathbf{W}$ and the magnetic polarizability $\boldsymbol{\alpha}_m$ tensors.

I. INTRODUCTION

There are numerous contributions to the problem of predicting the effective properties of inhomogeneous materials [1–3] and reviews appear regularly on this topic. Here we focus the discussion on transport properties [1], such as the viscosity of suspensions, dielectric constant, refractive index, thermal conductivity, and related physical properties of mixtures. Much of the research has been limited to the classical case of spherical particle suspensions and composites and a few other particle shapes that allow analytical treatment [4–20]. Even for suspensions of hard spheres in fluids, rigorous results are limited to the first couple of virial coefficients [4–20] and analytical bounds on the effective properties at higher volume fractions [21, 22]. Recent progress has been made in the numerical calculation of transport properties by finite element [23] and Brownian dynamics methods [24], which is associated with the advent of sufficiently powerful computational resources and faster computational algorithms such as the conjugate gradient method [25]. These new numerical results are very helpful in testing theoretical ideas about the effective properties of mixtures.

Much of the previous research, even the more recent numerical work, has focused on mixtures involving simple shapes, such as spheres, since the overall goal has usually been the development of a theory applicable at high volume fractions of suspended matter. The choice of simple particle shapes is motivated by the existence of the relatively few exact analytical results at lower concentrations, which provide a benchmark test for the numerical calculations.

However, many real particles in fluid–solid suspensions and in solid composites are not well represented by these simple shapes. The present work allows for general centrosymmetric particle shape, while focusing solely on the dilute limit. This is a natural first step towards treating complex-shaped particle mixtures at higher concentrations. Originally, the research was motivated by theoretical arguments, described below, that suggested a relation between the leading order virial coefficients for the viscosity and conductivity of suspensions of conducting particles having *arbitrary* shape. We have gathered analytical results for these properties, which are scattered widely throughout the mathematical and technical literature and have calculated new results as necessary (analytically and by finite element methods) to obtain a wide range of shapes with which to check the conjectured relation. The results obtained should have independent interest in the problem of developing a more realistic description of the properties of mixtures in terms of a more faithful description of the mixture components.

In Section II we summarize classical results for the conductivity virial expansion and discuss the relation between the leading virial coefficient, called the intrinsic conductivity $[\sigma]$ and certain functionals of particle shape that arise in other physical contexts—the electric polarizability, magnetic polarizability, and virtual mass. All these functionals of shape involve solving the Navier–Stokes or the Laplace equations on the exterior of a body with various boundary conditions [26, 27]. Exact results for these functionals are summarized for simple particle shapes to illustrate the general effect of particle anisotropy.

Section III summarizes classical results for the viscosity virial expansion of suspensions of particles and the virial expansion for the shear modulus of an elastic material with inclusions. An angular preaveraging approximation is invoked, as in previous calculations for the translational friction of a Brownian particle [27], to relate the intrinsic viscosity $[\eta]$ to the intrinsic conductivity $[\sigma]_\infty$ for highly conducting inclusions. The intrinsic viscosity is the leading order virial coefficient for the viscosity of a dilute mixture. Examination of exact results for $[\eta]$ and $[\sigma]_\infty$ show that the conjectured relation is not exact, but is a rather good approximation. An extensive range of particle shapes is considered in this comparison. In Section IV we pursue the universality of the $[\eta]-[\sigma]_\infty$ relation for a variety of complicated shapes using finite element methods. The conjectured relation between $[\eta]$ and $[\sigma]_\infty$ is found to hold to a good approximation ($\pm 5\%$) for all shapes considered.

More general results are possible in $d=2$ due to general conformal mapping methods. We exploit a mathematical identity of Pólya [26, 28], which implies that $[\sigma]$ for conducting and nonconducting inclusions is *exactly* related to the "transfinite diameter" C_L of the inclusions. This relation is useful since this quantity is fundamental in classical conformal mapping theory and, consequently, this property has been extensively investigated [26, 29, 30]. All previously known exact results for $[\sigma]_\infty$ in $d=2$, plus many new results, are obtained from our new relation expressing $[\sigma]_\infty$ purely in terms of the *geometrical* quantities, C_L and the particle area. Numerical results for $[\eta]$ in $d=2$ are obtained using finite element methods. Again the predicted approximate relation between $[\eta]$ and $[\sigma]_\infty$ is well confirmed. Analytical results for ellipses show that the conjectured relation is actually *exact* at all aspect ratios and we conjecture that $[\eta]$ equals $[\sigma]_\infty$ for all shapes in two dimensions.

Section VI considers the case of flat "plate-like" objects, which requires special analytic treatment. Plate-like particles are intermediate in their properties between the three- and two-dimensional (3D and 2D) cases and these problems tend to be especially difficult analytically. Numerical finite element calculations, however, can be carried out using methods similar to those used for other shapes (see Appendix E). This case has important applications in the scattering of sound waves and electromagnetic radiation through apertures and many numerical results have accumulated in the technical literature. We summarize these connections since they provide the source of predictions for $[\eta]$ through our proposed relation between $[\eta]$ and $[\sigma]_\infty$.

II. POLARIZABILITY, INTRINSIC CONDUCTIVITY, AND VIRTUAL MASS

Maxwell [4a] first considered the classic problem of the conductivity σ of a particle suspension in which the suspended particles have a different conductivity σ_p than the suspending medium σ_0. He recognized that the change in conductivity reflected the average dipole moment induced by the particles on the suspending medium in response to an applied field. For a dilute suspension of hard spheres the effect is the simple additive sum of the effects caused by the individual particle dipoles. The effective conductivity σ of the dilute mixture then equals,

$$\sigma/\sigma_0 = 1 + [3(\Delta_\sigma - 1)/(\Delta_\sigma + 2)]\phi + O(\phi^2), \qquad \Delta_\sigma \equiv \sigma_p/\sigma_0 \quad (2.1a)$$

where Δ_σ is the "relative conductivity" and ϕ is the volume fraction of suspended particles. Exact results that go beyond this classic result are limited, however. There are effective medium calculations [3] that attempt to extend the "virial expansion" [Eq. (2.1a)] to higher powers of ϕ. Sangani [5] recently generalized Maxwell's calculation for spherical particles to d dimensions. The second virial coefficient for σ/σ_0 was calculated by Levine and McQuarrie [6] for conducting spheres ($\Delta_\sigma \to \infty$), while Jefferey [7] treated the case of arbitrary Δ_σ.

The virial expansion [Eq. (2.1a)] has been verified experimentally for dilute suspensions of numerous substances. For example, the leading order virial coefficient for conducting spheres equals 3. This value has been observed by Voet for nearly spherical iron particles (diameter = 10 μm) in linseed and mineral oils [31]. Emulsions of salt water in fuel oil and mercury drops in different oils have also been found to be consistent with Eq. (2.1a) where Δ_σ is large [32]. The corresponding prediction for insulating suspended spheres (Δ_σ near 0) has been observed for suspensions of glass beads and sand particles in salt solutions [33] and for gas bubbles in salt solutions [34]. The virial coefficient notably changes sign and equals $-\frac{3}{2}$ in the insulating spherical particle suspension. Good agreement with Eq. (2.1a) has also been observed in fluidized beds where Δ_σ was tuned over a range of values [35].

At higher concentrations the independent particle approximation of the dilute regime no longer holds, but the leading order virial coefficient still plays a primary role in theoretical estimates of the high concentration variation of transport properties. In the simple effective medium theories of Bruggeman [36] and Brinkman [37], for example, the resummation of the virial expansion for an arbitrary transport property P of a suspension is quite generally given by,

$$P/P_0 \approx (1-\phi)^m \tag{2.1b}$$

where P_0 is the property for a pure suspending medium. Consistency at low concentration requires that the "critical exponent" m has the same magnitude as the leading order virial coefficient. This simple prediction, which is derived on the basis of very simplistic reasoning, is often in remarkably good agreement with observations in physically important systems [31–33,38–41]. For example, Archie's law for the conductivity of rocks saturated with salt water [39, 41] follows directly from Eqs. (2.1a) and (2.1b). The corresponding Brinkman–Roscoe [37, 38] result has also been cited often as a useful description of the viscosity of concentrated suspensions. We mention these approximate calculations only to illustrate the primary role of the leading order virial coefficient in developing a theoretical description of mixture properties at higher concentrations. The following development is restricted to the low concentration regime, where such uncontrolled approximations are unnecessary.

The practically important inverse problem of determining the volume fraction of a suspension of complicated shaped particles from electrical measurements motivated the generalization of Eq. (2.1a) to particles having arbitrary shape. Fricke [8] treated the case of ellipsoidal particles and utilized a Clausius–Mosotti style [42] effective medium theory to approximate the higher concentration regime. These effective medium calculations are exact in the dilute regime where they reduce to a virial expansion of the form for Eq. (2.1a).

The low concentration σ virial expansion of randomly oriented and arbitrarily shaped particles equals [43, 44]:

$$\sigma/\sigma_0 = 1 + [\sigma]\phi + O(\phi^2) \tag{2.2a}$$

$$[\sigma] \equiv \lim_{\phi\to 0^+} (\sigma - \sigma_0)/(\sigma_0\phi) \tag{2.2b}$$

where $[\sigma]$ is the "intrinsic conductivity." The magnitude of $[\sigma]$ can be rapidly varying for extended or flat particles depending on the relative conductivity Δ_σ, so that the effect of adding a given amount of material to a suspension can be greatly dependent on particle shape and composition. It is often convenient to define virials such as $[\sigma]$ in terms of a number concentration when the suspended particles have zero volume, as in the cases of needles, plates, and idealized random walk chains, for example.

The "polarizability" $\boldsymbol{\alpha}$ describes the average dipole moment induced on a particle in an applied field (electric or magnetic) and the calculation of the virial coefficient $[\sigma]$ therefore requires the determination of $\boldsymbol{\alpha}$ or at least an average of the matrix elements defining $\boldsymbol{\alpha}$ (see below). The quantity $\boldsymbol{\alpha}$ is a second rank tensor [4b,44] that depends on particle orientation, shape, size, and the ratio of the particle property to the matrix property for the property that is being considered [see Eqs. (2.1a) and (2.5b)]. The average polarizability $\langle \alpha \rangle$, which is $1/d$ times the trace of the polarizability tensor, is a scalar that is invariant under particle rotations [45, 46]. The polarizability has the units of volume so that the ratio of $\langle \alpha \rangle$ and the particle volume V_p is a scale invariant functional of particle shape and Δ_σ. Calculation of $\langle \alpha \rangle$ is often easier than the full polarizability tensor $\boldsymbol{\alpha}$, since any three orthogonal directions can be chosen for the field directions in the calculation of $\langle \alpha \rangle$. Equivalently, we can angularly average $\boldsymbol{\alpha}$ over all orientation angles with uniform probability [45, 46] to obtain $\langle \alpha \rangle$. In some applications it is useful to orient the suspended particles, in which case the effective conductivity $\boldsymbol{\sigma}$ of the composite becomes explicitly dependent on $\boldsymbol{\alpha}$ [9]. Historically, the anisotropic case was found to be important in the design of microwave lenses and other artificial dielectrics where large scale conducting elements are arrayed in an insulating matrix [47–50]. The anisotropic situation is also encountered in the optical properties of sheared anisotropic particle suspensions [51]. In this chapter, we emphasize the average polarizability $\langle \alpha \rangle$, which is relevant to suspensions in which the particle orientation is completely random.

In an electrostatic context the polarizability describes how the charges of a body of dielectric constant ε_p, embedded in a medium having a dielectric constant ε_0, are distorted in response to an applied electric field

[42, 52]. The distorted charge distribution gives rise to a dipolar field that reacts upon the applied field, thereby modifying the net effective field in the proximity of the body. This connection between conductivity and the dielectric constant is natural since Eqs. (2.1) and (2.2) also describe the dielectric constant of suspensions of particles with a relative dielectric constant $\Delta_\varepsilon = \varepsilon_p/\varepsilon_0$. Moreover, these equations apply equally well to the magnetic permeability, diffusion coefficient (see Appendix A), and the thermal conductivity of dilute suspensions, where the magnetic field, concentration gradient, and the temperature gradient are the corresponding "fields" [1, 2, 21d].

Although simple in principle, the calculation of the polarizability tensor for objects of general shape is a mathematical problem of notorious difficulty. Indeed, the ellipsoid [8, 52] is the only shape for which exact analytic results have been obtained as a function of Δ_σ. There have been recent numerical calculations of the polarizability tensor for other objects in relation to Rayleigh scattering (e.g., radar) applications [53, 54]. The situation is better for limiting values of the relative conductivity Δ_σ where the polarizability tensor $\boldsymbol{\alpha}(\Delta_\sigma)$ simplifies. For highly conducting (superconducting) inclusions, the polarizability tensor reduces to the electric polarizability $\boldsymbol{\alpha}_e$,

$$\lim_{\Delta_\sigma \to \infty} \boldsymbol{\alpha}(\Delta_\sigma) \equiv \boldsymbol{\alpha}_e \tag{2.3a}$$

and $[\sigma]$ for randomly oriented inclusions, having a much higher conductivity than the matrix, equals,

$$[\sigma(\Delta_\sigma \to \infty)] \equiv [\sigma]_\infty = \langle \alpha_e \rangle / V_p \tag{2.3b}$$

where $\langle \boldsymbol{\alpha}_e \rangle$ denotes the average electric polarizability tensor. The case of insulating inclusions in a conducting medium corresponds formally to $\Delta_\sigma \to 0^+$, so that we have

$$\lim_{\Delta_\sigma \to 0} \boldsymbol{\alpha}(\Delta_\sigma) \equiv \boldsymbol{\alpha}_m \tag{2.4a}$$

$$[\sigma(\Delta_\sigma \to 0^+)] \equiv [\sigma]_0 = \langle \alpha_m \rangle / V_p \tag{2.4b}$$

where $\boldsymbol{\alpha}_m$ is the magnetic polarizability with $\langle \alpha_m \rangle$ as the corresponding average. In the $\Delta_\sigma \to 0^+$ and $\Delta_\sigma \to \infty$ limits $[\sigma]$ is simply a functional of particle shape and spatial dimension. Many specific examples are given below. We note that $[\sigma]$ is rather insensitive to particle shape when the conductivity of the particles is similar to the embedding medium ($\Delta_\sigma \approx 1$). In this limit $[\sigma]$ equals [42b],

$$[\sigma] = (\Delta_\sigma - 1) + O[(\Delta_\sigma - 1)^2] \tag{2.4c}$$

which is completely independent of particle shape in leading order.

The limiting relations [Eqs. (2.3b) and (2.4b)], connecting the intrinsic conductivity to the electric and magnetic polarizabilities of a conductor, can be appreciated from a more general relation between the generalized electric **E** and magnetic **H** field polarizability tensors, $\boldsymbol{\alpha}(\mathbf{E})$ and $\boldsymbol{\alpha}(\mathbf{H})$, which allows a unified discussion of the response of complicated shaped objects to both electrostatic and magnetostatic fields. Senior [55] has proven the validity of the general relations,

$$\boldsymbol{\alpha}(\mathbf{E}) = \mathbf{X}(\Delta_\varepsilon), \qquad \boldsymbol{\alpha}(\mathbf{H}) = -\mathbf{X}(\Delta_\mu) \tag{2.5a}$$

where Δ_ε and Δ_μ are the relative dielectric constant and magnetic permeability,

$$\Delta_\varepsilon = \varepsilon_p/\varepsilon_0, \qquad \Delta_\mu = \mu_p/\mu_0 \tag{2.5b}$$

and **X** is the *same* function for *both* electric and magnetic fields. A perfect conductor ($\Delta_\varepsilon \to \infty$) is "magnetically impermeable" ($\Delta_\mu \to 0^+$), while a perfect insulator (($\Delta_\varepsilon \to 0^+$) is formally a "magnetic conductor" ($\Delta_\mu \to \infty$) [50, 56]. The relevance of the magnetic polarizability in describing the insulating limit ($\Delta_\varepsilon \to 0^+$) of the intrinsic conductivity is thus apparent.

Keller et al. [57] recently emphasized the *equivalence* of the magnetic polarizability tensor $\boldsymbol{\alpha}_m$ and the hydrodynamic effective mass tensor **M** describing a particle translating through an inviscid, irrotational, and incompressible ("perfect") liquid. The parameter **M** is equal to the particle volume V_p plus the "added mass" or "virtual mass" **W** associated with the kinetic energy imparted to the fluid from the particle motion,

$$\boldsymbol{\alpha}_m = -\mathbf{M}, \qquad \mathbf{M} = V_p\mathbf{I} + \mathbf{W} \tag{2.6}$$

where **I** is the identity matrix. The minus sign indicates that each matrix element is multiplied by -1. (Sometimes these tensors $\boldsymbol{\alpha}_m$ and **M** are defined to have the same sign.) The fundamental relation Eq. (2.6) was indicated earlier by Kelvin [58] and implicitly by others [59].

Almost all hydrodynamic calculations of **M** assume that the particle density is much higher than the surrounding fluid. Birkhoff and co-workers [60–63] showed that the dipolar field disturbance induced by the motion of particles having comparable density to the fluid medium involve the general polarizability tensor **X** in Eq. (2.5a), where the field permittivity parameter corresponding to Δ_σ is related to the relative density of the particle and the fluid. This generalized relation implies that

M of low density objects (e.g., air bubbles in water) corresponds to $\boldsymbol{\alpha}_e$ rather than $\boldsymbol{\alpha}_m$ [60]. In the following discussion we restrict ourselves to the more conventional case in which the moving object is assumed to have a much higher density than the fluid. The terms effective mass and virtual mass will then always imply the Keller–Kelvin relation [Eq. (2.6)]. A *nonperturbative* generalization of Eq. (2.6) to finite concentrations is discussed in Appendix A.

It should be mentioned that the hydrodynamic applications of **M** are not limited to transient hydrodynamic phenomena associated with the forces on accelerating particles in a fluid medium [64,65]. The presence of a body in a converging stream of an inviscid fluid, such as air in a wind tunnel impinging on an aircraft model, gives rise to a force on the body that is determined by the **M** tensor [**M** equals $(3V_p/2)\mathbf{I}$ for a sphere and the angular average of **M** for near-spherical and slender particles usually differs little from the sphere value; see below] and the pressure gradient in the channel. The force is in the direction of the pressure gradient. Thomson (Lord Kelvin) [66] and Taylor [67] showed that this "buoyancy drag force" is obtained even in non-simply connected spaces, like porous media. The Kelvin–Taylor result for drag forces in "perfect" fluids is a natural counterpart to the Stokes drag force [64] on slowly translating particles in viscous fluids, where a shape functional similar to **M** arises [see Eq. (3.10)].

The importance of Eqs. (2.3b), (2.4b), and (2.6) derives from the extensive mathematical and technical literature relating to the calculation of $\boldsymbol{\alpha}_e$ and **M** [6, 26, 46, 52, 68–76]. These functionals of object shape are naturally encountered in the solution of the Laplace equation on the exterior of regions of various shapes. Consequently, these shape functionals have attracted a mathematical interest quite apart from technical applications [26, 46]. For example, it has rigorously been shown that $\langle \alpha_e \rangle$ and $\langle M \rangle$ achieve their *absolute minima* for a circle and sphere in $d = 2,3$ dimensions [26, 77] of all objects having a finite area or volume, respectively. (Presumably, a hypersphere minimizes these functionals in d dimensions, $d \geq 2$.) Numerical illustrations of this sphere minimization property are presented below. It is also known (implicitly from the work of Keller and Mendelson [78]) that $\langle \alpha_e \rangle = -\langle \alpha_m \rangle$ in $d = 2$ and that these shape functionals have a general geometric interpretation in terms of conformal mapping, as will be discussed in Section V below.

Payne and Weinstein [79] proved that some components of $\boldsymbol{\alpha}_e$ and **M** for certain regions (having reflection symmetry in $d = 2$ or axisymmetric particles in $d = 3$) are related to the capacity C of the region. The capacity C is another shape functional related to solving the Laplace equation on the exterior of a particle. Numerous applications of this quantity are summarized in [80], which also describes a probabilistic method for calculating C by hitting a region of arbitrary shape with Brownian paths launched from an enclosing surface. This development makes the Payne–Weinstein relation attractive for the calculation of $\boldsymbol{\alpha}_e$ and $\boldsymbol{\alpha}_m$.

Finally, we mention that exact calculations of $\boldsymbol{\alpha}_e$ and $\mathbf{M}$ can be made for certain object boundaries and associated coordinate systems for which the Laplace equation is separable [46] and for regions related to the separable boundary cases by Kelvin inversion [4, 52b]. This leads to quite a few intrinsic conductivity results for objects with interesting shapes that are useful in checking numerical methods applicable to more generally shaped regions. Some results of this kind are summarized below.

The technology literature is also a rich source of results for $\boldsymbol{\alpha}_e$, $\boldsymbol{\alpha}_m$, and $\mathbf{M}$. Apart from the relation to transport coefficients like σ, ε, μ, and D, mentioned above, the magnetic and electric polarization tensors have fundamental interest because they *completely determine* [55–57] the scattering of electromagnetic waves having long wavelengths relative to the (metallic) scattering object size, this is, Rayleigh scattering [81]. It is a crucial application to discriminate object shape to the maximum extent possible from long wavelength radiation like radar and, needless to say, the technical literature reflects a preoccupation with objects having the shapes of missiles and space vehicles [53, 68]. Weather radar applications are also important [82]. We also note that scattering of long wavelength sound waves from hard obstacles is determined by $\mathbf{M}$ and the particle volume V_p [83–86] and that $\boldsymbol{\alpha}_e$ and $\mathbf{M}$ are also fundamental in the scattering of electromagnetic and sound waves through apertures [72, 87–91]. These applications especially require the calculation of $\boldsymbol{\alpha}_e$ and $\mathbf{M}$ for plate-shaped objects [92].

Next, we tabulate the components of the polarization tensors $\boldsymbol{\alpha}_e$ and $\boldsymbol{\alpha}_m$ for ellipsoids, since this information is important in the comparisons below with intrinsic viscosity data. These tabulations should give the reader some feeling for the magnitudes involved and illustrate some of the general ideas stated above.

The ellipsoid provides the simplest example of an object having variable shape. Appendix B summarizes the necessary mathematical and numerical computations involved in calculating $\boldsymbol{\alpha}_e$ (ellipsoid) for a range of principal axis radii ratios. In Figs. 2.1 and 2.2 we present the polarizability results for ellipsoids of revolution per unit particle volume (which actually have closed-form analytical solutions, see Appendix B). Enough points have been calculated to make the graphs of these quantities appear as smooth curves. The abscissa x denotes the length of the ellipsoid along the symmetry direction relative to the axis length

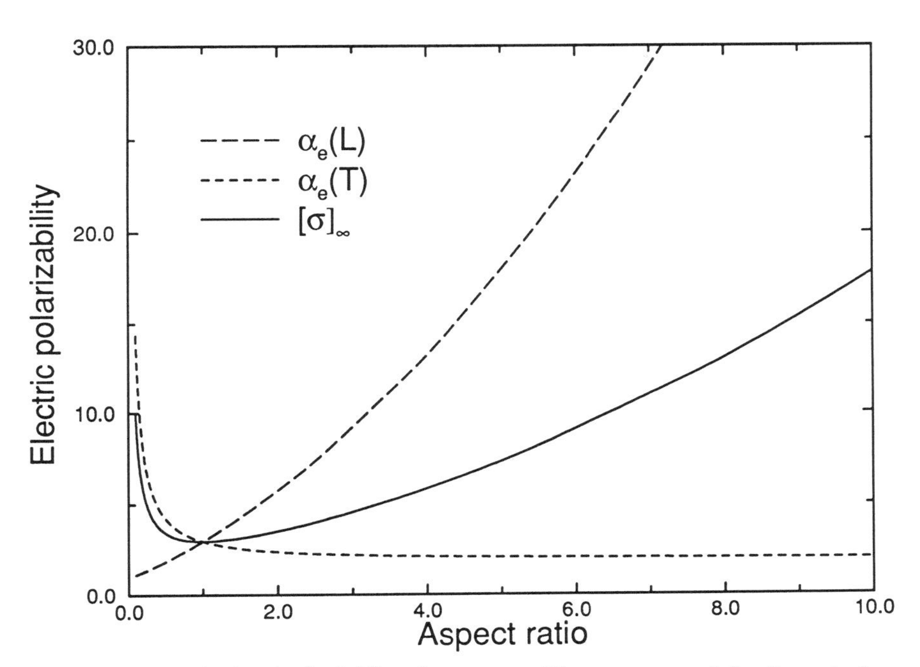

Figure 2.1. The longitudinal (L) and transverse (T) components of the dimensionless (normalized by the particle volume) electric polarizability tensor $\boldsymbol{\alpha}_e$ for ellipsoids of revolution and the average of these components, the intrinsic conductivity of a conductor $[\sigma]_\infty$.

normal to the symmetry axis (aspect ratio). The component of $\boldsymbol{\alpha}_e$ and $\boldsymbol{\alpha}_m$ along the symmetry axis is denoted by L and the component normal to the symmetry axis is denoted by T. The average polarizabilities, $\langle \boldsymbol{\alpha}_e \rangle / V_p = [\sigma]_\infty$ and $\langle \alpha_m \rangle / V_p = [\sigma]_0$, are also shown. These quantities are invariant under particle rotations and are functionals of particle shape only. A tabulation of this numerical data for ellipsoids of revolution is given in Table I. We give the data in the table in the dimensionless form $\boldsymbol{\alpha}/V_p$, since the results are then independent of the absolute particle size. A more general tabulation for the triaxial case is given separately for $\boldsymbol{\alpha}_e$ and $\boldsymbol{\alpha}_m$ in Tables II and III for a range of the two principal axis ratios. All reported digits shown in Tables I–III are significant.

We observe in Figs. 2.1 and 2.2 that the averages $\langle \alpha_e \rangle / V_p = [\sigma]_\infty$ and $\langle \alpha_m \rangle / V_p = [\sigma]_0$ obtain absolute minima for $x = 1$. This accords with the exact results [26, 77] mentioned above, which indicate that this minimum

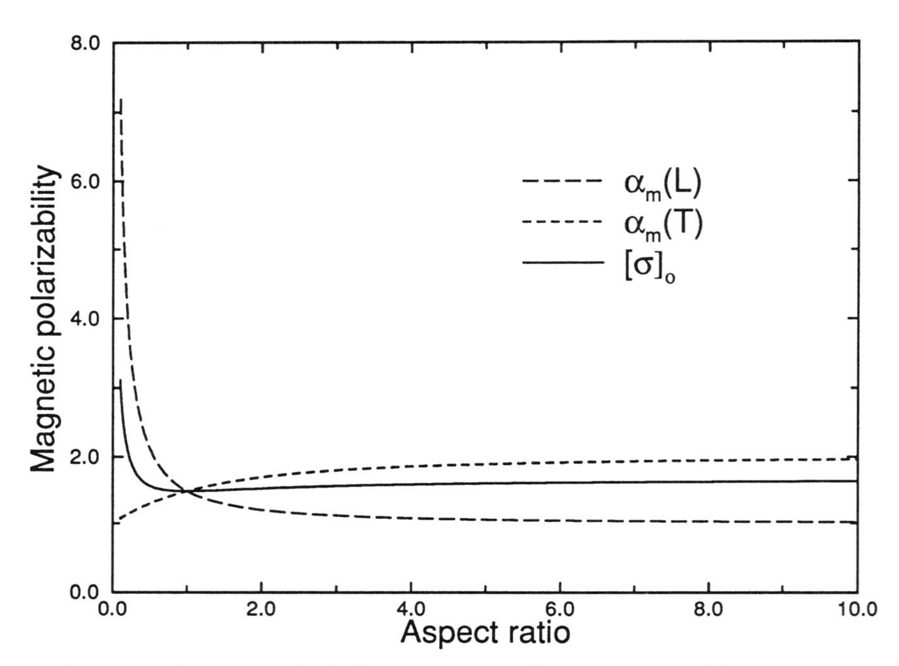

Figure 2.2. The longitudinal (L) and transverse (T) components of the dimensionless (normalized by the particle volume) magnetic polarizability tensor $\boldsymbol{\alpha}_m$ for ellipsoids of revolution and the average of these components, the intrinsic conductivity of an insulator $[\sigma]_0$.

is achieved in the case of a sphere for all objects having a given finite volume. These virial coefficients are observed to be quite sensitive to the aspect ratio x in the approach to the disk limit, but there are also significant effects of particle asymmetry on the virial coefficients for highly *conducting* needlelike ($x \gg 1$) particles. Needlelike *non-conducting* inclusions lead to remarkably little change in the intrinsic conductivity. Thus, we can understand the general experimental observation that nonconducting asymmetric inclusions in a conducting medium often lead to nearly the same intrinsic conductivity [93], except in the case of platelet shaped particles [94]. This implies that the most economical means of making a medium more insulating is through the introduction of a small concentration of nonconducting plate-like particles. Figure 2.2 also suggests that the polarizability of very irregular nonconducting objects, such as alkane chains and other nonconducting polymers, should

TABLE I
Polarizability Components for Ellipsoids of Revolution, Electric and Magnetic ($\boldsymbol{\alpha}_e$,$\boldsymbol{\alpha}_m$).

Aspect Ratio	Prolate					
	α_e (L)V_p	α_e(T)/V_p	$[\sigma]_\infty$	α_m(L)/V_p	α_m(T)/V_p	$[\sigma]_0$
2	5.7616	2.4200	3.5339	1.2100	1.7042	1.5395
3	9.1988	2.2439	4.5622	1.1220	1.8039	1.5766
4	13.2613	2.1631	5.8625	1.0816	1.8598	1.6004
5	17.9144	2.1182	7.3836	1.0591	1.8943	1.6159
6	23.1323	2.0904	9.1043	1.0452	1.9171	1.6265
7	28.8946	2.0717	11.0127	1.0358	1.9331	1.6340
8	35.1848	2.0585	13.1006	1.0293	1.9447	1.6396
9	41.9890	2.0488	15.3622	1.0244	1.9535	1.6438
10	49.2954	2.0414	17.7927	1.0207	1.9602	1.6471
20	148.168	2.0136	50.7320	1.0068	1.9866	1.6600
30	290.342	2.0069	98.1188	1.0035	1.9931	1.6632
40	472.623	2.0042	158.877	1.0021	1.9958	1.6646
50	693.013	2.0029	232.339	1.0014	1.9971	1.6652
60	950.083	2.0021	318.029	1.0011	1.9979	1.6656
70	1,242.74	2.0016	415.581	1.0008	1.9984	1.6659
80	1,570.10	2.0013	524.701	1.0006	1.9987	1.6660
90	1,931.43	2.0010	645.147	1.0005	1.9990	1.6661
100	2,326.12	2.0009	776.710	1.0004	1.9991	1.6662
200	8,013.36	2.0002	2,672.45	1.0001	1.9998	1.6665
300	16,675.8	2.0001	5,559.95	1.0001	1.9999	1.6666
400	28,145.8	2.0001	9,383.29	1.0000	1.9999	1.6666
500	42,316.9	2.0000	14,106.9	1.0000	2.0000	1.6666
600	59,112.3	2.0000	19,705.4	1.0000	2.0000	1.6666
700	78,472.2	2.0000	26,158.7	1.0000	2.0000	1.6667
800	100,348.0	2.0000	33,450.8	1.0000	2.0000	1.6667
900	124,700.0	2.0000	41,568.2	1.0000	2.0000	1.6667
1000	151,494.0	2.0000	50,499.4	1.0000	2.0000	1.6667

Reciprocal Aspect Ratio	Oblate					
	α_e(L)/V_p	α_e(T)/V_p	$[\sigma]_\infty$	α_m(L)/V_p	α_m(T)/V_p	$[\sigma]_0$
2	1.8968	4.2301	3.4524	2.1151	1.3096	1.5781
3	1.5738	5.4853	4.1815	2.7426	1.2230	1.7295
4	1.4212	6.7486	4.9728	3.3743	1.1740	1.9074
5	1.3325	8.0155	5.7878	4.0078	1.1425	2.0976
6	1.2746	9.2844	6.6145	4.6422	1.1207	2.2945
7	1.2338	10.5544	7.4476	5.2772	1.1047	2.4955
8	1.2036	11.8253	8.2847	5.9126	1.0924	2.6991
9	1.1802	13.0966	9.1245	6.5483	1.0827	2.9045
10	1.1617	14.3683	9.9661	7.1841	1.0748	3.1112
20	1.0797	27.0935	18.4222	13.5467	1.0383	5.2078
30	1.0529	39.8234	26.8999	19.9117	1.0258	7.3211
40	1.0396	52.5545	35.3829	26.2773	1.0194	9.4387
50	1.0316	65.2862	43.8680	32.6431	1.0156	11.5581
60	1.0263	78.0181	52.3541	39.0090	1.0130	13.6783
70	1.0225	90.7501	60.8409	45.3750	1.0111	15.7991
80	1.0197	103.482	69.3280	51.7411	1.0098	17.9202
90	1.0175	116.214	77.8154	58.1072	1.0087	20.0415
100	1.0158	128.946	86.3030	64.4733	1.0078	22.1630
200	1.0079	256.269	171.182	128.134	1.0039	43.3809
300	1.0052	383.593	256.064	191.796	1.0026	64.6007
400	1.0039	510.917	340.946	255.458	1.0020	85.8209
500	1.0031	638.241	425.828	319.120	1.0016	107.041
600	1.0026	765.565	510.711	382.782	1.0013	128.261
700	1.0022	892.889	595.593	446.444	1.0011	149.482
800	1.0020	1,020.21	680.476	510.106	1.0010	170.702
900	1.0017	1,147.53	765.358	573.768	1.0009	191.923
1000	1.0016	1,274.86	850.241	637.430	1.0008	213.144

TABLE II
Components of the Electric Polarizability Tensor $\boldsymbol{\alpha}_e$ for Triaxial Ellipsoids

a_1	a_2	a_3	$\alpha_e(11)/V_p$	$\alpha_e(22)/V_p$	$\alpha_e(33)/V_p$	$[\sigma]_\infty$
1	2	3	1.7345	3.7432	6.3979	3.9585
1	2	4	1.6587	3.5115	8.9007	4.6903
1	2	5	1.6162	3.3791	11.721	5.5723
1	2	6	1.5894	3.2950	14.847	6.5774
1	2	7	1.5714	3.2375	18.268	7.6924
1	2	8	1.5586	3.1962	21.974	8.9098
1	2	9	1.5490	3.1654	25.959	10.224
1	2	10	1.5418	3.1417	30.215	11.633
1	2	20	1.5140	3.0490	86.743	30.435
1	2	50	1.5030	3.0110	390.06	131.52
1	2	100	1.5009	3.0034	1,283.9	429.47
1	2	300	1.5001	3.0005	9,015.0	3006.5
1	2	500	1.5001	3.0002	22,717.0	7574.0
1	2	1,000	1.5000	3.0001	80,706.0	26903.0
1	3	4	1.4979	5.0419	7.4586	4.6662
1	3	5	1.4548	4.7842	9.6534	5.2975
1	3	6	1.4275	4.6178	12.061	6.0356
1	3	7	1.4089	4.5027	14.676	6.8626
1	3	8	1.3956	4.4190	17.491	7.7688
1	3	9	1.3857	4.3560	20.503	8.7483
1	3	10	1.3781	4.3070	23.706	9.7970
1	3	20	1.3486	4.1107	65.575	23.678
1	3	50	1.3367	4.0258	285.32	96.894
1	3	100	1.3344	4.0081	923.91	309.75
1	3	300	1.3335	4.0012	6,377.3	2127.5
1	3	500	1.3334	4.0005	15,980.0	5328.5
1	3	1,000	1.3333	4.0001	56,424.0	18809.0
1	4	5	1.3772	6.3308	8.6255	5.4445
1	4	6	1.3492	6.0582	10.667	6.0250
1	4	7	1.3300	5.8680	12.869	6.6892
1	4	8	1.3162	5.7286	15.228	7.4244
1	4	9	1.3059	5.6227	17.739	8.2227
1	4	10	1.2979	5.5400	20.400	9.0794
1	4	20	1.2667	5.2013	54.705	20.391
1	4	50	1.2538	5.0486	231.29	79.199
1	4	100	1.2512	5.0155	738.28	248.18
1	4	300	1.2502	5.0023	5,020.9	1675.7
1	4	500	1.2501	5.0009	12,520.0	4175.6
1	4	1,000	1.2500	5.0003	43,979.0	14661.0
1	5	6	1.3038	7.6145	9.8349	6.2511
1	5	7	1.2841	7.3327	11.785	6.8008
1	5	8	1.2698	7.1249	13.864	7.4198
1	5	9	1.2591	6.9661	16.069	8.0983
1	5	10	1.2508	6.8414	18.397	8.8300
1	5	20	1.2180	6.3225	48.043	18.528
1	5	50	1.2041	6.0802	198.00	68.430
1	5	100	1.2013	6.0260	623.89	210.37
1	5	300	1.2002	6.0040	4,186.8	1,398.0
1	5	500	1.2001	6.0016	10,396.0	3,467.7
1	5	1,000	1.2000	6.0005	36,349.0	12,118.0
1	6	7	1.2544	8.8954	11.065	7.0718
1	6	8	1.2397	8.6071	12.955	7.6009
1	6	9	1.2287	8.3860	14.952	8.1892
1	6	10	1.2201	8.2115	17.055	8.8289
1	6	20	1.1859	7.4754	43.525	17.395
1	6	50	1.1711	7.1214	175.29	61.195

Table II (*Continued*)

a_1	a_2	a_3	$\alpha_e(11)/V_p$	$\alpha_e(22)/V_p$	$\alpha_e(33)/V_p$	$[\sigma]_\infty$
1	6	100	1.1681	7.0399	545.81	184.67
1	6	300	1.1669	7.0062	3,618.4	1,208.8
1	6	500	1.1667	7.0025	8,950.0	2,986.0
1	6	1,000	1.1667	7.0007	31,163.0	10,390.0
1	7	8	1.2188	10.174	12.308	7.9005
1	7	9	1.2075	9.8815	14.155	8.4148
1	7	10	1.1987	9.6498	16.094	8.9809
1	7	20	1.1632	8.6607	40.253	16.692
1	7	50	1.1477	8.1728	158.74	56.021
1	7	100	1.1444	8.0576	488.85	166.01
1	7	300	1.1431	8.0091	3,204.4	1,071.2
1	7	500	1.1429	8.0037	7,897.7	2,635.6
1	7	1,000	1.1429	8.0011	27,395.0	9,134.8
1	8	9	1.1920	11.452	13.558	8.7343
1	8	10	1.1830	11.155	15.373	9.2375
1	8	20	1.1464	9.8790	37.772	16.265
1	8	50	1.1301	9.2348	146.10	52.157
1	8	100	1.1266	9.0792	445.32	151.84
1	8	300	1.1252	9.0126	2,888.4	966.19
1	8	500	1.1251	9.0052	7,095.0	2,368.3
1	8	1,000	1.1250	9.0015	24,524.0	8,178.2
1	9	10	1.1710	12.728	14.814	9.5714
1	9	20	1.1335	11.130	35.824	16.029
1	9	50	1.1165	10.307	136.12	49.182
1	9	100	1.1128	10.105	410.89	140.70
1	9	300	1.1114	10.017	2,638.5	883.23
1	9	500	1.1112	10.007	6,460.9	2,157.3
1	9	1,000	1.1111	10.002	22,259.0	7,423.5
1	10	20	1.1233	12.415	34.255	15.931
1	10	50	1.1057	11.392	128.01	46.839
1	10	100	1.1018	11.135	382.90	131.71
1	10	300	1.1003	11.022	2,435.6	815.92
1	10	500	1.1001	11.009	5,946.4	1,986.1
1	10	1,000	1.1000	11.002	20,423.0	6,811.8
1	20	50	1.0579	22.911	89.896	37.955
1	20	100	1.0527	21.715	250.13	90.969
1	20	300	1.0504	21.128	1,473.0	498.42
1	20	500	1.0502	21.055	3,510.9	1,177.6
1	20	1,000	1.0501	21.016	11,763.0	3,928.3
1	50	100	1.0243	57.057	161.17	73.084
1	50	300	1.0208	52.284	819.03	290.78
1	50	500	1.0203	51.582	1,860.8	637.82
1	50	1,000	1.0201	51.188	5,938.5	1,996.9
1	100	300	1.0112	107.87	568.97	225.95
1	100	500	1.0105	104.30	1,226.9	444.09
1	100	1000	1.0102	102.14	3,708.3	1,270.5
1	300	500	1.0043	345.37	744.03	363.47
1	300	1,000	1.0037	318.56	1,984.9	768.18
1	500	1,000	1.0024	559.34	1,589.2	716.52
1	1,000	500	1.0024	1589.0	559.36	716.47

TABLE III
Components of the Magnetic Polarizability Tensor $\boldsymbol{\alpha}_m$ for Triaxial Ellipsoids

a_1	a_2	a_3	$\alpha_m(11)/V_p$	$\alpha_m(22)/V_p$	$\alpha_m(33)/V_p$	$[\sigma]_0$
1	2	3	2.3615	1.3645	1.1852	1.6371
1	2	4	2.5181	1.3981	1.1265	1.6809
1	2	5	2.6229	1.4203	1.0932	1.7122
1	2	6	2.6965	1.4357	1.0722	1.7348
1	2	7	2.7501	1.4469	1.0579	1.7516
1	2	8	2.7903	1.4553	1.0476	1.7644
1	2	9	2.8213	1.4618	1.0400	1.7744
1	2	10	2.8457	1.4669	1.0342	1.7823
1	2	20	2.9455	1.4880	1.0116	1.8151
1	2	50	2.9879	1.4972	1.0025	1.8292
1	2	100	2.9963	1.4991	1.0007	1.8321
1	2	300	2.9994	1.4998	1.0001	1.8332
1	2	500	2.9997	1.4999	1.0000	1.8333
1	2	1000	2.9999	1.4999	1.0000	1.8333
1	3	4	3.0083	1.2474	1.1548	1.8035
1	3	5	3.1988	1.2642	1.1155	1.8595
1	3	6	3.3393	1.2764	1.0904	1.9020
1	3	7	3.4455	1.2854	1.0731	1.9347
1	3	8	3.5277	1.2924	1.0606	1.9603
1	3	9	3.5926	1.2979	1.0512	1.9806
1	3	10	3.6447	1.3023	1.0440	1.9971
1	3	20	3.8682	1.3214	1.0154	2.0684
1	3	50	3.9698	1.3304	1.0035	2.1013
1	3	100	3.9907	1.3324	1.0010	2.1081
1	3	300	3.9986	1.3332	1.0001	2.1107
1	3	500	3.9994	1.3332	1.0000	2.1109
1	3	1000	3.9998	1.3333	1.0000	2.1111
1	4	5	3.6510	1.1875	1.1311	1.9899
1	4	6	3.8638	1.1976	1.1034	2.0550
1	4	7	4.0303	1.2054	1.0842	2.1067
1	4	8	4.1626	1.2114	1.0702	2.1481
1	4	9	4.2694	1.2163	1.0597	2.1818
1	4	10	4.3568	1.2202	1.0515	2.2095
1	4	20	4.7497	1.2380	1.0186	2.3354
1	4	50	4.9407	1.2469	1.0043	2.3974
1	4	100	4.9816	1.2490	1.0013	2.4107
1	4	300	4.9973	1.2498	1.0001	2.4158
1	4	500	4.9989	1.2499	1.0000	2.4163
1	4	1000	4.9997	1.2499	1.0000	2.4166
1	5	6	4.2917	1.1511	1.1131	2.1854
1	5	7	4.5202	1.1579	1.0927	2.2570
1	5	8	4.7063	1.1632	1.0777	2.3158
1	5	9	4.8595	1.1676	1.0663	2.3645
1	5	10	4.9869	1.1711	1.0574	2.4052
1	5	20	5.5872	1.1878	1.0212	2.5988
1	5	50	5.8990	1.1968	1.0050	2.7003
1	5	100	5.9683	1.1989	1.0016	2.7230
1	5	300	5.9953	1.1998	1.0002	2.7318
1	5	500	5.9981	1.1999	1.0000	2.7327
1	5	1,000	5.9994	1.1999	1.0000	2.7332
1	6	7	4.9312	1.1266	1.0993	2.3857
1	6	8	5.1714	1.1314	1.0836	2.4622
1	6	9	5.3727	1.1353	1.0716	2.5266
1	6	10	5.5428	1.1386	1.0622	2.5813
1	6	20	6.3796	1.1544	1.0235	2.8525
1	6	50	6.8434	1.1633	1.0057	3.0042

Table III (*Continued*)

a_1	a_2	a_3	$\alpha_m(11)/V_p$	$\alpha_m(22)/V_p$	$\alpha_m(33)/V_p$	$[\sigma]_0$
1	6	100	6.9502	1.1655	1.0018	3.0392
1	6	300	6.9926	1.1664	1.0002	3.0531
1	6	500	6.9970	1.1665	1.0001	3.0546
1	6	1,000	6.9991	1.1666	1.0000	3.0553
1	7	8	5.5700	1.1090	1.0884	2.5891
1	7	9	5.8192	1.1125	1.0760	2.6693
1	7	10	6.0327	1.1156	1.0662	2.7382
1	7	20	7.1272	1.1305	1.0254	3.0944
1	7	50	7.7726	1.1394	1.0063	3.3061
1	7	100	7.9269	1.1416	1.0020	3.3569
1	7	300	7.9890	1.1426	1.0003	3.3774
1	7	500	7.9956	1.1427	1.0001	3.3795
1	7	1,000	7.9987	1.1428	1.0000	3.3805
1	8	9	6.2083	1.0956	1.0796	2.7945
1	8	10	6.4647	1.0984	1.0695	2.8776
1	8	20	7.8308	1.1126	1.0271	3.3236
1	8	50	8.6858	1.1214	1.0068	3.6047
1	8	100	8.8978	1.1237	1.0022	3.6746
1	8	300	8.9846	1.1248	1.0003	3.7032
1	8	500	8.9937	1.1249	1.0001	3.7063
1	8	1,000	8.9982	1.1249	1.0000	3.7077
1	9	10	6.8463	1.0852	1.0723	3.0013
1	9	20	8.4920	1.0987	1.0287	3.5398
1	9	50	9.5822	1.1074	1.0074	3.8990
1	9	100	9.8625	1.1098	1.0024	3.9916
1	9	300	9.9790	1.1109	1.0003	4.0301
1	9	500	9.9915	1.1110	1.0001	4.0342
1	9	1,000	9.9975	1.1110	1.0000	4.0362
1	10	20	9.1126	1.0876	1.0300	3.7434
1	10	50	10.461	1.0962	1.0078	4.1885
1	10	100	10.820	1.0986	1.0026	4.3073
1	10	300	10.972	1.0997	1.0004	4.3576
1	10	500	10.988	1.0999	1.0001	4.3630
1	10	1,000	10.996	1.0999	1.0000	4.3656
1	20	50	18.258	1.0456	1.0112	6.7718
1	20	100	19.980	1.0482	1.0040	7.3444
1	20	300	20.829	1.0496	1.0006	7.6267
1	20	500	20.929	1.0498	1.0002	7.6599
1	20	1,000	20.979	1.0499	1.0000	7.6764
1	50	100	42.139	1.0178	1.0062	14.721
1	50	300	49.147	1.0194	1.0012	17.056
1	50	500	50.191	1.0197	1.0005	17.404
1	50	1,000	50.751	1.0199	1.0001	17.590
1	100	300	90.683	1.0093	1.0017	30.898
1	100	500	96.135	1.0096	1.0008	32.715
1	100	1,000	99.404	1.0098	1.0002	33.805
1	300	500	235.88	1.0029	1.0013	79.295
1	300	1,000	274.51	1.0031	1.0005	92.172
1	500	1,000	413.72	1.0017	1.0006	138.58

increase in simple proportion to volume (molecular weight) for homologous molecular series, since the change in shape will not appreciably affect the intrinsic conductivity. This effect is observed in gas-phase polarizability estimates on normal alkanes, based on dielectric constant and refractive index measurements [43]. Corresponding gas-phase measurements on conjugated polymeric systems, on the other hand, exhibit a rapidly increasing polarizability with molecular weight [43], in accord with the calculations of $\boldsymbol{\alpha}_e$ and the simplistic view of such polymers as "conductors." The variation of the colors of dyes with molecular weight [95] and certain attractive forces between long chain molecules [96] can be similarly understood using this kind of picture of insulating and conducting polymers and geometrical estimates of the polarizability.

There are other shapes for which $\boldsymbol{\alpha}_e$ and $\boldsymbol{\alpha}_m$ can be determined exactly. Most of these additional exact results are summarized by Schiffer and Szegö [46], but are not well known in the physical science literature. (This is probably due to the rather complicated mathematical form of these exact analytical results.) As an example, we indicate exact $\boldsymbol{\alpha}_e$ results, per unit particle volume, for a torus in Fig. 2.3, where the symbol L again denotes the axis of symmetry. The abscissa is the ratio of the overall torus radius b to the radius a of the tube forming the body of the torus itself. For example, the limit $a \to 0$ for a fixed torus radius b gives an infinitely thin wire ring. Table IV tabulates the corresponding numerical values of the polarizability components used in Fig. 2.3, based on previous tabulations of the equivalent of these numbers by Belovitch and Boersma [97]. The present tabulation is given in a dimensionless form that avoids the problem of the choice of units. Further tabulations of analytic results for the electric and magnetic polarizability tensors, corresponding to other shapes [46], will require careful numerical work.

III. INTRINSIC VISCOSITY AND ITS RELATION TO INTRINSIC CONDUCTIVITY

An increase of the viscosity of suspensions and the shear modulus of solids is generally observed upon adding rigid particles to the medium. The introduction of rigid inclusions perturbs the stress field of the sheared pure medium since locally the field lines cannot penetrate the hard inclusions. There is evidently a qualitative analogy with the electrical conduction problem in a suspension of highly conducting particles where the electric field lines are similarly screened from the interiors of the conducting particles. Many authors have commented on the mathematical resemblance between electrical polarization and linearized flow theory calculations [7, 98, 99], which follow as a consequence of this physical

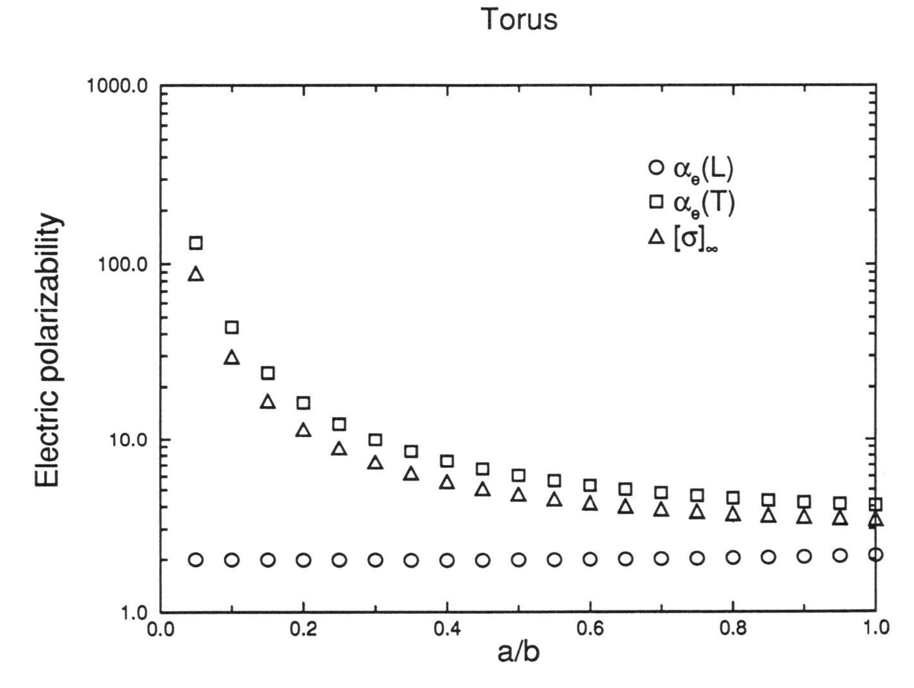

Figure 2.3. The longitudinal (L) and transverse (T) components of the dimensionless (normalized by the particle volume) electric polarizability tensor $\boldsymbol{\alpha}_e$ for tori and the average of these components, the intrinsic conductivity $[\sigma]_\infty$, plotted versus the ratio a/b where b is the overall radius of the torus and a is the radius of the torus tube. The ratio $a/b = 1$ corresponds to a torus without a hole.

analogy. In the following we develop an *approximate* relation between the electrical conductivity and suspension viscosity problems.

Einstein [10a], as part of his investigation of the molecular nature of matter, first calculated the incremental increase of the viscosity η of a dilute hard sphere suspension:

$$\eta/\eta_0 = 1 + (\tfrac{5}{2})\phi + O(\phi^2) \tag{3.1}$$

where η_0 is the pure solution viscosity. (Notably, his original calculation did not give the correct $\frac{5}{2}$ coefficient [10a] in Eq. (3.1) and was later corrected in light of experimental observations by Bancelin [100].) The leading virial coefficient is the intrinsic viscosity $[\eta]$, defined by

$$[\eta] \equiv \lim_{\phi \to 0^+} (\eta - \eta_0)/(\eta_0 \phi) \tag{3.2}$$

TABLE IV
Electric Polarizability Components for Torus

a/b	$\alpha_e(L)/V_p$	$\alpha_e(T)/V_p$	$[\sigma]_\infty$
0.05	1.998	131.851	88.567
0.10	1.995	43.641	29.759
0.15	1.991	24.051	16.698
0.20	1.986	16.282	11.517
0.25	1.983	12.313	8.870
0.30	1.980	9.971	7.307
0.35	1.978	8.455	6.296
0.40	1.977	7.409	5.598
0.45	1.978	6.651	5.093
0.50	1.980	6.083	4.716
0.55	1.985	5.645	4.425
0.60	1.991	5.300	4.197
0.65	1.999	5.024	4.016
0.70	2.008	4.799	3.869
0.75	2.020	4.614	3.750
0.80	2.033	4.461	3.652
0.85	2.047	4.333	3.571
0.90	2.063	4.226	3.505
0.95	2.079	4.135	3.450
0.96	2.083	4.118	3.440
0.97	2.086	4.103	3.431
0.98	2.090	4.087	3.422
0.99	2.094	4.073	3.413
1.00	2.097	4.058	3.405

Experiments on nearly spherical particles in low-concentration suspensions $[\phi < O(1\%)]$ commonly yield a value of $[\eta] \approx 2.7$, which is slightly higher [101] than the Einstein estimate $[\eta] = 2.5$. This small deviation is often ascribed to small particle asphericity or particle clustering [101] and, at any rate, the revised Einstein result [Eq. (3.1)] is a good approximation.

Rayleigh [102], Goodier [17], and Hill and Power [103] pointed out a fundamental analogy between the hydrodynamics of suspensions and the elastostatics of incompressible solids with rigid inclusions, which implies that Einstein's virial expansion for the viscosity of hard sphere suspensions also describes the shear modulus G virial expansion [17, 19],

$$G/G_0 = 1 + (\tfrac{5}{2})\phi + O(\phi^2) \tag{3.3}$$

for an elastic continuum of modulus G_0 containing stiff *spherical* inclusions at low concentration. The intrinsic shear modulus $[G]$ is defined by the limit,

$$[G] \equiv \lim_{\phi \to 0^+} (G - G_0)/(G_0 \phi) \tag{3.4}$$

For compressible spherical particles $[G]$ depends on the Poisson ratio ν of the particle [20],

$$[G] = (\tfrac{5}{2})(1 - 2\Delta\nu)/(1 - 10\Delta\nu/3), \qquad \Delta\nu = \nu - \tfrac{1}{2} \tag{3.5}$$

where $\nu \to \frac{1}{2}$ in the incompressible limit. Derivation of Eq. (3.5) assumes that the matrix material always remains in contact with the inclusion ("sticks") under deformation. A tabulation of experiments from a variety of sources indicates that the shear modulus and suspension viscosity have a common concentration dependence [104],

$$G(\phi)/G_0 \approx \eta(\phi)/\eta_0 \tag{3.6}$$

for nearly spherical, rigid inclusions in an incompressible elastic matrix and a Newtonian fluid, respectively. This behavior is consistent with the simple incompressibility assumption ($\nu = \frac{1}{2}$) and the viscoelastic and elastostatic analogy of Rayleigh [102]. It is emphasized that Eq. (3.6) is observed to hold *regardless of the concentration* of the suspended matter! We note that the simple relation between intrinsic viscosity and intrinsic shear modulus is limited to *spherical* particles and an incompressible suspending medium (see Appendix C).

Experience also indicates that the addition of "softer" materials to liquids and solids does not generally increase the viscosity and shear modulus. This physical situation is analogous to the addition of insulating material to a conducting medium [42a], since the inclusions are "permeable" to the shear-induced stress field lines in the suspending fluid or solid medium. In the extreme case, where the particle inclusions are highly deformable and the matrix is incompressible, $[G]$ becomes [18]

$$[G]_{d=3} = -\tfrac{5}{3}, \qquad [G]_{d=2} = -2 \tag{3.7}$$

so that the solid becomes *softer* with an increasing volume fraction of soft inclusions. The magnitude of $[G]$ for holes is comparable to $[\sigma]_0$ for an insulator in a conducting matrix [see Eq. (2.1)].

The introduction of liquid drops into another viscous fluid or a solid introduces some important additional features. In this case, momentum can propagate into the interior of the droplet and induce internal circulation within the droplet, so that the dissipation is altered from the hard sphere case. In many physical circumstances surface tension or internal pressure tends to make the drop resist deformation, however. Taylor [15] showed that the intrinsic viscosity of idealized *indeformable* liquid drops of viscosity η_{drop} equals

$$[\eta] = 1 + \tfrac{3}{2}[z_\eta/(1+z_\eta)], \qquad z_\eta = \eta_{\text{drop}}/\eta_0 \tag{3.8}$$

Note that $[\eta]$ reduces to 1 in the "bubble" limit $z_\eta \to 0^+$, rather than becoming negative. Experiments on liquid drops suspended in another liquid are often consistent with Eq. (3.8), although there can be complications with surface tension effects (impurities and small droplet size [105]), which can invalidate Eq. (3.8). In the complementary idealized case, where the spherical membrane surrounding the droplet is highly deformable, it is found that [106]

$$[\eta] = -\tfrac{5}{3}[1 - \tfrac{5}{3}z_\eta/(1+\tfrac{2}{3}z_\eta)] \tag{3.9}$$

which reduces to the hole limit $-\frac{5}{3}$ for the elastic problem [Eq. (3.7)] for $z_\eta \to 0$ and the *hard sphere* result ($z_\eta \to \infty$) of Einstein. Equation (3.9), which is comparable to Maxwell's formula [Eq. (2.1a)] for electrical conductivity, has been found to be a reasonable idealization for suspensions of red blood cells and other deformable particles [106]. The rest of this chapter considers only *rigid* particles. The brief discussion above was meant only to illustrate some of the complications that can arise when considering real particle mixtures.

Despite the fundamental importance of $[\eta]$ in determination of molecular shape [107], there are few analytical calculations of $[\eta]$ corresponding to nonspherical objects. Onsager [11] long ago calculated asymptotic results for long hard prolate ellipsoids, and these results were later generalized by Saito [12a] to analytical estimates for arbitrary aspect ratios. Kirkwood and Riseman [108] and Debye and Bueche [109] estimated $[\eta]$ for random coil polymer chains, but these calculations involved uncontrolled approximations. Rallison [13] and Haber and Brenner [14] recently obtained exact results for *triaxial* ellipsoids. The formalism required to treat the triaxial ellipsoid case is quite sophisticated and treatment of these more general shapes is necessarily complicated. The reason for the limited progress in calculating $[\eta]$, relative to $[\sigma]$, is simple: Solution of the steady-state Navier–Stokes equation on the exterior of the hard particles is a significantly more difficult technical problem than the corresponding solution of the Laplace equation.

Recently, Hubbard and Douglas [27] observed an interesting relation between hydrodynamic and electrostatic problems that suggests a route for developing a direct approximate relation between $[\eta]$ and $[\sigma]$. They observed that the *angular average* of the Green's function for the steady-state free space Navier–Stokes equation equals the Green's function of the free space Laplacian [27]. From this observation and the physical *angular averaging* associated with the Brownian particle diffusion process, they deduced that the scalar translational friction coefficient of arbitrarily shaped rigid paricles approximately equals

$$f_{\mathrm{T}} \approx 6\pi\eta C \tag{3.10}$$

where C is the electrostatic capacitance. (The parameter C is the Newtonian capacity as opposed to the logarithmic capacity discussed in Section V. The units of C are chosen so that a sphere of radius R has a capacitance $C = R$.) The capacitance C governs the far field decay of the solution of Laplace's equation where the solution equals 1 on the boundary and approaches zero at great distances from the boundary [26, 80]. Equation (3.10), which is consistent (within $\sim 1\%$ accuracy) with exactly known values of f_{T}, serves as an explicit connection between hydrodynamic and electrostatic problems. Direct comparisons of the average stress and electrostatic (or thermal) dipole coefficients [110–112] in the calculation of $[\eta]$ and $[\sigma]_\infty$, respectively, suggests that $[\eta]$ is simply proportional to $[\sigma]_\infty$ within angular averaging. In other words, it seems reasonable to preangularly average the steady-state Navier–Stokes Green's function so that the hydrodynamic problem reduces to the solution of the Laplace equation on the exterior of the particle as in the former calculations relating translational friction and capacity [27]. This procedure seems reasonable for a dilute particle suspension of randomly oriented particles, since $[\eta]$ is then an invariant under suspension rotations. In this chapter we are interested in checking the numerical accuracy of this relation. The existence of small numerical discrepancies in exact analytical results, described below, show that this relation is not exact, but rather a very good approximation for objects having diverse shapes.

The constant of proportionality between $[\eta]$ and $[\sigma]_\infty$ can be fixed by exact calculations for sphere suspensions [5, 113] in d spatial dimensions

$$[\eta] \approx [(d+2)/(2d)][\sigma]_\infty \tag{3.11}$$

We choose the sphere case to determine the proportionality constant since the preaveraging argument for the Oseen tensor leads to exact results for spheres. Of course, this is a rather trivial case and other shapes must be considered to check the conjectured relation [Eq. (3.11)] [114].

Further motivation of the approximation [Eq. (3.11)], derives from calculations by Kanwal [115], which show an *exact* relation between the rotational friction coefficient f_R and $\boldsymbol{\alpha}_e$ for a certain class of bodies

$$f_R(T) = 2\alpha_e(T)\eta_0 \tag{3.12}$$

corresponding to the rotation of a body of revolution, having an otherwise arbitrary profile, about its axis of symmetry. The parameter $\alpha_e(T)$ is the polarizability component normal to the axis of symmetry. Exact $f_R(T)$ results for a variety of complex-shaped particles can be directly obtained from Eq. (3.12) and from tabulations of $\alpha_e(T)$ [46].

Riseman and Kirkwood [116] noted that a proportionality relation should exist between the rotational friction coefficient and $[\eta]$, and this observation is consistent with the approximation in Eqs. (3.11) and (3.12). The rotational friction coefficient becomes difficult to measure and to calculate for nonsymmetric objects and for flexible objects so we do not pursue this connection further.

Brenner [117] developed the necessary mathematical machinery for calculating $[\eta]$ for rigid axisymmetric particles. It is useful to utilize this formalism to obtain some exact results that can be tested against Eq. (3.11). The particle shape made from two touching spheres of radius a is an interesting test case. Exact calculation, using the formalism of Brenner [117] and associated results for the stress dipole due previously to Wakiya [118], gives an *exact* value for the intrinsic viscosity of two touching and rigidly joined spheres

$$[\eta] = 3.4496\cdots \tag{3.13}$$

(We note that the value of $[\eta]$ given on p. 263 of [117] is incorrect.) An exact calculation of $\boldsymbol{\alpha}_e$ (and thus implicitly $[\sigma]_\infty$) for touching spheres is summarized by Schiffer and Szegö [46]. The electrical polarizability components along the symmetry axis $\alpha_e(L)$ and normal to the symmetry axis $\alpha_e(T)$ equal

$$\alpha_e(L) = 16\pi a^3 \zeta(3), \qquad \alpha_e(T) = 6\pi a^3 \zeta(3) \tag{3.14a}$$

where ζ is the Riemann zeta function, that is $\zeta(3) = 1.20206\cdots$.The intrinsic conductivity $[\sigma]_\infty$ for touching spheres is then

$$[\sigma]_\infty = 7\zeta(3)/2 = 4.2072\cdots, \qquad V_p = 8\pi a^3/3 \tag{3.14b}$$

Equations (3.13) and (3.14) imply that the ratio $[\eta]/[\sigma]$ for touching spheres equals

$$[\eta]/[\sigma]_\infty = 0.820\cdots \tag{3.15}$$

which agrees well with the estimate from Eq. (3.11)

$$[\eta]/[\sigma]_\infty = \tfrac{5}{6} = 0.833\cdots \tag{3.16}$$

We also compare the exact result to recent bead model calculations of $[\eta]$ for touching spheres by de la Torre and Bloomfield [119]. They find $[\eta] = 3.493$ for touching spheres, which is accurate to within 1% in comparison with the exact result [Eq. (3.13)].

Exact polarizability results are also known for the disk and needle limits of an ellipsoid of revolution. For a disk of radius a the polarizability components [46] and $[\sigma]_\infty$ in number density units equal

$$\alpha_e(L) = 0, \qquad \alpha_e(T) = 16a^3/3, \qquad [\sigma]_\infty = 32a^3/9 \tag{3.17}$$

and from the formalism of Brenner [117] we can also obtain an exact calculation of the intrinsic viscosity of a disk as

$$[\eta] = 128a^3/45 \tag{3.18}$$

which is also given in number density units. This result is probably known but we could not find a reference to it. For a disk we then obtain the exact ratio

$$[\eta]/[\sigma]_\infty = 0.8 \tag{3.19}$$

which is rather close to the approximation [Eq. (3.11)].

In the opposite *needle limit* ($x \rightarrow \infty$), corresponding to an extended prolate ellipsoid, the asymptotic scaling of $[\sigma]_\infty$ with x can be deduced analytically

$$[\sigma]_\infty \sim \tfrac{1}{3}x^2/\log(x) \tag{3.20}$$

where x is the ratio of the semimajor axis length to the semiminor axis length. Onsager [11] calculated the corresponding asymptotic prolate ellipsoid result for $[\eta]$ as

$$[\eta] \sim \tfrac{4}{15}x^2/\log(x) \tag{3.21}$$

which is consistent with more general calculations given later by Saito [12]. The exact limiting ratio for $[\eta]/[\sigma]_\infty$ for a *needle* ($x \rightarrow \infty$) then equals

$$[\eta]/[\sigma]_\infty = 0.8 \tag{3.22}$$

which is *identical* to the ratio obtained for a disk. The ratio $[\eta]/[\sigma]_\infty$ is thus found to be *nearly invariant* for a significant range of particle shapes, as expected from Eq. (3.11). In Fig. 2.4a we plot log $[\eta]$ versus log $[\sigma]_\infty$ for a wide range of aspect ratios for prolate ellipsoids of revolution, while Fig. 2.4b shows a similar graph for oblate ellipsoids of revolution, where the abscissa is now the inverse of the aspect ratio. The straight line is a fit that gives an average value of 0.8 for the intrinsic viscosity/conductivity ratio. Table V gives the numerical data shown in Fig. 2.4, where the $[\eta]$ results are taken from the original tabulation of Scheraga [120a].

The information required to obtain $[\eta]$ for triaxial ellipsoids is also known, although this information is rather inaccessible because of the complicated mathematical formalism that these calculations involve. The necessary formulas for the components of the electric polarizability [52a] are summarized in Appendix B and a summary of the necessary results of Haber and Brenner [14] for $[\eta]$ are provided in Appendix C. Tabulations

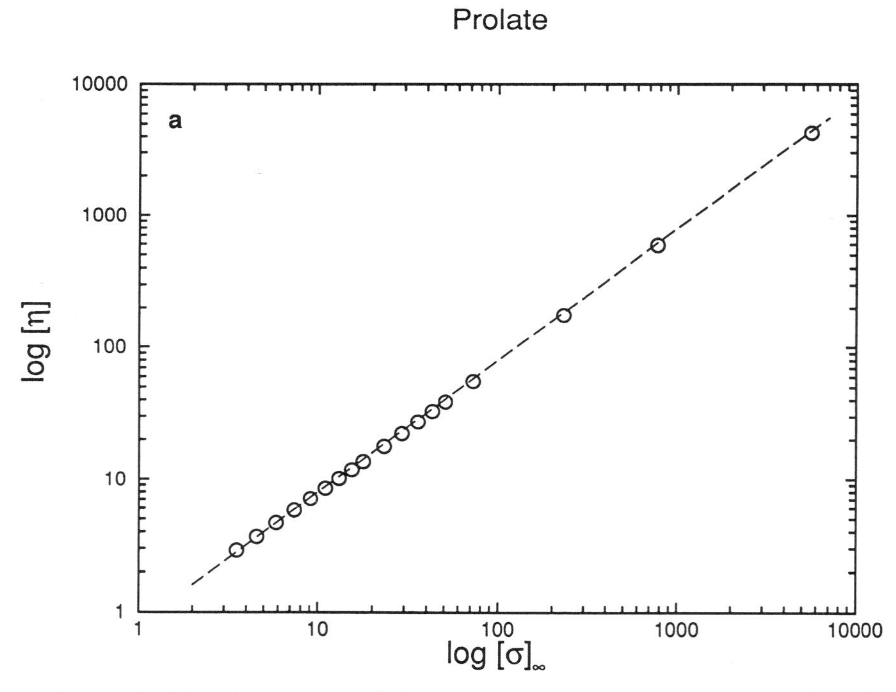

Figure 2.4. The intrinsic viscosity $[\eta]$ versus the intrinsic conductivity $[\sigma]_\infty$ for ellipsoids of revolution. The scales are logarithmic (base 10): (a) prolate and (b) oblate.

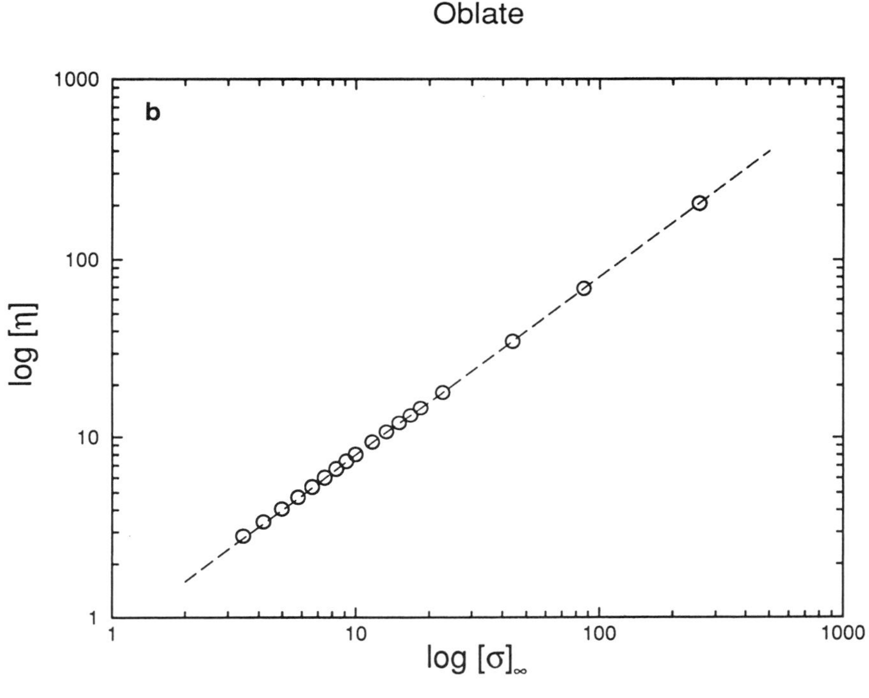

Figure 2.4. (*Continued*)

of these virial coefficients, which should be useful in applications, are given in Table VI.

We observe from Tables V and VI that all the ellipsoid data is nearly consistent with the ratio given in Eq. (3.16)

$$[\eta]/[\sigma]_\infty = 0.79 \pm 0.04 \tag{3.23}$$

so that $[\eta]/[\sigma]_\infty$ is an *invariant* to within a 5% accuracy. The angular averaging approximation is not as accurate for $[\eta]$ as in previous applications to f_T [27, 80], but Eq. (3.16) is sufficiently accurate for many practical applications since measurement and numerical calculation errors are often comparable to the 5% inaccuracy indicated by Eq. (3.23).

Equation (3.23) also holds for the spherical *dumbbell* at arbitrary separations. The dumbbell is defined by two identical spheres connected by a straight wire of zero thickness and fixed length. In the calculation of the polarizability the spheres are uncharged and the wire has zero

TABLE V
Intrinsic Viscosity and Intrinsic Conductivity for Ellipsoids of Revolution

Aspect Ratio $(x, 1/x)$	Prolate			Oblate		
	$[\sigma]_\infty$	$[\eta]$	$[\eta]/[\sigma]_\infty$	$[\sigma]_\infty$	$[\eta]$	$[\eta]/[\sigma]_\infty$
2	3.5339	2.908	0.82	3.4524	2.854	0.83
3	4.5622	3.685	0.81	4.1815	3.431	0.82
4	5.8625	4.663	0.80	4.9728	4.059	0.82
5	7.3836	5.806	0.79	5.7878	4.708	0.81
6	9.1043	7.099	0.78	6.6145	5.367	0.81
7	11.013	8.533	0.77	7.4476	6.032	0.81
8	13.101	10.10	0.77	8.2847	6.700	0.81
9	15.362	11.80	0.77	9.1245	7.371	0.81
10	17.793	13.63	0.77	9.9661	8.043	0.81
12	23.146	17.67	0.76	11.653	9.391	0.81
14	29.134	22.19	0.76	13.343	10.74	0.80
16	35.738	27.18	0.76	15.035	12.10	0.80
18	42.942	32.63	0.76	16.728	13.45	0.80
20	50.732	38.53	0.76	18.422	14.80	0.80
25	72.701	55.19	0.76	22.660	18.19	0.80
50	232.34	176.8	0.76	43.868	35.16	0.80
100	776.71	593.7	0.76	86.303	69.10	0.80
300	5,560.0	4,279.0	0.77	256.06	204.9	0.80

electrical resistance, while for the intrinsic viscosity calculation the wire has a vanishing hydrodynmic resistance. Brenner [117] summarized the information required to calculate $[\eta]$ for a dumbbell and an exact calculation of $[\sigma]_\infty$ for the dumbbell is given in Appendix D.

We define the quantity r_p to be the ratio of the distance between the centers of the spheres to their diameters, which completely characterizes the shape of the spherical particle dumbbell. It is then found that the value of $[\eta]$ is approximately quadratic in r_{p}. A useful approximate formula for $[\eta]$, covering the range $1 < r_{\mathrm{p}} < 10$, is given by

$$[\eta] = 2.5 + (1.037 - 0.3196\lambda) r_{\mathrm{p}}^2$$
$$\lambda = (r_{\mathrm{p}}/3)/(1 + r_{\mathrm{p}}/3) \tag{3.24}$$

which holds to about a 3% accuracy. Exact results for $[\eta]$ are tabulated in Table VII and the asymptotic variation of $[\eta]$ for a dumbbell at large separation equals

$$[\eta] \sim \tfrac{3}{4} r_{\mathrm{p}}^2, \qquad r_{\mathrm{p}} \to \infty \tag{3.25}$$

TABLE VI
Intrinsic Viscosity and Intrinsic Conductivity for Triaxial Ellipsoids

a_1	a_2	a_3	$[\sigma]_\infty$	$[\eta]$	$[\eta]/[\sigma]_\infty$
1	2	3	3.9585	3.2454	0.82
1	2	4	4.6903	3.8027	0.81
1	2	5	5.5723	4.4704	0.80
1	2	6	6.5774	5.2291	0.80
1	2	7	7.6924	6.0695	0.79
1	2	8	8.9098	6.9864	0.78
1	2	9	10.225	7.9765	0.78
1	2	10	11.633	9.0372	0.78
1	2	20	30.436	23.227	0.76
1	2	50	131.53	100.02	0.76
1	2	100	429.47	327.72	0.76
1	2	300	3,006.5	2,309.9	0.77
1	2	500	7,574.0	5,835.8	0.77
1	2	1,000	26,904.0	20,799.0	0.77
1	3	4	4.6662	3.8090	0.82
1	3	5	5.2975	4.2941	0.81
1	3	6	6.0356	4.8572	0.80
1	3	7	6.8626	5.4855	0.80
1	3	8	7.7688	6.1722	0.79
1	3	9	8.7483	6.9133	0.79
1	3	10	9.7970	7.7061	0.79
1	3	20	23.678	18.194	0.77
1	3	50	96.894	73.776	0.76
1	3	100	309.75	236.28	0.76
1	3	300	2,127.6	1,633.2	0.77
1	3	500	5,328.6	4,102.3	0.77
1	3	1,000	18,810.0	14,531.0	0.77
1	4	5	5.4445	4.4297	0.81
1	4	6	6.0250	4.8788	0.81
1	4	7	6.6892	5.3889	0.81
1	4	8	7.4244	5.9506	0.80
1	4	9	8.2227	6.5587	0.80
1	4	10	9.0794	7.2099	0.79
1	4	20	20.391	15.777	0.77
1	4	50	79.199	60.416	0.76
1	4	100	248.18	189.35	0.76
1	4	300	1,675.7	1,285.7	0.77
1	4	500	4,175.7	3,212.9	0.77
1	4	1,000	14,662.0	11,321.0	0.77
1	5	6	6.2511	5.0734	0.81
1	5	7	6.8008	5.5010	0.81
1	5	8	7.4198	5.9787	0.81
1	5	9	8.0983	6.4997	0.80
1	5	10	8.8300	7.0595	0.80
1	5	20	18.528	14.427	0.78
1	5	50	68.430	52.314	0.76
1	5	100	210.37	160.57	0.76
1	5	300	1,398.0	1,072.2	0.77
1	5	500	3,467.8	2,666.9	0.77
1	5	1,000	12,119.0	9,352.9	0.77
1	6	7	7.0718	5.7287	0.81
1	6	8	7.6009	6.1419	0.81
1	6	9	8.1892	6.5978	0.81
1	6	10	8.8289	7.0910	0.80
1	6	20	17.396	13.622	0.78
1	6	50	61.196	46.893	0.77

Table VI (*Continued*)

a_1	a_2	a_3	$[\sigma]_\infty$	$[\eta]$	$[\eta]/[\sigma]_\infty$
1	6	100	184.67	141.03	0.76
1	6	300	1,208.9	926.83	0.77
1	6	500	2,986.1	2,295.6	0.77
1	6	1,000	10,391.0	8,015.9	0.77
1	7	8	7.9005	6.3907	0.81
1	7	9	8.4148	6.7935	0.81
1	7	10	8.9809	7.2337	0.81
1	7	20	16.693	13.136	0.79
1	7	50	56.021	43.030	0.77
1	7	100	166.02	126.88	0.76
1	7	300	1,071.2	821.07	0.77
1	7	500	2,635.6	2,025.5	0.77
1	7	1,000	9,134.8	7,044.6	0.77
1	8	9	8.7343	7.0568	0.81
1	8	10	9.2375	7.4520	0.81
1	8	20	16.266	12.855	0.79
1	8	50	52.157	40.159	0.77
1	8	100	151.85	116.13	0.76
1	8	300	966.19	740.43	0.77
1	8	500	2,368.4	1,819.5	0.77
1	8	1,000	8,178.3	6,304.8	0.77
1	9	10	9.5714	7.7258	0.81
1	9	20	16.030	12.714	0.79
1	9	50	49.182	37.959	0.77
1	9	100	140.70	107.70	0.77
1	9	300	883.23	676.75	0.77
1	9	500	2,157.4	1,657.0	0.77
1	9	1,000	7,423.6	5,721.3	0.77
1	10	20	15.931	12.675	0.80
1	10	50	46.839	36.235	0.77
1	10	100	131.71	100.92	0.77
1	10	300	815.93	625.11	0.77
1	10	500	1,986.2	1,525.1	0.77
1	10	1,000	6,811.8	5,248.3	0.77
1	20	50	37.956	29.929	0.79
1	20	100	90.969	70.422	0.77
1	20	300	498.42	381.99	0.77
1	20	500	1,177.7	903.15	0.77
1	20	1,000	3,928.4	3,020.6	0.77
1	50	100	73.084	57.942	0.79
1	50	300	290.78	224.55	0.77
1	50	500	637.83	490.03	0.77
1	50	1,000	1,996.9	1,532.4	0.77
1	100	300	225.95	177.13	0.78
1	100	500	444.09	344.09	0.77
1	100	1,000	1,270.5	976.50	0.77
1	300	500	363.47	289.19	0.80
1	300	1,000	768.19	600.59	0.78
1	500	1,000	716.53	567.67	0.79

TABLE VII
Selected Values of the Intrinsic Viscosity and Intrinsic Conductivity for the Spherical Dumbbell

r_p	$[\sigma]_\infty$	$[\eta]$	$[\eta]/[\sigma]_\infty$
1.0000	4.2072	3.4496	0.82
1.0201	4.2707	3.4980	0.82
1.1276	4.6180	3.6754	0.80
1.5431	6.0912	4.8914	0.80
3.7622	19.176	14.756	0.77
6.1323	43.856	33.308	0.76
10.0677	109.603	82.690	0.75
$r_p \to \infty$	$3r_p^2/4$	r_p^2	0.75

Simha [120b] previously indicated a quadratic dependence of $[\eta]$ on r_p in the $r_p \to \infty$ limit, but his widely cited value for the prefactor, $\frac{3}{2}$, is not correct. The origin of this discrepancy is not clear, but we note that Simha [120b] ignored hydrodynamic interactions.

Schiffer and Szegö [46] previously summarized exact results for the electric polarizability of two *separated* spheres without the connecting wire. The generation of a large dipole in separated spheres, however, requires the electrical connection and the calculation of $[\sigma]_\infty$ in the case where there is a connecting wire is given in Appendix D. A tabulation of these new results for $[\sigma]_\infty$, along with the dimensionless polarizability components (normalized by the particle volume), is given in Table VII. These results are shown graphically in Fig. 2.5. It is hard to imagine a geometry more representative of a dipole. For large separations $[\sigma]_\infty$ is simply proportional to r_p^2

$$[\sigma]_\infty \sim r_p^2, \qquad r_p \to \infty \tag{3.26}$$

so that we have the asymptotic result

$$[\eta]/[\sigma]_\infty \sim 0.75, \qquad r_p \to \infty \tag{3.27}$$

It is interesting that the dumbbell accords with Eq. (3.23) even in the extreme limit of infinite separation.

We also mention some results for the intrinsic conductivity of insulating dumbbells. From the results of Schiffer and Szegö [46] for the effective mass **M** of touching spheres and Eq. (2.6) we have

$$\alpha_m(L) = -9\zeta(3)/8 \qquad \text{(touching spheres)} \tag{3.28}$$

and by finite element methods we calculate the other component

TABLE VIII
Electric Polarizability Components for Dumbbell

r_p	$\alpha_e(L)/V_p$	$\alpha_e(T)/V_p$	$[\sigma]_\infty$
1.005	2.70774	7.25347	4.22298
1.050	2.73443	7.62957	4.36615
1.100	2.76139	8.06009	4.52763
1.150	2.78562	8.50399	4.69174
1.200	2.80726	8.96133	4.85862
1.250	2.82651	9.43222	5.02841
1.300	2.84358	9.91671	5.20129
1.350	2.85869	10.4149	5.37743
1.400	2.87207	10.9268	5.55700
1.450	2.88393	11.4526	5.74015
1.500	2.89444	11.9923	5.92705
1.750	2.93207	14.9010	6.92172
2.000	2.95400	18.1651	8.02437
2.250	2.96749	21.7895	9.24149
2.500	2.97622	25.7778	10.5767
2.750	2.98209	30.1325	12.0322
3.000	2.98618	34.8556	13.6093
3.250	2.98912	39.9485	15.3089
3.500	2.99128	45.4122	17.1316
4.000	2.99415	57.4550	21.1478
4.500	2.99589	70.9883	25.6600
5.000	2.99700	86.0150	30.6697
5.500	2.99775	102.537	36.1774
6.000	2.99826	120.555	42.1838
6.500	2.99864	140.070	48.6892
7.000	2.99891	161.084	55.6938
7.500	2.99911	183.595	63.1977
8.000	2.99927	207.605	71.2011
8.500	2.99939	233.114	79.7041
9.000	2.99949	260.121	88.7067
9.500	2.99956	288.628	98.2091
10.000	2.99963	318.634	108.211
20.000	2.99995	1,233.70	413.231
30.000	2.99999	2,748.71	918.237
40.000	2.99999	4,863.72	1,623.24
50.000	3.00000	7,578.73	2,528.24
100.000	3.00000	30,153.7	10,053.2
150.000	3.00000	67,728.7	22,578.2
200.000	3.00000	120,304.0	40,103.2
250.000	3.00000	187,879.0	62,628.2
300.000	3.00000	270,454.0	90,153.2
350.000	3.00000	368,029.0	122,678.0
400.000	3.00000	480,604.0	160,203.0
450.000	3.00000	608,179.0	202,728.0
500.000	3.00000	750,754.0	250,253.0

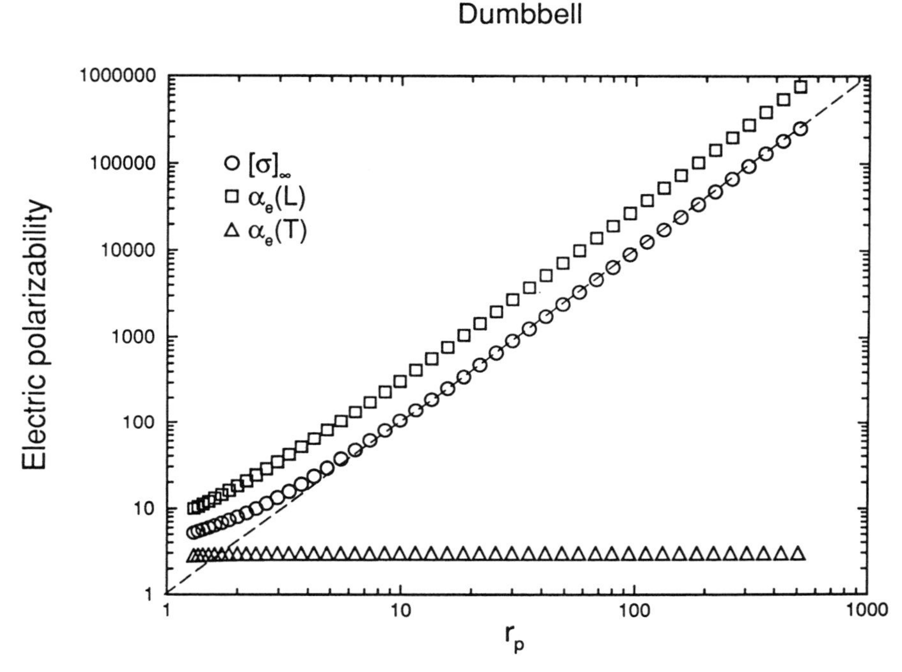

Figure 2.5. Longitudinal (L) and transverse (T) components of the dimensionless (normalized by the particle volume) electric polarizability tensor $\boldsymbol{\alpha}_e$ for the spherical particle dumbbell, along with the average of these components, the intrinsic conductivity $[\sigma]_\infty$.

$$\alpha_m(T) \approx -41\zeta(3)/30 \qquad \text{(touching spheres)} \tag{3.29}$$

which has long defied exact analytical calculation [46]. [The closed form estimate in Eq. (3.29) is based on the assumption that $\alpha_m(T)$ is proportional to $\zeta(3)$, as in Eq. (3.28), in combination with accurate numerical estimates of $\alpha_m(T)$.] We note that the known value of $\alpha_m(L)$ is given by our finite element method to an accuracy of better than 1%, so we expect that the corresponding value for the unknown $\alpha_m(T)$ should be correct within the same tolerance. (From previous experience, the finite element method used here is always more accurate for conducting matrix–insulating particle problems than for conducting matrix–superconducting particle problems.) We then obtain the intrinsic conductivity of the insulating doublet of spheres

$$[\sigma]_0 \approx -(463/360)\zeta(3) = -1.55\cdots \tag{3.30}$$

This value is only slightly different than the sphere result $[\sigma]_0 = -\frac{3}{2}$ of Maxwell [4a] [see Eq. (2.1)]. At large distances between the spheres, the dumbbell should approach the sphere value, so the spatial variation of $[\sigma]_0$ has limited interest in comparison with the conducting dumbbell case. We note that in the insulating case and long slender bodies, $\alpha_m(L)/V_p$ tends to approach -1. This result can be derived from slender body theory using results of Miles [120c] and the known relation between $\alpha_m(L)$ and $\alpha_m(T)$. The corresponding value of $\alpha_m(T)/V_p$ approaches -2, so that $[\sigma]_0$ obeys the general relation

$$[\sigma]_0 = -\tfrac{5}{3} \qquad \text{(slender body)} \tag{3.31}$$

Even in this extreme limit the deviation of $[\sigma]_0$ from the sphere value is unimpressive. The variation of $[\sigma]_0$ with shape is more interesting for flat bodies (see Fig. 2.2). We return to a discussion of flat bodies in Section VI.

As a final point, we mention that exact calculation of $[\eta]$ for other shapes is possible, in principle. Exact results for $\boldsymbol{\alpha}_e$ and $\mathbf{M}$ are known for the lens, bowl, spindle and other shapes [46]. Calculation of $[\eta]$ involves similar (albeit more complicated) mathematics.

IV. NUMERICAL INVESTIGATION OF $[\eta]/[\sigma]_\infty$ RATIO ($d = 3$)

Further examples of the approximate invariance of $[\eta]/[\sigma]_\infty$ for a variety of shapes are given in this section based on numerical finite element computations in combination with partial analytic results for $[\eta]$ and $[\sigma]_\infty$. All of the results obtained are consistent with Eq. (3.23).

The analogy of the elastostatic and hydrodynamic problems of fluid suspensions and solid composites [17, 102, 103], mentioned in Section III, indicates that a modification of existing finite element programs for calculating the effective elastic properties of composite bodies can be made to also obtain $[\eta]$. This modification and the variational principle for obtaining the Stokes' equation on which it is based, are described in Appendix E for particles with orthorhombic symmetry or higher (triaxial ellipsoids have orthorhombic symmetry). We note that Brenner's work [117] was essential in checking the consistency of this generalization, especially in the case of anisotropic elastic stiffness and viscosity tensors. We also utilize a similar finite element program for the calculation of $[\sigma]_\infty$. This finite element method is also described in Appendix E. All particles were represented by a cubical digital image, so that the elements were cubes arrayed on a simple cubic lattice. A standard lattice of size 104^3 was used, which was the largest that would fit in the memory of the computer available to us and which would allow reasonable running times. Even so, the total CPU time used to compute the results in this chapter was about 2000 h on a CONVEX 3820 supercomputer.

In these numerical calculations, arbitrary shapes had to be represented by collections of pixels. Because of the overall computational cell size limit, a compromise had to be taken between using enough pixels to give a good representation of the particle, and keeping the particle small compared to the overall unit cell, so as to keep the volume fraction small enough to be in the linear regime in concentration. The size and complexity of the objects that could be treated in this fashion is necessarily limited, but a good approximation to a wide range of physically interesting objects could still be obtained.

Periodic boundaries were used in all simulations to reduce the importance of finite size edge effects. Since a cubic cell was always used, in reality all computations were really for simple cubic periodic arrays of the object considered. Exact calculations exist for the intrinsic conductivity and viscosity of rigid spheres arranged on a simple cubic lattice. This example can then be used to illustrate the effect of finite resolution, as described above, and finite system size on the accuracy of the computations.

Zuzovsky and Brenner carried out computations for the effective conductivity of cubic arrays of spheres embedded in a matrix [121a], which are very useful for comparison with our numerical data. For the particular case where the spheres were *superconducting* and the matrix was an ordinary conductor of unit conductivity, they developed an accurate formula for the effective conductivity σ of the composite medium. Subtracting one from the effective conductivity, and dividing by the sphere volume fraction ϕ, gives their prediction for the effective *intrinsic conductivity* at any sphere volume fraction

$$[\sigma]_\infty = 3[1 - \phi - 1.306\phi^{10/3}/(1 - 0.407\phi^{7/3}) - 0.022\phi^{14/3} + O(\phi^6)]^{-1} \tag{4.1}$$

where $\phi = \pi(d/L_c)^3/6$, d = sphere diameter, and L_c = size of cubic unit cell. Actually, this quantity is only equal to the true intrinsic conductivity in the limit where ϕ is small enough so that the expansion in Eq. (2.2b) is applicable. Equation (4.1), however, provides a useful way to represent our numerical conductivity data.

Nunan and Keller [121b] computed the components of the viscosity tensor of the simple cubic array of rigid spheres in a fluid. There are two independent components for this symmetry (there would be three independent elastic components, but incompressibility reduces these to two), defined by Nunan and Keller as two functions of ϕ, p and q. With the use of their exact numerical results, they were able to show that an analytic expansion given by Zuzovsky et al. [121c] was accurate to within 0.2% up to $\phi = 0.13$ ($d/L_c = 0.63$) for simple cubic sphere packings. This analytic expression equals

$$p = 2.5\phi[1 - (1 - 60b)\phi + 12a\phi^{5/3} + \mathrm{O}(\phi^{7/3})]^{-1} \tag{4.2a}$$

$$q = 2.5\phi[1 - (1 + 40b)\phi - 8a\phi^{5/3} + \mathrm{O}(\phi^{7/3})]^{-1} \tag{4.2b}$$

where $a = 0.2857$ and $b = -0.04655$. In terms of p and q, the rotationally averaged intrinsic viscosity $[\eta] = (\eta/\eta_0 - 1)/\phi$ at any volume fraction ϕ is given by

$$[\eta] = (2p + 3q)/5 \tag{4.3}$$

Figure 2.6 shows the finite element results for periodic arrays of spheres, along with the exact results, Eqs. (4.1) and (4.3). First, consider the results for the intrinsic viscosity (circles). At small values of d/L_c, the numerical results are well above the exact result. This result is due to not having enough pixels to represent the spherical shape. For example, a sphere with a diameter of five pixels (a pixel is considered to be part of the sphere if its center lies within a radius of the center) does not look much like a smooth continuum sphere. In fact, all the finite element results shown in Fig. 2.6 have been rotationally averaged, as they have cubic symmetry. As d/L_c increases, allowing each sphere to be represented by more pixels, resolution improves, and the numerical points approach the exact curve. There is a region, around $d/L_c = 0.4$, where the numerical results are essentially exact, but the value of $[\eta]$ has not changed much from the $d/L_c \to 0^+$ limit. This is the region in which we have tried to run all the simulations: d/L_c high enough to give good particle shape resolution, but not high enough so that ϕ is out of the linear regime. Obviously, for higher aspect ratio particles, it is harder to stay in this range of d/L_c.

Now consider the results for $[\sigma]_\infty$ shown in Fig. 2.6. The comparison between exact formula and numerical (finite element) results is similar, except that the numerical results are consistently about 5–8% above the exact curve. This error is possibly larger for larger aspect ratio particles.

Judging by the results for spheres shown in Fig. 2.6, we can expect that both the numerically computed intrinsic conductivity and viscosity will be systemically high, with the ratio $[\eta]/[\sigma]_\infty$ probably somewhat low, as the intrinsic conductivity seems to overshoot the true result slightly more than

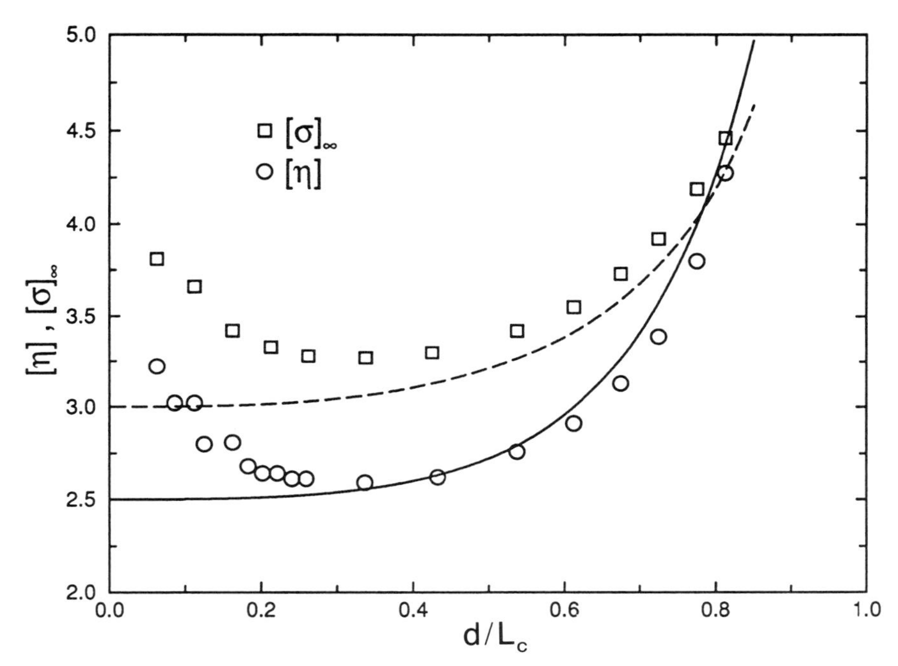

Figure 2.6. Numerical and analytical results for the intrinsic conductivity $[\sigma]_\infty$ and intrinsic viscosity $[\eta]$ for simple cubic arrays of spheres (superconducting and rigid) as a function of d/L_c, the ratio of the sphere diameter d to the sphere center spacing L_c. The length L is also defined as the size of the computational cell.

the intrinsic viscosity. Slightly increasing all the ratios involving a numerical computation of $[\sigma]_\infty$ in Tables IX–XI would improve the agreement with Eq. (3.16).

We have run other tests on the intrinsic viscosity using nonspherical shapes. Having exact results for nontrivial shapes is crucial in order to check the accuracy of numerical simulations. We computed $[\eta]$ for a spherical dumbbell with $r_\mathrm{p} = 1.526$, giving $[\eta] \approx 4.9$, as compared to the exact value $[\eta] = 4.89$ ($r_\mathrm{p} = 1.543$). An ellipsoid of revolution with an aspect ratio of 3 gave $[\eta] = 3.91$, which is 6.1% higher than the exact value of 3.685. The numerical result for two touching spheres was 3.62, about 5% larger than the exact result of $[\eta] = 3.45$ given in Eq. (3.13).

The important physical example of a right circular cylinder is considered next. Very precise analytical calculations of the polarizability $\boldsymbol{\alpha}_\mathrm{e}$ have

TABLE IX
Intrinsic Viscosity and Intrinsic Conductivity for Cylinders

Dimensions	Exact $[\sigma]_\infty$	Numerical $[\sigma]_\infty$	Numerical $[\eta]$	Numerical / *Exact* $[\eta]/[\sigma]_\infty$	Numerical / *Numerical* $[\eta]/[\sigma]_\infty$
$1 \times \frac{1}{2}$	4.106	4.32	3.45	0.84	0.80
1×1	3.401	3.56	2.90	0.85	0.81
1×2	3.622	3.79	3.08	0.85	0.81
1×4	4.704	4.93	3.93	0.84	0.80

TABLE X
Intrinsic Viscosity and Intrinsic Conductivity for Rectangular Parallelipipeds with Two Equal Edges

Aspect Ratio	Prolate			Oblate		
$(x, 1/x)$	$[\sigma]_\infty$	$[\eta]$	$[\eta]/[\sigma]_\infty$	$[\sigma]_\infty$	$[\eta]$	$[\eta]/[\sigma]_\infty$
1	3.72	3.05	0.82	3.72	3.05	0.82
2	4.22	3.41	0.81	4.15	3.36	0.81
3	5.21	4.13	0.79	4.84	3.88	0.80
4	6.38	4.98	0.78	5.58	4.44	0.80
5	7.72	5.96	0.77	6.34	5.02	0.79
10	16.4	12.3	0.75	10.2	7.92	0.78
∞				$1.11a^3$ [a]	$0.85a^3$ [a]	0.77

[a] Number density units.

TABLE XI
Intrinsic Viscosity and Intrinsic Conductivity for Various Shapes in Three Dimensions

Shape	m	$[\sigma]_\infty$	$[\eta]$	$[\eta]/[\sigma]_\infty$
Sponge	15/27	8.74	7.16	0.82
Sponge	21/27	27.1	22.0	0.81
Sponge	23/27	55.0	44.7	0.81
Sponge	25/27	192	156	0.81
Sponge	33/35	311	255	0.82
Dice ($r_p = 2.0$)		7.44	5.84	0.78
Jack		4.50	3.68	0.82
Square Ring	23/25	127	98.7	0.77
Square hollow tube	5/6	16.7	13.3	0.80

been made for the cylinder [122a and b]. Values of $[\sigma]_\infty$ calculated from these results are given in Table IX. Finite element calculations of $[\eta]$ and $[\sigma]_\infty$ for several aspect ratios, which have comparable accuracy to the sphere and touching sphere test cases, are also given in Table IX. The ratio $[\eta]/[\sigma]_\infty$ obtained from this combination of numerical and analytical calculations accords rather well with Eq. (3.16) and the general correlation Eq. (3.23). Note the difference between the columns marked Numerical/Exact and Numerical/Numerical giving the results for the ratio $[\eta]/[\sigma]_\infty$. (The term *numerical* refers to an estimate obtained by finite element calculation while *exact* refers to analytic results. The exact results often involve a nontrivial numerical evaluation of the integral expressions that define the analytic results, however.) Using the finite element estimates for the intrinsic conductivity instead of the exact results gave a value somewhat closer to the prediction of Eq. (3.16), since similar systematic computational errors for $[\eta]$ and $[\sigma]_\infty$ probably compensate.

Next, we examine the case of rectangular parallelepipeds which is summarized in Table X. Simulation results closely parallel the exact analytical calculations for the ellipsoid of revolution case discussed in Section III. Again the ratio $[\eta]/[\sigma]_\infty$ is shown to be nearly invariant with respect to shape. The oblate result for a very large aspect ratio (marked ∞ in Table X) corresponds to the case of a square plate and such objects are discussed more fully in Section VI. The value for $[\sigma]_\infty$ is approximately 7% higher than the best known experimental value (see Section VI) and we expect that our estimate of $[\eta]$ is too large by about the same amount. The $[\eta]/[\sigma]_\infty$ ratio tends to decrease as the aspect ratio increases. These results parallel the analytic results for ellipsoids of revolution in the prolate and oblate limits.

Next, we illustrate a simple means to increase $[\eta]$ and $[\sigma]_\infty$ to large values without making a very extended or flat object. We consider a cube of unit edge length, in which a square *channel* is cut through the center of each face, which passes completely through the cube. A picture of one such object is shown in Fig. 2.7a. The parameter m is taken to be the edge length of the cutout face in units of the cube edge length. We obtain a rigid cubic wire frame when m approaches 1. Notice that cutting out the center, which makes the particle more spongelike, has a very large effect on $[\eta]$ and $[\sigma]_\infty$, as seen in Table XI. It would be interesting to push the effect to the extreme in a different way by generating a Menger sponge [123] *fractal* by a repeated decimation of the cube at different scales so that $[\eta]$ and $[\sigma]_\infty$ would diverge in a characteristic fashion related to the fractal dimension of the sponge. The memory capacity of the computer was not large enough to allow us to consider more than one or two generations of such an iteratively constructed *diffuse* object, so we presently confine ourselves to the first generation wire frame structure shown in Fig. 2.7a. The rapid increase of $[\eta]$ and $[\sigma]_\infty$ when large holes are cut out is noted. In the limit where m goes to one the intrinsic conductivity and viscosity appear to scale roughly quadratically in $(1-m)$.

In a similar vein we also consider a flat square "ring" where the length of each side is 25 and $m = 23/25$. The effect on the virials is large, as in the "sponge" case (see Table XI). Results for a square cross-section tube of width one third of the side length are given in Table XI where m of the square face equals 5/6. In this case a less pronounced effect is found.

We consider other strategies of particle modification in Table XI. For example, instead of decimating the structure we introduce protuberances onto our object. Specifically, we poke three rectangular parallelipipeds orthogonally through a sphere (see Fig. 2.7b) to create a "jack-like"

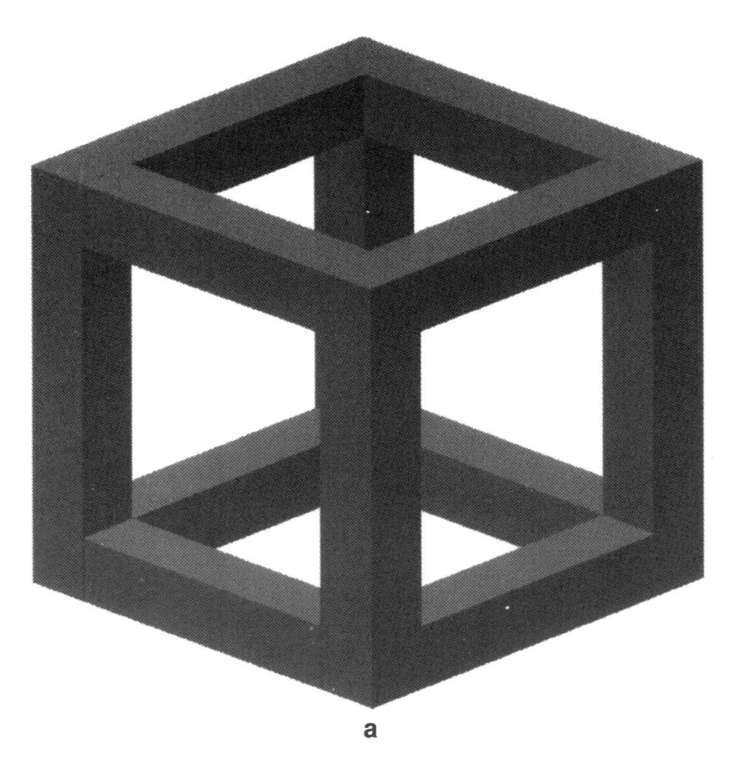

Figure 2.7. Model irregularly shaped objects considered by finite element computations. (a) sponge ($m = 0.6$), (b) jack-like object, and (c) dice.

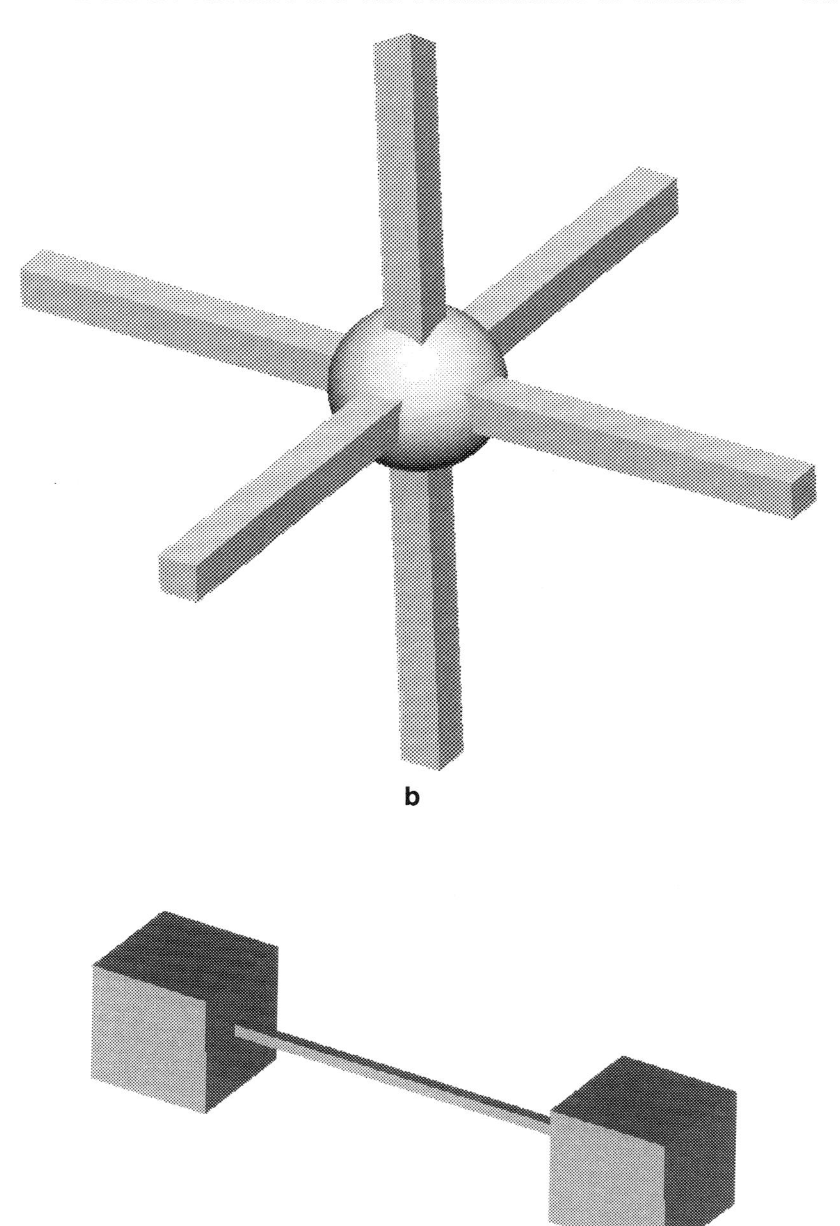

Figure 2.7. (*Continued*)

structure. Taking the width of these parallelipipeds as a unit of length, we take their length as 15 units and the sphere diameter as 9. The increase of $[\eta]$ and $[\sigma]_\infty$ is not as dramatic as the "sponge" case, but the effect is still appreciable. We next consider the case of separated and aligned cubes (dice), connected by a rigid, conducting wire of vanishing thickness to maintain object connectivity (see Fig. 2.7c). The results in Table XI show that such a tethering of *bulky* groups has a very large effect on the values of $[\eta]$ and $[\sigma]_\infty$. The values of $[\eta]$ for dice are similar to $[\eta]$ values for the spherical particle dumbbell (see Table VII) at comparable separations r_p.

V. INTRINSIC VISCOSITY AND CONDUCTIVITY IN d = 2

Calculation of the virial coefficients $[\eta]$, $[\sigma]_\infty$, $[\sigma]_0$, and $\langle M \rangle$ for a circle reveals a remarkable degeneracy in $d = 2$

$$[\eta] = [\sigma]_\infty = -[\sigma]_0 = \langle M \rangle / A = 2 \tag{5.1a}$$

where A is the circle area. We also have corresponding elastostatic results for hard and soft circular inclusions, $[G(\text{hard})] = -[G(\text{soft})] = 2$. The equality $[\sigma]_\infty = -[\sigma]_0$ in Eq. (5.1a) follows from the Keller–Mendelson inversion theorem [78]. Note that this relation holds for regions of *arbitrary* shape. The intrinsic viscosity result in Eq. (5.1a) is due to Brady [113] who found $[\eta] = (d + 2)/2$ for hyperspheres. It is easy to show that $[\sigma]_0$ and $[\sigma]_\infty$ for hyperspheres equal

$$[\sigma]_0 = -d/(d-1), \qquad [\sigma]_\infty = d, \qquad d \geq 2 \tag{5.1b}$$

so that the equality in $d = 2$ is evidently a rather special occurrence. This can be seen in Fig. 2.8, where $-[\sigma]_0$ and $[\sigma]_\infty$ are plotted versus dimensionality d. The relation $\langle M \rangle / A = -[\sigma]_0$ follows from the Keller–Kelvin relation [Eq. (2.6)], which is not restricted to $d = 2$ and holds for regions of arbitrary shape. We note that $[\eta] \approx 2$ actually has been measured in a quasi-two-dimensional film [124].

For objects of noncircular shape the degeneracy of these shape functionals reveals itself in the general relations

$$[\sigma]_\infty = -[\sigma]_0 = \langle M \rangle / A \tag{5.2}$$

and, moreover, Eq. (3.11) implies the approximation

$$[\sigma]_\infty \approx [\eta] \tag{5.3a}$$

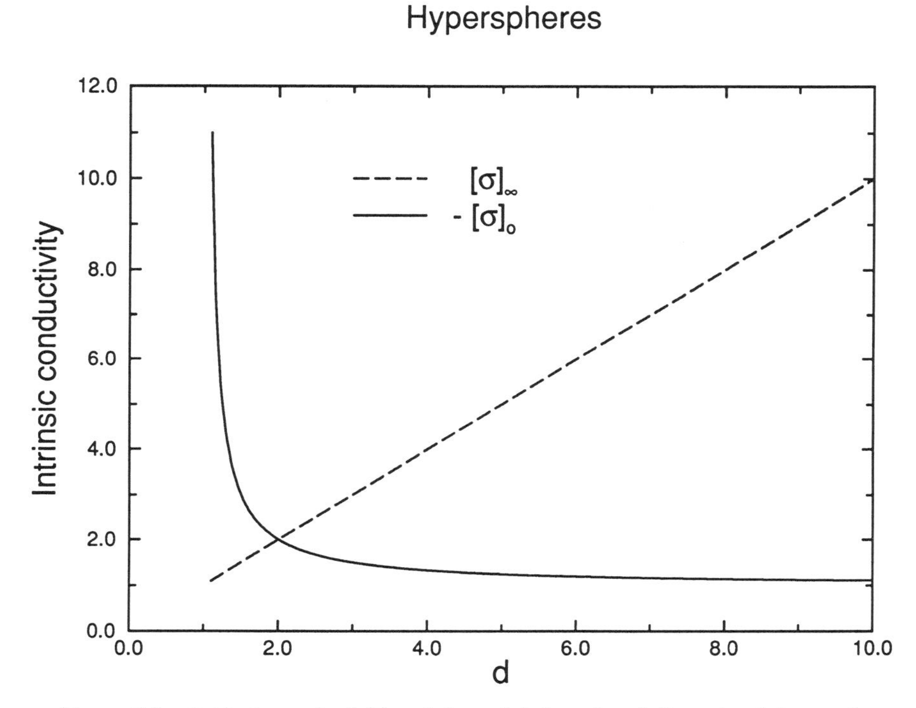

Figure 2.8. Intrinsic conductivities, $[\sigma]_0$ and $[\sigma]_\infty$, of a d-dimensional hypersphere versus dimensionality d.

In Appendix F we show that Eq. (5.3a) is *exact* for ellipses. The dependence of $[\sigma]_\infty$ and $[\eta]$ on aspect ratio $x = $ (semi-majoraxis)/(semi-minor axis) is given by

$$[\sigma]_\infty = [\eta] = (1 + x)^2/(2x) \tag{5.3b}$$

It seems entirely possible that the chain of particle property equalities in Eq. (5.1) could be exact for objects having *arbitrary* shape in $d = 2$. In this section we further examine the conjecture [Eq. (5.3a)] numerically for other shapes. General agreement is found, within numerical error, in accord with our conjecture. We also provide many new, exact results for $[\sigma]_0$, $[\sigma]_\infty$, and $\langle M \rangle / A$ that derive from the recognition of the chain of equalities [Eq. (5.2)] discussed above.

Pólya [28] proved rigorously that $\langle M \rangle / A$ is minimized for circular regions of all regions having finite area. Recognition of Eq. (5.2) then implies the physical result

$$[\sigma]_\infty = -[\sigma]_0 = \langle M \rangle / A \geq 2 \tag{5.4}$$

with equality obtained *uniquely* for the circular region. Pólya, in fact, proved much more in the process of deriving this fundamental isoperimetric inequality. He showed that $\langle M \rangle / A$ is given *exactly* by

$$\langle M \rangle / A = 2A_c / A, \qquad A_c = \pi C_L^2 \tag{5.5}$$

for regions having arbitrary shape, where C_L is variously termed the "transfinite diameter", "outer radius", or "logarithmic capacity", and A_c is defined here as the conformal area. (The area A and perimeter P of regions having general shape [30b] satisfy the important isoperimetric inequalities $A \leq A_c$, $P \geq 2\pi C_L$, and we note that $P \approx 2\pi C_L$ is often a reasonable approximation for objects having a modest shape irregularity [26].)

The transfinite diameter C_L is a basic measure of the average *size* of a bounded plane set, and can be defined in a variety of equivalent ways [29,30]. The parameter C_L, for example, is defined as the conformally invariant magnitude of Dirichlet's integral associated with the exterior of the region defining the particle [29]. The equivalent transfinite diameter can be expressed in terms of the Euclidean metric defining the distance between points in the set [30]. Perhaps the most useful definition of C_L involves the purely geometrical construction of mapping the exterior of a region Ω having an arbitrary but simply connected shape and finite area onto a circular region in such a fashion that the points at a large distance from Ω are asymptotically unaffected by the transformation [28]. The radius of this uniquely defined transformed circular region equals C_L. This transformation is basically the content of the Riemann mapping theorem [30c]. The origin of the outer radius terminology is thus apparent.

The invariance of C_L under conformal transformations is very convenient in the numerical computation of C_L and thus $[\sigma]_\infty$ and the other shape functionals of physical interest. It would be useful to have a program to calculate C_L for any conceivable bounded (compact) plane set, so that the physical consequences of shape variations could be explained easily. In Fig. 2.9 we indicate the results [125,126] obtained using "state of the art" conformal transformation methodology. The intrinsic conductivities of regions a) and b) in Fig. 2.9 were calculated using the energy integral definition of logarithmic capacity (see [29]) for certain spiral shaped lines

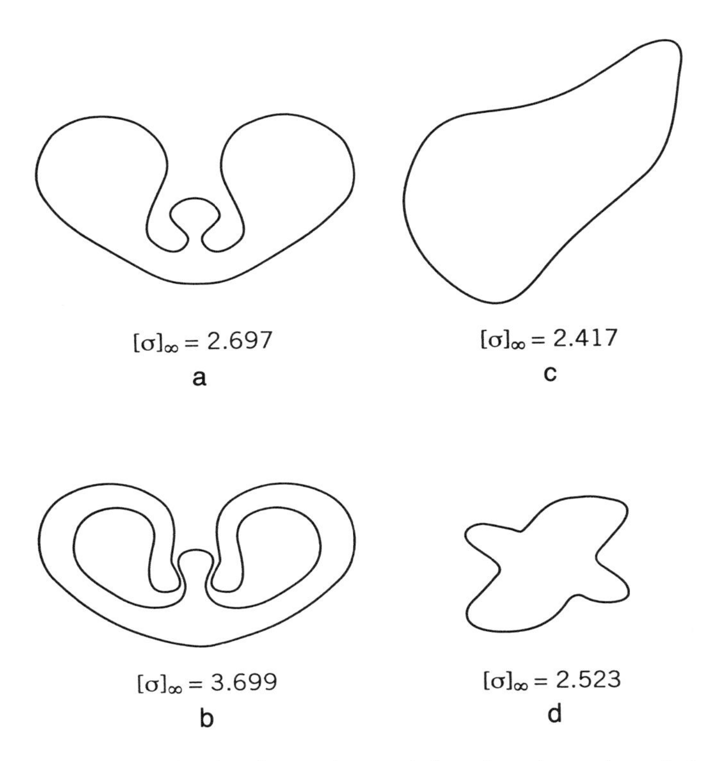

Figure 2.9. Various closed regions and numerical conformal mapping calculations for $[\sigma]_\infty$ corresponding to these regions. Numerical calculations were performed for us by McFadden [126].

that were then mapped conformally onto the indicated regions. By virtue of the invariance of C_L under comformal transformation, the C_L values for the original linear curves (not shown) equals C_L for the indicated regions in Fig. 2.9 and accurate estimates of C_L are obtained in this fashion. Any smooth nonintersecting curve in the plane can be mapped to its corresponding closed region in this fashion so that a large class of C_L calculations becomes possible for rather intricately shaped regions. Note the large value of $[\sigma]_\infty$ in Fig. 2.9c that results from the screening of the interior of the region in much the same fashion as the *sponge* shaped objects discused in the previous section. In c and d of Fig. 2.9 an alternative method of calculating C_L by conformally mapping the indicated regions onto a circular region of radius C_L numerically [126] is illustrated. This method is powerful, provided the region does not have sharp corners. (Other methods exist for polygonally shaped boundaries that are amenable to full analytic treatment and some of these results are discussed below.) We conclude that although many regions can be treated by these conformal techniques, there is still no simple and general method that can be adapted to calculate C_L for arbitrarily shaped regions. This remains an important mathematical challenge.

We mention that McKean [80c] summarized a formal algorithm for calculating C_L that has the generality we seek. This method involves hitting the 2D domains with random walks. This approach is similar to the technique implemented recently [80a and b] to calculate the Newtonian capacity [see Eq. (3.10)] in $d = 3$. Implementation of this algorithm should allow the numerical calculation of C_L for any bounded plane set.

Since C_L is a central object in the harmonic analysis of two dimensions [26, 28–30], there exist extensive tabulations of C_L [26, 29]. We may combine this information with Eq. (5.5) to obtain numerous *exact* results for the intrinsic conductivity. Table XII gives a sampling of some of these results. Further tabulation of expressions for C_L are given by Pólya and Szegö [26] and Landkof [29].

Pólya's isoperimetric inequality, $[\sigma]_\infty \geq 2$ [see Eq. (5.4)], is illustrated nicely by the case of the symmetric n-gon. Table XIII shows results for $[\sigma]_\infty$ as a function of n. The circle result is recovered in the limit $n \to \infty$.

TABLE XII
Logarithmic Capacity C_L Formulas for Variously Shaped Regions in Two Dimensions.

Region type	Logarithmic Capacity C_L	Intrinsic Conductivity $[\sigma]_\infty$
1. Circle of radius a	a	2
2. Ellipse, Axes of length a and b	$(a+b)/2$	$(a+b)^2/2ab$
3. Square of side length a	$\dfrac{\Gamma^2(\frac{1}{4})a}{4\pi^{3/2}} \approx 0.59017a$	2.1884
4. Triangles:		
(a) Equilateral triangle, Height h	$\dfrac{\Gamma^3(\frac{1}{3})h}{4\pi^2} \approx 0.4870h$	2.5809
(b) Right triangle, Long side of length h	$(5^{5/12}\Gamma^3(\frac{1}{3})/10^{10/3}\pi^2)h \simeq 0.37791h$	3.1090
(c) Isosceles right triangle, Long side of length h	$\dfrac{3^{3/4}\Gamma^2(\frac{1}{4})a}{4(2\pi)^{3/2}} \simeq 0.47563h$	2.8431
5. Hexagon of side length a	$[3\Gamma^3(\frac{1}{3})/2^{8/3}\pi^2]a \simeq 0.92032a$	2.0486
6. Symmetric n-gon, Side length a	$[n\Gamma(1+1/n)/2^{1+2/n}\sqrt{\pi}\,\Gamma(\frac{1}{2}+1/n)]a$	$[\tan(\pi/n)/2\pi n]\Gamma^4(1/n)/\Gamma^2(2/n)$
7. Semicircle of radius a	$(4/3^{3/2})a$	$(\frac{4}{3})^3 = 2.3704$

TABLE XIII
Intrinsic Conductivity for Regular n-gons in Two Dimensions

Shape	n	$[\sigma]_\infty$
Triangle	3	$2.5811\cdots$
Square	4	$2.1884\cdots$
Pentagon	5	$2.0878\cdots$
Hexagon	6	$2.0486\cdots$
Octagon	8	$2.0197\cdots$
Circle	∞	2

This beautiful result and others in Table XIII were recently rediscovered by Thorpe [127]. The domain fuctional $[\sigma]_\infty$ for n-gons tends to increase as the symmetry of the n-gon [26] is reduced by shape deformation. For example, the intrinsic conductivity of the equilateral triangle and the square are minimal for all triangles and quadrilaterals, respectively [26,128]. A systematic investigation of the variation of $[\sigma]_\infty$ with symmetry would be interesting, since ample evidence indicates that $[\sigma]_\infty$ is smaller for regions of higher symmetry.

Results for C_L and $[\sigma]_\infty$ are also given for linelike regions in Table XII. Such resuls are special to $d = 2$ since it is well known that the *Newtonian capacity* C [see Eq. (3.10)] vanishes for any finite length differentiable curve [129] in $d = 3$. Thus, we can expect $[\sigma]_\infty$ and $[\eta]$, as well as the friction coefficient of any smooth curve in $d = 3$, to equal zero. The finiteness for the capacity of Brownian paths [130] in three dimensions owes itself to the fact that such curves are *typically* nondifferentiable and this property of Brownian paths has numerous implications for polymer physics and phase transition theory.

We next turn to a test of the prediction Eq. (3.11), relating $[\eta]$ and $[\sigma]_\infty$ for nonelliptical shaped regions. Finite element calculations of $[\eta]$ for objects of various shapes are indicated in Table XIV. Our numerical estimate for a circle is $[\eta] = 1.96$, which is 2% *lower* than the exact value. Checks against the exact result [Eq. (5.3a)] for ellipses show that the finite element computations of $[\eta]$ in $d = 2$ tend to be slightly *lower* than the exact values, in contrast to the numerical calculations in $d = 3$, which tend to be slighly *higher* than the exact values. For example, the result for $[\eta]$ for an ellipse of aspect ratio $101/21 = 4.81$ is 3.33, 5.1% lower than predicted by Eq. (5.3b).

Exact calculations of **M** for many interesting shapes are available in the hydrodynamic literature because of important aerodynamic and ship dynamics applications [63, 69–71] and we include the corresponding exact $[\sigma]_\infty$ and numerical $[\eta]$ results for a parabolic lens, spherical lens,

TABLE XIV
Intrinsic Viscosity and Intrinsic Conductivity for Various Shapes in Two Dimensions

Shape	Aspect Ratio	$[\sigma]_\infty$	$[\eta]$	$[\eta]/[\sigma]_\infty$
Rectangle	1	2.19	2.15	0.98
	2	2.40	2.33	0.97
	3	2.76	2.65	0.96
	4	3.15	3.01	0.96
	5	3.55	3.37	0.95
	8	4.76	4.51	0.95
	12	6.38	6.03	0.95
	14	7.18	6.78	0.94
	16	7.99	7.52	0.94
	18	8.79	8.25	0.94
	19	9.20	8.64	0.94
	20	9.60	8.94	0.93
Parabolic lens		2.287	2.21	0.97
Spherical lens		3.236	3.19	0.99
Two touching circles	$r_p = 1$	$\pi^2/4$	2.38	0.96

touching circles, and a rectangle of varying aspect ratio x in Table XIV. Thorpe [127] gives a recent discussion of $[\sigma]_\infty$ for the rectangle. We also mention calculations of $\langle M \rangle / A$ and thus $[\sigma]_\infty$ for the exotic starlike hypocycloid and teardrop-like shapes calculated by Wrinch [70], which are difficult to approximate by finite element methods and even the numerical conformal mapping methods. These examples provide a good challenge for any numerical method for calculating C_L.

Looking more closely at the results for the $[\eta]/[\sigma]_\infty$ ratio for rectangles in Table XIV, we see that for the square this ratio is about 2% less than 1. The error in $[\eta]$ for the square should be comparable to that for the circle and so the actual value of this ratio could easily be one—allowing for a 2% error on the low side for the intrinsic viscosity as in the circle case. Consider now the rectangle with an aspect ratio of 5. If the error in the intrinsic viscosity is similar to that of the ellipse with an aspect ratio of 5, then this value should be about 5% lower, which would make the true value of the ratio 1. Higher resolution should improve the computation, if indeed the deviation from Eq. (5.3a) is only due to finite resolution error. To test this, we recomputed the intrinsic viscosity for a rectangle of aspect ratio 5 using approximately double the resolution (consuming 45 h of supercomputer time). The ratio in Table XIV changed from 0.95 to 0.97, as expected. Also, the error in the intrinsic viscosity is expected to gradually increase with the aspect ratio at a given resolution, as described above for the $d = 3$ case. Considering this finite resolution error, the ratio of the intrinsic conductivity and viscosity for rectangles is consistent with Eq. (5.3a) being *exact*. Results for the other shapes listed in Table XIV are also consistent with this error analysis and Eq. (5.3a). We also recall that Eq. (5.3a) is exact in the case of an ellipse [see Eq. (5.3b)], so on the basis of this analytical and numerical evidence we conjecture that $[\eta] = [\sigma]_\infty$ in $d = 2$.

VI. INTRINSIC VISCOSITY AND THE POLARIZABILITY OF PLATES

Plates occupy an intermediate position between objects extended three dimensionally and objects in three dimensions that can be confined to a plane. The loss of the extension of the body in the dimension normal to the object plane has the effect of decreasing the number of nonzero tensor components for $\boldsymbol{\alpha}_e$ and $\boldsymbol{\alpha}_m$. We noticed in Section II that the tensor components for nearly flat tend to exhibit rapid variation when the thickness is varied. This makes such object shapes useful for effectively modifying the properties of a medium.

The polarization tensors of plates are crucial in the description of long wavelength scattering of electromagnetic and pressure waves through apertures [88–91]. This connection was apparently first noticed by Rayleigh [81a], but the practical significance of this connection was appreciated more recently because of the difficulty of calculations of the polarization tensor components for plates having general shapes. The magnetic polarizability $\boldsymbol{\alpha}_m$ (or the mathematical equivalent $\mathbf{M}$) is also important in the description of the flow of viscous fluid through screens [131]. The technological literature is a good source of results regarding the polarizability of plate-like regions.

The theoretical impetus for calculating the polarizability tensors of plates came from the needs of a developing microwave technology [47–50]. Bethe [88a] calculated the magnetic and electric polarizability of circular plates in his classical theory of the diffraction of electromagnetic radiation by a hole small compared with the incident wavelength and later [89b] he gave results for elliptic plates. Cohn [49, 132] made electrolytic-tank measurements of the polarizability components of plates of numerous shapes (rectangular slots, rounded slots, rosettes, dumbbells, crossed slots, etc.) to provide this important technological information. The electric polarizability of an aperture of general shape was measured by simply cutting out a nonconducting material of the given shape and suspending the object with thin nonconducting wires in an electrolytic solution between two electrodes coplanar with the inclusion. Similarly, the magnetic polarization was measured by suspending a metallic inclusion normal to the electrode surfaces [132]. These experimental measurements of the electric and magnetic polarizability of plates have had many important applications.

Recently, numerical solutions of integral equations defining the electric and magnetic polarizability tensors have been obtained for a wide variety of shapes [72, 73]. These calculations have confirmed the accuracy of Cohn's measurements and a general correlation of the magnetic and electrical aperture polarizations have been obtained for certain families of objects [72, 73]. Significant progress has recently been made by Fabrikant [75] who developed an *analytical* technique for calculating the magnetic and electric polarizability tensors of plates that compares very well with previous numerical and experimental results. Fabrikant treats polygons, rectangles, the rhombus, a circular sector, and other shapes, and the method can apparently be applied to regions having very general shapes.

The utilization of these important theoretical developments requires the recognition of the relation between *aperture polarizabilities* $\boldsymbol{\alpha}_e$(apert.) and $\boldsymbol{\alpha}_m$(apert.), used in the technical literature, and the ordinary polarizabilities, $\boldsymbol{\alpha}_e$ and $\boldsymbol{\alpha}_m$, discussed above. Babinet's principle [73,89] implies that these quantities are related by the definitions

$$\boldsymbol{\alpha}_e(\text{apert.}) = -\boldsymbol{\alpha}_m/4, \qquad \boldsymbol{\alpha}_m(\text{apert.}) = \boldsymbol{\alpha}_e/4 \tag{6.1}$$

Simple physical considerations show that $\boldsymbol{\alpha}_m$ for a flat plate is effectively a scalar, having only one nonzero component. Correspondingly, the component of $\boldsymbol{\alpha}_e$ for a metal plate normal to an applied field vanishes, since there is no way to separate charges in this case. However, there are nonvanishing components to $\boldsymbol{\alpha}_e$ when the plate is aligned along the field direction. From these observations we immediately obtain numerous results for $\boldsymbol{\alpha}_e$ and $\boldsymbol{\alpha}_m$ from the literature of electromagnetic and sound scattering through apertures.

For example, we can obtain the magnetic polarizability of an elliptic plate from Bethe's formula for the electric aperture polarizability of an ellipse (plate is normal to field direction)

$$\alpha_m(T) = (4\pi ab^2/3)/E(e) \tag{6.2}$$

where a and b are semimajor and semiminor axes, e is the eccentricity, $e = [1 - (b/a)^2]^{1/2}$, and E is a complete elliptic function of the second kind (T and L denote the transverse and longitudinal components of the polarizability tensor as defined in Sect. 3). The average $\langle \alpha_m \rangle$ of a plate generally equals

$$\langle \alpha_m \rangle = \tfrac{1}{3}\alpha_m(T) \tag{6.3}$$

since there is only one nonzero component, as mentioned above. Note that Eqs. (6.2) and (6.3) reduce to Eq. (3.17) for the circular disk. Numerous other examples follow along these lines from recent numerical calculations [72, 73] and analytical calculations by Fabrikant [75].

Results for the electric polarizability can be similarly obtained, although these results tend to have a more complicated mathematical description. We mention the important experimental estimate [49] of $\alpha_e(L)$ for a square plate having a side length a,

$$\alpha_e(L) = 1.02a^3 \text{ (expt.)} \tag{6.4}$$

placed parallel to the applied field. There are two components of the electric polarizability for an asymmetric rectangular plate, of course. The estimate Eq. (6.4) was obtained from Cohn's electrolytic tank measurements and this value has been confirmed by more recent numerical [72, 73] and analytical studies [75] to an accuracy on the order of 1%. Our finite element technique gives $\alpha_e(L) = 1.09a^3$ for the square plate, which is apparently too high by the usual 5–6% in $d = 3$. This result is listed in Table X.

These quantitative estimates of $\boldsymbol{\alpha}_e$ and $\boldsymbol{\alpha}_m$ for plates from this variety of sources is very useful in combination with the approximate invariant relation [Eq. (3.16)]. In Section III we mentioned the exact result

$$[\eta] = \tfrac{4}{5}[\sigma]_\infty \tag{6.5}$$

for circular plates. Combination of the extensive plate estimates with Eq. (6.5) then yields predictions for $[\eta]$ that can be checked against experiment. Further efforts are needed on the difficult problem of analytically calculating $[\eta]$ for arbitrarily shaped plate-like regions, which certainly is not going to be any more tractable than the electrostatic and magnetostatic analogs. We note that our finite element technique could be used to estimate the polarizabilities, intrinsic conductivity, and intrinsic viscosity of arbitrarily shaped objects, and that the method is not limited to large values of the relative conductivity Δ_σ. The numerical calculations are actually *faster and more accurate* when Δ_σ is not large.

VII. CONCLUSIONS AND SUMMARY

There are many physical processes for which the solution of the Laplace equation on the exterior of a body of general shape is central to the theoretical description. Previous papers [80] discussed the exterior Dirichlet problem for the Laplace equation and the calculation of the capacity C, which is the shape functional associated with this problem. This chapter discusses other functionals of shape associated with the solution of the Laplace and the Navier–Stokes equations on the exterior of objects having general shape. These functionals [26] include the electric and magnetic polarizabilities, the hydrodynamic virtual mass, and the intrinsic viscosity.

New numerical and analytical results for these shape functionals, along with values from a large literature, were obtained to check a proposed relation between shape functionals associated with the Laplace equation, namely, the electric polarizability and a shape functional associated with the Navier–Stokes equation, the intrinsic viscosity. Our new approximate relation is a natural generalization of a result of Hubbard and Douglas [27] approximately relating the translational friction coefficient (Navier–Stokes equation) to the electrostatic capacity (Laplace equation). These relations between hydrodynamic theory and electromagnetic theory complement the classical relation between the *effective mass* $\mathbf{M}$ of *perfect* fluids and the magnetic polarizability $\boldsymbol{\alpha}_m$ observed by Keller et al. and Kelvin [57,58].

Exact and numerical results confirm that the intrinsic viscosity $[\eta]$ is proportional to the intrinsic conductivity $[\sigma]_\infty$ of conducting particles of arbitrary shape to within about a 5% approximation,

$$[\eta] \approx (0.79 \pm 0.04)[\sigma]_\infty \tag{7.1a}$$

in three dimensions. In two dimensions we find $[\eta]$ exactly equals $[\sigma]_\infty$ for ellipses. Data for other shapes, allowing for a usual underestimation of $[\eta]$ by 2–5% in our $d = 2$ finite element method, are in general agreement with this conjectured approximation. On the basis of this evidence, we conjecture that $[\eta] = [\sigma]_\infty$ for all shapes in $d = 2$. Further exact calcuations of $[\eta]$ for some of the shapes discussed would be useful in developing a proof (or disproof) of this conjecture. All of our findings agree well with the predictions of our angular averaging approximation

$$[\eta] \approx [(d+2)/(2d)][\sigma]_\infty \tag{7.2a}$$

$$[\eta] \approx 0.83[\sigma]_\infty, \qquad d = 3 \tag{7.2b}$$

$$[\eta] \approx [\sigma]_\infty, \qquad d = 2 \tag{7.2c}$$

Although our primary goal in this chapter was to test relation (7.2), the tabulated values of $[\sigma]_\infty$, $[\sigma]_0$, and $[\eta]$ for numerous shapes and the discussion of the general shape parameters that affect these quantities should find wide application. Inevitably, this information becomes important when we attempt to resume the virial expansions, such as Eq. (2.1), to provide a useful description of suspensions of arbitrarily shaped particles in a matrix at high concentrations.

APPENDIX A
VIRTUAL MASS AND THE ACOUSTIC INDEX OF REFRACTION

It seems likely that the fundamental relation [Eq. (2.6)] was recognized much earlier since the mathematical equivalence between the flow of inviscid fluids and electrical conduction is well known [58,64,102]. Recently, Brown [133a] pointed out a nonperturbative generalization of Eq. (2.6). He showed that the conductivity σ of a nonconducting rigid matrix filled with a conducting fluid of conductivity σ_0 is related to the change in the average *effective mass* (effective fluid density) of the corresponding inviscid fluid at arbitrary volume fractions. Johnson and Sen [134] showed that this analogy also implies that the acoustic index of refraction n of an ideal fluid in a rigid matrix equals

$$n^2 = (1 - \phi)(\sigma_0/\sigma) \tag{A.1}$$

where $1 - \phi$ is the *porosity*. This relation, like so many others, was apparently known to Rayleigh [135]. Acoustic index of refraction measurements (fourth sound) for superfluid He^4 in a porous medium [136] and salt water in a sintered glass bead pack [24d] confirm Eq. (A.1) to a good approximation. Recently, there has been a nonperturbative generalization of Eq. (2.2a) [137]

$$\sigma/\sigma_0 = (1 - \phi)/t_E \tag{A.2}$$

where the *electrical tortuosity* α_E is defined as $t_E = D_0/D_p$, where D_p is the diffusion coefficient of particles in the fluid region such that the diffusing particles obey a *reflecting* boundary condition when they encounter the matrix. The parameter D_0 is the diffusion coefficient of the particle in the fluid in the absence of the rigid matrix.

The diffusion coefficient observed in a macroscopic diffusion measurement on a porous medium with insulating rigid inclusions, however, is *not equal* to the pore diffusion coefficient, D_p. Rather, the diffusion coefficient D measured in a macroscopic measurement is related to D_p as

$$D = D_p/(1 - \phi) \tag{A.3}$$

so that Eq. (A.2) reduces to the generalized Einstein relation [138].

$$\sigma/\sigma_0 = D/D_0 \tag{A.4}$$

For low concentrations ($\phi \to 0^+$) the nonperturbative relation [Eq. (A.4)] reduces to the known virial expansion [16a, 16d] for fixed hard sphere inclusions having a reflecting boundary condition

$$\sigma/\sigma_0 = D/D_0 = 1 + [\sigma]_0 + O(\phi^2), \qquad [\sigma]_0 = -\langle M \rangle / V_p \tag{A.5a}$$

$$\langle M \rangle / V_p = \tfrac{3}{2} \qquad (d = 3;\ \text{sphere}) \tag{A.5b}$$

From a conductivity standpoint Eq. (A.5a) corresponds to insulating inclusions. Equation (A.5a) shows that the average effective mass $\langle M \rangle$ of the rigid inclusions determines the leading concentration dependence of the diffusion coefficient in a porous medium in the absence of particle interaction. We note that the insensitivity of $[\sigma]_0$ to particle shape for extended particles (not platelets) means that as particles aggregate at higher concentration then Eq. (A.5) should remain a good approximation. Experiment, indeed, shows that the leading linear concentration dependence in Eq. (A.5) holds to a good approximation over a wide concentration range [33, 34].

APPENDIX B
POLARIZATION FORMALISM FOR ELLIPSOIDS

Stratton [52a] defines a set of numbers A_i, which often arise in the discussion of the properties of ellipsoids. These numbers are defined by the integrals (Stratton does not include prefactor $a_1 a_2 a_3$):

$$A_i = a_1 a_2 a_3 \int_0^\infty dx[(a_i^2 + x)R(x)]^{-1} \tag{B.1}$$

where

$$R(x) = [(a_1^2 + x)(a_2^2 + x)(a_3^2 + x)]^{1/2} \tag{B.2}$$

and the constants a_1, a_2, and a_3 are the ellipsoid semiaxes lengths with $V_p = 4\pi a_1 a_2 a_3 / 3$. The A_i parameters obey the simple sum rule,

$$A_1 + A_2 + A_3 = 2 \tag{B.3}$$

Equation (B.3) is useful since it reduces the integrals that need to be computed for a general ellipsoid from three to two. Various combinations and ratios of these numbers are required for the intrinsic viscosity calculation and are given in Appendix C. Note that Stratton uses the notation A_i for these quantities, while Haber and Brenner [14] and

Scheraga [120a] use α_i. We followed Stratton's notation to avoid confusion with our notation for the polarizability.

The A_i integrals can be expressed as combinations of the standard elliptic integrals. However, for numerical purposes, it is just as simple to evaluate the integrals directly using Gauss–Legendre quadrature. It is first useful to transform the integrals by letting $x = \tan^2(\theta)$, so that $0 < \theta < \pi/2$. An integration mesh can be easily set up and enough points chosen to achieve convergence. For the smaller values of the ratios a_3/a_1 and a_3/a_2 (on the order of 50 or less) we needed less than 100 points in the quadrature mesh to achieve 5–6 significant figure accuracy, while for the highest values of these ratios studied (on the order 10,00) we needed 10000 points in the integration mesh to achieve the same accuracy.

For ellipsoids of revolution we have $a_1 = a_2$ so that $a_3/a_1 = a_3/a_2 = x$, the aspect ratio and the number of integrals needed reduces to 1. In this case we have $A_1 = A_2 = (2 - A_3)/2$ and A_3 is given by

$$A_3 = [x/(x^2 - 1)][B - 2/x] \tag{B.4}$$

where for prolate ellipsoids of revolution $(x > 1)$

$$B = [(x^2 - 1)]^{-1/2} \ln \{[x + (x^2 + 1)^{1/2}]/[x - (x^2 + 1)^{1/2}]\} \tag{B.5}$$

and for oblate ellipsoids of revolution $(x < 1)$

$$B = 2\cos^{-1}(x)/(x^2 - 1)^{1/2} \tag{B.6}$$

With the A_i functions defined for any ellipsoid the formulas in Stratton can be easily evaluated to compute the polarizability for any choice of matrix conductivity σ_0 and particle conductivity σ_p. For the case of a highly conducting ellipsoidal inclusion, $\sigma_p \gg \sigma_0$, the components of α_e are defined as

$$\alpha_e(i)/V_p = 2/A_i \tag{B.7}$$

In the case of an insulating ellipsoidal inclusion the components of $\boldsymbol{\alpha}_m$ are given by

$$\alpha_m(i)/V_p = (A_i/2 - 1)^{-1} \tag{B.8}$$

and the intrinsic conductivity equals

$$[\sigma] = (\alpha_1 + \alpha_2 + \alpha_3)/3 \tag{B.9}$$

APPENDIX C
INTRINSIC VISCOSITY FORMALISM FOR ELLIPSOIDS AND OTHER SHAPES

Haber and Brenner worked out the isotropically averaged intrinsic viscosity for ellipsoidal particles [14]. This formalism is valid for any choice of the axis lengths a_1, a_2, and a_3. Values of the A_i functions are required, as well as auxiliary functions that can be defined in terms of the A_i, denoted A'_i and A''_i. They are given as follows (i, j, k are permuted cyclically):

$$A'_i = a_j a_k (A_j - A_k)/(a_k^2 - a_j^2) \tag{C.1}$$

$$A''_i = (a_j^2 A_j - a_k^2 A_k)/(a_j^2 - a_k^2) \tag{C.2}$$

[The expression for A'_i is opposite in sign from that given in Haber and Brenner [14], and this typographical error is corrected in Eq. (C.1).] The following quantities are then formed:

$$Q_i = \tfrac{4}{5} A''_i/(A''_1 A''_2 + A''_2 A''_3 + A''_1 A''_3) \tag{C.3}$$

$$q_i = a_j a_k (A_j + A_k)/[A'_i(a_j^2 A_j + a_k^2 A_k)] \tag{C.4}$$

The intrinsic viscosity can be expressed in terms of these constants

$$[\eta] = \tfrac{2}{15}(Q_1 + Q_2 + Q_3) + \tfrac{2}{2}(q_1 + q_2 + q_3) \tag{C.5}$$

This equation is tabulated in Tables V and VI.

The intrinsic viscosity for anisotropic particles is in general a fourth rank tensor, with the same symmetries as the elastic stiffness tensor [117] for the given symmetry of the particle. For example, the intrinsic viscosity for an aligned triaxial ellipsoid is a fourth rank tensor having the same symmetry as the elastic stiffness tensor for an orthorhombic crystal [117, 139]. Formally, this can be stated using either the stress to define the effective viscosity of a single-particle suspension, or using the energy dissipation rate [14]. In the following, the discussion of Haber and Brenner is followed [14] in presenting both methods.

Consider an isolated particle, immersed at rest in a homogeneous, incompressible fluid with viscosity η_0. We restrict consideration to particles that have a center of symmetry. The traceless rate of strain field is denoted $\mathbf{S}$, such that if the particle were not there, $\mathbf{S} = \mathbf{S}_0$ would be a uniform traceless rate of strain tensor. Far away from the particle, the velocity fields must go to $\mathbf{v} = \mathbf{S}_0 \mathbf{r}$ and the average of $\mathbf{S}$ over all space, denoted $\langle S \rangle$, equals $\mathbf{S}_0$ [14]. Solving Stokes' equations for the fluid velocity and pressure fields due to the presence of the rigid particle and then averaging the viscous stress $\boldsymbol{\sigma}_s$ over the volume of the sample gives

$$\boldsymbol{\sigma}_s = 2\eta_0 \, [\mathbf{I} + \phi \boldsymbol{\Psi}] \colon \mathbf{S}_0 \tag{C.6}$$

where ϕ is the volume fraction of the particles. This result is limited to a sufficiently dilute suspension so that each particle can be considered to be independent of the others. The parameter Ψ_{ijkl} is a fourth rank tensor of the same symmetry as the particle. If the calculation is carried out in a periodic array of particles, as was done in the finite element work explained in Appendix E, then Ψ_{ijkl} has whichever symmetry is lower, the particle or the array. For example, when considering a cubic array of spheres, Ψ_{ijkl} has cubic symmetry and is not isotropic like the sphere.

The isotropically averaged intrinsic viscosity that we have studied in this chapter is obtained by calculating the isotropic average $\langle \Psi_{ijkl} \rangle$, using the standard definition of rotational tensor averaging [140] and then using Eq. (3.2) to obtain

$$[\eta] = \langle \Psi_{ijkl} \rangle \tag{C.7}$$

We are operating under the assumption of "overwhelming Brownian motion" [117], so that the applied shear is weak compared to the Brownian motion of the particle, so that all orientations are equally probable. In this case, the quantity Ψ_{ijkl} is calculated when holding the particle positionally and orientationally fixed in space. The rotational average of Ψ_{ijkl} then incorporates the fact that all orientations are equally probable. The opposite case would be the strong applied shear case, where Brownian motion can be ignored, and the particle has anisotropic orientation due to the applied shear field. For an anisotropic particle this would correspond to the elastic case, since in the elastic case a rigid particle maintains its shape but can orient itself with the applied field.

Since the intrinsic viscosity tensor has the same symmetries as the equivalent elastic tensor, the averaging procedure discussed above is exactly the same as taking the Voigt average [141] of the elastic stiffness tensor for a polycrystalline sample. This procedure has been worked out for every crystal symmetry of interest and results are available in the literature. For example, all the particles considered numerically in this chapter have tetragonal or higher symmetry. Tetragonal symmetry means square symmetry in the cross-sectional plane, with the third direction being different (a rectangular parallelipiped, with $a = b \neq c$, has tetragonal symmetry). The angularly averaged shear modulus $\langle G \rangle$ is obtained from the elastic stiffness tensor C_{ijkl} (tetragonal symmetry) by

$$\langle G \rangle = \tfrac{1}{30}[M_G + 3C_{1111} - 3C_{1122} + 12C_{2323} + 6C_{1212})] \tag{C.8a}$$

$$M_G \equiv C_{1111} + C_{1122} + 2C_{3333} - 4C_{1133} \tag{C.8b}$$

To obtain the rotationally averaged intrinsic viscosity, simply substitute Ψ_{ijkl} for C_{ijkl}, and multiply by a factor of 2, because the elastic tensor averages are defined for a system of notation such that the shear strain is twice that used for the rate of shear strain in fluid mechanics [142,143]. One should note that in [14] and [117] a factor of $\frac{5}{2}$ is often taken out of the definition of Ψ_{ijkl} so that $\langle \Psi_{ijkl} \rangle$ is then normalized to equal 1 for a spherical particle.

The intrinsic viscosity can also be defined via the energy-dissipation rate. Since this is the way that the finite-element simulations were done and since it offers an easy way to do the averages numerically, we present this method as well. We again follow the discussion of [14].

For the same case considered above, it is found that the rotationally averaged rate of energy dissipation $\langle E \rangle$ is given by

$$\langle E \rangle = 2\eta_0[1 + [\eta]\phi]S^0_{ij}S^0_{ij} \tag{C.9}$$

For tetragonal symmetry and higher it is possible to select the terms of $\mathbf{S}_0$ so that the energy dissipation rate E [143]

$$E = \tfrac{1}{4}S^0_{ij}\eta_{ijkl}S^0_{kl} \tag{C.10}$$

gives the terms needed to form the average Eq. (C.8). One choice for S^0_{ij}, which gives the correct combination of terms for tetragonal symmetry, is the following:

$$S_{11} = [2 + \sqrt{3}]^{1/2}/\sqrt{30} \tag{C.11a}$$

$$S_{22} = -[2 - \sqrt{3}]^{1/2}/\sqrt{30} \tag{C.11b}$$

$$S_{33} = -(S_{11} + S_{22})/\sqrt{30} \tag{C.11c}$$

$$S_{13} = S_{23} = S_{12} = 1/\sqrt{5} \tag{C.11d}$$

Computing the energy for this applied rate of strain tensor gives precisely the combination of terms in Eq. (C.8). By subtracting the original energy from that found without a particle being present, the rotationally averaged intrinsic viscosity can be read off from the numerical results.

APPENDIX D
EXACT SOLUTION OF THE POLARIZATION OF A DUMBBELL

It was not previously noticed that all the mathematical apparatus necessary for computing the components of $\boldsymbol{\alpha}_e$ for the spherical particle dumbbell had been in place since the publication of a paper by Davis [144]. The problem is the following: Take two highly conducting spheres of the same radius connected by a very thin conducting wire so that both spheres in this dumbbell are at the same potential. Apply an electric field that is uniform at infinity and solve for the potential. Once this is done, calculate the normal component of the electric field right at the surface of the dumbbell, thus giving the charge density Σ (charge/unit area) on the dumbbell surface. The dipole moment of the dumbbell is then calculated from the average of $\Sigma\,\mathbf{r}$ over the surface and divided by the dumbbell volume.

Davis [144] gives the solution for the potential when the spheres are both held at zero potential and this solution is all that is required for solving our dumbbell problem. In calculating the polarization there are two difficult integrals that arise, which fortunately have been tabulated by Apelblat [145]. The required integrals are given by,

$$\int_{-1}^{1} \frac{P_n(x)\,dx}{(\cosh(a)-x)^{3/2}} = \frac{2^{3/2}\,e^{-(n+1/2)a}}{\sinh(a)} \tag{D.1}$$

$$\int_{-1}^{1} \frac{P_n(x)\,dx}{[\cosh(a)-x]^{5/2}} = \frac{2^{3/2}\,e^{-(n+1/2)a}[2n+1+2\coth(a)]}{3\sinh^2(a)} \tag{D.2}$$

The polarizability of the dumbbell then equals

$$\alpha_e(T)/V_p = 12(r_p-1)^3[1+2/(r_p-1)]^{3/2}Z_2 \tag{D.3}$$

$$\alpha_e(L)/V_p = 6(r_p-1)^3[1+2/(r_p-1)]^{3/2}(4S_2+4S_1+S_0) \tag{D.4}$$

$$[\sigma]_\infty = (2\alpha_e(T)/V_p + \alpha_e(L)/V_p)/3 \tag{D.5}$$

where r_p is the ratio of the center–center separation to the sphere diameter and the constants Z_2 and S_j are defined by

$$Z_2 = \sum_{n=1}^{\infty} \frac{n(n+1)}{\{\exp[(2n+1)\mu_1]+1\}} \tag{D.6}$$

$$S_j = \sum_{n=0}^{\infty} \frac{n^j}{\{\exp[(2n+1)\mu_1]-1\}} \tag{D.7}$$

where $\cosh(\mu_1) = r_p$. These infinite sums are rapidly converging so that numerical computation is straightforward. In the limit $r_p \to \infty$ only the first term of S_0 is important and we find the simple limiting behavior

$$[\sigma]_\infty \sim r_p^2, \qquad r_p \to \infty \tag{D.8}$$

The electric polarizability for two spheres without the wire [46] is quite different in general. For $r_p \to \infty$ the intrinsic conductivity $[\sigma]_\infty$ of the untethered spheres approaches 3 (the single sphere result) and for nearly touching spheres ($r_p \to 1$), $[\sigma]_\infty$ approaches $7\zeta(3)/2$ [see Eq. (3.14)].

APPENDIX E
FINITE ELEMENT METHOD COMPUTATION OF $[\eta]$ AND $[\sigma]_\infty$

Finite element methods [146] are well suited to find the extrema of *energy functionals*. In the physical cases considered in this chapter, where variational principles exist from which the relevant equations can be derived, the energy functional to be minimized is the actual energy of the system, where perhaps Lagrange multiplier terms have been added to enforce constraints. The appropriate functional for elastic problems is the elastic energy [146]

$$U(\text{elastostatic}) = \frac{1}{2} \int \varepsilon_{ij} C_{ijkl} \varepsilon_{kl} d^3 r \tag{E.1}$$

where ε_{ij} is the strain tensor and C_{ijkl} is the elastic stiffness tensor and the appropriate functional for the conductivity problem [2] is the dissipated electrical power

$$U\ (\text{electrostatic}) = \frac{1}{2} \int E_i \sigma_{ij} E_j d^3 r \tag{E.2}$$

where E_i is the electric field vector and σ_{ij} is the conductivity tensor.

The elastic-fluid mapping mentioned previously [17, 102, 103, 147] implies that a similar variational principle exists for the linearized fluid problem (Stokes equation), where only terms linear in the velocities are retained. This fact has been known for some time, being first discussed by Helmholz [148] and proven and elaborated on by Korteweg [149]. Millikan [150], however, proved that if one is restricted to action integrals involving only fluid velocities and first-order spatial derivatives, there will be no variational principle whose Euler equations will give the full Navier–Stokes equations.

Two recent discussions of the variational principle for viscous fluids were given by de Veubeke [151] and Keller et al. [152], where the pressure is introduced as a Lagrange multiplier [151]:

$$\delta\left[\int \{p(\nabla\cdot u) - \frac{1}{4}\eta\theta_{ij}\theta_{ij}\right]d^3r = 0 \tag{E.3}$$

$$\theta_{ij} = \frac{1}{2}\left(\frac{\partial u_j}{\partial x_i} + \frac{\partial u_i}{\partial x_j}\right) \tag{E.4}$$

$$F = -\tfrac{1}{4}\eta\theta_{ij}\theta_{ij} \tag{E.5}$$

where F is the rate of energy dissipation, $\mathbf{u}$ is the fluid velocity, and p is the pressure. Carrying out the variation indicated in Eq. (E.2) gives the linearized steady-state Navier–Stokes equation in the creeping flow limit, commonly called the Stokes equation. To enforce the incompressibility condition $\boldsymbol{\nabla}\cdot\mathbf{u} = 0$ requires an extra effort on the part of the would-be solver of Eq. (E.3).

In the finite element solution used in this chapter we used a formulation of the problem described by Zienkiewicz [153] in which the pressure is ignored and the incompressibility condition is only approximately maintained via a "penalty method" [146]. A term is added to the energy dissipation F of the form $\frac{1}{2}\beta(\boldsymbol{\nabla}\cdot\mathbf{u})^2$, where taking a large *penalty parameter* ($\beta \rightarrow \infty$) corresponds to the incompressibile limit ($\boldsymbol{\nabla}\cdot\mathbf{u} = 0$). This method works extremely well, although run times become longer as the value of β is increased. Finite values of β imply some degree of compressibility of the simulated fluid.

However, it turns out that the intrinsic viscosity, or more specifically, the extra dissipated energy terms in Eq. (C.9) from Appendix C do not depend (under steady-state conditions) on the compressibility of the fluid, which means that the intrinsic viscosity does not depend on the compressibility either [154]. This allowed run times to be considerably shortened, as much smaller values of β could be used with *no change* in the final results. The argument goes as follows.

The stress tensor of a compressible isotropic fluid is a function of both the shear viscosity η and the bulk viscosity ξ [143]. The total stress tensor σ_{ij} is the sum of an ideal fluid pressure component and a viscosity component

$$\sigma_{ij} = -p\delta_{ij} + 2\eta\hat{\theta}_{ij} \tag{E.6a}$$

where p is the macroscopic pressure and the viscosity contribution to the stress tensor $\hat{\theta}_{ij}$ reflects the explicit dependence on η and ξ [143],

$$\hat{\theta}_{ij} = \left(\theta_{ij} - \frac{\delta_{ij}}{3}\frac{\partial u_k}{\partial x_k}\right) + \left(\frac{\xi}{2\eta}\delta_{ij}\frac{\partial u_k}{\partial x_k}\right) \tag{E.6b}$$

For an incompressible fluid ($\nabla \cdot \mathbf{u} = 0$), the viscous stress tensor $\hat{\theta}_{ij}$ evidently reduces to θ_{ij} in Eq. (E.4). The compressibility related terms in Eq. (E.4) can be absorbed into a redefinition of the pressure P_{comp}, which then varies with the local spatial position,

$$\sigma_{ij} = -P_{\text{comp}}\delta_{ij} + 2\eta\theta_{ij} \tag{E.7a}$$

We then observe that the integral of the gradient of P_{comp} over any closed circuit C in the fluid equals zero

$$\oint_C \nabla P_{\text{comp}} \cdot d\mathbf{s} = 0 \tag{E.7b}$$

since P_{comp} is a single-valued function of position. Consequently, there is no extra dissipation [154] associated with the fluid compressibility ($\nabla \cdot \mathbf{u} \neq 0$). Notably, Eq. (E.7b) is independent of the surface boundary condition.

Finally, in actually carrying out the finite element computation for the viscosity, the "no-slip" boundary condition at a fluid–solid boundary is enforced by holding the velocity at all computational nodes either at the boundary or inside the particle fixed at zero during the conjugate gradient [25] cycles that find the minimum energy. To treat a plate geometry one simply holds the velocities of all the nodes in the appropriate planar region fixed at zero. In the analogous electrical problem one holds the voltage constant at the same nodes as for the fluid problem.

APPENDIX F ELLIPSE TRANSPORT VIRIAL COEFFICIENTS IN $d = 2$

The intrinsic conductivity and viscosity for an ellipse in a 2D fluid can be obtained from the triaxial ellipsoid results in three dimensions by taking appropriate limits and averaging. The average required to obtain virial coefficients in $d = 2$ involves rotationally averaging around the z axis of the ellipsoid and letting a_3 (the semiaxis length in the z direction) approach infinity. We then need to evaluate A_1, A_2, and A_3 from Appendix B in the limit $a_3 \to \infty$. From Eqs. (B.1) and (B.2) along with the limit $a_3 \to \infty$ we find

$$A_1 \rightarrow 2a_2/(a_1 + a_2) \tag{F.1}$$

$$A_2 \rightarrow 2a_1/(a_1 + a_2) \tag{F.2}$$

$$A_3 \rightarrow 0 \tag{F.3}$$

and the A_i sum rule in $d = 2$ becomes,

$$A_1 + A_2 = 2 \tag{F.4}$$

The rotational average in $d = 2$ for the polarization is then

$$\langle \alpha_e \rangle = (\alpha_{11} + \alpha_{22})/2 \tag{F.5}$$

since the second-rank polarization tensor is diagonal for orthorhombic symmetry and higher. The intrinsic conductivity then equals

$$[\sigma]_\infty = (a_1 + a_2)^2/(2a_1a_2), \qquad d = 2 \tag{F.6}$$

which agrees with Garboczi et al. [155].

The intrinsic viscosity in $d = 2$ is found in the same way as the average shear modulus by rotationally averaging an elastic stiffness tensor having rectangular symmetry around the z-axis:

$$\langle G \rangle = [C_{1111} + C_{2222} + 4C_{1212} - 2C_{1122}]/8 \tag{F.7}$$

After inserting the appropriate combinations of A_1, A_2, and A_3 into Haber and Brenner's formalism given in Appendix C and [14], the rotationally averaged intrinsic viscosity becomes

$$[\eta] = \tfrac{5}{8}[Q_1 + Q_2 + 4Q_3] \tag{F.8}$$

where the Q_i values are defined in [14]. This reduces to

$$[\eta] = (a_1 + a_2)^2/(2a_1a_2) \tag{F.9}$$

which is exactly equal to the intrinsic conductivity [see Eq. (F.6)].

ACKNOWLEDGMENTS

We would like to thank Dr. Geoffrey McFadden of the Computational and Applied Mathematics laboratory at NIST for his numerical calculations of the logarithmic capacity of regions having complicated shape (Section V), Dr. Joe Hubbard for his contribution relating to understanding the role of solvent compressibility on transport properties (Appendix E), and Dr. Holly Rushmeier for generating Fig. 2.7 for us. We also thank Dr. Howard Brenner for useful discussions, encouragement, and for supplying important references.

REFERENCES

1. G. K. Batchelor, *Annu. Rev. Fluid Mech.*, **6**, 227 (1974).
2. S. Torquato, *Appl. Mech. Rev.*, **44**, 37 (1991).
3. D. J. Bergman, *Phys. Rep.*, **9**, 377 (1978).
4. (a) J. C. Maxwell, *A Treatise on Electricity and Magnetism* (Dover, N.Y., 1954), (b) J. B. Keller, *Philips Res. Rep.* **30**, 83 (1975).
5. A. S. Sangani, *Soc. Ind. Appl. Math., J. Appl. Math.*, **50**, 64 (1990).
6. H. B. Levine and D.A. McQuarrie, *J. Chem. Phys.* **49**, 4181 (1968).
7. D. J. Jeffrey, *Proc. R. Soc. London A* **335**, 355 (1973).
8. (a) H. Fricke, *Phys. Rev.* **24**, 575 (1924); (b) H. Fricke and S. Morse, *Phys. Res.*, **25**, 361 (1925).
9. (a) A. Rocha and A. Acrivos, *Q. J. Mech. Appl. Math.*, **26**, 217 (1973); (b) H.-S. Chen and A. Acrivos, *Proc. R. Soc. London A* , **349**, 261 (1976).
10. (a) A. Einstein, *Ann. Phys.* **19**, 289 (1906); **34**, 591 (1911). (b) The second-order calculation is developed by G. K. Batchelor and J. T. Green, *J. Fluid Mech.*, **56**, 401 (1972).
11. L. Onsager, *Phys. Rev.* **40**, 1028 (1932).
12. (a) N. Saito, *J. Phys. Soc. Jpn.*, **6**, 297 (1951). (b) R. Simha *J. Phys. Chem.*,**44**, 25 (1940); See Ref. 14 for a discussion of technical difficulties in this historically important calculation.
13. J. M. Rallison, *J. Fluid Mech.*, **84**, 237 (1978).
14. S. Haber and H. Brenner, *J. Coll. Int. Sci.*, **97**, 496 (1984).
15. G. I. Taylor, *Proc. R. Soc. London A*, **138**, 41 (1932).
16. (a) J. R. Lebenhaft and K. Kapral, *J. Stat. Phys.*, **20**, 25 (1979), (b) M. Fixman, *J. Phys. Chem.* **88**, 6472 (1984), (c) S. Sridharan and R. Cukier, *J. Phys. Chem.* **88**, 1237 (1984). (d) J. H. Wang, *J. Amer. Chem. Soc.*,**76**, 4755 (1954).
17. J. N. Goodier, *J. Appl. Mech. A*, **55**, 39 (1933).
18. (a) J. K. MacKenzie, *Proc. Phys. Soc. London*, **B 63**, 361 (1950); (b) J. M. Dewey, *J. Appl. Phys.*, **18**, 579 (1947); (c) J. D. Eshelby, *Proc. R. Soc. London A*, **241**, 376 (1957); (d) L. J. Walpole, *Q. J. Mech. Appl. Math.*, **25**, 153 (1972).
19. E. Guth, *J. Appl. Phys.*, **16**, 20 (1945).
20. (a) H. Froehlich and R. Sack, *Proc. R. Soc. London A*, **185**, 415 (1946), (b) R. Roscoe, *J. Fluid Mech.* **28**, 273 (1967).
21. (a) Z. Hashin, *Bull. Res. Council Isr.*, **C5**, 46 (1955), (b) Z. Hashin and S. Shtrikman, *J. Appl. Phys.*, **33**, 3125 (1962), (c) S. Praeger, *Physica*, **29**, 129 (1963); *J. Chem. Phys.*, **50**, 4305 (1969), (d) Z. Hashin, *J. Appl. Mech.*, **50**, 481 (1983).
22. (a) G. W. Milton, *J. Appl. Phys.*, **52**, 5286 (1981); *Phys. Rev. Lett.*, **46**, 542 (1981); (b) R. Lipton, *J. Mech. Phys. Solids*, **41**, 809 (1993), (c) J. D. Beasley and S. Torquato, *Phys. Fluids A*, **1**, 199 (1989).
23. E. J. Garboczi and A. R. Day, *J. Mech. Phys. Solids*, in press.
24. (a) S. Torquato and I. C. Kim, *Appl. Phys. Lett.*, **55**, 1847 (1989); (b) S. B. Lee, I. C. Kim, C. A. Miller, and S. Torquato, *Phys. Rev. B*, **39**, 11833 (1989); (c) L. M. Schwartz and J. R. Banavar, *Phys. Rev. B*, **39**, 11965 (1989); (d) K. A. Akanni, J. W. Evans, and I. S. Abramson, *Chem. Eng. Sci.*, **42**, 1945 (1987).
25. W. H. Press, B. P. Flannery, S. A. Teukolsky, and W. T. Vetterling, *Numerical Recipes*, Cambridge University Press, Cambridge, UK, 1989.

26. G. Pólya and G. Szegö, *Isoperimetric Inequalities in Mathematical Physics*, Princeton University Press, Princeton, NJ, 1951.

27. (a) J. B. Hubbard and J. F. Douglas, *Phys. Rev. E*, **47**, R-2983 (1993); (b) see [80].

28. G. Pólya, *Proc. Nat. Acad. Sci.*, **33**, 218 (1947).

29. N. S. Landkof, *Foundations of Modern Potential Theory*, Springer-Verlag, NY, 1972.

30. (a) G. Szegö, "Relation Between Different Capacity Concepts," in *Proceedings of the Conference on Differential Equations*, J. B. Diaz and L. E. Payne, Eds., University of Maryland Bookstore, College Park, MD, 1956, p. 139; (b) M. Schiffer, *Proc. Camb. Philos. Soc.* **37**, 373 (1941); (c) L. V. Ahlfors, *Conformal Invariants: Topics in Geometric Function Theory*, McGraw-Hill, NY, 1973.

31. A. Voet, *J. Phys. Chem.*, **51**, 1037 (1947).

32. (a) C. Pearce, *Br. J. Appl. Phys.*, **6**, 113 (1955); (b) R. Guillien, *Ann. Phys. (Paris)*, **16**, 205 (1941).

33. R. E. De La Rue and C. W. Tobias, *J. Electrochem. Soc.*, **106**, 827 (1959).

34. L. Sigrist and O. Dossenbach, *J. Appl. Electrochem.*, **10**, 223 (1980).

35. (a) C. R. Turner, *Chem. Engr. Sci.* **31**, 487 (1976); (b) A strong variation of thermal conductivity has been observed in metal filled plastics [D. L. Cullen, M. S. Zawojski, and A. L. Holbrook, *Plastics Engr. J.*, **44** (1), 37 (1988)]. This important example deserves quantitative examination.

36. D. A. G. Bruggeman, *Ann. Phys. (Leipzig)*, **24**, 636 (1935).

37. H.C. Brinkman, *J. Chem. Phys.*, **20**, 571 (1952).

38. (a) R. Roscoe, *Br. J. Appl. Phys.*, **3**, 267 (1952); (b) I. M. Krieger, *Adv. Coll. Int. Sci.*, **3**, 111 (1972); (c) A. B. Metzner, *J. Rheol.*, **29**, 739 (1985); (d) J. N. Goodwin and R. H. Ottewill, *Faraday Trans.*, **87**, 357 (1991).

39. G. E. Archie, *Trans. AIME*, **146**, 54 (1942).

40. P. N. Sen, C. Scala, and M. H. Cohen, *Geophysics*, **46**, 781 (1981).

41. P. D. Jackson, D. T. Smith, and P. N. Stanford, *Geophysics*, **43**, 1250 (1978).

42. (a) C. J. Bottcher, *Theory of Electric Polarization*, Elsevier, Amsterdam, 1952; (b) L.D. Landau and E. M. Lifshitz, *Electrodynamics of Continuous Media*, Pergamon, New York, 1960.

43. W. F. Brown, *J. Chem. Phys.*, **23**, 114 (1955).

44. N. Hill, W. E. Vaughn, A. H. Price, and M. Davis, *Dielectric Properties and Molecular Behavior*, Van Nostrand Reinhold, NY, 1969.

45. S. Szegö, *Duke Math. J.*, **16**, 209 (1949).

46. M. Schiffer and G. Szegö, *Trans. Am. Math. Soc.*, **67**, 130 (1949).

47. W. E. Kock, *Bell. Syst. Tech. J.*, **27**, 58 (1948).

48. G. Estrin, *J. Appl. Phys.*, **21**, 667 (1950).

49. S. Cohn, *J. Appl. Phys.*, **20**, 257 (1949); **22**, 628 (1951).

50. R. E. Collin, *Field Theory of Guided Waves*, McGraw-Hill, NY, 1960.

51. (a) P. Debye, *J. Appl. Phys.*, **15**, 338 (1944); (b) A. Peterlin, "Streaming and Stress Birefringence," in *Rheology*, Vol. 1, F. R. Eirich, Ed., Academic, New York, 1956.

52. (a) J. Stratton, *Electromagnetic Theory*, McGraw-Hill, NY, 1941; (b) G. Dassios and R. E. Kleinman, *SIAM Rev.*, **31**, 565 (1989); (c) L. Eyges, *Ann. Phys.*, **90**, 266 (1975).

53. (a) R. E. Kleinman and T. B. A. Senior, *Radio Sci.*, **7**, 937 (1972); (b) T. B. A. Senior, *Radio Sci.*, **17**, 741 (1982).

54. K. M. Siegel, *Proc. IEEE*, **51**, 232 (1962).

55. T. B. A. Senior, *Radio Sci.*, **11**, 477 (1976).

56. D. F. Herrick and T. B. A. Senior, *IEEE Trans. Ant. Prop.*, **25**, 590 (1977).

57. (a) J. B. Keller, R. E. Kleinman, and T. B. A. Senior, *J. Inst. Math. Appl.*, **9**, 14 (1972); (b) K. S. J. Lee, *Radio Sci.*, **22**, 1235 (1987).

58. W. M. Thomson (Lord Kelvin), *Reprints on Electricity and Magnetism*, MacMillan, London, 1884. See Sections 16 and 17 and Article 32, Section 633 for a discussion of the "hydrokinetic analogy" behind the **M**-magnetic polarizability relation.

59. A. B. Bassett, *Philos. Mag.*, **16**, 286 (1885).

60. G. Birkhoff in *Studies in Math. and Mech.: Essays in Honor of Richard Von Mises*, Academic, NY, 1954, pp. 88–96.

61. G. Birkhoff and E. H. Zarantonello, *Jets, Wakes, and Cavities*, Academic, NY, 1957, Section 6.7.

62. G. Birkhoff, *Q. Appl. Math.*, **10**, 81 (1952); **11**, 109 (1953).

63. G. Birkhoff, *Hydrodynamics: A Study of Logic, Fact, and Similitude*, Princeton University Press, Princeton, NJ, 1960, Chap. 6.

64. (a) H. Lamb, *Hydrodynamics*, Dover, NY, 1945; (b) M. Thompson, *Theoretical Hydrodynamics*, MacMillan, London, 1962.

65. G. I. Taylor, *Proc. R. Soc. A*, **120**, 13 (1928).

66. W. M. Thomson ('Lord Kelvin'), *Philos. Mag.*, **45**, 332 (1873).

67. G. I. Taylor, *Proc. R. Soc. A*, **120**, 260 (1928).

68. R. E. Kleinman and T. B. A. Senior, *IEEE Trans. Aerospace Elect. Sys.*, **11**, 672 (1975).

69. E. H. Kennard, *Irrotational Flow of Frictionless Fluids Mostly of Invariable Density*, Department of Navy Report 2299, Feb. 1967, David Taylor Model Basin, Washington, DC.

70. (a) D. M. Wrinch, *Philos. Mag.*, **48**, 1089 (1924); (b) W. G. Bickley, *Proc. London Math. Soc.*, **37**, 82 (1934).

71. (a) J. L. Taylor, *Phil. Mag.*, **9**, 160 (1930); (b) E. B. Moullin, *Proc. Phil. Soc.*, **29**, 400 (1928); (c) X. Cai and G.B. Wallis, *Phys. Fluids A*, **5**, 1614 (1993).

72. F. De Meulenaere and J. Van Bladel, *IEEE Trans. Ant. Propag.*, **25**, 198 (1977).

73. (a) E. Avras and R.F. Harrington, *IEEE Trans. Ant. Propag.*, **31**, 719 (1983); (b) N. A. McDonald, *IEEE Trans. Microwave Th. Techn.*, **35**, 20 (1987); (c) R. L. Gluckstein, R. Li, and R.K. Cooper, *IEEE Trans. Microwave Th. Techn.*, **38**, 186 (1990).

74. L. Eyges and P. Gianino, *IEEE Trans. Ant. Prop.*, **27**, 557 (1979).

75. (a) V. I. Fabrikant, *J. Phys. A*, **20**, 323 (1987); (b) V. I. Fabrikant, *J. Sound and Vibration*, **121**, 1 (1988); (c) V. I. Fabrikant, *J. Acoust. Soc.*, **80**, 1438 (1986).

76. M. Fixman, *J. Chem. Phys.*, **75**, 4040 (1981).

77. M. Schiffer, *C. R. Acad. Sci. (Paris)*, **244**, 3118 (1957).

78. (a) J. B. Keller, *J. Math. Phys.*, **5**, 548 (1964); (b) K. S. Mendelson, *J. Appl. Phys.*, **46**, 917 (1975).

79. (a) L. E. Payne and A. Weinstein, *Pacific J. Math.*, **2**, 633 (1952); (b) L. E. Payne, *Q. Appl. Math.*, **10**, 197 (1952); *Soc. Ind. Appl. Math. Rev.*, **9**, 453 (1967); (c) A. Weinstein, *Bull. Am. Math. Soc.*, **59**, 20 (1953); (d) See also [53], [55], and [57] for a review of Payne and Weinstein's results.

80. (a) H.-X. Zhou, A. Szabo, J. F. Douglas, and J. B. Hubbard, *J. Chem. Phys.*, **100**, R-3821 (1994); (b) J. F. Douglas, H.-X. Zhou, and J. B. Hubbard, *Phys. Rev. E*, **49**, 5319 (1994); (c) McKean, *J. Math. Kyoto Univ.*, **4**, 617 (1965).

81. (a) J. W. Strutt (Lord Rayleigh), *Philos. Mag.*, **44**, 28 (1897); (b) V. Twersky, *Appl. Opt.* **3**, 1150 (1964); (c) A. F. Stevenson, *J. Appl. Opt.*, **24**, 1134 (1953); (d) P. Debye, *J. Phys. Coll. Chem.*, **51**, 18 (1947); (e) J. W. Crispin and A. L. Maffett, *IEEE*, **53**, 833 (1965). See the rest of the issue for other theoretical and experimental contributions to radar (Rayleigh) scattering; (f) H. Brysk, R. E. Hiatt, V. H. Weston, and K. M. Siegel, *Can. J. Phys.*, **37**, 675 (1959).

82. (a) N. R. Labrum, *J. Appl. Phys.*, **23**, 1324 (1952); (b) W. Zhang, *IEEE Trans. Ant. Propag.*, **42**, 347 (1994).

83. J. W. Strutt (Lord Rayleigh), *Philos. Mag. Ser.*, 4 **41**, 447 (1871); *Ser. 6* **35**, 373 (1918); *Proc. London Math. Soc.*, **4**, 253 (1872).

84. J. Van Bladel, *J. Acoust. Soc. Am.*, **44**, 1069 (1968).

85. T. B. A. Senior, *J. Acoust. Soc. Am.*, **53**, 742 (1973).

86. D. S. Jones, *Proc. R. Soc. Edin. A*, **83**, 245 (1979).

87. J. W. Strutt (Lord Rayleigh), *Philos. Mag.*, **43**, 259 (1897); see also [82a].

88. (a) H. A. Bethe, *Phys. Rev.*, **66**, 163 (1944); (b) H. A. Bethe, "Lumped Constants for Small Irises," M. I. T. Radiation Laboratory Reports. 43–22 (1943); (c) H. Levine and J. Schwinger, *Phys. Rev.*, **74**, 958 (1948); *Comm. Pure Appl. Math.*, **3**, 355 (1950); (d) D. L. Jaggard and C. H. Papas, *Appl. Phys.*, **15**, 21 (1978); (e) T. Wang, J. R. Mautz, and R. F. Harrington, *Radio Sci.*, **22**, 1289 (1979).

89. C. J. Bouwkamp, *Rep. Prog. Phys.*, **17**, 35 (1954).

90. J. Van Bladel, *Proc. IEEE*, **17**, 1098 (1970).

91. J. Van Bladel, *Radio Sci.*, **14**, 319 (1979).

92. (a) T. B. A. Senior and D. J. Ahlgren, *IEEE Trans. Ant. Propag.*, **21**, 134 (1973); (b) T. B. A. Senior and T. M. Willis III, *IEEE Trans. Ant. Propag.*, **30**, 1271 (1982).

93. (a) R. L. Hamilton and O. K. Crosser, *I.E.C. Fundamentals*, **1**, 187 (1962); (b) J. Francl and W. D. Kingery, *J. Am. Ceram. Soc.*, **37**, 99 (1954).

94. K. S. Mendelson and H. H. Cohen, *Geophysics*, **47**, 257 (1982). See also [42].

95. (a) T. Murai, *Prog. Theor. Phys.*, **27**, 899 (1962); (b) G. Araki and T. Murai, *Prog. Theor. Phys.*, **8**, 639 (1952); (c) H. Kuhn, *J. Chem. Phys.*, **17**, 1198 (1949).

96. H. Sternlicht, *J. Chem. Phys.*, **40**, 1175 (1964).

97. (a) V. Belovitch and J. Boersma, *Philips J. Res.*, **38**, 79 (1983); (b) T. Miloh, G. Waisman, and D. Weihs, *J. Engr. Math.*, **12**, 1 (1978). This work calculates the magnetic polarizability (added or virtual mass) of tori. The intrinsic conductivity $[\sigma]_0$ of insulating toroidal particles is never greater than 12% of the spherical particle value or less than the spherical particle value so that $[\sigma]_0$ is rather insensitive to shape.

98. M. Fixman, *J. Chem Phys.*, **76**, 6124 (1982).

99. (a) K. F. Freed and S. F. Edwards, *J. Chem. Phys.*, **62**, 4032 (1975); (b) G. J. Kynch, *Br. J. Phys.*, **3**, S-5 (1954); *Proc. R. Soc. London A*, **237**, 90 (1956).

100. T. F. Ford, *J. Phys. Chem.*, **64**, 1168 (1960). This work references and discusses Bancelin's data and the interchange between Einstein and Bancelin, and is more accessible than Bancelin's original paper.
101. (a) D. J. Jeffrey and A. Acrivos, *A.I. Chem. E.*, **22**, 417 (1976); (b) R. Rutgers, *Rheol. Acta*, **2**, 305 (1964).
102. J. W. Strutt (Lord Rayleigh), *Theory of Sound, Vol. 2* (Dover, N.Y., 1945).
103. R. Hill and G. Power, *Q. J. Mech. and Appl. Math.*, **9**, 313 (1956).
104. (a) J. S. Chang, E. B. Christiansen, and A. D. Baer, *J. Appl. Polym. Sci.*, **15**, 2007 (1971); (b) N.J. Mills, *J. Appl. Polym. Sci.*, **15**, 2791 (1971).
105. W. N. Bond, *Philos. Mag.*, **4**, 889 (1927); **5**, 794 (1928).
106. C. Brenner, *Can. J. Chem. Engr.*, **53**, 126 (1975).
107. J. W. Mehl, J. L. Oncley, and R. Simha, *Nature* (London), **92**, 132 (1940).
108. J. Kirkwood and J. Riseman, *J. Chem. Phys.*, **16**, 565 (1948).
109. P. Debye and A. M. Bueche, *J. Chem. Phys.*, **16**, 573 (1948).
110. S. Kim and S.-Y. Lu., *Int. J. Multiphase Flow*, **10**, 113 (1984).
111. G. K. Batchelor, *J. Fluid Mech.*, **41**, 545 (1970). See Eq. (5.11) of this work.
112. E. J. Hinch, *J. Fluid Mech.*, **54**, 423 (1972).
113. J. F. Brady, *Int. J. Multiphase Flow*, **10**, 113 (1984).
114. G. Bossis, A. Meunier, and J. F. Brady, *J. Chem. Phys.*, **94**, 5064 (1991). The results of this work suggest to us that $[\eta]$ of flocculated aggregates, held together by tenuous forces, should be approximately proportional to $[\sigma]_0$, rather than $[\sigma]_\infty$. This possibility deserves further consideration.
115. R. P. Kanwal, *J. Fluid Mech.*, **10**, 17 (1960).
116. J. Riseman and J. G. Kirkwood, *J. Chem. Phys.*, **17**, 442 (1949).
117. H. Brenner, *Int. J. Multiphase Flow*, **1**, 195 (1974).
118. S. Wakiya, *J. Phys. Soc. Jpn.*, **31**, 1581 (1972).
119. G. de la Torre and V. Bloomfield, *Q. Rev. Biophys.*, **14**, 81 (1981).
120. (a) H. A. Scheraga, *J. Chem. Phys.*, **23**, 1526 (1955); (b) R. Simha, *J. Res. N.B.S.*, **42**, 409 (1949); (c) J. W. Miles, *J. Appl. Phys.*, **38**, 192 (1967).
121. (a) M. Zuzovsky and H. Brenner, *J. Appl. Math. Phys.*, **28**, 979 (1977). See especially Eq. (77); (b) K. C. Nunan and J. B. Keller, *J. Fluid Mech.*, **142**, 269 (1984); (c) M. Zuzovsky, P.M. Adler, and H. Brenner, *Phys. Fluids* **26**, 1714 (1983).
122. (a) T. T. Taylor, *J. Res. NBS* **64**, 135 (1960); (b) P. K. Wang, *IEEE Trans. Ant. Prop.*, **32**, 956 (1984); (c) M. Fixman, *J. Chem. Phys.*, **75**, 4040 (1981).
123. B. B. Mandelbrot, *The Fractal Geometry of Nature*, W. H. Freeman and Co., San Francisco, CA, 1982.
124. M. Belzons, R. Blanc, and C. Carmoin, *C. R. Acad. Sci. (Paris)*, **292**, 68 (1981).
125. F. Bauer, P. Garabedian, D. Korn, and A. Jameson, *Lect. Notes Econ. Math. Sys.*, **108**, 1975.
126. G. B. McFadden, P. W. Voorhees, R. F. Boisvert, and D. I. Meiron, *J. Sci. Comp. 1*, **117** (1986). The methods of [125] and [126] were adapted to the calculation of the logarithmic capacity by G. B. McFadden of the Computational and Applied Mathmatics Laboratory at NIST.
127. (a) M. F. Thorpe, *Proc. R. Soc. London A*, **437**, 215 (1992); (b) J. H. Hetherington and M. F. Thorpe, *Proc. R. Soc. London A*, **438**, 591 (1992).
128. G. Pólya and G. Szegö, *Am. J. Math.*, **68**, 1 (1945).

129. O. D. Kellogg, *Foundations of Potential Theory*, (Springer-Verlag, Berlin, 1929, see p. 331.

130. (a) A. Dvoretsky, P. Erdös, and S. Kakutani, *Acta Sci. Math. (Szeged)* **12**, 75 (1950); (b) S. J. Taylor, *Proc. Camp. Phil. Soc.*, **51**, 265 (1955).

131. H. Hasimoto, *J. Phys. Soc. Jpn.*, **13**, 633 (1958).

132. (a) S. Cohn, *Proc. I.R.E.*, **39**, 1416 (1951); (b) S. Cohn, *Proc. I.R.E.*, **40**, 1069 (1952).

133. (a) R. J. S. Brown, *Geophysics*, **45**, 1269 (1980); (b) M. A. Biot, *J. Acoust. Soc. Am.*, **28**, 168 (1956).

134. D. L. Johnson and P. N. Sen, *Phys. Rev. B*, **24**, 2486 (1981).

135. J. W. Strutt (Lord Rayleigh), *Philos. Mag.*, **4**, 481 (1892).

136. D. L. Johnson, T. J. Plona, C. Scala, F. Pasierb, and H. Kojima, *Phys. Rev. Lett.*, **49**, 1840 (1982).

137. J. G. Berryman, *Phys. Rev. B*, **27**, 7789 (1983).

138. A. Atkinson and A. K. Nickerson, *J. Mat. Sci.*, **19**, 3068 (1984).

139. J. F. Nye, *Physical properties of crystals: Their representation by tensors and matrices*, Oxford University Press, London, 1957.

140. G. Arfken, *Mathematical Methods for Physicists*, Academic, New York, 1970, p. 121.

141. J. P. Watt, *J. Appl. Phys.*, **51**, 1525 (1980).

142. L. D. Landau and E. M. Lifshitz, *Theory of Elasticity*, Pergamon Press, New York, 1970.

143. L. D. Landau and E. M. Lifshitz, *Fluid Mechanics*, Pergamon Press, New York, 1959.

144. M. H. Davis, *Q. J. Mech. Appl. Math.*, **17**, 499 (1964).

145. A. Apelblat, *Physical Sciences Data*, Vol. 13, Elsevier, Amsterdam, 1983, p. 56.

146. R. D. Cook, D. S. Malkus, and M. E. Plesha, *Concepts and Applications of Finite Element Analysis*, Wiley, New York, 1989.

147. R. M. Christensen, *Mechanics of Composite Materials*, Krieger Publishing Co., Malabar, FL, 1991.

148. H. Helmholz, *Wiss. Abh.*, **1**, 223. See also the collected works of Helmholz.

149. D. J. Korteweg, *Philos. Mag.*, **16**, 112 (1883).

150. C.B. Millikan, *Philos. Mag.*, **7**, 641 (1929).

151. B. Fraeijs de Veubeke, "Variational principles in fluid mechanics and finite element applications," in *Progress in Numerical Fluid Dynamics Lecture Notes in Physics*, J. Ehlers, K. Hepp, H. A. Weidenmuller, and J. Zittartz, Eds., Springer-Verlag, Heidelberg, 1975. Vol. 41, pp 226–259

152. (a) J. B. Keller, L.A. Rubenfeld, and J.E. Molyneux, *J. Fluid Mech.*, **30**, 97 (1967); (b) R. Skalak, *J. Fluid Mech.*, **42**, 527 (1951).

153. O. C. Zienkiewicz, "Viscous, incompressible flow with special reference to non-Newtonian (plastic) fluids," in *Finite Elements in Fluids—Volume I: Viscous Flow and Hydrodynamics*, R. H. Gallagher, J. T. Oden, C. Taylor, and O. C. Zienkiewicz, Eds., Wiley, London, 1975, pp. 25–56.

154. P. J. Stiles and J. B. Hubbard, *Chem. Phys. Lett.*, **105**, 655 (1984).

155. E. J. Garboczi, M. F. Thorpe, M. S. DeVries, and A. R. Day, *Phys. Rev. A*, **43**, 6473 (1991).

THE PHENOMENOLOGICAL AND STATISTICAL THERMODYNAMICS OF NONUNIFORM SYSTEMS

ELIGIUSZ WAJNRYB,* ANDRZEJ R. ALTENBERGER, AND JOHN S. DAHLER

Department of Chemical Engineering and Materials Science, University of Minnesota, Minneapolis, MN 55455

CONTENTS

* Permanent address: Institute of Fundamental Technological Research, Swietokrzyska 21, 00-049 Warszawa, Poland.

Advances in Chemical Physics, *Volume XCI*, Edited by I. Prigogine and Stuart A. Rice.
ISBN 0-471-12002-2

I. INTRODUCTION

The objectives of the research presented here are to extend conventional, global thermodynamics to nonuniform systems and, at the same time, to construct a complementary statistical mechanical theory. To accomplish this we shall adhere to a minimalist approach, seeking at all times to invoke as few additional postulates, assumptions, and approximations as possible.

Although few if any systematic investigations of this topic appear to have been made, the literature abounds with equations pertaining to local thermodynamic properties of nonuniform systems. These equations have been obtained both by statistical mechanical (kinetic theory) and continuum mechanical arguments. Common to most of these presentations are the (often tacit) assumptions that there are well-defined local counterparts to all the standard thermodynamic variables, for example, internal energy, entropy, temperature, pressure, and heat capacities, and that these local variables and/or their differentials and derivatives are related by the obvious local analogs of the equations of conventional, global thermodynamics. One of our aims here is to investigate the extent to which it is possible to demonstrate that these expectations are indeed "natural consequences" of the global theory. We shall see that this involves something beyond the simple act of replacing intensive variables such as the temperature and pressure with corresponding scalar fields and the obvious observation that one can define the "point density' $m(\mathbf{r})$ of an arbitrary extensive variable $M(\Omega)$ (associated with a sample of volume Ω) as the limit of the quotient $M(\Omega_i)/\Omega_i$ with Ω_i a sequence of nested volumes converging on the point $\mathbf{r}$.

Furthermore, it will be necessary to contend with the fact that global thermodynamics almost invariably is formulated in terms of macroscopic, quasiuniform systems and the (measurable) properties that can be associated with them, whereas a thermodynamic field theory necessarily involves properties that can vary continuously from point to point. Because of the associated uneven distributions of particles and energy, property variations within one small volume element or "cell" of a nonuniform system will be correlated with those in others. It nevertheless is reasonable to expect, given the perspective of global thermodynamics, that there exists a minimal set of independent observables in terms of which all other thermodynamic observables are determined. Although the selection of this basis set is to some extent arbitrary, once a proper choice has been made, all other quantities presumably can be treated as functionals of the basic field variables. The number of independent variables or thermodynamic degrees of freedoms then will be determined by the number of constraints imposed upon the system. For example, it is expected that an inhomogeneous system composed of c distinct chemical species can be characterized by a set of c particle-density fields $\{n_\alpha(\mathbf{r}); \alpha = 1, \ldots, c\}$ and an energy density $\varepsilon(\mathbf{r})$. There then will be $c+1$ degrees of freedom. All other thermodynamic variables such as entropy, temperature, pressure and species chemical potentials are to be treated as functionals of these basic thermodynamic fields. An important part of our task is to devise a systematic means for constructing a thermodynamic field theory of this sort, which conforms as closely as possible to the well-established global theory. There is, however, a distinct possibility, in this context, that the definitions and/or "ground rules" of global thermodynamics may not provide adequate guidance in all cases about how to proceed to the limit of continuously varying field variables. The strategy we adopt in contending with this difficulty is to strive for a seamless match between the thermodynamic theory and the complementary, micromechanically based statistical mechanical theory.

When a macroscopic system is in thermodynamic equilibrium with its surroundings, intensive properties, such as temperature and the chemical potentials, generally are expected to be constant throughout the system. However, due to random, spontaneous exchanges of particles and energy among the adjacent "cells" or subsystems into which the total system can be arbitrarily subdivided, the local instantaneous values of the particle and energy density fields invariably will differ from their corresponding stoichiometric averages. Because of the functional relationships that exist among all of the thermodynamic variables there will be corresponding fluctuations of the other fields as well. (It generally is accepted that the fluctuations of a system in thermodynamic equilibrium can be represented as a stationary, random process.)

The spatial correlations and time or frequency dependent spectra of local thermodynamic quantities can be observed using modern techniques of light and neutron scattering spectroscopy [1]. To account for the experimental observations one must extend the standard classical formulation of thermodynamic fluctuation theory [2–6] to include the behavior of inhomogeneous systems. To the best of our knowledge the first attempt to generalize the classical theory of Landau and Lifshitz [6] was due to Schofield [7, 8] who produced a formalism that incorporated correlation functions of wavevector (position) dependent thermodynamic quantities. The dynamics of local fluctuations has since been studied by a number of authors (see [9] and [10] for reviews). However, much less attention has been devoted to the task of developing a systematic equilibrium theory of local fluctuations and to the formulation of the equilibrium thermodynamics of inhomogeneous systems. Instead, most studies in this field have addressed specific applications of the theory of inhomogeneous fluids such as surface phenomena [11–13], capillary waves [14, 15], crystallization, and the formation of ordered structures [16–19], and nucleation [20, 21].

As we shall see, an important feature of the theory of thermodynamic fluctuations is its applicability not only to fluctuations of mechanical quantities, which can as well be treated using various Gibbs equilibrium ensembles, but also to fluctuations of thermal quantities such as temperature, entropy, and chemical potentials, which cannot be treated as random variables within the formalism of generalized equilibrium ensembles. The problem of defining the fluctuations of these thermal quantities has been controversial in the past and it would seem that a consensus still has not been reached [22]. One aim of the present paper is to propose a generalization and extension of the Schofield theory of local fluctuations, which will serve to lessen the distinction between fluctuations of mechanical and thermal properties.

The format of this chapter is as follows: Section II is devoted to the systematic development of a thermodynamic field theory, beginning with a brief review of global thermodynamics patterned after that presented by Callen [4]. This theory and its representation in terms of Massieu–Planck functions is then used as a model for drafting an analogous theory for nonuniform systems. Once this has been accomplished, the utility of the resulting field theory is demonstrated by establishing a number of illustrative results. In Section III the information theory definition of entropy is used to construct a generalized Gibbs distribution, conditioned by prescribed compositional and energy fields that differ from those characteristic of equilibrium. The statistical thermodynamics associated with this distribution is compared with the phenomenological theory of Section II. Among the results obtained from this comparison are correlation-integral formulas for many of the quantities encountered in Section II. This section concludes with an examination of fluctuations for the important case of macroscopically uniform states of equilibrium.

II. THE THERMODYNAMICS OF NONUNIFORM SYSTEMS

Although the systems considered in most treatments of conventional thermodynamics can have extraordinarily varied and complicated microstructures, the thermodynamic variables descriptive of these systems are macroscopic and global; the values assigned to them (resulting from measurements or theory) are identified as macroscopic characteristics of the entire bulk sample. Transfers from one such macroscopic system to another are imagined to produce corresponding changes of these global variables and coefficients descriptive of these transfer processes (be they of heat, volume, or matter) are themselves taken to be global variables, for example, heat capacities C_p and C_v, the coefficients κ_T and α_p of isothermal compressibility and thermal expansion, respectively, and analogous mass transfer functions. However, it is certainly conceivable that as the systems among which these transfers occur become more and more numerous and individually smaller, the transfer processes will require a nonlocal description characterized by coefficients that depend on both donors and acceptors. Some molecular theories previously have produced descriptions of this sort and, as we shall see, nonlocality can as well be viewed as a natural outcome of the thermodynamic approach.

A. Global Thermodynamics

There is a number of equally sound formulations of global thermodynamics, any one of which could serve as the basis for the task at hand. However, in the interests of brevity and conceptual clarity we adopt a postulatory formulation patterned on that presented in Callen's excellent treatise [4] in place of a more intuitively appealing, phenomenological approach. Furthermore, a formalism based on the Massieu–Planck function is better suited to our purposes than the entropy (or internal energy) formalism because the former is so closely related to the generalized grand-canonical ensemble theory that we subsequently develop in Section III.

The first of Callen's postulates is the existence of a special class of states, called equilibrium states, which are determined by the intrinsic characteristics of a macroscopic system and to which this system tends to evolve in the presence of a static environment. These states are totally characterized by the values of a "complete set" of extensive variables $X_1, X_2, \ldots, X_r$, which, for a $p\Omega$-system are known to consist of the internal energy E, volume Ω, and mole numbers $N_1, N_2, \ldots, N_c$. It is understood, in this context, that the internal energy is a measurable property of a macroscopic system [4]. The adjective "extensive" is used here to identify quantities so defined that the property value for the total system is the sum of corresponding values for all possible partitions of the system into macroscopic subsystems. This implies that the interactions among these subsystems are weak and short ranged in comparison with the dimensions of the subsystems themselves. In addition to this extensivity, the properties of a system in a state of thermodynamic equilibrium usually are assumed to be spatially uniform despite the presence of external forces, including those associated with walls that may confine the system. Thus, the interactions between the walls and the system also must be weak and short ranged.

The second postulate is of the existence of the entropy S, a function of the extensive variables of a composite system that is defined for all *equilibrium states*. This entropy is an extensive function that assumes its maximum value in the absence of "internal constraints," namely, constraints that prevent flows of energy, volume, and matter among constituent subsystems. Furthermore, the entropy is continuous and differentiable and increases monotonically with the internal energy.

The relation

$$S = S(X_1, X_2, \ldots, X_r) \tag{2.1}$$

giving the entropy as a function of the extensive variables, is called the *fundamental equation* or the *fundamental relation*. The entropy function so defined is characteristic in the sense that it contains every statement that thermodynamics can make about the system. Associated with each extensive variable X_i is the corresponding intensive variable

$$P_i = \left(\frac{\partial S}{\partial X_i}\right)_{\{X_{j \neq i}\}} \tag{2.2}$$

which itself can be expressed as an order zero homogeneous function of the variables $\{X_i\}$. However, because of this homogeneity, P_i also can be expressed as a function of $r-1$ intensive variables. Thus, for example, if $X_1 = \Omega$, then $P_i = P_i(X_2/\Omega, X_3/\Omega, \ldots, X_r/\Omega)$. Since the internal energy is included among the variables $\{X_i\}$, the assumed monotonicity of S insures that the inverse of the absolute temperature $T^{-1} = \partial S/\partial E$ is nonnegative.

The relationships

$$P_i = P_i(X_1, X_2, \ldots, X_r) \qquad i = 1, \ldots, r \tag{2.3}$$

are called equations of state. These are not independent since, due to the assumed extensivity (first-order homogeneity) of the entropy, it follows that

$$S = \sum_{i=1}^{r} X_i P_i \tag{2.4}$$

From this and Eq. (2.2), which implies that $dS = \Sigma_{i=1}^{r} P_i dX_i$, one then obtains the Gibbs–Duhem equation

$$\sum_{i=1}^{r} X_i dP_i = 0 \tag{2.5}$$

We now introduce the so-called Massieu–Planck functions, which are Legendre transforms of the fundamental equation (2.1) defined according to the prescription

$$\Phi_k = S - \sum_{i=1}^{k} P_i X_i \qquad k < r \tag{2.6}$$

By invoking Eq. (2.4) this can be expressed in the alternative form

$$\Phi_k = \sum_{i=k+1}^{r} P_i X_i \tag{2.7}$$

Then by combining this with the Gibbs–Duhem equation (2.5) we obtain for the differential of Φ_k the formula

$$d\Phi_k = -\sum_{i=1}^{k} X_i dP_i + \sum_{i=k+1}^{r} P_i dX_i \tag{2.8}$$

The Massieu–Planck function Φ_k, expressed in terms of its natural variables $P_1, P_2, \ldots, P_k, X_{k+1}, \ldots, X_r$ constitutes a fundamental relation

$$\Phi_k = \Phi_k(P_1, \ldots, P_k; X_{k+1}, \ldots, X_r) \tag{2.9}$$

which contains exactly the same information about the system as the entropic fundamental equation (2.1). In particular, the expressions

$$\frac{\partial \Phi_k}{\partial P_i} = -X_i \qquad i = 1, \ldots, k \tag{2.10a}$$

$$\frac{\partial \Phi_k}{\partial X_i} = P_i \qquad i = k+1, \ldots, r \tag{2.10b}$$

obtained from Eq. (2.8) provide equations of state

$$X_i = X_i(P_1, \ldots, P_k; X_{k+1}, \ldots, X_r) \qquad i = 1, \ldots, k \tag{2.11a}$$

$$P_i = P_i(P_1, \ldots, P_k; X_{k+1}, \ldots, X_r) \qquad i = k+1, \ldots, r \tag{2.11b}$$

which are equivalent to Eq. (2.3) of the entropy representation.

The development that follows requires a generalization of the theory that has just been presented. The need for this extension arises when our knowledge of the entropy function is restricted to its dependence on only a subset of the extensive variables, for example, on $X_1, X_2, \ldots, X_\ell$, with $\ell < r$, for fixed values of those that remain, namely, $\mathring{X}_{\ell+1}, \ldots, \mathring{X}_r$. This situation can be expressed by a relation

$$\tilde{S} = \tilde{S}(X_1, \ldots, X_\ell; \mathring{X}_{\ell+1}, \ldots, \mathring{X}_r) \tag{2.12}$$

which is analogous but not equivalent to the fundamental equation (2.1). The difference is that Eq. (2.12) does not provide a complete thermodynamic statement about the system; since the information contained in the function $\tilde{S}$ is limited to a fixed set of values for the variables $X_{\ell+1}, \ldots, X_r$, one cannot perform the differentiations of $\tilde{S}$ needed to produce the corresponding equations of state for the quantities P_i with $i = \ell + 1, \ldots, r$.

To complete the information that is lacking from Eq. (2.12) we can specify the missing intensive variables [zero-order homogeneous functions of the variables $X_1, \ldots, X_\ell$; $\mathring{X}_{\ell+1}, \ldots, \mathring{X}_r$] by the independently prescribed equations of state

$$\tilde{P}_i = \tilde{P}_i(X_1, \ldots, X_\ell; \mathring{X}_{\ell+1}, \ldots, \mathring{X}_r) \qquad i = \ell + 1, \ldots, r \tag{2.13}$$

The other intensive variables are obtained, as before, by differentiating the entropy;

$$P_i = \frac{\partial}{\partial X_i} \tilde{S}(X_1, \ldots, X_\ell; \mathring{X}_{\ell+1}, \ldots, \mathring{X}_r) \qquad i = 1, \ldots, \ell \tag{2.14}$$

However, the intensive variables given by Eqs. (2.13) cannot be selected arbitrarily; to insure the extensivity of the entropy the functions $\tilde{P}_i$ for $i = \ell + 1, \ldots, r$ must be so chosen that they satisfy the equation

$$\begin{aligned} \tilde{S}(X_1, \ldots, X_\ell; \mathring{X}_{\ell+1}, \ldots, \mathring{X}_r) = {} & \sum_{i=1}^{\ell} X_i P_i(X_1, \ldots, X_\ell; \mathring{X}_{\ell+1}, \ldots, \mathring{X}_r) \\ & + \sum_{i=\ell+1}^{r} \mathring{X}_i \tilde{P}_i(X_1, \ldots, X_\ell; \tilde{X}_{\ell+1}, \ldots, \mathring{X}_r) \end{aligned} \tag{2.15}$$

Since $\tilde{S}$ of Eq. (2.12) involves only ℓ independent variables, we see from Eq. (2.14) that

$$d\tilde{S} = \sum_{i=1}^{\ell} P_i dX_i \tag{2.16}$$

while from Eq. (2.15) it follows that

$$d\tilde{S} = \sum_{i=1}^{\ell} [P_i dX_i + X_i dP_i] + \sum_{i=\ell+1}^{r} \mathring{X}_i d\tilde{P}_i \tag{2.17}$$

By equating the two formulas [Eqs. (2.16) and (2.17)] for $d\tilde{S}$ we obtain the relation

$$\sum_{i=1}^{\ell} X_i dP_i + \sum_{i=\ell+1}^{r} \mathring{X}_i d\tilde{P}_i = 0 \tag{2.18}$$

which, although formally the same as the previously derived Gibbs–Duhem equation (2.5), contains the two differently defined sets of intensive variables, $\{P_i\}$ and $\{\tilde{P}_i\}$.

The extensivity condition [Eq. (2.15)] can be interpreted as a constraint (identity) on the intensive variables $\tilde{P}_i$. By using Eq. (2.14) we recast this identity as

$$\begin{aligned}\tilde{S}(X_1, \ldots, X_\ell; \mathring{X}_{\ell+1}, \ldots, \mathring{X}_r) - \sum_{i=1}^{\ell} X_i \frac{\partial \tilde{S}}{\partial X_i} \\ = \sum_{i=\ell+1}^{r} \mathring{X}_i \tilde{P}_i(X_1, \ldots, X_\ell; \mathring{X}_{\ell+1}, \ldots, \mathring{X}_r)\end{aligned} \tag{2.19}$$

The set of coupled equations (2.14) is then inverted to produce formulas

$$X_i = X_i(P_1, \ldots, P_\ell; \mathring{X}_{\ell+1}, \ldots, \mathring{X}_r) \qquad i = 1, \ldots, \ell \tag{2.20}$$

for the extensive variables $\{X_i; i = 1, \ldots, \ell\}$. These, in turn, permit us to rewrite the identity [Eq. (2.19)] in the form

$$\tilde{S} - \sum_{i=1}^{\ell} X_i P_i = \sum_{i=\ell+1}^{r} \mathring{X}_i \tilde{P}_i(P_1, \ldots, P_\ell; \mathring{X}_{\ell+1}, \ldots, \mathring{X}_r) \tag{2.21}$$

The resulting Massieu–Planck function

$$\Phi = \Phi(P_1, \ldots, P_\ell; \mathring{X}_{\ell+1}, \ldots, \mathring{X}_r)$$
$$\equiv \sum_{i=\ell+1}^{r} \mathring{X}_i \tilde{P}_i(P_1, \ldots, P_\ell; \mathring{X}_{l+1}, \ldots, \mathring{X}_r) \qquad (2.22)$$

is the Legendre transform of $\tilde{S}$ with respect to the "available" extensive properties $X_1, \ldots, X_\ell$.

The equations of state [cf. Eqs. (2.13) and (2.20)]

$$\tilde{P}_i = \tilde{P}_i(P_1, \ldots, P_\ell; \mathring{X}_{\ell+1}, \ldots, \mathring{X}_r) \qquad i = l+1, \ldots, \ell \qquad (2.23)$$

contain all the thermodynamic information about the system. In particular, the extensive variables $\{X_i; i = 1, \ldots, \ell\}$ are related to the Massieu–Planck function by the equations of state

$$X_i = -\frac{\partial}{\partial P_i} \Phi(P_1, \ldots, P_\ell; \mathring{X}_{\ell+1}, \ldots, \mathring{X}_r) \qquad i = 1, \ldots, \ell \quad (2.24)$$

and the entropy $\tilde{S}(X_1, \ldots, X_\ell; \mathring{X}_{\ell+1}, \ldots, \mathring{X}_r)$ is obtained by constructing the Legendre transform of Φ with respect to the variables $P_1, \ldots, P_\ell$.

Thus, we have two equivalent descriptions of a thermodynamic system characterized by ℓ available or "free" extensive variables $X_1, \ldots, X_\ell$ and by another $r - \ell$ "fixed" variables with the assigned values $\mathring{X}_{\ell+1}, \ldots, \mathring{X}_r$. The entropy and Massieu–Planck representations applicable to this situation require not only a fundamental equation, Eq. (2.15) on the one hand and (2.22) on the other, but $r - \ell$ equations of state for the intensive variables $\tilde{P}_i$, conjugate to the fixed extensive properties.

B. Application of Formalism to a Composite System

In our search for a thermodynamic description of the continuously varying field variables of a nonuniform system, it is natural to adopt a procedure according to which a macroscopic system is partitioned successively into ever finer sets of subsystems. Accordingly, we consider here a single-component macroscopic system the volume of which is partitioned into cells by rigid, permeable and energy conducting walls. There are k of these cells, the ith of which has a volume $\mathring{\Omega}_i$; $\Omega = \Sigma_{i=1}^{k} \mathring{\Omega}_i$. The entropy function of this system may be written as

$$S = S(E_1, N_1, \mathring{\Omega}_1; \ldots; E_k, N_k, \mathring{\Omega}_k) \qquad (2.25)$$

and the equations of state for the $\tilde{P}_i$-type intensive variables conjugate to the cell volumes are of the form [cf. Eq. (2.13)]

$$\phi_i = \phi_i(E_1, N_1, \mathring{\Omega}_1; \ldots ; E_k, N_k, \mathring{\Omega}_k) \qquad i = 1, 2, \ldots, k \tag{2.26}$$

These are related by the standard formulas $\phi_i = p_i / k_B T_i$ to the pressures and temperatures of the cells. The other equations of state, for the P_i-type variables, are derivable from the entropy function of Eq. (2.25). These are written as

$$\beta_i = \frac{1}{k_B} \frac{\partial S}{\partial E_i} \qquad i = 1, \ldots, k \tag{2.27a}$$

$$\nu_i = \frac{1}{k_B} \frac{\partial S}{\partial N_i} \qquad i = 1, \ldots, k \tag{2.27b}$$

where, in anticipation of the subsequent statistical mechanical development, we have inserted the Boltzmann constant k_B. The parameters β_i and ν_i are related by the standard formulas $\beta_i = 1/k_B T_i$ and $\nu_i = -\mu_i / k_B T_i$ to the absolute temperature T_i and chemical potential μ_i.

The functions ϕ_i must, of course, be so defined that [cf. Eq. (2.15)]

$$\begin{aligned} S &= \sum_{i=1}^{k} \left[E_i \frac{\partial S}{\partial E_i} + N_i \frac{\partial S}{\partial N_i} + \mathring{\Omega}_i k_B \phi_i \right] \\ &= k_B \sum_{i=1}^{k} [E_i \beta_i + N_i \nu_i + \mathring{\Omega}_i \phi_i] \end{aligned} \tag{2.28}$$

This, in turn, can be rewritten as

$$S = k_B \sum_{i=1}^{k} \mathring{\Omega}_i [\varepsilon_i \beta_i + n_i \nu_i + \phi_i] \equiv \sum_{i=1}^{k} \mathring{\Omega}_i \sigma_i \tag{2.29}$$

with $\varepsilon_i = E_i / \mathring{\Omega}_i$ and $n_i = N_i / \mathring{\Omega}_i$ denoting the densities of energy and molecules associated with cell i and where

$$\sigma_i = k_B(\varepsilon_i \beta_i + n_i \nu_i + \phi_i) \tag{2.30}$$

is the corresponding density of entropy. To complete the picture we note that the Gibbs–Duhem relation, Eq. (2.18), appropriate to this situation is given by

$$\sum_{i=1}^{k} [E_i d\beta_i + N_i d\nu_i + \mathring{\Omega}_i d\phi_i] = \sum_{i=1}^{k} \mathring{\Omega}_i [\varepsilon_i d\beta_i + n_i d\nu_i + d\phi_i] = 0 \tag{2.31}$$

and that the appropriate form of Eq. (2.16) is

$$dS = \sum_{i=1}^{k} \left[\frac{\partial S}{\partial E_i} dE_i + \frac{\partial S}{\partial N_i} dN_i \right]$$

$$= \sum_{i=1}^{k} \mathring{\Omega}_i k_B(\beta_i d\varepsilon_i + \nu_i dn_i) \equiv \sum_{i=1}^{k} \mathring{\Omega}_i ds_i \tag{2.32}$$

with

$$ds_i = k_B(\beta_i d\varepsilon_i + \nu_i dn_i) \tag{2.33}$$

Finally, the Massieu–Planck function of the composite system can be written in the form [cf. Eq. (2.22)]

$$\Phi = \Phi(\beta_1, \nu_1, \mathring{\Omega}_1; \ldots ; \beta_k, \nu_k, \mathring{\Omega}_k) \equiv \sum_{i=1}^{k} \mathring{\Omega}_i \phi_i(\beta_1, \nu_1, \mathring{\Omega}_1; \ldots ; \beta_k, \nu_k, \mathring{\Omega}_k) \tag{2.34}$$

and the equations of state for the $\tilde{P}_i$-type variables conjugate to the fixed variables $\mathring{\Omega}_1, \ldots, \mathring{\Omega}_k$ become [cf. Eq. (2.23)]

$$\phi_i = \phi_i(\beta_1, \nu_1, \mathring{\Omega}_1; \ldots ; \beta_k, \nu_k, \mathring{\Omega}_k) \qquad i = 1, \ldots, k \tag{2.35}$$

Last of all are the equations [corresponding to Eq. (2.24)]

$$E_i = -\frac{\partial \Phi}{\partial \beta_i} \qquad i = 1, \ldots, k \tag{2.36a}$$

$$N_i = -\frac{\partial \Phi}{\partial \nu_i} \qquad i = 1, \ldots, k \tag{2.36b}$$

which relate the extensive variables E_i and N_i to derivatives of the Massieu–Planck function.

By assigning fixed volumes to the cells we have called attention to the fact that the volume is distinct from other extensive variables. This is because of the unique role it plays in defining the densities of other extensive variables. The corresponding intensive variables, the "pressures" ϕ_i, are equally unique. They are $\tilde{P}_i$-type variables that are not derivable from the entropy function of the constrained system nor are they derivable from the corresponding Massieu–Planck function.

If the subsystems are truly macroscopic, it then follows from the assumed additivity of the extensive variables that the function ϕ_i will be dependent only on the thermodynamic coordinates of cell i, that is,

$\phi_i = \phi_i(E_i, N_i, \mathring{\Omega}_i)$ or equivalently $\phi_i = \phi_i(\beta_i, \nu_i, \mathring{\Omega}_i)$. Under these circumstances the global Gibbs–Duhem equation (2.31) is separately satisfied in each cell, that is, $\varepsilon_i d\beta_i + n_i d\nu_i + d\phi_i = 0$ for $i = 1, 2, \ldots, k$ and ds_i, given by Eq. (2.33), is equal to the differential of the entropy density σ_i; $ds_i = d\sigma_i$. In fact, each of these equalities implies the other.

However, as the volume elements $\mathring{\Omega}_i$ become progressively smaller a stage invariably will be reached beyond which the assumptions of macroscopic thermodynamics no longer are satisfied. Although it may remain possible to assign values to the energy and material contents of individual cells, interactions among the material contents of neighboring cells inevitably grow so strong that the concept of independent thermodynamic states becomes untenable. [The cell dimensions at which this "breakdown" occurs does, of course, depend on the density of the macroscopic, composite system.] Nevertheless, if one wishes to construct a thermodynamic field theory he must move beyond these limits of conventional macroscopic thermodynamics into a realm where the only available source for guidance is statistical mechanics.

From our own statistical mechanical analysis of the situation, presented in Section III, emerges a conclusion that bears directly on the issue facing us here. In particular, it is found that the statistical mechanical analogs of the $\tilde{P}_i$-type variables ϕ_i are *nonlocal* functionals of the temperature and chemical potential fields, that is, of the field variables analogous to the thermodynamic intensive properties β_i and ν_i. This means that to obtain a thermodynamic field theory that is consistent with statistical mechanics one must replace the macroscopic equations of state $\phi_i = \phi_i(\beta_i, \nu_i, \mathring{\Omega}_i)$ with the more general and "nonlocal" relationships $\phi_i = \phi_i(\beta_1, \nu_1, \mathring{\Omega}_1; \ldots; \beta_k, \nu_k, \mathring{\Omega}_k)$ of Eq. (2.35). Once this has been done, it follows that

$$d\phi_i = \sum_{j=1}^{k} \left[\frac{\partial \phi_i}{\partial \beta_j} d\beta_j + \frac{\partial \phi_i}{\partial \nu_j} d\nu_j \right] \tag{2.37}$$

and so it becomes abundantly clear that the Gibbs–Duhem relation (2.31) for the composite system no longer can be replaced with a set of "local Gibbs–Duhem equations," $\varepsilon_i d\beta_i + n_i d\nu_i + d\phi_i = 0$, specific to the individual cells. Coincident with this is the conclusion, that when the cells are sufficiently reduced in size, the quantity ds_i defined by Eq. (2.33) ceases to be identifiable with the differential of the entropy density σ_i.

C. Continuous Generalization of the Massieu–Planck Representation

The conclusions of the preceding sections are now adopted as the bases for constructing an extension of thermodynamics that will be applicable to systems characterized by continuously varying temperature and chemical potential fields. In the Massieu–Planck representation of the previous section the quantities β_i and ν_i for $i = 1, \ldots, k$ were identified as the basic intensive variables of a composite system, defined by partitioning a macroscopic system into a finite set of k cells. When these cells become sufficiently small their contents no longer can be expected to satisfy the criteria demanded of macroscopic thermodynamic systems. It is at this point, as we verge on the transition from finite to infinitesimal cells (and from macroscopic to microscopic thermodynamics), that something new is injected into the theory, namely, the assumption that as the cells become vanishingly small the pressurelike, $\tilde{P}_i$-type intensive properties ϕ_i become nonlocally dependent on the variables $\{\beta_\ell, \nu_\ell; \ell = 1, \ldots, k\}$. The rationale for this assumption is our assertion that the resulting thermodynamic field theory will prove to be consistent with the (still to be demonstrated) statistical mechanical theory and, conversely, that the failure to adopt this assumption will result in contradictory thermodynamic and statistical mechanical theories of nonuniform systems.

Once the decision to make this assumption has been made we then proceed without further reference to statistical mechanics or the atomistic fine structure of the system. Thus, as the cell sizes are imagined to diminish without limit the discrete labels $i = 1, \ldots, k$ are replaced with position vectors $\mathbf{r}$ of points within the supposedly continuous system and the variable sets $\{\beta_i, \nu_i; i = 1, \ldots, k\}$ are replaced with the corresponding functions or "field variables" $\beta(\mathbf{r})$ and $\nu(\mathbf{r})$. Furthermore, the equations of state $\phi_i = \phi_i(\beta_1, \nu_1, \mathring{\Omega}_1; \ldots; \beta_k, \nu_k, \mathring{\Omega}_k)$ are reexpressed as

$$\phi(\mathbf{r}) = \phi(\mathbf{r}; \{\beta(\mathbf{r}'), \nu(\mathbf{r}')\}) \tag{2.38}$$

Here, the argument of $\phi(\mathbf{r})$ refers to the field point at which the function is evaluated while the symbols $\{\beta(\mathbf{r}'), \nu(\mathbf{r}')\}$ indicate the fields upon which $\phi(\mathbf{r})$ is functionally dependent.

Before proceeding further let us slightly generalize the analysis by shifting attention to a multicomponent system consisting of a number c of distinct species, labeled by lower case Greek symbols. Thus, the number density or concentration field of species α will be denoted by the symbol $n_\alpha(\mathbf{r})$; the associated conjugate field is $\nu_\alpha(\mathbf{r})$. The aggregates of these variables are written as $\mathbf{n}(\mathbf{r})$ and $\boldsymbol{\nu}(\mathbf{r})$, respectively. The Massieu–Planck function for the system then can be written in the form

$$\Phi = \Phi(\Omega; \{\beta(\mathbf{r}), \boldsymbol{\nu}(\mathbf{r})\}) \tag{2.39}$$

and the equation of state, Eq. (2.38), becomes

$$\phi(\mathbf{r}) = \phi(\mathbf{r}; \{\beta(\mathbf{r}'), \boldsymbol{\nu}(\mathbf{r}')\}) \tag{2.40}$$

These are related by the identity

$$\Phi(\Omega; \{\beta(\mathbf{r}), \boldsymbol{\nu}(\mathbf{r})\}) = \int_\Omega d\mathbf{r}\,\Phi(\mathbf{r}; \{\beta, \boldsymbol{\nu}\}) \tag{2.41}$$

which is the continuous analog of Eq. (2.34).

Next come the temperature, chemical potential, and pressure fields defined by the formulas $T(\mathbf{r}) = 1/k_B\beta(\mathbf{r})$, $\mu_\alpha(\mathbf{r}) = -\nu_\alpha(\mathbf{r})/\beta(\mathbf{r})$, and $p(\mathbf{r}) = \phi(\mathbf{r})/\beta(\mathbf{r})$, respectively, and in terms of which Eq. (2.40) may be written equivalently as

$$p(\mathbf{r}) = p(\mathbf{r}; \{T(\mathbf{r}'), \boldsymbol{\mu}(\mathbf{r}')\}) \tag{2.42}$$

From either of the functionals ϕ or p [cf., Eqs. (2.40) and (2.42), respectively] one can obtain all conceivable thermodynamic information about the system such as, the internal energy and species density fields [cf. Eqs. (2.36) (2.41)]

$$\varepsilon(\mathbf{r}) = -\left.\frac{\delta\Phi}{\delta\beta(\mathbf{r})}\right|_{\boldsymbol{\nu}} = -\int_\Omega d\mathbf{r}' \left.\frac{\delta\phi(\mathbf{r}')}{\delta\beta(\mathbf{r})}\right|_{\boldsymbol{\nu}} \tag{2.43a}$$

and

$$n_\alpha(\mathbf{r}) = -\left.\frac{\delta\Phi}{\delta\nu_\alpha(\mathbf{r})}\right|_{\beta,\boldsymbol{\nu}'_\alpha} = -\int_\Omega d\mathbf{r}' \left.\frac{\delta\phi(\mathbf{r}')}{\delta\nu_\alpha(\mathbf{r})}\right|_{\beta,\boldsymbol{\nu}'_\alpha} \tag{2.43b}$$

for $\alpha = 1, \ldots, c$. The symbol $\boldsymbol{\nu}'_\alpha$ appearing here denotes the set of all variables included in $\boldsymbol{\nu}$, *except* for $\nu_\alpha(\mathbf{r})$.

Our identification of $p(\mathbf{r})$ as a bearer of all thermodynamic information is the field analog of the familiar fact that the equation of state $p = p(T, \boldsymbol{\mu})$ of global thermodynamics is a fundamental relation equivalent in content to $S = S(\Omega, E, \mathbf{N})$ or $E = E(S, \Omega, \mathbf{N})$. Here $\boldsymbol{\mu} = \{\mu_\alpha; \alpha = 1, \ldots, c\}$ denotes the entire set of species chemical potentials and $\mathbf{N} = \{N_\alpha; \alpha = 1, \ldots, c\}$ the associated set of particle numbers.

The entropy of the system is obtained, as follows, by constructing the Legendre transform of the Massieu–Planck function [cf. Eqs. (2.21),

(2.22), (2.34), and (2.41)]

$$S/k_{\mathrm{B}} = \Phi - \int_{\Omega} d\mathbf{r} \left[\beta(\mathbf{r}) \frac{\delta\Phi}{\delta\beta(\mathbf{r})}\bigg|_{\boldsymbol{\nu}} + \sum_{\alpha=1}^{c} \nu_{\alpha}(\mathbf{r}) \frac{\delta\Phi}{\delta\nu_{\alpha}(\mathbf{r})}\bigg|_{\beta,\boldsymbol{\nu}'_{\alpha}} \right]$$

$$= \Phi + \int_{\Omega} d\mathbf{r} \left[\beta(\mathbf{r})\varepsilon(\mathbf{r}) + \sum_{\alpha=1}^{c} \nu_{\alpha}(\mathbf{r}) n_{\alpha}(\mathbf{r}) \right]$$

$$= \int_{\Omega} d\mathbf{r} \left[\beta(\mathbf{r})\varepsilon(\mathbf{r}) + \sum_{\alpha=1}^{c} \nu_{\alpha}(\mathbf{r}) n_{\alpha}(\mathbf{r}) + \phi(\mathbf{r}) \right] \tag{2.44}$$

This can be written as

$$S = \int_{\Omega} d\mathbf{r} \sigma(\mathbf{r}) \tag{2.45}$$

with

$$\sigma(\mathbf{r}) = k_{\mathrm{B}} \left[\beta(\mathbf{r})\varepsilon(\mathbf{r}) + \sum_{\alpha=1}^{c} \nu_{\alpha}(\mathbf{r}) n_{\alpha}(\mathbf{r}) + \phi(\mathbf{r}) \right] \tag{2.46}$$

denoting the entropy density field. Furthermore, by performing a functional variation of S, using the second line of Eq. (2.44) and Eqs. (2.43a and b), it easily can be shown that

$$\delta S = \int_{\Omega} d\mathbf{r} \delta s(\mathbf{r}) \tag{2.47}$$

wherein

$$\delta s(\mathbf{r}) = k_{\mathrm{B}} \left[\beta(\mathbf{r}) \delta\varepsilon(\mathbf{r}) + \sum_{\alpha=1}^{c} \nu_{\alpha}(\mathbf{r}) \delta n_{\alpha}(\mathbf{r}) \right] \tag{2.48}$$

An alternative expression for δS is produced by constructing the variation of the formula [Eq. (2.45)] with $\sigma(\mathbf{r})$ given by Eq. (2.46). From this expression and Eqs. (2.47) and (2.48) one then obtains an equation

$$\int_{\Omega} d\mathbf{r} \left[\varepsilon(\mathbf{r}) \delta\beta(\mathbf{r}) + \sum_{\alpha=1}^{c} n_{\alpha}(r) \delta\nu_{\alpha}(\mathbf{r}) + \delta\phi(\mathbf{r}) \right] = 0 \tag{2.49}$$

which immediately can be identified with the continuous analog of the Gibbs–Duhem relation (2.31) for a composite system. Because the functional dependencies of $\phi(\mathbf{r})$ on the fields β and $\boldsymbol{\nu}$ are nonlocal, the integrand of Eq. (2.49) is *not* equal to zero at each point $\mathbf{r}$. Thus, according to the assumption made at the beginning of this section

$$\delta\phi(\mathbf{r}) = \int_\Omega d\mathbf{r}' \left[\left.\frac{\delta\phi(\mathbf{r})}{\delta\beta(\mathbf{r}')}\right|_{\boldsymbol{\nu}} \delta\beta(\mathbf{r}') + \sum_{\alpha=1}^{c} \left.\frac{\delta\phi(\mathbf{r})}{\delta\nu_\alpha(\mathbf{r}')}\right|_{\beta,\boldsymbol{\nu}'_\alpha} \delta\nu_\alpha(\mathbf{r}') \right] \tag{2.50}$$

This is the continuum analog of Eq. (2.37). Consistent with these observations is the conclusion that the difference

$$\begin{aligned} \delta\sigma(\mathbf{r}) - \delta s(\mathbf{r}) &= k_B \left[\varepsilon(\mathbf{r})\delta\beta(\mathbf{r}) + \sum_{\alpha=1}^{c} n_\alpha(\mathbf{r})\delta\nu_\alpha(\mathbf{r}) + \delta\phi(\mathbf{r}) \right] \\ &= -\frac{1}{T(\mathbf{r})} \left[\sigma(\mathbf{r})\delta T(\mathbf{r}) + \sum_{\alpha=1}^{c} n_\alpha(\mathbf{r})\delta\mu_\alpha(\mathbf{r}) - \delta p(\mathbf{r}) \right] \end{aligned} \tag{2.51}$$

does not equal zero.

The quantity $\delta\sigma(\mathbf{r})$ is the first variation of the functional $\sigma(\mathbf{r})$ defined by Eq. (2.46) but there is no functional $s(\mathbf{r})$ for which $\delta s(\mathbf{r})$, defined by Eq. (2.48), is the variation. The integrals of both $\delta\sigma(\mathbf{r})$ and $\delta s(\mathbf{r})$ are, however, equal to δS. Although we could have proceeded without introducing $\delta s(\mathbf{r})$, there are at least two reasons for doing so; (1) the formula (2.48) is the precise field analog of the global thermodynamic expression $dS = T^{-1}(dE - \Sigma_{\alpha=1}^{c} \mu_\alpha dN_\alpha)$ and; (2) even if we had not introduced $\delta s(\mathbf{r})$, it is likely that alert readers would have done so. Some of these then possibly might not have noticed the distinction between $\delta s(\mathbf{r})$ and $\delta\sigma(\mathbf{r})$, thereby incorrectly concluding that the integrand of Eq. (2.49) is everywhere zero.

Having thus established the basic connections between conventional, macroscopic thermodynamics and an analogous thermodynamic field theory based on the fundamental equations (2.39)–(2.42), we are now poised to investigate further consequences of this theory. Thus, Eqs. (2.46), (2.48), and (2.49) can be written more familiarly as

$$\sigma(\mathbf{r}) = \frac{1}{T(\mathbf{r})} \left[\varepsilon(\mathbf{r}) + p(\mathbf{r}) - \sum_{\alpha=1}^{c} n_\alpha(\mathbf{r})\mu_\alpha(\mathbf{r}) \right] \tag{2.52}$$

$$\delta s(\mathbf{r}) = \frac{1}{T(\mathbf{r})} \left[\delta\varepsilon(\mathbf{r}) - \sum_{\alpha=1}^{c} \mu_\alpha(\mathbf{r})\delta n_\alpha(\mathbf{r}) \right] \tag{2.53}$$

and

$$\int_{\Omega} d\mathbf{r} \frac{1}{T(\mathbf{r})} \left[\sigma(\mathbf{r})\delta T(\mathbf{r}) + \sum_{\alpha=1}^{c} n_{\alpha}(\mathbf{r})\delta\mu_{\alpha}(\mathbf{r}) - \delta p(\mathbf{r}) \right] = 0 \tag{2.54}$$

respectively. Furthermore, Eq. (2.50) then becomes

$$\delta p(\mathbf{r}) = \int_{\Omega} d\mathbf{r}' \left[\left. \frac{\delta p(\mathbf{r})}{\delta T(\mathbf{r}')} \right|_{\boldsymbol{\mu}} \delta T(\mathbf{r}') + \sum_{\alpha=1}^{c} \left. \frac{\delta p(\mathbf{r})}{\delta \mu_{\alpha}(\mathbf{r}')} \right|_{T,\boldsymbol{\mu}'_{\alpha}} \delta\mu_{\alpha}(\mathbf{r}') \right] \tag{2.55}$$

with

$$\left. \frac{\delta p(\mathbf{r})}{\delta T(\mathbf{r}')} \right|_{\boldsymbol{\mu}} = \frac{p(\mathbf{r})}{T(\mathbf{r})} \delta(\mathbf{r} - \mathbf{r}') - \frac{T(\mathbf{r})}{T^2(\mathbf{r}')} \left\{ \left. \frac{\delta\phi(\mathbf{r})}{\delta\beta(\mathbf{r}')} \right|_{\boldsymbol{\nu}} - \sum_{\alpha=1}^{c} \mu_{\alpha}(\mathbf{r}') \left. \frac{\delta\phi(\mathbf{r})}{\delta\nu_{\alpha}(\mathbf{r}')} \right|_{\beta,\boldsymbol{\nu}'_{\alpha}} \right\} \tag{2.56}$$

and

$$\left. \frac{\delta p(\mathbf{r})}{\delta \mu_{\alpha}(\mathbf{r}')} \right|_{T,\boldsymbol{\mu}'_{\alpha}} = - \frac{T(\mathbf{r})}{T(\mathbf{r}')} \left. \frac{\delta\phi(\mathbf{r})}{\delta\nu_{\alpha}(\mathbf{r}')} \right|_{T,\boldsymbol{\nu}'_{\alpha}} \tag{2.57}$$

It should be recognized that the values of field variables, such as $T(\mathbf{r})$ and $p(\mathbf{r})$, appearing in these and other formulas for functional derivatives are those appropriate to an implicit reference state, for which the derivatives are to be evaluated. In the event that this reference state is uniform, point functions, such as $T(\mathbf{r})$ and $\mu_{\alpha}(\mathbf{r})$, will have constant values and quantities, such as $\delta p(\mathbf{r})/\delta T(\mathbf{r}')|_{\boldsymbol{\mu}}$, will depend only on the separation $|\mathbf{r} - \mathbf{r}'|$ of the two points $\mathbf{r}$ and $\mathbf{r}'$.

From Eq. (2.43) together with the connections [Eqs. (2.46), (2.56) and (2.57)], it can be shown that

$$\sigma(\mathbf{r}) = \int_{\Omega} d\mathbf{r}' \frac{T(\mathbf{r})}{T(\mathbf{r}')} \left. \frac{\delta p(\mathbf{r}')}{\delta T(\mathbf{r})} \right|_{\boldsymbol{\mu}} \tag{2.58}$$

and

$$n_{\alpha}(\mathbf{r}) = \int_{\Omega} d\mathbf{r}' \frac{T(\mathbf{r})}{T(\mathbf{r}')} \left. \frac{\delta p(\mathbf{r}')}{\delta \mu_{\alpha}(\mathbf{r})} \right|_{T,\boldsymbol{\mu}'_{\alpha}} \tag{2.59}$$

These are the local, field analogs of the familiar thermodynamic relations $S/\Omega = (\partial p/\partial T)_{\boldsymbol{\mu}}$ and $N_{\alpha}/\Omega = (\partial p/\partial \mu_{\alpha})_{T,\boldsymbol{\mu}'_{\alpha}}$, respectively.

It now proves convenient to introduce the "blurred" entropy density and particle densities

$$\tilde{\sigma}(\mathbf{r}, \mathbf{r}') = \left.\frac{\delta p(\mathbf{r})}{\delta T(\mathbf{r}')}\right|_{\boldsymbol{\mu}} \tag{2.60}$$

and

$$\tilde{n}_\alpha(\mathbf{r}, \mathbf{r}') = \left.\frac{\delta p(\mathbf{r})}{\delta \mu_\alpha(\mathbf{r}')}\right|_{T, \boldsymbol{\mu}'_\alpha} \tag{2.61}$$

in terms of which Eq. (2.55) can be rewritten as

$$\delta p(\mathbf{r}) = \int_\Omega d\mathbf{r}' \left[\tilde{\sigma}(\mathbf{r}, \mathbf{r}')\delta T(\mathbf{r}') + \sum_{\alpha=1}^{c} \tilde{n}_\alpha(\mathbf{r}, \mathbf{r}')\delta\mu_\alpha(\mathbf{r}') \right] \tag{2.62}$$

In addition to the field variables already considered we introduce the total particle density

$$n(\mathbf{r}) = \sum_{\alpha=1}^{c} n_\alpha(\mathbf{r}) \tag{2.63}$$

and the $c-1$ mole fraction fields

$$x_\alpha(\mathbf{r}) = n_\alpha(\mathbf{r})/n(\mathbf{r}) \qquad \alpha = 2, 3, \ldots, c \tag{2.64}$$

The collection of field variables for which variations have now been defined consists of the $4c+6$ quantities $\delta\beta$, δT, $\delta\varepsilon$, δn, δp, δs, $\delta\sigma$; $\delta\nu_\alpha$, $\delta\mu_\alpha$, δn_α for $\alpha = 1, \ldots, c$ and δx_α for $\alpha = 2, \ldots, c$. There are $3c+5$ functional relations connecting these variables. These consist of the three previously derived equations,

$$\delta s(r) = \frac{1}{T(\mathbf{r})}\left[\delta\varepsilon(\mathbf{r}) - \sum_{\alpha=1}^{c} \mu_\alpha(\mathbf{r})\delta n_\alpha(\mathbf{r}) \right] \tag{2.65}$$

$$\delta\sigma(\mathbf{r}) = \delta s(\mathbf{r}) - \frac{1}{T(\mathbf{r})}\left[\sigma(\mathbf{r})\delta T(\mathbf{r}) + \sum_{\alpha=1}^{c} n_\alpha(\mathbf{r})\delta\mu_\alpha(\mathbf{r}) - \delta p(\mathbf{r}) \right] \tag{2.66}$$

and

$$\delta p(\mathbf{r}) = \int_\Omega d\mathbf{r}' \left[\tilde{\sigma}(\mathbf{r}, \mathbf{r}')\delta T(\mathbf{r}') + \sum_{\alpha=1}^{c} \tilde{n}_\alpha(\mathbf{r}, \mathbf{r}')\delta\mu_\alpha(\mathbf{r}') \right] \tag{2.67}$$

the $2c + 1$ rather obvious relations

$$\delta\beta(\mathbf{r}) = -\frac{1}{k_B T^2(\mathbf{r})}\,\delta T(\mathbf{r}) \tag{2.68}$$

$$\delta n(\mathbf{r}) = \sum_{\alpha=1}^{c} \delta n_\alpha(\mathbf{r}) \tag{2.69}$$

$$\delta\nu_\alpha(\mathbf{r}) = \frac{\mu_\alpha(\mathbf{r})}{k_B T^2(\mathbf{r})}\,\delta T(\mathbf{r}) - \frac{1}{k_B T(\mathbf{r})}\,\delta\mu_\alpha(\mathbf{r}) \qquad \alpha = 1, \ldots, c \tag{2.70}$$

$$\delta x_\alpha(\mathbf{r}) = \frac{1}{n(\mathbf{r})}\,\delta n_\alpha(\mathbf{r}) - \frac{n_\alpha(\mathbf{r})}{n^2(\mathbf{r})}\,\delta n(\mathbf{r}) \qquad \alpha = 2, \ldots, c \tag{2.71}$$

and the $c + 1$ linear connections

$$\delta\varepsilon(\mathbf{r}) = -\int_\Omega d\mathbf{r}' \left[\mathscr{G}_{00}(\mathbf{r}, \mathbf{r}')\delta\beta(\mathbf{r}') + \sum_{\gamma=1}^{c} \mathscr{G}_{0\gamma}(\mathbf{r}, \mathbf{r}')\delta\nu_\gamma(\mathbf{r}') \right] \tag{2.72}$$

$$\delta n_\alpha(\mathbf{r}) = -\int_\Omega d\mathbf{r}' \left[\mathscr{G}_{\alpha 0}(\mathbf{r}, \mathbf{r}')\delta\beta(\mathbf{r}') + \sum_{\gamma=1}^{c} \mathscr{G}_{\alpha\gamma}(\mathbf{r}, \mathbf{r}')\delta\nu_\gamma(\mathbf{r}') \right] \tag{2.73}$$

which follow directly from Eq. (2.43). The quantities $\mathscr{G}_{ij}$ appearing in these last equations are the second-order functional derivatives of the Massieu–Planck function, defined by the formula

$$\mathscr{G}_{ij}(\mathbf{r}, \mathbf{r}') = \frac{\delta^2\Phi}{\delta\nu_i(\mathbf{r})\delta\nu_j(\mathbf{r}')} \qquad i, j = 0, 1, \ldots, c \tag{2.74}$$

with $\nu_0(\mathbf{r}) \equiv \beta(\mathbf{r})$. They satisfy the relations

$$\mathscr{G}_{ij}(\mathbf{r}, \mathbf{r}') = \mathscr{G}_{ji}(\mathbf{r}', \mathbf{r}) \tag{2.75}$$

and so form a symmetric matrix. The functions $\mathscr{G}_{ij}$ as well as the other coefficients occurring in the set of $3c + 5$ linear connections [Eqs. (2.65)–(2.73)] are "characteristic," in the sense that they are completely defined in terms of functional derivatives of Φ and/or $\phi(\mathbf{r})$.

The difference, $c + 1$, between the number of field variations and the linear constraints that they must satisfy can be identified as the number of "functional degrees of freedom." By this we mean that the functional variation of a selected field variable can be expressed in terms of the variations of an arbitrarily chosen set of $c + 1$ others. For example, one could choose to express $T(\mathbf{r})\delta s(\mathbf{r})$ in terms of δT, δp and $\{\delta x_\alpha; \alpha = 2, \ldots, c\}$ and so obtain a relationship

$$T(\mathbf{r})\delta s(\mathbf{r}) = \int_\Omega d\mathbf{r}' \left[C_p(\mathbf{r}, \mathbf{r}')\delta T(\mathbf{r}') + T(\mathbf{r}) \times \left\{ \left. \frac{\delta s(\mathbf{r})}{\delta p(\mathbf{r}')} \right|_{T,\mathbf{x}} \delta p(\mathbf{r}') + \sum_{\alpha=2}^{c} \left. \frac{\delta s(\mathbf{r})}{\delta x_\alpha(\mathbf{r}')} \right|_{T,p,\mathbf{x}'_\alpha} \delta x_\alpha(\mathbf{r}') \right\} \right] \tag{2.76}$$

in which the quantity $T(\mathbf{r})\delta s(\mathbf{r})/\delta T(\mathbf{r}')|_{p,\mathbf{x}}$ has been identified as a generalized heat capacity (per unit volume) at constant pressure.

By using the connections [Eqs. (2.65), (2.68), and (2.70)] one can transform Eqs. (2.72) and (2.73) into the alternative set

$$\delta s(\mathbf{r}) = \int_\Omega d\mathbf{r}' \left[\mathcal{K}_{00}(\mathbf{r}, \mathbf{r}') \frac{\delta T(\mathbf{r}')}{k_B T(\mathbf{r}')} + \sum_{\gamma=1}^{c} \mathcal{K}_{0\gamma}(\mathbf{r}, \mathbf{r}') \frac{\delta \mu_\gamma(\mathbf{r}')}{k_B T(\mathbf{r}')} \right] \tag{2.77}$$

$$\delta n_\alpha(\mathbf{r}) = \int_\Omega d\mathbf{r}' \left[\mathcal{K}_{\alpha 0}(\mathbf{r}, \mathbf{r}') \frac{\delta T(\mathbf{r}')}{k_B T(\mathbf{r}')} + \sum_{\gamma=1}^{c} \mathcal{K}_{\alpha\gamma}(\mathbf{r}, \mathbf{r}') \frac{\delta \mu_\gamma(\mathbf{r}')}{k_B T(\mathbf{r}')} \right] \tag{2.78}$$

with coefficients $\mathcal{K}_{ij}(\mathbf{r}, \mathbf{r}')$ that likewise form a symmetric matrix:

$$\mathcal{K}_{00}(\mathbf{r}, \mathbf{r}') = k_B^2 \sum_{i=0}^{c} \sum_{j=0}^{c} \nu_i(\mathbf{r}) \mathcal{G}_{ij}(\mathbf{r}, \mathbf{r}') \nu_j(\mathbf{r}') \tag{2.79}$$

$$\mathcal{K}_{0\alpha}(\mathbf{r}, \mathbf{r}') = k_B \sum_{i=0}^{c} \nu_i(\mathbf{r}) \mathcal{G}_{i\alpha}(\mathbf{r}, \mathbf{r}') \tag{2.80}$$

$$\mathcal{K}_{\alpha 0}(\mathbf{r}, \mathbf{r}') = k_B \sum_{j=0}^{c} \mathcal{G}_{\alpha j}(\mathbf{r}, \mathbf{r}') \nu_j(\mathbf{r}') \tag{2.81}$$

$$\mathcal{K}_{\alpha\beta}(\mathbf{r}, \mathbf{r}') = \mathcal{G}_{\alpha\beta}(\mathbf{r}, \mathbf{r}') \tag{2.82}$$

To proceed further with the thermodynamic theory we introduce the matrix inverse of $\{\mathcal{G}_{\alpha\beta}(\mathbf{r}, \mathbf{r}'); \alpha, \beta = 1, 2, \ldots, c\}$, so defined that

$$\sum_{\gamma=1}^{c} \int_\Omega d\mathbf{r}'' \mathcal{G}^{-1}_{\alpha\gamma}(\mathbf{r}, \mathbf{r}'') \mathcal{G}_{\gamma\beta}(\mathbf{r}'', \mathbf{r}') = \delta_{\alpha\beta} \delta(\mathbf{r} - \mathbf{r}') \tag{2.83}$$

Although the existence of this inverse matrix must be assumed at this point in the development, it is demonstrated in Section III.B that the analogous statistical mechanical matrix does indeed exist. The result of acting upon Eq. (2.73) with the inverse operator [i.e., multiply with $\mathcal{G}^{-1}_{\beta\alpha}(\mathbf{r}'', \mathbf{r})$, sum on α, and integrate over the domain of $\mathbf{r}$] is the formula

$$\delta\nu_\alpha(\mathbf{r}) = \int_\Omega d\mathbf{r}' \left[\left.\frac{\delta\nu_\alpha(\mathbf{r})}{\delta\beta(\mathbf{r}')}\right|_{\mathbf{n}} \delta\beta(\mathbf{r}') + \sum_{\gamma=1}^{c} \left.\frac{\delta\nu_\alpha(\mathbf{r})}{\delta n_\gamma(\mathbf{r}')}\right|_{\beta,\mathbf{n}'_\gamma} \delta n_\gamma(\mathbf{r}') \right] \tag{2.84}$$

wherein

$$\left.\frac{\delta\nu_\alpha(\mathbf{r})}{\delta\beta(\mathbf{r}')}\right|_{\mathbf{n}} = -\sum_{\gamma=1}^{c} \int_\Omega d\mathbf{r}'' \mathcal{G}^{-1}_{\alpha\gamma}(\mathbf{r}, \mathbf{r}'') \mathcal{G}_{\gamma 0}(\mathbf{r}'', \mathbf{r}') \tag{2.85}$$

and

$$\left.\frac{\delta\nu_\alpha(\mathbf{r})}{\delta\gamma(\mathbf{r}')}\right|_{\beta,\mathbf{n}'_\gamma} = -\mathcal{G}^{-1}_{\alpha\gamma}(\mathbf{r}, \mathbf{r}') \tag{2.86}$$

The linear connections [Eqs. (2.68) and (2.70)] enable us to transform Eq. (2.84) into the corresponding formula

$$\delta\mu_\alpha(\mathbf{r}) = \int_\Omega d\mathbf{r}' \left[\left.\frac{\delta\mu_\alpha(\mathbf{r})}{\delta T(\mathbf{r}')}\right|_{\mathbf{n}} \delta T(\mathbf{r}') + \sum_{\gamma=1}^{c} \left.\frac{\delta\mu_\alpha(\mathbf{r})}{\delta n_\gamma(\mathbf{r}')}\right|_{T,\mathbf{n}'_\gamma} \delta n_\gamma(\mathbf{r}') \right] \tag{2.87}$$

with

$$\left.\frac{\delta\mu_\alpha(\mathbf{r})}{\delta T(\mathbf{r}')}\right|_{\mathbf{n}} = \frac{1}{T(\mathbf{r}')} \Bigg[\mu_\alpha(\mathbf{r})\delta(\mathbf{r} - \mathbf{r}') - \frac{T(\mathbf{r})}{T(\mathbf{r}')} \sum_{\gamma=1}^{c} \int_\Omega d\mathbf{r}'' \mathcal{G}^{-1}_{\alpha\gamma}(\mathbf{r}, \mathbf{r}'') \mathcal{G}_{\gamma 0}(\mathbf{r}'', \mathbf{r}') \Bigg] \tag{2.88}$$

$$\left.\frac{\delta\mu_\alpha(\mathbf{r})}{\delta n_\gamma(\mathbf{r}')}\right|_{T,\mathbf{n}'_\gamma} = k_B T(\mathbf{r}) \mathcal{G}^{-1}_{\alpha\gamma}(\mathbf{r}, \mathbf{r}') \tag{2.89}$$

D. The Pressure Representation

The basic set of thermodynamic variables used throughout much of the preceding development has consisted of the energy density and the particle densities of the c independent components. Although the energy density field then was replaced with the temperature and later by the entropy density, we continued to describe the composition of the system in terms of the full set of c particle density fields. However, it sometimes is desirable to select the thermodynamic pressure field as one of the basic observables, along with the temperature and an appropriately chosen set of $c-1$ compositional variables. The arguments for doing this become particularly cogent when the solution under consideration contains a special component (which we designate by $\alpha = 1$) that plays the role of a dispersing agent or solvent.

The first step in transforming to the representation in which the temperature, pressure, and solute mole fractions appear as the basic thermodynamic degrees of freedom, is to rewrite the variation of the chemical potential given by Eq. (2.87) in terms of the variations of the total particle density $n(\mathbf{r}) = \Sigma_{\alpha=1}^{c} n_\alpha(\mathbf{r})$ and the $c-1$ mole fraction fields $\{x_\alpha(\mathbf{r}) = n_\alpha(\mathbf{r})/n(\mathbf{r});\ \alpha = 2, 3, \ldots, c\}$.

$$\delta\mu_\alpha(\mathbf{r}) = \int_\Omega d\mathbf{r}' \left\{ \left.\frac{\delta\mu_\alpha(\mathbf{r})}{\delta T(\mathbf{r}')}\right|_{n,\mathbf{x}} \delta T(\mathbf{r}') + \left.\frac{\delta\mu_\alpha(\mathbf{r})}{\delta n(\mathbf{r}')}\right|_{T,\mathbf{x}} \delta n(\mathbf{r}') + \sum_{\gamma=2}^{c} \left.\frac{\delta\mu_\alpha(\mathbf{r})}{\delta x_\gamma(\mathbf{r}')}\right|_{n,T,\mathbf{x}'_\gamma} \delta x_\gamma(\mathbf{r}') \right\} \tag{2.90}$$

The functional derivatives appearing in this formula are given by the following expressions:

$$\left.\frac{\delta\mu_\alpha(\mathbf{r})}{\delta T(\mathbf{r}')}\right|_{n,\mathbf{x}} = \left.\frac{\delta\mu_\alpha(\mathbf{r})}{\delta T(\mathbf{r}')}\right|_{\mathbf{n}} \tag{2.91}$$

$$\left.\frac{\delta\mu_\alpha(\mathbf{r})}{\delta n(\mathbf{r}')}\right|_{T,\mathbf{x}} = \sum_{\gamma=1}^{c} \left.\frac{\delta\mu_\alpha(\mathbf{r})}{\delta n_\gamma(\mathbf{r}')}\right|_{T,\mathbf{n}'_\gamma} x_\gamma(\mathbf{r}') = k_B T(\mathbf{r}) \sum_{\gamma=1}^{c} \mathscr{G}_{\alpha\gamma}^{-1}(\mathbf{r},\mathbf{r}') x_\gamma(\mathbf{r}') \tag{2.92}$$

and

$$\begin{aligned}\left.\frac{\delta\mu_\alpha(\mathbf{r})}{\delta x_\beta(\mathbf{r}')}\right|_{T,n,\mathbf{x}'_\beta} &= n(\mathbf{r}') \left\{ \left.\frac{\delta\mu_\alpha(\mathbf{r})}{\delta n_\beta(\mathbf{r}')}\right|_{T,\mathbf{n}'_\beta} - \left.\frac{\delta\mu_\alpha(\mathbf{r})}{\delta n_1(\mathbf{r}')}\right|_{T,\mathbf{n}'_1} \right\} \\ &= k_B T(\mathbf{r}) n(\mathbf{r}') [\mathscr{G}_{\alpha\beta}^{-1}(\mathbf{r},\mathbf{r}') - \mathscr{G}_{\alpha 1}^{-1}(\mathbf{r},\mathbf{r}')]\end{aligned} \tag{2.93}$$

Next, by combining Eqs. (2.67), (2.91)–(2.93) we find that

$$\delta p(\mathbf{r}) = \int_\Omega d\mathbf{r}' \left\{ \left.\frac{\delta p(\mathbf{r})}{\delta T(\mathbf{r}')}\right|_{n,\mathbf{x}} \delta T(\mathbf{r}') + \left.\frac{\delta p(\mathbf{r})}{\delta n(\mathbf{r}')}\right|_{T,\mathbf{x}} \delta n(\mathbf{r}') + \sum_{\gamma=2}^{c} \left.\frac{\delta p(\mathbf{r})}{\delta x_\gamma(\mathbf{r}')}\right|_{T,n,\mathbf{x}'_\gamma} \delta x_\gamma(\mathbf{r}') \right\} \tag{2.94}$$

with

$$\left.\frac{\delta p(\mathbf{r})}{\delta T(\mathbf{r}')}\right|_{n,\mathbf{x}} = \tilde{\sigma}(\mathbf{r}, \mathbf{r}') + \sum_{\alpha=1}^{c} \int_{\Omega} d\mathbf{r}'' \tilde{n}_{\alpha}(\mathbf{r}, \mathbf{r}'') \left.\frac{\delta \mu_{\alpha}(\mathbf{r}'')}{\delta T(\mathbf{r}')}\right|_{\mathbf{n}} \tag{2.95}$$

$$\begin{aligned}\left.\frac{\delta p(\mathbf{r})}{\delta n(\mathbf{r}')}\right|_{T,\mathbf{x}} &= \sum_{\alpha=1}^{c} \int_{\Omega} d\mathbf{r}'' \tilde{n}_{\alpha}(\mathbf{r}, \mathbf{r}'') \left.\frac{\delta \mu_{\alpha}(\mathbf{r}'')}{\delta n(\mathbf{r}')}\right|_{T,\mathbf{x}} \\ &= \sum_{\alpha=1}^{c} \sum_{\beta=1}^{c} \int_{\Omega} d\mathbf{r}'' \tilde{n}_{\alpha}(\mathbf{r}, \mathbf{r}'') k_{\mathrm{B}} T(\mathbf{r}'') \mathcal{G}_{\alpha\beta}^{-1}(\mathbf{r}', \mathbf{r}'') x_{\beta}(\mathbf{r}'')\end{aligned} \tag{2.96}$$

and

$$\begin{aligned}\left.\frac{\delta p(\mathbf{r})}{\delta x_{\alpha}(\mathbf{r}')}\right|_{T,n,\mathbf{x}'_{\alpha}} &= \sum_{\beta=1}^{c} \int_{\Omega} d\mathbf{r}'' \tilde{n}_{\beta}(\mathbf{r}, \mathbf{r}'') \left.\frac{\delta \mu_{\beta}(\mathbf{r}'')}{\delta x_{\alpha}(\mathbf{r}')}\right|_{T,n,\mathbf{x}'_{\alpha}} \\ &= n(\mathbf{r}') \sum_{\beta=1}^{c} \int_{\Omega} d\mathbf{r}'' \tilde{n}_{\beta}(\mathbf{r}, \mathbf{r}'') k_{\mathrm{B}} T(\mathbf{r}'') [\mathcal{G}_{\beta\alpha}^{-1}(\mathbf{r}'', \mathbf{r}') - \mathcal{G}_{\beta 1}^{-1}(\mathbf{r}'', \mathbf{r}')]\end{aligned} \tag{2.97}$$

As a final example, we examine the connection between the two heat capacity functions,

$$C_p(\mathbf{r}, \mathbf{r}') = T(\mathbf{r}) \left.\frac{\delta s(\mathbf{r})}{\delta T(\mathbf{r}')}\right|_{p,\mathbf{x}} \tag{2.98}$$

and [cf., Eq. (2.53)]

$$C_v(\mathbf{r}, \mathbf{r}') = T(\mathbf{r}) \left.\frac{\delta s(\mathbf{r})}{\delta T(\mathbf{r}')}\right|_{\mathbf{n}} = \left.\frac{\delta \varepsilon(\mathbf{r})}{\delta T(\mathbf{r}')}\right|_{\mathbf{n}} \tag{2.99}$$

With $\delta s(\mathbf{r})$ expressed as a linear functional of δT, δn, and $\delta \mathbf{x}$, the formula (2.98) can be rewritten as

$$C_p(\mathbf{r}, \mathbf{r}') = T(\mathbf{r}) \left.\frac{\delta s(\mathbf{r})}{\delta T(\mathbf{r}')}\right|_{n,\mathbf{x}} + T(\mathbf{r}) \int_{\Omega} d\mathbf{r}'' \left.\frac{\delta s(\mathbf{r})}{\delta n(\mathbf{r}'')}\right|_{T,\mathbf{x}} \left.\frac{\delta n(\mathbf{r}'')}{\delta T(\mathbf{r}')}\right|_{p,\mathbf{x}} \tag{2.100}$$

where

$$\left.\frac{\delta s(\mathbf{r})}{\delta n(\mathbf{r}'')}\right|_{T,\mathbf{x}} = \sum_{\gamma=1}^{c} \left.\frac{\delta s(\mathbf{r})}{\delta n_\gamma(\mathbf{r}'')}\right|_{T,\mathbf{n}'_\gamma} x_\gamma(\mathbf{r}'')$$

$$= -\frac{T(\mathbf{r})}{T(\mathbf{r}'')n(\mathbf{r}'')} \sum_{\gamma=1}^{c} n_\gamma(\mathbf{r}'') \left.\frac{\delta \mu_\gamma(\mathbf{r}'')}{\delta T(\mathbf{r})}\right|_{n,\mathbf{x}} \tag{2.101}$$

To obtain the last of these expressions use has been made of the Maxwell relation

$$\frac{T(\mathbf{r})}{T(\mathbf{r}')} \left.\frac{\delta \mu_\alpha(\mathbf{r}')}{\delta T(\mathbf{r})}\right|_{\mathbf{n}} = -\left.\frac{\delta s(\mathbf{r})}{\delta n_\alpha(\mathbf{r}')}\right|_{T,\mathbf{n}'_\alpha} \tag{2.102}$$

which, together with the two other equalities,

$$\frac{T(\mathbf{r})}{T(\mathbf{r}')} \left.\frac{\delta \mu_\alpha(\mathbf{r}')}{\delta n_\beta(\mathbf{r})}\right|_{T,\mathbf{n}'_\beta} = \left.\frac{\delta \mu_\beta(\mathbf{r})}{\delta n_\alpha(\mathbf{r}')}\right|_{T,\mathbf{n}'_\alpha} \tag{2.103}$$

and

$$\frac{T(\mathbf{r})}{T(\mathbf{r}')} \left.\frac{\delta s(\mathbf{r}')}{\delta T(\mathbf{r})}\right|_{\mathbf{n}} = \left.\frac{\delta s(\mathbf{r})}{\delta T(\mathbf{r}')}\right|_{\mathbf{n}} \tag{2.104}$$

can be extracted from Eqs. (2.77) and (2.78). By combining Eqs. (2.100) and (2.101) with the formula [cf. Eq. (2.67)]

$$\left.\frac{\delta p(\mathbf{r})}{\delta T(\mathbf{r}')}\right|_{n,\mathbf{x}} = \tilde{\sigma}(\mathbf{r},\mathbf{r}') + \sum_{\gamma=1}^{c} \int_\Omega d\mathbf{r}''\, \tilde{n}_\gamma(\mathbf{r}',\mathbf{r}'') \left.\frac{\delta \mu_\gamma(\mathbf{r}'')}{\delta T(\mathbf{r})}\right|_{n,\mathbf{x}} \tag{2.105}$$

we obtain the formula

$$\begin{aligned} C_p(\mathbf{r},\mathbf{r}') - C_v(\mathbf{r},\mathbf{r}') &= T^2(\mathbf{r}) \int_\Omega d\mathbf{r}'' \int_\Omega d\mathbf{r}'''\alpha_p(\mathbf{r}'',\mathbf{r}) \frac{\kappa_T^{-1}(\mathbf{r}''',\mathbf{r}'')}{T(\mathbf{r}''')} \alpha_p(\mathbf{r}''',\mathbf{r}') \\ &\quad - T^2(\mathbf{r}) \int_\Omega d\mathbf{r}'' \frac{1}{T(\mathbf{r}'')} \{(\tilde{\sigma}(\mathbf{r}'',\mathbf{r}) \\ &\quad + \sum_{\gamma=1}^{c} \left[\int_\Omega d\mathbf{r}'''\tilde{n}_\gamma(\mathbf{r}'',\mathbf{r}''') \left.\frac{\delta \mu_\gamma(\mathbf{r}''')}{\delta T(\mathbf{r})}\right|_{n,\mathbf{x}} \right. \\ &\quad \left. - n_\gamma(\mathbf{r}'') \left.\frac{\delta \mu_\gamma(\mathbf{r}'')}{\delta T(\mathbf{r})}\right|_{n,\mathbf{x}} \right] \Bigg\} \alpha_p(\mathbf{r}'',\mathbf{r}') \end{aligned} \tag{2.106}$$

The quantities

$$\alpha_p(\mathbf{r}, \mathbf{r}') = -\frac{1}{n(\mathbf{r})} \frac{\delta n(\mathbf{r})}{\delta T(\mathbf{r}')}\bigg|_{p,\mathbf{x}} \tag{2.107}$$

$$\kappa_T^{-1}(\mathbf{r}, \mathbf{r}') = n(\mathbf{r}') \frac{\delta p(\mathbf{r})}{\delta n(\mathbf{r}')}\bigg|_{T,\mathbf{x}} \tag{2.108}$$

appearing in Eq. (2.106) can be identified as the local thermodynamic analogs of the isobaric thermal expansivity and the bulk modulus (or inverse of the isothermal compressibility), respectively.

The first term on the right-hand side of Eq. (2.106) closely resembles the familiar result of global thermodynamics but the additional terms are neither familiar nor obviously negligible. A partial explanation of this is that the heat capacities commonly used in conventional thermodynamics are not defined on a volumetric basis, as has been done here, but on a molar or particle basis [23]. Thus, to establish correspondence with the classical results we should work in terms of the modified heat capacity functions

$$\begin{aligned}\tilde{C}_p(\mathbf{r}, \mathbf{r}') &= T(\mathbf{r}) \frac{\delta}{\delta T(\mathbf{r}')} \left\{ \frac{\sigma(\mathbf{r})}{n(\mathbf{r})} \right\}\bigg|_{p,\mathbf{x}} \\ &= \frac{1}{n(\mathbf{r})} \{ C_p(\mathbf{r}, \mathbf{r}') + T(\mathbf{r})\sigma(\mathbf{r})\alpha_p(\mathbf{r}, \mathbf{r}') \\ &\quad + T(\mathbf{r}) \left[\frac{\delta \sigma(\mathbf{r})}{\delta T(\mathbf{r}')}\bigg|_{p,\mathbf{x}} - \frac{\delta s(\mathbf{r})}{\delta T(\mathbf{r}')}\bigg|_{p,\mathbf{x}} \right] \}\end{aligned} \tag{2.109}$$

and

$$\begin{aligned}\tilde{C}_v(\mathbf{r}, \mathbf{r}') &= T(\mathbf{r}) \frac{\delta}{\delta T(\mathbf{r}')} \left\{ \frac{\sigma(\mathbf{r})}{n(\mathbf{r})} \right\}\bigg|_{\mathbf{n}} \\ &= \frac{1}{n(\mathbf{r})} \left\{ C_v(\mathbf{r}, \mathbf{r}') + T(\mathbf{r}) \left[\frac{\delta \sigma(\mathbf{r})}{\delta T(\mathbf{r}')}\bigg|_{\mathbf{n}} - \frac{\delta s(\mathbf{r})}{\delta T(\mathbf{r}')}\bigg|_{\mathbf{n}} \right] \right\}\end{aligned} \tag{2.110}$$

From these formulas and Eq. (2.106) it then follows that

$$\tilde{C}_p(\mathbf{r},\mathbf{r}') - \tilde{C}_v(\mathbf{r},\mathbf{r}') = \frac{T^2(\mathbf{r})}{n(\mathbf{r})} \int_\Omega d\mathbf{r}'' \int_\Omega d\mathbf{r}''' \alpha_p(\mathbf{r}'',\mathbf{r}) \times \frac{\kappa_T^{-1}(\mathbf{r}''',\mathbf{r}'')}{T(\mathbf{r}''')} \alpha_p(\mathbf{r}''',\mathbf{r}') + \Delta\tilde{C}(\mathbf{r},\mathbf{r}') \tag{2.111}$$

with

$$\begin{aligned}\Delta\tilde{C}(\mathbf{r},\mathbf{r}') &= \frac{T(\mathbf{r})}{n(\mathbf{r})}[\ell_{p,\mathbf{x}}(\mathbf{r},\mathbf{r}') - \ell_{\mathbf{n}}(\mathbf{r},\mathbf{r}')] \\ &\quad - \frac{T^2(\mathbf{r})}{n(\mathbf{r})} \int_\Omega d\mathbf{r}'' \frac{1}{T(\mathbf{r}'')} \sum_{\gamma=1}^{c} \left[\left[\int_\Omega d\mathbf{r}''' \tilde{n}_\gamma(\mathbf{r}'',\mathbf{r}''') \frac{\delta\mu_\gamma(\mathbf{r}''')}{\delta T(\mathbf{r})} \right|_{n,\mathbf{x}} \\ &\quad - n_\gamma(\mathbf{r}'') \frac{\delta\mu_\gamma(\mathbf{r}'')}{\delta T(\mathbf{r})} \bigg|_{n,\mathbf{x}} \right] \alpha_p(\mathbf{r}'',\mathbf{r}') + \frac{T(\mathbf{r})}{n(\mathbf{r})} \Big[\sigma(\mathbf{r})\alpha_p(\mathbf{r},\mathbf{r}') \\ &\quad - T(\mathbf{r}) \int_\Omega d\mathbf{r}'' \frac{1}{T(\mathbf{r}'')} \tilde{\sigma}(\mathbf{r}'',\mathbf{r})\alpha_p(\mathbf{r}'',\mathbf{r}') \Big]\end{aligned} \tag{2.112}$$

and where

$$\ell_y(\mathbf{r},\mathbf{r}') = \frac{\delta\sigma(\mathbf{r})}{\delta T(\mathbf{r}')}\bigg|_y - \frac{\delta s(\mathbf{r})}{\delta T(\mathbf{r}')}\bigg|_y \tag{2.113}$$

The first term on the right-hand side of Eq. (2.111) is an obvious generalization of what appears in the well-known relationship, $\tilde{C}_p - \tilde{C}_v = (T/n)(\alpha_p^2/\kappa_T)$, of global thermodynamics. The second term is so constituted that its integral $\int_\Omega d\mathbf{r}\, \Delta\tilde{C}(\mathbf{r},\mathbf{r}')$ vanishes when the system is in a homogeneous state of equilibrium. A stronger condition than this is established in the Section III.D.

E. The Osmotic Representation

A second case of interest is the representation in which the chemical potential of one of the components, the "solvent," enters as one of the basic thermodynamic variables. In situations of this sort the value of the solvent potential can be controlled by placing the solution in contact with a reservoir of pure solvent, across a semipermeable membrane through which only it can pass. When the conditions are satisfied for this "osmotic equilibrium," a natural choice of basic thermodynamic fields consists of the temperature, the solvent ($\alpha = 1$) chemical potential, and the particle densities of the $c-1$ solute species. This set of basic fields is particularly well adapted to the treatment of colloidal and macromolecular solutions where the solvent frequently is treated as a continuum in which the solute particles are dispersed [24]. A closely analogous model serves as the basis of Debye–Hückel type theories of electrolyte solutions [25].

From the constitutive relationship [Eq. (2.78)] we obtain the following formulas for the chemical potential variations of the solute ($\alpha \geq 2$) species;

$$\delta\mu_\alpha(\mathbf{r}) = \int_\Omega d\mathbf{r}' \left\{ \left.\frac{\delta\mu_\alpha(\mathbf{r})}{\delta T(\mathbf{r}')}\right|_{\mu_1,\mathbf{n}'_1} \delta T(\mathbf{r}') + \left.\frac{\delta\mu_\alpha(\mathbf{r})}{\delta\mu_1(\mathbf{r}')}\right|_{T,\mathbf{n}'_1} \delta\mu_1(\mathbf{r}') + \sum_{\beta=2}^{c} \left.\frac{\delta\mu_\alpha(\mathbf{r})}{\delta n_\beta(\mathbf{r}')}\right|_{T,\mu_1,\mathbf{n}''_{1,\beta}} \delta n_\beta(\mathbf{r}') \right\} \tag{2.114}$$

wherein

$$\left.\frac{\delta\mu_\alpha(\mathbf{r})}{\delta T(\mathbf{r}')}\right|_{\mu_1,\mathbf{n}'_1} = \frac{\mu_\alpha(\mathbf{r})}{T(\mathbf{r})}\,\delta(\mathbf{r}-\mathbf{r}') - \frac{T(\mathbf{r})}{T(\mathbf{r}')^2} \sum_{\beta=2}^{c} \int_\Omega d\mathbf{r}''\mathscr{G}^{\mathrm{eff}^{-1}}_{\alpha\beta}(\mathbf{r},\mathbf{r}'')[\mathscr{G}_{\beta 0}(\mathbf{r}'',\mathbf{r}') - \mathscr{G}_{\beta 1}(\mathbf{r}'',\mathbf{r}')\mu_1(\mathbf{r}')] \tag{2.115}$$

$$\left.\frac{\delta\mu_\alpha(\mathbf{r})}{\delta\mu_1(\mathbf{r}')}\right|_{T,\mathbf{n}'_1} = -\frac{T(\mathbf{r})}{T(\mathbf{r}')} \sum_{\beta=2}^{c} \int_\Omega d\mathbf{r}''\mathscr{G}^{\mathrm{eff}^{-1}}_{\alpha\beta}(\mathbf{r},\mathbf{r}'')\mathscr{G}_{\beta 1}(\mathbf{r}'',\mathbf{r}') \tag{2.116}$$

and

$$\left.\frac{\delta\mu_\alpha(\mathbf{r})}{\delta n_\beta(\mathbf{r}'')}\right|_{T,\mu_1,\mathbf{n}''_{1,\beta}} = k_{\mathrm{B}} T(\mathbf{r})\mathscr{G}^{\mathrm{eff}^{-1}}_{\alpha\beta}(\mathbf{r},\mathbf{r}') \tag{2.117}$$

The symbol $\mathbf{n}''_{\alpha,\beta}$ denotes the set of all variables included in $\mathbf{n}$, *except* for $n_\alpha(\mathbf{r})$ and $n_\beta(\mathbf{r})$. The "effective" inverse kernels $\mathscr{G}^{\mathrm{eff}^{-1}}_{\alpha\beta}(\mathbf{r},\mathbf{r}')$ are defined by the relationships

$$\sum_{\gamma=2}^{c} \int_\Omega d\mathbf{r}''\mathscr{G}^{\mathrm{eff}^{-1}}_{\alpha\gamma}(\mathbf{r},\mathbf{r}'')\mathscr{G}_{\gamma\beta}(\mathbf{r}'',\mathbf{r}') = \delta_{\alpha\beta}\delta(\mathbf{r}-\mathbf{r}') \tag{2.118}$$

which, although closely analogous to Eq. (2.83), are applicable exclusively to the solute species, that is, to $\alpha, \beta \geq 2$.

When Eq. (2.114) is inserted into the fluctuational equation of state

[Eq. (2.67)], written as

$$\delta p(\mathbf{r}) = \int_\Omega d\mathbf{r}' \left[\tilde{\sigma}(\mathbf{r}, \mathbf{r}')\delta T(\mathbf{r}') + \tilde{n}_1(\mathbf{r}, \mathbf{r}')\delta\mu_1(\mathbf{r}') + \sum_{\alpha=2}^{c} \tilde{n}_\alpha(\mathbf{r}, \mathbf{r}')\delta\mu_\alpha(\mathbf{r}') \right] \tag{2.119}$$

it is found that

$$\begin{aligned} \left.\frac{\delta p(\mathbf{r})}{\delta n_\alpha(\mathbf{r}')}\right|_{T,\mu_1,\mathbf{n}''_{1,\alpha}} &= \sum_{\beta=2}^{c} \int_\Omega d\mathbf{r}'' \tilde{n}_\beta(\mathbf{r}, \mathbf{r}'') \left.\frac{\delta \mu_\beta(\mathbf{r}'')}{\delta n_\alpha(\mathbf{r}')}\right|_{T,\mu_1,\mathbf{n}''_{1,\alpha}} \\ &= \sum_{\beta=2}^{c} \int_\Omega d\mathbf{r}'' \tilde{n}_\beta(\mathbf{r}, \mathbf{r}'') k_B T(\mathbf{r}'') \mathscr{G}^{\text{eff}^{-1}}_{\beta\alpha}(\mathbf{r}'', \mathbf{r}') \end{aligned} \tag{2.120}$$

To obtain the second of these expressions use has been made of Eq. (2.117).

We return to these equations and their pressure representation counterparts in Section III.E.

III. THE STATISTICAL MECHANICS OF NONUNIFORM SYSTEMS

Now that a thermodynamic field theory has been fashioned for inhomogeneous systems, we turn to the task of constructing a complementary statistical mechanical formalism. The principal value of such a theory is to be found in the microscopic interpretation it provides for the thermodynamic properties and in the associated potential for generating numerical estimates of these properties.

A. The Constrained Grand Canonical Ensemble

From the point of view of statistical mechanics, the most convenient set of basic thermodynamic observables consists of the energy density field and the particle densities of all components. These thermodynamic variables can be identified with the ensemble average values of the corresponding microscopic fields, namely,

$$\hat{n}_\alpha(\mathbf{r}) = \sum_{i=1}^{N_\alpha} \delta(\mathbf{r} - \mathbf{r}_{i\alpha}) \qquad \alpha = 1, 2, \ldots, c \tag{3.1}$$

and

$$\hat{\varepsilon} = \sum_{\alpha=1}^{c} \sum_{i=1}^{N_\alpha} H_{i\alpha} \delta(\mathbf{r} - \mathbf{r}_{i\alpha}) \tag{3.2}$$

wherein

$$H_{i\alpha} = \frac{1}{2m_\alpha} \mathbf{p}_{i\alpha}^2 + v_\alpha(\mathbf{r}_{i\alpha}) + \frac{1}{2} \sum_{\beta=1}^{c} \sum_{j=1}^{N_\beta}{}' u_{\alpha\beta}(|\mathbf{r}_{i\alpha} - \mathbf{r}_{j\beta}|) \tag{3.3}$$

Here $u_{\alpha\beta}$ denotes the interaction energy of two particles, one of species α and another of species β. The potential, v_α, of the conservative external fields of force acting on a species-α particle, includes a part that confines these particles to the interior of a region of volume Ω. The prime (′) in Eq. (3.3) indicates that the sums do not include the term with $\beta = \alpha$ and $j = i$. Finally, by $\mathbf{r}_{i\alpha}$ and $\mathbf{p}_{i\alpha}$ we denote the position and momentum of particle i of species α.

The microscopic variables $\{\hat{n}_\alpha(\mathbf{r});\ \alpha = 1, 2, \ldots, c\}$ and $\hat{\varepsilon}(\mathbf{r})$ depend not only on the field point $\mathbf{r}$ but upon the entire collection of particle locations and momenta. This set will be denoted by the symbol $\Gamma_{\{N_\alpha\}}$. In the case of an open system the particle numbers $\{N_\alpha\}$ and the phase space variables $\Gamma_{\{N_\alpha\}}$ are random quantities. The averages of the fields defined by Eqs. (3.1) and (3.2) are then defined by the formulas

$$n_\alpha(\mathbf{r}) = \langle \hat{n}_\alpha(\mathbf{r}) \rangle$$

and

$$\varepsilon(\mathbf{r}) = \langle \hat{\varepsilon}(\mathbf{r}) \rangle \tag{3.4}$$

where the pointed brackets denote the averaging operation

$$\langle \hat{\phi} \rangle = \sum_{\{N_\alpha\}} \prod_{\alpha=1}^{c} \left(\frac{1}{h^{3N_\alpha} N_\alpha!} \right) \int d\Gamma_{\{N_\alpha\}} \psi(\Gamma_{\{N_\alpha\}}) \hat{\phi} \tag{3.5}$$

Here h is the Planck constant and $\psi(\Gamma_{\{N_\alpha\}})$ is a probability density associated with the particle numbers $\{N_\alpha\}$ and the corresponding set of phase space variables $\Gamma_{\{N_\alpha\}}$.

According to information theory [26, 27], the entropy of this system is to be identified with the Boltzmann–Shannon functional,

$$S = -k_B \langle \ln \psi(\Gamma_{\{N_\alpha\}}) \rangle \tag{3.6}$$

of the set of distribution functions $\psi(\Gamma_{\{N_\alpha\}})$. The basic premise of information theory is that the least biased estimates of these functions are those that maximize the functional Eq. (3.6), *subject* to constraints reflecting the observer's information or presumptions about the basic thermodynamic observables. This principle of the theory appears to correspond to the Clausius formulation of the second law of thermodynamics [28] according to which all microscopic processes lead naturally to the largest value of the entropy that is consistent with the given physical constraints upon the system. In this context, the distribution functions $\psi_{eq}(\Gamma_{\{N_\alpha\}})$ associated with a state of *thermodynamic equilibrium*, are those corresponding to a situation for which the number of constraints is the least possible. This minimal set of constraints consists of the specification of the average number of particles of each kind and of the average total energy of the system.

More restrictive constraints lead to nonequilibrium, constrained Gibbs ensembles. An example of this is provided here where we shall adopt the nonminimal set of constraints [Eq. (3.4)] to treat systems with prescribed compositional and energetic fields that differ from those characteristic of equilibrium. The corresponding nonequilibrium distribution functions are obtained by using a variational calculus generalization of the conventional Lagrange method of undetermined multipliers [29]. Thus, we seek the unconstrained extremum of the auxiliary functional

$$\mathcal{S} = S - k_B \int_\Omega d\mathbf{r} \left[\beta(\mathbf{r})\varepsilon(\mathbf{r}) + \sum_{\alpha=1}^{c} \nu_\alpha(\mathbf{r}) n_\alpha(\mathbf{r}) \right] \tag{3.7}$$

wherein $\beta(\mathbf{r})$ and $\{\nu_\alpha(\mathbf{r}); \alpha = 1, 2, \ldots, c\}$ are fields that play the roles of Lagrange multipliers associated with $\varepsilon(\mathbf{r})$ and $\{n_\alpha(\mathbf{r})\}$, respectively. According to Eq. (3.4) these, in turn, are linear functionals of $\psi(\Gamma_{\{N_\alpha\}})$.

The nonequilibrium, constrained Gibbs distribution resulting from this extremalization procedure may be written in the form

$$\psi(\Gamma_{\{N_\alpha\}}) = \Xi^{-1}[\Omega; \{\beta, \boldsymbol{\nu}\}] \exp\left[-\int_\Omega d\mathbf{r} \left\{ \beta(\mathbf{r})\hat{\varepsilon}(\mathbf{r}) + \sum_{\alpha=1}^{c} \nu_\alpha(\mathbf{r}) \hat{n}_\alpha(\mathbf{r}) \right\} \right] \tag{3.8}$$

with

$$\Xi[\Omega; \{\beta, \boldsymbol{\nu}\}] = \sum_{\{N_\alpha\}} \prod_{\alpha=1}^{c} \left(\frac{1}{h^{3N_\alpha} N_\alpha!}\right) \int d\Gamma_{\{N_\alpha\}} \times \exp\left[-\int_\Omega d\mathbf{r}\left\{\beta(\mathbf{r})\hat{\varepsilon}(\mathbf{r}) + \sum_{\alpha=1}^{c} \nu_\alpha(\mathbf{r})\hat{n}_\alpha(\mathbf{r})\right\}\right] \tag{3.9}$$

identified as the grand-canonical Gibbs partition function of the constrained ensemble; it is a functional of the Lagrange multiplier fields $\beta(\mathbf{r})$ and $\{\nu_\alpha(\mathbf{r})\}$.

It is important to note that the conventional *equilibrium* Gibbs grand canonical distribution can be identified with the special case of the constrained, *nonequilibrium* distribution [Eq. (3.8)] corresponding to spatially uniform Lagrange multiplier fields, that is, to the special case where

$$\begin{aligned} \beta(\mathbf{r}) &\to \beta_{\mathrm{eq}} = 1/k_{\mathrm{B}} T_{\mathrm{eq}} \\ \nu_\alpha(\mathbf{r}) &\to \nu_{\alpha,\mathrm{eq}} = -\beta_{\mathrm{eq}} \mu_{\alpha,\mathrm{eq}} \end{aligned} \tag{3.10}$$

Under these conditions $\ln \Xi$ can be identified with $\beta_{\mathrm{eq}} p \Omega$.

It follows directly from the definition [Eq. (3.9)] of the nonequilibrium grand partition function that

$$\left.\frac{\delta \ln \Xi[\Omega; \{\beta, \boldsymbol{\nu}\}]}{\delta \beta(\mathbf{r})}\right|_{\boldsymbol{\nu}} = -\varepsilon(\mathbf{r}) \tag{3.11}$$

and

$$\left.\frac{\delta \ln \Xi[\Omega; \{\beta, \boldsymbol{\nu}\}]}{\delta \nu_\alpha(\mathbf{r})}\right|_{\beta, \boldsymbol{\nu}'_\alpha} = -n_\alpha(\mathbf{r}) \tag{3.12}$$

These relationships insure a 1:1 correspondence between the fields $\varepsilon(\mathbf{r})$, $\{n_\alpha(\mathbf{r})\}$ and the corresponding Lagrange multipliers. In principle, the latter can be expressed as functionals of the prescribed energy and particle density profiles associated with the constraints [Eq. (3.4)]. Attention is drawn to the close resemblance between these last two equations and the thermodynamic field theory formulas Eqs. (2.42) and (2.43) in which the Massieu–Planck function Φ appears in place of $\ln \Xi$. This suggests that we proceed a step further by introducing a function

$\psi(\mathbf{r})$ so defined that

$$\ln \Xi[\Omega; \{\beta, \boldsymbol{\nu}\}] = \int_{\Omega} d\mathbf{r}\, \psi(\mathbf{r}) \tag{3.13}$$

This "density of ln Ξ" is the statistical counterpart of the thermodynamic field $\phi(\mathbf{r})$ of Eq. (2.38).

By inserting the distribution function of Eq. (3.8) into the entropy functional defined by Eq. (3.6) and invoking the relationship (3.13) one immediately concludes that

$$S = \int_{\Omega} d\mathbf{r}\, \sigma(\mathbf{r}) \tag{3.14}$$

with

$$\sigma(\mathbf{r}) = k_{\mathrm{B}}\left[\beta(\mathbf{r})\varepsilon(\mathbf{r}) + \sum_{\alpha=1}^{c} \nu_{\alpha}(\mathbf{r}) n_{\alpha}(\mathbf{r}) + \psi(\mathbf{r})\right] \tag{3.15}$$

The second of these equations would be formally identical with the previously obtained thermodynamic relation [Eq. (2.46)] if $\psi(\mathbf{r})$ were replaced with the thermodynamic field $\phi(\mathbf{r})$.

Next, by combining Eqs. (3.11)–(3.15) we obtain the following expression for the first-order functional derivatives of the entropy:

$$\left.\frac{\delta S}{\delta \varepsilon(\mathbf{r})}\right|_{\mathbf{n}} = k_{\mathrm{B}}\beta(\mathbf{r}) \tag{3.16}$$

$$\left.\frac{\delta S}{\delta n_{\alpha}(\mathbf{r})}\right|_{\beta, \mathbf{n}'_{\alpha}} = k_{\mathrm{B}}\nu_{\alpha}(\mathbf{r}) \tag{3.17}$$

These, in turn, permit us to write the variation of entropy and its density as

$$\delta S = \int_{\Omega} d\mathbf{r}\, \delta s(\mathbf{r}) \tag{3.18}$$

and

$$\delta s(\mathbf{r}) = k_{\mathrm{B}}\left[\beta(\mathbf{r})\delta\varepsilon(\mathbf{r}) + \sum_{\alpha=1}^{c} \nu_{\alpha}(\mathbf{r})\delta n_{\alpha}(\mathbf{r})\right] \tag{3.19}$$

respectively. The second of these is formally identical with Eq. (2.48) of the previous section. However, the quantities $\varepsilon(\mathbf{r})$ and $n_{\alpha}(\mathbf{r})$ appearing here are ensemble averages of microscopic fields, whereas those of Section II.C are phenomenological thermodynamic field variables. Furthermore, the function $\psi(\mathbf{r})$ of Eq. (3.13) has been defined here in terms of the generalized grand partition function $\Xi[\Omega; \{\beta, \boldsymbol{\nu}\}]$ while the counterpart Massieu–Planck function Φ of Section II.C and its density $\phi(\mathbf{r})$ are "typical" thermodynamic variables, in the sense that their functional dependencies must be determined experimentally.

In Section II.C it was argued that $\phi(\mathbf{r})$ must be a nonlocal functional of the fundamental thermodynamic variables β and $\boldsymbol{\nu}$ in order for the resulting thermodynamic field theory to be consistent with statistical mechanics. In the present context this requires a demonstration that $\psi(\mathbf{r})$, the statistical mechanical analog of $\psi(\mathbf{r})$, is a nonlocal functional of the corresponding Lagrange multiplier fields β and $\boldsymbol{\nu}$. To establish this we consider a system identical with that treated above except for the replacement of each pair potential function $u_{\alpha\beta}(r)$ with $\lambda u_{\alpha\beta}(r)$, where λ denotes a dimensionless parameter. Subscripts λ are then used to distinguish the properties of this modified system from those of the actual system, for which $\lambda = 1$. The microscopic energy density of this system, $\hat{\varepsilon}_\lambda(\mathbf{r})$, can be written as the sum of kinetic and external-field potential terms, which are independent of λ and a pair-interaction term $\lambda\hat{\varepsilon}_{\text{INT}}(\mathbf{r})$ with

$$\hat{\varepsilon}_{\text{INT}}(\mathbf{r}) = \frac{1}{2}\sum_{\alpha=1}^{c}\sum_{\beta=1}^{c}\int d\mathbf{r}'\, u_{\alpha\beta}(|\mathbf{r}-\mathbf{r}'|)\hat{n}_{\alpha\beta}(\mathbf{r},\mathbf{r}') \tag{3.20}$$

and

$$\hat{n}_{\alpha\beta}(\mathbf{r},\mathbf{r}') = \sum_{i=1}^{N_\alpha}\sum_{j=1}^{N_\beta}{}' \,\delta(\mathbf{r}-\mathbf{r}_{i\alpha})\delta(\mathbf{r}'-\mathbf{r}_{j\beta}) \tag{3.21}$$

It then readily can be verified that

$$\begin{aligned}\frac{\partial \ln \Xi_\lambda}{\partial\lambda} &= -\left\langle \frac{\partial}{\partial\lambda}\int_\Omega d\mathbf{r}\,\beta(\mathbf{r})\hat{\varepsilon}_\lambda(\mathbf{r})\right\rangle_\lambda \\ &= -\int_\Omega d\mathbf{r}\,\beta(\mathbf{r})\langle\hat{\varepsilon}_{\text{INT}}(\mathbf{r})\rangle_\lambda \end{aligned}\tag{3.22}$$

and

$$\ln \Xi_{\lambda=1} = \ln \Xi_{\lambda=0} - \int_\Omega d\mathbf{r}\,\beta(\mathbf{r})\int_0^1 d\lambda\,\langle\hat{\varepsilon}_{\text{INT}}(\mathbf{r})\rangle_\lambda \tag{3.23}$$

The quantity $\Xi_{\lambda=0}$ is the generalized grand partition function of an ideal gas, namely, a system devoid of particle interactions. It can be calculated without difficulty; the result is given by the formula

$$\Xi_{\lambda=0}[\Omega; \{\beta, \boldsymbol{\nu}\}] = \exp\left[\int_{\Omega} d\mathbf{r} \sum_{\alpha=1}^{c} n_{\alpha 0}(\mathbf{r})\right] \tag{3.24}$$

wherein

$$n_{\alpha 0}(\mathbf{r}) = \left\{\frac{e^{-\beta(\mathbf{r})v_\alpha(\mathbf{r})}}{\lambda_\alpha^3(\mathbf{r})}\right\} e^{-\nu_\alpha(\mathbf{r})} \tag{3.25}$$

with $\lambda_\alpha(\mathbf{r}) = h[\beta(\mathbf{r})/2\pi m_\alpha]^{1/2}$ is the number density of the ideal gas. Consequently, Eq. (3.23) can be rewritten as

$$\ln \Xi[\Omega; \{\beta, \boldsymbol{\nu}\}] = \int_{\Omega} d\mathbf{r} \sum_{\alpha=1}^{c} \left[n_{\alpha 0}(\mathbf{r}) - \frac{1}{2} \sum_{\beta=1}^{c} \int_{\Omega} d\mathbf{r}' \beta(\mathbf{r}) u_{\alpha\beta}(|\mathbf{r} - \mathbf{r}'|) \times \int_0^1 d\lambda\, n_{\alpha\beta}(\mathbf{r}, \mathbf{r}'; \lambda)\right] \tag{3.26}$$

where, since

$$\langle \hat{n}_{\alpha\beta}(\mathbf{r}, \mathbf{r}')\rangle_\lambda = \langle \delta\hat{n}_\alpha(\mathbf{r})\delta\hat{n}_\beta(\mathbf{r}')\rangle_\lambda + \langle \hat{n}_\alpha(\mathbf{r})\rangle_\lambda \langle \hat{n}_\beta(\mathbf{r}')\rangle_\lambda - \delta_{\alpha\beta}\delta(\mathbf{r} - \mathbf{r}')\langle \hat{n}_\alpha(\mathbf{r})\rangle_\lambda$$

it follows that

$$n_{\alpha\beta}(\mathbf{r}, \mathbf{r}'; \lambda) = \frac{\delta^2 \ln \Xi_\lambda}{\delta\nu_\alpha(\mathbf{r})\delta\nu_\beta(\mathbf{r}')} + \frac{\delta \ln \Xi_\lambda}{\delta\nu_\alpha(\mathbf{r})} \frac{\delta \ln \Xi_\lambda}{\delta\nu_\beta(\mathbf{r}')} + \delta_{\alpha\beta}\delta(\mathbf{r} - \mathbf{r}') \frac{\delta \ln \Xi_\lambda}{\delta\nu_\alpha(\mathbf{r})} \tag{3.27}$$

The density $\Sigma_\alpha n_{\alpha 0}(\mathbf{r})$ of $\ln \Xi_{\lambda=0}$ is clearly a local functional of β and $\boldsymbol{\nu}$ but the density of $\ln \Xi_{\lambda=1} = \ln \Xi$ is not; the pair distribution function $n_{\alpha\beta}(\mathbf{r}, \mathbf{r}'; \lambda)$, defined as the ensemble average of the many-body variable $\hat{n}_{\alpha\beta}(\mathbf{r}, \mathbf{r}')$, depends in principle on $\beta(\mathbf{r}'')$ and $\{\nu_\alpha(\mathbf{r}''); \alpha = 1, \ldots, c\}$ for all $\mathbf{r}'' \in \Omega$. However, in reality, this range is limited to points $\mathbf{r}''$ lying within microscopic correlation lengths from $\mathbf{r}$ and $\mathbf{r}'$. Thus, while the "degree of nonlocality" certainly is limited, it nevertheless is finite. This nonlocality is a direct consequence of the particle interactions; it vanishes in the ideal gas limit of $\lambda \rightarrow 0$.

The integrand appearing on the right-hand side of Eq. (3.26) can be identified with $\psi(\mathbf{r})$, the density of $\ln \Xi$ defined by Eq. (3.13). However, this definition is not unique since, for example, the integrand factor of $\beta(\mathbf{r})$ can be replaced by $\beta(\mathbf{r}')$ without affecting the validity of Eq. (3.26).

We later conduct (in Section III.D) a more thorough examination of the quantity $\psi(\mathbf{r})$ and its connection to the thermodynamic pressure. Our sole objective here has been to establish that $\psi(\mathbf{r})$ is indeed a nonlocal functional of the Lagrange multiplier fields β and $\boldsymbol{\nu}$.

The fields encountered thus far in the development can be divided into two categories; mechanical and thermal. By the term "mechanical" we refer to fields, such as the densities of energy and particles, which can be defined as ensemble averages of many-body microscopic variables, such as $\hat{\varepsilon}(\mathbf{r})$ and $\hat{n}_\alpha(\mathbf{r})$. Any "thermal" characteristics of the average fields must originate from the distribution function of the phase space variables, $\Gamma_{\{N_\alpha\}}$. Among the thermal fields we include objects such as the Lagrange multipliers and the entropy, for which there are no immediately obvious microscopic counterparts. These thermal variables arise either as parameters of the distribution function or, as in the case of entropy [see Eq. (3.6)], are defined in terms of the distribution function. Thus, thermal fields are related not so much to the mechanical properties of the system as they are to its statistics.

It is both natural and convenient to adopt the definition

$$\delta\hat{a}(\mathbf{r}) = \hat{a}(\mathbf{r}) - \langle \hat{a}(\mathbf{r}) \rangle_{\text{ref}} = \hat{a}(\mathbf{r}) - a_{\text{ref}}(\mathbf{r}) \tag{3.28}$$

for the microscopic fluctuation of an arbitrary thermodynamic field, $a(\mathbf{r})$, to which $\hat{a}(\mathbf{r})$ is the corresponding many-body mechanical variable. In particular, the microscopic fluctuations of the particle and energy densities are

$$\begin{aligned} \delta\hat{n}_\alpha(\mathbf{r}) &= \hat{n}_\alpha(\mathbf{r}) - n_{\alpha,\text{ref}}(\mathbf{r}) \\ \delta\hat{\varepsilon}(\mathbf{r}) &= \hat{\varepsilon}(\mathbf{r}) - \varepsilon_{\text{ref}}(\mathbf{r}) \end{aligned} \tag{3.29}$$

The subscripts symbols "ref" appearing in these formulas refer to a *reference ensemble* which may, but need not, correspond to a macroscopic state of thermodynamic equilibrium. In accordance with the notation used in the preceding thermodynamic theory, we rarely shall retain these subscripts, relying instead on an implicit identification of field variables with the reference state.

The variations $\{\delta n_\alpha(\mathbf{r}) = \langle \delta\hat{n}_\alpha(\mathbf{r}) \rangle\}$ and $\delta\varepsilon(\mathbf{r}) = \langle \delta\hat{\varepsilon}(\mathbf{r}) \rangle$ are then to be interpreted as the average values of the microscopic fields $\{\delta\hat{n}_\alpha(\mathbf{r})\}$ and $\delta\hat{\varepsilon}(\mathbf{r})$ associated with a nonequilibrium Gibbs distribution. The basic assumption of thermodynamic fluctuation theory is that every instantaneous profile of energy and/or of the particle densities also can be interpreted as an average profile associated with some constrained grand canonical ensemble. This assumption is similar in spirit but wider in scope than the "instantaneous entropy hypothesis" used in the Greene–Callen theory of global thermodynamic fluctuations [30]. Later in our development we shall also encounter a close analog of the Onsager regression hypothesis, according to which (in the present context) there is to every variation $\delta a(\mathbf{r})$ of a mechanical *or* thermal field a corresponding microscopic field $\delta \hat{a}(\mathbf{r})$, which can be represented as a linear combination of well-defined *mechanical* fields.

B. Fluctuations of Mechanical Fields

To each microscopic *mechanical* field $\hat{a}(\mathbf{r})$ there is a corresponding thermodynamic field

$$a(\mathbf{r}) = \langle \hat{a}(\mathbf{r}) \rangle \tag{3.30}$$

averaged over the constrained Gibbs grand canonical ensemble. This average is a functional of the Lagrange multiplier fields $\beta(\mathbf{r})$ and $\{\nu_\alpha(\mathbf{r})\}$. Its first-order total variation is given by the expression

$$\delta a(\mathbf{r}) = \int_\Omega d\mathbf{r}' \left\{ \frac{\delta a(\mathbf{r})}{\delta \beta(\mathbf{r}')}\bigg|_{\boldsymbol{\nu}} \delta\beta(\mathbf{r}') + \sum_{\alpha=1}^{c} \frac{\delta a(\mathbf{r})}{\delta \nu_\alpha(\mathbf{r}')}\bigg|_{\beta,\boldsymbol{\nu}'_\alpha} \delta\nu_\alpha(\mathbf{r}') \right\} \tag{3.31}$$

Thus, $\delta a(\mathbf{r})$ is a linear functional of the fields $\delta\beta(\mathbf{r})$ and $\{\delta\nu_\alpha(\mathbf{r})\}$. We again call attention to the fact that the functional derivatives, such as those appearing in Eq. (3.31), depend on the fields $\beta(\mathbf{r})$ and $\{\nu_\alpha(\mathbf{r})\}$ of a reference ensemble. By using the relationships

$$\frac{\delta \ln \psi(\Gamma_{\{N_\alpha\}})}{\delta \beta(\mathbf{r})} = -\,\delta\hat{\varepsilon}(\mathbf{r}) \tag{3.32}$$

$$\frac{\delta \ln \psi(\Gamma_{\{N_\alpha\}})}{\delta \nu_\alpha(\mathbf{r})} = -\,\delta\hat{n}_\alpha(\mathbf{r}) \tag{3.33}$$

these derivatives can be expressed as correlation functions of appropriate microscopic fields, namely,

$$\frac{\delta a(\mathbf{r})}{\delta \beta(\mathbf{r}')}\bigg|_{\boldsymbol{\nu}} = -\,\langle \delta\hat{a}(\mathbf{r})\delta\hat{\varepsilon}(\mathbf{r}')\rangle_{\text{ref}} \tag{3.34}$$

$$\frac{\delta a(\mathbf{r})}{\delta \nu_\alpha(\mathbf{r}')}\bigg|_{\beta,\boldsymbol{\nu}'_\alpha} = -\,\langle \delta\hat{a}(\mathbf{r})\delta\hat{n}_\alpha(\mathbf{r}')\rangle_{\text{ref}} \tag{3.35}$$

with $\delta\hat{a}(\mathbf{r})$ defined by Eq. (3.28).

The Eq. (3.31) also can be expressed as a functional of the variations of temperature and of the chemical potentials. Thus, by substituting the relationships (2.68) and (2.71) into Eq. (3.31), we find that

$$\delta a(\mathbf{r}) = \int_\Omega d\mathbf{r}' \left\{ \frac{\delta a(\mathbf{r})}{\delta T(\mathbf{r}')}\bigg|_{\boldsymbol{\mu}} \delta T(\mathbf{r}') + \sum_{\alpha=1}^{c} \frac{\delta a(\mathbf{r})}{\delta \mu_\alpha(\mathbf{r}')}\bigg|_{T,\boldsymbol{\mu}'_\alpha} \delta\mu_\alpha(\mathbf{r}') \right\} \tag{3.36}$$

The functional derivatives occurring here can be written as the correlation functions

$$\frac{\delta a(\mathbf{r})}{\delta T(\mathbf{r}')}\bigg|_{\boldsymbol{\mu}} = \frac{1}{k_B T(\mathbf{r}')} \langle \delta\hat{a}(\mathbf{r})\delta\hat{s}(\mathbf{r}')\rangle_{\text{ref}} \tag{3.37}$$

$$\frac{\delta a(\mathbf{r})}{\delta \mu_\alpha(\mathbf{r}')}\bigg|_{T,\boldsymbol{\mu}'_\alpha} = \frac{1}{k_B T(\mathbf{r}')} \langle \delta\hat{a}(\mathbf{r})\delta\hat{n}(\mathbf{r}')\rangle_{\text{ref}} \tag{3.38}$$

with

$$\delta\hat{s}(\mathbf{r}) = \frac{1}{T(\mathbf{r})} \left\{ \delta\hat{\varepsilon}(\mathbf{r}) - \sum_{\alpha=1}^{c} \mu_\alpha(\mathbf{r})\delta\hat{n}_\alpha(\mathbf{r}) \right\} \tag{3.39}$$

identified as the microscopic fluctuation of entropy, corresponding to the averaged expression [Eq. (3.19)]. Indeed, $\delta\hat{s}(\mathbf{r})$ as given by Eq. (3.39) is the first of the thermal fields for which we now have found a representation in terms of micromechanical variables (and quantities characteristic of the reference state).

The pairs of equations (3.34), (3.35) and (3.37), (3.38) provide the basic connections among the functional derivatives of thermodynamic fields and reference state correlation functions of related microscopic fluctuations. In particular, with $a(\mathbf{r})$ identified first with $n_\alpha(\mathbf{r})$ and then with $\varepsilon(\mathbf{r})$ we obtain the formulas

$$-\frac{\delta n_\alpha(\mathbf{r})}{\delta \nu_\gamma(\mathbf{r}')}\bigg|_{\beta,\boldsymbol{\nu}'_\gamma} = k_B T(\mathbf{r}') \frac{\delta n_\alpha(\mathbf{r})}{\delta \mu_\gamma(\mathbf{r}')}\bigg|_{T,\boldsymbol{\mu}'_\gamma} = \langle \delta\hat{n}_\alpha(\mathbf{r})\delta\hat{n}_\gamma(\mathbf{r}')\rangle_{\text{ref}} \equiv G_{\alpha\gamma}(\mathbf{r},\mathbf{r}') \tag{3.40}$$

and

$$-\frac{\delta \varepsilon(\mathbf{r})}{\delta \beta(\mathbf{r}')}\bigg|_{\boldsymbol{\nu}} = k_B T^2(\mathbf{r}') \left\{ \frac{\delta \varepsilon(\mathbf{r})}{\delta T(\mathbf{r}')}\bigg|_{\boldsymbol{\mu}} + \sum_{\gamma=1}^{c} \frac{\delta \varepsilon(\mathbf{r})}{\delta \mu_\gamma(\mathbf{r}')}\bigg|_{T,\boldsymbol{\mu}'_\gamma} \frac{\mu_\gamma(\mathbf{r}')}{T(\mathbf{r}')} \right\}$$
$$= \langle \delta\hat{\varepsilon}(\mathbf{r})\delta\hat{\varepsilon}(\mathbf{r}')\rangle_{\text{ref}} \equiv G_{00}(\mathbf{r},\mathbf{r}') \tag{3.41}$$

The first of these can be expressed in terms of the irreducible pair correlation function, $h_{\alpha\beta}(\mathbf{r}, \mathbf{r}') = g_{\alpha\beta}(\mathbf{r}, \mathbf{r}') - 1$, with $g_{\alpha\beta}(\mathbf{r}, \mathbf{r}')$ denoting the generalization of the radial distribution function appropriate to a nonuniform system:

$$G_{\alpha\beta}(\mathbf{r}, \mathbf{r}') = n_\alpha(\mathbf{r}) n_\beta(\mathbf{r}') h_{\alpha\beta}(\mathbf{r}, \mathbf{r}') + n_\alpha(\mathbf{r}) \delta_{\alpha\beta} \delta(\mathbf{r} - \mathbf{r}') \tag{3.42}$$

The formulas (3.40) and (3.41) are local fluctuation theory counterparts of the relationships

$$\langle \delta N_\alpha \delta N_\beta \rangle_{\mathrm{eq}} = k_{\mathrm{B}} T_{\mathrm{eq}} \left. \frac{\partial N_\alpha}{\partial \mu_\beta} \right|_{T, \Omega, \boldsymbol{\mu}'_\beta} \tag{3.43}$$

and

$$\langle (\delta E)^2 \rangle_{\mathrm{eq}} = k_{\mathrm{B}} T^2_{\mathrm{eq}} \left. \frac{\partial E}{\partial T} \right|_{\{\boldsymbol{\mu}/T\}, \Omega} \tag{3.44}$$

familiar from the classical fluctuation theory of extensive mechanical variables [6].

To complete the set of correlation functions $G_{\alpha\beta}$ and G_{00} we include

$$- \left. \frac{\delta n_\alpha(\mathbf{r})}{\delta \beta(\mathbf{r}')} \right|_{\boldsymbol{\nu}} = \langle \delta \hat{n}_\alpha(\mathbf{r}) \delta \hat{\varepsilon}(\mathbf{r}') \rangle_{\mathrm{ref}} \equiv G_{\alpha 0}(\mathbf{r}, \mathbf{r}') \tag{3.45}$$

$$- \left. \frac{\delta \varepsilon(\mathbf{r})}{\delta n_\alpha(\mathbf{r}')} \right|_{\beta, n'_\alpha} = \langle \delta \hat{\varepsilon}(\mathbf{r}) \delta \hat{n}_\alpha(\mathbf{r}') \rangle_{\mathrm{ref}} \equiv G_{0\alpha}(\mathbf{r}, \mathbf{r}') \tag{3.46}$$

the latter of which also can be identified with $G_{\alpha 0}(\mathbf{r}', \mathbf{r})$. Here and in Eq. (3.41) the index $\alpha = 0$ is used to indicate the energy and all others, $\alpha > 0$, to label components of the fluid mixture.

By combining the relationships (3.40), (3.41), (3.45), and (3.46) with Eq. (3.31) we obtain a set of constitutive equations

$$- \langle \delta \hat{\varepsilon}(\mathbf{r}) \rangle = - \delta \varepsilon(\mathbf{r}) = \int_\Omega d\mathbf{r}' \left[G_{00}(\mathbf{r}, \mathbf{r}') \delta \beta(\mathbf{r}') + \sum_{\gamma=1}^{c} G_{0\gamma}(\mathbf{r}, \mathbf{r}') \delta \nu_\gamma(\mathbf{r}') \right] \tag{3.47}$$

$$- \langle \delta \hat{n}_\alpha(\mathbf{r}) \rangle = - \delta n_\alpha(\mathbf{r}) = \int_\Omega d\mathbf{r}' \left[G_{\alpha 0}(\mathbf{r}, \mathbf{r}') \delta \beta(\mathbf{r}') + \sum_{\gamma=1}^{c} G_{\alpha\gamma}(\mathbf{r}, \mathbf{r}') \delta \nu_\gamma(\mathbf{r}') \right] \tag{3.48}$$

connecting the variations of the mechanical fields $\delta\varepsilon(\mathbf{r})$ and $\{\delta n_\alpha(\mathbf{r})\}$ with variations of the Lagrange multiplier fields $\beta(\mathbf{r})$ and $\{\delta\nu_\alpha(\mathbf{r})\}$. These equations are, of course, completely analogous to the thermodynamic field theory equations (2.72) and (2.73) that contain in place of the correlation integrals $G_{ij}(\mathbf{r}, \mathbf{r}')$ the corresponding second-order functional derivatives $\mathcal{G}_{ij}(\mathbf{r}, \mathbf{r}')$, of the Massieu–Planck function. This illustrates the utility of the statistical mechanical theory that identifies thermodynamic quantities that are both difficult to measure and difficult to interpret with physically significant correlation functions that are susceptible to microscopic theoretical analysis and to computer simulations as well.

Corresponding to Eqs. (3.47) and (3.48) is a similar set of relations connecting the variations of the particle densities and entropy density [defined as the average of Eq. (3.39)] to variations of the temperature and chemical potentials, as defined by Eqs. (2.68) and (2.71). These relations are identical in form to the previously derived expressions (2.77) and (2.78) but contain in place of the thermodynamic quantities $\mathcal{K}_{ij}(\mathbf{r}, \mathbf{r}')$, defined by Eqs. (2.79)–(2.82), the corresponding correlation functions

$$
\begin{aligned}
K_{00}(\mathbf{r}, \mathbf{r}') &= \langle \delta\hat{s}(\mathbf{r})\delta\hat{s}(\mathbf{r}')\rangle_{\text{ref}} \\
K_{\alpha 0}(\mathbf{r}, \mathbf{r}') &= \langle \delta\hat{n}_\alpha(\mathbf{r})\delta\hat{s}(\mathbf{r}')\rangle_{\text{ref}} \\
K_{0\alpha}(\mathbf{r}, \mathbf{r}') &= \langle \delta\hat{s}(\mathbf{r})\delta\hat{n}_\alpha(\mathbf{r}')\rangle_{\text{ref}} \\
K_{\alpha\beta}(\mathbf{r}, \mathbf{r}') &= \langle \delta\hat{n}_\alpha(\mathbf{r})\delta\hat{n}_\beta(\mathbf{r}')\rangle_{\text{ref}}
\end{aligned}
\tag{3.49}
$$

To pursue the connection between the thermodynamic and statistical mechanical theories further we now introduce inverse elements $G^{-1}_{\alpha\beta}(\mathbf{r}, \mathbf{r}')$ through the defining equations [cf. Eq. (2.83)]

$$
\sum_{\gamma=1}^{c} \int_\Omega d\mathbf{r}'' G^{-1}_{\alpha\gamma}(\mathbf{r}, \mathbf{r}'') G_{\gamma\beta}(\mathbf{r}'', \mathbf{r}') = \delta_{\alpha\beta}\delta(\mathbf{r} - \mathbf{r}') \tag{3.50}
$$

What distinguishes the situation here from the earlier thermodynamic analysis is that the van Hove functions $G_{\alpha\beta}(\mathbf{r}, \mathbf{r}')$ are familiar, well-understood quantities that are related by Eq. (3.42) to the functions $h_{\alpha\beta}(\mathbf{r}, \mathbf{r}')$. These, in turn, have been subjects of innumerable theoretical studies and computer simulations.

By combining Eqs. (3.42) with (3.50) one finds that the inverse functions $G^{-1}_{\alpha\beta}(\mathbf{r}, \mathbf{r}')$ are related to the direct correlation functions $C_{\alpha\beta}(\mathbf{r}, \mathbf{r}')$ [12] by the formulas

$$
G^{-1}_{\alpha\beta}(\mathbf{r}, \mathbf{r}') = \frac{1}{n_\alpha(\mathbf{r})}\delta_{\alpha\beta}\delta(\mathbf{r} - \mathbf{r}') - C_{\alpha\beta}(\mathbf{r}, \mathbf{r}') \tag{3.51}
$$

and that the latter are connected to the irreducible pair correlation functions $h_{\alpha\beta}(\mathbf{r}, \mathbf{r}')$ by the Ornstein–Zernike equations appropriate to an inhomogeneous system. Using these inverse functions one again obtains the relationship (2.84), together with the statistical mechanical counterparts of the thermodynamic relations (2.85)–(2.89), in each of which the quantities G_{ij} $(i, j = 0, 1, \ldots, c)$ and $G^{-1}_{\alpha\gamma}$ appear in place of the corresponding $\mathcal{G}_{ij}$ and $\mathcal{G}^{-1}_{\alpha\gamma}$. From these statistical mechanical analogs of Eqs. (2.84)–(2.89) and Eq. (3.31) we then obtain the "mixed basis" representation for $\delta a(\mathbf{r})$, namely,

$$\delta a(\mathbf{r}) = \int_\Omega d\mathbf{r}' \left\{ \frac{\delta a(\mathbf{r})}{\delta\beta(\mathbf{r}')}\bigg|_{\mathbf{n}} \delta\beta(\mathbf{r}') + \sum_{\gamma=1}^{c} \frac{\delta a(\mathbf{r})}{\delta n_\gamma(\mathbf{r}')}\bigg|_{\beta,\mathbf{n}'_\gamma} \delta n_\gamma(\mathbf{r}') \right\} \tag{3.52}$$

with

$$\frac{\delta a(\mathbf{r})}{\delta\beta(\mathbf{r}')}\bigg|_{\mathbf{n}} = -\langle \delta\hat{a}(\mathbf{r})[1 - P_n]\delta\hat{\varepsilon}(\mathbf{r}')\rangle_{\text{ref}} \tag{3.53}$$

and

$$\frac{\delta a(\mathbf{r})}{\delta n_\alpha(\mathbf{r}')}\bigg|_{\beta,\mathbf{n}'_\alpha} = \sum_{\gamma=1}^{c} \int_\Omega d\mathbf{r}'' \langle \delta\hat{a}(\mathbf{r})\delta\hat{n}_\gamma(\mathbf{r}'')\rangle_{\text{ref}} G^{-1}_{\gamma\alpha}(\mathbf{r}'', \mathbf{r}') \tag{3.54}$$

The symbol P_n appearing in Eq. (3.53) is the projection operator onto the subspace spanned by the microscopic fluctuations of the particle density fields. Its action on an arbitrary many-body function $\hat{X}$ is defined by the formula

$$P_n\hat{X} = \sum_{\alpha=1}^{c} \sum_{\beta=1}^{c} \int_\Omega d\mathbf{r}' \int_\Omega d\mathbf{r}'' \delta\hat{n}_\alpha(\mathbf{r}') G^{-1}_{\alpha\beta}(\mathbf{r}', \mathbf{r}'') \langle \delta\hat{n}_\beta(\mathbf{r}'')\hat{X}\rangle_{\text{ref}} \tag{3.55}$$

With $a(\mathbf{r})$ chosen equal to $\varepsilon(\mathbf{r})$, Eq. (3.53) yields the following formulas for the generalized heat capacity at constant volume,

$$\begin{aligned} C_v(\mathbf{r}, \mathbf{r}') &= \frac{\delta\varepsilon(\mathbf{r})}{\delta T(\mathbf{r}')}\bigg|_{\mathbf{n}} = \frac{1}{k_B T^2(\mathbf{r}')} \langle \delta\hat{\varepsilon}(\mathbf{r})[1 - P_n]\delta\hat{\varepsilon}(\mathbf{r}')\rangle_{\text{ref}} \\ &= \frac{1}{k_B} \frac{T(\mathbf{r})}{T(\mathbf{r}')} \langle \delta\hat{s}(\mathbf{r})[1 - P_n]\delta\hat{s}(\mathbf{r}')\rangle_{\text{ref}} \end{aligned} \tag{3.56}$$

C. Fluctuations of Thermal Fields

Neither the Lagrange multiplier fields $\beta(\mathbf{r})$ and $\{\nu_\alpha(\mathbf{r})\}$ nor such closely related thermal variables as the temperature, chemical potentials, and entropy density have direct mechanical analogs. In particular, within the context of generalized ensemble theory, the Lagrange multiplier fields are parameters of the distribution function that enter the theory as nonrandom functions of location. However, one might anticipate that local variations of these fields, produced by fluctuations of the particle and energy density fields, could be expressed as linear functionals of the spatial variations of these basic mechanical fields. In fact, one example of this is the microscopic entropy fluctuation field $\delta\hat{s}(\mathbf{r})$, which we already have succeeded in representing as a linear combination of the microscopic fluctuations of the energy and particle densities. Our immediate objective here is to identify analogous microscopic mechanical representations of the Lagrange multiplier fields. The way to accomplish this is illustrated by the procedure we used in Section III.A to obtain the formula (3.39) for $\delta\hat{s}(\mathbf{r})$ from Eq. (3.19). Thus, with $\delta\varepsilon(\mathbf{r})$ and $\delta n_\alpha(\mathbf{r})$ identified as ensemble averages of the microscopic variables $\delta\hat{\varepsilon}(\mathbf{r})$ and $\delta\hat{n}_\alpha(\mathbf{r})$, respectively, we simply defined $\delta\hat{s}(\mathbf{r})$ as that linear combination of $\delta\hat{\varepsilon}(\mathbf{r})$ and the $\delta\hat{n}_\alpha(\mathbf{r})$ values, which produces the field equation (3.39) when averaged. Proceeding in this same way we can replace each of the basic thermodynamic field relations (2.65)–(2.73) with the corresponding linear connection among variations of the local microscopic fields. A few manipulations are then required to solve these equations for the individual microscopic field variations in terms of the variables $\delta\hat{\varepsilon}$ and $\delta\hat{n}_\alpha$. Two examples of this will now be presented.

When δa in Eq. (3.52) is chosen equal to $\delta\varepsilon$ and use is made of Eqs. (3.53)–(3.55), we obtain an expression,

$$\delta\varepsilon(\mathbf{r}) = \int_\Omega d\mathbf{r}'\left\{\frac{\delta\varepsilon(\mathbf{r})}{\delta\beta(\mathbf{r}')}\bigg|_{\mathbf{n}}\delta\beta(\mathbf{r}') + \sum_{\alpha=1}^{c}\frac{\delta\varepsilon(\mathbf{r})}{\delta n_\alpha(\mathbf{r}')}\bigg|_{\beta,\mathbf{n}'_\alpha}\delta n_\alpha(\mathbf{r}')\right\}$$

$$= \int_\Omega d\mathbf{r}'\Big\{-k_{\mathrm{B}}T^2(\mathbf{r}')C_v(\mathbf{r},\mathbf{r}')\delta\beta(\mathbf{r}')$$

$$+ \sum_{\alpha=1}^{c}\sum_{\beta=1}^{c}\int_\Omega d\mathbf{r}''\langle\delta\hat{\varepsilon}(\mathbf{r})\delta\hat{n}_\beta(\mathbf{r}'')\rangle G^{-1}_{\beta\alpha}(\mathbf{r}'',\mathbf{r}')\delta n_\alpha(\mathbf{r}')\Big\}$$

which, with the help of Eq. (3.55) and the identity $\delta\varepsilon(\mathbf{r}) = \langle\delta\hat{\varepsilon}(\mathbf{r})\rangle$, can be rewritten as

$$\langle\delta\hat{\varepsilon}(\mathbf{r})\rangle = -k_{\mathrm{B}}\int_\Omega d\mathbf{r}'T^2(\mathbf{r}')C_v(\mathbf{r},\mathbf{r}')\delta\beta(\mathbf{r}') + \langle P_N\delta\hat{\varepsilon}(\mathbf{r})\rangle$$

The result of acting upon this equation with the inverse of $C_v(\mathbf{r},\mathbf{r}')$, defined by

$$\int_{\Omega} d\mathbf{r}'' C_v^{-1}(\mathbf{r}, \mathbf{r}'') C_v(\mathbf{r}'', \mathbf{r}') = \delta(\mathbf{r} - \mathbf{r}') \tag{3.57}$$

is the formula

$$\delta\beta(\mathbf{r}) = -\frac{1}{k_B T^2(\mathbf{r})} \left\langle \int_{\Omega} d\mathbf{r}' C_v^{-1}(\mathbf{r}, \mathbf{r}')[1 - P_n]\delta\hat{\varepsilon}(\mathbf{r}') \right\rangle \tag{3.58}$$

This relationship enables us to identify a microscopic mechanical variable,

$$\delta\hat{\beta}(\mathbf{r}) = -\frac{1}{k_B T^2(\mathbf{r})} \int_{\Omega} d\mathbf{r}' C_v^{-1}(\mathbf{r}, \mathbf{r}')[1 - P_n]\delta\hat{\varepsilon}(\mathbf{r}') \tag{3.59}$$

the ensemble average of which is the variation $\delta\beta(\mathbf{r})$ of a thermal field!

A similar analysis, based on the substitution of Eq. (3.58) into Eq. (2.84), leads to the conclusion that

$$\delta\hat{\nu}_\alpha(\mathbf{r}) = -\sum_{\gamma=1}^{c} \int_{\Omega} d\mathbf{r}' G_{\alpha\gamma}^{-1}(\mathbf{r}, \mathbf{r}')[1 - P_\varepsilon]\delta\hat{n}_\gamma(\mathbf{r}') \tag{3.60}$$

with

$$P_\varepsilon \hat{X} = \int_{\Omega} d\mathbf{r}' \int_{\Omega} d\mathbf{r}'' \langle \delta\hat{\varepsilon}(\mathbf{r}')\hat{X}\rangle_{\text{ref}} \frac{C_v^{-1}(\mathbf{r}', \mathbf{r}'')}{k_B T^2(\mathbf{r}')} [1 - P_n]\delta\hat{\varepsilon}(\mathbf{r}'') \tag{3.61}$$

is a many-body mechanical variable, so defined that $\delta\nu_\alpha(\mathbf{r}) = \langle \delta\hat{\nu}_\alpha(\mathbf{r})\rangle$. This completes our search for microscopic mechanical fields associated with the Lagrange multipliers $\beta(\mathbf{r})$ and $\{\nu_\alpha(\mathbf{r})\}$. The way is now open for examining fluctuations of these and other thermal fields.

From Eqs. (3.59) and (3.60) it immediately follows that fluctuations of the fields $\hat{\beta}(\mathbf{r})$ and $\{\hat{\nu}_\alpha(\mathbf{r})\}$ are orthogonal to fluctuations of the particle and energy densities, respectively, that is

$$\langle \delta\hat{\beta}(\mathbf{r})\delta\hat{n}_\alpha(\mathbf{r}')\rangle_{\text{ref}} = 0 \tag{3.62}$$

and

$$\langle \delta\hat{\nu}_\alpha(\mathbf{r})\delta\hat{\varepsilon}(\mathbf{r}')\rangle_{\text{ref}} = 0 \tag{3.63}$$

Microscopic fluctuations of the temperature now can be defined, using

Eqs. (3.56) and (3.59), by the formula

$$\delta\hat{T}(\mathbf{r}) = \int_{\Omega} d\mathbf{r}' C_v^{-1}(\mathbf{r}, \mathbf{r}')\delta\hat{\varepsilon}_{\perp}(\mathbf{r}') \tag{3.64}$$

Here the field

$$\delta\hat{\varepsilon}_{\perp}(\mathbf{r}) = [1 - P_n]\delta\hat{\varepsilon}(\mathbf{r}) \tag{3.65}$$

can be identified as that part of the microscopic energy fluctuation $\delta\hat{\varepsilon}(\mathbf{r})$ that is orthogonal to the subspace spanned by the particle density fluctuations $\{\delta\hat{n}_\alpha(\mathbf{r})\}$. It is then easily verified [cf. Eqs. (3.56) and (3.65)] that

$$\langle \delta\hat{\varepsilon}_{\perp}(\mathbf{r})\delta\hat{\varepsilon}_{\perp}(\mathbf{r}')\rangle_{\text{ref}} = \langle \delta\hat{\varepsilon}(\mathbf{r})[1 - P_n]\delta\hat{\varepsilon}(\mathbf{r}')\rangle_{\text{ref}} = k_B T^2(\mathbf{r}')C_v(\mathbf{r}, \mathbf{r}') \tag{3.66}$$

A consequence of this last formula is the relationship

$$\langle \delta\hat{T}(\mathbf{r})\delta\hat{T}(\mathbf{r}')\rangle_{\text{ref}} = k_B T^2(\mathbf{r}')C_v^{-1}(\mathbf{r}', \mathbf{r}) \tag{3.67}$$

which can be recognized as the generalization to nonuniform systems of a well-known result of the classical theory of global equilibrium fluctuations [6].

The formula (3.64) is precisely the same as one proposed by Schofield [7, 8], who argued that the fluctuation of temperature should be identified with that part of the fluctuation of energy that is not caused by fluctuations of composition and/or density. The heat capacity formulas (3.56) and (3.66) likewise are identical with Schofield's findings.

Since it is now established that the fluctuations of temperature and particle densities are statistically independent [see Eq. (3.62)], these variables can be used as a complete set of basic thermodynamic fields. The thermodynamic "state space" is then spanned by a pair of orthogonal projections, P_n defined by Eq. (3.55) and P_T, the action of which is

$$P_T\hat{X} = \int_{\Omega} d\mathbf{r}' \int_{\Omega} d\mathbf{r}''\delta\hat{T}(\mathbf{r}') \frac{C_v(\mathbf{r}', \mathbf{r}'')}{k_B T^2(\mathbf{r}')} \langle \delta\hat{T}(\mathbf{r}'')\hat{X}\rangle_{\text{ref}} \tag{3.68}$$

The microscopic equilibrium fluctuations of an arbitrary field $\hat{a}(\mathbf{r})$ therefore can be expressed as linear combinations of mutually orthogonal temperature and particle density fluctuations:

$$\delta\hat{a}(\mathbf{r}) = [P_T + P_n]\delta\hat{a}(\mathbf{r})$$

$$= \int_\Omega d\mathbf{r}' \left\{ \frac{\delta a(\mathbf{r})}{\delta T(\mathbf{r}')}\bigg|_{\mathbf{n}} \delta\hat{T}(\mathbf{r}') + \sum_{\alpha=1}^{c} \frac{\delta a(\mathbf{r})}{\delta n_\alpha(\mathbf{r}')}\bigg|_{T,\mathbf{n}'_\alpha} \delta\hat{n}_\alpha(\mathbf{r}') \right\} \tag{3.69}$$

From Eqs. (3.59) and (3.60) it immediately follows that

$$\langle \delta\hat{\beta}(\mathbf{r})\delta\hat{\varepsilon}(\mathbf{r}')\rangle_{\text{ref}} = -\,\delta(\mathbf{r}-\mathbf{r}') \tag{3.70}$$

$$\langle \delta\hat{\nu}_\alpha(\mathbf{r})\delta\hat{n}_\beta(\mathbf{r}')\rangle_{\text{ref}} = -\,\delta_{\alpha\beta}\delta(\mathbf{r}-\mathbf{r}') \tag{3.71}$$

Thus, the values of these two correlation functions are independent from the material properties of the system. Together with the (statistical mechanical counterparts of the) relations (2.85) and (2.86), the expression (3.67), and the definition (3.57), the two formulas (3.70) and (3.71) imply that

$$\langle \delta\hat{\beta}(\mathbf{r})\delta\hat{\beta}(\mathbf{r}')\rangle_{\text{ref}} = \frac{1}{k_B T^2(\mathbf{r})} C_v^{-1}(\mathbf{r},\mathbf{r}') = -\frac{\delta\beta(\mathbf{r})}{\delta\varepsilon(\mathbf{r}')}\bigg|_{\mathbf{n}} \tag{3.72}$$

$$\langle \delta\hat{\nu}_\alpha(\mathbf{r})\delta\hat{\nu}_\beta(\mathbf{r}')\rangle_{\text{ref}} = G^{-1}_{\alpha\beta}(\mathbf{r},\mathbf{r}') = -\frac{\delta\nu_\alpha(\mathbf{r})}{\delta n_\beta(\mathbf{r}')}\bigg|_{\beta,\mathbf{n}'_\beta} \tag{3.73}$$

and

$$\langle \delta\hat{\nu}_\alpha(\mathbf{r})\delta\hat{\beta}(\mathbf{r}')\rangle_{\text{ref}} = \frac{1}{k_B T^2(\mathbf{r})} \int_\Omega d\mathbf{r}'' C_v^{-1}(\mathbf{r}',\mathbf{r}'') \frac{\delta\nu_\alpha(\mathbf{r})}{\delta\beta(\mathbf{r}'')}\bigg|_{\mathbf{n}}$$

$$= -\frac{\delta\nu_\alpha(\mathbf{r})}{\delta\varepsilon(\mathbf{r}')}\bigg|_{\mathbf{n}} = -\frac{\delta\beta(\mathbf{r})}{\delta n_\alpha(\mathbf{r}')}\bigg|_{\beta,\mathbf{n}'_\alpha} \tag{3.74}$$

To obtain yet another set of correlation functions we first combine Eq. (3.39) with Eq. (3.64) to produce the formula

$$\delta\hat{T}(\mathbf{r}) = \int_\Omega d\mathbf{r}' T(\mathbf{r}') C_v^{-1}(\mathbf{r},\mathbf{r}')(1-P_n)\delta\hat{s}(\mathbf{r}') \tag{3.75}$$

Next, by acting on Eq. (2.78) with the inverse of the correlation function $K_{\alpha\beta} = G_{\alpha\beta}$ we obtain a formula for $\delta\mu_\alpha(\mathbf{r})$ that can be identified with the ensemble average of the microscopic field

$$\delta\hat{\mu}_\alpha(\mathbf{r}) = k_\mathrm{B}T(\mathbf{r}) \sum_{\gamma=1}^{c} \int_\Omega d\mathbf{r}' G^{-1}_{\alpha\gamma}(\mathbf{r}, \mathbf{r}')(1 - P_s)\delta\hat{n}_\gamma(\mathbf{r}') \tag{3.76}$$

The quantity P_s appearing here is a projection operator so defined that

$$P_s\hat{X} = \frac{1}{k_\mathrm{B}} \int_\Omega d\mathbf{r}' \int_\Omega d\mathbf{r}'' \frac{T(\mathbf{r}'')}{T(\mathbf{r}')} C_v^{-1}(\mathbf{r}', \mathbf{r}'')\langle \delta\hat{s}(\mathbf{r}')\hat{X}\rangle_{\mathrm{ref}}(1 - P_n)\delta\hat{s}(\mathbf{r}'') \tag{3.77}$$

From the relationships (3.75) and (3.76) we then obtain the formulas

$$\langle \delta\hat{s}(\mathbf{r})\delta\hat{T}(\mathbf{r}')\rangle_{\mathrm{ref}} = k_\mathrm{B}T(\mathbf{r})\delta(\mathbf{r} - \mathbf{r}') \tag{3.78}$$

$$\langle \delta\hat{n}_\alpha(\mathbf{r})\delta\hat{\mu}_\beta(\mathbf{r}')\rangle_{\mathrm{ref}} = k_\mathrm{B}T(\mathbf{r})\delta_{\alpha\beta}\delta(\mathbf{r} - \mathbf{r}') \tag{3.79}$$

and

$$\langle \delta\hat{s}(\mathbf{r})\delta\hat{\mu}_\alpha(\mathbf{r}')\rangle_{\mathrm{ref}} = \langle \delta\hat{T}(\mathbf{r})\delta\hat{n}_\alpha(\mathbf{r}')\rangle_{\mathrm{ref}} = 0 \tag{3.80}$$

The one remaining thermal field to be considered is the statistical mechanical analog of the thermodynamic pressure, $p(\mathbf{r}) = \phi(\mathbf{r})/\beta(\mathbf{r})$. Equation (2.62) provides a linear connection between the fluctuation of this variable and the fluctuations of the thermodynamic temperature and chemical potentials. The statistical mechanical counterpart of Eq. (2.62) is identical in form, differing only because it is formulated in terms of $\psi(\mathbf{r})$, the density of $\ln \Xi$, instead of $\phi(\mathbf{r})$. By identifying the quantities $\delta T(\mathbf{r})$ and $\delta\mu_\alpha(\mathbf{r})$ in Eq. (2.62) with the ensemble averages of the "mechanical variables" $\delta\hat{T}(\mathbf{r})$ and $\delta\hat{\mu}_\alpha(\mathbf{r})$, defined by Eqs. (3.75) and (3.78), respectively, we arrive at the definition

$$\delta\hat{p}(\mathbf{r}) = \int_\Omega d\mathbf{r}' \left[\tilde{\sigma}(\mathbf{r}, \mathbf{r}')\delta\hat{T}(\mathbf{r}') + \sum_{\alpha=1}^{c} \tilde{n}_\alpha(\mathbf{r}, \mathbf{r}')\delta\hat{\mu}_\alpha(\mathbf{r}') \right] \tag{3.81}$$

for the microscopic fluctuation of pressure.

From this formula and the relationships [cf. (2.58)–(2.61)] $\sigma(\mathbf{r}) = \int_\Omega d\mathbf{r}'[T(\mathbf{r})/T(\mathbf{r}')]\tilde{\sigma}(\mathbf{r}', \mathbf{r})$ and $n_\alpha(\mathbf{r}) = \int_\Omega d\mathbf{r}'[T(\mathbf{r})/T(\mathbf{r}')]\tilde{n}_\alpha(\mathbf{r}', \mathbf{r})$ it follows that

$$\int_\Omega d\mathbf{r} \frac{1}{T(\mathbf{r})} \left[\sigma(\mathbf{r})\delta\hat{T}(\mathbf{r}) + \sum_{\alpha=1}^{c} n_\alpha(\mathbf{r})\delta\hat{\mu}_\alpha(r) - \delta\hat{p}(\mathbf{r}) \right] = 0 \tag{3.82}$$

We see that Eq. (3.82) is the microfield analog of the Gibbs–Duhem relation (2.47). Furthermore, the result of combining Eq. (3.82) with the microfield counterpart of Eq. (2.49) is the equality

$$\int_\Omega d\mathbf{r}\, \delta\hat{s}(\mathbf{r}) = \int_\Omega d\mathbf{r}\, \delta\hat{\sigma}(\mathbf{r}) \equiv \delta\hat{S} \tag{3.83}$$

Thus, while $\delta s(\mathbf{r}) \neq \delta\sigma(\mathbf{r})$ and $\delta\hat{s}(\mathbf{r}) \neq \delta\hat{\sigma}(\mathbf{r})$, the corresponding global variations of these two quantities are indeed equal.

The definition (3.81) and Eqs. (3.78)–(3.80) lead directly to the relationships

$$\langle \delta\hat{p}(\mathbf{r})\delta\hat{s}(\mathbf{r}')\rangle_{\text{ref}} = k_{\text{B}}T(\mathbf{r}')\tilde{\sigma}(\mathbf{r},\mathbf{r}') = k_{\text{B}}T(\mathbf{r}')\left.\frac{\delta p(\mathbf{r})}{\delta T(\mathbf{r}')}\right|_{\boldsymbol{\mu}} \tag{3.84}$$

and

$$\langle \delta\hat{p}(\mathbf{r})\delta\hat{n}_\alpha(\mathbf{r}')\rangle_{\text{ref}} = k_{\text{B}}T(\mathbf{r}')\tilde{n}_\alpha(\mathbf{r},\mathbf{r}') = k_{\text{B}}T(\mathbf{r}')\left.\frac{\delta p(\mathbf{r})}{\delta \mu_\alpha(\mathbf{r}')}\right|_{T,\boldsymbol{\mu}'_\alpha} \tag{3.85}$$

The first of these is a surprise because Landau's global thermodynamic theory [6] predicts that the fluctuations of pressure and entropy are independent. However, Landau's conclusion applies to the molar entropy or the entropy per molecule, whereas $\delta\hat{s}(\mathbf{r}')$ in Eq. (3.84) is the fluctuation of the entropy density, that is, the entropy per unit volume. Thus, we proceed as in Section II.D by introducing the entropy per molecule $\eta(\mathbf{r}) = \sigma(\mathbf{r})/n(\mathbf{r})$ and the corresponding first-order functional variation

$$\delta\hat{\eta}(\mathbf{r}) = \frac{1}{n(\mathbf{r})^2}\,[n(\mathbf{r})\delta\hat{\sigma}(\mathbf{r}) - \sigma(\mathbf{r})\delta\hat{n}(\mathbf{r})] \tag{3.86}$$

of the associated microscopic field. The result of combining this expression with the definition (3.81) of $\delta\hat{p}(\mathbf{r})$ and the connection $\delta\hat{\sigma} = \delta\hat{s} - T^{-1}[\sigma\delta\hat{T} + \Sigma_\alpha n_\alpha\delta\hat{\mu}_\alpha - \delta\hat{p}]$ of Eq. (2.66) is the formula

$$\begin{aligned}\langle \delta\hat{p}(\mathbf{r})\delta\hat{\eta}(\mathbf{r}')\rangle_{\text{ref}} &= \frac{k_{\text{B}}T(\mathbf{r}')}{n(\mathbf{r}')^2}\{n(\mathbf{r}')\tilde{\sigma}(\mathbf{r},\mathbf{r}') - \sigma(\mathbf{r}')\tilde{n}(\mathbf{r},\mathbf{r}')\} - \frac{1}{n(\mathbf{r}')T(\mathbf{r}')}\\ &\quad\times\int_\Omega d\mathbf{r}''\Big\{[\delta(\mathbf{r}''-\mathbf{r}')\sigma(\mathbf{r}'') - \tilde{\sigma}(\mathbf{r}',\mathbf{r}'')]\langle\delta\hat{p}(\mathbf{r})\delta\hat{T}(\mathbf{r}'')\rangle_{\text{ref}}\\ &\quad+\sum_{\alpha=1}^{c}[\delta(\mathbf{r}''-\mathbf{r}')n_\alpha(\mathbf{r}'') - \tilde{n}_\alpha(\mathbf{r}',\mathbf{r}'')]\langle\delta\hat{p}(\mathbf{r})\delta\hat{\mu}_\alpha(\mathbf{r}'')\rangle_{\text{ref}}\Big\}\end{aligned} \tag{3.87}$$

To obtain this result use also has been made of the correlation integral equations (3.84) and (3.85).

From Eq. (3.87) it is readily verified that, for a spatially uniform state of reference

$$\left\langle \delta\hat{p}(\mathbf{r}) \int_{\Omega} d\mathbf{r}' \delta\hat{n}(\mathbf{r}') \right\rangle_{\text{uniform}} = 0 \tag{3.88}$$

This agrees with Landau's result for the global fluctuations of pressure and entropy *per molecule*.

Next on the agenda is a brief consideration of formulas connected with the pressure and osmotic representations of Sections II.D and II.E. The statistical mechanical formula for the functional derivative $\delta p(\mathbf{r})/\delta n(\mathbf{r}')|_{T,\mathbf{x}}$ differs from the thermodynamic formula (2.96) in that the function $\mathcal{G}^{-1}_{\alpha\beta}(\mathbf{r}', \mathbf{r}'')$ is replaced with the inverse of the density–density correlation function $G_{\alpha\beta}(\mathbf{r}, \mathbf{r}')$. This inverse is related by Eq. (3.51) to the direct correlation function $C_{\alpha\beta}(\mathbf{r}, \mathbf{r}')$. The result of joining these equations is the formula

$$\left.\frac{\delta p(\mathbf{r})}{\delta n(\mathbf{r}')}\right|_{T,\mathbf{x}} = k_{\mathrm{B}} T(\mathbf{r}') \left[\frac{\tilde{n}(\mathbf{r}, \mathbf{r}')}{n(\mathbf{r}')} - n(\mathbf{r}) C_{nn}(\mathbf{r}, \mathbf{r}') \right] \tag{3.89}$$

wherein

$$\tilde{n}(\mathbf{r}, \mathbf{r}') = \sum_{\alpha=1}^{c} \tilde{n}_{\alpha}(\mathbf{r}, \mathbf{r}') \tag{3.90}$$

and

$$C_{nn}(\mathbf{r}, \mathbf{r}') = \sum_{\alpha=1}^{c} \sum_{\beta=1}^{c} \int_{\Omega} d\mathbf{r}'' \frac{T(\mathbf{r}'')}{T(\mathbf{r}')} \frac{\tilde{n}_{\alpha}(\mathbf{r}, \mathbf{r}'')}{n(\mathbf{r})} C_{\alpha\beta}(\mathbf{r}'', \mathbf{r}') x_{\beta}(\mathbf{r}') \tag{3.91}$$

The relationship (3.89) closely resembles the usual compressibility equation of state. The precise connection between the two will be established in Section III.E.

In conclusion, we consider the statistical mechanical counterpart of the thermodynamic equation (2.120), in which the inverse element $G^{\mathrm{eff}^{-1}}_{\beta\alpha}(\mathbf{r}'', \mathbf{r}')$ appears in place of $\mathcal{G}^{\mathrm{eff}^{-1}}_{\beta\alpha}(\mathbf{r}'', \mathbf{r}')$. Associated with this inverse operation, which is restricted to the subspace spanned by the particle densities of the *solute* species, is a set of "effective" direct correlation functions $C^{\mathrm{eff}}_{\alpha\beta}(\mathbf{r}, \mathbf{r}')$, to which the kernels $G^{\mathrm{eff}^{-1}}_{\alpha\beta}(\mathbf{r}, \mathbf{r}')$ are related as follows:

$$G_{\alpha\beta}^{\text{eff}^{-1}}(\mathbf{r}, \mathbf{r}') = \frac{\delta_{\alpha\beta}\delta(\mathbf{r} - \mathbf{r}')}{n_\alpha(\mathbf{r})} - C_{\alpha\beta}^{\text{eff}}(\mathbf{r}, \mathbf{r}') \tag{3.92}$$

These effective direct correlation functions first were introduced by Adelman [31] as part of his theory of fluid mixtures. It can be shown directly from the defining relation

$$\sum_{\gamma=2}^{c} \int_\Omega d\mathbf{r}'' G_{\alpha\gamma}^{\text{eff}^{-1}}(\mathbf{r}, \mathbf{r}'') G_{\gamma\beta}(\mathbf{r}'', \mathbf{r}') = \delta_{\alpha\beta}\delta(\mathbf{r} - \mathbf{r}') \tag{3.93}$$

and Eq. (3.42) that these functions are related to the irreducible pair correlation functions $h_{\alpha\beta}(\mathbf{r}, \mathbf{r}')$ by the "osmotic" Ornstein–Zernike equations

$$h_{\alpha\beta}(\mathbf{r}, \mathbf{r}') = C_{\alpha\beta}^{\text{eff}}(\mathbf{r}, \mathbf{r}') + \sum_{\gamma=2}^{c} \int_\Omega d\mathbf{r}'' C_{\alpha\gamma}^{\text{eff}}(\mathbf{r}, \mathbf{r}'') n_\gamma(\mathbf{r}'') h_{\gamma\beta}(\mathbf{r}'', \mathbf{r}') \tag{3.94}$$

which first were proposed by Adelman and Deutch [32].

By inserting Eq. (3.92) into the counterpart of Eq. (2.120) we obtain at last a formula

$$\left.\frac{\delta p(\mathbf{r})}{\delta n_\alpha(\mathbf{r}')}\right|_{T, \mu_1, \mathbf{n}'_\alpha} = k_\text{B} T(\mathbf{r}') \left[\frac{\tilde{n}_\alpha(\mathbf{r}, \mathbf{r}')}{n_\alpha(\mathbf{r}')} - \sum_{\beta=2}^{c} \int_\Omega d\mathbf{r}'' \tilde{n}_\beta(\mathbf{r}, \mathbf{r}'') \frac{T(\mathbf{r}'')}{T(\mathbf{r}')} C_{\beta\alpha}^{\text{eff}}(\mathbf{r}'', \mathbf{r}') \right] \tag{3.95}$$

which is the generalization to a nonuniform fluid of Adelman's formula [31] for the osmotic pressure.

Situations where the osmotic representation is most appropriate also are those for which it is natural to introduce effective, solvent-averaged potentials of interaction among the solute particles. In these cases the only role of the solvent is its effect upon the quasicontinuum, macroscopic parameters that appear in the effective solute interaction potentials. If the solution is so dilute that the solute particles are widely dispersed, most observable properties of the system will depend only on the asymptotic limits of the irreducible pair and direct correlation functions, the latter of which is given by

$$C_{\alpha\beta}^{\text{eff}}(\mathbf{r}, \mathbf{r}') \approx -\frac{1}{k_\text{B} T} u_{\alpha\beta}^{\text{eff}}(|\mathbf{r} - \mathbf{r}'|) \qquad |\mathbf{r} - \mathbf{r}'| \to \infty \tag{3.96}$$

Once the choice of $u_{\alpha\beta}^{\text{eff}}(|\mathbf{r} - \mathbf{r}'|)$ has been made, the corresponding pair correlation functions, consistent with the approximation (3.96), can be obtained by solving Eq. (3.94). For example, the long-ranged part of the effective potential of interaction of two ions dissolved in a solvent characterized by a dielectric constant ε is given by the formula

$$u_{\alpha\beta}^{\text{eff}}(|\mathbf{r}|) = \frac{q_\alpha q_\beta}{\varepsilon|\mathbf{r}|} \tag{3.97}$$

where q_γ denotes the electric charge of a species-γ ion. The solution of Eq. (3.94) obtained by using Eq. (3.97) in Eq. (3.96) is the well-known Debye–Hückel (DH) function

$$h_{\alpha\beta}^{\text{DH}}(|\mathbf{r} - \mathbf{r}'|) = -\frac{1}{\varepsilon k_{\text{B}} T} \frac{q_\alpha q_\beta}{|\mathbf{r} - \mathbf{r}'|} \exp(-\kappa_{\text{D}}|\mathbf{r} - \mathbf{r}'|) \tag{3.98}$$

with $\kappa_{\text{D}} = [(4\pi/\varepsilon k_{\text{B}} T) \Sigma_{\alpha=1}^{c} n_\alpha q_\alpha^2]^{1/2}$ equal to the inverse of the characteristic Debye length. Adelman [31] has shown how the effective interactions also can be determined from information about the correlations between the solvent and solute species.

D. The Thermodynamic Pressure and Its Relationship to the Mechanical Pressure Tensor

It is appropriate at this point to examine further the pressure field $p(\mathbf{r})$ that we previously have identified with $\phi(\mathbf{r})/\beta(\mathbf{r})$ in Section II.C and/or with the corresponding statistical mechanical quantity $\psi(\mathbf{r})/\beta(\mathbf{r})$ of Section III.A. According to conventional statistical mechanics the pressure associated with a uniform state of thermodynamic equilibrium is related by the formula

$$\beta p = \frac{\partial}{\partial \Omega} \ln \Xi[\Omega; \beta, \boldsymbol{\nu}] \tag{3.99}$$

to the volume derivative of the grand partition function. Consider now a simply connected region of volume $\Omega_{\mathbf{R}}$, centered on a point $\mathbf{R}$ that lies within a *non*uniform system. It is reasonable in light of Eq. (3.99) to identify the pressure at this point with the limit

$$\bar{p}(\mathbf{R}) = \beta(\mathbf{R})^{-1} \lim_{\Omega_{\mathbf{R}} \to 0} \frac{\partial}{\partial \Omega_{\mathbf{R}}} \ln \Xi_{\mathbf{R}} \tag{3.100}$$

wherein $\Xi_{\mathbf{R}} = \Xi[\Omega_{\mathbf{R}}; \{\beta, \boldsymbol{\nu}\}]$. The limit indicated must not, of course, be interpreted literally, in the strict mathematical sense. Thus, we conclude here for much the same reasons as given earlier in Sections II.B and II.C that the sequence of nested volume elements implied by the formula

(3.100) must not sink beneath a critical size of order ℓ^3 with the length ℓ so chosen that the subsystem occupying this cell satisfies the criteria of macroscopic thermodynamics. Consequently, the function $\bar{p}(\mathbf{R})$ defined by the formula (3.100) is a coarse-grained average.

We intend to use the definition (3.100) together with the previously postulated existence of a density function $\psi(\mathbf{r})$ that satisfies the equation

$$\ln \Xi[\Omega; \{\beta, \boldsymbol{\nu}\}] = \int_{\Omega} d\mathbf{r}\, \psi(\mathbf{r}; \{\beta, \boldsymbol{\nu}\}) \tag{3.101}$$

Therefore, a minimum requirement upon the length scale ℓ is that it greatly exceeds the ranges of the pair potential functions $u_{\alpha\beta}(|\mathbf{r} - \mathbf{r}'|)$ of the constituent particle species.

To proceed further some means must be found for constructing the volume derivatives appearing in the formula (3.100). For this purpose we introduce the affine transformation

$$\mathbf{r} \rightarrow \mathbf{r}' = \mathbf{r}_0 + (\Omega'/\Omega)^{1/3}(\mathbf{r} - \mathbf{r}_0)$$

or

$$\mathbf{r}' - \mathbf{r}_0 = (\Omega'/\Omega)^{1/3}(\mathbf{r} - \mathbf{r}_0) \tag{3.102}$$

the geometric significance of which is shown in Fig. 3.1. The location of $\mathbf{r}_0$, the "center of the transformation," is arbitrary; it may, but need not,

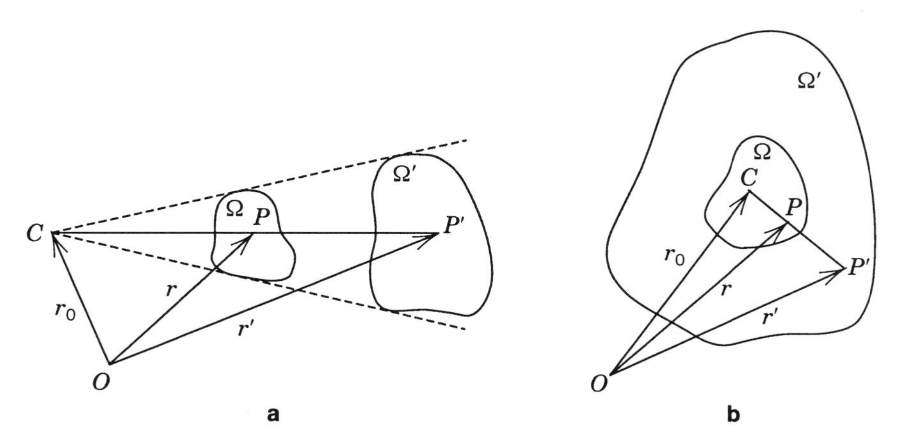

Figure 3.1. Geometric depiction of the transformation (3.102). In panel (a) the center C lies outside of the volume elements Ω and Ω', whereas in (b) it is internal. The symbol O denotes the origin of an inertial frame.

be situated near the field point $\mathbf{R}$ of Eq. (3.100). Consider then an integral

$$I(\Omega) = \int_{\Omega} d\mathbf{r}\, \chi(\mathbf{r}) \tag{3.103}$$

the volume derivative of which is given by the expression

$$\frac{\partial I(\Omega)}{\partial \Omega} = \lim_{\Omega' \to \Omega} \frac{I(\Omega') - I(\Omega)}{\Omega' - \Omega} \tag{3.104}$$

with $I(\Omega')$ defined, in accordance with Eq. (3.102), by the formula

$$I(\Omega') = \int_{\Omega} d\mathbf{r}' \chi(\mathbf{r}') = \frac{\Omega'}{\Omega} \int_{\Omega} d\mathbf{r}\, \chi\left[\mathbf{r}_0 + \left(\frac{\Omega'}{\Omega}\right)^{1/3} (\mathbf{r} - \mathbf{r}_0)\right] \tag{3.105}$$

and where Ω' is the image [cf. Fig. 3.1] of the volume Ω under the transformation Eq. (3.102). It is important to recognize that the functions $\partial I(\Omega)/\partial\Omega$ and $I(\Omega')$, defined by Eqs. (3.104) and (3.105), depend on the choice of $\mathbf{r}_0$. From these formulas we conclude that

$$\begin{aligned}
\frac{\partial I(\Omega)}{\partial \Omega} &= \left[\frac{\partial I(\Omega')}{\partial \Omega'}\right]_{\Omega'=\Omega} = \frac{\partial}{\partial \Omega'} \left\{\frac{\Omega'}{\Omega} \int_{\Omega} d\mathbf{r}\, \chi\left[\mathbf{r}_0 + \left(\frac{\Omega'}{\Omega}\right)^{1/3} (\mathbf{r} - \mathbf{r}_0)\right]\right\}_{\Omega'=\Omega} \\
&= \frac{1}{\Omega} \int_{\Omega} d\mathbf{r} \left\{\chi(\mathbf{r}) + \frac{1}{3} (\mathbf{r} - \mathbf{r}_0) \cdot \nabla \chi(\mathbf{r})\right\} \\
&= \frac{1}{\Omega} \int_{\Omega} d\mathbf{r} \frac{1}{3} \nabla \cdot [(\mathbf{r} - \mathbf{r}_0) \chi(\mathbf{r})]
\end{aligned} \tag{3.106}$$

The result of applying this to the integral $\int_{\Omega} d\mathbf{r}\, \psi(\mathbf{r})$ is the formula

$$\frac{\partial}{\partial \Omega} \int_{\Omega} d\mathbf{r}\, \psi(\mathbf{r}) = \frac{1}{\Omega} \int_{\Omega} d\mathbf{r}\, \psi(\mathbf{r}) + \frac{1}{3\Omega} \int_{\Omega} d\mathbf{r} (\mathbf{r} - \mathbf{r}_0) \cdot \nabla \psi(\mathbf{r}) \tag{3.107}$$

When Ω is identified with $\Omega_{\mathbf{R}}$ and $\mathbf{r}_0$ with $\mathbf{R}$, this leads to the result

$$\begin{aligned}
\lim_{\Omega_{\mathbf{R}} \to \ell^3} \left\{\frac{\partial}{\partial \Omega_{\mathbf{R}}} \int_{\Omega_{\mathbf{R}}} d\mathbf{r}\, \psi(\mathbf{r})\right\} &= \lim_{\Omega_{\mathbf{R}} \to \ell^3} \left\{\frac{1}{\Omega_{\mathbf{R}}} \int_{\Omega_{\mathbf{R}}} d\mathbf{r}\, \psi(\mathbf{r}) \right. \\
&\qquad \left. + \frac{1}{3\Omega_R} \int_{\Omega_{\mathbf{R}}} d\mathbf{r} (\mathbf{r} - \mathbf{R}) \cdot \nabla \psi(\mathbf{r})\right\} \\
&\equiv \bar{\psi}(\mathbf{R}) + 0(\ell)
\end{aligned} \tag{3.108}$$

with

$$\bar{\psi}(\mathbf{R}) = \frac{1}{\omega} \int_{\omega} d\mathbf{r}\, \psi(\mathbf{r}) \tag{3.109}$$

denoting the average of the field $\psi(\mathbf{r})$ over the interior of a region, centered on $\mathbf{R}$, with a volume ω of order ℓ^3. A comparison of Eqs. (3.100) and (3.108) establishes the equality of the coarse-grained pressure $\bar{p}(\mathbf{R})$ with the quotient $\bar{\psi}(\mathbf{R})/\beta(\mathbf{r})$ of the "ω average of $\psi(\mathbf{r})$" and $\beta(\mathbf{r})$.

To use the formula (3.106) for computing the volume derivative of $\ln \Xi[\Omega; \{\beta, \boldsymbol{\nu}\}]$ we write Eq. (3.9) as

$$\begin{aligned}
\Xi[\Omega'; \{\beta, \boldsymbol{\nu}\}] = \sum_{\{N_\alpha\}} &\left\{ \prod_{\alpha=1}^{c} \frac{(\Omega'/\Omega)^{N_\alpha}}{N_\alpha!} \right\} \int_{\Omega} \cdots \int_{\Omega} \\
&\times \left\{ \prod_{\alpha=1}^{c} \prod_{i=1}^{N_\alpha} \frac{d\mathbf{r}_{i\alpha}}{\lambda_\alpha^3[\mathbf{r}_0 + (\Omega'/\Omega)^{1/3}(\mathbf{r}_{i\alpha} - \mathbf{r}_0)]} \right\} \\
&\times \exp\left\{ -\frac{1}{2} \sum_{\alpha} \sum_{\beta} \sum_{i} \sum_{j}{}' u_{\alpha\beta}\left[\left(\frac{\Omega'}{\Omega}\right)^{1/3} |\mathbf{r}_{i\alpha} - \mathbf{r}_{j\beta}| \right] \right. \\
&\times \frac{1}{2}\left(\beta\left[\mathbf{r}_0 + \left(\frac{\Omega'}{\Omega}\right)^{1/3} (\mathbf{r}_{i\alpha} - \mathbf{r}_0) \right] \right. \\
&\left. + \beta\left[\mathbf{r}_0 + \left(\frac{\Omega'}{\Omega}\right)^{1/3} (\mathbf{r}_{j\beta} - \mathbf{r}_0) \right] \right) \\
&\left. - \sum_{\alpha} \sum_{j} \omega_\alpha\left[\mathbf{r}_0 + \left(\frac{\Omega'}{\Omega}\right)^{1/3} (\mathbf{r}_{i\alpha} - \mathbf{r}_0) \right] \right\}
\end{aligned} \tag{3.110}$$

with λ_α appearing in Eq. (3.25) and where $\omega_\alpha = \beta v_\alpha + \nu_\alpha$. It then can be verified that

$$\begin{aligned}
\frac{\partial}{\partial \Omega} \ln \Xi[\Omega; \{\beta, \boldsymbol{\nu}\}] = \frac{1}{\Omega} \sum_{\alpha} &\left[\langle N_\alpha \rangle - \frac{1}{3} \left\langle \sum_{i=1}^{N_\alpha} (\mathbf{r}_{i\alpha} - \mathbf{r}_0) \cdot \{ \nabla_{\mathbf{r}_{i\alpha}} \omega_\alpha(\mathbf{r}_{i\alpha}) \right.\right. \\
&\left.\left. + 3\nabla_{\mathbf{r}_{i\alpha}} \ln \lambda_\alpha(\mathbf{r}_{i\alpha}) \} \right\rangle \right] \\
&+ \frac{1}{3\Omega} \left\langle -\frac{1}{2} \sum_{\alpha} \sum_{\beta} \sum_{i} \sum_{j}{}' \mathbf{r}_{i\alpha,j\beta} \cdot \frac{\partial u_{\alpha\beta}(|\mathbf{r}_{i\alpha,j\beta}|)}{\partial \mathbf{r}_{i\alpha,j\beta}} \right. \\
&\left. \times \frac{1}{2} \{ \beta(\mathbf{r}_{i\alpha}) + \beta(\mathbf{r}_{j\beta}) \} \right\rangle
\end{aligned}$$

$$-\frac{1}{3\Omega}\left\langle \frac{1}{2}\sum_{\alpha}\sum_{\beta}\sum_{i}\sum_{j}{}' u_{\alpha\beta}(|\mathbf{r}_{i\alpha,j\beta}|)(\mathbf{r}_{i\alpha}-\mathbf{r}_0)\cdot\nabla_{\mathbf{r}_{i\alpha}}\beta(\mathbf{r}_{i\alpha})\right\rangle \tag{3.111}$$

with $\mathbf{r}_{i\alpha,j\beta}=\mathbf{r}_{i\alpha}-\mathbf{r}_{j\beta}$ and where

$$\sum_{\alpha}\langle N_\alpha\rangle=\int_\Omega d\mathbf{r}\,\beta(\mathbf{r})\left\langle \frac{1}{3}\operatorname{tr}\hat{\mathbf{p}}_{\mathrm{KIN}}(\mathbf{r})\right\rangle \tag{3.112}$$

$$\left\langle \sum_{i=1}^{N_\alpha}(\mathbf{r}_{i\alpha}-\mathbf{r}_0)\cdot\{\nabla_{\mathbf{r}_{i\alpha}}\omega_\alpha(\mathbf{r}_{i\alpha})+3\nabla_{\mathbf{r}_{i\alpha}}\ln\lambda_\alpha(\mathbf{r}_{i\alpha})\}\right\rangle$$

$$=\int_\Omega d\mathbf{r}(\mathbf{r}-\mathbf{r}_0)\cdot\{\langle\hat{n}_\alpha(\mathbf{r})\rangle\nabla\omega_\alpha(\mathbf{r})+\langle\hat{\varepsilon}_{\alpha,\mathrm{KIN}}(\mathbf{r})\rangle\nabla\beta(\mathbf{r})\} \tag{3.113}$$

$$\left\langle \frac{1}{2}\sum_{\alpha}\sum_{\beta}\sum_{i}\sum_{j}{}' u_{\alpha\beta}(|\mathbf{r}_{i\alpha,j\beta}|)(\mathbf{r}_{i\alpha}-\mathbf{r}_0)\cdot\nabla_{\mathbf{r}_{i\alpha}}\beta(\mathbf{r}_{i\alpha})\right\rangle$$

$$=\int_\Omega d\mathbf{r}\langle\hat{\varepsilon}_{\mathrm{INT}}(\mathbf{r})\rangle(\mathbf{r}-\mathbf{r}_0)\cdot\nabla\beta(\mathbf{r}) \tag{3.114}$$

and

$$\left\langle -\frac{1}{2}\sum_{\alpha}\sum_{\beta}\sum_{i}\sum_{j}{}' \mathbf{r}_{i\alpha,j\beta}\cdot\frac{\partial u_{\alpha\beta}(|\mathbf{r}_{i\alpha,j\beta}|)}{\partial\mathbf{r}_{i\alpha,j\beta}}\frac{1}{2}\{\beta(\mathbf{r}_{i\alpha})+\beta(\mathbf{r}_{j\beta})\}\right\rangle$$

$$=\int_\Omega d\mathbf{r}\,\beta(\mathbf{r})\langle\operatorname{tr}\hat{\mathbf{p}}^{(0)}(\mathbf{r})\rangle \tag{3.115}$$

The microfields appearing in these last four formulas are defined as follows:

$$\hat{\mathbf{p}}_{\mathrm{KIN}}(\mathbf{r})=\sum_{\alpha}\sum_{i}\frac{\mathbf{p}_{i\alpha}\mathbf{p}_{i\alpha}}{m_\alpha}\delta(\mathbf{r}_{i\alpha}-\mathbf{r}) \tag{3.116}$$

$$\hat{\varepsilon}_{\alpha,\mathrm{KIN}}(\mathbf{r})=\sum_{i}\frac{\mathbf{p}_{i\alpha}^2}{2m_\alpha}\delta(\mathbf{r}_{i\alpha}-\mathbf{r}) \tag{3.117}$$

$$\hat{\varepsilon}_{\mathrm{INT}}(\mathbf{r})=\frac{1}{2}\sum_{\alpha}\sum_{\beta}\int_\Omega d\mathbf{r}'\,\hat{n}_{\alpha\beta}(\mathbf{r},\mathbf{r}')u_{\alpha\beta}(|\mathbf{r}-\mathbf{r}'|) \tag{3.118}$$

and

$$\hat{\mathbf{p}}_{\mathrm{INT}}^{(0)}(\mathbf{r}) = -\frac{1}{2}\sum_{\alpha}\sum_{\beta}\int_{\Omega} d\mathbf{r}'\,\frac{1}{2}\{\hat{n}_{\alpha\beta}(\mathbf{r},\mathbf{r}') + \hat{n}_{\alpha\beta}(\mathbf{r}',\mathbf{r})\}(\mathbf{r}-\mathbf{r})\frac{\partial u_{\alpha\beta}(|\mathbf{r}-\mathbf{r}'|)}{\partial(\mathbf{r}-\mathbf{r}')} \tag{3.119}$$

In terms of these quantities Eq. (3.111) assumes the more compact and physically interpretable form

$$\frac{\partial}{\partial\Omega}\ln\Xi[\Omega;\{\beta,\boldsymbol{\nu}\}] = \frac{1}{\Omega}\int_{\Omega} d\mathbf{r}\,\beta(\mathbf{r})\left\langle\frac{1}{3}\,\mathrm{tr}\,\hat{\mathbf{p}}^{(0)}\right\rangle$$
$$-\frac{1}{3\Omega}\int_{\Omega} d\mathbf{r}(\mathbf{r}-\mathbf{r}_0)\cdot\left\{\varepsilon(\mathbf{r})\nabla\beta(\mathbf{r}) + \sum_{\alpha} n_{\alpha}(\mathbf{r})\nabla\omega_{\alpha}(\mathbf{r})\right\} \tag{3.120}$$

with $\hat{\mathbf{p}}^{(0)} = \hat{\mathbf{p}}_{\mathrm{KIN}} + \hat{\mathbf{p}}_{\mathrm{INT}}^{(0)}$ and $\varepsilon(\mathbf{r}) = \langle\hat{\varepsilon}_{\mathrm{KIN}}(\mathbf{r}) + \hat{\varepsilon}_{\mathrm{INT}}(\mathbf{r})\rangle$.

Because of the postulated equality, Eq. (3.101), the two expressions (3.107) and (3.120) must be equal for an arbitrary choice of the transformation center C and for arbitrary, but not too small values of the volume Ω. Therefore, we conclude that

$$\bar{\psi}(\mathbf{r}) = \beta(\mathbf{r})\langle\tfrac{1}{3}\,\mathrm{tr}\,\hat{\mathbf{p}}^{(0)}(\mathbf{r})\rangle \tag{3.121}$$

and

$$\nabla\bar{\psi}(\mathbf{r}) + [\bar{\varepsilon}(\mathbf{r})\nabla\beta(\mathbf{r}) + \sum_{\alpha}\bar{n}_{\alpha}(\mathbf{r})\nabla\omega_{\alpha}(\mathbf{r})] = 0 \tag{3.122}$$

The second of these is the spatial version of the "local Gibbs–Duhem relation," which would have resulted from Eq. (2.49) had it not been for the nonlocal character of $\psi(\mathbf{r})$. Thus, Eq. (3.122) is what one naturally might have expected by analogy with global thermodynamics.

From Eq. (3.121) we conclude that the coarse-grained, local thermodynamic pressure is related by the formula

$$\bar{p}(\mathbf{r}) = \tfrac{1}{3}\,\mathrm{tr}\langle\hat{\mathbf{p}}^{(0)}(\mathbf{r})\rangle \tag{3.123}$$

to the microscopic tensor $\hat{\mathbf{p}}^{(0)}(\mathbf{r})$. The commonly accepted formula for the microscopic interactional contribution to the fluid pressure tensor is not

$\hat{\mathbf{p}}_{\text{INT}}^{(0)}$ of Eq. (3.121) [33] but

$$\hat{\mathbf{p}}_{\text{INT}}(\mathbf{r}) = -\frac{1}{2}\sum_{\alpha}\sum_{\beta}\sum_{i}\sum_{j}{}' \mathbf{r}_{i\alpha,j\beta}\frac{\partial u_{\alpha\beta}(|\mathbf{r}_{i\alpha,j\beta}|)}{\partial \mathbf{r}_{i\alpha,j\beta}}\int_0^1 da\, \delta(\mathbf{r}-\mathbf{r}_{i\alpha}+a\mathbf{r}_{i\alpha,j\beta}) \tag{3.124}$$

However, Eq. (3.124) becomes the same as Eq. (3.121) when the integral of the delta function is replaced with $\frac{1}{2}[\delta(\mathbf{r}-\mathbf{r}_{i\alpha})+\delta(\mathbf{r}-\mathbf{r}_{j\beta})]$, the average of the integrand values for $a=0$ and $a=1$. This replacement has no effect upon averages computed for a uniform state and it is expected to be of little consequence unless there are significant fractional changes of local thermodynamic state over distances on the order of ℓ. Under these conditions the local *average* pressure is a *function* of the local average thermodynamic state, given by precisely the same formula,

$$\bar{p}(\mathbf{r}) = \tfrac{1}{3}\,\mathrm{tr}\langle \hat{\mathbf{p}}_{\text{KIN}}(\mathbf{r}) + \hat{\mathbf{p}}_{\text{INT}}(\mathbf{r})\rangle \tag{3.125}$$

as one would expect on the basis of a micromechanical calculation of the rate of momentum transfer. It nevertheless would be incorrect to identify this coarse-grained average, $\bar{p}(\mathbf{r})$, with the nonlocal thermodynamic pressure $p(\mathbf{r}) = \psi(\mathbf{r})/\beta(\mathbf{r})$.

The "ω average of a field variable," exemplified by $\bar{\psi}(\mathbf{r})$ defined by Eq. (3.109), is a concept deserving of further consideration. Indeed, this sort of averaging can even be made a useful part of the thermodynamic theory of Section II by assuming that the functional derivatives $\delta\phi(\mathbf{r})/\delta\beta(\mathbf{r}')|_{\boldsymbol{\nu}}$ and $\delta\phi(\mathbf{r})/\delta\nu_\alpha(\mathbf{r}')|_{\beta,\boldsymbol{\nu}'_\alpha}$ are finite ranged, that is, that the magnitudes of these two-point functions are negligibly small for values of $|\mathbf{r}-\mathbf{r}'|$ in excess of a characteristic length ℓ. This range ℓ cannot be predicted by the thermodynamic theory. However, by borrowing from statistical mechanics one can anticipate that the order of magnitude of ℓ will be that of a typical equilibrium correlation function of microscopic variables, for example, $\ell \sim 5$–20 Å for a small-molecule fluid. Then, with $S_{\mathbf{r}'}$ denoting a sphere of radius ℓ centered on the point $\mathbf{r}'$ it follows from Eq. (2.43) that

$$\int_{S_{\mathbf{r}'}} d\mathbf{r}\,\frac{\delta\phi(\mathbf{r})}{\delta\beta(\mathbf{r}')}\bigg|_{\boldsymbol{\nu}} = -\varepsilon(\mathbf{r}') \tag{3.126a}$$

$$\int_{\Omega-S_{\mathbf{r}'}} d\mathbf{r}\,\frac{\delta\phi(\mathbf{r})}{\delta\beta(\mathbf{r}')}\bigg|_{\boldsymbol{\nu}} = 0 \tag{3.126b}$$

Similar relations apply to the functional derivatives $\delta\phi(\mathbf{r})/\delta\nu_\alpha(\mathbf{r}')|_{\beta,\boldsymbol{\nu}'_\alpha}$.

We now integrate Eq. (2.50) over a region ω, the volume of which is significantly greater than $S_{\mathbf{r}'}$ but generally much smaller than Ω. In particular, the volume of ω is taken to be just large enough that its contents are capable of satisfying the criteria demanded of a macroscopic thermodynamic system. The integral of Eq. (2.50) then can be written as

$$\int_\omega d\mathbf{r}\, \delta\phi(\mathbf{r}) = \int_\Omega d\mathbf{r}'\, \delta\beta(\mathbf{r}') \int_\omega d\mathbf{r}\left[\frac{\delta\phi(\mathbf{r})}{\delta\beta(\mathbf{r}')}\right]\Bigg|_{\boldsymbol{\nu}} + \sum_{\alpha=1}^{c} \int_\Omega d\mathbf{r}'\, \delta\nu_\alpha(\mathbf{r}') \int_\omega d\mathbf{r}\left[\frac{\delta\phi(\mathbf{r})}{\delta\nu_\alpha(\mathbf{r}')}\right]\Bigg|_{\beta,\boldsymbol{\nu}'_\alpha} \tag{3.127}$$

where, according to Eq. (3.126)

$$\int_\omega d\mathbf{r}\frac{\delta\phi(\mathbf{r})}{\delta\beta(\mathbf{r}')}\Bigg|_{\boldsymbol{\nu}} = \begin{bmatrix} -\varepsilon(\mathbf{r}')\,; & \mathbf{r}'\in\omega \\ 0\,; & \mathbf{r}'\notin\omega \end{bmatrix} \tag{3.128}$$

A similar conclusion applies to the integral of $\delta\phi(\mathbf{r})/\delta\nu_\alpha(\mathbf{r}')|_{\beta,\boldsymbol{\nu}'_\alpha}$. Consequently, one concludes from Eqs. (3.127) and (3.128) that

$$\int_\omega d\mathbf{r}\left[\varepsilon(\mathbf{r})\delta\beta(\mathbf{r}) + \sum_{a=1}^{c} n_\alpha(\mathbf{r})\delta\nu_\alpha(\mathbf{r}) + \delta\phi(\mathbf{r})\right] = 0$$

or

$$\int_\omega d\mathbf{r}\frac{1}{T(\mathbf{r})}\left[\sigma(\mathbf{r})\delta T(\mathbf{r}) + \sum_{\alpha=1}^{c} n_\alpha(\mathbf{r})\delta\mu_\alpha(\mathbf{r}) - \delta p(\mathbf{r})\right] = 0 \tag{3.129}$$

Although the integrand of Eq. (3.129) is the same as that of Eq. (2.49), the domain of integration ω is much smaller than Ω. Indeed, if the integrand factors are interpreted as *averages* over domains with linear dimensions of the order of $\omega^{1/3} \gg \ell$, Eq. (3.129) can be replaced with the "local Gibbs–Duhem relation"

$$\bar{\varepsilon}(\mathbf{r})\delta\bar{\beta}(\mathbf{r}) + \sum_{\alpha=1}^{c} \bar{n}_\alpha(\mathbf{r})\delta\bar{\nu}_\alpha(\mathbf{r}) + \delta\bar{\phi}(\mathbf{r}) = 0 \tag{3.130}$$

or, equivalently

$$\bar{\sigma}(\mathbf{r})\delta\bar{T}(\mathbf{r}) + \sum_{\alpha=1}^{c} \bar{n}_\alpha(\mathbf{r})\delta\bar{\mu}_\alpha(\mathbf{r}) - \delta\bar{p}(\mathbf{r}) = 0 \tag{3.131}$$

Other conclusions of a similar nature can be obtained by first rewriting Eqs. (2.72) and (2.73) as

$$\int_{\Omega} d\mathbf{r}\left[\delta\varepsilon(\mathbf{r}) + g_{00}(\mathbf{r})\delta\beta(\mathbf{r}) + \sum_{\beta=1}^{c} g_{0\beta}(\mathbf{r})\delta\nu_{\beta}(\mathbf{r})\right] = 0 \qquad (3.132)$$

and

$$\int_{\Omega} d\mathbf{r}\left[\delta n_{\alpha}(\mathbf{r}) + g_{00}(\mathbf{r})\delta\beta(\mathbf{r}) + \sum_{\beta=1}^{c} g_{\alpha\beta}(\mathbf{r})\delta\nu_{\beta}(\mathbf{r})\right] = 0 \qquad (3.133)$$

respectively, with $g_{ij}(\mathbf{r}) = \int_{\Omega} d\mathbf{r}'\mathcal{G}_{ij}(\mathbf{r}', \mathbf{r})$. Arguments similar to those used to obtain Eq. (3.129) then lead to the conclusion that the relations Eqs. (3.132) and (3.133) remain valid when the domains of integration Ω are replaced with the much smaller $\omega \ll \Omega$. This same line of reasoning enables us to conclude that for a homogeneous state of equilibrium the integrals of the function $\Delta\tilde{C}(\mathbf{r}, \mathbf{r}')$, defined by Eq. (2.112), and the correlation integral $\langle\delta\hat{p}(\mathbf{r})\delta\hat{\eta}(\mathbf{r}')\rangle$ of Eq. (3.87) both will vanish when extended over ω-sized domains.

Finally, by combining Eqs. (2.66) and (3.129) we are led to the conclusion that

$$\int_{\omega} d\mathbf{r}\, \delta\sigma(\mathbf{r}) = \int_{\omega} d\mathbf{r}\, \delta s(\mathbf{r}) \qquad (3.134)$$

Thus, even though $\delta\sigma(\mathbf{r})$ and $\delta s(\mathbf{r})$ cannot be equated to one another, their averages over ω-sized domains are numerically equal. These examples establish the rather comforting conclusion that most, if not all, of the nonlocal aspects of the thermodynamic and statistical mechanical theories presented above can be dispensed with by adopting the coarse-graining (ω-averaging) approximation. The costs of the resulting simplicity include an obvious lack of formal rigor and the requirement that one maintain a degree of skepticism concerning the applicability of this theory to distances on the order of molecular correlation lengths.

E. Fluctuations in Uniform Systems

In most laboratory experiments the size of the system under investigation is so small that long-ranged external fields acting on the particles may be treated as uniform. Furthermore, the conditions of measurement often can be so arranged that the influence of container walls is negligible. Under these circumstances the fluid properties are expected to be translationally and rotationally invariant. Equilibrium correlation functions such as $G_{\alpha\beta}(\mathbf{r}, \mathbf{r}')$ then will depend only on the separation between the two field points, so that $G_{\alpha\beta}(\mathbf{r}, \mathbf{r}') \rightarrow G_{\alpha\beta}(|\mathbf{r} - \mathbf{r}'|)$. Similarly, the average of a quantity that depends on a single field point will assume a constant, uniform value throughout the system, equal to the corresponding stoichiometric value, for example, $n_\alpha(\mathbf{r}) \rightarrow N_\alpha/\Omega$ and $\sigma(\mathbf{r}) \rightarrow S/\Omega$.

It then is convenient to introduce the Fourier representative of the fluctuating field $a(\mathbf{r})$ by means of its relationship,

$$a(\mathbf{r}) = \frac{1}{\Omega} \sum_{\mathbf{k}} e^{i\mathbf{k}\cdot\mathbf{r}} a(\mathbf{k}) \tag{3.135}$$

to the Fourier expansion coefficient

$$a(\mathbf{k}) = \int_\Omega d\mathbf{r}\, e^{-i\mathbf{k}\cdot\mathbf{r}} a(\mathbf{r}) \tag{3.136}$$

The summation in Eq. (3.135) extends over all wavevectors that are compatible with the container to which the system is confined. Associated with this Fourier picture are the representations

$$\delta(\mathbf{r} - \mathbf{r}') = \frac{1}{\Omega} \sum_{\mathbf{k}} e^{i\mathbf{k}\cdot(\mathbf{r}-\mathbf{r}')} \tag{3.137}$$

and

$$\delta^{Kr}(\mathbf{k} - \mathbf{k}') = \frac{1}{\Omega} \int_\Omega d\mathbf{r}\, e^{-i(\mathbf{k}-\mathbf{k}')\cdot\mathbf{r}} \tag{3.138}$$

of the Dirac and Kronecker deltas, respectively.

Due to the translational and rotational invariance of the system, Fourier representations of the correlation functions assume special forms. To illustrate this we consider the correlation function,

$$\mathscr{G}_{fg}(\mathbf{r}, \mathbf{r}') = \langle \delta\hat{f}(\mathbf{r}) \delta\hat{g}(\mathbf{r}') \rangle_{\text{eq}} \tag{3.139}$$

of the fluctuations of two microscopic fields $\hat{f}(\mathbf{r})$ and $\hat{g}(\mathbf{r})$. Because of the invariance property $\mathscr{G}_{fg}(\mathbf{r}, \mathbf{r}') = \mathscr{G}_{fg}(|\mathbf{r} - \mathbf{r}'|)$ it follows directly that

$$\begin{aligned} \mathscr{G}_{fg}(\mathbf{k}, \mathbf{k}') &\equiv \int_\Omega d\mathbf{r}\, e^{-i\mathbf{k}\cdot\mathbf{r}} \int_\Omega d\mathbf{r}'\, e^{-i\mathbf{k}'\cdot\mathbf{r}'} \mathscr{G}_{fg}(\mathbf{r}, \mathbf{r}') \\ &= \langle \delta\hat{f}(\mathbf{k}) \delta\hat{g}(\mathbf{k}') \rangle_{\text{eq}} = \Omega \delta^{Kr}(\mathbf{k} + \mathbf{k}') \mathscr{G}_{fg}(|\mathbf{k}|) \end{aligned} \tag{3.140}$$

where

$$\mathscr{G}_{fg}(|\mathbf{k}|) = \int_{\Omega} d\mathbf{r}\, e^{-i\mathbf{k}\cdot\mathbf{r}} \mathscr{G}_{fg}(|\mathbf{r}|)$$
$$= \frac{1}{\Omega} \int_{\Omega} d\mathbf{r} \int_{\Omega} d\mathbf{r}'\, e^{-i\mathbf{k}\cdot(\mathbf{r}-\mathbf{r}')} \mathscr{G}_{fg}(|\mathbf{r}-\mathbf{r}'|)$$
$$= \frac{1}{\Omega} \langle \delta\hat{f}(\mathbf{k})\delta\hat{g}(-\mathbf{k})\rangle_{\text{eq}} \tag{3.141}$$

Consequently, we obtain the relationships

$$\mathscr{G}_{fg}(\mathbf{k}, \mathbf{k}') = \delta^{Kr}(\mathbf{k}+\mathbf{k}')\langle \delta\hat{f}(\mathbf{k})\delta\hat{g}(-\mathbf{k})\rangle_{\text{eq}} \tag{3.142}$$

and

$$\mathscr{G}_{fg}(\mathbf{r}, \mathbf{r}') = \frac{1}{\Omega} \sum_{\mathbf{k}} e^{i\mathbf{k}\cdot(\mathbf{r}-\mathbf{r}')} \mathscr{G}_{fg}(|\mathbf{k}|) \tag{3.143}$$

The next step is to establish connections between the formulas of the local fluctuation theory, presented above, and those of the conventional, "global" statistical thermodynamic theory. It is useful for this purpose to use the formula

$$\bar{a} = \frac{1}{\Omega} \int_{\Omega} d\mathbf{r}\, a(\mathbf{r}) = \frac{1}{\Omega} \lim_{k\to 0} a(\mathbf{k}) \tag{3.144}$$

to define the volume average or "stoichiometric value" of an arbitrary field $a(\mathbf{r})$. We then identify two fields $a(\mathbf{r})$ and $b(\mathbf{r})$, so related to $\hat{f}(\mathbf{r})$ and $\hat{g}(\mathbf{r})$, respectively, that

$$\frac{\delta a(\mathbf{r})}{\delta b(\mathbf{r}')} = \langle \delta\hat{f}(\mathbf{r})\delta\hat{g}(\mathbf{r}')\rangle_{\text{eq}} = \mathscr{G}_{fg}(|\mathbf{r}-\mathbf{r}'|) \tag{3.145}$$

This formula is, in fact, typical of many previously encountered in this Chapter. It can be written in the alternative form

$$\delta a(\mathbf{r}) = \int_{\Omega} d\mathbf{r}' \mathscr{G}_{fg}(|\mathbf{r}-\mathbf{r}'|)\delta b(\mathbf{r}') \tag{3.146}$$

as a linear integral relationship between the variations of the two fields $a(\mathbf{r})$ and $b(\mathbf{r}')$. The Fourier representation of Eq. (3.146), given by

$$\delta a(\mathbf{k}) = \frac{1}{\Omega} \langle \delta\hat{f}(\mathbf{k})\delta\hat{g}(-\mathbf{k})\rangle_{\text{eq}} \delta b(\mathbf{k}) \tag{3.147}$$

can be rearranged into the form

$$\frac{\delta a(\mathbf{k})}{\delta b(\mathbf{k})} = \frac{1}{\Omega} \langle \delta\hat{f}(\mathbf{k}) \delta\hat{g}(-\mathbf{k}) \rangle_{\text{eq}} \tag{3.148}$$

The long wavelength limit of this last expression then becomes

$$\lim_{\mathbf{k}\to 0} \frac{\delta a(\mathbf{k})}{\delta b(\mathbf{k})} = \lim_{\Delta\bar{b}\to 0} \frac{\bar{a}[b_{\text{eq}} + \Delta\bar{b}] - \bar{a}[b_{\text{eq}}]}{\Delta\bar{b}} = \frac{\partial\bar{a}}{\partial\bar{b}} \tag{3.149}$$

with $\bar{a}$ and $\bar{b}$ denoting the stoichiometric values of the fields $a(\mathbf{r})$ and $b(\mathbf{r})$, respectively. Thus, we obtain the connection,

$$\left.\frac{\partial\bar{a}}{\partial\bar{b}}\right|_{\Omega} = \frac{1}{\Omega} \lim_{\mathbf{k}\to 0} \langle \delta\hat{f}(\mathbf{k}) \delta\hat{g}(-\mathbf{k}) \rangle_{\text{eq}} \tag{3.150}$$

that we were seeking. [The overbars indicating stoichiometric values are henceforth deleted, except occasionally for emphasis.]

As an example, the Fourier representation of the functional derivative [Eq. (3.40)] can be written as

$$\begin{aligned} k_{\text{B}}T \left.\frac{\delta n_\alpha(\mathbf{k})}{\delta\mu_\beta(\mathbf{k})}\right|_{T,\boldsymbol{\mu}'_\beta} &= \frac{1}{\Omega} \langle \delta\hat{n}_\alpha(\mathbf{k}) \delta\hat{n}_\beta(-\mathbf{k}) \rangle_{\text{eq}} \\ &= n_\alpha n_\beta h_{\alpha\beta}(\mathbf{k}) + n_\alpha \delta_{\alpha\beta} \end{aligned} \tag{3.151}$$

which, according to Eq. (3.150), implies the "global" relationship

$$\begin{aligned} k_{\text{B}}T \left.\frac{\partial n_\alpha}{\partial\mu_\beta}\right|_{T,\Omega,\boldsymbol{\mu}'_\beta} &= \frac{1}{\Omega} \langle \delta\hat{N}_\alpha \delta\hat{N}_\beta \rangle_{\text{eq}} \\ &= n_\alpha n_\beta \int_\Omega d\mathbf{r}\, h_{\alpha\beta}(|\mathbf{r}|) + n_\alpha \delta_{\alpha\beta} \end{aligned} \tag{3.152}$$

wherein

$$\delta\hat{N}_\alpha = \hat{N}_\alpha - \langle \hat{N}_\alpha \rangle_{\text{eq}} = \int_\Omega d\mathbf{r}\, \delta\hat{n}_\alpha(\mathbf{r}) \tag{3.153}$$

Similarly, the Fourier representation of the local compressibility formula (3.89) is

$$\left.\frac{\delta p(\mathbf{k})}{\delta n(\mathbf{k})}\right|_{T,\mathbf{x}} = k_{\text{B}}T \left[\frac{\hat{n}(k)}{n} - nC_{nn}(\mathbf{k}) \right] \tag{3.154}$$

with

$$C_{nn}(k) = \sum_{\alpha=1}^{c} \sum_{\beta=1}^{c} \frac{\tilde{n}_\alpha(k)}{n} C_{\alpha\beta}(k) x_\beta \tag{3.155}$$

In the long wavelength limit [cf Eq. (2.59)]

$$\lim_{\mathbf{k}\to 0} \tilde{n}_\alpha(k) = \left.\frac{\partial \bar{p}}{\partial \bar{\mu}_\alpha}\right|_{T,\bar{\boldsymbol{\mu}}'_\alpha} = n_\alpha \tag{3.156}$$

and Eq. (3.154) reduces to the standard compressibility equation of state

$$\left.\frac{\partial \bar{p}}{\partial n}\right|_{T,\mathbf{x}} = \frac{1}{n\bar{\kappa}_T} = k_B T\left[1 - n\int_\Omega d\mathbf{r}\, C_{nn}(|\mathbf{r}|)\right] \tag{3.157}$$

To invert the formula (3.154) we introduce the pair-space correlation function $h_{nn}(\mathbf{r}, \mathbf{r}')$, defined in terms of $C_{nn}(\mathbf{r}, \mathbf{r}')$ by the Ornstein–Zernike type equation

$$h_{nn}(\mathbf{r}, \mathbf{r}') = C_{nn}(\mathbf{r}, \mathbf{r}') + \int_\Omega d\mathbf{r}'' C_{nn}(\mathbf{r}, \mathbf{r}'') n(\mathbf{r}'') h_{nn}(\mathbf{r}'', \mathbf{r}') \tag{3.158}$$

In the case of a uniform system the Fourier transform of this can be solved to yield an expression $C_{nn}(k) = h_{nn}(k)/[1 + nh_{nn}(k)]$, which when substituted into Eq. (3.154) produces the formula

$$\begin{aligned}\left.\frac{\delta n(k)}{\delta p(k)}\right|_{T,\mathbf{x}} &= \frac{1}{k_B T}[1 + nh_{nn}(k)]\left[\frac{\tilde{n}(k)}{n} + \{\tilde{n}(k) - n\}h_{nn}(k)\right]^{-1} \\ &\equiv n\kappa_T(k)\end{aligned} \tag{3.159}$$

In the long-wavelength limit Eq. (3.159) reduces to the familiar expression [34],

$$\bar{\kappa}_T = \frac{1}{nk_B T}\left[1 + n\int_\Omega d\mathbf{r}\, h_{nn}(\mathbf{r})\right] \tag{3.160}$$

for the isothermal compressibility. This can be rewritten in the form

$$k_B T n\bar{\kappa}_T = |\mathbf{G}(\mathbf{k}=0)| \Big/ \left\{n \sum_{\alpha=1}^{c} \sum_{\beta=1}^{c} x_\alpha x_\beta |\mathbf{G}(\mathbf{k}=0)|_{\alpha\beta}\right\} \tag{3.161}$$

derived by Kirkwood and Buff [35]. Here $\mathbf{G}(\mathbf{k})$ denotes a $c \times c$ dimensional matrix, the elements of which are the Fourier transforms of the particle-density correlation functions, namely,

$$G_{\alpha\beta}(|\mathbf{k}|) = n[x_\alpha \delta_{\alpha\beta} + x_\alpha x_\beta n h_{\alpha\beta}(|\mathbf{k}|)] \tag{3.162}$$

and $|\mathbf{G}|_{\alpha\beta}$ is the cofactor of $G_{\alpha\beta}(|\mathbf{k}|)$ in the determinant $|\mathbf{G}(\mathbf{k})|$.

Comparable to these are the equations [cf. Eqs. (3.95) and (3.94)]

$$\left.\frac{\delta p(\mathbf{k})}{\delta n_\alpha(\mathbf{k})}\right|_{T,\boldsymbol{\mu}_1,\mathbf{n}'_\alpha} = k_\mathrm{B} T \left[\frac{\tilde{n}_\alpha(k)}{n_\alpha} - \sum_{\beta=2}^{c} \tilde{n}_\beta(k) C_{\beta\alpha}(k)\right] \tag{3.163}$$

and

$$h_{\alpha\beta}(k) = C^{\mathrm{eff}}_{\alpha\beta}(k) + \sum_{\alpha\gamma}^{c} C^{\mathrm{eff}}_{\alpha\gamma}(k) n_\gamma h_{\gamma\beta}(k) \tag{3.164}$$

of the osmotic representation.

As a final example, we consider the heat capacity at constant volume, defined by Eq. (3.56). The Fourier representation of this quantity, for a uniform fluid, is given by the expression

$$C_v(\mathbf{k}) = \left.\frac{\delta \varepsilon(\mathbf{k})}{\delta T(\mathbf{k})}\right|_{\mathbf{n}} = \frac{1}{k_\mathrm{B} T^2 \Omega} \langle \delta\hat{\varepsilon}_\perp(\mathbf{k}) \delta\hat{\varepsilon}_\perp(-\mathbf{k})\rangle_{\mathrm{eq}} \tag{3.165}$$

where

$$\delta\hat{\varepsilon}_\perp(\mathbf{k}) = \delta\hat{\varepsilon}(\mathbf{k}) - \sum_{\beta=1}^{c} \left.\frac{\delta\varepsilon(\mathbf{k})}{\delta n_\beta(\mathbf{k})}\right|_T \delta\hat{n}_\beta(\mathbf{k}) \tag{3.166}$$

with $\delta\varepsilon(\mathbf{k})/\delta n_\beta(\mathbf{k})|_T = \Sigma_\alpha\, G^{-1}_{\beta\alpha}(\mathbf{k}) G_{\alpha 0}(\mathbf{k})$, denotes the Fourier transform of the component of the microscopic energy density that is orthogonal to the fluctuations of the particle densities. In the long wavelength limit this expression reduces to the following formula for the heat capacity (per unit volume) of an open system:

$$\begin{aligned} C_v &= \lim_{\mathbf{k}\to 0} C_v(\mathbf{k}) = \left.\frac{\partial \varepsilon}{\partial T}\right|_{\mathbf{n}} = 4\pi \int_0^\infty dr\, r^2 C_v(r) \\ &= \frac{1}{k_\mathrm{B} T^2 \Omega} \langle \delta\hat{E}^2_\perp \rangle_{\mathrm{eq}} \end{aligned} \tag{3.167}$$

Here,

$$\delta\hat{E}_\perp = \delta\hat{E} - \sum_{\beta=1}^{c} \left.\frac{\partial\varepsilon}{\partial n_\beta}\right|_{T,\Omega} \delta\hat{N}_\beta \tag{3.168}$$

with

$$\delta \hat{E} = \hat{E} - \langle \hat{E} \rangle_{\text{eq}} = \int_{\Omega} d\mathbf{r}\, \delta \hat{\varepsilon}(\mathbf{r}) \tag{3.169}$$

is that part of the fluctuation of energy that is not related to the particle number fluctuations $\{\delta \hat{N}_\gamma\}$, defined by Eq. (3.153).

The wavevector dependence of thermodynamic quantities is often called spatial dispersion. This dispersion becomes experimentally observable when the linear extension of the system is comparable to the wavelength of the probe beam. Only then are the fractional deviations of the energy and particle densities sufficiently large to be detected. Since the range of particle correlations depends on the range of their mutual interactions, it is not surprising that spatial dispersion most often is taken into account in plasmas. By "plasmas" we mean systems of particles with long-ranged potentials of interaction that are inversely proportional to the particle separation.

As the simplest example, let us consider a multicomponent gravitational plasma composed of species $\alpha = 1, 2, \ldots, c$, each characterized by a particle mass m_α and interacting with one another according to the Newtonian formula

$$u_{\alpha\beta}(|\mathbf{r}|) = -\frac{G m_\alpha m_\beta}{r} \tag{3.170}$$

where G denotes the gravitational constant. From the asymptotic Abé formula [36] Eq. (3.96), connecting the direct corelation function $C_{\alpha\beta}(|\mathbf{r} - \mathbf{r}'|)$ to the potential energy of interaction, we obtain the corresponding Fourier representation formula

$$C_{\alpha\beta}(k) = \frac{G m_\alpha m_\beta}{k_{\mathrm{B}} T} \frac{4\pi}{k^2} \qquad k \to 0 \tag{3.171}$$

Then, by using the Ornstein–Zernike equations, as described in the preceding section, we obtain for the irreducible pair correlation functions the asymptotic ($k \to 0$) formulas

$$h_{\alpha\beta}(\mathbf{k}) = \frac{4\pi G m_\alpha m_\beta}{k_{\mathrm{B}} T} \frac{1}{k^2 - \kappa_G^2} \tag{3.172}$$

The quantity

$$\kappa_G^2 = \frac{4\pi G}{k_B T} \sum_{\gamma=1}^{c} m_\gamma^2 n_\gamma \tag{3.173}$$

appearing in this expression is the inverse square of a length characteristic of the homogeneous gravitational plasma. It is analogous in many respects to the well-known Debye parameter κ_D, to which we previously have alluded in connection with Eq. (3.99). However, in contrast to the familiar "screened Coulomb" functions of the Debye–Hückel theory, the coordinate space transforms of Eq. (3.172) are the functions.

$$h_{\alpha\beta}(|\mathbf{r}-\mathbf{r}'|) = \frac{G m_\alpha m_\beta}{k_B T} \frac{\cos(\kappa_G |\mathbf{r}-\mathbf{r}'|)}{|\mathbf{r}-\mathbf{r}'|} \tag{3.174}$$

Thus, in place of the exponential damping characteristic of an electrically neutral charged particle plasma, the gravitational plasma exhibits spatial *oscillations* on a scale fixed by the length κ_G^{-1}.

Cosmological analyses of galactic spatial correlations [37, 38] invariably are restricted to a density correlation function,

$$\chi(r) = \frac{1}{n^2} \sum_{\alpha=1}^{c} \sum_{\beta=1}^{c} n_\alpha n_\beta h_{\alpha\beta}(|\mathbf{r}|) = \frac{G}{k_B T} \left(\frac{\rho}{n}\right)^2 \frac{\cos(\kappa_G r)}{r} \tag{3.175}$$

that is averaged over all observed (or inferred) galaxies, regardless of mass. The quantity $\rho = \Sigma_{\alpha=1}^{c} m_\alpha n_\alpha$ appearing in this last formula is the average mass density of the gravitational plasma, that is, the density of that part of the universe associated with galaxies. In terms of $\chi(r)$, the probability density for finding two galaxies with a separation r is given by the formula

$$P(r) = n[1 + \chi(r)] \tag{3.176}$$

Similarly, the static form factor for the universe is

$$S(\mathbf{k}) = 1 + n\chi(k) = 1 + n \frac{4\pi G}{k_B T} \left(\frac{\rho}{n}\right)^2 \frac{1}{k^2 - \kappa_G^2} \tag{3.177}$$

with $\chi(k)$ denoting the Fourier transform of $\chi(r)$.

The asymptotic form of the correlation function given by Eq. (3.175) is valid for arbitrary concentrations, in the limit of large intergalactic separations. The only restriction (see, e.g., Hansen and McDonald [34]) is that the particle separations be so large that the values of irreducible pair correlation functions are significantly less than unity. Consequently, the asymptotic formulas (3.174) and (3.175) are expected to be accurate when the average particle separation satisfies the condition $r \gtrsim \kappa_G^{-1}$. Because these formulas have been obtained without reference to galactic substructure (indeed, the galaxies have been treated as mass points), the only scale length that appears is κ_G^{-1}. At separations much smaller than κ_G^{-1} the distribution functions may be distorted by "short-ranged" interactions characteristic of galactic substructure. Furthermore, it is by no means certain that the expanding universe can be treated as a system in a homogeneous state of equilibrium.

Two distinct asymptotic power laws $[\chi(r) \propto r^{-\gamma}]$ can be associated with the correlation function $\chi(r)$ of Eq. (3.175). For separations such that $\kappa_G r \ll 1$

$$\ln \chi(r) \approx \ln\left[\frac{G\kappa_G}{k_B T}\left(\frac{\rho}{n}\right)^2\right] - \ln(\kappa_G r) \qquad \kappa_G r \ll 1 \tag{3.178}$$

From this we obtain the power law exponent $\gamma_1 = 1$. However, when the value of $\kappa_G r$ lies near unity, the correlation function can be represented by the series

$$\ln \chi(r) = \ln\left[\frac{G\kappa_G}{k_B T}\left(\frac{\rho}{n}\right)^2\right] - 0.62 - 2.56\ln(\kappa_G r)$$
$$- 2.49[\ln(\kappa_G r)]^2 - 0.50[\ln(\kappa_G r)]^3 - \cdots$$

from which we extract the asymptotic approximation

$$\ln \chi(r) \approx \ln\left[\frac{G\kappa_G}{k_B T}\left(\frac{\rho}{n}\right)^2 e^{-0.62}\right] - 2.56\ln(\kappa_G r) \qquad \kappa_G r \approx 1 \tag{3.179}$$

with a power law exponent $\gamma_2 = 2.56$. The arithmetic average of these two, namely, $\gamma = \frac{1}{2}(\gamma_1 + \gamma_2) = 1.78$, is very nearly equal to the observationally determined value of 1.77 ± 0.04 that Groth and Peebles [39] obtained using the Zwicky galactic catalog. A final item connected with the gravitational plasma is the wavevector dependent compressibility, given by the formula.

$$\frac{\kappa_T(k)}{\kappa_T^0} = \frac{k^2}{k^2 - n\dfrac{4\pi G}{k_B T}\left(\dfrac{\rho}{n}\right)^2} \tag{3.180}$$

Here $\kappa_T^0 = (nk_B T)^{-1}$ denotes the compressibility of an ideal gas. A more detailed comparison of observations with the theoretical formula (3.175) has been presented elsewhere [40].

In the case of an electrically neutral ionic plasma [see the interaction potential Eq. (3.92)], the analog of Eq. (3.172) is the Debye–Hückel (DH) function

$$h_{\alpha\beta}^{\mathrm{DH}}(k) = -\frac{4\pi}{\varepsilon k_{\mathrm{B}} T} \frac{q_\alpha q_\beta}{k^2 + \kappa_{\mathrm{D}}^2} \tag{3.181}$$

and the corresponding coordinate space function is $h_{\alpha\beta}^{\mathrm{DH}}(r)$, given by Eq. (3.99). Due to the electroneutrality of the plasma, its compressibility is the same as that of an ideal gas with the same density, unless short-ranged interactions are taken into account.

The quantity of interest in the case of a neutral charged-particle plasma is the autocorrelation function of the charge density fluctuation field

$$\delta\hat{q}(\mathbf{r}) = \sum_{\alpha=1}^{c} q_\alpha \delta\hat{n}_\alpha(\mathbf{r}) \tag{3.182}$$

The Fourier representation of this function is found to be

$$\langle |\delta\hat{q}(\mathbf{k})|^2 \rangle = \frac{\varepsilon k_{\mathrm{B}} T}{4\pi} \frac{\kappa_{\mathrm{D}}^2}{k^2 + \kappa_{\mathrm{D}}^2} \tag{3.183}$$

The asymptotic formulas (3.172) and (3.181) [or Eqs. (3.174) and (3.99)] for the irreducible equilibrium pair correlation functions are direct consequences of the functional form, $u_{\alpha\beta}(r) \propto r^{-1}$, of the plasma interaction potential. However, Ornstein and Zernike [41, 42] noticed that a similar asymptotic behavior was displayed by the pair correlation functions of more ordinary fluids, formed from particles with relatively short-ranged interactions similar, for example, to the Lennard–Jones potential. In an effort to understand how this might come about, let us consider the particle density autocorrelation function for a single-component fluid, namely,

$$G_{11}(|\mathbf{k}|) = \langle |\delta\hat{n}_1(\mathbf{k})|^2 \rangle_{\mathrm{eq}} = \frac{k_{\mathrm{B}} T}{\left.\dfrac{\delta\mu_1(\mathbf{k})}{\delta n_1(\mathbf{k})}\right|_T} \tag{3.184}$$

The denominator factor from this expression can be written in the form

$$\left.\frac{\delta\mu_1(\mathbf{k})}{\delta n_1(\mathbf{k})}\right|_T = \left.\frac{\partial\mu_1}{\partial n_1}\right|_T \{1+\Gamma(k)\} \tag{3.185}$$

with

$$\Gamma(k) = \frac{4\pi n_1 \int_0^\infty dr\, r^2 C_{11}(r)\left[1-\dfrac{\sin kr}{kr}\right]}{1-4\pi n_1 \int_0^\infty dr\, r^2 C_{11}(r)} \tag{3.186}$$

and where

$$\left.\frac{\partial\mu_1}{\partial n_1}\right|_T = \frac{k_B T}{n_1}\left[1-4\pi n_1 \int_0^\infty dr\, r^2 C_{11}(r)\right] = \frac{1}{n_1^2 \bar{\kappa}_T} \tag{3.187}$$

is the standard statistical mechanical formula for the density derivative of the chemical potential of a homogeneous, equilibrium fluid. The symbol $\bar{\kappa}_T$ in Eq. (3.187) stands for the corresponding isothermal compressibility.

For an equilibrium state of the system to be mechanically stable it is both necessary and sufficient that the corresponding value of the derivative $\partial\mu_1/\partial\mu_1|_T$ be positive definite. The boundary between these and the unstable states, for which the derivative is negative valued, defines the spinodal curve,

$$\left.\frac{\partial\mu_1}{\partial n_1}\right|_T = 0 \tag{3.188}$$

Thus, as the state of the system approaches the spinodal curve the compressibility becomes unbounded but the direct correlation function remains an integrable function, related to fluid density by the formula

$$\frac{1}{n_1} = 4\pi \int_0^\infty dr\, r^2 C_{11}(r) \tag{3.189}$$

The spinodal curve of a single-component system consists of two "branches," separated by the region of unstable states. These meet at the critical point where, in addition to Eq. (3.188), $\partial^2\mu_1/\partial n_1^2|_T = 0$ and $\partial^3\mu_1/\partial n_1^3|_T > 0$.

The direct correlation function is a dimensionless functional of the pair interaction potential $u_{11}(|\mathbf{r}|)$ and this, in turn, invariably can be represented as a function of a dimensionless particle separation $\rho = |\mathbf{r}|/\sigma$ and an energy scale parameter u_0. Here σ is a natural microscopic unit of length that gauges the range of the potential $u_{11}(|\mathbf{r}|)$; typically, it is the particle separation at which the pair potential exhibits the minimum value, $-u_0$. All other characteristic lengths of the fluid can be expressed in terms of σ.

The direct correlation function then is expected to depend on the reduced interparticle separation $\rho = |\mathbf{r}|/\sigma$, the reduced interaction energy amplitude βu_0, and the reduced particle density $n_1\sigma^3$. Consequently, the function $\Gamma(k)$ defined by Eq. (3.186), is dependent on the dimensionless wavenumber $k\sigma$ and the parameters βu_0 and $n_1\sigma^3$. By thus expressing all functional dependences in terms of dimensionless variables we are able to state precisely what is meant by the small wavenumber limit of the expression (3.184). It is the range of values of the wavenumber $k = |\mathbf{k}|$ for which $k\sigma \ll 1$. The physical implication of this condition is that the linear dimensions of the regions over which density fluctuations are localized must greatly exceed the range σ of the particle interactions. In this long wavelength limit

$$\lim_{\mathbf{k}\to 0} \frac{\Gamma(k)}{k^2} = \xi_n^2 = \frac{2\pi}{3} n_1 \frac{\int_0^\infty dr\, r^4 C_{11}(r)}{1 - 4\pi n_1 \int_0^\infty dr\, r^2 C_{11}(r)}$$

$$= \sigma^2 \left\{ \frac{n_1^* \int_0^\infty d\rho\, \rho^4 C_{11}^*(\rho)}{1 - 6n_1^* \int_0^\infty d\rho\, \rho^2 C_{11}^*(\rho)} \right\} \tag{3.190}$$

wherein $n_1^* = 2\pi n_1 \sigma^3/3$ and $C_{11}^*(\rho) = C_{11}(r)$. The quantity ξ_n defined by the formula (3.186) is a microscopic correlation length, characteristic of the particle distribution in a uniform fluid at equilibrium.

From Eqs. (3.184) and (3.190) we see that the long wavelength limit of the density autocorrelation function $G_{11}(|\mathbf{k}|)$ is given by the Lorentzian,

$$G_{11}(|k|) \underset{k\to 0}{\approx} \frac{k_B T}{\left.\dfrac{\partial \mu_1}{\partial n_1}\right|_T} \frac{1}{1 + k^2 \xi_n^2} = \frac{k_B T}{\left.\dfrac{\partial \mu_1}{\partial n_1}\right|_T \xi_n^2} \frac{1}{k^2 + \xi_n^{-2}} \tag{3.191}$$

It should be noted that for mechanically stable states of the fluid the algebraic sign of ξ_n^2, as defined by Eq. (3.190), is the same as that of the integral

$$\zeta = \frac{2\pi}{3} \int_0^\infty dr\, r^4 C_{11}(r) \tag{3.192}$$

As the particle separation increases beyond the value σ, it generally is expected that the direct correlation function will decay to zero more rapidly than does the corresponding irreducible pair correlation function

$h_{11}(r)$. Furthermore, the direct correlation function invariably is negative valued at small interparticle separations because of short-ranged repulsive interactions and it usually exhibits a positive valued peak at some intermediate separation, due to attractive long-ranged interactions. Particles with purely repulsive interactions such as those, for example, in a hard-sphere fluid, have negative definite direct correlation functions.

An example of an experimentally determined direct correlation function has been reported, for the case of argon, in a review article by Pings [43]. This system then was treated theoretically by Woodhead-Galloway et al. [44], using model potentials consisting of hard-sphere cores plus attractive long-ranged tails. They found that the inclusion of an attractive contribution to the potential was essential if the integral ζ was to be positive valued. It also was discovered that the values of this integral depended significantly upon density and that even when long-ranged interactions were included, the sign of ζ became negative at sufficiently high densities, indicating that under these conditions the correlations were dominated by the repulsive interactions.

The coordinate-space representation for the asymptotic form of the density autocorrelation function can be obtained by Fourier inversion of the formula (3.191). The functional form of the result does, of course, depend on the algebraic sign of ξ_n^2. When this quantity is negative valued, as it is when the repulsive interactions are dominant, the Fourier inverse of Eq. (3.191) is given by an expression

$$G_{11}(r) \underset{r/\sigma\to\infty}{\approx} \left\{ -\frac{k_B T}{(\partial \mu_1/\partial n_1)|_T} \frac{2\pi^2}{|\xi_n^2|} \right\} \frac{\cos(r/|\xi_n|)}{r} \tag{3.193}$$

which closely resembles the formula (3.175) for a gravitational plasma. On the other hand, when $\xi_n^2 > 0$, Fourier inversion of Eq. (3.191) leads to the familiar Ornstein–Zernike formula,

$$G_{11}(r) \underset{r/\sigma\to\infty}{\approx} \left\{ \frac{k_B T}{(\partial \mu_1/\partial n_1)|_T \, 4\pi\xi_n^2} \right\} \frac{e^{-r/\xi_n}}{r} \tag{3.194}$$

Thus, the asymptotic form of the density autocorrelation function depends very much on the pair potential and, as Woodhead-Galloway's previously cited study indicates, by the interplay between the attractive and repulsive parts of this potential.

This usual description of critical fluctuations (cf. [6]), based on the formula (3.194), hinges upon the observation [see Eq. (3.190)] that at the critical point $\xi_n^{-1} \to 0$, whereas $\xi_n^2\, \partial\mu_1/\partial n_1|_T$ remains finite. Therefore, at the critical point the screening factor $\exp(-r/\xi_n)$ disappears from Eq.

(3.194) and the density autocorrelation function becomes proportional to r^{-1} in the asymptotic limit. It then is concluded that the isothermal compressibility, as given by the formula $\bar{\kappa}_T = [n_1^2 k_B T]^{-1} \int d\mathbf{r} G_{11}(r)$, becomes unbounded at the critical point because the integral of $G_{11}(r)$ diverges.

In this context it is of interest to consider an alternative asymptotic formula for the density autocorrelation function that can be obtained by writing Eq. (3.185) in the form

$$\ln \frac{\delta\mu_1(k)}{\delta n_1(k)}\bigg|_T - \ln \frac{\partial\mu_1}{\partial n_1}\bigg|_T = \ln\{1 + \Gamma(k)\} \approx \xi_n^2 k^2 + \cdots \tag{3.195}$$

When this is substituted into Eq. (3.184) one obtains the Gaussian asymptotic formula

$$G_{11}(k) \approx \frac{k_B T}{(\partial\mu_1/\partial n_1)|_T} e^{-\xi_n^2 k^2} = G_{11}(k=0)e^{-\xi_n^2 k^2} \tag{3.196}$$

in place of the Lorentzian Eq. (3.191). The coordinate-space counterpart of Eq. (3.196) is the Gaussian,

$$G_{11}(r) \approx G_{11}(k=0) \frac{e^{-r^2/4\xi_n^2}}{(4\pi\xi_n^2)^{3/2}} \tag{3.197}$$

the integral of which equals $G_{11}(k=0) = k_B T/\partial\mu_1/\partial n_1|_T$, for all finite and positive values of ξ_n^2. Consequently, the divergence of the compressibility at the critical point is not attributable, in this case, to the analytic form of the autocorrelation function in the asymptotic limit: $G_{11}(k=0)$ and $\bar{\kappa}_T$ become unbounded at the critical point simply because the condition (3.189) is satisfied by the density and temperature dependent direct correlation function $C_{11}(r)$.

According to the modern theory of critical phenomena [45] the asymptotic form of the particle density-fluctuation autocorrelation function (more generally, the autocorrelation function of the order parameter) should be

$$G_{11}(r) \underset{r/\sigma\to\infty}{=} a(T)\left(\frac{\sigma}{r}\right)^{\eta} \frac{e^{-r/\xi}}{r^{d-2}} \tag{3.198}$$

Here, d is the dimension of the space, η is the "Fisher critical index," and $a(T)$ denotes an unspecified analytic function of temperature. This relationship (with $0 \leq \eta \leq 2$) is identified as a generalization of the

Ornstein–Zernike formula (3.194), the latter of which is considered to be the result of the so-called mean-field or Landau–Ginzburg approach to the theory of phase transitions.

At the present time there appears to be no compelling experimental evidence that η differs from zero. Thus, as examples of critical exponents obtained from experimental data [46], the values of η reported for xenon and helium are $\eta_{\mathrm{Xe}} = 0.1 \pm 0.1$ and $\eta_{\mathrm{He}^4} = 0.02 \pm 0.05$, respectively. Because the error estimates are comparable in magnitude to the values of the indexes themselves, these data do not appear to exclude the possibility that η is zero. Furthermore, theoretical calculations based upon renormalization group methods are nonconvergent and so could be judged as inconclusive. This situation presumably will be resolved when further, more sensitive experiments are conducted.

The importance of the correlation functions of particle density and energy fluctuations is that they can be related directly to experimentally measurable quantities such as the intensity distributions of scattered light and neutron beams. In particular, random fluctuations of the temperature and particle densities produce local variations of the dielectric permittivity of the system, thereby altering its optical properties. In consequence of this it is natural to introduce the local dielectric permittivity, $\varepsilon(\mathbf{r})$, treated as a functional of the temperature and particle density fields of a single-component system. Then, according to Eq. (3.69), we can write the expression

$$\delta\hat{\varepsilon}(\mathbf{k}) = \left.\frac{\delta\varepsilon(\mathbf{k})}{\delta T(\mathbf{k})}\right|_{n_1} \delta\hat{T}(\mathbf{k}) + \left.\frac{\delta\varepsilon(\mathbf{k})}{\delta n_1(k)}\right|_{T} \delta\hat{n}_1(\mathbf{k}) \tag{3.199}$$

for the Fourier transform of the fluctuation of the microscopic dielectric permittivity field of the fluid. The Einstein–Smoluchowski theory of quasielastic light scattering [47–50] provides a direct connection between the intensity of the scattered light and the autocorrelation function of $\delta\hat{\varepsilon}(\mathbf{k})$, namely,

$$\begin{aligned} I(k) \propto \langle \delta\hat{\varepsilon}(\mathbf{k})\delta\hat{\varepsilon}(-\mathbf{k})\rangle_{\mathrm{eq}} &= \left|\frac{\delta\varepsilon(\mathbf{k})}{\delta T(\mathbf{k})}\right|^2_{n_1} \langle|\delta\hat{T}(\mathbf{k})|^2\rangle_{\mathrm{eq}} \\ &+ \left|\frac{\delta\varepsilon(\mathbf{k})}{\delta n_1(\mathbf{k})}\right|^2_{T} \langle|\delta\hat{n}_1(\mathbf{k})|^2\rangle_{\mathrm{eq}} \end{aligned} \tag{3.200}$$

The asymptotic form of the temperature fluctuation autocorrelation function appearing in Eq. (3.200) can be gotten from the formula (3.67), the Fourier representative of which is

$$\langle |\delta \hat{T}(\mathbf{k})|^2 \rangle_{\text{eq}} = k_{\text{B}}T/C_v(\mathbf{k}) \tag{3.201}$$

The long wavelength limit of the wavevector dependent heat capacity $C_v(\mathbf{k})$ is given by the formula

$$C_v(\mathbf{k}) \approx C_v[1 + k^2\xi_T^2] \tag{3.202}$$

with C_v defined by Eq. (3.167) and where

$$\xi_T^2 = -\int_0^\infty dr\, r^4 C_v(r) \Big/ 6 \int_0^\infty dr\, r^2 C_v(r) \tag{3.203}$$

By combining these results with Eqs. (3.191) and (3.200) we then obtain the formula

$$I(k) \propto \left\{ \left| \frac{\partial \varepsilon}{\partial T} \right|^2_{n_1} \frac{k_{\text{B}}T^2}{C_v} \right\} \frac{1}{1 + k^2\xi_T^2} + \left\{ \left| \frac{\partial \varepsilon}{\partial n_1} \right|^2_T \frac{k_{\text{B}}T}{(\partial \mu_1/\partial n_1)|_T} \right\} \frac{1}{1 + k^2\xi_n^2} \tag{3.204}$$

which characterizes the spectrum of the scattered light in the asymptotic limit, $k\sigma \ll 1$. Although the spectrum given by Eq. (3.204) is a superposition of contributions from fluctuations of temperature and particle density, the first of these will be negligible unless the fluid is composed of polar molecules; only then does $\varepsilon(\mathbf{r})$ exhibit a significant dependence upon temperature. Thus, the inverse of the scattering intensity for a nonpolar fluid is expected to be a linear function of k^2 with a slope equal to ξ_n^2. And, indeed, there is ample experimental evidence in support of this functional dependence on k^2 [50]. The few instances where deviations from linearity are observed (in the long-wavelength regime) may have been caused by multiple-scattering events. Those for which negative slopes were observed could have been due to negative values of ξ_n^2.

This brings us, somewhat belatedly, to a consideration of the algebraic signs of the terms appearing in Eq. (3.204). For globally stable states of the system both $\partial \mu_1/\partial n_1|_T$ and the heat capacity C_v are positive definite. Therefore, the factors enclosed by the bracket symbols $\{\cdots\}$ are expected to be positive. As previously argued, the sign of ξ_n^2 will be positive or negative depending on whether the correlations of density fluctuations are dominated by attractive or repulsive interactions, respectively. Finally, we come to the algebraic sign of ξ_T^2. Because the denominator factor of the definition (3.203) is directly proportional to the bulk heat capacity C_v, the sign of ξ_T^2 is the opposite of that of the integral $\int_0^\infty dr\, r^4 C_v(r)$. Strong evidence is presented in the following section that this integral is negative valued and $\xi_T^2 > 0$. Consequently, the correlation function of the temperature fluctuations can be expected to exhibit a long-ranged behavior,

$$\langle \delta\hat{T}(\mathbf{r})\delta\hat{T}(\mathbf{r}')\rangle_{\text{eq}} \approx \frac{k_B T^2}{4\pi C_v \xi_T^2} \frac{e^{-|\mathbf{r}-\mathbf{r}'|/\xi_T}}{|\mathbf{r}-\mathbf{r}'|} \tag{3.205}$$

analogous to the density fluctuations of a neutral ionic plasma.

Before closing this section let us briefly examine the case of particles with pair interactions consisting of a long-ranged part inversely proportional to the interparticle separation plus a short-ranged part representative of their impenetrable "cores." In our previous considerations we first studied particles for which the asymptotic correlations were dominated by long-ranged interactions and then progressed to systems with correlations determined exclusively by short-ranged interactions. In the first case the direct correlation function was nonintegrable while in the second it was assumed that this function not only was integrable but that it also possessed finite higher order moments. The correlation lengths characteristic of the plasmalike systems were found to depend on temperature in a very simple way and were not divergent for any finite value of T. In contrast to this, the correlation lengths associated with short-ranged (SR) forces became unbounded as the system approached a stability limit. A simple example of a system that can exhibit both of these behaviors is a one-component gravitational plasma composed of particles interacting with the pair potential

$$u_{11}(r) = -\frac{Gm_1^2}{r} + u_{11}^{\text{SR}}(\text{r}) \tag{3.206}$$

Here $u_{11}^{\text{SR}}(r)$ is a short-ranged potential, such as a Lennard–Jones, square-well, or hard-sphere interaction.

The direct correlation function for this system can be written as the sum of the asymptotic, long-ranged contribution characteristic of the gravitational interaction plus a short-ranged, integrable part:

$$C_{11}(r) = \frac{Gm_1^2}{k_B T}\frac{1}{r} + C_{11}^{\text{SR}}(r) \tag{3.207}$$

The Fourier representation of Eq. (3.207) is

$$C_{11}(k) = \frac{4\pi G m_1^2}{k_B T} \frac{1}{k^2} + C_{11}^{SR}(k) \tag{3.208}$$

By introducing the last of these into the Ornstein–Zernike relationship and taking the long wavelength limit we obtain for the irreducible pair correlation function the asymptotic formula

$$h_{11}(k) \underset{k\sigma \gg 1}{=} \left\{ \left(\frac{4\pi G m_1^2}{k_B T} \right) \frac{k_B T}{n_1 (\partial \mu_1^{SR} / \partial n_1)_T} \right\} \frac{1}{k^2 - \dfrac{k_B T \kappa_G^2}{n_1 (\partial \mu_1^{SR} / \partial n_1)_T}} \tag{3.209}$$

The quantity,

$$\left. \frac{\partial \mu_1^{SR}}{\partial n_1} \right|_T \equiv \frac{k_B T}{n_1} [1 - n_1 C_{11}^{SR}(k=0)] \tag{3.210}$$

appearing in Eq. (3.209) is the contribution from the short-ranged interactions to the density derivative of the chemical potential.

The formula (3.209) is similar to Eq. (3.172) but it contains a factor $\partial \mu_1^{SR} / \partial n_1 |_T$ that can vanish at a spinodal-type stability limit and serves to renormalize the gravitational screening constant $\kappa_G = (4\pi G m_1^2 n_1 / k_B T)^{1/2}$. Thus, just as one expects, it is the short-ranged interactions that control whatever phase transitions may occur while the long-ranged gravitational interactions play the dominant role in fixing the asymptotic form of the pair correlation function.

F. Stability of Equilibrium States

Here we shall consider the response of a system at equilibrium to small, specified variations of the basic thermodynamic fields. The initial state will be stable with respect to the displacement

$$[\varepsilon(\mathbf{r}), \{n_\alpha(\mathbf{r})\}] \rightarrow [\varepsilon(\mathbf{r}) + \delta\varepsilon(\mathbf{r}), \{n_\alpha(\mathbf{r}) + \delta n_\alpha(\mathbf{r})\}]$$

provided that it is accompanied by a decrease of entropy. A state of the system is said to be *internally* stable with respect to the variations $[\delta\varepsilon(\mathbf{r}), \{\delta n_\alpha(\mathbf{r})\}]$ if; (1) the internal energy and contents of the entire system remain constant during the process and; (2) if the total entropy of the system diminishes. The limit of this internal stability is reached when the second variation of the entropy vanishes.

The change of entropy caused by variations of the energy and particle density fields can be written as

$$\Delta S = S[\varepsilon_{\text{eq}}(\mathbf{r}) + \delta\varepsilon(\mathbf{r}), \{n_{\alpha,\text{eq}}(\mathbf{r}) + \delta n_\alpha(\mathbf{r})\}] - S[\varepsilon_{\text{eq}}(\mathbf{r}), \{n_{\alpha,\text{eq}}(\mathbf{r})\}]$$
$$= \int_\Omega d\mathbf{r}\, \delta s(\mathbf{r}) + \frac{1}{2}\int_\Omega d\mathbf{r}\, \delta^2 s(\mathbf{r}) + \cdots \tag{3.211}$$

We henceforth restrict our considerations to internal stability. The first term on the right-hand side of Eq. (3.211) is then equal to zero because of the energy and particle conservation conditions

$$\int_\Omega d\mathbf{r}\, \delta\varepsilon(\mathbf{r}) = 0$$
$$\int_\Omega d\mathbf{r}\, \delta n_\alpha(\mathbf{r}) = 0 \tag{3.212}$$

appropriate to an isolated and closed system. Consequently, the criterion for stability with respect to infinitesimal variations of the basic fields becomes

$$\int_\Omega d\mathbf{r}\, \delta^2 s(\mathbf{r}) \le 0 \tag{3.213}$$

The equality sign in this expression identifies the stability limit.

Because of Eq. (3.19), this stability criterion can be written in an alternative form

$$\int_\Omega d\mathbf{r}\left[\delta\beta(\mathbf{r})\delta\varepsilon(\mathbf{r}) + \sum_{\alpha=1}^{c} \delta\nu_\alpha(\mathbf{r})\delta n_\alpha(\mathbf{r})\right] \le 0 \tag{3.214}$$

which, with the help of Eqs. (3.47) and (3.48), then becomes

$$\int_\Omega d\mathbf{r} \int_\Omega d\mathbf{r}'\left[G_{00}(\mathbf{r},\mathbf{r}')\delta\beta(\mathbf{r})\delta\beta(\mathbf{r}') + 2\sum_{\alpha=1}^{c} G_{0\alpha}(\mathbf{r},\mathbf{r}')\delta\beta(\mathbf{r})\delta\nu_\alpha(\mathbf{r}')\right.$$
$$\left. + \sum_{\alpha=1}^{c}\sum_{\beta=1}^{c} G_{\alpha\beta}(\mathbf{r},\mathbf{r}')\delta\nu_\alpha(\mathbf{r})\delta\nu_\beta(\mathbf{r}')\right] \ge 0 \tag{3.215}$$

More convenient than this is the equivalent formula

$$\int_{\Omega} d\mathbf{r} \int_{\Omega} d\mathbf{r}' \left[\frac{C_v(\mathbf{r}, \mathbf{r}')}{k_B T^2} \delta T(\mathbf{r}) \delta T(\mathbf{r}') + \frac{1}{k_B T} \sum_{\alpha=1}^{c} \sum_{\beta=1}^{c} \times \frac{\delta \mu_\alpha(\mathbf{r})}{\delta n_\beta(\mathbf{r}')} \bigg|_{T, \mathbf{n}'_\beta} \delta n_\alpha(\mathbf{r}) \delta n_\beta(\mathbf{r}') \right] \geq 0 \tag{3.216}$$

obtained by expressing the $\delta \nu_\alpha$ values in terms of $\delta \beta$ and the δn_α values, according to Eqs. (3.47) and (3.48), and using the definition (3.56) of C_v.

An equilibrium state of the system will be internally stable if and only if this criterion, Eq. (3.216), is satisfied for arbitrary variations of the temperature and particle density fields. Since these fields are independent it follows from Eq. (3.216) that the two relationships

$$\int_{\Omega} d\mathbf{r} \int_{\Omega} d\mathbf{r}' \, C_v(\mathbf{r}, \mathbf{r}') \delta T(\mathbf{r}) \delta T(\mathbf{r}') \geq 0 \tag{3.217}$$

and

$$\sum_{\alpha=1}^{c} \sum_{\beta=1}^{c} \int_{\Omega} d\mathbf{r} \int_{\Omega} \frac{d\mu_\alpha(\mathbf{r})}{\delta n_\beta(\mathbf{r}')} \bigg|_{T} \delta n_\alpha(\mathbf{r}) \delta n_\beta(\mathbf{r}') \geq 0 \tag{3.218}$$

must be satisfied separately. The first of these pertains to the thermal stability of the system while the second combines the conditions for mechanical stability and for stability with respect to particle diffusion.

If the initial state is one of homogeneous equilibrium, the stability criterion (3.216) can be replaced by its more useful Fourier transform,

$$\sum_{\mathbf{k}} \left\{ \frac{C_v(\mathbf{k})}{k_B T^2} \delta T(\mathbf{k}) \delta T(-\mathbf{k}) + \frac{1}{k_B T} \sum_{\alpha=1}^{c} \sum_{\beta=1}^{c} \frac{\delta \mu_\alpha(\mathbf{k})}{\delta n_\beta(\mathbf{k})} \bigg|_{T, \mathbf{n}'_\beta} \delta n_\alpha(\mathbf{k}) \delta n_\beta(-\mathbf{k}) \right\} \geq 0 \tag{3.219}$$

Now, consider a spatially periodic temperature fluctuation field, $\delta T(\mathbf{k})$, characterized by a single wavevector $\mathbf{k}$. Then it follows from Eq. (3.217) that the system will be stable with respect to this fluctuation if and only if the corresponding value of the wavevector dependent heat capacity is nonnegative, that is, if $C_v(|\mathbf{k}|) \geq 0$. However, according to Eq. (3.167),

$$C_v(|\mathbf{k}|) = \frac{1}{k_B T^2 \Omega} \langle |\delta \hat{\varepsilon}_\perp(\mathbf{k})|^2 \rangle_{\text{eq}}$$

and so $C_v(|\mathbf{k}|)$ is, in fact, nonnegative for *all* values of the wavevector $\mathbf{k}$. From this we conclude that systems in states of homogeneous equilibrium are stable with respect to thermal fluctuations of *all* wavelengths.

A further conclusion can be drawn by combining this last observation with the previously derived formula

$$C_v(|\mathbf{k}|) = C_v[1 + k^2\xi_T^2 + O(k^4)]$$

Thus, if the quantity ξ_T^2 were negative valued there then would be an associated range of long wavelengths for which $C_v(|\mathbf{k}|)$ could be negative, provided of course that $C_v > 0$. This implies that $\xi_T^2 > 0$ for all globally stable, homogeneous states of equilibrium.

The formation of a spatially periodic distribution of the particles is subject to the stability criterion

$$\sum_{\mathbf{k}} \left[\sum_{\alpha=1}^{c} \sum_{\beta=1}^{c} \left. \frac{\delta\mu_\alpha(\mathbf{k})}{\delta n_\beta(\mathbf{k})} \right|_{T,\mathbf{n}'_\beta} \delta n_\alpha(\mathbf{k}) \delta n_\beta(-\mathbf{k}) \right] \geq 0 \tag{3.220}$$

Conditions for stability with respect to other isothermal processes can be obtained as special cases of Eq. (3.220). For example, stability with respect to displacements from equilibrium that are both isothermal and isobaric is conditioned by the requirement that

$$\sum_{\mathbf{k}} \left[\sum_{\alpha=2}^{c} \sum_{\beta=2}^{c} \left\{ \left. \frac{\delta\mu_\alpha(\mathbf{k})}{\delta x_\beta(\mathbf{k})} \right|_{T,p,\mathbf{x}'_\beta} - \left. \frac{\delta\mu_1(\mathbf{k})}{\delta x_\beta(\mathbf{k})} \right|_{T,p,\mathbf{x}'_\beta} + \Delta\mu_{\alpha\beta}(\mathbf{k}) \right\} \right.$$
$$\left. \delta x_\alpha(\mathbf{k}) \delta x_\beta(-\mathbf{k}) \right] \geq 0 \tag{3.221}$$

with

$$\left. \frac{\delta\mu_\alpha(\mathbf{k})}{\delta x_\beta(\mathbf{k})} \right|_{T,p,\mathbf{x}'_\beta} = \left. \frac{\delta\mu_\alpha(\mathbf{k})}{\delta x_\beta(\mathbf{k})} \right|_{T,n,\mathbf{x}'_\beta} + \left. \frac{\delta\mu_\alpha(\mathbf{k})}{\delta n(\mathbf{k})} \right|_{T,\mathbf{x}} \left. \frac{\delta n(\mathbf{k})}{\delta x_\beta(\mathbf{k})} \right|_{T,p,\mathbf{x}'_\beta} \tag{3.222}$$

and where [using the Fourier transform, $\delta p(\mathbf{k}) = \tilde{\sigma}(\mathbf{k})\delta T(\mathbf{k}) + \Sigma_{\alpha=1}^{c} \tilde{n}_\alpha(\mathbf{k})\delta\mu_\alpha(\mathbf{k})$, of Eq. (2.62)]

$$\Delta\mu_{\alpha\beta}(\mathbf{k}) = \left\{ \frac{1}{n} \left. \frac{\delta n(\mathbf{k})}{\delta x_\alpha(\mathbf{k})} \right|_{T,p,\mathbf{x}'_\alpha} \right\} \left\{ \sum_{\gamma=1}^{c} [n_\gamma - \tilde{n}_\gamma(\mathbf{k})] \left. \frac{\delta\mu_\gamma(\mathbf{k})}{\delta x_\beta(\mathbf{k})} \right|_{T,p,\mathbf{x}'_\beta} \right\} \tag{3.223}$$

In the long-wavelength ($k \to 0$) limit $\tilde{n}_\gamma(\mathbf{k}) \to n_\gamma$, and $\Delta\mu_{\alpha\beta}(\mathbf{k}) \to 0$ so that the criterion (3.221) reduces to the global thermodynamic condition for stability with respect to diffusion.

Stability with respect to a displacement from equilibrium of the total particle (number) density, with the temperature and composition held fixed, is conditioned by the requirement that

$$\begin{aligned}
&\sum_{\mathbf{k}} \left[\sum_{\alpha=1}^{c} \sum_{\beta=1}^{c} \left. \frac{\delta\mu_\alpha(\mathbf{k})}{\delta n_\beta(\mathbf{k})} \right|_{T,\mathbf{n}'_\beta} x_\alpha x_\beta \right] \delta n(\mathbf{k})\, \delta n(-\mathbf{k}) \\
&\quad = \frac{1}{n} \sum_{\mathbf{k}} \left\{ \sum_{\alpha=1}^{c} \left. \frac{\delta\mu_\alpha(\mathbf{k})}{\delta n(\mathbf{k})} \right|_{T,\mathbf{x}} n_\alpha \right\} |\delta n(\mathbf{k})|^2 \\
&\quad = \sum_{\mathbf{k}} \left\{ \frac{1}{n} \left. \frac{\delta p(\mathbf{k})}{\delta n(\mathbf{k})} \right|_{T,\mathbf{x}} + \frac{1}{n} \sum_{\alpha=1}^{c} [n_\alpha - \tilde{n}_\alpha(\mathbf{k})] \left. \frac{\delta\mu_\alpha(\mathbf{k})}{\delta n(\mathbf{k})} \right|_{T,\mathbf{x}} \right\} |\delta n(\mathbf{k})|^2 \\
&\quad \geq 0
\end{aligned} \tag{3.224}$$

From this we conclude that a system will be stable with respect to the formation of a spatially periodic displacement from equilibrium of the total density, $\delta n(\mathbf{k})$, provided that the corresponding value of $n\delta p(\mathbf{k})/\delta n(\mathbf{k})_{T,\mathbf{x}} + n \; \Sigma_{\alpha=1}^{c} \; [n_\alpha - \tilde{n}_\alpha(\mathbf{k})]\delta\mu_\alpha(\mathbf{k})/\delta n(\mathbf{k})|_{T,\mathbf{x}} \equiv \kappa_T^{-1}(\mathbf{k}) + \Delta\kappa_T^{-1}(\mathbf{k})$ is positive definite. The system will be stable with respect to arbitrary fluctuations of density provided that this condition is met for all values of k. The function $\Delta\kappa_T^{-1}(\mathbf{k})$, like $\Delta\mu_{\alpha\beta}(\mathbf{k})$ of Eq. (3.223), vanishes in the long-wavelength limit, in which case we recover the global stability condition $\bar{\kappa}_T \geq 0$.

Since $\kappa_T^{-1}(\mathbf{k}) + \Delta\kappa_T^{-1}(\mathbf{k})$ need not be positive under all circumstances, we expect there to be spatial inhomogeneities of density with respect to which certain states of equilibrium are *un*stable. Indeed, there surely must be instabilities of this sort that can be identified with the onset of the transition from the homogeneous, isotropic fluid to an ordered, crystalline solid. Wavevectors $\mathbf{k}_c$ characteristic of spatially periodic density fluctuations with respect to which the homogeneous fluid state is unstable satisfy the equation [see Eq. (3.154)]

$$\begin{aligned}
&nk_{\mathrm{B}}T[1 - nC_{nn}(k_c)] + \sum_{\alpha} [\tilde{n}_\alpha(k_c) - n_\alpha] \\
&\qquad \times \left[k_{\mathrm{B}}T - n \left. \frac{\delta\mu_\alpha(k_c)}{\delta n(k_c)} \right|_{T,\mathbf{x}} \right] = 0
\end{aligned} \tag{3.225}$$

By discarding the second term (the summation) of Eq. (3.225) we obtain the stability limit of a supercooled liquid,

$$\kappa_T^{-1}(k_c) = nk_B T[1 - nC_{nn}(k_c)] = 0 \tag{3.226}$$

which Schneider et al. [51] and Lovett [19] previously identified with the onset of an ordered, crystalline phase.

By expanding $C_{nn}(k_c)$ in a Taylor series and retaining only the first and second terms we can obtain from Eq. (3.226) an estimate of the smallest value of k_c for which the fluid phase reaches its limit of stability. The result of this calculation is

$$k_c^{-2} = -\xi_n^2 \tag{3.227}$$

with ξ_n^2 denoting the multicomponent analog of the quantity defined by Eq. (3.190). According to this admittedly approximate result, the fluid will be stable with respect to density fluctuations provided that $\xi_n^2 > 0$, for then there is no real-valued wavevector that satisfies the truncated, long-wavelength limit of Eq. (3.226). Under these same conditions, possible only if the particle interactions include attractive parts, the asymptotic form of the density autocorrelation function is given by the familiar Ornstein–Zernike formula (3.194). On the other hand when $\xi_n^2 < 0$, as it is when the repulsive interactions are dominant, the prediction of Eq. (3.227) is that the fluid will be unstable with respect to periodic density fluctuations characterized by a real valued wavenumber $k_c = |\xi_n|^{-1}$. In this case the asymptotic behavior of the density fluctuation autocorrelation function is the damped oscillatory function given by Eq. (3.193).

With the second term on the left-hand side of Eq. (3.225) included, the formula (3.227) is replaced by

$$k_c^{-2} = -\left[\xi_n^2 + \sum_\alpha \tilde{n}_\alpha^{(2)} \{\bar{v}_\alpha - k_B T \bar{\kappa}_T\}\right] \tag{3.228}$$

with $\bar{v}_\alpha = (\partial \mu_\alpha / \partial p)_{T,\mathbf{x}}$ denoting the partial molecular volume and where $\tilde{n}_\alpha^{(2)}$ is the coefficient of $-k^2$ in the expansion

$$\begin{aligned}\tilde{n}_\alpha(k) &= \frac{1}{\Omega} \int_\Omega d\mathbf{r} \int_\Omega d\mathbf{r}'\, e^{-i\mathbf{k}\cdot(\mathbf{r}-\mathbf{r}')}\, \tilde{n}_\alpha(\mathbf{r}, \mathbf{r}')_{\mathrm{eq}} \\ &= n_\alpha - \tilde{n}_\alpha^{(2)} k^2 + O(k^4)\end{aligned} \tag{3.229}$$

An estimate of the corrective term in Eq. (3.228) can be made using the approximation $\hat{p}(\mathbf{r}) \doteq \hat{p}_{\mathrm{HY}}(\mathbf{r}) \equiv \frac{1}{3}\, \mathrm{tr}[\hat{\mathbf{p}}_{\mathrm{KIN}}(\mathbf{r}) + \hat{\mathbf{p}}_{\mathrm{INT}}(\mathbf{r})]$ of Eq. (3.125).

Thus, from this approximation and Eq. (3.38) it follows that

$$\tilde{n}_\alpha(\mathbf{r},\mathbf{r}') = \frac{\delta p(\mathbf{r})}{\delta \mu_\alpha(\mathbf{r}')}\bigg|_{T,\boldsymbol{\mu}'_\alpha} = \frac{1}{k_\mathrm{B}T}\langle \delta\hat{p}_\mathrm{HY}(\mathbf{r})\delta\hat{n}_\alpha(\mathbf{r}')\rangle \tag{3.230}$$

and

$$\tilde{n}^{(2)}_\alpha = \frac{1}{6k_\mathrm{B}T\Omega}\int_\Omega d\mathbf{r}\int_\Omega d\mathbf{r}'|\mathbf{r}-\mathbf{r}'|^2\langle \delta\hat{p}_\mathrm{HY}(\mathbf{r})\delta\hat{n}_\alpha(\mathbf{r}')\rangle_\mathrm{eq} \tag{3.231}$$

The contribution to $\tilde{n}^{(2)}_\alpha$ from $\delta\hat{p}_\mathrm{KIN}$ is $\frac{2}{3}\pi n_\alpha \sum_\beta n_\beta \int_0^\infty dr\, r^4 h_{\alpha\beta}(r)$; the contribution from $\delta\hat{p}_\mathrm{INT}$ is easily formulated but difficult to evaluate because it depends on integrals of three-particle spatial correlation functions.

IV. CLOSING REMARKS

Situations frequently are encountered that call for extensions of conventional, global thermodynamics to systems with intensive properties that vary continuously from point to point. The natural response in such cases is to associate with each of these properties a corresponding thermodynamic field. Thus, for example, it is eminently reasonable to introduce fields $T(\mathbf{r})$ and $p(\mathbf{r})$ corresponding to the temperature T and pressure p, respectively, each of which is a well-defined property of a spatially uniform, macroscopic system. Questions then arise about the relationships connecting the variations of these and other thermodynamic field variables.

One purpose of the theory presented here has been to provide answers to questions of this sort. Thus, we have found that some of these relationships such as the connection

$$\delta\varepsilon(\mathbf{r}) = T(\mathbf{r})\delta s(\mathbf{r}) + \sum_{\alpha=1}^{c} \mu_\alpha(\mathbf{r})\delta n_\alpha(\mathbf{r}) \tag{4.1}$$

between the variations of (the densities of) internal energy, entropy, and the species particle numbers, are precise analogs of the relationships among the corresponding intensive variables of a uniform fluid. However, there are other connections such as

$$\delta\varepsilon(\mathbf{r}) = \int_\Omega d\mathbf{r}'\left[C_v(\mathbf{r},\mathbf{r}')\delta T(\mathbf{r}') + \sum_{\alpha=1}^{c}\frac{\delta\varepsilon(\mathbf{r})}{\delta n_\alpha(\mathbf{r}')}\bigg|_{T,\mathbf{n}'_\alpha}\delta n_\alpha(\mathbf{r}')\right] \tag{4.2}$$

which incorporate functions with two spatial arguments, such as the heat capacity $C_v(\mathbf{r}, \mathbf{r}')$. These two-point functions invariably can be identified as functional derivatives, for example, $C_v(\mathbf{r}, \mathbf{r}') = \delta\varepsilon(\mathbf{r})/\delta T(\mathbf{r}')|_{\mathbf{n}}$, which are not singular or "local," that is, not proportional to $\delta(\mathbf{r} - \mathbf{r}')$. [In this context notice from Eq. (4.1) that the functional derivative $\delta\varepsilon(\mathbf{r})/\delta s(\mathbf{r}')|_{\mathbf{n}} = T(\mathbf{r})\delta(\mathbf{r} - \mathbf{r}')$ *is* singular.]

We also have discovered that to achieve consistency between the thermodynamic field theory and a corresponding statistical mechanical theory it is necessary to assume that the pressure is a nonlocal functional of the temperature and chemical potential fields, that is,

$$\delta p(\mathbf{r}) = \int_{\Omega} d\mathbf{r}' \left[\tilde{\sigma}(\mathbf{r}, \mathbf{r}')\delta T(\mathbf{r}') + \sum_{\alpha=1}^{c} \tilde{n}_\alpha(\mathbf{r}, \mathbf{r}')\delta\mu_\alpha(\mathbf{r}') \right] \tag{4.3}$$

with $\tilde{\sigma}(\mathbf{r}, \mathbf{r}') = \delta p(\mathbf{r})/\delta T(\mathbf{r}')|_{\boldsymbol{\mu}}$ and $\tilde{n}_\alpha(\mathbf{r}, \mathbf{r}') = \delta p(\mathbf{r})/\delta\mu_\alpha(\mathbf{r}')|_{T,\boldsymbol{\mu}'_\alpha}$ denoting nonsingular functional derivatives. As a consequence of this assumed nonlocality, we were *un*able to establish the local field analog,

$$\sigma(\mathbf{r})\delta T(\mathbf{r}) + \sum_{\alpha=1}^{c} n_\alpha(\mathbf{r})\delta\mu_\alpha(\mathbf{r}) - \delta p(\mathbf{r}) = 0 \tag{4.4}$$

of the global Gibbs–Duhem relation. These discrepancies between the field theory and one's intuitive expectations were shown to disappear if the quantities involved in Eqs. (4.3) and (4.4) were replaced by averages extended over regions with linear dimensions equal or greater than the range of appropriate many-body correlation functions. Similar conclusions were found to apply to other discrepancies of this same sort.

Important outputs of the statistical mechanical theory are the correlation function formulas that it provides for the functional derivatives of the thermodynamic field theory. Thus, for example, the heat capacity (per unit volume) at constant volume is related by the expression

$$C_v(\mathbf{r}, \mathbf{r}') = \frac{1}{k_B T^2(\mathbf{r})} \langle \delta\tilde{\varepsilon}_\perp(\mathbf{r})\delta\hat{\varepsilon}_\perp(\mathbf{r}')\rangle_{\text{ref}} \tag{4.5}$$

to that portion of the micromechanical energy fluctuation (about a reference state) that is orthogonal to fluctuations of the corresponding micromechanical compositional fluctuations $\delta\hat{n}_\alpha(\mathbf{r})$. Analogous correlation integral formulas were obtained as the statistical mechanical counterparts to other functional derivatives.

In addition to these we have demonstrated a general procedure for generating micro*mechanical* formulas corresponding to fluctuations of *thermal* variables, such as temperature, entropy, and the chemical potentials. These formulas permit the construction of correlation integral expressions such as

$$\langle \delta \hat{T}(\mathbf{r}) \delta \hat{n}_\alpha(\mathbf{r}') \rangle_{\text{ref}} = 0 \tag{4.6}$$

and

$$\langle \delta \hat{T}(\mathbf{r}) \delta \hat{T}(\mathbf{r}') \rangle_{\text{ref}} = k_{\text{B}} T^2(\mathbf{r}') C_v^{-1}(\mathbf{r}', \mathbf{r}) \tag{4.7}$$

involving thermal as well as micromechanical variables.

In Section III.E we addressed the important topic of fluctuations in macroscopically uniform systems, first by establishing the connections between the long-wavelength limits of (the Fourier transforms of) certain correlation integrals and the partial derivatives of global thermodynamics and then by considering several examples of thermodynamic quantities that display experimentally observable spatial dispersion. A number of conclusions were then reached concerning the effects upon these quantities of long- and short-ranged interactions. This chapter concluded with a brief examination of stability with respect to spatial inhomogeneities.

ACKNOWLEDGMENTS

The research reported here has been supported by grants from the Theoretical and Computational Chemistry Program and from the International Program Division of the National Science Foundation.

REFERENCES

1. See, for example, R. Pecora, Ed., *Dynamic Light Scattering*, Plenum, New York, 1985.
2. R. F. Greene and H. B. Callen, *Phys. Rev.*, **83**, 1231 (1951).
3. H. B. Callen and R. F. Greene, *Phys. Rev.*, **86**, 702 (1952).
4. H. B. Callen, *Thermodynamics*, Chap. 15, Wiley, New York, 1962.
5. L. Tisza and P. M. Quay, *Ann. Phys.*, **25**, 48 (1963).
6. L. D. Landau and E. M. Lifshitz, *Statistical Physics*, Chap. 12, Pergamon, London, 1958.
7. P. Schofield, *Proc. Phys. Soc.*, **88**, 149 (1966).
8. P. Schofield, in *Physics of Simple Liquids*, H. N. V. Temperley, J. S. Rowlinson, and G. S. Rushbrooke, Eds., Chap. 13, North-Holland, Amsterdam, 1968.
9. J. Keizer, *Statistical Thermodynamics of Non-Equilibrium Processes*, Springer, New York, 1987.
10. A. R. Altenberger and J. S. Dahler, in *Advances in Thermodynamics*, Vol. 6, S. Sieniutycz and P. Salamon, Eds., Taylor & Francis, New York, 1991.
11. R. Defay, I. Prigogine, A. Bellemans, and D. H. Everett, *Surface Tension and Adsorption*, Longmans, London, 1966.

12. R. Evans, *Adv. Phys.*, **28**, 143 (1979).
13. P. Schofield and J. R. Henderson, *Proc. R. Soc. London* **A379**, 231 (1982); **A380**, 211 (1982).
14. J. K. Percus, *Int. J. Quantum Chem. Quantum Chem. Symp.* **16**, 33 (1982) and an article in *The Liquid State of Matter*, E. W. Montroll and J. L. Lebowitz, Eds., North-Holland, Amsterdam, 1982.
15. M. S. Jhon, J. S. Dahler, and R. C. Desai, *Adv. Chem. Phys.*, **44**, 279 (1981).
16. F. F. Abraham, *Phys. Rep.*, **53**, 93 (1979).
17. A. R. Altenberger, J. Chem. Phys., **76**, 1473 (1982).
18. M. Baus, *Mol. Phys.*, **50**, 543 (1983).
19. R. Lovett, *J. Chem. Phys.*, **66**, 1225 (1977).
20. J. D. Gunton and M. Droz, *Introduction to the Theory of Metastable and Unstable States*, Springer, Berlin, 1983.
21. R. Lovett, P. Ortoleva, and J. Ross, *J. Chem. Phys.*, **69**, 947 (1978).
22. See, for example, H. J. Kreuzer, *Nonequilibrium Thermodynamics and its Statistical Foundations*, Chap. 1, Clarendon, Oxford, 1981.
23. H. B. Callen, *Thermodynamics*, Chap. 15, Wiley, New York, 1962, p. 55.
24. See, for example, K. S. Schmitz, *Dynamic Light Scattering by Macromolecules*, Chap. 2, Academic, Boston, 1990.
25. H. L. Friedman, *A Course of Statistical Mechanics*, Prentice-Hall, Englewood Cliffs, NJ, 1985.
26. D. N. Zubarev, *Non-Equilibrium Statistical Thermodynamics*, Chap. 1, Consultants Bureau, New York, 1974.
27. R. S. Ingarden, *Fortsh. Phys.*, **13**, 755 (1965).
28. M. S. Longair, *Theoretical Concepts in Physics*, Cambridge University, Cambridge, 1986, p. 145.
29. L. E. Elsgole, *Variational Calculus*, Chap. 4, Polish Scientific Publ., Warsaw, Poland, 1960.
30. H. B. Callen, *Thermodynamics*, Chap. 3, Wiley, New York, 1962.
31. S. A. Adelman, *J. Chem. Phys.*, **64**, 724 (1976); *Chem. Phys. Lett.*, **38**, 567 (1976).
32. S. A. Adelman and J. M. Deutsch, *Adv. Chem. Phys.*, **31**, 103 (1975).
33. Our assertion that the formula (3.124) is "commonly accepted" is not strictly correct, inasmuch as P. Schofield and J. R. Henderson have shown in [13] that the procedures commonly used to establish this formula are not unique. However, there do exist additional conditions that one reasonably can impose upon $\hat{p}_{\mathrm{INT}}(\mathbf{r})$, which establish the uniqueness of the formula (3.124). This will be demonstrated in a paper we currently are preparing for publication.
34. J. P. Hansen and J. R. McDonald, *Theory of Simple Liquids*, Chap. 5, Academic, London, 1990.
35. J. G. Kirkwood and F. P. Buff, *J. Chem. Phys.*, **19**, 774 (1951).
36. R. Abe, *Prog. Theor. Phys.*, **19**, 57, 407 (1958).
37. G. Börner, *The Early Universe*, Chaps. 10, 12, Springer, Berlin, 1988.
38. W. C. Saslaw, *Gravitational Physics of Stellar and Galactic Systems*, Chap. 27, Cambridge University Press, Cambridge, 1985.
39. E. J. Groth and P. J. E. Peebles, *Ap. J.*, **217**, 385 (1977).

40. A. R. Altenberger and J. S. Dahler, *Ap. J.*, **421**, L9 (1994).
41. L. S. Ornstein and F. Zernike, *Phys. Z.*, **19**, 139 (1918).
42. M. E. Fisher, *J. Math. Phys.*, **5**, 944 (1964).
43. C. J. Pings, "Structure of Simple Liquids by X-Ray Diffraction," Chap. 10 in *Physics of Simple Liquids*, H. N. V. Temperley, J. S. Rowlinson and G. S. Rushbrooke, Eds., Wiley, New York, 1968.
44. J. Woodhead-Galloway, T. Gaskell, and N. H. March, *J. Phys. C, Ser. 2*, 1, 271 (1968).
45. J. J. Binney, N. J. Dowrick, A. J. Fisher, and M. E. J. Newman, *The Theory of Critical Phenomena*, Oxford University Press, Oxford, 1992.
46. J. J. Binney, N. J. Dowrick, A. J. Fisher, and M. E. J. Newman, *The Theory of Critical Phenomena*, Oxford University Press, Oxford, 1992, p. 22.
47. M. Smoluchowski, *Ann. Phys.*, **25**, 205 (1908); *Bull. Int. Ac. Pol. Sci. L (A)*, 493 (1911).
48. A. Einstein, *Ann. Phys.*, **33**, 1275 (1910).
49. J. Kocinski and L. Wojtczak, *Theory of Critical Scattering. An Introduction*, Elsevier-PWN, Warsaw, Poland, 1978.
50. See, for example, D. McIntyre and J. V. Sengers, *Physics of Simple Liquids*, Chap. 11, H. N. V. Temperley, J. S. Rowlinson, and G. S. Rushbrooke, Eds., Wiley, New York, 1968.
51. T. Schneider, R. Brout, H. Thomas, and J. Feder, *Phys. Rev. Lett.*, **25**, 1423 (1970).

NONLINEAR DIELECTRIC AND KERR EFFECT RELAXATION IN ALTERNATING FIELDS

J.L. DEJARDIN AND G. DEBIAIS

Groupe de Physique Théorique, Université de Perpignan, 66860 Perpignan Cédex, France

CONTENTS

Advances in Chemical Physics, Volume XCI, Edited by I. Prigogine and Stuart A. Rice.
ISBN 0-471-12002-2

I. INTRODUCTION

The first investigator to observe and give a precise description of the random movement of microparticles in suspension, namely, spores in water, was the English botanist Robert Brown [1] in 1828 (see also Nelson in [2]). This movement, which is named after him, increases when the temperature increases. It is ceaseless, totally irregular, and composed of rotations and translations. Molecules in a fluid show an identical agitation movement. The effects observed on a macroscopic scale are not solely resultants of these irregular movements. To put it more exactly, these average effects must be described by a statistical theory.

Numerous works have described in a succinct fashion the evolution of our knowledge and research in this area. Recently, Coffey [3] documented the history of the Brownian motion in detail. To illustrate our approach it seems necessary to give an account of the main steps in the historical development of the subject.

The first theory is that of Einstein in 1905 developed in a subsequent book [4] (see also Risken in [5]). This theory begins with a one-dimensional diffusion equation which does not restrict the generality of the problem. We consider f particles per unit volume situated in a volume element lying between x and $x + dx$, and then another equivalent volume element situated at x'. The probability that one particle in the element x at time t enters the volume element x' at time $t' = t + \tau$ depends on the distance $x' - x$ and the time interval τ. This transition probability is denoted by $\phi(x' - x, \tau)$. The particles in the volume element x' at time $t + \tau$ are necessarily in another volume element, or eventually in the same element at time t, and so we obtain the Chapman–Kolmogorov equation

$$f(x', t + \tau) = \int_{-\infty}^{+\infty} f(x, t)\phi(x' - x, \tau)\, dx \tag{4.1}$$

Putting $X = x - x'$, and knowing that positive and negative displacements are equally likely to occur, one has

$$dX = dx \qquad \phi(X, \tau) = \phi(-X, \tau)$$

so that Eq. (4.1) becomes

$$f(x', t + \tau) = \int_{-\infty}^{+\infty} f(x' + X, t)\phi(X, \tau)\, dX \tag{4.2}$$

If τ and X are small enough, we can expand each side of Eq. (4.2) in a Taylor series

$$\begin{aligned} f(x', t) + \tau \frac{\partial f}{\partial t} + (\tau^2/2!) \frac{\partial^2 f}{\partial t^2} + \cdots \\ = \int_{-\infty}^{+\infty} \left[f(x', t) + X \frac{\partial f}{\partial x} + (X^2/2!) \frac{\partial^2 f}{\partial x^2} + \cdots \right] \phi(X, \tau) dX \end{aligned} \tag{4.3}$$

Taking account of the definition of the function ϕ, simple mathematical considerations give

$$\begin{aligned} &\int_{-\infty}^{+\infty} \phi(X, \tau) dX = 1\,, \qquad \text{(sum over all the probabilities)} \\ &\int_{-\infty}^{+\infty} \phi(X, \tau) X\, dX = \langle X \rangle = 0 \qquad \text{(equal probability of } X\text{)} \end{aligned} \tag{4.4}$$

The expression

$$\langle X^2 \rangle = \int_{-\infty}^{+\infty} \phi(X, \tau) X^2 dX$$

represents the mean-square displacement.

We assume that higher order terms such as $\langle X^4 \rangle$ and $\langle X^6 \rangle$, in the right-hand side of Eq. (4.3) are all of second order in τ, so that Eq. (4.3) may be written

$$\tau \frac{\partial f}{\partial t} = 1/2 \langle X^2 \rangle \frac{\partial^2 f}{\partial x^2} \tag{4.5}$$

By analogy, one remarks that Eq. (4.5) is a diffusion equation very similar to that of heat conduction where the density of particles f is equivalent to the temperature. From this, it is possible to define a diffusion coefficient (D) of the particles that is

$$D = \frac{\langle X^2 \rangle}{2\tau} \tag{4.6}$$

We also suppose that the particles have an equilibrium distribution of the Maxwell–Boltzmann type

$$f = f_0 \exp(-V/kT) \tag{4.7}$$

where

$V = Fx$ is the potential energy of the particles
$F = 6\pi\eta a v = \xi_t v$ is the vicous force given by Stokes's law for translation (η is the viscosity coefficient, a and v are the radius and the velocity of a particle, respectively)
k is the Boltzmann constant and T is the absolute temperature.

One thus deduces that

$$D = \frac{kT}{6\pi\eta a}$$

from which Einstein's formula is obtained

$$\frac{\langle X^2 \rangle}{\tau} = \frac{kT}{3\pi\eta a} = \frac{2kT}{\xi_t} \tag{4.8}$$

Some time after the formulation of this theory, Langevin [5, 6] in 1908, exactly calculated the mean value of the square displacement by supposing that the particle had an equation of motion

$$m \frac{d^2 x}{dt^2} + \xi_t \frac{dx}{dt} = F(t) \tag{4.9}$$

where $F(t)$ is the fluctuating force due to the translational Brownian motion and $x(t)$ is a random variable. The force F is assumed to have no dependence on $x(t)$ and to vary much quicker than $x(t)$.

Equation (4.9) can be written

$$\frac{m}{2} \frac{d}{dt} \left(\frac{d(x^2)}{dt} \right) - m \left(\frac{dx}{dt} \right)^2 = -\frac{\xi_t}{2} \frac{d(x^2)}{dt} + Fx \tag{4.10}$$

and so, averaging for each particle

$$\frac{m}{2}\frac{d}{dt}\left(\frac{d(\overline{x^2})}{dt}\right)-m\overline{\left(\frac{dx}{dt}\right)^2}=-\frac{\xi_t}{2}\frac{d(\overline{x^2})}{dt}+\overline{Fx} \tag{4.11}$$

The quantity $\overline{Fx}$ tends to zero due to the totally irregular variation of F. On taking account of the equipartition theorem, which results from a Maxwellian velocity distribution, one has

$$\frac{1}{2}m\overline{\left(\frac{dx}{dt}\right)^2}=\frac{1}{2}kT \tag{4.12}$$

so that Eq. (4.11) becomes

$$\frac{m}{2}\frac{d}{dt}\left(\frac{d(\overline{x^2})}{dt}\right)+\frac{\xi_t}{2}\frac{d(\overline{x^2})}{dt}=kT \tag{4.13}$$

On putting

$$u=\frac{d(\overline{x^2})}{dt}$$

the solution of Eq. (4.13) may be written

$$u=\frac{2kT}{\xi_t}+C\exp\left(-\frac{\xi_t t}{m}\right) \tag{4.14}$$

where C is a constant of integration.

If one assumes that $t \gg m/\xi_t$, this condition allows us to neglect the mass in comparison with the friction. Hence, Eq. (4.14) reduces to $u=2kT/\xi_t$. Integrating over a time interval ranging from 0 to τ, we obtain

$$\overline{x^2}-\overline{x_0^2}=\frac{2kT}{\xi_t}\tau \tag{4.15}$$

On using the initial condition $x_0=0$ at time $t=0$, one also has

$$\overline{(\Delta x)^2}=\frac{2kT}{\xi_t}\tau \tag{4.16}$$

whose relation is identical to that of Einstein.

At the same time, in 1906, a fundamental new contribution to the theory was made by Von Smoluchowski [6]. He considered the problem of the Brownian movement of a particle under the influence of an external force $K(x)$ and showed that if such a force acts on an assembly of particles, the distribution function obeys the following equation:

$$\frac{\partial f(x,t)}{\partial t} = -\frac{\partial}{\partial x}\left[\frac{K(x)}{\xi_t} f(x,t)\right] + \frac{kT}{\xi_t}\frac{\partial^2 f(x,t)}{\partial x^2} \tag{4.17}$$

Smoluchowski originally developed this equation in the case of an elastic force (harmonic oscillator) such that $K(x) = -cx$.

Debye [7] employed this equation in 1913 in an attempt to explain the abnormal dispersion of the electric permittivity observed in polar liquids at radio frequencies (microwaves and far-infrared). He adapted the Smoluchowski equation to his problem by replacing the coordinate x by the angular variable θ, then he replaced the translational friction coefficient ξ_t by the rotational coefficient ξ, and the force $K(x)$ by $-\partial V(\theta,t)/\partial\theta$, with $V(\theta,t)$ being the potential energy of the considered particle. The relevant equation in the noninertial limit is

$$\frac{\partial f(\theta,t)}{\partial t} = \frac{\partial}{\partial\theta}\left[\frac{1}{\xi}\frac{\partial V(\theta,t)}{\partial\theta} f(\theta,t) + \frac{kT}{\xi}\frac{\partial f(\theta,t)}{\partial\theta}\right] \tag{4.18}$$

This equation corresponds to the Langevin equation

$$\xi\frac{\partial\theta(t)}{\partial t} + \frac{\partial V(\theta,t)}{\partial\theta} = \lambda(t) \tag{4.19}$$

where $\lambda(t)$ is the white-noise driving torque due to the Brownian motion. It is generally assumed that $\lambda(t)$ is a centered Gaussian random variable that satisfies the following conditions:

$$\overline{\lambda(t)} = 0$$

$$\overline{\lambda(t)\lambda(t')} = 2kT\xi\delta(t-t')$$

where the overbar indicates a statistical average, and δ is the Dirac delta function. In this theory, one remarks that inertia is ignored. Physically speaking, Eq. (4.18) is appropriate for a two-dimensional (2D) medium. The molecules carry a permanent dipolar moment μ, and are supposed to be noninteracting with each other, with the sole exception of the perturbation due to the external electric field $E(t)$. The function $f(\theta,t)$ represents the distribution function that describes the number of dipoles whose axes are contained in an angle element $d\theta$ on the circumference of a circle and θ is the angle that makes the orientation of the dipole with the direction of the field. If we neglect the induced dipole moment, the potential energy is given by

$$V(\theta, t) = -\mu E(t) \cos \theta$$

Later, Debye [7] generalized this equation to the sphere model, that is an extent to three-dimensional (3D) space. For the evolution of the probability density function he obtained

$$\frac{\partial f(\theta, t)}{\partial t} = \frac{kT}{\xi \sin \theta} \frac{\partial}{\partial \theta} \left[\sin \theta \left(\frac{f(\theta, t)}{kT} \frac{\partial V(\theta, t)}{\partial \theta} + \frac{\partial f(\theta, t)}{\partial \theta} \right) \right] \tag{4.20}$$

The well-known Debye relation is derived from this work, which describes the dispersion of the dielectric constant as a function of the angular frequency of the alternating electric field

$$\boldsymbol{\varepsilon} = \varepsilon_\infty + \frac{S}{1 + i\omega\tau} \tag{4.21}$$

where τ is the Debye relaxation time equal to ξ/kT in 2D space and $\xi/2kT$ in 3D space, $S = \varepsilon_0 - \varepsilon_\infty$, ε_0 is the dielectric constant at very low frequencies, and ε_∞ is the dielectric constant at high frequencies resulting from the electronic polarization.

To take account of the absorption properties of the liquid medium, the dielectric constant is written in a complex form: $\boldsymbol{\varepsilon} = \varepsilon' - i\varepsilon''$. When applied to a frequency region lying from low frequencies to microwaves, its success has been enormous due, in particular, to the fact that Eq. (4.21) simply represents the Fourier transform of an exponentially decaying function. Unfortunately, its application to very high frequencies and the visible region predicts, for example, black water, which constitutes evidence of its limitation. The origin of these difficulties [8, 9] arises from the consideration that time is assumed to be greater than m/ξ_t, and thus Einstein's relation is not differentiable at time $t = 0$ since $(\Delta x^2)^{0.5}$ is proportional to $(t)^{0.5}$.

Much progress was made between 1918 and 1930 by Uhlenbeck and Ornstein [4, 6]. Their argument rests on statistical calculations based on the Langevin equation. By taking account of inertia, they showed that the correct mean-square displacement is given by

$$\langle (\Delta x)^2 \rangle = \frac{2kTm}{\xi_t^2} \left[\frac{\xi_t t}{m} - 1 + \exp\left(-\frac{\xi_t t}{m} \right) \right] \tag{4.22}$$

This equation is differentiable at the origin, and allows us to find Einstein's relation for $t \gg m/\xi_t$ again. When $t \ll m/\xi_t$, one obtains a dependence on the mass such that

$$\langle(\Delta x)^2\rangle = \frac{kTt^2}{m} \tag{4.23}$$

Uhlenbeck and Ornstein made use of Klein's [10] (1921) important diffusion equation

$$\frac{\partial W}{\partial t} + v\,\frac{\partial W}{\partial x} - \frac{1}{m}\frac{\partial V}{\partial x}\frac{\partial W}{\partial v} = \frac{\xi_t}{m}\frac{\partial}{\partial v}\left[vW + \frac{kT}{m}\frac{\partial W}{\partial v}\right] \tag{4.24}$$

In this equation, W is the distribution function in phase space. It depends on the position x and the velocity v at time t. The distribution function $f(x, t)$ in configuration space can then be calculated by integrating over all the possible velocities

$$f(x, t) = \int_{-\infty}^{+\infty} W(x, v, t)\, dv \tag{4.25}$$

The corresponding Langevin equation thus takes the most complete form

$$m\,\frac{d^2x}{dt^2} + \xi_t\,\frac{dx}{dt} + \frac{\partial V}{\partial x} = F(t) \tag{4.26}$$

where $F(t)$ is a stochastic force arising from the interaction of the particle with neighbors.

Solutions of Klein's equation may be easily obtained in two particular cases. The first case was considered by Uhlenbeck and Orstein. It corresponds to $V = 0$, that is, the free Brownian particle. The second case is defined by $V = \gamma x^2/2$. Thus we have the harmonic oscillator. For other forms of the potential V, solution of this equation becomes extremely difficult.

Equation (4.24) was reconsidered in 1940 by Kramers [6] who studied the escape of Brownian particles over a potential barrier. It is usually known as the Kramers equation, the Fokker–Planck–Kramers equation, or the generalized Liouville equation [2, 8]. Transposed to rotation in 2D space, it becomes

$$\frac{\partial W}{\partial t} + \bar{\omega}\,\frac{\partial W}{\partial \theta} - \frac{1}{I}\frac{\partial V}{\partial \theta}\frac{\partial W}{\partial \bar{\omega}} = \frac{\xi_t}{I}\frac{\partial}{\partial \bar{\omega}}\left(\bar{\omega}W + \frac{kT}{I}\frac{\partial W}{\partial \bar{\omega}}\right) \tag{4.27}$$

where $\bar{\omega} = \partial\theta/\partial t$ is the angular velocity of the particle, $W = W(\theta, \bar{\omega}, t)$, and I is the moment of inertia of the particle about its rotation axis.

The Kramers equation was applied in 1955 by Gross [11] (see also

Evans et al. in [12]) and in 1957 by Sack [13] to study rotational Brownian motion of an assembly of noninteracting dipolar molecules. This study corresponds to that of Debye, but seeks to include inertia. These authors resulted in a relation identical to that already obtained by Rocard [14] in 1933 after a less rigorous analysis (see also Coffey et al. in [15]). This equation is a modification of the Debye equation leading to a new definition of the dielectric constant such that

$$\boldsymbol{\varepsilon} = \varepsilon_\infty + \frac{S}{1 + i\omega\tau - (\omega^2 I\tau/\xi)} \tag{4.28}$$

It predicts negligible absorption at very high frequencies, and so transparent water in this region, which is much in conformity with experiment.

Faced with the difficulties encountered for solving the Kramers equation when the potential V is of any form, Brinkman [16] and Sack [17] showed that, for small inertial effects, the Fokker–Planck–Kramers equation could be put into a Smoluchowski equation form that is easier to manipulate and that contains the supplementary inertial term

$$\frac{m}{\xi_t}\frac{\partial^2 f}{\partial t^2} \quad \text{for translation}$$

or

$$\frac{I}{\xi}\frac{\partial^2 f}{\partial t^2} \quad \text{for rotation} \tag{4.29}$$

This equation is the so-called modified Smoluchowski equation (MSE). It was employed by Alexiewicz [18], and then by Coffey and McGoldrick [19] to calculate the expected values of the first two Legendre polynomials, namely, $\langle P_1(\cos\theta)\rangle$ and $\langle P_2(\cos\theta)\rangle$ appropriate to dielectric and Kerr effect relaxation, respectively. At the same time, this MSE was severely criticized by several authors [20–24], when applied, in particular, to nonlinear responses.

Coffey et al. [19, 25, 26], reconsidering the Kramers equation, deduced a method from perturbation theory for the solution of this equation for a plane rotator. This consists of separating the variables in the distribution function W, and then by developing the part relevant to angular velocity in a series of Weber functions of degree n and the spatial part in a Fourier series of degree p. These functions lead to a set of differential difference equations called Brinkman equations. Noting that it is possible to uncouple the dependence between n and p by considering only matrices of the form $n \times n$, these authors deduced that the limiting value $n = 2$ is equivalent to the MSE, and it is necessary in order to treat nonlinear problems to carry the calculations up to $n = 3$. Hence, they give interesting results concerning the linear response of the dielectric relaxation and the nonlinear response of the Kerr effect relaxation; that is to say concerning those molecules that present a permanent dipole moment and are not polarized by the orienting field.

Very recently, Déjardin [27] carried out a calculation in two dimensions for an ac field superimposed on a dc field. His method allows one to obtain equations of the Brinkman type easier, and he includes the induced dipole moment in his results.

In conclusion, it is necessary to mention some recent work of Coffey [28] who showed that inertial effects for molecules free to rotate in 3D space may be obtained in the case of dielectric relaxation by averaging the Langevin equation directly. This method may be applied to an assembly of needlelike rotators by choosing as variables the Hermite polynomials in the angular velocities and the associated Legendre functions of order 1 in the angular variables. An extension of this work to electric birefringence in 3D has been given by Coffey et al. [29] to yield the after-effect solution, that is, the response following the removal of a constant field.

II. INERTIAL EFFECTS IN DIELECTRIC RELAXATION

A. Influence of Inertia on the Dielectric Properties of Polar Liquids

The influence of inertia in dielectric relaxation was originally studied in the 1950s. It is only during the last few years, however, that such a study has begun in relation to the dynamic Kerr effect. It is clear that inertial effects are often masked by friction or viscosity, and, in general, they are neither important nor apparent, say in very high-frequency regions (e.g., microwave and infrared regions). These regions are usually difficult enough to access experimentally for dielectric relaxation and indeed they are more difficult for birefringence. The effects of electric conduction and electromagnetic radiation constitute severe obstacles to the study of these phenomena and so require high-quality and convenient apparatus. In addition, the mathematical treatment of Kramers equation in the general case has until quite recently been inaccessible.

Thus, experimentally, one usually has been contented with the description of dielectric relaxation provided by the Debye equation, which described the observed phenomena from very low frequencies to high frequencies of the order of 10^9 Hz well in most cases. Some aberrations have been reported, but they usually included experimental errors. In the case of the dynamic Kerr effect, the experimental region did not exceed 10^7 Hz and inertial effects were not existent. Moreover, the theoretical treatment excluding inertia (Smoluchowski diffusion equation) was not as easy an exercise.

In this chapter we wish to highlight the importance of including these effects and to show how that inclusion produces some very interesting situations while considering the limits of applicability of the Debye relations. We wish also to justify the original solution method of the Smoluchowski equation we recently proposed and we shall give further details on it.

We refer to the work of Coffey and McGoldrick who included inertia in the theory of dielectric relaxation by two different methods. The first method consists in a perturbation treatment of the Brinkman equations, which allows us to describe the dielectric linear response of noninteracting dipolar molecules. These molecules are disks that are free to rotate in two dimensions. The second method utilizes the MSE proposed by Sack [17] and concerns molecules of spherical symmetry that can rotate in three dimensions [19, 30]. The results obtained are identical and are totally rigorous in dielectric relaxation.

In this section, we limit ourselves to a consideration of the dielectric constant. We present its expression including inertia, which has the same form as that used in condensed matter physics. We shall describe methods of determining the parameters underlying these effects when they are accessible, and shall interpret some diagrams, such as the Cole–Cole diagrams $\varepsilon''(\varepsilon')$. Our purpose is to demonstrate how the Debye model is a limiting case of this new expression and inertial effects are not significant except in particular cases, which we propose to emphasize, using a typical example.

The solution of the MSE for the expected value of the first Legendre polynomial [19] yields

$$\langle P_1(\cos\theta)\rangle(t) = \frac{\mu}{3kT}\left[E_0 \frac{\left(1-\dfrac{\omega^2\tau}{\beta}\right)\cos\omega t + \omega\tau\sin\omega t}{\left(1-\dfrac{\omega^2\tau}{\beta}\right)^2 + (\omega\tau)^2}\right] \tag{4.30}$$

where $\beta = \xi/I$, is the reciprocal of the friction time. The other symbols have the meaning used in the Introduction.

The real and imaginary parts of the electric susceptibility, χ' and χ'', may be deduced from the relation

$$N\mu\langle\cos\theta\rangle(t) = \chi' E_0\cos\omega t + \chi'' E_0\sin\omega t \tag{4.31}$$

where N designates the number of dipoles per unit volume.

Thus, introducing the dielectric constant at very high frequencies, ε_∞, which results from the electronic polarization, the real and imaginary parts of the complex dielectric constant may be written

$$\varepsilon' = \varepsilon_\infty + \frac{N\mu^2}{3kT}\left[\frac{\left(1-\frac{\omega^2\tau}{\beta}\right)}{\left(1-\frac{\omega^2\tau}{\beta}\right)^2+(\omega\tau)^2}\right]$$
$$\varepsilon'' = \frac{N\mu^2}{3kT}\left[\frac{\omega\tau}{\left(1-\frac{\omega^2\tau}{\beta}\right)^2+(\omega\tau)^2}\right] \tag{4.32}$$

On setting

$$S = \frac{N\mu^2}{3kT} \qquad \omega_0^2 = \frac{2kT}{I} = \frac{\beta}{\tau} \qquad \gamma = \omega_0\tau = \frac{\xi}{\sqrt{2IkT}} \tag{4.33}$$

these become

$$\varepsilon' = \varepsilon_\infty + \frac{S\omega_0^2(\omega_0^2-\omega^2)}{(\omega_0^2-\omega^2)^2+\gamma^2\omega_0^2\omega^2} \tag{4.34a}$$

$$\varepsilon'' = \frac{S\omega_0^3\gamma\omega}{(\omega_0^2-\omega^2)^2+\gamma^2\omega_0^2\omega^2} \tag{4.34b}$$

This formulation is strictly identical to that used for the dielectric constant in the physics of condensed matter, and is also useful in liquid crystals and gel states. It may be established by starting from the Maxwell equations and the classical theory of oscillators, as discussed in Appendix A. For example, ω_0 is the angular frequency of an oscillator, S is the oscillator strength, that is its own contribution to the static dielectric constant, and γ is an adimensional damping constant which, given a temperature, reflects the duality between friction (ξ) and inertia (I). In the physics of condensed matter, these three parameters have an equivalent definition. However, they may be regarded as transposed to rectilinear vibrations of dipoles rather than to rotational ones as is the case for the liquid state.

Equations (4.33) may be extended to an ensemble of n oscillators in order to take account of the fact that each oscillator contributes linearly to the dielectric constant of the medium, as may be seen in Appendix A as well. This extension may of course be written in a complex form [see Eq. (A.12)]

$$\varepsilon = \varepsilon_\infty + \sum_{i=1}^{n} \frac{S_i \omega_i^2}{\omega_i^2 - \omega^2 + i\gamma_i \omega_i \omega} \tag{4.35}$$

Equation (4.35) again is identical to that given by Murthy and Vij [31] in their interpretation of absorption phenomena in liquid crystals of 4-*n*-alkyl-4′-cyano biphenyls in the submillimeter wave range. The approach of these authors, briefly summarized, is to use the superposition principle appropriate to the linear response and to suppose that the liquid under consideration is composed of a number of molecular rotations that are not coupled to each other. These movements may then be described by a linear combination of damped harmonic oscillators. Each one can be represented, in the absence of the field, by an equation of the Langevin type

$$I \frac{d^2\phi}{dt^2} + \xi \frac{d\phi}{dt} + V_0 \phi = 0 \tag{4.36}$$

where ϕ is the angle between the dipole axis and the reference direction that is the direction of the applied field, and V_0 is a constant.

Thus, their theory is in fact an adaptation to rotation of rectilinear oscillators. It is worth noting that the convergence of the results found from the oscillator theory and that based upon a rotational diffusion equation is remarkable. By comparison of both methods, one is able to express the constant V_0 in terms of $2kT$. This allows us to obtain (as seen in Section II.C) a number of microscopic parameters. Furthermore, the conditions of validity of such an equation have been well described by Sack and Coffey, whereas Murthy and Vij did not give such limits.

The equations we have given above are limited to small inertial effects. Sack [13], by exactly solving the Liouville equation, found the solution for the macroscopic polarization of an assembly of dipolar molecules rotating in 2D space, in the form of a continued fraction. On limiting himself to the second convergent (small inertia) of this continued fraction, he found that the solution corresponds to that obtained from the MSE. He introduced a parameter that he called γ and that we designate by g' to eliminate any confusion with our damping constant γ

$$g' = \frac{kT}{I\beta^2} = \frac{I}{\xi\tau'} \tag{4.37}$$

where $\tau' = \xi/kT$ is the Debye relaxation time in 2D space.

Sack estimated that the parameter g' should not exceed about 0.05 in order that the expression based on the second convergent should be valid. Coffey et al. [26] on the other hand, used the value 0.16 in discussing the linear dielectric response. In passing to the 3D case, we should note that τ' becomes τ, which is twice smaller, and g' is therefore changed into g, two times greater, such that

$$g = \frac{I}{\xi\tau} = 2g' \leq 0{,}32$$

Hence, the condition for validity of γ is the following

$$\gamma = \frac{1}{\sqrt{g}} \geq 1{,}75$$

Writing ε' and ε'' in the form

$$\begin{aligned}
\varepsilon' &= \varepsilon_\infty + \frac{S\left[1\Big/\left(1-\dfrac{\omega^2}{\omega_0^2}\right)\right]}{1+\left[\gamma^2\omega_0^2\omega^2\Big/\omega_0^4\left(1-\dfrac{\omega^2}{\omega_0^2}\right)^2\right]} \\
\varepsilon'' &= \frac{S\left[\gamma\omega_0^3\omega\Big/\omega_0^4\left(1-\dfrac{\omega^2}{\omega_0^2}\right)\right]}{1+\left[\gamma^2\omega_0^2\omega^2\Big/\omega_0^4\left(1-\dfrac{\omega^2}{\omega_0^2}\right)^2\right]}
\end{aligned} \tag{4.38}$$

it is clear that these relations are similar to those of Debye

$$\varepsilon' = \varepsilon_\infty + \frac{S}{1+\omega^2\tau^2}, \qquad \varepsilon'' = \frac{S\omega\tau}{1+\omega^2\tau^2} \tag{4.39}$$

when $(\omega^2/\omega_0^2) \rightarrow 0$, or at least, $(\omega^2/\omega_0^2) \ll 1$.

To satisfy this last condition, and to fix ideas, one is able in practice to take

$$\frac{\omega^2}{\omega_0^2} < \frac{1}{100} \qquad \text{or} \qquad \omega < \frac{\omega_0}{10}$$

from which one deduces that

$$\gamma > 10\,\omega\tau \tag{4.40}$$

Thus, when ω is greater than $\omega_0/10$, inertial effects will appear unless they are masked by frictional effects. The condition $\omega < \omega_0/10$ is, in effect, a necessary condition in order to observe a Debye-type mechanism, but it is not sufficient. It is necessary that the region investigated satisfies not only this condition but on the other hand contains a significant and characteristic variation of ε' and ε'', that is to say, in the region of absorption. If this is so, the quantity $\omega\tau$ is close to 1, because it is then well known, following Debye, that the equality $\omega\tau = 1$ gives the maximum of $\varepsilon''(\omega)$ and approximately the point of inflexion of the curve $\varepsilon'(\omega)$. We can then deduce that the condition $\gamma > 10$ reduces to Debye-type behavior. For negligible inertial effects, the definition $\gamma = \xi/(2kTI)^{0.5}$ shows that γ may even tend to infinity.

To illustrate and confirm these estimates, we considered a damped oscillator whose natural frequency is arbitrarily chosen, $N_0 = \omega_0/2\pi = 10^7$ Hz, with $\varepsilon_0 = 100$ and $\varepsilon_\infty = 50$ ($S = \varepsilon_0 - \varepsilon_\infty = 50$). Figures 4.1 and 4.2 represent the curves of dispersion of ε' and ε'' as a function of frequency following the formulation given by Eqs. (4.34), γ successively takes on the values 20 (case a), 5 (case b), and 1.75 (case c). The dispersion curves that correspond to the Debye relations [Eq. (4.39)] with $\tau = \gamma/\omega_0$ are represented in the same graphs (dashed lines). It appears that both curves

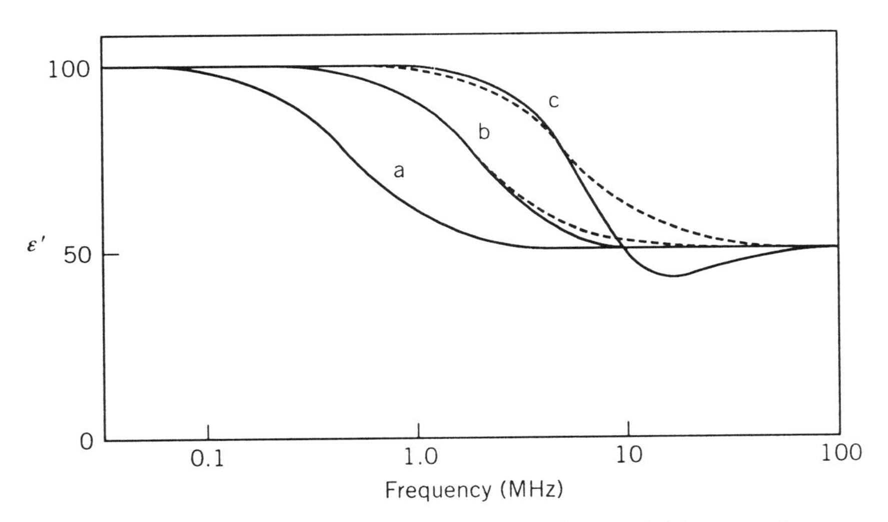

Figure 4.1. Variation of the real part ε' of the complex permittivity versus frequency for various values of γ. (a): $\gamma = 20$; (b): $\gamma = 5$; (c): $\gamma = 1.75$ (ε_0 and ε_∞ have been chosen arbitrarily equal to 100 and 50, respectively).

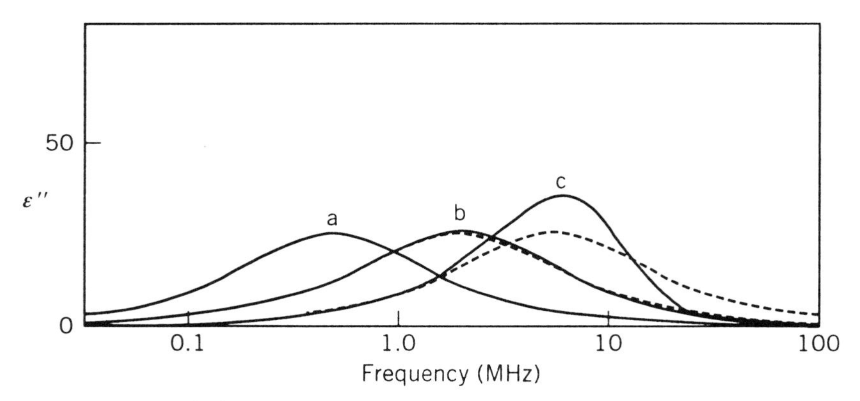

Figure 4.2. Variation of the imaginary part ε'' of the complex permittivity versus frequency for various values of γ. (a): $\gamma = 20$; (b): $\gamma = 5$; (c): $\gamma = 1.75$ (ε_0 and ε_∞ have been chosen arbitrarily equal to 100 and 50, respectively).

coincide for $\gamma = 20$. At high frequencies, starting from 10^6 Hz, differences appear for other γ values, a small difference for $\gamma = 5$, and a large one for $\gamma = 1.75$.

The Cole–Cole diagrams illustrated in Fig. 4.3 confirm these results. For $\gamma = 20$, we observe a semicircle, typical of the Debye response, but

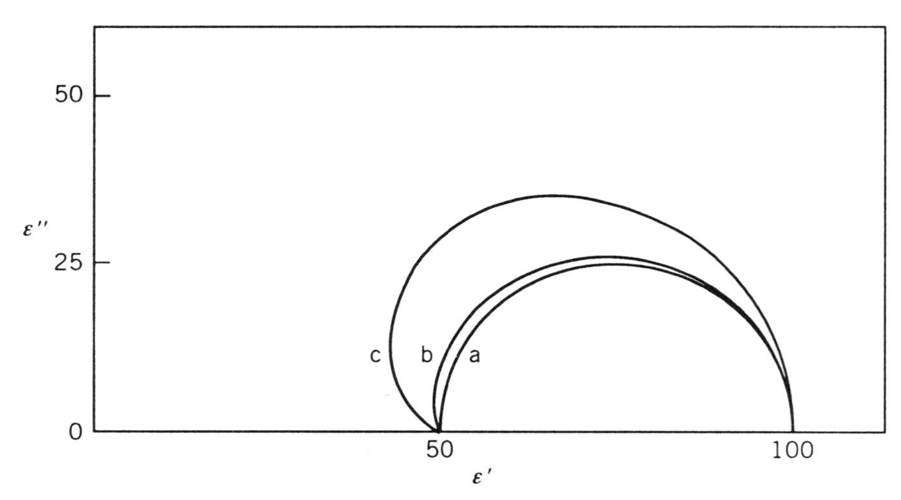

Figure 4.3. Cole–Cole diagrams obtained for various values of γ. (a): $\gamma = 20$; (b): $\gamma = 5$; (c): $\gamma = 1.75$ (ε_0 and ε_∞ have been chosen arbitrarily equal to 100 and 50, respectively).

this is progressively increased and deformed in the high-frequency direction as γ decreases.

The variation of other physical quantities is also very useful in order to reveal inertial effects. We think particularly of the phase angle $\varphi' = \tan^{-1}[\varepsilon''/(\varepsilon' - \varepsilon_\infty)]$, and eventually of the phase angle $\varphi = \tan^{-1}(\varepsilon''/\varepsilon')$, which are represented by any point on the Cole–Cole diagram, as shown in Fig. 4.4. In effect, by drawing the Cole–Cole diagrams and in accordance with the theoretical relations, one can easily see that φ' evolves for increasing frequencies from 0 to $\pi/2$, which is characteristic of pure Debye behavior. On the other hand, once inertial effects manifest themselves, φ' evolves from 0 to π, as clearly indicated in case c ($\gamma = 1.75$). As for the usual phase angle φ, it starts from 0, passes through a maximum linked to the top of the Cole–Cole diagram, the value of which depends on all the oscillator parameters (S, ω_0, γ) and returns to 0. It is less interesting than φ' because it is more difficult to exploit.

Figures 4.5 and 4.6 show the respective dependences of φ' and φ on the frequency and clearly show how sensitive the variations of φ' are. In effect, the curves corresponding to cases b and c show, as expected, a well-defined progression to the limit π. On the contrary, the Cole–Cole diagram does not reveal any inertial effects in case a ($\gamma = 20$), whereas the relevant phase angle, having been $\pi/2$, exceeds this value at very high frequencies.

It is necessary to remark that the experimental accuracy in the

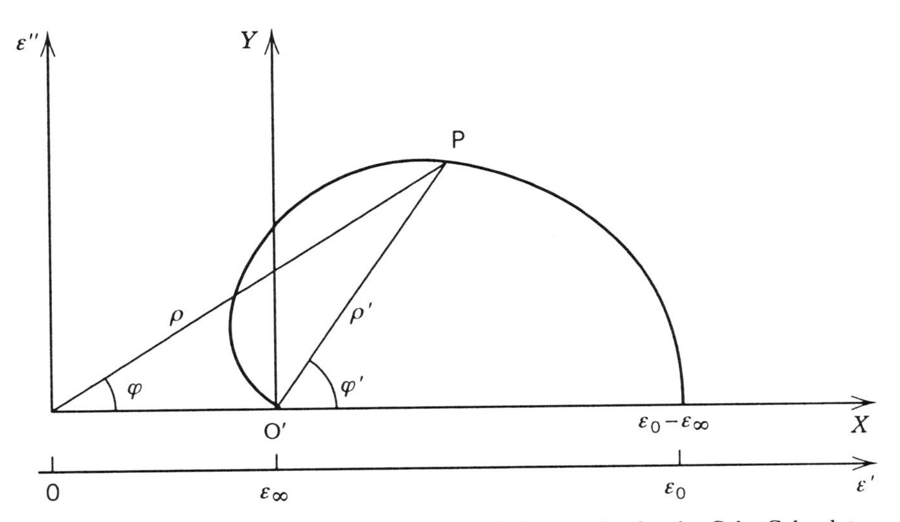

Figure 4.4. Definition of reference axes and phase angles for the Cole–Cole plot.

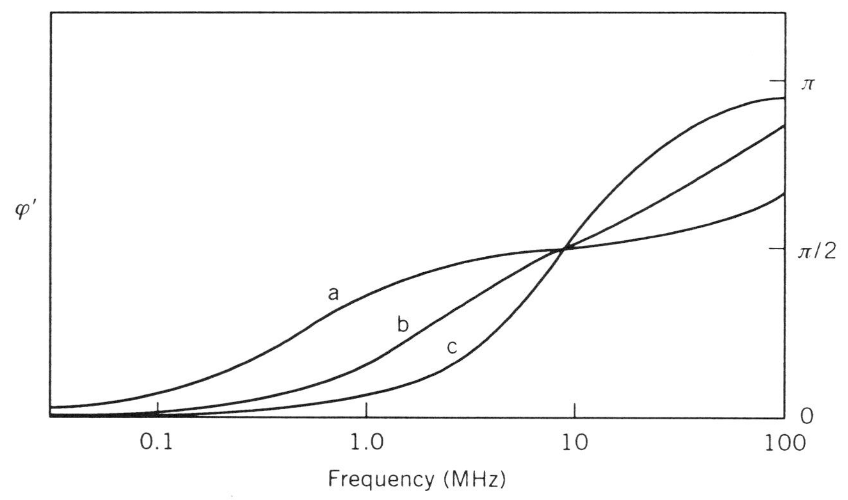

Figure 4.5. Variation of the phase angle φ' versus frequency for various values of γ. (a): $\gamma = 20$; (b): $\gamma = 5$; (c): $\gamma = 1.75$ (the natural frequency has been chosen arbitrarily equal to 10^7Hz).

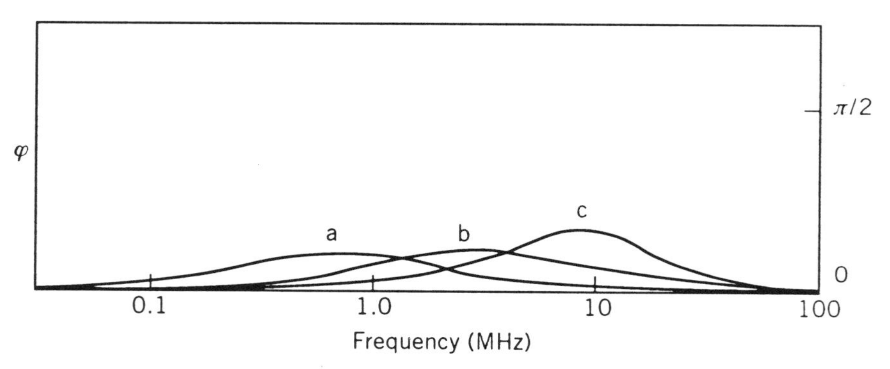

Figure 4.6. Variation of the phase angle φ versus frequency for various values of γ. (a): $\gamma = 20$; (b): $\gamma = 5$; (c): $\gamma = 1.75$ (the natural frequency has been chosen arbitrarily equal to 10^7Hz).

measurement of the phase angle that often exceeds 15% does not take into account the large sensitivity of the phase angle to inertial effects.

B. Determination of Oscillator Parameters

By using the Cole–Cole diagrams for the analysis of experimental results, we shall determine their Cartesian and polar equations. We refer everything to the origin O′ defined by $\varepsilon' = \varepsilon_\infty$, $\varepsilon'' = 0$, and make the change of variables $X = \varepsilon' - \varepsilon_\infty$, $Y = \varepsilon''$, as shown in Fig. 4.4.

By eliminating the angular frequency ω in Eqs. (4.34), we obtain the Cartesian equation (see Appendix B)

$$4A^2(-A + \sqrt{\Delta})^2 + \gamma^2 B^2(2A^2 + B^2 - 2A\sqrt{\Delta}) = 2S\gamma\omega_0 B^2(-A + \sqrt{\Delta}) \tag{4.41}$$

where

$$A = X\gamma\omega_0 \qquad B = 2Y\omega_0 \qquad \Delta = A^2 + B^2$$

We also find the polar equation

$$\rho' = \frac{S}{2}\left(\cos\varphi' \pm \sqrt{\cos^2\varphi' + \frac{4}{\gamma^2}\sin^2\varphi'}\right) \tag{4.42}$$

where

$$\rho' = (X^2 + Y^2)^{0.5} \qquad \varphi' = \tan^{-1}(Y/X)$$

Simple calculations obtained from Eqs. (4.34), (4.41), and (4.42) yield interesting results, which are presented in Table I, and allow one to determine the characteristic parameters of the oscillator quickly. The angular frequency ω' is given by

$$\omega' = \omega_0\left(\frac{\sqrt{(\gamma^2 - 2)^2 + 12} - (\gamma^2 - 2)}{6}\right)^{1/2} \tag{4.43}$$

which defines the maximum M of ε''. One sees that the limit of this equation is $\omega_0/\gamma = \tau^{-1}$ when $\gamma \to +\infty$. One regains the Debye result $\omega\tau = 1$, and then M tends to $S/2$. For small γ values, ω' becomes the

TABLE I
Variation of ε', ε'', and φ' versus ω

ω	ε'	ε''	φ'
0	$\varepsilon_0 = S + \varepsilon_\infty$	0	0
ω_0	ε_∞	S/γ	$\pi/2$
ω'	Close to inflection point	M (Maximum)	$\pi/2 < \varphi' < \pi$
∞	ε_∞	0	π

angular frequency of the oscillator, and M goes to infinity. It is also interesting to remark that the phase angle $\pi/2$ corresponds strictly to ω_0. These results reflect the behavior of a (more or less) damped oscillator.

Experiment shows that the Cole–Cole diagrams are in most cases not very different from the semicircular shape. When the plot passes above the semicircle, we know that this is due to small inertial effects. On the contrary, curves situated below have been interpreted for many years by empirical expressions derived from those of Debye, describing circles the center of which would be below the ε' axis [32]. These curves result, in fact, in a multiplication or a distribution of relaxation times where several oscillators contribute to the dielectric response. It seems that the generalized complex relation [Eqs. (4.34)] provides a good description of the phenomenon. Limiting ourselves to the Debye behavior, that is high γ values, the diagrams d, e, f in Fig. 4.7 correspond to a number of oscillators i equal to 3, 4, and 5, respectively, with identical S_i contributions and taking always $\varepsilon_0 = 100$, $\varepsilon_\infty = 50$. We see that the resulting curves are lowered in proportion to the number i and their general shape may be regarded as arcs of circles in the case of a homogeneous distribution of relaxation times. When the contributions S_i of the oscillators are not equal, for example, for the curve g, where $i = 5$, one has an asymmetry often observed in practice.

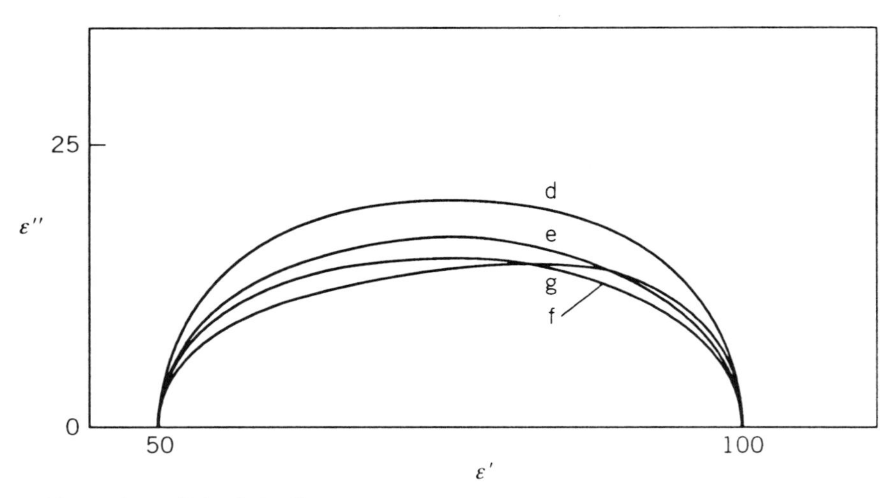

Figure 4.7. Cole–Cole diagrams obtained from the generalized Debye relation for different values of i (number of oscillators) and equal S_i (except (g)). (d): $i = 2$; (e): $i = 4$; (f): $i = 5$; (g): $i = 5$, but unequal S_i.

C. Application to Deoxyhemoglobin S: An Attempt to Explain Aggregation from Molecular Inertia

Inertial effects have often been studied and compared with experiment by a number of authors who have made measurements of dielectric parameters or other quantities, such as the absorption coefficient and the dielectric loss. For instance, we mention the interesting works of Morita et al. [33] and Coffey et al. [34].

We have a slightly different objective. We have tried to apply the previous theoretical considerations to the dielectric properties of an appropriately chosen polar liquid. In effect, we wish to measure moments of inertia and then to link the variation of the damping coefficient γ with the physical state of the liquid studied. The problem was to find experimental data for evolving dielectric constants, showing the progressive influence of inertia. This led us to the study of biological molecules. We have therefore considered deoxygenated hemoglobin S for which Déjardin [35, 36] proposed a mathematical model of kinetics of polymerization where the notion of inertial force is introduced to explain aggregation of blood cells. Hemoglobin S is an abnormal hemoglobin that leads to sickle-cell anemia, a plague that is prevalent in African countries. When oxygen pressure is low, the molecules arrange themselves in such a way as to form polymers due to hydrophobic interactions. This arrangement is, in the limit, comparable to the formation of the nematic phase of paracrystalline gels [37].

Dispersion measurements of the dielectric constant found by Delalic et al. [38] on deoxygenated hemoglobin S, regarded at different stages of aggregation, allow us to test our theory and to give an interpretation of their results.

Starting from the modulus of the complex dielectric constant measured by these authors, we shall calculate the real and imaginary parts of this quantity using the classical Kramers–Kronig method. We shall explain why the critical relaxation frequency seems to move towards high frequencies with polymerization, and why the Debye theory predicts the contrary. In addition, our analysis allows us to determine and characterize three principal aggregates whose respective contributions to the dielectric properties of the ensemble may be measured. Furthermore, we shall verify that in the condensed state of matter, the lowering of the coefficient γ is accompanied by an arrangement and some structuring of the medium studied.

1. *Kramers–Kronig Relations*

The modulus ρ of the complex dielectric constant $\boldsymbol{\varepsilon}$ is defined by

$$\rho = (\varepsilon'^2 + \varepsilon''^2)^{0.5} \tag{4.44}$$

where

$$\boldsymbol{\varepsilon} = \rho \exp(-i\varphi) \qquad \varphi = \tan^{-1}(\varepsilon''/\varepsilon')$$

Knowing only the modulus ρ, it is essential to calculate either the phase angle φ to obtain $\varepsilon' = \rho \cos \varphi$ and $\varepsilon'' = \rho \sin \varphi$, or eventually φ', according to the method of investigation retained.

To calculate φ, we appeal to the mathematical relations useful for the inversion of the Kramers–Kronig formulas. These allow one to determine the real part of a complex function $G(\omega)$, which depends on the angular frequency ω starting from the imaginary part, and inversely, ω being known over a large frequency interval. The conditions for their application are well known [32, 39, 40]. We recall them briefly here

- The function must not have any singularities on the real frequency axis.
- The integral of $G(\omega)/\omega$ over an infinite semicircle in the upper one half of the complex plane in ω must tend to zero.
- The real part of the function $G(\omega)$ is an even function of the angular frequency ω and the imaginary part is an odd function.
- The Boltzmann superposition principle must apply.

It is clear that the complex dielectric constant satisfies those conditions. In particular, ε' is an even function of the frequency and ε'' is an odd function. Thus from Eq. (4.44) we can write

$$\ln \boldsymbol{\varepsilon} = \ln \rho - i\varphi \tag{4.45}$$

The function $\ln \boldsymbol{\varepsilon}$ satisfies the same conditions imposed on it as $\boldsymbol{\varepsilon}$, in accordance with Bode [41], and we may deduce the imaginary part from the real part as follows:

$$\varphi(\omega_c) = -\frac{1}{\pi} \mathscr{P} \int_0^\infty \frac{d \ln \rho}{d\omega} \ln\left(\frac{\omega + \omega_c}{\omega - \omega_c}\right) d\omega \tag{4.46}$$

The symbol $\mathscr{P}$ is the Cauchy principal value of the integral and ω_c denote the angular frequency under consideration.

Experimentally, the dielectric constant is not known over a frequency interval lying from 0 to $+\infty$, but in a limited domain (ω_1, ω_2). So, the calculation of the phase angle may be split in three parts

$$\varphi = \int_0^{\omega_1} + \int_{\omega_1}^{\omega_2} + \int_{\omega_2}^{\infty}$$

or (4.47)

$$\varphi = \varphi_1 + \varphi_2 + \varphi_3$$

The integration between the limits ω_1 and ω_2 gives φ_2. This restriction is compensated for by the correction terms φ_2 and φ_3 the contribution of which is not negligible. These terms may be estimated by means of either a mathematical or a semiempirical law, and many methods and relations have been proposed. The experience that we acquire in solid physics [42] leads us to use the expressions that Ruiz-Pérez [43] employed in a development of mean values. The advantages, in particular, allow us to obtain a self-consistent approximation to all the intervals studied (ω_1, ω_2). These expressions are as follows:

$$\varphi_1(\omega_c) = C[(\omega_1 + \omega_c)\ln(\omega_1 + \omega_c) + (\omega_c - \omega_1) \times \ln(\omega_c - \omega_1) - 2\omega_c \ln \omega_c] \tag{4.48a}$$

$$\varphi_3(\omega_c) = D\left[\frac{\omega_2 + \omega_c}{\omega_2\omega_c}\ln\frac{\omega_2 + \omega_c}{\omega_2\omega_c} + \frac{\omega_2 - \omega_c}{\omega_2\omega_c}\ln\frac{\omega_2 - \omega_c}{\omega_2\omega_c} - \frac{2}{\omega_c}\ln\frac{1}{\omega_c}\right] \tag{4.48b}$$

In correction terms, C and D are two constants that must be evaluated first. To do this, it suffices to prescribe for the calculated total phase φ, a certain or a priori known value, and this can be done by choosing two distinct angular frequencies ω' and ω''. One therefore obtains two equations allowing one to get C and D. Generally, if this procedure is possible, the frequencies ω' and ω'' are chosen at the extremities of the spectrum outside the absorption region and corresponding to values of φ, which are negligible or zero. These corrections when applied to φ_2 are successful over a wide frequency range despite the lack of experimental data at very low or high frequencies, which may cause difficult integration conditions.

The method used in such a fashion produces an error in the phase of only about 3%. This supposes that, knowing the modulus to within an accuracy of 4 or 5%, φ may be deduced to a precision of 7 or 8%. This should be compared with the accuracy of the experimental results that can vary between 12 and 20%. The theoretical and practical details of the Kramers–Kronig relations are explained in Appendix C.

2. *Use of the Method and Results*

As a first step, we examine the dispersion curves of ε' and ε'' obtained using the preceding relations and the corresponding Cole–Cole diagrams. These clearly show the presence of many constituents because of the numerous arcs of circle that are revealed in the diagrams. The methods of investigation that we have just given exhibit certain difficulties in their application. For example, we know that the phase φ' does not exhibit inertial effects in the spectral domain studied for an oscillator, except at very high frequencies. The presence of a number of oscillators may run the risk of diminishing or masking these inertial effects by superposition and we have avoided elucidating this quantity.

So, we compare the preceding curves with those obtained starting from a simple model that in first approximation consists in using the Debye relations with several components. This procedure suggests a systematic best-fit method akin to that already used in studying the infrared (IR) responses of thin layers of mercury sulfide [44]. This entails having a knowledge of ε_∞ and the adjustment of two parameters, S_i and τ_i, where i designates the number of constituents. If these results are not satisfactory, we recommence the best-fit procedure with the aid of the oscillator relations [Eqs. (4.34)], which now require adjustment of three oscillator parameters S_i, ω_i, and γ_i, where the first results accommodate the second ones. Thus, it is possible in the latter case to have access to the moment of inertia.

Bear in mind that it is illusory and not useful to use Eqs. (4.34) in the case of quasi-Debye behavior; that is, when γ is greater than 10, because the two theories now lead to confused dielectric responses. Indeed, if the relaxation time could be determined easily (e.g., by the maximum of ε'' when $\omega\tau = 1$) it becomes more difficult to find the pair of parameters (ω_i, γ_i) that are given by the relation $\tau_i = \gamma_i/\omega_i$. To make the last remark more rigorous, consider the following numerical example: An oscillator characterized by $\omega_1 = 10^8$ rad s^{-1} and $\gamma_1 = 10^3$ presents a response that is practically indistinguishable from the one characterized by $\omega_2 = 2 \times 10^6$ rad s^{-1} and $\gamma_2 = 20$.

Following the definitions that we have given, one sees that the determination of these parameters leads to important information concerning the medium. That is, starting from the relaxation time τ, one can deduce the friction coefficient ξ from $\xi = 2kT\tau$. From the fundamental frequency N_0, if this is measurable, one may calculate the moment of inertia I, according to the formula $I = 2\,kT/(2\pi N_0)^2$.

The experimental measurements of Delalic et al. [38], which we considered, are measurements of the dielectric constants on suspensions of red blood cells containing hemoglobin S in concentrated phosphate buffer (28.6 g dL^{-1}) at 37 °C. The first measurement is made immediately after deoxygenation, the second after an elapsed time of 4 h, the third after 7 h, the fourth after 9 h and the fifth after 21 h. The state of the medium corresponding to the first measurement is called state 1 (or S1), and goes successively up to state 5 (or S5). In Fig. 4.8, the values of ε' and ε'' as functions of frequency are represented by the open round dots (○) and the solid circles (●), respectively. They are calculated from experimental points by the Kramers–Kronig method for S1. Figure 4.9 shows the Cole–Cole diagram corresponding to them. Figures 4.10–17 similarly illustrate the other states. On fitting all these graphs, the results that we obtained by successive refinements are pictured in the same Figs. 4.8–4.10 (full lines). The values of all the parameters of interest are grouped in Table II. We note that we write S_i in the form of contributions C_i (expressed in percentage) to the total oscillator force $S = \varepsilon_0 - \varepsilon_\infty$, namely, $S_i = C_i S$, where $\Sigma\, C_i = 1$.

Four main constituents are found in all, in variable proportions, in accordance with the state of polymerization. Constituent 1 may be considered as the monomeric phase, the three others correspond to aggregation forms, each one characterized by a natural frequency re-

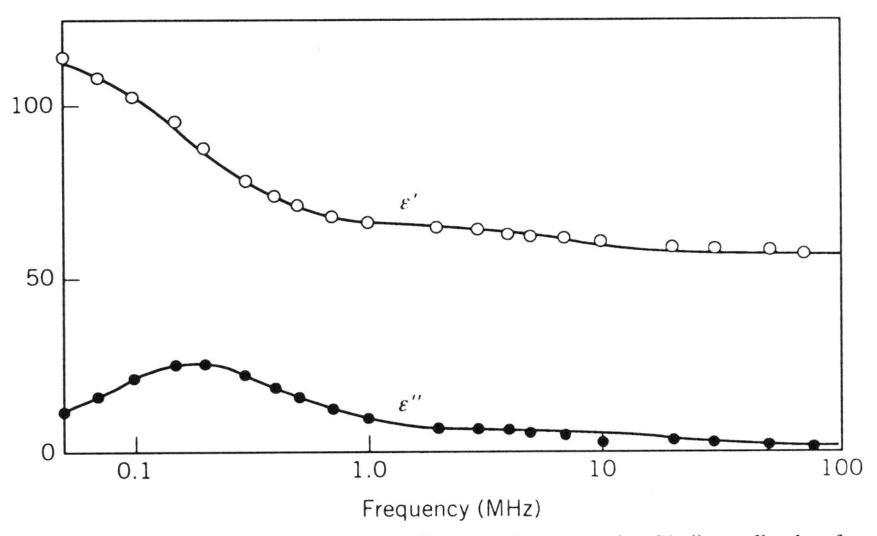

Figure 4.8. Dispersion plots ε' and ε'' versus frequency for S1 (immediately after deoxygenation). ○, ●: our treatment by Kramers–Kronig; full lines: best-fit procedure.

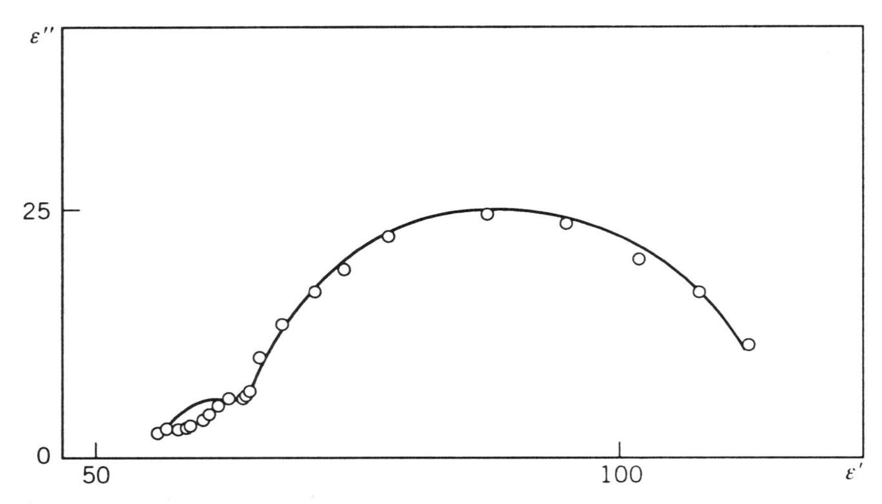

Figure 4.9. Cole–Cole plot for S1. ○: our treatment by Kramers–Kronig; full line: best-fit procedure.

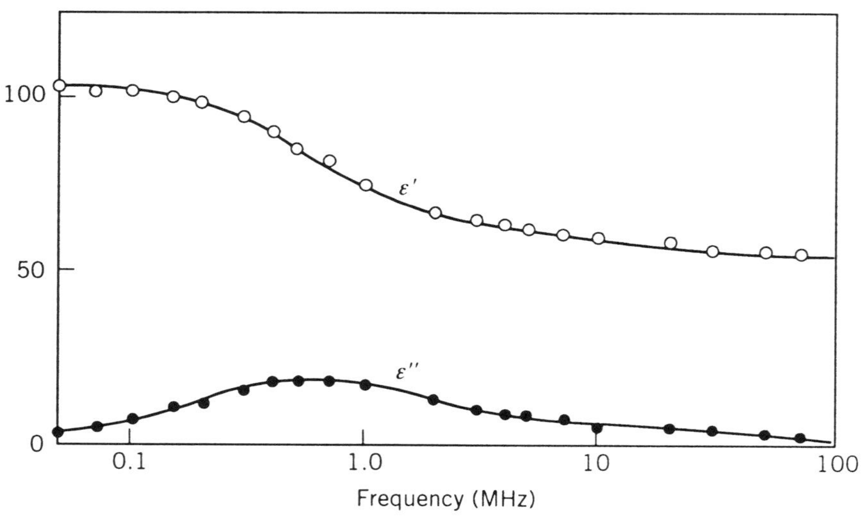

Figure 4.10. Dispersion plots ε' and ε'' versus frequency for S2 (four hours after deoxygenation). ○, ●: our treatment by Kramers–Kronig; full lines: best-fit procedure.

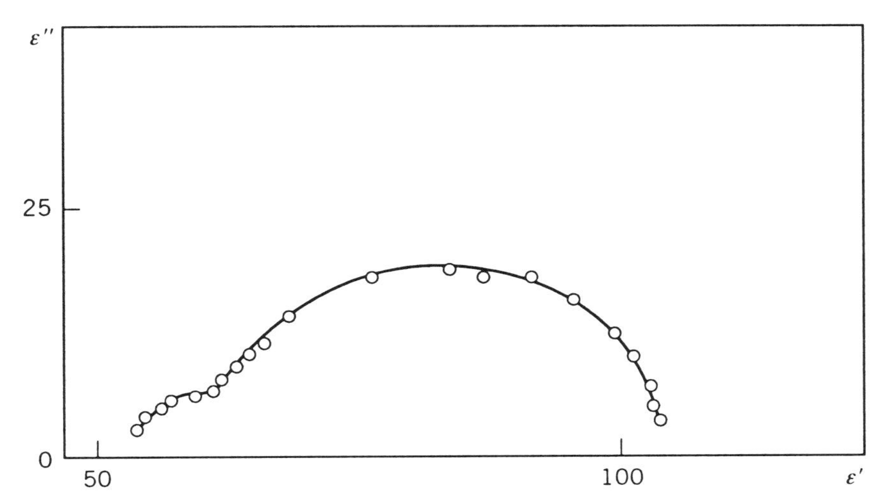

Figure 4.11. Cole–Cole plot for S2. ○: our treatment by Kramers–Kronig; full line: best-fit procedure.

markably constant all the time. The calculation of the moments of inertia of these aggregates gives the following results:

$$I_2 = 4.42 \times 10^{-31}\ \text{g cm}^2 \qquad I_3 = 9.04 \times 10^{-29}\ \text{g cm}^2$$
$$I_4 = 3.00 \times 10^{-27}\ \text{g cm}^2$$

The monomeric form is obviously lighter than the other constituents and has a Debye-type behavior. So, we need to determine nothing but the relaxation times. To get more thorough information, we searched the literature [37] for the radius of the equivalent sphere of the monomer, $r_1 = 62$ Å, along with the molecular weight, $M = 64{,}000$. We deduce the moment of inertia I_1 from the relation $I = 0.4\,Mr^2/\mathcal{N}$, where $\mathcal{N}$ is Avogadro's number,

$$I_1 = 1.63 \times 10^{-32}\ \text{g cm}^2$$

then the natural frequency

$$N_1 = 364\ \text{MHz}$$

and finally the evolution of γ with the different states of polymerization.

By using inverse reasoning, and assuming the three aggregates assimi-

TABLE II
Time Evolution of Typical Dielectric Parameters of Deoxyhemoglobin S

Parameter[a]	S1(0 h)	S2(4 h)	S3(7 h)	S4(9 h)	S5(21 h)
ε_0	114	104	99	94	92
ε_∞	56	54	52	53	47
$S = \varepsilon_0 - \varepsilon_\infty$	58	50	47	41	45
C_1	0.85	0.48	0.11	0	0
S_1	49.3	24.0	5.2		
N_1(MHz)	364	364	364		
γ_1	2000	900	600		
τ_1(ns)	875	393	262		
ξ_1(J s)	748×10^{-29}	337×10^{-29}	225×10^{-29}		
C_2	0.15	0.18	0.20	0.20	0.21
S_2	8.7	9.0	9.4	8.2	9.45
N_2(MHz)	70	70	70	70	70
γ_2	9.0	4.7	3.5	3	3
τ_2(ns)	20.5	10.7	8.0	6.8	6.8
ξ_2(J s)	175×10^{-30}	915×10^{-31}	685×10^{-31}	582×10^{-31}	582×10^{-31}
C_3	0	0.34	0.51	0.55	0.54
S_3		17.0	24.0	22.6	24.3
N_3(MHz)		5	5	4.9	4.8
γ_3		4.8	2.1	2.3	2.2
τ_3(ns)		153	66.8	74.7	72.9
ξ_3(J s)		131×10^{-29}	572×10^{-30}	640×10^{-30}	625×10^{-30}
C_4	0	0	0.18	0.25	0.25
S_4			8.4	10.2	11.25
N_4(MHz)			0.8	0.85	0.85
γ_4			1.7	1.6	1.6
τ_4(ns)			338	300	300
ξ_4(J s)			290×10^{-29}	256×10^{-29}	256×10^{-29}

[a] Units are in parentheses.

lated to spheres, we find that

$$r_2 = 120 \text{ Å} \qquad r_3 = 348 \text{ Å} \qquad r_4 = 700 \text{ Å}$$

that is, shapes that are larger and larger. In fact, it is known that only the smallest aggregate is approximately spherical and that the others are fibers built from an initial nucleus. It is remarkable to find that $r_2 \approx 2r_1$, which supposes the association of $n = (r_2/r_1)^3 = 8$ monomeric spheres. On the other hand, if this small aggregate sustains the nucleation mechanism, which is an assumption, we can find the length of the fibers.

Owing to the difficulties encountered in certain fits, it is possible that polydispersity is still more important. One can, however, consider that the three types of aggregates are average representations in accordance

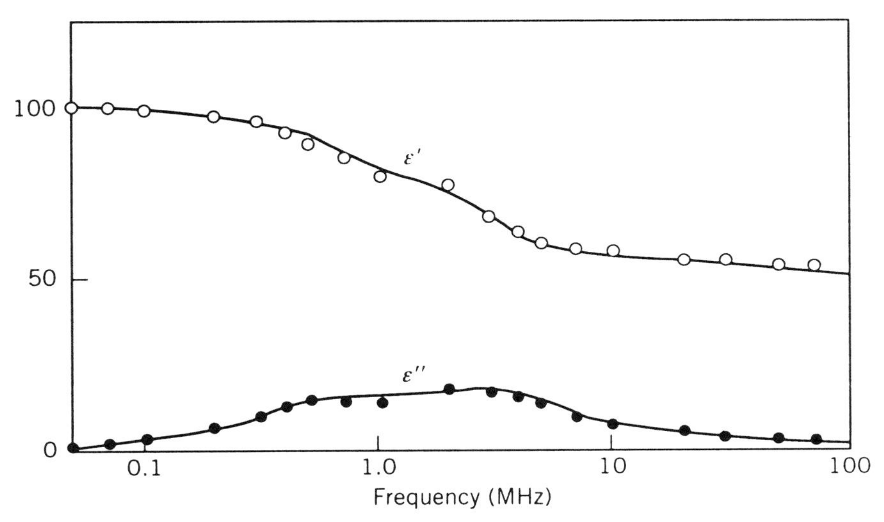

Figure 4.12. Dispersion plots ε' and ε'' versus frequency for S3 (seven hours after deoxygenation). ○, ●: our treatment by Kramers–Kronig; full lines: best-fit procedure.

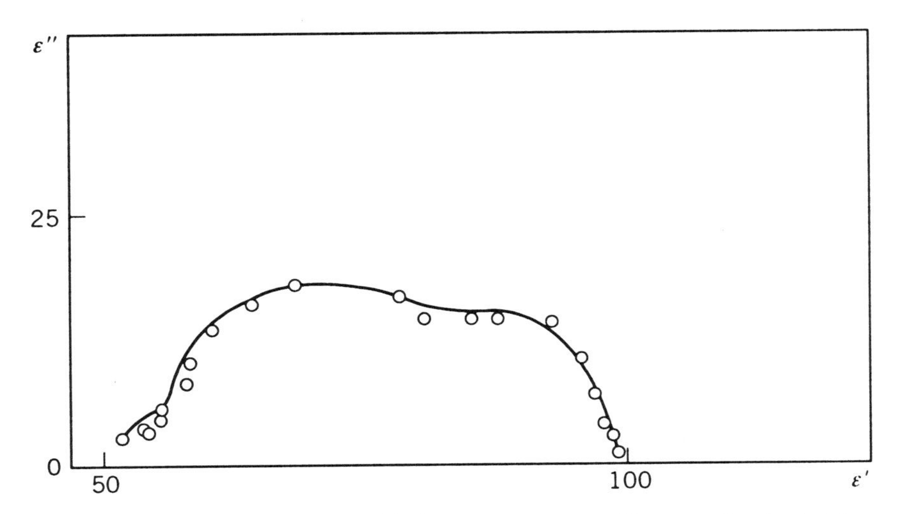

Figure 4.13. Cole–Cole plot for S3. ○: our treatment by Kramers–Kronig; full line: best fit procedure.

with reality. It would not be serious to still consider other oscillators because 12 parameters must already be adjusted to S3.

With regard to the damping coefficient γ, or the friction coefficient ξ, we note that each diminishes with time before later tending to a stable value that may be verified for the four constituents. The explanation of

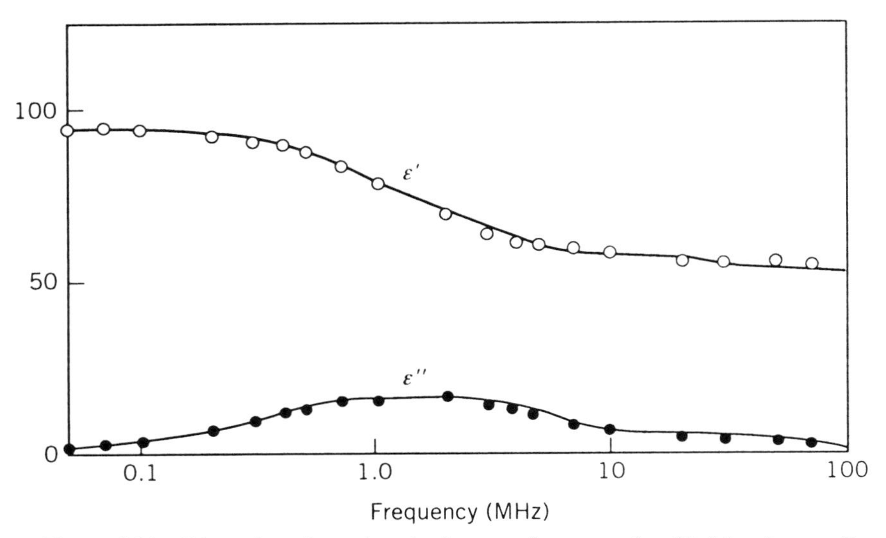

Figure 4.14. Dispersion plots ε' and ε'' versus frequency for S4 (nine hours after deoxygenation). ○, ●: our treatment by Kramers–Kronig; full lines: best-fit procedure.

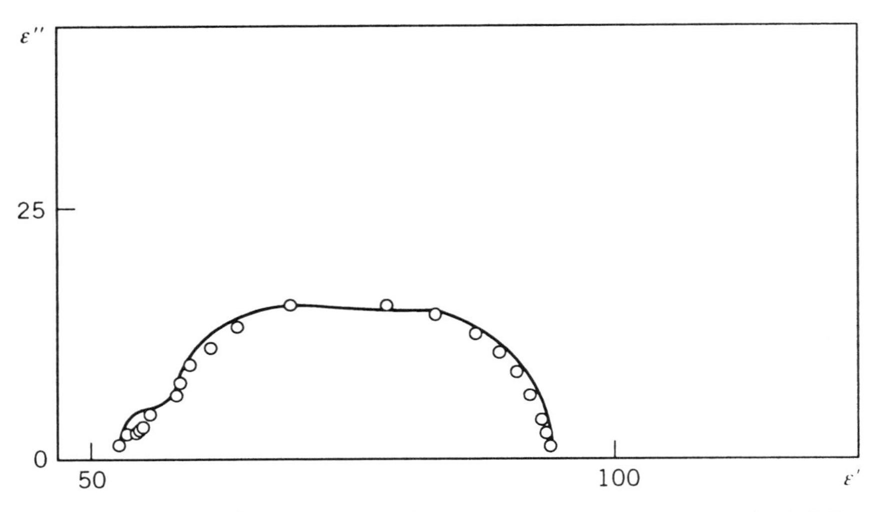

Figure 4.15. Cole–Cole plot for S4. ○: our treatment by Kramers–Kronig; full line: best-fit procedure.

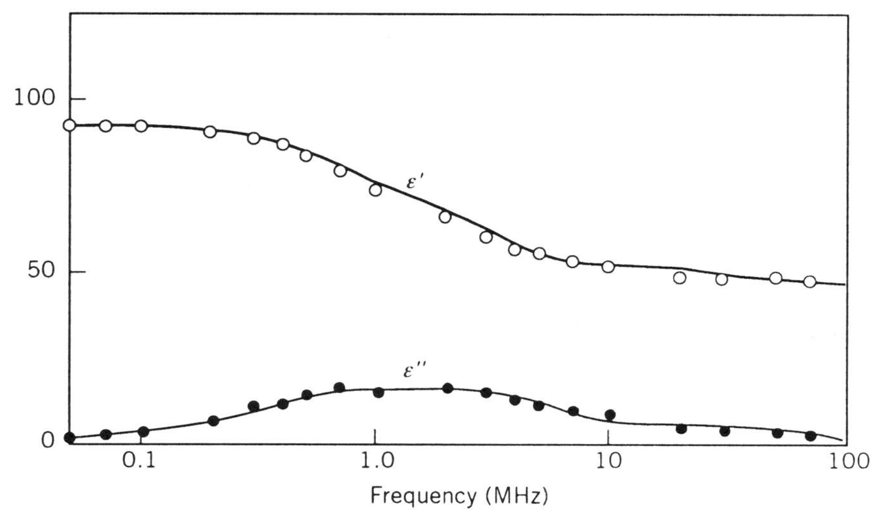

Figure 4.16. Dispersion plots ε' and ε'' versus frequency for S5 (twenty one hours after deoxygenation). ○, ●: our treatment by Kramers–Kronig; full lines: best-fit procedure.

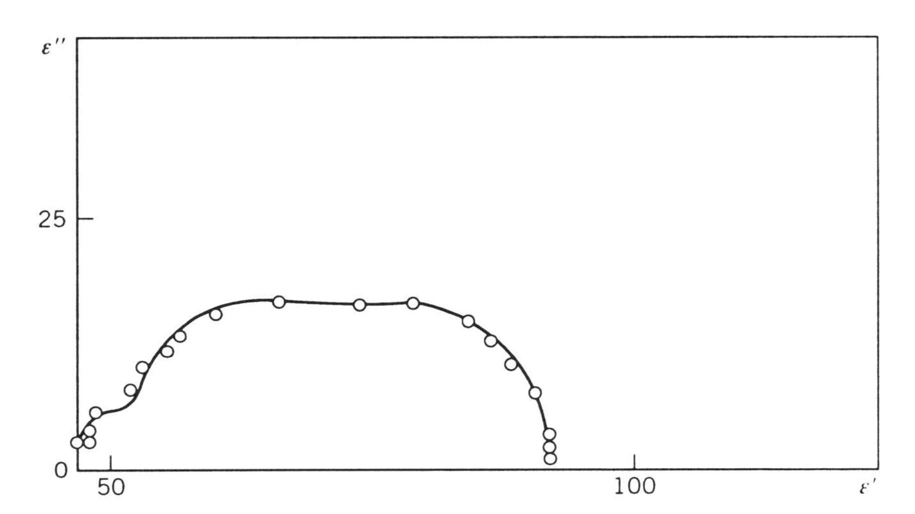

Figure 4.17. Cole–Cole plot for S5. ○: our treatment by Kramers–Kronig; full line: best-fit procedure.

this seems to be in the decrease of the concentration of the solution due to the process of polymerization, or more exactly to the diminution of the global density N of oscillators. This is well described by the systematic lowering of the whole oscillator force S, which depends on N. The reduction of the hindrance of solution allows one to understand that the rotation–vibration movements are made easier, and that the friction and damping coefficients logically diminish. Equivalently, there is a lessening in the time constant and hence a shift to high frequencies of the critical frequencies. Thus, the appearance of inertial effects may explain this evolution. Limiting oneself to these aggregates alone, one may see that for a given state, the friction coefficient ξ increases with size or inertia, which is normal since ξ is, for example, proportional to r^3 in the case of a sphere. What appears less normal is the increased value of ξ for the monomer that is much more than that of the aggregate. Therefore it seems that strong hydrophobic type interactions sustained by the monomers cause them to aggregate. This suffices to explain the constraint of their movement, which adds to the natural friction in the solvent, and leads to the high value of ξ.

Finally, the evolving state of the medium with time may be interpreted by taking account of all the previous considerations. Initially, the monomeric phase is largely prevalent since its contribution C_1 to S is 85%, and the remaining 15% are small aggregates that are probably formed during the deoxygenation process. Four hours afterwards, the contribution is not more than 48% for monomers and 18% for small aggregates. One notices the appearance of mean-sized aggregates with $C_3 = 34\%$. After 7 h, we see very big aggregates. After 9 h, the monomeric phase becomes exhausted, and the general situation no longer evolves because few variations are recorded after 21 h. We remark that the percentage of C_2 is almost stable because, following our hypothesis, the small aggregates serve to form the others that are bigger and whose mean contributions and characteristics are known.

To sum up, polydispersity seems to involve, at first, an increase of inertia. According to our estimates, which were deduced from measurements, the increase of size, and consequently that of friction, is compensated by the disencumbering of the medium in which the oscillators evolve, and it results in a slight lowering of the friction coefficient. The damping coefficient is contributed to by both of these effects and decreases rapidly. It is therefore necessary to take account of inertial effects in order to interpret correctly the dielectric response of hemoglobin S, and the information that can be extracted from them is very fruitful for a better understanding of molecular relaxation mechanisms. We found that the minimum value of γ equals 1.6 corresponding to S5 and the biggest aggregate. For this state, which experimentalists attribute to a state very close to gel, the mean value of γ is 2.2. These numerical values are in accordance with our estimates based on the assumptions of Coffey who predicted that the limit of validity of the oscillator relations was given by $\gamma = 1.75$. This confirms the idea that this last value reflects border between the liquid state and condensed matter.

In this section, we attempted to show how to define the limits of validity of the classical Debye theory applied to dispersion and absorption measurements of the dielectric constant when inertial effects are taken into account. In this way, we introduced the notion of oscillators, which allows us to obtain useful information about a number of macroscopic parameters that characterize the liquid state. Indeed, inertial effects are always present, and the frequency domain that allows their observation is in relation to the molecular size. This result means that such observations are possible over a widely varying frequency range lying from very low frequencies up to the IR region. However, their influence, as we have seen, is often masked at a given temperature by the frictional effects of the molecules in the solvent. The study of hemoglobin S shows that inertial effects do not appear clearly in customary experimental domains, except for a physical state that is close to the gel state. The Debye relations may suffice in a large majority of cases, but information is a bit less complete. It is in the study of the phase angle φ' that inertial effects reveal themselves most clearly because when they are included the phase angle at high frequencies passes from value $\pi/2$ to value π, which is twice that with no inertial effects. Thus, it is necessary to greatly enlarge the region of investigation at high frequencies. This is difficult experimentally and bad precision in φ' could reduce these interesting prospects. Nevertheless, use of Kramers–Kronig relations may serve to compensate for this handicap in so far as this method does no require phase measurements. When many constituents are present in the medium, the study becomes very intricate. A systematic fit of all the experimental curves is therefore needed according to a technique often used in the physics of condensed matter.

D. Nonlinear Dielectric Response Including the Induced Polarization of Polar Molecules

The theory of electric polarization in dielectric fluids contributes greatly to the understanding of electrooptical relaxation phenomena. The usual treatment of this problem is due to Debye [7], who proposed an extension of the work of Einstein [4] to study the rotational Brownian motion of polar molecules and account for their anomalous dispersion. The linear response corresponding to a pure alternating electric field leads to the well-known Cole–Cole diagrams, which are perfect semicircles, and to phase angles between in-phase and out-of-phase components of the susceptibility whose asymptotic values (very high-frequency region) tend to $\pi/2$. Such a response is independent of the electric field strength. Later, Coffey and Paranjape [45] proposed an extension of the Debye model to obtain the field and frequency dependence of the relative permittivity of polar liquids in the case of two different electric fields, either (a) a strong alternating field $E(t) = E_0 \cos \omega t$ or (b) a strong dc constant field superimposed on a weak alternating field in the same direction $E(t) = E_c + E_0 \cos \omega t$. Their calculation provides analytic expressions for the dielectric relaxation (DR) as far as terms in E^3. This section is devoted to development of the nonlinear behavior of this response at a double level, both resulting from the effect of a double electric stimulus [case (b)] and from the influence of not only permanent but also induced dipolar moments of molecules. But any collective effect is neglected, which is the case of a very dilute medium where the assembly pertains to a closed system undergoing independent rotational motions. As we shall see in Section III, the influence of the permanent and induced dipole moments is usually described in Kerr effect theories. We shall confine ourselves to the steady-state DR, which is obtained by solving the rotational diffusion equation (Smoluchowski equation) by means of a general perturbative method. This latter method, derived from Morita's work [46], has been used to obtain the nonlinear response in Kerr effect relaxation. Hereupon, we shall employ a similar approach to obtain (for the stationary regime) the harmonic components in DR valid up to the third order of the electric field [47].

1. Expansion of the Orientational Distribution Function Up To Third Order in the Electric Field

The Smoluchowski equation, also called the rotational diffusion equation, allows one to calculate the orientational distribution function $f(u, t)$ of a molecule subjected to the action of an external electric field. By assuming that the molecules of the dielectric liquid are equivalent to small rigid ellipsoids of rotation whose geometric axes coincide with those of the electric polarizability tensor and that they do not interact with each other (dilute solution), one may treat the rotational motion of a single molecule. Hence, the orientational distribution function $f(u, t)$ obeys the following differential equation written in terms of Liouville-type operators [48]

$$\frac{\partial f(u, t)}{\partial t} = [\hat{A}_0(u) + \gamma(t)\hat{A}_\gamma(u) + \beta(t)\hat{A}_\beta(u)]f(u, t) \tag{4.49}$$

where

$$\hat{A}_0(u) = D\left[(1-u^2)\frac{\partial^2}{\partial u^2} - 2u\frac{\partial}{\partial u}\right] = D\Delta_u \tag{4.50}$$

is the unperturbed Laplacian operator, independent of the electric field $E(t)$, and

$$\hat{A}_\gamma(u) = D\left[2u - (1-u^2)\frac{\partial}{\partial u}\right] \tag{4.51}$$

$$\hat{A}_\beta(u) = D\left[(3u^2-1) - u(1-u^2)\frac{\partial}{\partial u}\right] \tag{4.52}$$

are two perturbed operators due to the presence of the electric field and related to the effects of permanent and induced dipole moments on the molecule, respectively.

In Eqs. (4.49)–(4.52), the angular variable is u equal to $\cos\theta$, where θ is the angle the molecule symmetry axis makes with the direction of $E(t)$. The parameter D is the rotary diffusion constant. The parameters $\gamma(t)$ and $\beta(t)$ are time-dependent parts of the orientational potential energy of the molecule in the field with respect to the thermal energy, defined as follows:

$$\gamma(t) = \frac{\mu}{kT}E(t) \qquad \beta(t) = \frac{\Delta\alpha}{kT}E^2(t) \tag{4.53}$$

where μ is the permanent moment along the symmetry axis of the molecule (symmetric top) and $\Delta\alpha = \alpha_{\parallel} - \alpha_{\perp}$ is the difference between the principal electric polarizabilities parallel and perpendicular to the axis of symmetry, respectively. Now, on setting

$$P = \frac{\Delta\alpha}{\mu^2}kT \qquad \hat{A}'_\beta(u) = P\hat{A}_\beta(u) \tag{4.54}$$

Eq. (4.49) becomes

$$\frac{\partial f(u,t)}{\partial t} = [\hat{A}_0(u) + eE(t)\hat{A}_\gamma(u) + e^2E^2(t)\hat{A}'_\beta(u)]f(u,t) \tag{4.55}$$

where e is a small perturbation parameter equal to μ/kT.

It is very simple to solve Eq. (4.55) when $E(t)$ is zero. The corresponding solution for this unperturbed state is called by Morita [46] the conditional-probability-density function and must satisfy

$$\frac{\partial g(u, u', t)}{\partial t} = \hat{A}_0(u) g(u, u', t) \tag{4.56}$$

Since $\hat{A}_0(u)$ is a Laplacian operator, $g(u, u', t)$ may be expanded in a series of Legendre polynomials

$$g(u, u', t) = \sum_n e^{-n(n+1)Dt} P_n(u) P_n(u') \tag{4.57}$$

with the following closure relation:

$$\sum_n P_n(u) P_n(u') = \delta(u - u') \tag{4.58}$$

where $\delta(u - u')$ is the Dirac delta function.

Actually, Eq. (4.57) agrees with the equation obtained by Déjardin [49] in an earlier work, with $u' = 0$, for rise transients following the sudden application of a rectangular electric field. Now, we may deduce the formal expression of the orientational distribution function $f(u, t)$, namely,

$$\begin{aligned} f(u, t) = {} & \int f(u_0, 0) g(u, u_0, t)\, du_0 \\ & + e \int \int_0^t g(u, u_0, t) \hat{A}_\gamma(u_0) E(t_0) f(u_0, t_0) du_0 dt_0 \\ & + e^2 \int \int_0^t g(u, u_0, t) \hat{A}'_\beta(u_0) E^2(t_0) f(u_0, t_0) du_0 dt_0 \end{aligned} \tag{4.59}$$

So, by expanding $f(u, t)$ in a power series of the small parameter e up to terms of third order, we find that

$$\begin{aligned} f(u, t) = {} & f(u, 0) + e f_{\gamma 1}(u, t) + e^2 [f_{\gamma 2}(u, t) + f_{\beta 2}(u, t)] \\ & + e^3 [f_{\gamma 3}(u, t) + f_{\beta 2, \gamma 1}(u, t)] \cdots \end{aligned} \tag{4.60}$$

where

$$f_{\gamma 1}(u, t) = \int_0^t \mathbf{g} \mathbf{f}_\gamma (t - t') E(t') dt' \tag{4.60a}$$

$$f_{\gamma 2}(u, t) = \int_0^t \int_0^{t_1} \mathbf{g} \mathbf{A}_\gamma (t - t_1) \mathbf{f}_\gamma (t_1 - t_2) E(t_1) E(t_2) dt_1 dt_2 \tag{4.60b}$$

$$f_{\beta 2}(u,t) = \int_0^t \mathbf{g}\mathbf{f}_{\boldsymbol{\beta}}(t-t')E^2(t')dt' \tag{4.60c}$$

$$f_{\gamma 3}(u,t) = \int_0^t \int_0^{t_1} \int_0^{t_2} \mathbf{g}\mathbf{A}_{\boldsymbol{\gamma}}(t-t_1)\mathbf{A}_{\boldsymbol{\gamma}}(t_1-t_2)\mathbf{f}_{\boldsymbol{\gamma}}(t_2-t_3) \times E(t_1)E(t_2)E(t_3)dt_1dt_2dt_3 \tag{4.60d}$$

$$f_{\beta 2,\gamma 1}(u,t) = \frac{1}{2}\left[\int_0^t \int_0^{t_1} \mathbf{f}_{\boldsymbol{\beta}}(t-t_1)E^2(t_1)\mathbf{f}_{\boldsymbol{\gamma}}(t_1-t_2)E(t_2)dt_1dt_2 + \int_0^t \int_0^{t_1} \mathbf{f}_{\boldsymbol{\gamma}}(t-t_1)E(t_1)\mathbf{f}_{\boldsymbol{\beta}}(t_1-t_2)E^2(t_2)dt_1dt_2\right] \tag{4.60e}$$

The subscripts 1–3 stand for the order of the electric field $E(t)$.

In Eqs. (4.60a)–(4.60b), $\mathbf{g}$ represents a row matrix the elements of which are the Legendre polynomials from $i = 1$ to $i = n$, while $\mathbf{f}_{\boldsymbol{\gamma}}$ and $\mathbf{f}_{\boldsymbol{\beta}}$ are column relaxation matrices with the jth term given by

$$f_\gamma^{(j)} = e^{-j(j+1)Dt} \int P_j(u)\hat{A}_\gamma(u)f(u,0)du$$

$$f_\beta^{(j)} = e^{-j(j+1)Dt} \int P_j(u)\hat{A}_\beta'(u)f(u,0)du \qquad j = 1, \ldots, n$$

$\mathbf{A}_{\boldsymbol{\gamma}}$ is a square matrix in which the ij element is equal to

$$A_\gamma^{(ij)} = e^{-i(i+1)Dt} \int P_i(u)\hat{A}_\gamma(u)P_j(u)du$$

Note that no $\mathbf{A}_{\boldsymbol{\beta}}$ square matrix appears in our theoretical analysis because in the development of $f(u,t)$ terms higher than e^3 are discarded. Moreover, it is obvious from Eq. (4.60) that the electric polarization is given as a series of odd powers of e. Also, it is interesting to remark that the last term in the right-hand side of this equation gives rise to a coupling effect between the induced dipole moment and the permanent dipole moment; this is clearly a new nonlinear effect that is not present in the Kerr effect relaxation for the second-order response where $f_{\gamma 2}$ and $f_{\beta 2}$ are, in effect, nonlinear functions but provide separated contributions in γ and β (even powers of e). We must mention here that Benoit, in his study of Kerr effect relaxation from rectangular pulses [50], finds a coupling of permanent dipoles and polarizabilities (actually cross terms) for the electric birefringence only and not for the polarizability density. The same remark applies to the works by Cole [51] and also by Watanabe and Morita [52] for the Kerr effect. So, the nonlinear DR response due to the application of a double external stimulus and to the consideration of permanent dipoles and polarizabilities involves such a coupling effect and, to our knowledge, has not been elucidated yet.

In a first step, we shall carry out the general expressions of the nonlinear DR responses. Then, we shall present analytic expressions of these responses that may be useful in experimental investigations, obtained by assuming that the amplitude E_0 of the cosine field is much less than that of the constant field E_c. We shall illustrate all these equations from numerous typical plots, such as dispersion and Cole–Cole diagrams, together with the variation of phase angle against frequency.

2. *Expressions for the Nonlinear DR Responses for the Stationary Regime*

For the stationary regime of the electric polarization $P(t)$, that is, when the system has removed all the transient effects, we need to calculate $f_{\gamma 1}(u,t)$, $f_{\gamma 3}(u,t)$, and $f_{\beta 2,\gamma 1}(u,t)$. On assuming that the double external electric field $E(t)$ has been applied at time $t=0$, we find that

$$\begin{aligned} P(t) &= \lim_{t\to\infty} [f_{\gamma 1}(u,t) + f_{\gamma 3}(u,t) + f_{\beta 2,\gamma 1}(u,t)] \\ &= f^{\infty}_{\gamma 1} + f^{\infty}_{\gamma 3} + f^{\infty}_{\beta 2,\gamma 1} \end{aligned} \tag{4.61}$$

We shall now evaluate each term of the right-hand side of Eq. (4.61). The results are as follows:

$$\begin{aligned} f^{\infty}_{\gamma 1} = \mathbf{g}E_c \int_0^{\infty} \mathbf{f}_{\gamma}(t)dt + \mathbf{g}E_0\bigg[\left(\int_0^{\infty} \mathbf{f}_{\gamma}(t_1) \cos \omega t_1 dt_1 \right) \cos \omega t \\ + \left(\int_0^{\infty} \mathbf{f}_{\gamma}(t_1) \sin \omega t_1 dt_1 \right) \sin \omega t \bigg] \end{aligned} \tag{4.62}$$

which represents the first-order nonlinear response, and

$$
\begin{aligned}
f^{\infty}_{\gamma 3} = {} & \mathbf{g}E_c^3\left(\int_0^{\infty} \mathbf{A}_{\gamma}(t)dt\right)^2 \int_0^{\infty} \mathbf{f}_{\gamma}(t)dt \\
& + \mathbf{g}E_0E_c^2 \int_0^{\infty} \mathbf{f}_{\gamma}(t)dt\left[\left(\int\int \mathbf{A}_{\gamma}(t_1)\boldsymbol{A}_{\gamma}(t_2)\cos\omega(t_1+t_2)dt_1dt_2\right)\cos\omega t\right. \\
& \left. + \left(\int\int \mathbf{A}_{\gamma}(t_1)\mathbf{A}_{\gamma}(t_2)\sin\omega(t_1+t_2)dt_1dt_2\right)\sin\omega t\right] \\
& + \mathbf{g}E_0E_c^2 \int\int \mathbf{A}_{\gamma}(t_1)\mathbf{f}_{\gamma}(t_2)dt_1dt_2\left[\left(\int_0^{\infty} \mathbf{A}_{\gamma}(t_1)\cos\omega t_1 dt_1\right)\cos\omega t\right. \\
& \left. + \left(\int_0^{\infty} \mathbf{A}_{\gamma}(t_1)\sin\omega t_1 dt_1\right)\sin\omega t\right] \\
& + \mathbf{g}E_0E_c^2\left\{\left[\int\int\int \mathbf{A}_{\gamma}(t_1)\mathbf{A}_{\gamma}(t_2)\mathbf{f}_{\gamma}(t_3)\cos\omega(t_1+t_2+t_3)dt_1dt_2dt_3\right]\cos\omega t\right. \\
& \left. + \left[\int\int\int \mathbf{A}_{\gamma}(t_1)\mathbf{A}_{\gamma}(t_2)\mathbf{f}_{\gamma}(t_3)\sin\omega(t_1+t_2+t_3)dt_1dt_2dt_3\right]\sin\omega t\right\} \\
& + \mathbf{g}E_cE_0^2\left\{\int_0^{\infty} \mathbf{A}_{\gamma}(t)dt \int_0^{\infty} \mathbf{A}_{\gamma}(t)\cos\omega t\, dt \int_0^{\infty} \mathbf{f}_{\gamma}(t)dt\right. \\
& + \int_0^{\infty} \mathbf{f}_{\gamma}(t)dt\left[\left(\int\int \mathbf{A}_{\gamma}(t_1)\mathbf{A}_{\gamma}(t_2)\cos\omega(2t_1+t_2)dt_1dt_2\right)\cos 2\omega t\right. \\
& \left.\left. + \left(\int\int \mathbf{A}_{\gamma}(t_1)\mathbf{A}_{\gamma}(t_2)\sin\omega(2t_1+t_2)dt_1dt_2\right)\sin 2\omega t\right]\right\} \\
& + \mathbf{g}E_cE_0^2 \int\int\int \mathbf{A}_{\gamma}(t_1)\mathbf{A}_{\gamma}(t_2)\mathbf{f}_{\gamma}(t_3)\cos\omega(t-t_1-t_2) \\
& \times \cos\omega(t-t_1-t_2-t_3)dt_1dt_2dt_3 \\
& + \mathbf{g}E_cE_0^2 \int\int\int \mathbf{A}_{\gamma}(t_1)\mathbf{A}_{\gamma}(t_2)\mathbf{f}_{\gamma}(t_3)\cos\omega(t-t_1) \\
& \times \cos\omega(t-t_1-t_2-t_3)dt_1dt_2dt_3 \\
& + \mathbf{g}E_0^3 \int\int\int \mathbf{A}_{\gamma}(t_1)\mathbf{A}_{\gamma}(t_2)\mathbf{f}_{\gamma}(t_3)\cos\omega(t-t_1) \\
& \times \cos\omega(t-t_1-t_2)\cos\omega(t-t_1-t_2-t_3)dt_1dt_2dt_3 \qquad (4.63)
\end{aligned}
$$

which is the third-order nonlinear response due to the permanent moments only. We do not give explicit developments of the last three terms in Eq. (4.63) for reasons of space, but it can be seen that they provide harmonic contributions in 2ω, 2ω, and 3ω, respectively. It suffices to expand the product of cosines appearing in each expression.

The last term in Eq. (4.61) is also a third-order nonlinear dielectric response arising from the joint contribution of permanent and induced dipole moments. This yields

$$
\begin{aligned}
f^{\infty}_{\beta 2,\gamma 1} = {} & \mathbf{g}\left(E_c^3 + \frac{1}{2} E_c E_0^2\right) \int_0^{\infty} \mathbf{f}_{\gamma}(t)dt \int_0^{\infty} \mathbf{f}_{\beta}(t)\, dt \\
& + \mathbf{g}E_c^2 E_0 \int_0^{\infty} \mathbf{f}_{\gamma}(t)dt \left[\left(\int_0^{\infty} \mathbf{f}_{\beta}(t_1) \cos \omega t_1 dt_1\right) \cos \omega t \right. \\
& \left. + \left(\int_0^{\infty} \mathbf{f}_{\beta}(t_1) \sin \omega t_1 dt_1\right) \sin \omega t\right] \\
& + \mathbf{g}E_c^2 E_0 \left\{\left[\int\int \mathbf{f}_{\gamma}(t_1)\mathbf{f}_{\beta}(t_2) \cos \omega(t_1 + t_2)\, dt_1 dt_2\right] \cos \omega t \right. \\
& \left. + \left[\int\int \mathbf{f}_{\gamma}(t_1)\mathbf{f}_{\beta}(t_2) \sin \omega(t_1 + t_2)\, dt_1 dt_2\right] \sin \omega t\right\} \\
& + \frac{1}{2}\mathbf{g}\left(E_c^2 E_0 + \frac{1}{2} E_0^3\right)\left\{\left[\int\int \mathbf{f}_{\beta}(t_1)\mathbf{f}_{\gamma}(t_2) \right.\right. \\
& \left. \times \cos \omega(t_1 + t_2)dt_1 dt_2\right] \cos \omega t \\
& \left. + \left[\int\int \mathbf{f}_{\beta}(t_1)\mathbf{f}_{\gamma}(t_2) \sin \omega(t_1 + t_2)dt_1 dt_2\right] \sin \omega t\right\} \\
& + \frac{1}{2}\mathbf{g}\left(E_c^2 E_0 + \frac{1}{2} E_0^3\right) \int_0^{\infty} \mathbf{f}_{\beta}(t)dt \left[\left(\int_0^{\infty} \mathbf{f}_{\gamma}(t_1) \cos \omega t_1 dt_1\right) \cos \omega t \right. \\
& \left. + \left(\int_0^{\infty} \mathbf{f}_{\gamma}(t_1) \sin \omega t_1 dt_1\right) \sin \omega t\right] \\
& + \frac{1}{4}\mathbf{g}E_c E_0^2 \int_0^{\infty} \mathbf{f}_{\gamma}(t)dt \left[\left(\int_0^{\infty} \mathbf{f}_{\beta}(t_1) \cos 2\omega t_1 dt_1\right) \cos 2\omega t \right. \\
& \left. + \left(\int_0^{\infty} \mathbf{f}_{\beta}(t_1) \sin 2\omega t_1 dt_1\right) \sin 2\omega t\right]
\end{aligned}
$$

$$
\begin{aligned}
&+\frac{1}{4}\mathbf{g}E_cE_0^2\left\{\left[\int\int \mathbf{f}_\gamma(t_1)\mathbf{f}_\beta(t_2)\cos 2\omega(t_1+t_2)dt_1dt_2\right]\cos 2\omega t\right.\\
&\left.+\left[\int\int \mathbf{f}_\gamma(t_1)\mathbf{f}_\beta(t_2)\sin 2\omega(t_1+t_2)dt_1dt_2\right]\sin 2\omega t\right\}\\
&+\frac{1}{2}\mathbf{g}E_cE_0^2\left\{\int_0^\infty \mathbf{f}_\beta(t)dt\int_0^\infty \mathbf{f}_\gamma(t)\cos\omega t dt\right.\\
&+\left[\int\int \mathbf{f}_\beta(t_1)\mathbf{f}_\gamma(t_2)\cos\omega(2t_1+t_2)dt_1dt_2\right]\cos 2\omega t\\
&\left.+\left[\int\int \mathbf{f}_\beta(t_1)\mathbf{f}_\gamma(t_2)\sin\omega(2t_1+t_2)dt_1dt_2\right]\sin 2\omega t\right\}\\
&+\frac{1}{2}\mathbf{g}E_cE_0^2\left\{\int_0^\infty \mathbf{f}_\gamma(t)dt\int_0^\infty \mathbf{f}_\beta(t)\cos\omega t dt\right.\\
&+\left[\int\int \mathbf{f}_\gamma(t_1)\mathbf{f}_\beta(t_2)\cos\omega(2t_1+t_2)dt_1dt_2\right]\cos 2\omega t\\
&\left.+\left[\int\int \mathbf{f}_\gamma(t_1)\mathbf{f}_\beta(t_2)\sin\omega(2t_1+t_2)dt_1dt_2\right]\sin 2\omega t\right\}\\
&+\frac{1}{8}\mathbf{g}E_0^3\left\{\left[\int\int \mathbf{f}_\beta(t_1)\mathbf{f}_\gamma(t_2)\cos\omega(t_1-t_2)dt_1dt_2\right]\cos\omega t\right.\\
&\left.+\left[\int\int \mathbf{f}_\beta(t_1)\mathbf{f}_\gamma(t_2)\sin\omega(t_1-t_2)dt_1dt_2\right]\sin\omega t\right\}\\
&+\frac{1}{8}\mathbf{g}E_0^3\left\{\left[\int\int \mathbf{f}_\beta(t_1)\mathbf{f}_\gamma(t_2)\cos\omega(3t_1+t_2)dt_1dt_2\right]\cos 3\omega t\right.\\
&\left.+\left[\int\int \mathbf{f}_\beta(t_1)\mathbf{f}_\gamma(t_2)\sin\omega(3t_1+t_2)dt_1dt_2\right]\sin 3\omega t\right\}\\
&+\frac{1}{8}\mathbf{g}E_0^3\left\{\left[\int\int \mathbf{f}_\gamma(t_1)\mathbf{f}_\beta(t_2)\cos\omega(t_1+2t_2)dt_1dt_2\right]\cos\omega t\right.\\
&\left.+\left[\int\int \mathbf{f}_\gamma(t_1)\mathbf{f}_\beta(t_2)\sin\omega(t_1+2t_2)dt_1dt_2\right]\sin\omega t\right\}\\
&+\frac{1}{8}\mathbf{g}E_0^3\left\{\left[\int\int \mathbf{f}_\gamma(t_1)\mathbf{f}_\beta(t_2)\cos\omega(3t_1+2t_2)dt_1dt_2\right]\cos 3\omega t\right.\\
&\left.+\left[\int\int \mathbf{f}_\gamma(t_1)\mathbf{f}_\beta(t_2)\sin\omega(3t_1+2t_2)dt_1dt_2\right]\sin 3\omega t\right\}
\end{aligned}
\tag{4.64}
$$

Equations (4.62)–(4.64) show that the nonlinear DR response for terms in E^3 consists of three harmonic components varying at once, twice, and three times the circular frequency ω. We also find that three relaxation functions $[\mathbf{f}_\gamma(t), \mathbf{f}_\beta(t), \mathbf{A}_\gamma(t)]$ are necessary to describe this nonlinear phenomenon instead of one $[\mathbf{f}_\gamma(t)]$ in the linear regime. The most novel result arises from Eq. (4.64), where $\mathbf{f}_\gamma(t)$ and $\mathbf{f}_\beta(t)$ are coupled so that the nonlinearity causes a coupling between permanent and induced dipolar moments. We reiterate that this result is only apparent in dielectric relaxation (which is all that we are concerned with here) and not in Kerr effect for an expansion carried to third order terms.

Finally, we may write down the following simplified expression for the electric polarization:

$$P(t) = P_{st} + \sum_{j=1}^{3} [P_j'(\omega)\cos(j\omega t) + P_j''(\omega)\sin(j\omega t)] \tag{4.65}$$

where P_{st} is a time-independent but frequency-dependent component and P_j' and P_j'' are the real and imaginary parts of the nonlinear harmonic components of $P(t)$, respectively. We shall now stress on these $j\omega$ components arising from the application of a weak alternating field superimposed on a direct field in view of obtaining expressions easier to manipulate and to be checked by experiment.

3. *Analytic Expressions for the Nonlinear Susceptibilities*

We again consider the Smoluchowski equation, which is valid if inertial effects are fully ignored and written in the usual form. This is

$$\frac{1}{D}\frac{\partial f}{\partial t} = \frac{1}{\sin\theta}\frac{\partial}{\partial\theta}\left[\sin\theta\left(\frac{\partial f}{\partial\theta} + \frac{1}{kT}\frac{\partial V}{\partial\theta} f\right)\right] \tag{4.66}$$

where the symbols have the same meaning as previously. The orientational potential energy V is explicitly given by

$$V = -\mu E\cos\theta - \tfrac{1}{2}\Delta\alpha E^2\cos^2\theta \tag{4.67}$$

To solve Eq. (4.66), we use the definition of the ensemble average of the nth Legendre polynomial $P_n(u)$, namely,

$$\langle P_n(u)\rangle(t) = \int_{-1}^{+1} f(u,t)P_n(u)du \tag{4.68}$$

where $u = \cos\theta$. We also assume that the amplitude E_0 of the cosine field is much less than that of the dc field E_c.

Multiplying Eq. (4.66) by $P_n(u)$ and integrating over the variable u, we find that the $\langle P_n(u)\rangle(t)$ must satisfy an infinite hierarchy of coupled differential–difference equations [49, 52]

$$\begin{aligned}\frac{d}{d\tau}\langle P_n(u)\rangle(\tau) = &-n(n+1)\left[1 - \frac{\beta(\tau)}{(2n-1)(2n+3)}\right]\langle P_n(u)\rangle(\tau)\\ &+\gamma(\tau)\frac{n(n+1)}{2n+1}[\langle P_{n-1}(u)\rangle(\tau) - \langle P_{n+1}(u)\rangle(\tau)]\\ &+\beta(\tau)\left[\frac{(n-1)n(n+1)}{(2n-1)(2n+1)}\langle P_{n-2}(u)\rangle(\tau)\right.\\ &\left.-\frac{n(n+1)(n+2)}{(2n+1)(2n+3)}\langle P_{n+2}(u)\rangle(\tau)\right]\end{aligned} \tag{4.69}$$

where $\tau = Dt$ is a reduced variable, $\gamma(\tau) = (\mu/kT)E(\tau)$, $\beta(\tau) = (\Delta\alpha/kT)E^2(\tau)$. Since the electric polarization is proportional to $\langle P_1(u)\rangle(\tau)$, Eq. (4.69) is written as

$$\begin{aligned}\langle \dot{P}_1(u)\rangle(\tau) = &-2\left[1 - \frac{\beta(\tau)}{5}\right]\langle P_1(u)\rangle(\tau) + \frac{2}{3}\gamma(\tau)[1 - \langle P_2(u)\rangle(\tau)]\\ &-\frac{2}{5}\beta(\tau)\langle P_3(u)\rangle(\tau)\end{aligned} \tag{4.70}$$

If only terms of third order in the electric field are retained, one arrives at

$$\langle \dot{P}_1(u)\rangle(\tau) + 2\left[1 - \frac{\beta(\tau)}{5}\right]\langle P_1(u)\rangle(\tau) = \frac{2}{3}\gamma(\tau)[1 - \langle P_2(u)\rangle(\tau)] \tag{4.71}$$

In Eq. (4.71), we discarded the product $\beta(\tau)\langle P_3(u)\rangle(\tau)$ because this term would give a contribution at least in E^5.

The stationary solution of Eq. (4.71) may be obtained by assuming the system to be in equilibrium at $t = -\infty$, and considering its behavior a long time after the electric field has been switched on. Thus, one readily finds that

$$\begin{aligned}\langle P_1(u)\rangle(\tau) = &\int_{-\infty}^{\tau} \exp\left\{-2\int_{\tau'}^{\tau}\left[1 - \frac{\beta(\tau_1)}{5}\right]d\tau_1\right\}\\ &\times\left\{\frac{2}{3}\gamma(\tau')[1 - \langle P_2(u)\rangle(\tau')]\right\}d\tau'\end{aligned} \tag{4.72}$$

Expanding the exponential in the integral of Eq. (4.72) in Taylor's series, we obtain

$$\langle P_1(u)\rangle(\tau) = \int_{-\infty}^{\tau} e^{-2(\tau-\tau')} \left[1 + \frac{2}{5}\int_{\tau'}^{\tau} \beta(\tau_1)d\tau_1\right] \times \left\{\frac{2}{3}\gamma(\tau')[1 - \langle P_2(u)\rangle(\tau')]\right\} d\tau' \tag{4.73}$$

From the definition of $\beta(\tau)$,

$$\beta(\tau) = \frac{\Delta\alpha}{kT}\left[E_c^2 + 2E_cE_0\cos\omega'\tau + \frac{1}{2}E_0^2(1+\cos 2\omega'\tau)\right]$$

where $\omega' = \omega/D$, and on setting

$$\beta_0 = \frac{\Delta\alpha}{kT}E_0^2 \qquad \beta_c = \frac{\Delta\alpha}{kT}E_c^2 \qquad \beta_{0c} = \frac{\Delta\alpha}{kT}(2E_0E_c)$$

Eq. (4.73) may be explicitly rearranged to read as follows:

$$\begin{aligned}\langle P_1(u)\rangle(\tau) = \frac{2}{3}\int_{-\infty}^{\tau} e^{-2(\tau-\tau')} \Bigg\{1 + \frac{2}{5}\Bigg[\left(\frac{\beta_0}{2} + \beta_c\right)(\tau-\tau') \\ + \beta_{0c}\frac{\sin\omega'\tau - \sin\omega'\tau'}{\omega'} \\ + \frac{\beta_0}{2}\frac{\sin 2\omega'\tau - \sin 2\omega'\tau'}{2\omega'}\Bigg]\Bigg\} \\ \times \gamma(\tau')[1 - \langle P_2(u)\rangle(\tau')]d\tau'\end{aligned} \tag{4.74}$$

where

$$\gamma(\tau) = \frac{\mu E_c}{kT} + \frac{\mu E_0}{kT}\cos\omega'\tau = \gamma_c + \gamma_0\cos\omega'\tau$$

The last term on the right-hand side of Eq. (4.74) contains the ensemble average of the second Legendre polynomial, which is characteristic of the electric birefringence. So, one may retain only terms in E^2 of $\langle P_2(u)\rangle(\tau)$, since the product $\gamma(\tau)\langle P_2(u)\rangle(\tau)$ is then third order. Moreover, this term would be the only one in the case of pure polar molecules ($\Delta\alpha = 0$). Setting $n = 2$ in Eq. (4.69) and limiting ourselves up to second order in the electric field, one readily finds that $\langle P_2(u)\rangle(\tau)$ may be derived from the following expression

$$\langle P_2(u)\rangle(\tau) = \frac{4}{5}\int_{-\infty}^{\tau}\int_{-\infty}^{\tau_1} e^{-6(\tau-\tau')}\gamma(\tau')e^{-2(\tau'-\tau_1)}\gamma(\tau_1)d\tau' d\tau_1$$
$$+\frac{2}{5}\int_{-\infty}^{\tau} e^{-6(\tau-\tau')}\beta(\tau')d\tau' \qquad (4.75)$$

Furthermore, it is easy to see that the linear dielectric response reduces to

$$\langle P_1(u)\rangle_{\mathrm{L}}(\tau) = \frac{2}{3}\int_{-\infty}^{\tau} e^{-2(\tau-\tau')}\gamma(\tau')d\tau' \qquad (4.76)$$

where the subscript "L" stands for linear.

Applying the Laplace transform to Eq. (4.74) and after some algebraic operations checked on a computer, we find that the three harmonic components of the electric polarization may be expressed in the following compact form (with ω or ω')

$$\mathbf{P}_j(\omega) = P_j'(\omega) - iP_j''(\omega) \qquad (4.77)$$

where $\mathbf{P}_j(\omega)$ is the complex electric polarization of rank j, and

$$P_1'(\omega') = \frac{\gamma_0}{3(1+\omega'^2/4)} - \gamma_0^3\frac{(3-13\omega'^2/36) + P(-6+5\omega'^2/9+7\omega'^4/72)}{180(1+\omega'^2/4)^2(1+\omega'^2/9)}$$
$$-\gamma_0\gamma_c^2\frac{(3+\omega'^2/36-\omega'^4/72) - P(6+7\omega'^2/18+\omega'^4/18)}{45(1+\omega'^2/4)^2(1+\omega'^2/36)} \qquad (4.78a)$$

$$P_1''(\omega') = \omega'\left[\frac{\gamma_0}{6(1+\omega'^2/4)}\right.$$
$$-\gamma_0^3\frac{\left(\frac{7}{3}+\omega'^2/36\right)+P\left(-\frac{25}{6}-7\omega'^2/72+\omega'^4/36\right)}{180(1+\omega'^2/4)^2(1+\omega'^2/9)}$$
$$\left.-\gamma_0\gamma_c^2\frac{\left(\frac{14}{3}+25\omega'^2/36+\omega'^4/144\right)-P\left(\frac{25}{3}+8\omega'^2/9+5\omega'^4/144\right)}{90(1+\omega'^2/4)^2(1+\omega'^2/36)}\right] \qquad (4.78b)$$

$$P_2'(\omega') = -\gamma_0^2\gamma_c \times \frac{(3 - 17\omega'^2/12 - 31\omega'^4/216 - \omega'^6/216) + P(-6 + 4\omega'^2/3 + 29\omega'^4/216 - \omega'^6/216)}{90(1+\omega'^2)(1+\omega'^2/4)(1+\omega'^2/9)(1+\omega'^2/36)} \tag{4.79a}$$

$$P_2''(\omega') = -\gamma_0^2\gamma_c\omega' \times \frac{(\frac{14}{3} + 11\omega'^2/27 + \omega'^4/54) - P(\frac{25}{3} + 65\omega'^2/54 + 29\omega'^4/432 + \omega'^6/432)}{90(1+\omega'^2)(1+\omega'^2/4)(1+\omega'^2/9)(1+\omega'^2/36)} \tag{4.79b}$$

$$P_3'(\omega') = -\gamma_0^3 \frac{(1 - 5\omega'^2/12) + P(-2 + 5\omega'^2/3 + \omega'^4/8)}{180(1+\omega'^2/4)(1+\omega'^2 9)(1+9\omega'^2/4)} \tag{4.80a}$$

$$P_3''(\omega') = -\gamma_0^3\omega' \frac{(\frac{14}{3} - \omega'^2/2) - P(\frac{25}{3} + 5\omega'^2/12)}{360(1+\omega'^2/4)(1+\omega'^2/9)(1+9\omega'^2/4)} \tag{4.80b}$$

From Eq. (4.65), it should be noted that there exists a dc component $\langle P_1(u)\rangle_{\text{st}}$, which is only due to the presence of the constant field E_c. Its expression is given by

$$\langle P_1(u)\rangle_{\text{st}} = \frac{\gamma_c}{3}\left(1 + \frac{2\beta_c + \beta_0 - \gamma_c^2}{15}\right) - \gamma_0^2\gamma_c \frac{3 + 7\omega'^2/36 + P(-1 + 5\omega'^2/12)}{90(1+\omega'^2/4)(1+\omega'^2/36)} \tag{4.81}$$

The same reason also applies to the harmonic term in 2ω. The appearance of this term varying at twice the fundamental frequency of the alternating field in DR is a direct consequence of the nonlinear nature of the response when a unidirectional field is simultaneously impressed on. Furthermore we remark the coupling effect of interactive fields E_0 and E_c together with cross-terms between permanent and induced dipole moments. On setting $P = 0$ and $\gamma_c = 0$ in Eqs. (4.78) and (4.80), the results so obtained are in full agreement with those of Coffey and Paranjape [45]. To study the frequency behavior of the nonlinear complex components of the polarization $\mathbf{P}_j (j = 1, 2, 3)$, we shall present them in a form normalized to unity, so that

$$X_j(\omega') = \frac{P_j'(\omega')}{P_j'(0)} \qquad Y_j(\omega') = \frac{P_j''(\omega')}{P_j'(0)} \tag{4.82}$$

where

$$\begin{aligned} P_1'(0) &= \frac{1}{3}\gamma_0\left[1 - \left(\frac{\gamma_0^2}{5}\right)\left(\frac{1}{4} + \gamma_c^2/\gamma_0^2\right)(-2P+1)\right] \\ P_2'(0) &= -\frac{1}{30}\gamma_0^2\gamma_c(-2P+1) \\ P_3'(0) &= -(\gamma_0^3/180)(-2P+1) \end{aligned} \tag{4.83}$$

are the normalization coefficients. By inspection of Eq. (4.83), it is immediately seen that the proportionality factor is $(-2P+1)$ instead of $(P+1)$ in the Kerr effect (see Section III).

To a have a complete description about this dielectric relaxation process, we shall also consider the evolution of phase angles Θ_j between the in-phase and out-of-phase j components [53], defined by

$$\Theta_j(\omega') = \tan^{-1}(P_j''/P_j') = \tan^{-1}(Y_j/X_j)$$

4. *Dispersion and Cole–Cole Plots*

The frequency dependence of P_{st} equal to $\langle P_1(u)\rangle_{st}$ is shown in Fig. 4.18. We have, in fact, considered the variations of the second term in the right-hand side of Eq. (4.81) (noted P_{st}'), normalized to unity, since only this term is ω dependent. The plots show a similar behavior to the dispersion plots relative to the real parts $X_j(\omega)$. Figures 4.19–4.22 illustrate these latter together with the imaginary parts $Y_j(\omega)$ for various values of the parameter P. Given a value of P, each of them exhibits the characteristic differences between the harmonic components. The essential feature of these plots clearly depicts their resonant behavior. This may be explained as follows. Equations (4.77)–(4.80) indicate that the relaxation mechanisms occur with typical relaxation times. In particular, the ω component contains three distinct relaxation times, $1/2D$, $1/3D$, $1/6D$, instead of only one ($1/2D$) in the linear regime. The 2ω term is characterized by four relaxation times, $1/D$, $1/2D$, $1/3D$, and $1/6D$. The third harmonic (in 3ω), which is only due to the presence of the

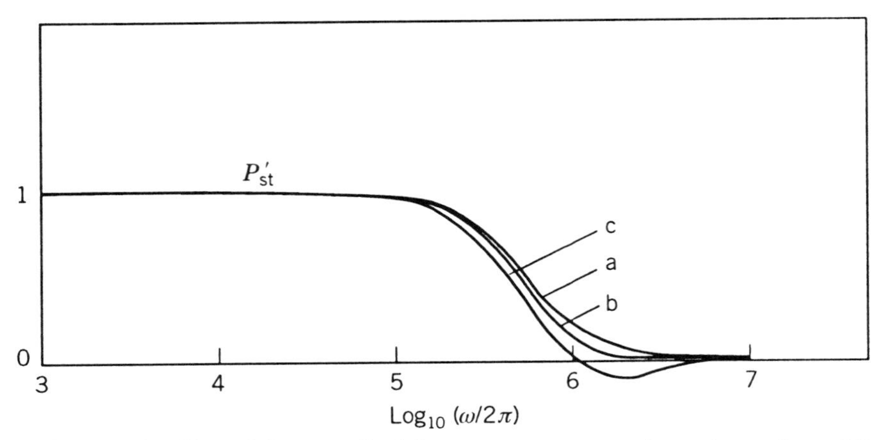

Figure 4.18. Plots of the normalized dc component versus ω for various values of P. (a): $P = 0$; (b): $P = -2$; (c): $P = 10$. (D has been chosen arbitrarily equal to $6 \times 10^6\ s^{-1}$).

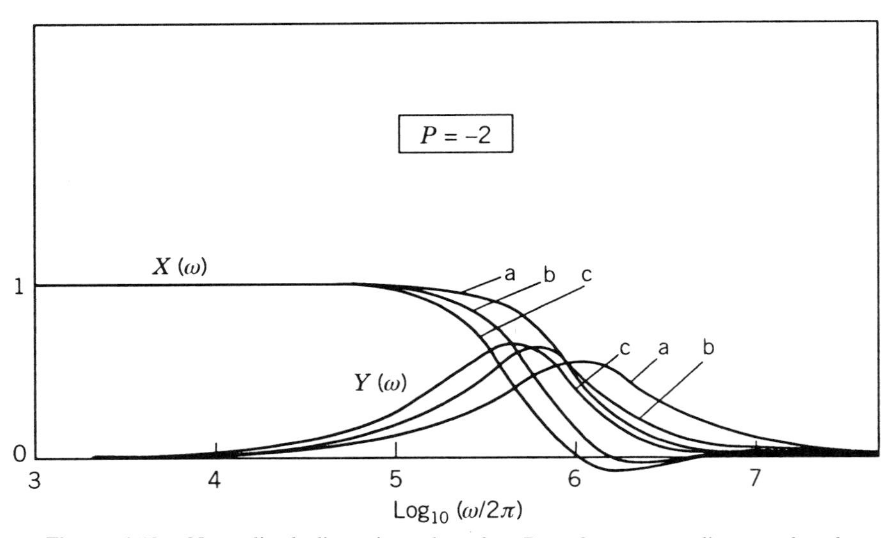

Figure 4.19. Normalized dispersion plots for $P = -2$ corresponding to the three harmonic terms. (a): ω-component; (b): 2ω-component; (c): 3ω-component. ($\gamma_0 = 0.1$; $\gamma_c = 1.83$; D has been chosen arbitrarily equal to $6 \times 10^6\ s^{-1}$).

alternating field shows the existence of three relaxation times, $3/2D$, $1/3D$, and $1/2D$. The combination of all these relaxation times arises from nonlinear effects. We may remark that the maximum of the imaginary parts $Y_j(\omega')$ is shifted on the left of the frequency scale when one passes from the fundamental to the third harmonic term. The

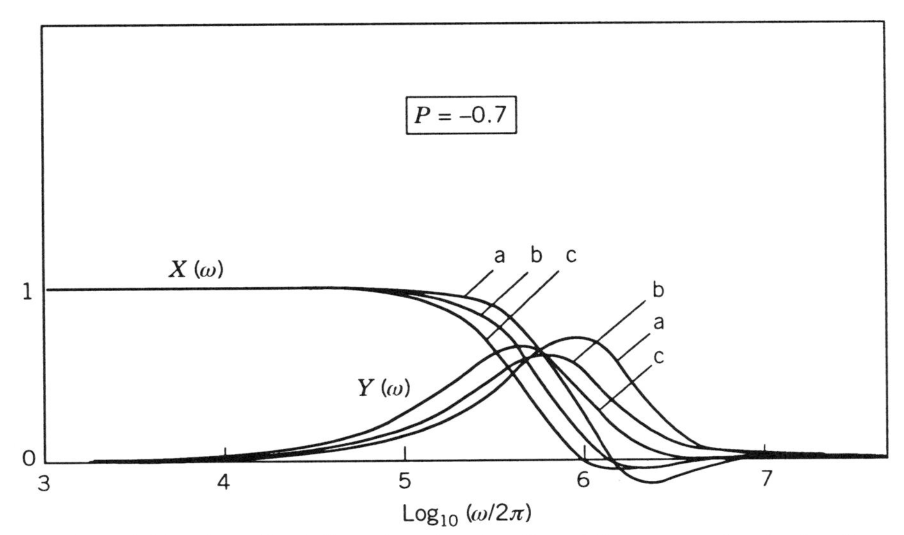

Figure 4.20. Normalized dispersion plots for $P = -0.7$ corresponding to the three harmonic terms. (a): ω-component; (b): 2ω-component; (c): 3ω-component. ($\gamma_0 = 0.1$; $\gamma_c = 1.83$; D has been chosen arbitrarily equal to $6 \times 10^6\ \mathrm{s}^{-1}$).

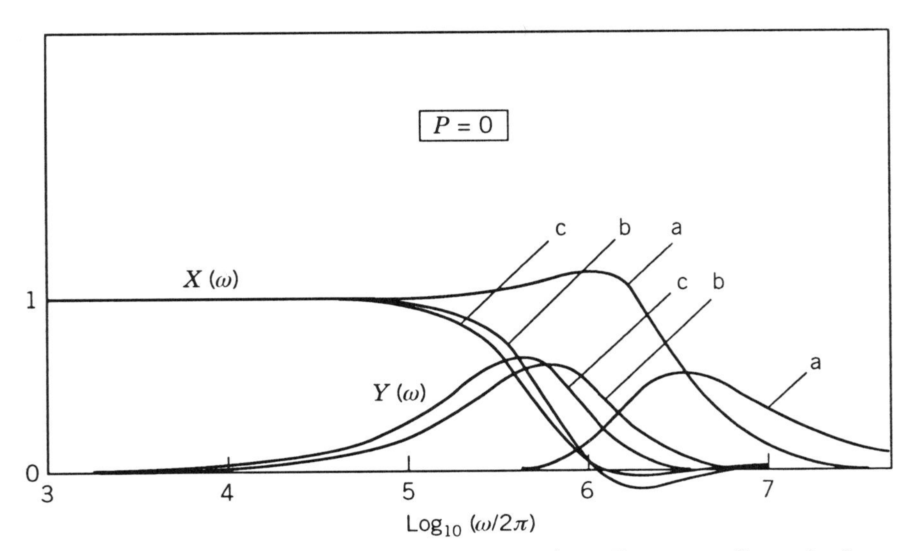

Figure 4.21. Normalized dispersion plots for $P = 0$ ($\Delta\alpha + 0$) corresponding to the three harmonic terms. (a): ω-component; (b): 2ω-component; (c): 3ω-component. ($\gamma_0 = 0.1$; $\gamma_c = 1.83$; D has been chosen arbitrarily equal to $6 \times 10^6\ \mathrm{s}^{-1}$).

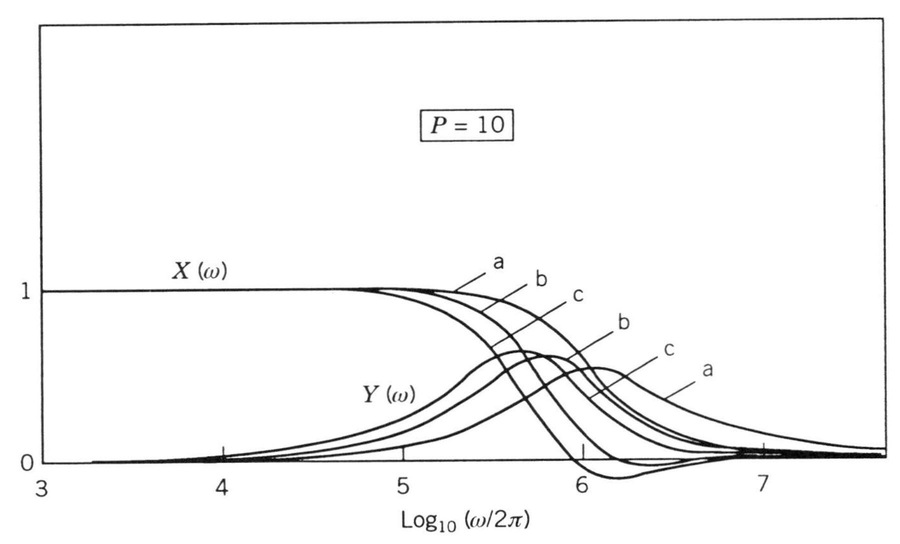

Figure 4.22. Normalized dispersion plots for $P = 10$ ($\mu \rightarrow 0$) corresponding to the three harmonic terms. (a): ω-component; (b): 2ω-component; (c): 3ω-component. ($\gamma_0 = 0.1$; $\gamma_c = 1.83$; D has been chosen arbitrarily equal to $6 \times 10^6\ \mathrm{s}^{-1}$).

dispersion plots of the first harmonic component obtained for $P = 0$ show how they differ from the usual Debye behavior. This reveals the nonlinear nature of the dielectric response. In particular, an overshoot is observed for $X_1(\omega)$ while $Y_1(\omega)$ has lost its even symmetry, the rise being faster than the decay. In the case $P = 10$, it is quite surprising to have Debye-like plots for this ω component.

Cole–Cole diagrams $Y_j(X_j)$ are presented in Figs. 4.23–4.26. They clearly show how they deviate from the semicircle obtained by Debye in the case of a linear dielectric response. The value of $P = 0$ corresponds to purely polar molecules ($\Delta\alpha = 0$), while $P = 10$ is relative to molecules where the induced moment over the permanent one ($\mu \approx 0$) prevails. If we look at the plot corresponding to the ω component for $P = 0$ (Fig. 4.25), we see that $Y_1(\omega)$ takes negative values in the very low-frequency region. As a consequence, $X_1(\omega)$ clearly exceeds the normalized value of 1, and the plot presents a flattened aspect. However, the maximum height attained by $Y_1(\omega)$ remains close to 0.5. This is at variance with the linear response where the semicircular Cole–Cole plot lies completely in the upper half-plane. Regarding the same component for $P = 10$, we see that

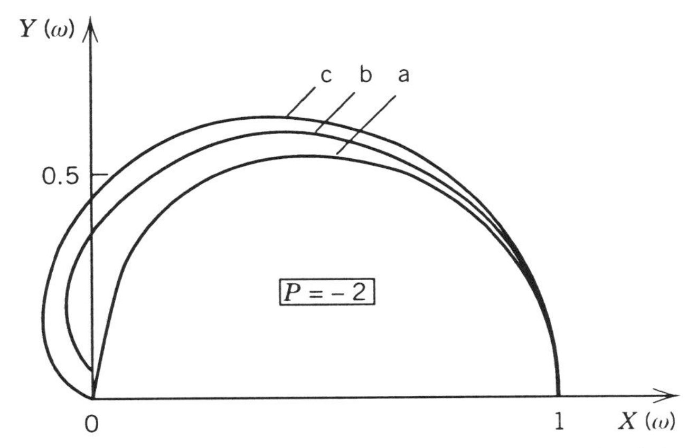

Figure 4.23. Plots of the imaginary part Y_j vs the real part X_j of the normalized dielectric relaxation functions for $P = -2$. (a): ω-component; (b): 2ω-component; (c): 3ω-component. ($\gamma_0 = 0.1$; $\gamma_c = 1.83$).

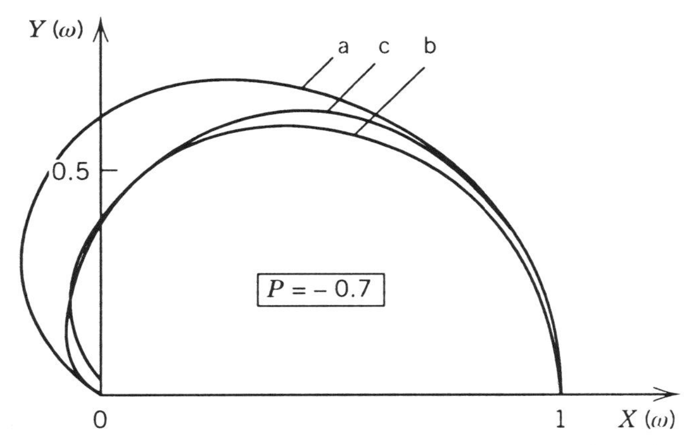

Figure 4.24. Plots of the imaginary part Y_j vs the real part X_j of the normalized dielectric relaxation functions for $P = -0.7$. (a): ω-component; (b): 2ω-component; (c): 3ω-component. ($\gamma_0 = 0.1$; $\gamma_c = 1.83$).

the skewed arc is rather deformed in the high-frequency domain, and the maximum of $Y_1(\omega)$ is less than 0.5.

5. *Asymptotic Values of Phase Angles (High-Frequency Domain)*

Since each harmonic term may always be separated experimentally, we now have a triple set of data on the liquid under consideration. This

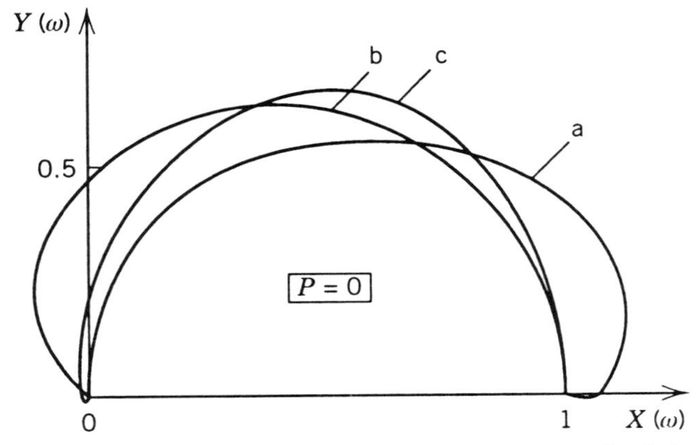

Figure 4.25. Plots of the imaginary part Y_j vs the real part X_j of the normalized dielectric relaxation functions for $P = 0$. (a): ω-component; (b): 2ω-component; (c): 3ω-component. ($\gamma_0 = 0.1$; $\gamma_c = 1.83$).

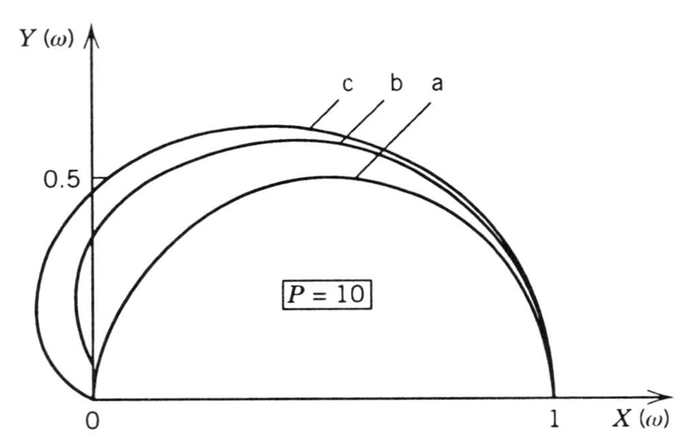

Figure 4.26. Plots of the imaginary part Y_j vs the real part X_j of the normalized dielectric relaxation functions for $P = 10$. (a): ω-component; (b): 2ω-component; (c): 3ω-component. ($\gamma_0 = 0.1$; $\gamma_c = 1.83$).

increases the accuracy with which molecular parameters (such as D and P) can be determined. In this way, the phase angles yield very interesting and complementary information. They allow us to evaluate the modulus of the complex polarization $\mathbf{P}_j$ by means of the Kramers–Kronig relations and to extract the harmonic components, since

$$P_j'(\omega) = |\mathbf{P}_j| \cos \Theta_j$$

$$P_j''(\omega) = |\mathbf{P}_j| \sin \Theta_j$$

We mention here that the components of the electric polarization $P(t)$ have been carried out by direct calculation of $\langle P_1(u) \rangle (t)$. These two quantities are in fact proportional to each other, so that we have rigorously

$$P(t) = N\mu \langle P_1(u) \rangle (t) = N\mu \left[\langle P_1(u) \rangle_{\text{st}} + \sum_{j=1}^{3} P_j' \cos(j\omega t) + P_j'' \sin(j\omega t) \right] \tag{4.84}$$

where N is the number of dipoles per unit volume.

Furthermore, each harmonic component of the electric polarization may be related to a complex susceptibility $\boldsymbol{\chi}_j$ defined by the following equations

$$\begin{aligned} P_1(t) &= \text{Re}[\boldsymbol{\chi}_1(\omega)] E_0 \exp(i\omega t) \\ P_2(t) &= \text{Re}[\boldsymbol{\chi}_2(\omega)] E_c E_0^2 \exp(2i\omega t) \\ P_3(t) &= \text{Re}[\boldsymbol{\chi}_3(\omega)] E_0^3 \exp(3i\omega t) \end{aligned} \tag{4.85}$$

where

$$\boldsymbol{\chi}_j(\omega) = \chi_j'(\omega) - i\chi_j''(\omega) \tag{4.86}$$

These considerations are important in the determination of the phase angles. So, if we apply the definition of the loss tangent tan δ_1 given by Scaife [54] to the first harmonic component, we obtain the following equation

$$\delta_1(\omega) = \tan^{-1} \frac{\chi_1''(\omega)}{\varepsilon_0[1 + \chi_1'(\omega)]} \tag{4.87}$$

in which

$$\chi_1'(\omega) = \frac{N\mu}{\gamma_0} P_1'(\omega) \qquad \chi_1''(\omega) = \frac{N\mu}{\gamma_0} P_1''(\omega)$$

In Eq. (4.87), $\delta_1(\omega)$ is also strongly dependent on P values and exactly represents the phase angle existing between the imaginary $\varepsilon''(\omega)$ and real $\varepsilon'(\omega)$ parts of the complex electric permittivity $\boldsymbol{\varepsilon}(\omega)$. If P is set equal to zero, this result agrees with that of Coffey and Paranjape. These authors introduced the loss tangent tan δ_0 obtained for very small values of E_0 and plotted the phase shift $(\delta_0 - \delta_1)$ against ω' for various values of E_c the amplitude of the constant field. They found that strong dc fields are necessary to observe appreciable phase shifts. Because of the difficulties that would result experimentally, we prefer to study the phase angles Θ_j arising from $\mathbf{P}_j$ or $\boldsymbol{\chi}_j$. As shown in Figs. 4.27–4.29, their variations with the frequency may be important. They indicate asymptotic values (high-frequency region) lying between $\pi/2$ and $3\pi/2$. In the linear regime where only the leading terms of the first harmonic remain [see Eqs. (4.78)], the high-frequency limit is $\pi/2$. From this, it seems that such shifts could be measurable by experiment to appreciate nonlinear effects in dielectric relaxation.

We shall now make some detailed comments about the phase plots obtained for each harmonic component of the electric polarization. In Fig. 4.27, it is seen how Θ_1 is very sensitive to negative P values close to

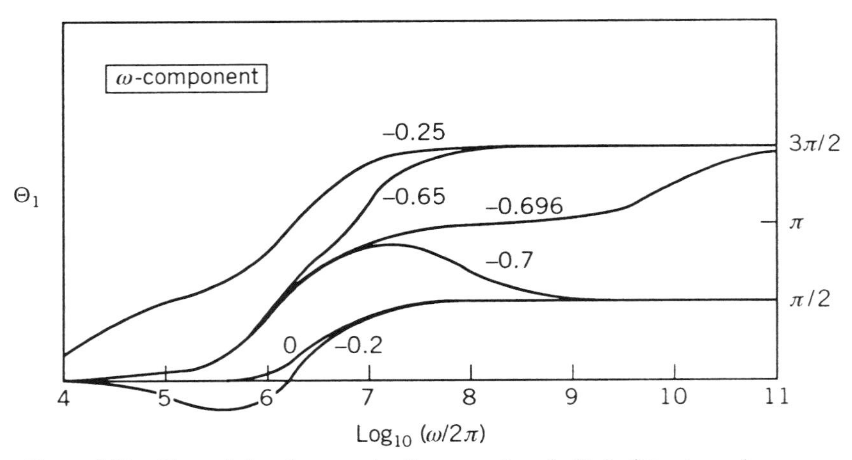

Figure 4.27. Plots of the phase angles Θ_1 versus $\log_{10}(\omega/2\pi)$. (Numbers above curves are for different values of P).

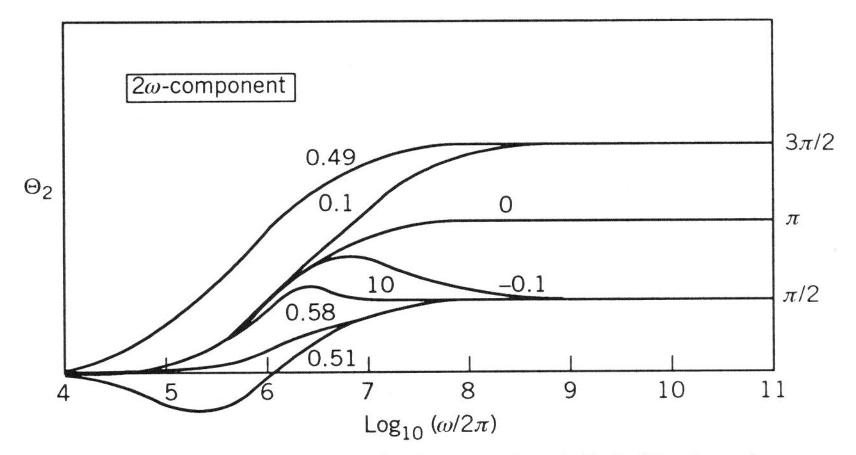

Figure 4.28. Plots of the phase angles Θ_2 versus $\log_{10}(\omega/2\pi)$. (Numbers above curves are for different values of P).

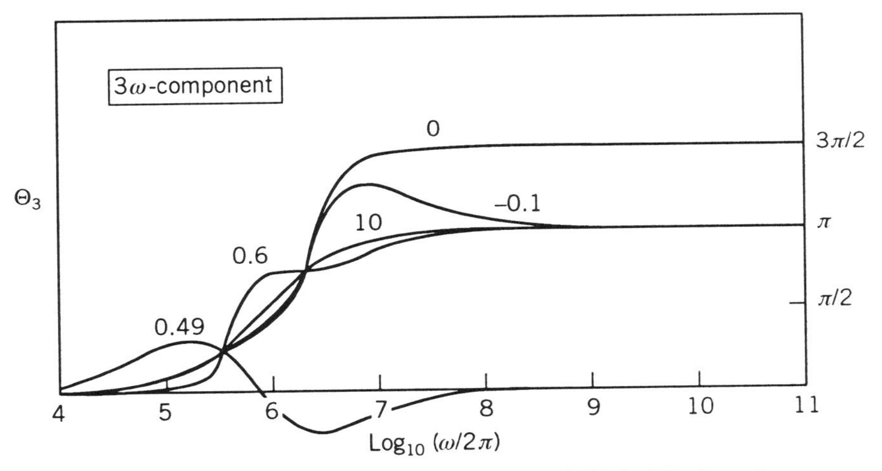

Figure 4.29. Plots of the phase angles Θ_3 versus $\log_{10}(\omega/2\pi)$. (Numbers above curves are for different values of P).

-0.7. For $P = -0.7$, the relevant curve goes through a maximum a little less than π before decaying to $\pi/2$. Very near this value ($P = -0.696$), the variations of Θ_1 are quite different. First, a plateau located at $\Theta_1 = \pi$ is attained in the frequency range comprised between 10^8 and 10^9 Hz, then Θ_1 increases up to $3\pi/2$ in the very high frequency region. Similar remarks may be made for $P = -0.2$ and -0.25 with some differences. In the first case, a minimum is obtained indicating that Θ_1 takes negative values for relatively low frequencies before passing again through zero and tending to $\pi/2$. By contrast, in the second case when $P = -0.25$, the asymptotic value of Θ_1 is $3\pi/2$.

The frequency evolution of Θ_1 for highly polar molecules ($P = 0$) from 0 to $\pi/2$ does not require any special comment. To conclude our discussion about this ω component, we may state that the anisotropy of the polarizability plays a great role in the variations of Θ_1. This is particularly true of the negative values of P. This means that we have situations such as $\Delta\alpha < 0$ or $\alpha_\perp > \alpha_\parallel$($\alpha_\perp$ and $\alpha_\parallel$ are the principal electric polarizabilities, perpendicular and parallel to the symmetry axis of the molecule, respectively). We mention here that the positive values of P yield changes in Θ_1 comparable to those obtained for $P = 0$.

Figure 4.28 pictures the variations of Θ_2 against frequency that are relevant to the second harmonic component. Because of the normalization factor proportional to $(1 - 2P)$, we have a discontinuity for $P = 0.5$. Hence, we consider it of interest to see how Θ_2 behaves according to whether $P \to 0.5^-$ or $P \to 0.5^+$. To illustrate this, we have chosen $P = 0.49$ and $P = 0.51$. By inspection of the corresponding curves, it should be noted that the limiting values of Θ_2 for infinitely high frequencies are not the same, being equal to $3\pi/2$ and $\pi/2$, respectively. The evolution of Θ_2 for $P = 0.49$ is regular while for $P = 0.51$, Θ_2 starts from zero, decays to a negative minimum, then increases to $\pi/2$.

Around $P = 0$, the plots of Θ_2 present some typical features. Indeed, the limiting value attained by Θ_2 for $P = -0.1$ is $\pi/2$. This value is multiplied by a factor 2 for $P = 0$ ($\Theta_2 = \pi$), and a factor 3 for $P = 0.1$ ($\Theta_2 = 3\pi/2$).

For anisotropically polarizable molecules ($P = 10$), $\Theta_2 \to \pi/2$, passing through a maximum that exceeds this asymptotic value.

Now, consider the plots presented in Fig. 4.29. Whatever the value of P, it is evident that these curves intersect in two characteristic points lying approximately in the low-frequency region, 3×10^5 and 2×10^6 Hz. Here too, we must consider the discontinuity arising from the normalizing factor. The case $P = 0.49$ provides curious behavior of the phase angle Θ_3. This situation has not occurred until now since this angle goes through two extrema, a maximum and a minimum. Consequently, Θ_3 starts from and arrives at zero.

For purely polar molecules, $\Theta_3 \to 3\pi/2$. In all the other cases save $P = 0.49$, Θ_3 tends to π. This result is very interesting from an experimental point of view in so far as it allows us to differentiate between the dipole moments (permanent or induced) in dielectric fluids.

III. THE DYNAMIC KERR EFFECT

An unperturbed fluid is always subjected to thermal energy that produces a disordered movement of the molecules, that is, Brownian motion. Initially isotropic fluid may become anisotropic under the action of an external force. Such a fluid acquires birefringent properties in consequence of the orientation of the molecules and has behavior similar to that of a uniaxial medium having optical axis along the direction of the force [55]. This latter can take various forms, such as a magnetic field that is the Cotton–Mouton effect, a velocity gradient that is the Maxwell effect, or an electric field that is called the Kerr effect. The theoretical work presented here concerns the dynamic Kerr effect, but it can be transferred easily to describe the other effects. This electrooptical effect is another way of investigating fluids and gives information complementary to that obtained by dielectric relaxation.

Kerr [56] was the first to discover the phenomenon of electric birefringence in 1875. He established the following law, which describes the induced birefringence Δn, which is proportional to the square of the electric field E

$$\Delta n = B\lambda E^2$$

where B is the Kerr constant, and λ is the wavelength in vacuo of the radiation that allows us to observe the phenomenon.

Since the discovery of the Kerr effect, numerous theoretical studies have been presented to describe the duality between the thermal energy of agitation and the potential energy of orientation of molecules in a dielectric liquid. The electric field may be applied in various forms, namely, the dc field, rectangular pulses, the ac field, and so on. Most of the experimental works, however, have largely corroborated the interest of this effect by measurements, particularly of permanent and induced dipole moments. In addition to the authors quoted in the Introduction, we mention Thurston and Bowling [57], Wegener [58], Filippini [59], as well as Morita and Watanabe [52], to which we shall refer in this section. The Laboratory of Applied Physics of the University of Perpignan has also made an important contribution to the study of this phenomenon [49, 60–63].

The domain of experimental investigation remains that of relatively low frequencies, rarely exceeding 10^9 Hz. Thus, inertial effects that, for a normal fluid, may be predicted to lie in the same regions as those of the dielectric constant, which is in the microwave region, are generally imperceptible to us, and up to the present time, we have been unable to find experimental data in the literature about these inertial effects, at least in the dynamic Kerr effect. Recently, a study in the optical Kerr effect [64], induced by the electric field of laser producing impulses of the order of some tens of femtoseconds on small molecules (carbon disulfide, chlorobenzene, or nitrobenzene) has shown why it is essential to include inertial effects and so, to leave the theory of rotational diffusion.

For these reasons (those working with frequencies less than 10^9 Hz) it seems interesting to consider only the rotational diffusion equation of Smoluchowski, which does not include inertia, and to propose a method of solving that equation for the orientation factor characterizing the electric birefringence. Just as in the previous work undertaken in the Laboratory of Applied Physics, we choose the electric field consisting of the superposition of an alternating field on a continuous field. This method, as we shall see in Section III.A.3, leads to double information about a same molecular parameter.

To delimitate the domain of validity of our study, we shall make the following assumptions:

- The fluid may be a liquid formed of molecules or macromolecules in dilute solution, or a pure organic liquid. It is understood that the molecular interactions in the solute or in the pure liquid, and those between solute and solvent may be neglected. So, the solvent like the pure liquid behaves as a continuous medium.
- Molecules of the liquid are small rigid ellipsoids of rotation whose geometric axes coincide with those of the electric polarizability tensor.
- Molecules are electrically anisotropic and carry a permanent dipole moment μ along the symmetry axis. We can also consider an induced dipole moment resulting from the displacement of movable charges under the action of the electric field and defined by

$$\mathbf{m} = \bar{\bar{\boldsymbol{\alpha}}}\mathbf{E} \tag{4.88}$$

  where $\bar{\bar{\boldsymbol{\alpha}}}$ is the polarizability tensor. This tensor may be diagonalized just like the tensor of dielectric constants (see Appendix C) and may be reduced to two principal values.

The birefringence induced in a fluid represents the difference between the optical indexes n' and n'' corresponding to directions parallel and perpendicular to the field, respectively. These indexes are known under the names of extraordinary $n' = n_e$ and ordinary $n'' = n_0$ refractive indexes. These are real quantities when the medium is not absorbing, but in general, we will have complex values when we take account of

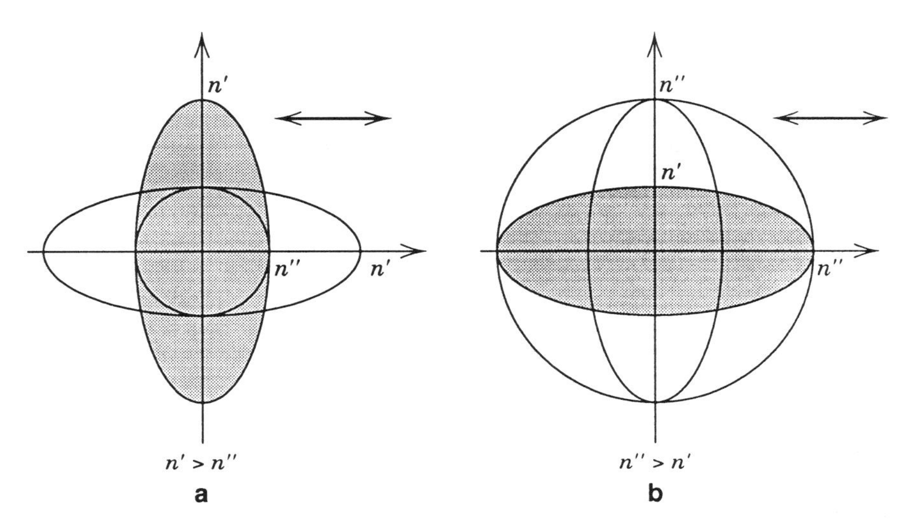

Figure 4.30. Index ellipsoids of a uniaxial medium. (a): $\Delta n > 0$ (prolate); (b): $\Delta n < 0$ (oblate). The arrows represent the optical axis.

absorption. Figure 4.30 shows the index ellipsoids that are typical of a uniaxial medium.

It is usual to put

$$\mathbf{n} = n - ik \tag{4.89}$$

where $\mathbf{n}$ denotes the complex refractive index, n is the refractive index, and k is the index of absorption.

Now, if the medium is birefringent, we can write

$$\begin{aligned} \Delta\mathbf{n} &= \mathbf{n}' - \mathbf{n}'' = (n' - ik') - (n'' - ik'') \\ &= \Delta n - i\,\Delta k \end{aligned} \tag{4.90}$$

The name birefringence may be ascribed to Δn, and we shall designate Δk as biabsorption. From Eq. (4.90) and regardless of the nature of the applied field, it follows that the Kerr constant necessarily has a complex value, namely,

$$\mathbf{B} = B' - iB'' \tag{4.91}$$

so that,

$$\Delta n = B'\lambda E^2 \qquad \Delta k = B''\lambda E^2 \tag{4.92}$$

A. Time-Dependent Behavior of Electric Birefringence Caused by an AC Field Superimposed on a DC Field

When an alternating electric field of amplitude E_0, such that

$$E(t) = E_0 \cos \omega t$$

is applied to an assembly of molecules, we know that the birefringence that results [57, 65] is given by

$$|\Delta \mathbf{n}| = \Delta n_{st}(\omega) + |\Delta n|_{alt} \cos(2\omega t - \theta) \tag{4.93}$$

where $\Delta n_{st}(\omega)$ is the stationary component, which is time independent, but frequency dependent, $|\Delta n|_{alt}$ is the modulus of the alternating component, and θ is the phase angle.

That equation, which has been verified experimentally, was established from linear response theory, which is when the Boltzmann superposition principle may be applied. This principle obtains when the applied field is small in comparison to the thermal energy. We mention the work of Bordewijk and co-workers [32] who showed that for a time dependent field, the birefringence function $\Delta n(t)$ is given by the correlation function $C_2(t)$ of the second Legendre polynomial following the law

$$\Delta n(t) \approx \{E^2(t) \otimes [-\dot{C}_2(t)]\}_{t=\infty} \tag{4.94}$$

where the symbol $\otimes$ designates a convolution product.

In what follows, we shall apply the nonlinear response theory developed by Morita and Watanabe [46, 66] to the rotational diffusion model, and we shall deduce expressions that describe Kerr effect relaxation. The main difference in comparison with the linear response theory lies in the fact that now two relaxation functions are needed, instead of only one, to describe the dynamic effect.

1. *Need for Two Relaxation Functions to Characterize the Birefringence Process*

We refer to our master equation [see Eq. (4.55) in Section II.D.1] written in terms of unperturbed $\hat{A}_0(u)$ and perturbed $\hat{A}_\gamma(u)$, $\hat{A}'_\beta(u)$ operators, which allows us to calculate the orientational distribution function $f(u, t)$. By inspection of the terms in the bracket of this equation, one may already note that the Kerr effect response when restricted to second order in the electric field strength is linear for only anisotropically polarizable molecules and nonlinear for pure dipole moments. The solution of Eq. (4.55) rests upon an original idea introduced by Morita [46], who showed that the nonlinear response resulting from a strong external perturbation may be obtained once the conditional-probability-density function $g(u, u', t)$, characteristic of an unperturbed state, is known [see Eq. (4.56)]. This theoretical approach enables us to determine the nonlinear responses of dielectric and Kerr effect relaxation up to any desired order in the electric field. Now, if we confine ourselves to the second-order nonlinear Kerr effect response, we may expand $f(u, t)$ in a power series of the small parameter $e = \mu/kT$. We find that

$$f(u,t) = f(u,0) + ef_{\gamma 1}(u,t) + e^2[f_{\gamma 2}(u,t) + f_{\beta 2}(u,t)] + \cdots \tag{4.95}$$

where $f_{\gamma 1}$, $f_{\gamma 2}$, and $f_{\beta 2}$ have already been expressed by Eqs. (4.60a)–(4.60c) in Section II. It is obvious that one only needs terms proportional to e^2 in so far as the electric birefringence is given as a series of even powers of e. As previously indicated, we consider an electric field applied at $t = 0$ to the dielectric liquid, and of the form:

$$E(t) = E_c + E_0 \cos \omega t$$

where E_c and E_0 are the respective amplitudes of the dc and ac fields, and ω is the angular frequency. Hence, on substituting the expression of $E(t)$ into Eqs. (4.60b), and (4.60c), the quantity $e^2[f_{\gamma 2} + f_{\beta 2}]$ yields the complete birefringence signal. One has

$$\begin{aligned} f_{\gamma 2} + f_{\beta 2} = {} & \mathbf{g}E_0^2\left[\int_0^t \int_0^{t_1} \mathbf{A}_\gamma(t - t_1)\mathbf{f}_\gamma(t_1 - t_2) \cos \omega t_1 \cos \omega t_2 dt_1 dt_2 \right. \\ & \left. + \int_0^t \mathbf{f}_\beta(t - t_1) \cos^2 \omega t_1 dt_1\right] + \mathbf{g}E_c^2\left[\int_0^t \int_0^{t_1} \mathbf{A}_\gamma(t - t_1)\mathbf{f}_\gamma(t_1 - t_2) dt_1 dt_2 \right. \\ & \left. + \int_0^t \mathbf{f}_\beta(t - t_1) dt_1\right] + \mathbf{g}E_cE_0\left[\int_0^t \int_0^{t_1} \mathbf{A}_\gamma(t - t_1)\mathbf{f}_\gamma(t_1 - t_2)(\cos \omega t_1 \right. \\ & \left. + \cos \omega t_2) dt_1 dt_2 + 2\int_0^t \mathbf{f}_\beta(t - t_1) \cos \omega t_1 dt_1\right] \end{aligned} \tag{4.96}$$

From Eq. (4.96), we immediately note that the leading term in brackets on the right-hand side of this equation is the nonlinear Kerr effect response for the pure alternating field. In the second bracket, we have the response due to the dc field only. The last bracket contains terms proportional to E_cE_0 (mutual action of both fields), which is clearly a nonlinear coupling effect in full accordance with previous works [66, 67]. Moreover, it is shown that two relaxation functions $\mathbf{A}(t)$ and $\mathbf{f}(t)$ are necessary to describe the nonlinear relaxation process of the electric birefringence (note that $\mathbf{f}(t)$ may represent $\mathbf{f}_\gamma(t)$ or $\mathbf{f}_\beta(t)$, which yields three relaxation functions). This may be explained as the action of the field interaction operators $\hat{A}_\gamma(u)$ and $\hat{A}'_\beta(u)$ on the perturbed distribution function $f(u, t)$. In the case of pure polar molecules, we have $\mathbf{f}_\beta(t) = 0$, while for only polarizable molecules $\mathbf{A}_\gamma(t) = 0$ and $\mathbf{f}_\gamma(t) = 0$.

2. *Analytic Expressions for the Steady-State Response*

The stationary regime corresponds to the physical situation where all the transients are removed. This occurs theoretically in the long-time limit $(t \to \infty)$ and allows us to get combinations of sinusoidal functions only since all the exponential terms of the form $\exp[-i(i+1)Dt]$ vanish.

On setting

$$f_2(t) = \lim_{t\to\infty} e^2[f_{\gamma 2}(u, t) + f_{\beta 2}(u, t)] \tag{4.97}$$

after some algebra (see Appendix D) we obtain

$$\begin{aligned}
f_2(t) = {} & \frac{1}{2}\mathbf{g}e^2E_0^2\left\{\int_0^\infty \mathbf{A}_\gamma(t)dt \int_0^\infty \mathbf{f}_\gamma(t)\cos\omega t\, dt\right.\\
& + \left[\int\!\!\int \mathbf{A}_\gamma(t_1)\mathbf{f}_\gamma(t_2)\cos\omega(2t_1 + t_2)dt_1dt_2\right]\cos 2\omega t\\
& \left. + \left[\int\!\!\int \mathbf{A}_\gamma(t_1)\mathbf{f}_\gamma(t_2)\sin\omega(2t_1 + t_2)dt_1dt_2\right]\sin 2\omega t\right\}\\
& + \frac{1}{2}\mathbf{g}e^2E_0^2\int_0^\infty \mathbf{f}_\beta(t)dt + \frac{1}{2}\mathbf{g}e^2E_0^2\left\{\left[\int_0^\infty \mathbf{f}_\beta(t_1)\cos 2\omega t_1 dt_1\right]\cos 2\omega t\right.\\
& \left. + \left[\int_0^\infty \mathbf{f}_\beta(t_1)\sin 2\omega t_1 dt_1\right]\sin 2\omega t\right\}\\
& + \mathbf{g}e^2E_c^2\left[\int_0^\infty \mathbf{A}_\gamma(t)dt \int_0^\infty \mathbf{f}_\gamma(t)dt + \int_0^\infty \mathbf{f}_\beta(t)dt\right]\\
& + \mathbf{g}e^2E_cE_0\int_0^\infty \mathbf{f}_\gamma(t)dt\left\{\left[\int_0^\infty \mathbf{A}_\gamma(t_1)\cos\omega t_1 dt_1 dt_2\right]\cos\omega t\right.\\
& \left. + \left[\int_0^\infty \mathbf{A}_\gamma(t_1)\sin\omega t_1 dt_1 dt_2\right]\sin 2\omega t\right\}\\
& + \mathbf{g}e^2E_cE_0\left\{\left[\int\!\!\int \mathbf{A}_\gamma(t_1)\mathbf{f}_\gamma(t_2)\cos\omega(t_1 + t_2)dt_1dt_2\right]\cos\omega t\right.
\end{aligned}$$

$$+\left[\iint \mathbf{A}_\gamma(t_1)\mathbf{f}_\gamma(t_2)\sin\omega(t_1+t_2)dt_1\,dt_2\right]\sin\omega t\Bigg\}$$

$$+2\,\mathbf{g}e^2E_cE_0\Bigg\{\left[\int_0^\infty \mathbf{f}_\beta(t_1)\cos\omega t_1 dt_1\right]\cos\omega t$$

$$+\left[\int_0^\infty \mathbf{f}_\beta(t_1)\sin\omega t_1 dt_1\right]\sin\omega t\Bigg\} \tag{4.98}$$

Some general comments may be drawn from this expression. First, we remark that it consists of three different components, namely a time-independent part and two distinct time-dependent parts varying at once and twice the circular frequency of the alternating field. On the other hand, we note that the leading time-independent term in the right-hand side of Eq. (4.98) is frequency dependent. This is the fact of permanent dipole moments only. Indeed, no frequency dependence is observed in the static terms relating to induced dipolar moments. Also, all these time-independent terms are proportional to either E_0^2 or E_c^2 and not to the product E_cE_0. Concerning the harmonic components, it is shown how the terms in 2ω arise from the ac field only while the terms in ω result from the conjugate action of both ac and dc biasing fields. Moreover, the second harmonics involve two matrices, such as $\mathbf{A}_\gamma\mathbf{f}_\gamma$ and $\mathbf{f}_\beta$, whereas the fundamental terms bring into play three matrices $\mathbf{A}_\gamma\mathbf{f}_\gamma$, $\mathbf{A}_\gamma$, and $\mathbf{f}_\beta$. On using the definition of the Fourier–Laplace transforms with respect to time, the nonlinear Kerr effect response may also be reexpressed as follows:

$$f_2(t)=\Delta n_{\text{st}}+\sum_{j=1}^{2}\Delta n_j(\omega)\cos(j\omega t)+\Delta k_j(\omega)\sin(j\omega t) \tag{4.99}$$

In this equation, Δn_{st} is the steady component, $\Delta n_j(\omega)$ and $\Delta k_j(\omega)$ are the harmonic components of the electric birefringence such that

$$\Delta n_{\text{st}}=\frac{1}{2}\,\mathbf{g}e^2E_0^2[A_\gamma(0)\mathcal{R}eF_\gamma(i\omega)+F_\beta(0)]$$

$$+\,\mathbf{g}E_c^2[A_\gamma(0)F_\gamma(0)+F_\beta(0)] \tag{4.99a}$$

$$\Delta n_1(\omega)-i\,\Delta k_1(\omega)=\mathbf{g}e^2E_cE_0[F_\gamma(0)A_\gamma(i\omega)+A_\gamma(i\omega)F_\gamma(i\omega)+2F_\beta(i\omega)] \tag{4.99b}$$

$$\Delta n_2(\omega)-i\,\Delta k_2(\omega)=\frac{1}{2}\,\mathbf{g}e^2E_0^2[A_\gamma(2i\omega)F_\gamma(i\omega)+F_\beta(2i\omega)] \tag{4.99c}$$

where $A_{\gamma}(s)$, $F_{\gamma}(s)$, and $F_{\beta}(s)$ are the Laplace transforms of $\mathbf{A}_{\gamma}(t)$, $\mathbf{F}_{\gamma}(t)$, and $\mathbf{f}_{\beta}(t)$, respectively, $\mathcal{Re}$ denotes "the real part of ", and s is the symbolic variable.

To get analytic expressions of Kerr effect relaxation easily tractable for experiment, we shall now consider the situation where E_0/E_c is much less than 1, which corresponds to the case of a weak ac field superimposed on a strong dc field. To do that, we may start from Eq. (4.69) which was established in Section II, where we set $n=2$. For the stationary regime, we find [68] that the ensemble average of the second Legendre polynomial up to the second approximation in the electric field is given by

$$\langle P_2(u)\rangle(t) = \langle P_2(u)\rangle_{\gamma}(t) + \langle P_2(u)\rangle_{\beta}(t) \tag{4.100}$$

where the subscripts γ and β stand for the contributions due to the permanent and induced dipole moments, respectively, with

$$\langle P_2(u)\rangle_{\gamma}(t) = \frac{4}{5} D^2 \int_0^{\infty} \int_0^{\infty} e^{-6D(t-t')} \gamma(t') e^{-2D(t'-t_1)} \gamma(t_1) dt' dt_1 \tag{4.101a}$$

$$\langle P_2(u)\rangle_{\beta}(t) = \frac{2}{5} D \int_0^{\infty} e^{-6D(t-t')} \beta(t') dt' \tag{4.101b}$$

These expressions may be rearranged in the form of convolution products (denoted by $\otimes$) to read

$$\begin{aligned} \langle P_2(u)\rangle_{\gamma}(t) &= \frac{4}{5} D^2 \left(\frac{\mu}{kT}\right)^2 \{e^{-6Dt} \otimes (E_0 \cos \omega t + E_c) \\ &\quad \times [e^{-2Dt} \otimes (E_0 \cos \omega t + E_c)]\}|_{t=\infty} \\ \langle P_2(u)\rangle_{\beta}(t) &= \frac{2}{5} D \left(\frac{\mu}{kT}\right)^2 P e^{-6Dt} \otimes \left[\left(\frac{1}{2} E_0^2 + E_c^2\right) \right. \\ &\quad \left. + \frac{1}{2} E_0^2 \cos 2\omega t + 2E_0 E_c \cos \omega t \right]|_{t=\infty} \end{aligned} \tag{4.102}$$

We remark that the birefringence signal resulting from the presence of the permanent dipole moments leads us to carry out a double convolution product where two different exponentials appear. Moreover, it is shown by a simple comparison that the functions $f_2(t)$ and $\langle P_2(u)\rangle(t)$ have the same physical meaning. On using the Laplace transform and the convolution theorem on these equations, and after inverting them into the time domain, we arrive at the final result

$$\langle P_2(u)\rangle(t) = \langle P_2(u)\rangle_{\text{st}} + \sum_{j=1}^{2} \Delta n_j(\omega)\cos(j\omega t) + \Delta k_j(\omega)\sin(j\omega t) \qquad (4.103)$$

where

$$\langle P_2(u)\rangle_{\text{st}} = \frac{1}{15}\langle E^2(t)\rangle\left(\frac{\mu}{kT}\right)^2\left\{P + \frac{1}{1+2(E_c/E_0)^2} \times\left[\frac{1}{1+(\omega/2D)^2} + 2\left(\frac{E_c}{E_0}\right)^2\right]\right\} \qquad (4.104)$$

and

$$\langle E^2(t)\rangle = \frac{E_0^2}{2} + E_c^2 \qquad (4.105)$$

$$\Delta n_1(\omega) = \frac{2}{15}E_0E_c\left(\frac{\mu}{kT}\right)^2\left\{\frac{P+\frac{1}{2}}{1+(\omega/6D)^2} + \frac{\frac{1}{2}[1-(\omega/6D)(\omega/2D)]}{[1+(\omega/6D)^2][1+(\omega/2D)^2]}\right\} \qquad (4.106a)$$

$$\Delta k_1(\omega) = \frac{2}{15}E_0E_c\left(\frac{\mu}{kT}\right)^2\left\{\frac{P+\frac{1}{2}}{1+(\omega/6D)^2}\left(\frac{\omega}{6D}\right)\right\} + \frac{\frac{1}{2}[(\omega/6D)+(\omega/2D)]}{[1+(\omega/6D)^2][1+(\omega/2D)^2]} \qquad (4.106b)$$

$$\Delta n_2(\omega) = \frac{1}{30}E_0^2\left(\frac{\mu}{kT}\right)^2\left\{\frac{P}{1+(\omega/3D)^2} + \frac{1-(\omega/2D)(\omega/3D)}{[1+(\omega/2D)^2][1+(\omega/3D)^2]}\right\} \qquad (4.107a)$$

$$\Delta k_2(\omega) = \frac{1}{30}E_0^2\left(\frac{\mu}{kT}\right)^2\left\{\frac{P}{1+(\omega/3D)^2}\left(\frac{\omega}{3D}\right) + \frac{(\omega/2D)+(\omega/3D)}{[1+(\omega/2D)^2][1+(\omega/3D)^2]}\right\} \qquad (4.107b)$$

The nonlinear response is clearly established by means of $\Delta n_1(\omega)$ and $\Delta k_1(\omega)$ with a coupling effect due to the interaction of the fields E_0 and E_c. This nonlinearity is particularly true of the limiting low-frequency value of $\langle P_2(u)\rangle_{st}$, which is different from $\Sigma_{j=1}^{2} \Delta n_j(0)$, whereas, in a pure ac field, $\lim_{\omega\to 0} \langle P_2(u)\rangle_{st}$, is strictly identical to $\Delta n_2(0)$. The angular braces in Eq. (4.105) represent a mean value over time and not an ensemble average. Equations (4.106) and (4.107) are in full agreement with previous results obtained by Thurston and Bowling [57] only for an alternating field ($j = 2$) and by Morita and Watanabe [69] for an alternating field combined with a unidirectional field ($j = 1, 2$).

3. *Cartesian and Polar Equations of Cole–Cole Diagrams*

It is well known that Cole–Cole plots are particularly used by experimentalists in the Debye theory of dielectric relaxation. They are perfect semicircles if linear response theory applies. When a dielectric medium is subjected to a periodic field of circular frequency ω in a fixed direction, it polarizes itself. This may be characterized by the electric induction **D** (or displacement) such that

$$\mathbf{D} = \varepsilon_0\mathbf{E} + \mathbf{P} = \boldsymbol{\varepsilon}(\omega)\mathbf{E} = [\varepsilon'(\omega) - i\varepsilon''(\omega)]\mathbf{E} \tag{4.108}$$

where $\boldsymbol{\varepsilon}$ is the complex electric permittivity and **P** is the electric polarization. We make use of a complex notation in order to take account of dispersion in the dielectric.

Hence, we have

$$\mathbf{P} = [\boldsymbol{\varepsilon}(\omega) - \varepsilon_0]\mathbf{E} = \varepsilon_0\boldsymbol{\chi}(\omega)\mathbf{E} = \varepsilon_0[\chi'(\omega) - i\chi''(\omega)]\mathbf{E} \tag{4.109}$$

where $\boldsymbol{\chi}$ is called the complex electric susceptibility. This means that the electric polarization is not, in general, in phase with the electric field and may always by split into two components, one in-phase and the other out-of-phase.

From Eqs. (4.108) and (4.109), we deduce that

$$\begin{aligned} \varepsilon'(\omega) - \varepsilon_0 &= \varepsilon_0\chi'(\omega) \\ \varepsilon''(\omega) &= \varepsilon_0\chi''(\omega) \end{aligned} \tag{4.110}$$

Let us recall that according to Maxwell's equations, the refractive index is related to the electric permittivity by

$$\mathbf{n}^2(\omega) = \boldsymbol{\varepsilon}(\omega)/\varepsilon_0 \tag{4.111}$$

Graphs of the imaginary part $\varepsilon''(\omega)$ against the real part $\varepsilon'(\omega)$ are named Cole–Cole plots. It is also possible to draw dispersion plots representing the evolution of ε' and ε'' against ω (see Section II.D).

From all these considerations, it appears interesting to employ similar curves with the harmonic components of the electric birefringence. More precisely, we shall study Cole–Cole-like diagrams deduced from the real X_j and imaginary Y_j parts of both normalized complex birefringence functions $\Delta\mathbf{N}_j = X_j - iY_j$. Furthermore, we shall show that, in the complex plane, the tip of the vector (X_j, Y_j) describes a set of quasiconchoids of circles whose shapes are strongly dependent on P values.

Therefore, Kerr functions of interest are defined as

$$X_j(\omega) = \frac{\Delta n_j(\omega)}{\Delta n_j(0)} \qquad Y_j(\omega) = \frac{\Delta k_j(\omega)}{\Delta n_j(0)} \tag{4.112}$$

where $\Delta n_j(0)$ is the value of $\Delta n_j(\omega)$ when $\omega \to 0$, namely,

$$\begin{aligned} \Delta n_1(0) &= \frac{2}{15} E_0 E_c \left(\frac{\mu}{kT}\right)^2 (P+1) \\ \Delta n_2(0) &= \frac{1}{30} E_0^2 \left(\frac{\mu}{kT}\right)^2 (P+1) \end{aligned} \tag{4.113}$$

Now, by eliminating the angular frequency ω in Eqs. (4.106) and (4.107) (parametric equations), we shall derive Cartesian and polar equations of the corresponding curves for both cases $j = 1$ and $j = 2$. For this purpose, it will be useful to set as a reduced frequency variable $r_j = j(\omega/6D)$. We first consider the case of a pure alternating field corresponding only to $j = 2$, which is the most common case.

1. Case $j = 2$

By a simple combination of parametric equations given by the components X_2 and Y_2, we have the following relationship:

$$[(3r_2^2/2) - 1]\, X_2 - \tfrac{5}{2}\, r_2 Y_2 + 1 = 0 \tag{4.114}$$

For different values of r_2, Eq. (4.114) is representative of a set of straight lines, whose envelope is an ellipse (nearly a circle) defined by

$$X_2^2 + \frac{25}{24} Y_2^2 - X_2 = 0 \tag{4.115}$$

The semiaxes along OX and OY are 0.5 and $(6/25)^{0.5}$, respectively. The latter value plays an important role in determining P and will be discussed in some detail below. The roots of Eq. (4.114), considered as a second-degree equation of r_2, are for a particular pair (X_2, Y_2)

$$r_{2+} = (3\,X_2)^{-1}(\tfrac{5}{2}\,Y_2 + \sqrt{\Delta_2})$$
$$r_{2-} = (3\,X_2)^{-1}(\tfrac{5}{2}\,Y_2 - \sqrt{\Delta_2}) \tag{4.116}$$

where

$$\Delta_2 = \frac{25}{4}\,Y_2^2 + 6\,X_2(X_2 - 1)$$

We shall see their importance, as well. Finally, after some tedious algebra, we find the analytical expression for $Y_2(X_2)$ curves.

$$X_2^2 + Y_2^2 - X_2(1 + C_2) + C_2 = Y_2^2/2\,X_2 + (Y_2/5\,X_2)\sqrt{\Delta_2} \tag{4.117}$$

where

$$C_2 = 3\,P/5(P + 1)$$

Because the solutions of Eq. (4.117) have a very complicated form, we have made use of polar coordinates (ρ_2', θ_2') after translating the origin from O to O_2', so that $\overline{OO_2'} = C_2$. The overbar above OO_2' indicates an algebraic length. If M is a point of the $Y_2(X_2)$ curve, by putting $\overline{OM} = \rho_2$ and $\overline{O_2'M} = \rho_2'$, we then have

$$X_2 = \rho_2 \cos\theta_2 = \rho_2' \cos\theta_2' + C_2$$
$$Y_2 = \rho_2 \sin\theta_2 = \rho_2' \sin\theta_2' \tag{4.118}$$

This is displayed in Fig. 4.31. Hence we obtain the following polar equation

$$[\rho_2' + (C_2 - 1)\cos\theta_2']^2 + [\rho_2' + (C_2 - 1)\cos\theta_2']\cos\theta_2' - \frac{6}{25}\sin^2\theta_2' = 0 \tag{4.119}$$

with physically significant solutions

$$\rho_{2+}' = (\tfrac{1}{2} - C_2)(\cos\theta_2') + K_2(\theta_1')$$
$$\rho_{2-}' = (\tfrac{1}{2} - C_2)(\cos\theta_2') - K_2(\theta_2') \tag{4.120}$$

where

$$K_2(\theta_2') = \tfrac{1}{2}(1 - \tfrac{1}{25}\sin^2\theta_2')^{0.5}$$

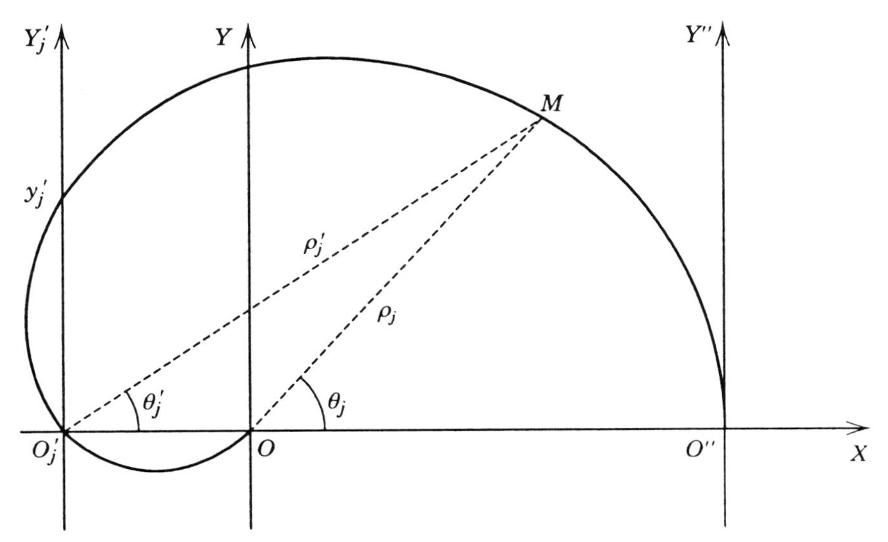

Figure 4.31. Reference frame for $Y_j(X_j)$ plots ($OO_j' = C_j$, $OO'' = 1$); $y_1' = (3/16)^{0.5}$, $y_2' = (6/25)^{0.5}$.

The classical equation of a conchoid being of the form $\rho_2' = 2\,R_2(\cos\theta_2') \pm K_2'$, where K_2' is a constant, it can be seen that Eqs. (4.120) represent quasiconchoids of circles with the origin at O_2'. Indeed, by identification, it is clear that $2\,R_2 = \frac{1}{2} - C_2$ is a constant and K_2' varies very slowly with respect to θ_2'. Moreover, ρ_{2+}' and ρ_{2-}' are identical solutions but with opposite directions of rotation for θ_2'. Note that the relative position of O_2' is strongly dependent on P values, as indicated in Fig. 4.32. For any value of P, C_2 is positive, except for $-1 < P < 0$, where C_2 becomes negative.

Typical results are illustrated in Figs. 4.33 and 4.34, for $P < -1$ and $P > -1$, respectively, owing to the discontinuity for $P = -1$ contained in the normalization factor [see Eq. (4.113)], which is proportional to $P + 1$. Numerous determinations of intersection conditions of $Y_2(X_2)$ curves with preferential axes OX, OY, $O_2'Y'$, and OO'' are presented in Table III. The coordinates of intersection points and the corresponding value of r_2 are obtained using parametric, Cartesian, or polar equations [cf. Eqs. (4.112), (4.117), and (4.119)].

Now, we shall consider for any value of P the variations of the phase angle θ_j (argument of the complex number $\Delta\mathbf{N}_j$) with the angular frequency ω. The determination of extrema of that function $\theta_j(\omega)$ together with its limiting values ($\omega \to 0$ and $\omega \to \infty$) are very important in

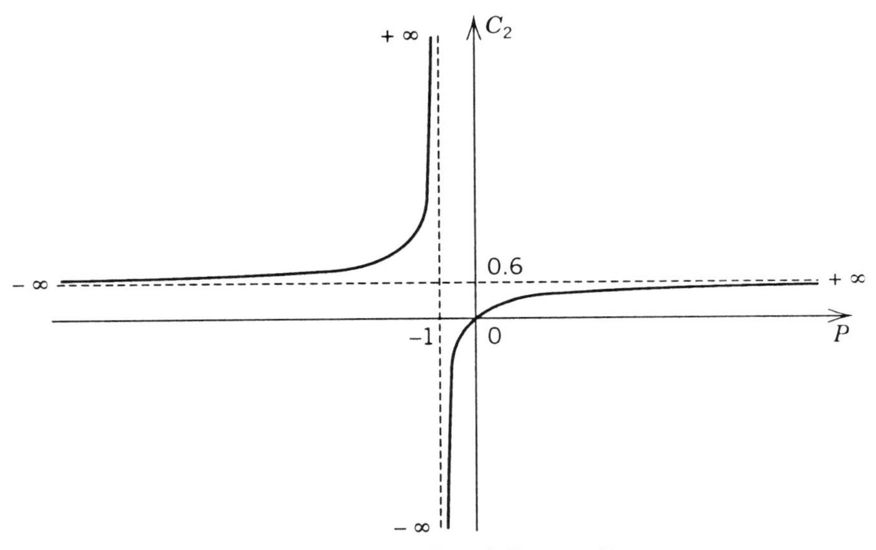

Figure 4.32. Plot of C_2 versus P.

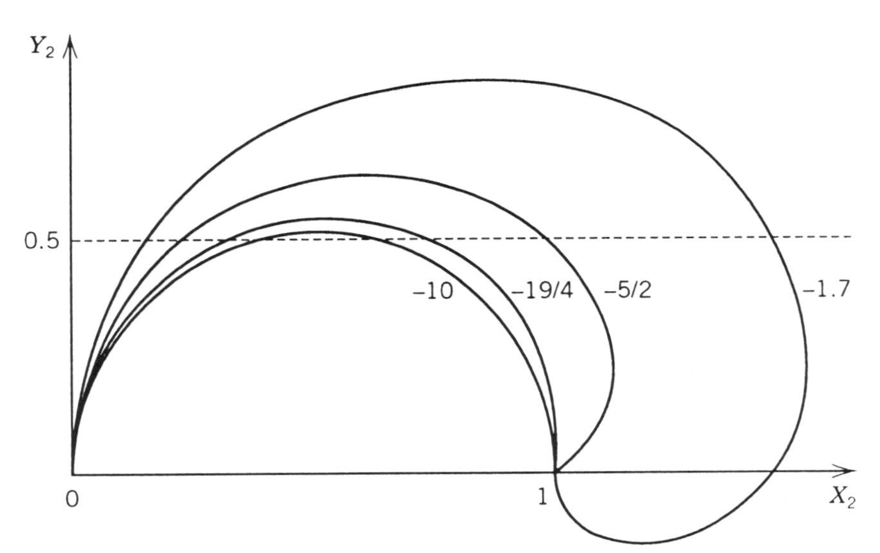

Figure 4.33. Plots of the imaginary part Y_2 versus the real part X_2 of the normalized birefringence function, for $P < -1$. Numbers above or below plots represent various values of P.

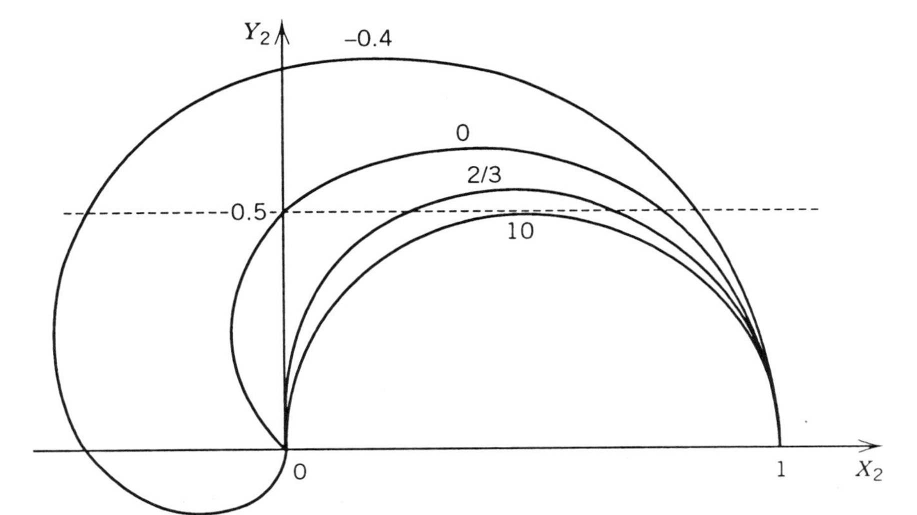

Figure 4.34. Plots of the imaginary part Y_2 versus the real part X_2 of the normalized birefringence function, for $P > -1$. Numbers above or below plots represent various values of P.

TABLE III
Determination of Coordinates of Intersection Points with Axes

Axis	Intersection Point (X_2, Y_2) Coordinates		Domain of Validity for P	r_2
OX	$X_2 = 1$	$Y_2 = 0$	$-\infty < P < +\infty$	0
	$X_2 = 0$	$Y_2 = 0$	$-\infty < P < +\infty$	$+\infty$
	$X_2 = C_2 = \dfrac{3P}{5(P+1)}$	$Y_2 = 0$	$-\dfrac{5}{2} < P < 0$	$\left(-\dfrac{4P+10}{9P}\right)^{1/2}$
OY	$Y_2 = \left[\dfrac{3(2-3P)}{25(P+1)}\right]^{1/2}$	$X_2 = 0$	$-1 < P < \dfrac{2}{3}$	$\left[\dfrac{4(P+1)}{3(2-3P)}\right]^{1/2}$
$O_2'Y'$	$Y_2 = \sqrt{6}/5$	$X_2 = C_2$	$-\infty < P < +\infty$	$\sqrt{2/3}$
$O''Y''$	$Y_2 = \dfrac{3}{5}\left[-\dfrac{4P+19}{9(P+1)}\right]^{1/2}$	$X_2 = 1$	$-\dfrac{19}{4} < P < -1$	$\left[-\dfrac{4P+19}{9(P+1)}\right]^{1/2}$

removing any ambiguity in using Kramers–Kronig relations from experimental results [9, 70].

The sum in the second term of Eq. (4.103) may be readily written in the form

$$\sum_{j=1}^{2} \{[\Delta n_j^2(\omega) + \Delta k_j^2(\omega)]^{0.5} \cos(j\omega t - \theta_j)\} \tag{4.121}$$

where θ_j is the phase angle between in-phase and out-of-phase j components (harmonic terms) of the electric birefringence, defined as

$$\theta_j(\omega) = \tan^{-1} \frac{\Delta k_j(\omega)}{\Delta n_j(\omega)} = \tan^{-1} \frac{Y_j(\omega)}{X_j(\omega)} \tag{4.122}$$

Here, we are concerned with the second harmonic component for which we have $r_2 = \omega/3D$. Hence, Eqs. (4.107) enable us to express θ_2 as a function of r_2 (or ω). From Eq. (4.122), we have

$$\theta_2 = \tan^{-1}(Y_2/X_2)$$

and therefore

$$\theta_2(r_2) = \tan^{-1} \left[r_2 \frac{(P-2)(1+\frac{9}{4}r_2^2) + \frac{9}{2}(1+r_2^2)}{(P-2)(1+\frac{9}{4}r_2^2) + 3(1+r_2^2)} \right] \tag{4.123}$$

To get the extrema of $\theta_2(r_2)$, if they exist, it suffices to differentiate θ_2 with respect to r_2. A fourth-degree equation is established, of the following form:

$$L_2(P)r_2^4 + M_2(P)r_2^2 + N_2(P) = 0 \tag{4.124}$$

where L_2, M_2, and N_2 are P-dependent coefficients such that

$$\begin{aligned} L_2(P) &= 27[\tfrac{3}{16}(P-2)^2 + \tfrac{5}{8}(P-2) + \tfrac{1}{2}] \\ M_2(P) &= 3[\tfrac{3}{2}(P-2)^2 + \tfrac{55}{8}(P-2) + 9] \\ N_2(P) &= (P-2)^2 + \tfrac{15}{2}(P-2) + \tfrac{27}{2} \end{aligned} \tag{4.125}$$

A mathematical analysis of signs of L_2, M_2, and N_2 for various values of P allows us to know the sign of roots $(r_2')^2$ and $(r_2'')^2$ in Eq. (4.114), which must be positive. In that way, we have found that only two extrema are physically possible: the first one is a minimum for $-\frac{5}{2} < P < -1$, the

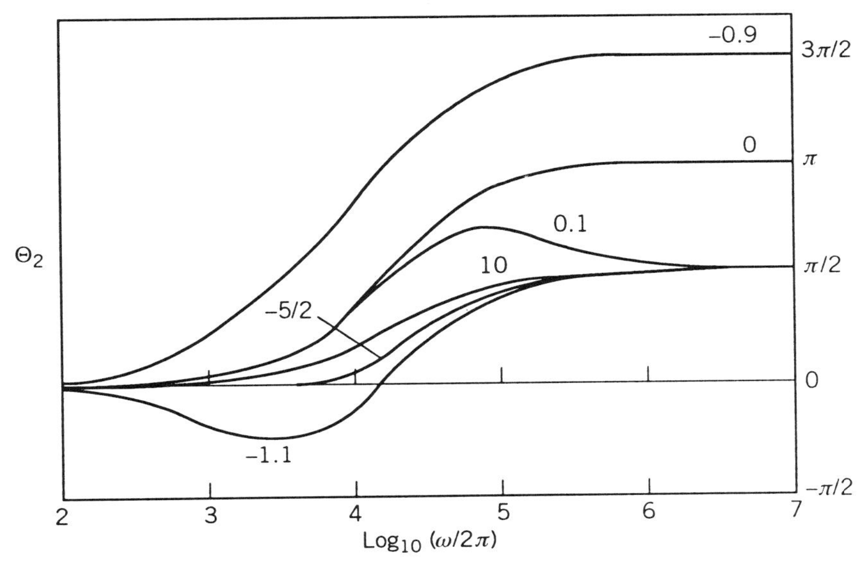

Figure 4.35. Plots of the phase angle θ_2 versus $\log_{10}(\omega/2\pi)$. Numbers above curves are for different values of P(D has been chosen arbitrarily equal to $4 \times 10^4\ \mathrm{s}^{-1}$).

second a maximum for $0 < P < \frac{2}{3}$, as shown in Fig. 4.35. In Table IV we show changes of θ_2 and its limiting values as a function of r_2 and P.

The detailed characteristics shown in Table III allow identification and interpretation without ambiguity of every diagram $Y_2(X_2)$, that is of every value of P between $-\frac{19}{4}$ and $\frac{2}{3}$. For $P > \frac{2}{3}$, the plots become semicircles, which are rigorously obtained for $P = 2$ and $+\infty$. In that latter domain, the maximum of Y_2 is always slightly less than or equal to 0.5. For $P < -\frac{19}{4}$, we have a similar situation as for $P > \frac{2}{3}$, but the maximum of Y_2 slightly exceeds the limiting value of 0.5. Determination of P remains easy for $-6 < P < 2$, where semicircles are skewed in a significant way towards OY or $O''Y''$ axes, and accuracy is close to that obtained with experimental measurements of birefringence components X_2 and Y_2. Outside that domain, a 10% uncertainty on X_2 and Y_2 may produce an error up to 50% because of the quasisemicircular shape of the diagrams. To refine this result, an improvement could be made regarding theoretical plots of the modulus of the birefringence function $\Delta \mathbf{N}_2$ or the phase θ_2 versus frequency, and comparing them with experiment. Furthermore, if the measurements are not accurate enough, the sign of P may still be in doubt, but this last point can be easily clarified by determining the sign of the nonnormalized birefringence $\Delta n_2(0)$ in the low-frequency range

TABLE IV
Variation of the Phase Angle θ_2 Versus r_2

$r_2 = 0$	θ_2 Values $r_2 = r_{2E}$ (Extremum)	$r_2 = \infty$	Intervals of Validity for P
0	↘ $-\frac{\pi}{2} < \theta_{2m} < 0$ ↗	$\frac{\pi}{2}$	$-\frac{5}{2} < P < -1$
0	↗	$\frac{\pi}{2}$	$P \geq \frac{2}{3}, P \leq -\frac{5}{2}$
0	↗ $\frac{\pi}{2} < \theta_{2m} < \pi$ ↘	$\frac{\pi}{2}$	$0 < P < \frac{2}{3}$
0	↗	π	$P = 0$
0	↗	$\frac{3\pi}{2}$	$-1 < P < 0$

$(\omega \to 0)$ and by comparing it to a sample liquid: if $\Delta n_2(0)$ has a positive value, $P > -1$; for negative values of $\Delta n_2(0)$, $P < -1$.

When P is determined, it becomes possible to obtain the relaxation time of the birefringence $\tau_2 = (6D)^{-1} = r_2/(2\omega)$ by considering the results presented in Table III. In any case, the coordinates $X_2 = C_2$, $Y_2 = (\frac{6}{25})^{0.5}$ form a point for which $r_2 = (\frac{2}{3})^{0.5}$ and, the frequency being known, determination of τ_2 is easy. In passing, we see that the position where $\theta_2' = \pi/2$ leads to a phase angle

$$\theta_2 = \tan^{-1}[\sqrt{6}(P+1)/3P] \tag{4.126}$$

Equation (4.126) indicates that for infinite values of P (negative or positive), a limiting value of θ_2 is attained, denoted by $\theta_2^{\lim}(\approx 0.22\pi)$, which may be convenient for the determination of the sign of P,

$$\begin{aligned} 0 < \theta_2 < \theta_2^{\lim} \quad &\text{for} \quad -\infty < P < -1 \\ \theta_2^{\lim} < \theta_2 < \pi/2 \quad &\text{for} \quad 0 < P < +\infty \\ \pi/2 < \theta_2 < \pi \quad &\text{for} \quad -1 < P < 0 \end{aligned} \tag{4.127}$$

The latter relations allow one (to a first estimate) to know in which domain P lies, and corroborate all the the previous results.

2. Case $j = 1$

As already indicated, this case occurs when a continuous field E_c is superposed on an alternating field $E_0 \cos(\omega t)$. Proceeding just as before, the main results are summarized below.

The basis equation [from Eqs. (4.106)] is as follows:

$$(3r_1^2 - 1)X_1 - 4r_1Y_1 + 1 = 0 \tag{4.128}$$

The Cartesian equation is

$$X_1^2 + Y_1^2 - X_1(1 + C_1) + C_1 = (Y_1^2/2X_1) + (Y_1/8X_1)\sqrt{\Delta_1} \tag{4.129}$$

where

$$C_1 = 3(P + \tfrac{1}{2})/[4(P + 1)] \tag{4.130}$$

$$\Delta_1 = 16\,Y_1^2 + 12\,X_1(X_1 - 1) \tag{4.131}$$

The polar equations ($\overline{OO_1'} = C_1$; see Fig. 4.31 for polar coordinates ρ_1', θ_1') are

$$\begin{aligned} \rho_{1+}' &= (\tfrac{1}{2} - C_1)(\cos\theta_1') + K_1(\theta_1') \\ \rho_{1-}' &= (\tfrac{1}{2} - C_1)(\cos\theta_1') - K_1(\theta_1') \end{aligned} \tag{4.132}$$

where

$$K_1(\theta_1') = \tfrac{1}{2}(1 - \tfrac{1}{4}\sin^2\theta_1')^{0.5}$$

The variation of θ_1 versus r_1 [from Eqs. (4.106)] is

$$\theta_1(r_1) = \tan^{-1}\left[r_1 \frac{(P + \frac{1}{4})(1 + 9r_1^2) + \frac{9}{4}(1 + r_1^2)}{(P + \frac{1}{4})(1 + 9r_1^2) + \frac{3}{4}(1 + r_1^2)} \right] \tag{4.133}$$

Calculation of the extrema of $\theta_1(r_1)$ leads to the following equation:

$$L_1(P)r_1^4 + M_1(P)r_1^2 + N_1(P) = 0 \tag{4.134}$$

where

$$\begin{aligned} L_1(P) &= 27[3(P + \tfrac{1}{4})^2 + (P + \tfrac{1}{4}) + \tfrac{1}{16}] \\ M_1(P) &= 3[6(P + \tfrac{1}{4})^2 + 2(P + \tfrac{1}{4}) + \tfrac{9}{8}] \\ N_1(P) &= (P + \tfrac{1}{4})^2 + 3(P + \tfrac{1}{4}) + \tfrac{27}{16} \end{aligned} \tag{4.135}$$

Equations (4.132) represent quasiconchoids of circles, with an origin at O'_1. Changes in θ'_j have a more sensitive effect on K_1 than on K_2; as a result, those conchoids are more flattened than for case $j = 2$.

The fundamental features of the first harmonic term are shown in Tables V and VI. Figure 4.36 illustrates how C_1 depends on P values: for any value of P, C_1 remains positive except for $-1 < P < -0.5$. Figures 4.37 and 4.38 show the plots of Y_1 against X_1 for $P < -1$ and $P > -1$, respectively.

TABLE V
Determination of Coordinates of Intersection Points with Axes

Axis	Intersection Point (X_1, Y_1) Coordinates		Domain of Validity for P	r_1
OX	$X_1 = 1$	$Y_1 = 0$	$-\infty < P < +\infty$	0
	$X_1 = 0$	$Y_1 = 0$	$-\infty < P < +\infty$	$+\infty$
	$X_1 = C_1 = \dfrac{3(P+\frac{1}{2})}{4(P+1)}$	$Y_1 = 0$	$-\frac{5}{2} < P < -\frac{1}{2}$	$\left[-\dfrac{P+\frac{5}{2}}{9(P+\frac{1}{2})}\right]^{1/2}$
OY	$Y_1 = \left[-\dfrac{3(3P+1)}{16(P+1)}\right]^{1/2}$	$X_1 = 0$	$-1 < P < -\frac{1}{3}$	$\left[-\dfrac{P+1}{3(3P+1)}\right]^{1/2}$
O'_1Y'	$Y_1 = \sqrt{3}/4$	$X_1 = C_1$	$-\infty < P < +\infty$	$\sqrt{1/3}$
$O''Y''$	$Y_1 = \frac{3}{4}\left[-\dfrac{P+7}{9(P+1)}\right]^{1/2}$	$X_1 = 1$	$-7 < P < -1$	$\left[-\dfrac{P+7}{9(P+1)}\right]^{1/2}$

TABLE VI
Variation of the Phase Angle θ_1 versus r_1

$r_1 = 0$	θ_1 values $r_1 = r_{1E}$(extremum)	$r_1 = \infty$	Intervals of Validity for P
0 ↘	$-\dfrac{\pi}{2} < \theta_{1\,m} < 0$ ↗	$\dfrac{\pi}{2}$	$-\frac{5}{2} < P < -1$
0	→	$\dfrac{\pi}{2}$	$P \geq -\frac{1}{3}, P \leq -\frac{5}{2}$
0 ↗	$\dfrac{\pi}{2} < \theta_{1\,m} < \pi$ ↘	$\dfrac{\pi}{2}$	$-\frac{1}{2} < P < -\frac{1}{3}$
0	→	π	$P = -\frac{1}{2}$
0	→	$\dfrac{3\pi}{2}$	$-1 < P < -\frac{1}{2}$

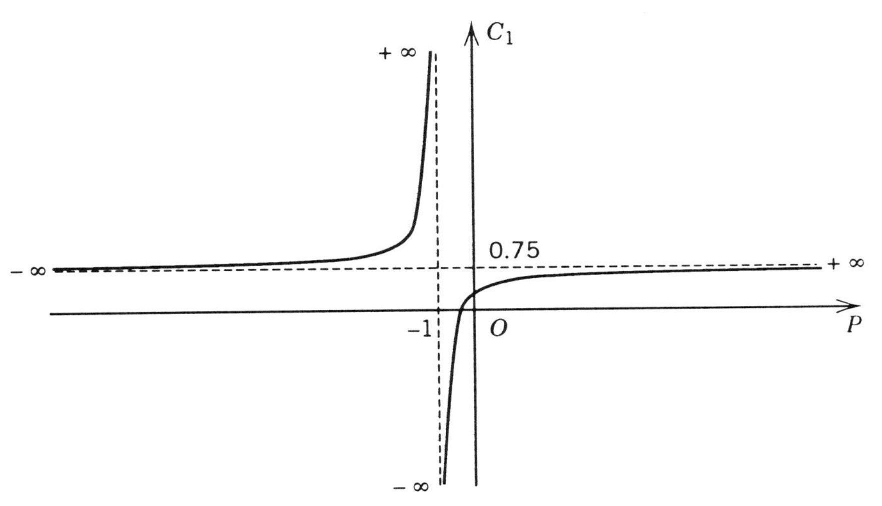

Figure 4.36. Plot of C_1 versus P.

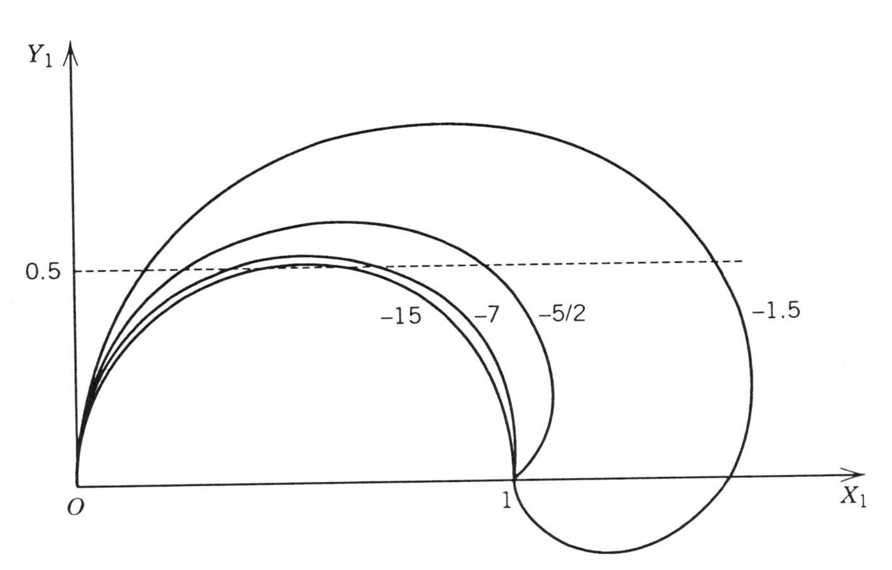

Figure 4.37. Plots of the imaginary part Y_1 versus the real part X_1 of the normalized birefringence function, for $P < -1$. Numbers above or below plots represent various values of P.

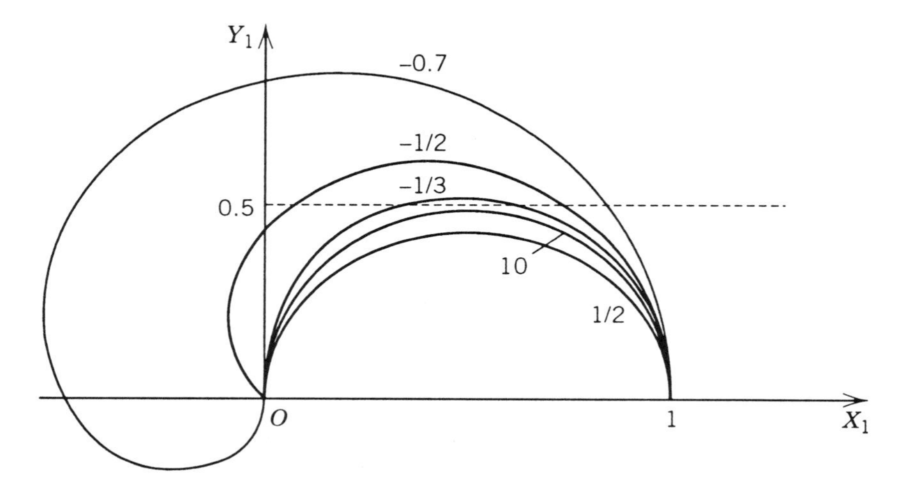

Figure 4.38. Plots of the imaginary part Y_1 versus the real part X_1 of the normalized birefringence function, for $P > -1$. Numbers above or below plots represent various values of P.

From Table V, it is seen that $Y_1(X_1)$ diagrams are well differentiated for $-7 < P < -\frac{1}{3}$. Comparing Figs. 4.34 and 4.38 for $-1 < P < 2$, the curves show responses that are different enough for both harmonic components, with deviations from the semicircles more pronounced for $j = 1$ than for $j = 2$. As P becomes larger than 1, the diagrams tend to semicircular shapes and the same trend is observed for $P < -7$. Three particular values of P lead to full semicircles, namely, $P = -\frac{1}{4}$ and $P = \pm\infty$. Outside those ranges, P values may present some uncertainty, and a similar approach, as that used before for $j = 2$, may be made in order to estimate these values. When the effective value of P is determined, the relaxation time $\tau_2 = r_1/\omega$ is easily found from the point $X_1 = C_1$, $Y_1 = (\frac{3}{16})^{0.5}$, where $r_1 = (\frac{1}{3})^{0.5}$. Plots of θ_1 as a function of frequency and P are drawn in Fig. 4.39. As shown previously, a supplementary indication for the sign of P is obtained from Eq. (4.133), that is,

$$\theta_1 = \tan^{-1}[\sqrt{3}(P + 1)/3(P + 0.5)] \tag{4.136}$$

For infinite values of P the limit of θ_1 is exactly $\pi/6$ and the following

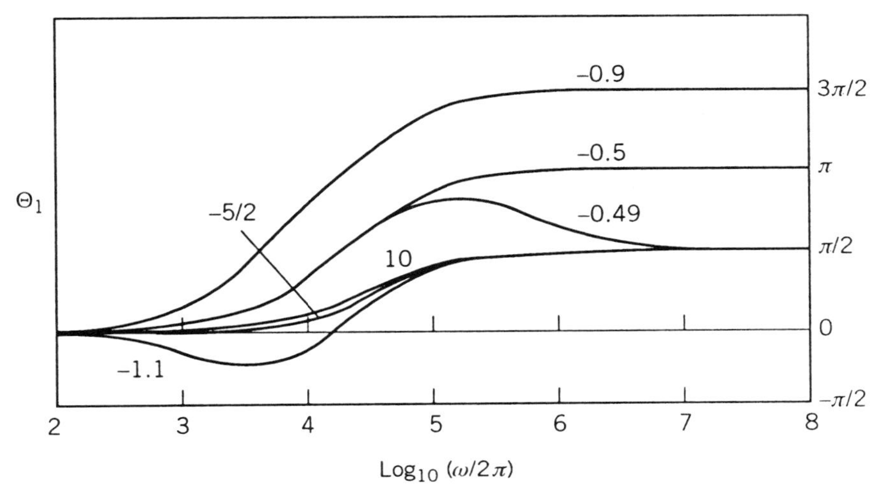

Figure 4.39. Plots of the phase angle θ_1 versus $\log_{10}(\omega/2\pi)$. Numbers above curves are for different values of $P(D$ has been chosen arbitrarily equal to $4 \times 10^4\,\mathrm{s}^{-1})$.

conditions can therefore be considered:

$$
\begin{aligned}
0 < \theta_2 < \theta_1^{\lim} \quad & \text{for} \quad -\infty < P < -1 \\
\theta_1^{\lim} < \theta_1 < \pi/2 \quad & \text{for} \quad 0.5 < P < +\infty \\
\pi/2 < \theta_1 < \pi \quad & \text{for} \quad -1 < P < 0.5
\end{aligned}
\tag{4.137}
$$

4. *Comments on the Evolution of Phase Angles with Frequency*

By studying both harmonic components of the electric birefringence, we have at our disposal a double set of information that allows identification and characterization of any monodisperse liquid system and is complementary to the Kramers–Kronig relations. By studying the variations of θ_j' from 0 to 2π, we deduced that all the quasiconchoids of circles present a symmetry axis (OX) whatever value the parameter P may have. As physically r_j is a positive quantity, its corresponding variations connected to θ_j' lie from 0 to $+\infty$. Moreover, it is seen from polar equations that when θ_j' is equal to $\pi/2$, Y_j has a constant value $[Y_2 = (\frac{6}{25})^{0.5}$, $Y_1 = (\frac{3}{16})^{0.5}]$ and $X_j = C_j$; this is a general result for any value of P. In other words, the first root ρ_{j+}' is associated with θ_j' lying between 0 and π, while the second root ρ_{j-}' describes the same plot but with a variation of θ_j' from π to 2π. Thus, we have kept only ρ_{j+}'. We also mention that one of the main advantages in normalizing the birefringence functions is to refer the origin of all phases θ_j to zero, for any P. All the $Y_j(X_j)$ curves start effectively from zero frequency where the coordinates in the (OXY) frame are $(1, 0)$, that is, for $\theta_j' = 0$ or $\theta_j = 0$. This fact is very interesting for the specification of the origin of experimental phase-angle measurements, which are sometimes known within the range $0 - \pi$, such as the phase angles carried out by the Kramers–Kronig relations. In the latter case, Tables IV and VI provide fruitful information for the calculation of correction terms based on the Kramers–Kronig method, and a better accuracy for the phase angle.

5. *Determination of Molecular Parameters from Nonlinear Transients*

This approach goes well beyond a purely mathematical exercise. We shall indeed show how transients of electric birefringence may provide some interesting information for the determination of the molecular parameters P and D. Moreover, an original method for measuring phase angles is proposed [71]. As before, we are always concerned with the simultaneous action of two electric fields (ac field plus dc bias field) and we restrict ourselves up to second order in the field strength.

On using Eqs. (4.101a) and (4.101b), the time-dependent behavior of the electric birefringence is given by

$$\langle P_2(u)\rangle(t) = \frac{6}{5} D \int_0^{\infty} e^{-6D(t-t')}\gamma(t')\langle P_1(u)\rangle(t')dt' + \frac{2}{5} D \int_0^{\infty} e^{-6D(t-t')}\beta(t')dt' \tag{4.138}$$

where

$$\langle P_1(u)\rangle(t) = \frac{2}{3} D \int_0^{\infty} e^{-2D(t-t')}\gamma(t')dt' \tag{4.139}$$

On writing $\gamma(t) = \gamma_c + \gamma_0 \cos \omega t$, with $\gamma_c = (\mu/kT)E_c$ and $\gamma_0 = (\mu/kT)E_0$, the Laplace transform of Eq. (4.139) leads to

$$\langle P_1(u)\rangle(t) = \frac{1}{3}\gamma_c(1 - e^{-2Dt}) + \frac{(\gamma_0/3)}{1 + (\omega/2D)^2}\left(\cos \omega t + \frac{\omega}{2D}\sin \omega t - e^{-2Dt}\right) \tag{4.140}$$

Equation (4.140) holds for electric polarization but, as seen from Eq. (4.138), it is also useful in obtaining $\langle P_2(u)\rangle(t)$. To evaluate the effect of the dc field on the ac field, we introduce the parameter $x_E = E_0/E_c$, and using reduced variables as done by Watanabe and Morita, we get the following results:

$$
\begin{aligned}
i_\gamma(\tau) = {} & (1 - e^{-6\tau})\left(\frac{4x_E^2}{4+\omega'^2} + 2\right) + 3(e^{-6\tau} - e^{-2\tau})\left(\frac{4x_E}{4+\omega'^2} + 1\right) \\
& + \frac{72x_E}{36+\omega'^2}\left[\left(1 + \frac{4}{4+\omega'^2}\right)\left(\cos\omega'\tau + \frac{\omega'}{6}\sin\omega'\tau - e^{-6\tau}\right)\right. \\
& \left. + \frac{1}{3}\frac{\omega'^2}{4+\omega'^2}\left(-\cos\omega'\tau + \frac{6}{\omega'}\sin\omega'\tau + e^{-6\tau}\right)\right] \\
& + \frac{48}{16+\omega'^2}\left(x_E + \frac{4x_E^2}{4+\omega'^2}\right)\left[e^{-6\tau} - e^{-2\tau}\left(\cos\omega'\tau + \frac{\omega'}{4}\sin\omega'\tau\right)\right] \\
& + \frac{36x_E^2}{(4+\omega'^2)(9+\omega'^2)}\left[\left(\cos 2\omega'\tau + \frac{\omega'}{3}\sin 2\omega'\tau - e^{-6\tau}\right)\right. \\
& \left. + \frac{\omega'^2}{6}\left(-\cos 2\omega'\tau + \frac{3}{\omega'}\sin 2\omega'\tau + e^{-6\tau}\right)\right]
\end{aligned}
\tag{4.141}
$$

$$
\begin{aligned}
i_\beta(\tau) = {} & (2 + x_E^2)(1 - e^{-6\tau}) + \frac{9x_E^2}{9+\omega'^2}\left(\cos 2\,\omega'\tau + \frac{\omega'}{3}\sin 2\omega'\tau - e^{-6\tau}\right) \\
& + \frac{144x_E}{36+\omega'^2}\left(\cos\omega'\tau + \frac{\omega'}{6}\sin\omega'\tau - e^{-6\tau}\right)
\end{aligned}
\tag{4.142}
$$

where $i_\gamma(t)$ and $i_\beta(t)$ are the Kerr responses due to the permanent dipole moments and the induced dipole moments, respectively. So, we may write the complete solution for the transients of the electric birefringence in the compact form

$$
\langle P_2(u)\rangle(\tau) = \frac{1}{30}\gamma_c^2[i_\gamma(\tau) + Pi_\beta(\tau)] \tag{4.143}
$$

with $\tau = Dt$, $\omega' = \omega/D$, and $P = (\Delta\alpha/\mu^2)kT$.

For high values of x_E, it is easy to see that Eq. (4.143) is in full agreement with previous work corresponding to the dynamic Kerr effect resulting from a pure sinusoidal field [52]. We shall now proceed to a wide discussion of the evolution of the ensemble averages $\langle P_2(u)\rangle(t)$ for various values of P, x_E, and ω' by means of some figures chosen to illustrate typical results. Figures 4.40a–h represent, in fact, the evolution

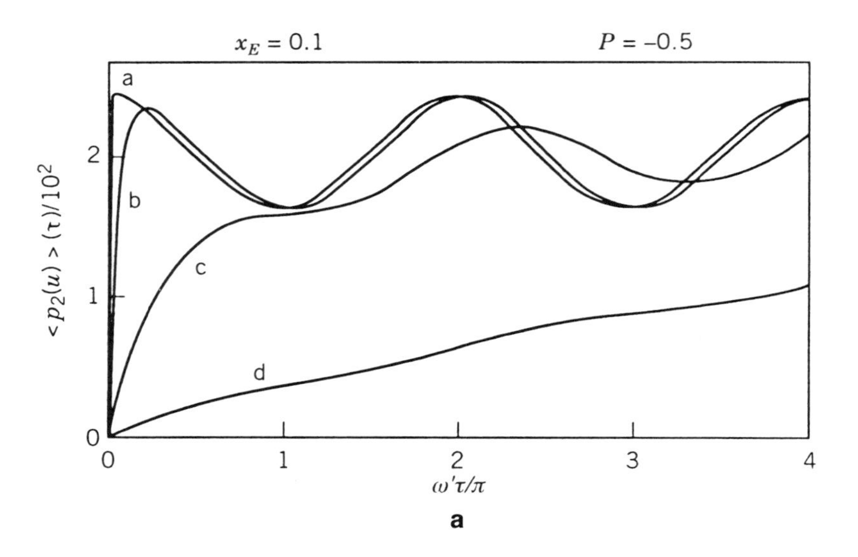

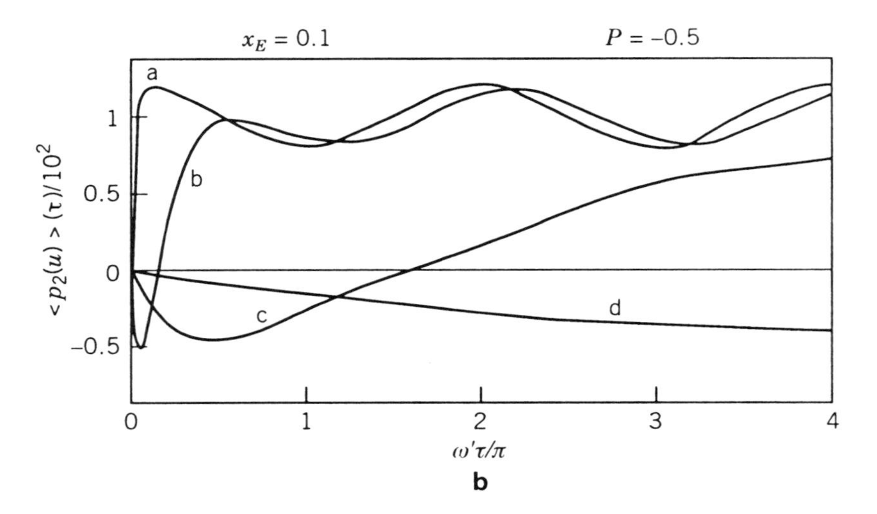

Figure 4.40. Plots of the nonnormalized transient birefringence signals for various values of ω'. a, b, c, d stand for $\omega' = 0.1$, 1, 10, and 100, respectively. (**a**): $x_E = 0.1$, $P = 100$; (**b**): $x_E = 0.1$, $P = -0.5$; (**c**): $x_E = 1$, $P = 100$; (**d**): $x_E = 1$, $P = -0.5$; (**e**): $x_E = 5$, $P = 100$; (**f**): $x_E = 5$, $P = -0.5$; (**g**): $x_E = 100$, $P = 100$; (**h**): $x_E = 100$, $P = -0.5$.

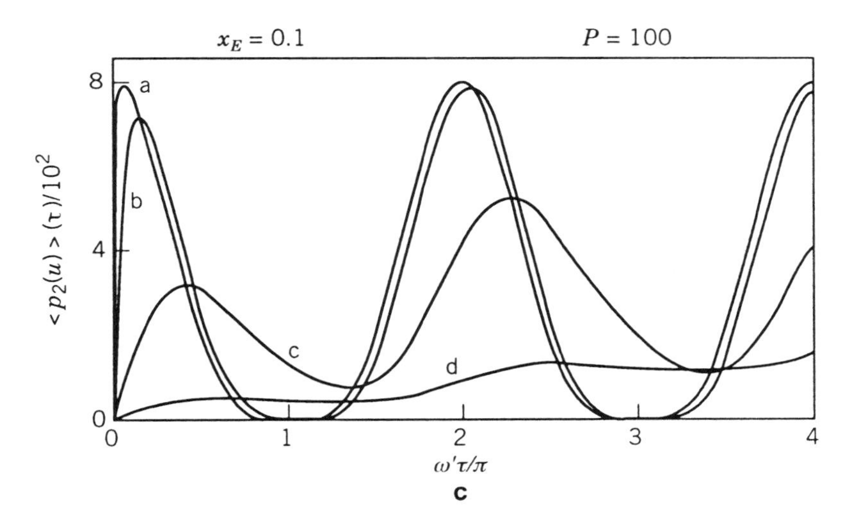

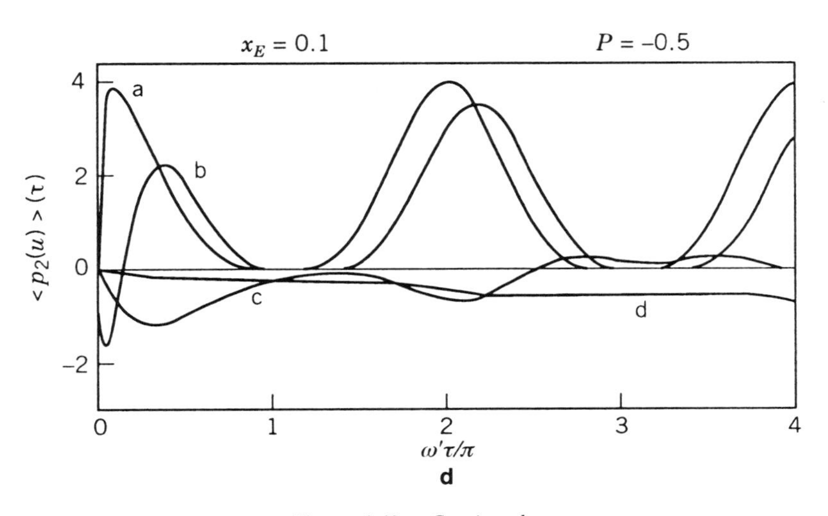

Figure 4.40. *Continued.*

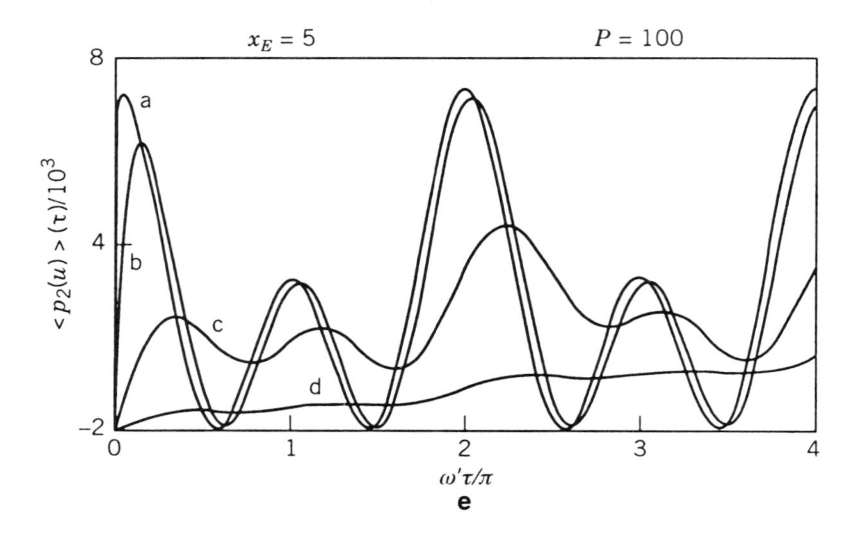

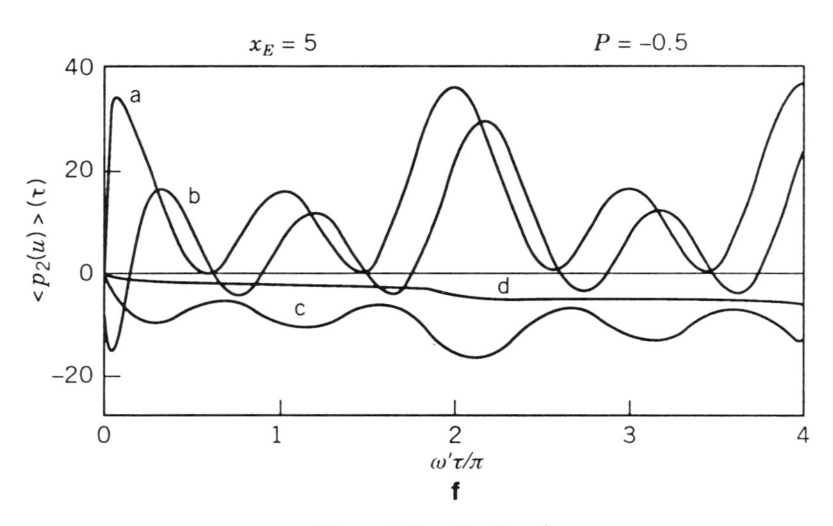

Figure 4.40. *Continued.*

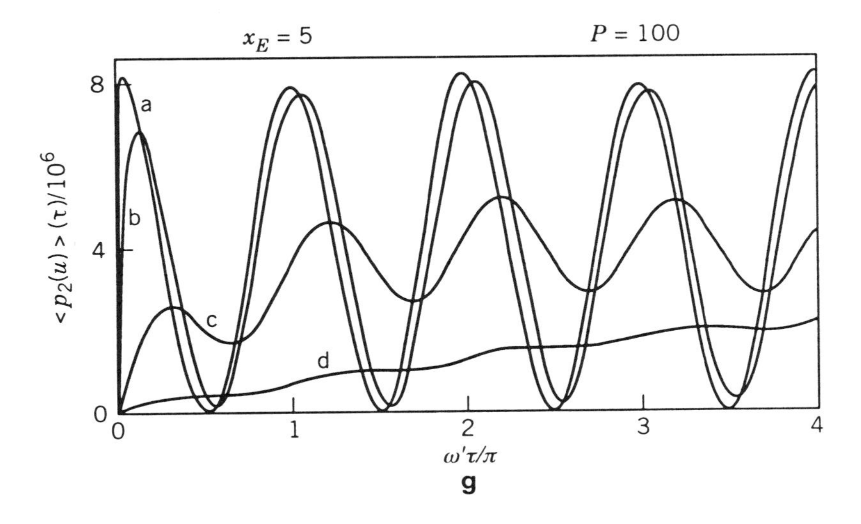

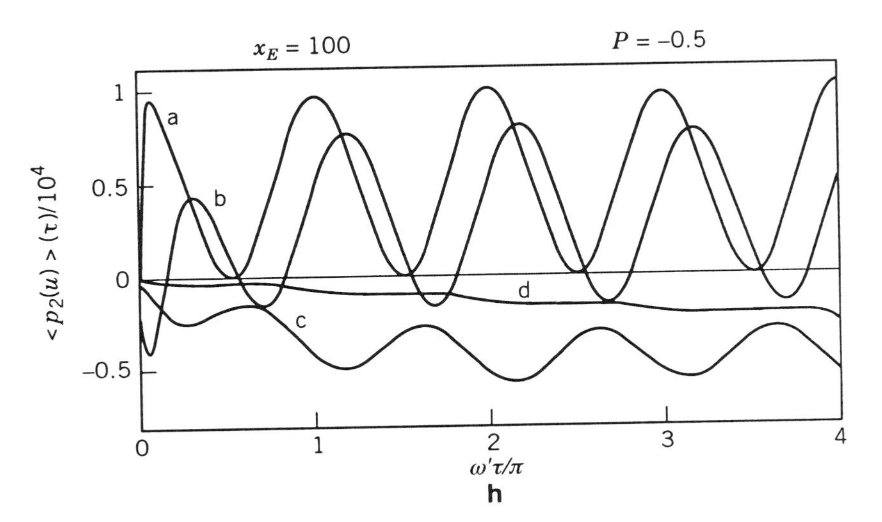

Figure 4.40. *Continued.*

with $\omega'\tau$ (or ωt) of the term in brackets in Eq. (4.143), such that

$$\langle p_2(u)\rangle(\tau) = [i_\gamma(\tau) + Pi_\beta(\tau)] \tag{4.144}$$

which simplifies the ordinate scale units. As P may vary from $-\infty$ to $+\infty$, several responses are possible and exhibit a large variety. In order to reduce the number of curves, only two particular values of P are considered.

1. $P = 100$. This situation consists in favoring $i_\beta(t)$ compared to $i_\gamma(t)$, which is the case where the induced dipolar moments of molecules prevail over the permanent dipolar moments. Note that for $P = -100$, the corresponding plots would give negative values of the birefringence signals and would be almost symmetric with respect to the abscissa axis.
2. $P = -0.5$. This value refers to molecules whose contribution due to the induced moments is less than that due to the permanent ones ($\Delta\alpha < 0$, polar molecules). In what follows, we shall examine how the phase angles θ_1 and θ_2 may be determined in this case from their respective variations with ω'.

Before going into further detail, some general comments may be drawn from Figs. 4.40a–h. Thus, it is evident that the amplitude of the ω component increases with decreasing x_E values. This merely arises from the fact that if the harmonic term in 2 ω is the only one existing for $x_E = 100$ (Figs. 4.40g and h), the harmonic term in ω predominates when $x_E = 0.1$ (Figs. 4.40a and b). What is happening there may be clearly understood if one remarks that for the lowest values of x_E ($E_0 \ll E_c$), the Kerr response varies approximately as $\cos(\omega t)$, while for a prevalence of the alternating field on the unidirectional bias field ($E_0 \gg E_c$), the resulting variation is effected as $\cos(2\omega t)$. For intermediate values of $x_E (x_E = 1,\ x_E = 5)$, both components are mixed and one obtains a periodic but not sinusoidal signal. In any case, we first and foremost observe the transient regime, which is essentially driven by exponential functions, followed by the steady state, the evolution of which is regular and periodic.

In all the figures presented, it is worth remarking that the steady-state electric birefringence attains its equilibrium value quickly enough for times approximately equal to π/ω in the low-frequency region and 10 π/ω in the high-frequency region. Consequently, although a hasty reading of curves could lead one to state the contrary, the stationary state is attained faster for high frequencies than for low frequencies (note, indeed, that the abscissa are expressed as ωt and not as t).

The problem now is the calculation of both constants P and D, which are characteristics of the molecular medium, and the determination of which is of essential interest. In this way, we may refer the reader to previous papers [48, 68] in which the stationary regime was considered, and where we have used a spectral analysis based on Kramers–Kronig relations for the determination of phases and Cole–Cole diagrams of some dielectric liquids. With our present theoretical treatment of the time-dependent behavior of the birefringence, we are able to work with only one frequency, which is interesting from an experimental point of view. On assuming that an entire experimental study could be at our disposal, one might always carry out Eq. (4.143) by a numerical analysis leading to the parameters P and D, which are the only unknown quantities in our problem. As usually done, we should act by successive refining cycles to obtain the best-fit procedure between theoretical and experimental plots. This solution is not, however, very easy to apply because of the complexity of $\langle p_2(u)\rangle(\tau)$. But the problem may be reduced to a very simple form if one considers the extreme limits of frequency range, that is, the transient behavior for low or high frequencies.

When ω' is small, that is, $\omega' < 1$ (or $\omega < D$), the corresponding transients are of short duration, and Eq. (4.144) may be written in the interval $0 < \omega'\tau < \pi/10$ as follows

$$\langle p_2(u)\rangle(\tau) \approx (1 + x_E)^2[2(P+1) + (1-2P)e^{-6\tau} - 3e^{-2\tau}] \quad (4.145)$$

We have verified that Eq. (4.145) gives very satisfactory results with a relative error less than 3% compared with the whole theoretical expression. For most common dielectric materials, we may take D as about 10^5 s^{-1} and ensure that Eq. (4.145) is correct for frequencies lying from 0 to 20 kHz.

In a similar way, for high values of ω', say $\omega' > 50$ (or $\omega > 50D$), we may express Eq. (4.144) as

$$\langle p_2(u)\rangle(\tau) \approx 2 + P(2 + x_E^2) + [1 - P(2 + x_E^2)]e^{-6\tau} - 3\,e^{-2\tau} \quad (4.146)$$

The latter equation may be applied with the same uncertainty, as well. Note that Eqs. (4.145) and (4.146) are valid whatever value x_E may have, and represent Kerr responses due to the sudden application of both fields until the system has reached its final stationary state.

In addition to this, we mention that it should be interesting to work with the lowest frequencies because experimental conditions then become more suitable and the birefringence magnitudes are more important than for high frequencies. This would provide some improvement in the precision of the numerical treatment of measurements.

At this stage of our quantitative description, one may object that it is not necessary to use superimposed fields to apply this method. This is true, but with the help of some curves obtained for various values of x_E, the determination of the molecular parameters P and D will be of better accuracy. Moreover, it is possible to find another approach where the influence of the coupled fields is more pronounced. To show that, we consider the maxima of the function $\langle p_2(u)\rangle(t)$ immediately after the establishment of the transients on taking into account two limiting cases, $x_E = 100$ and $x_E = 0.1$, which, as indicated before, favor the $2\,\omega$ component and the ω component, respectively.

For $x_E = 100$, these maxima are defined by

$$\cos(2\,\omega t - \theta_2) = 1 \qquad \text{or} \qquad \theta_2 = 2\,\omega t - 2\,k_2\pi \tag{4.147a}$$

while, for $x_E = 0.1$, one has

$$\cos(\omega t - \theta_1) = 1 \qquad \text{or} \qquad \theta_1 = \omega t - 2\,k_1\pi \tag{4.147b}$$

where k_1 and k_2 are integers.

It should be noted that this method is feasible for any value of x_E in so far as the ω and 2ω components may always be separated experimentally [3, 62]. So, a direct measurement of the phase angles θ_1 and θ_2 may be obtained regardless of any ambiguity since their values are included in an interval lying from $-\pi/2$ to $3\pi/2$. The variations of θ_j with angular frequency have been widely developed in Section III.A.3. We recall here their theoretical expressions in terms of ω' and P, namely,

$$\theta_1(\omega') = \tan^{-1}\left[\frac{\omega'}{6}\,\frac{(P+\frac{1}{4})(1+\frac{1}{4}\,\omega'^2)+\frac{9}{4}(1+\frac{1}{36}\,\omega'^2)}{(P+\frac{1}{4})(1+\frac{1}{4}\omega'^2)+\frac{3}{4}(1+\frac{1}{36}\,\omega'^2)}\right] \tag{4.148}$$

$$\theta_2(\omega') = \tan^{-1}\left[\frac{\omega'}{3}\,\frac{(P-2)(1+\frac{1}{4}\,\omega'^2)+\frac{9}{2}(1+\frac{1}{9}\,\omega'^2)}{(P-2)(1+\frac{1}{4}\,\omega'^2)+3(1+\frac{1}{9}\,\omega'^2 0}\right] \tag{4.149}$$

Consequently, it is shown how the dual set of information given by both components for the phase angles requires one to solve a set of two equations with two unknown quantities P and D, which are then readily deduced. This is really specific to the study with superimposed fields. We turn now to the particular case of $P = -0.5$. This case is interesting in so far as the corresponding phase angles θ_1 and θ_2 are large and tend to quite different values in the high-frequency region, π and $3\pi/2$, respectively.

Their behavior as a function of ω' is also fast enough at low and mid-frequencies. From Figs. 4.40a and b, it appears that the first maximum of plots a and b (corresponding to an increase of θ_1 with ω') is shifted in a more pronounced way for $P = -0.5$ than for $P = 100$. The same is true of θ_2 if one refers to Figs. 40g and h. Figure 4.41 portrays two plots of $\langle p_2(u) \rangle(\tau)$ for $\omega' = 10$ and $P = -0.5$, which allow us to measure θ_1 and θ_2. On taking D equal to $10^5\ s^{-1}$, this corresponds to a frequency of 1.6×10^5 Hz. One easily finds $\theta_1 = 0.82\ \pi$ and $\theta_2 = 1.26\ \pi$, values in excellent agreement with the limits we mentioned above. An estimate of experimental phase angles obtained by other techniques generally provides a relative precision of about 15 or 20%. Our method must lead to accuracies twice as small, at least. Hence, it is quite convenient to complement data extracted from the Kramers–Kronig relations.

In passing, we may remark in Fig. 4.41 that the birefringence signal is very sensitive to the influence of the induced dipole moment (negative here) and the alternating field strength. In a rapidly varying electric field, it is obvious that the permanent dipole moment can no longer follow any molecular reorientation dictated by this electric field.

6. *Application of the Kramers–Kronig Method*

All along in our study, we emphasized the importance of calculating phase angles and showed that this enables us to provide useful in-

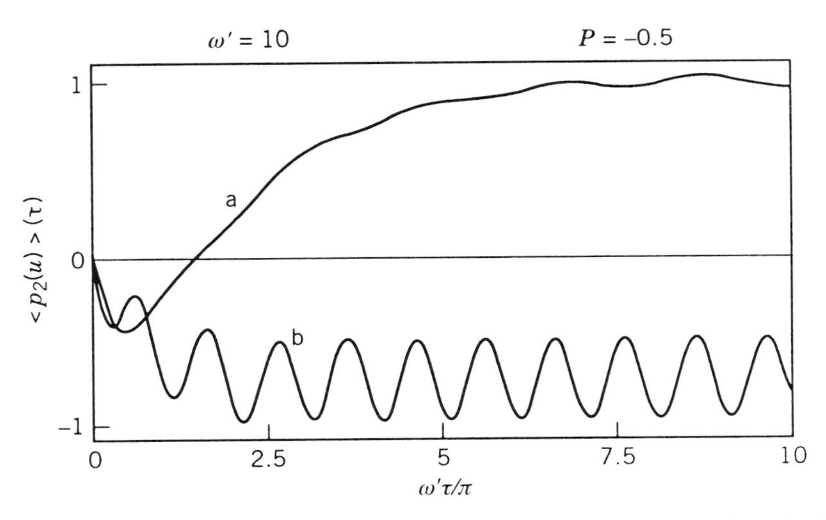

Figure 4.41. Plots of the nonnormalized transient birefringence signals for $\omega' = 10$, $P = -0.5$. (a): $x_E = 0.1$; (*b*): $x_E = 100$ (with a reduction factor of 6,000).

formation as much in electric birefringence as in dielectric relaxation. So, we applied the Kramers–Kronig relations to the dynamic Kerr effect to get a better interpretation of the characteristic diagrams of the phenomenon, and then to obtain results as interesting as those already described in dielectric relaxation. In particular, we shall demonstrate the usefulness of the Kramers–Kronig relations [48, 72, 73] as a technique for interpreting experimental birefringence data of liquids. To test that, we have chosen the measurements obtained from the alternating Kerr effect on linoleic acid (glycerol acetate) and pentachlorodiphenyl [63]. These relations are generally used in the study of dielectric relaxation of liquids characterized by a complex refractive index or a complex electric permittivity when submitted to the action of sinusoidally varying electric fields. In Section III.A.3, we likewise showed that the birefringence induced by the superposition of two fields, ac plus dc fields, in a liquid may be expressed in the form of a complex quantity such that

$$\Delta\mathbf{N}_j = X_j - iY_j \,, \qquad j = 1, 2 \tag{4.150}$$

where X_j and Y_j are the real and imaginary parts of the complex birefringence normalized to unity.

This equation may also be written as follows:

$$\Delta\mathbf{N}_j = |\Delta\mathbf{N}_j| \exp(-i\theta_j) \tag{4.151}$$

where

$$|\Delta\mathbf{N}_j| = (X_j^2 + Y_j^2)^{0.5} \tag{4.152a}$$

$$\theta_j = \tan^{-1}\frac{Y_j}{X_j} \tag{4.152b}$$

The Kramers–Kronig formulas allow one to calculate the real part of a complex quantity from its imaginary part and conversely. Applied to our birefringence problem, these are given by

$$X_j(\omega') = \frac{1}{\pi}\mathcal{P}\int_{-\infty}^{+\infty}\frac{Y_j(\omega)}{\omega-\omega'}\,d\omega \tag{4.153a}$$

$$Y_j(\omega') = -\frac{1}{\pi}\mathcal{P}\int_{-\infty}^{+\infty}\frac{X_j(\omega)}{\omega-\omega'}\,d\omega \tag{4.153b}$$

where $\mathcal{P}$ denotes the Cauchy principal value.

We shall not use Eqs. (4.153) in so far as the experimental procedure provides us with the modulus of $\Delta\mathbf{N}_j$. So, to extract the phase angle θ_j, we rewrite Eq. (4.151) as

$$\ln(\Delta\mathbf{N}_j) = \ln|\Delta\mathbf{N}_j| - i\theta_j \tag{4.154}$$

so that Kramers–Kronig relations become

$$\ln|\Delta\mathbf{N}_j|(\omega') = \frac{1}{\pi}\mathscr{P}\int_{-\infty}^{+\infty} \frac{\theta_j(\omega)}{\omega - \omega'}\,d\omega \tag{4.155a}$$

$$\theta_j(\omega') = -\frac{1}{\pi}\mathscr{P}\int_{-\infty}^{+\infty} \frac{\ln|\Delta\mathbf{N}_j|(\omega)}{\omega - \omega'}\,d\omega \tag{4.155b}$$

In writing Eqs. (4.155), it is obviously assumed that $\Delta\mathbf{N}_j(0)$ is different from zero. It is clear that this condition is always experimentally satisfied. At zero frequency, we get the static birefringence.

Before going further, it should be convenient to recall that the dispersion relations obey Boltzmann's superposition principle, which normally holds for linear responses. However, they may also be applied to the nonlinear regime [9, 66] when one restricts oneself to second-order nonlinear terms. This is exactly the limit we have fixed in our approach to the dynamic Kerr effect.

In practice, the range of integration is finite and lies in a frequency interval from ω_a to ω_b. Hence, we may split up θ_j into three parts as follows:

$$\theta_j^{(1)}(\omega') = -\frac{1}{\pi}\mathscr{P}\int_{-\infty}^{\omega_a} \frac{\ln|\Delta\mathbf{N}_j|(\omega)}{\omega - \omega'}\,d\omega \tag{4.156a}$$

$$\theta_j^{(2)}(\omega') = -\frac{1}{\pi}\mathscr{P}\int_{\omega_a}^{\omega_b} \frac{\ln|\Delta\mathbf{N}_j(\omega)}{\omega - \omega'}\,d\omega \tag{4.156b}$$

$$\theta_j^{(3)}(\omega') = -\frac{1}{\pi}\mathscr{P}\int_{\omega_b}^{+\infty} \frac{\ln|\Delta\mathbf{N}_j|(\omega)}{\omega - \omega'}\,d\omega \tag{4.156c}$$

$$\theta_j(\omega') = \theta_j^{(1)}(\omega') + \theta_j^{(2)}(\omega') + \theta_j^{(3)}(\omega') \tag{4.156d}$$

The contribution of $\theta_j^{(1)}$ and $\theta_j^{(3)}$ is far from negligible. We recall that these two terms are correction terms that can be evaluated by means of a mathematical or empirical law. Here again, we shall take the same correction expressions as those derived by Ruiz-Perez [43]. We have

$$\theta_j^{(1)}(\omega') = k_a[(\omega_a + \omega')\ln(\omega_a + \omega') + (\omega_a - \omega') \times \ln(\omega_a - \omega') - 2\omega' \ln \omega'] \tag{4.157a}$$

$$\theta_j^{(3)}(\omega') = k_b \left[\frac{\omega_b + \omega'}{\omega_b \omega'} \ln \frac{\omega_b + \omega'}{\omega_b \omega'} + \frac{\omega_b - \omega'}{\omega_b \omega'} \times \ln \frac{\omega_b - \omega'}{\omega_b \omega'} - \frac{2}{\omega'} \ln \frac{1}{\omega'} \right] \tag{4.157b}$$

The constants k_a and k_b are calculated by making an appropriate choice of two frequencies ω_1 and ω_2 comprised in the interval $[\omega_a, \omega_b]$ for which the phase angle θ_j is known either by experiment or by theoretical considerations. Then, it suffices to solve the following system

$$\begin{aligned} \theta_j(\omega_1) &= \theta_j^{(1)}(\omega_1) + \theta_j^{(2)}(\omega_1) + \theta_j^{(3)}(\omega_1) \\ \theta_j(\omega_2) &= \theta_j^{(1)}(\omega_2) + \theta_j^{(2)}(\omega_2) + \theta_j^{(3)}(\omega_2) \end{aligned} \tag{4.158}$$

The determination of the main part $\theta_j^{(2)}(\omega')$ may be calculated by cutting up the plot of $\ln|\Delta \mathbf{N}_j|(\omega)$ into $(m-1)$ straight segments. Denoting the slope of one of them by p_i, we have

$$\theta_j^{(2)}(\omega') = -\frac{1}{\pi} \left[\sum_{i=1}^{m-1} p_i \int_{\omega_i}^{\omega_{i+1}} \ln\left(\frac{\omega + \omega'}{\omega - \omega'}\right) d\omega \right] \tag{4.159}$$

We already mentioned that the birefringence functions $|\Delta \mathbf{N}_1|(\omega)$ and $|\Delta \mathbf{N}_2|(\omega)$ are normalized to unity. This implies that the phase angle θ_j vanishes as $\omega \rightarrow 0$. Hence, we can consider that $\theta_j(\omega_1)$ is very close to zero when ω_1 lies in the very low frequency region. By contrast, one needs to know, if possible, an experimental value (and only one) of θ_j in the high-frequency region to make a homogeneous correction all along the curve of θ_j against ω. The corrections obtained from $\theta_j^{(1)}$ and $\theta_j^{(3)}$ are revealed to be very sensitive and efficient even in difficult conditions of integration. This is particularly the case where the function $|\Delta \mathbf{N}_j|(\omega)$ may undergo important variations at the end of the frequency range. Finally, the accuracy of the present calculation improves as the number $(m-1)$ segments become larger.

The main advantage of the Kramers–Kronig technique over the experimental method is clearly established by comparison of the respective accuracies. If the modulus of $\Delta \mathbf{N}_j$ is experimentally measured with a relative error ranging from 3 to 5%, for increasing frequencies, the precision of the experimental phase lies in the range 12–20%. On using our method, we found that the phase angle calculated from $|\Delta \mathbf{N}_j|$ can be determined within an error of 7 or 8%, and is thus twice as small. As a consequence, a good alignment of calculated points is obtained, whereas experimental points are rather scattered. This is particularly evident in the Cole–Cole diagrams.

Use of Kramers–Kronig relations was applied to study the dynamic Kerr effect of linoleic acid (at 20 °C) and pentachlorodiphenyl (at 10 °C). From these investigations, we have shown that both liquids lead to very interesting results not at all apparent from the experimental procedure alone. Two meaningful conclusions arise from that treatment. First, experimental and calculated curves are well described on assuming the principle of additive responses. On the other hand, better information on molecular association of these dielectric materials is obtained (characterization of polydisperse systems). Moreover, our treatment of experimental data is relevant to both harmonic components, which appear when a dc field is superimposed on an ac field. Calculation is carried out with the modulus of $\Delta \mathbf{N}_j$ normalized to unity, that is divided by the static value of $|\Delta \mathbf{N}_j|$ obtained in a weak field and a low-frequency limiting value, which is without any influence on the results. The frequency range lies from 250 Hz to 3 MHz. For linoleic acid and pentachlorodiphenyl, 17 and 20 experimental points were to our hand, respectively. To increase the accuracy of this method, dispersion curves were adjusted to these points by successive interpolations and then cut up into 50 segments. As the initial part of the dispersion spectrum reduces greatly with frequency, the first correction term $\theta_j^{(1)}$ corresponding to an integration from 0 to $(2\pi \times 250)$ rad s^{-1} is nearly zero in all cases. This is confirmed by very small values calculated for $\theta_j^{(2)}$ at 250 Hz. On the contrary, $\theta_j^{(3)}$, which is the correction term occurring in the high-frequency domain is meaningful throughout the experimental range. It is determined by matching the definitive phase angle with an experimental phase angle at 2 or 3 MHz. Particular aspects and general information are now developed for each liquid.

1. Linoleic Acid

Figure 4.42 shows how the evolution of θ_1 versus ω, related to the first harmonic term, is in fair agreement with experiment. The same trend is observed for the second harmonic term θ_2 (see Fig. 4.43). A diagram of Y_1 against X_1 (Fig. 4.44), where $X_1 = |\Delta \mathbf{N}_1| \cos \theta_1$ and $Y_1 = |\Delta \mathbf{N}_1| \sin \theta_1$, is composed of two arcs, indicating that two constituents

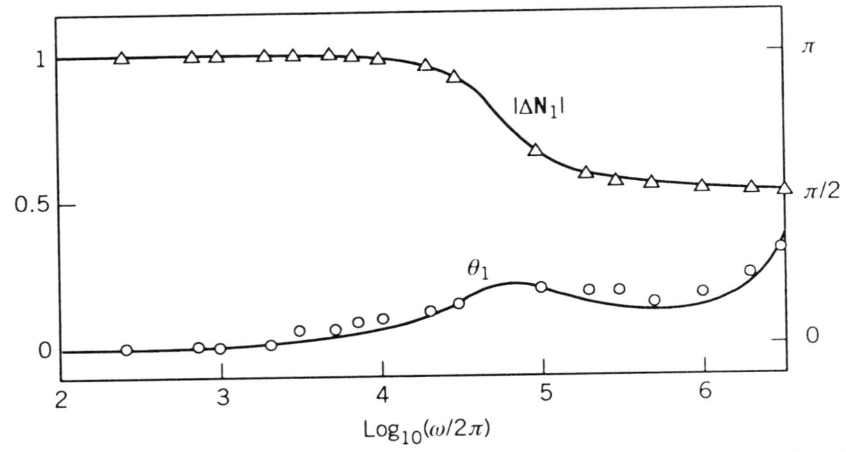

Figure 4.42. Plot of calculated θ_1 versus $\log_{10}(\omega/2\pi)$ for linoleic acid on using the Kramers–Kronig method. (Δ) and (o) are experimental points for $|\mathbf{\Delta N}_1|$ and θ_1, respectively, extracted from ref. 63.

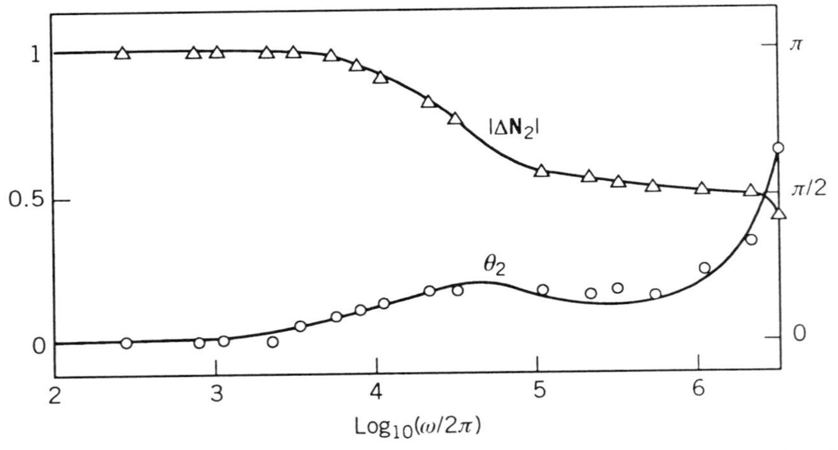

Figure 4.43. Plot of calculated θ_2 versus $\log_{10}(\omega/2\pi)$ for linoleic acid on using the Kramers–Kronig method. (Δ) and (o) are experimental points for $|\mathbf{\Delta N}_2|$ and θ_2, respectively, extracted from ref. 63.

exist in the medium where the birefringence contributions are 48 and 52%, respectively. That proves the existence of some polydispersity in this liquid. It is the same in Fig. 4.45, where we can see two maxima that are very far from each other; the first at 52 kHz and the second beyond 3 MHz. On considering the second harmonic component of the

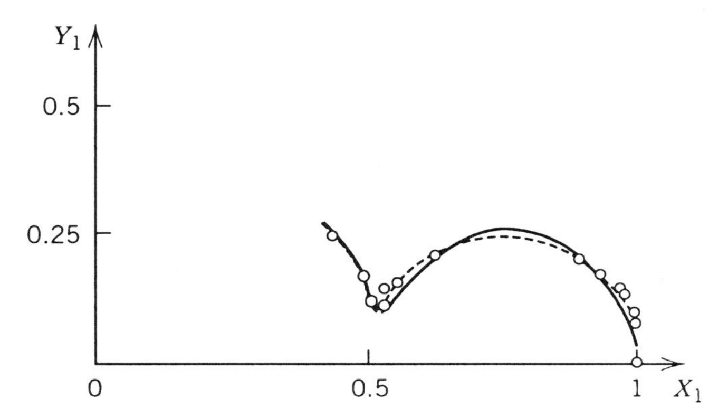

Figure 4.44. Cole–Cole diagram Y_1 versus X_1 for linoleic acid (full line). (o) are experimental points extracted from ref. 63. The broken line corresponds to our best-fit procedure.

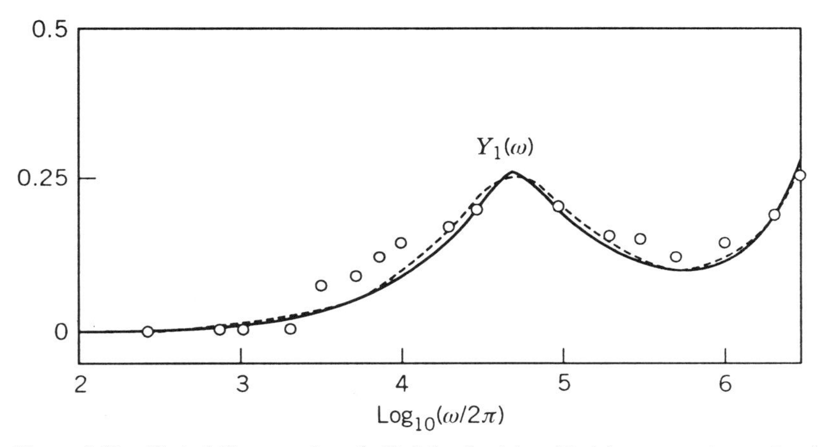

Figure 4.45. Plot of Y_1 versus $\log_{10}(\omega/2\pi)$ for linoleic acid. (o) are experimental points extracted from ref. 63. The broken line corresponds to our best-fit procedure.

electric birefringence, the Cole–Cole diagram exhibits a slight difference in the low-frequency region, as shown in Fig. 4.46. It is seen, indeed, that the corresponding arc is flattened, which could reveal the existence of a third constituent. This last assumption is supported by the plot of Y_2 versus ω in Fig. 4.47, where one sees a very broadened maximum, located at 30 kHz. Note that this third species was not apparent from measurements.

The high quality of our results, obtained by inversion of the dispersion relations, allows us to extract very refined information

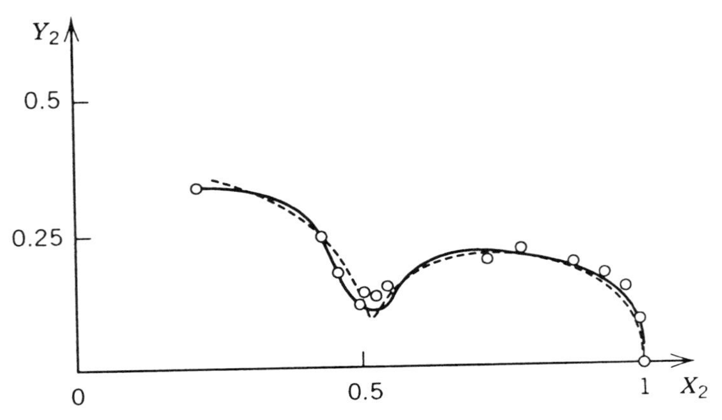

Figure 4.46. Cole–Cole diagram Y_2 versus X_2 for linoleic acid (full line). (o) are experimental points extracted from ref. 63. The broken line corresponds to our best-fit procedure.

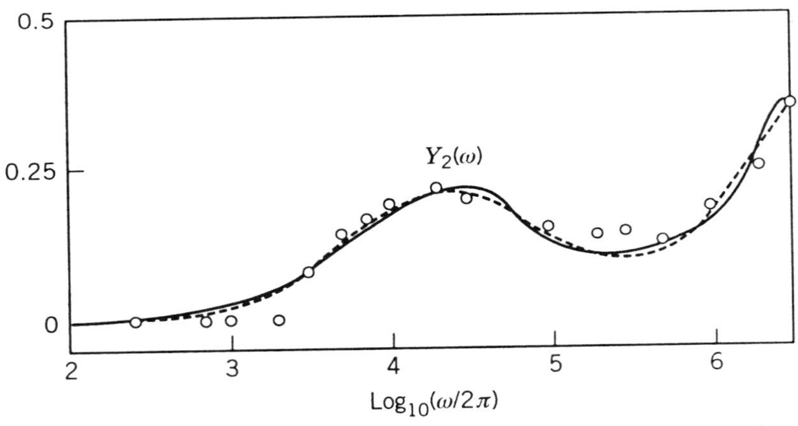

Figure 4.47. Plot of Y_2 versus $\log_{10}(\omega/2\pi)$ for linoleic. (o) are experimental points extracted from ref. 63. The broken line corresponds to our best-fit procedure.

about the molecular structure of liquids. We restrict ourselves to the determination of some characteristic physical parameters for which we shall attempt to give some qualitative interpretation. The polydisperse nature of a liquid appears clearly on the Cole–Cole diagrams and results from a careful analysis of experimental data by using this technique. So, for a pure liquid, these diagrams are characterized by only one arc. In the case of a polydisperse material, several arcs can be distinguished, each of them corresponding to a constituent of the medium. Assuming that these different species do not interact (or only weakly) with each other from an optical point of view, the Kerr functions obey the following model of simple additivity

$$|\Delta \mathbf{N}_j| = \sum_{h=1}^{n} C_j^{(h)} |\Delta \mathbf{N}_j^{(h)}| \qquad \sum_{h=1}^{n} C_j^{(h)} = 1 \tag{4.160}$$

where $|\Delta \mathbf{N}_j^{(h)}|$ is the modulus of the complex Kerr function of the hth species, considered as an individual molecular ensemble, n is the number of constituents and $C_j^{(h)}$ is a homogeneous weighting factor assigned to the hth species in its contribution to the whole birefringence. In the following, the particles of species h in the sample under consideration will have $P_j^{(h)}$ and $D_j^{(h)}$ as a ratio of dipolar moments and rotational diffusion constants, respectively. As a first step, minima of Cole–Cole type diagrams provide us with a good enough valuation of $C_j^{(h)}$ values whereas amplitudes of arcs give an indication of the $P_j^{(h)}$ values. Moreover, the $D_j^{(h)}$ constants [or the Kerr effect relaxation times $\tau_2^{(h)}$] can be derived from the frequency peaks of $Y_j(\omega)$ (maximum loss). Then, we proceed to the best-fit procedure on these two graphs by successive refining cycles.

Typical parameters having a physical significance are shown in Tables VII and VIII corresponding to both harmonic components. By a simple comparison, one can see that three constituents are apparent for the second harmonic and only two with the first one. However, this apparent discrepancy can be explained by the lack of experimental measurements for the first harmonic in a spectral region of the Cole–Cole diagram corresponding to the top of the larger arc and so, it is

TABLE VII
Determination of Birefringence Parameters (first harmonic, $j = 1$) for Linoleic Acid

h	$C_1^{(h)}$	$P_1^{(h)}$	$D_1^{(h)}(\mathrm{s}^{-1})$	$\tau_2^{(h)}(\mathrm{s})$
1	0.48	30	5.55×10^4	3×10^{-6}
2	0.52	−0.6	2.77×10^7	6×10^{-9}

TABLE VIII
Determination of Birefringence Parameters (second harmonic, $j = 2$) for Linoleic Acid

h	$C_2^{(h)}$	$P_2^{(h)}$	$D_2^{(h)}(\mathrm{s}^{-1})$	$\tau_2^{(h)}(\mathrm{s})$
1	0.20	30	2.31×10^4	7.2×10^{-6}
2	0.28	30	8.33×10^4	2.0×10^{-6}
3	0.52	−0.10	1.75×10^7	9.5×10^{-9}

possible that a constituent has not been perceived. This assessment is strengthened by the value of the $D_1^{(1)}$ intermediate between $D_2^{(1)}$ and $D_2^{(2)}$. Of the three constituents, the first one is bigger because it occurs at the lowest frequencies and is certainly related to a disordered aggregative form of the molecules in the liquid for which the net permanent dipolar moment is nearly zero, as indicated by the high value of $P_j^{(1)}$. The same approach can be made for the second constituent, a smaller aggregate. The third is related to isolated molecules, that is, monomeric forms of the nonassociated liquid. It should be noted that the $D_j^{(h)}$ values and therefore the $\tau_2^{(h)}$ values for the latter constituent are slightly different. This could originate at the onset of molecular association due, in part, to the long time (several hours) it takes to achieve birefringence measurements. Moreover, both harmonic components are not simultaneously determined, and the liquid is formed of aggregates with various sizes and shapes. So, evolution of the aggregation kinetics might happen in spite of good temperature control. This qualitative description should explain changes in $P^{(3)}$ from -0.6 for $P_1^{(3)}$ to -0.1 for $P_2^{(3)}$. To date we have been unable to give other quantitative reasons.

2. Pentachlorodiphenyl

On calculating $\theta_j^{(2)}$ for both harmonic terms, we found some disagreement in the high-frequency range by comparison with experiment. Whereas Bénet [63] finds a continuous increase of the phase angle with ω, we found a deep minimum close to 1 MHz, and correction with $\theta_j^{(3)}$ has not modified the intensity of that minimum. To clarify that discrepancy and to prove that our method does provide a good tool for analyzing the character of polydispersion processes, we referred to the work of Filippini [59]. He obtained using the same liquid at a different temperature (0 °C) and in a pure alternating field a minimum around 10 kHz, hence confirming our technique. In this way, we revised our corrective term $\theta_j^{(3)}$ in order to obtain a definitive phase angle in good accordance with experiment at the lowest frequencies and to get the asymptotic value of $\pi/2$ beyond 3 MHz by extrapolating the plots of θ_1 and θ_2. This is shown in Figs. 4.48 and 4.49.

In Figs. 4.50 and 4.51 we show typical Cole–Cole diagrams represented by a large arc constituted of three shoulders in their first part, followed by a small arc, which lead us to four apparent constituents for both harmonic terms. This last interpretation is illustrated by the plots of Y_1 and Y_2 against ω, as shown in Figs. 4.52 and 4.53, where the same shoulder series can be observed. The corre-

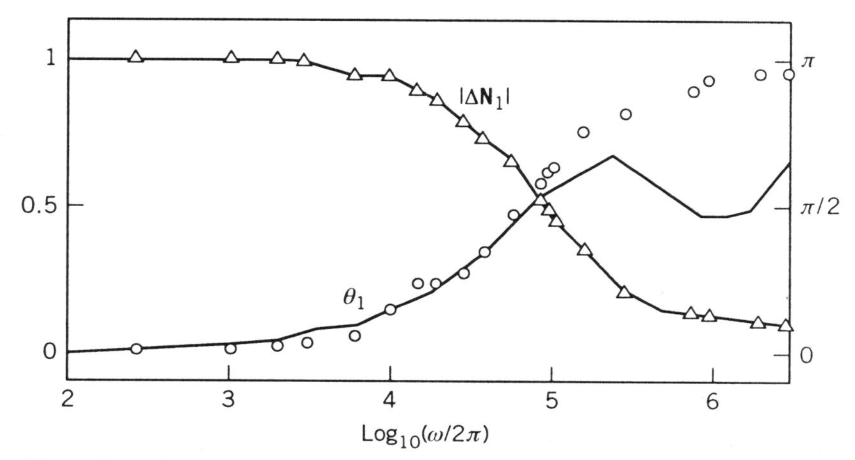

Figure 4.48. Plot of calculated θ_1 versus $\log_{10}(\omega/2\pi)$ for pentachlorodiphenyl on using the Kramers–Kronig method. (Δ) and (o) are experimental points for $|\mathbf{\Delta N}_1|$ and θ_1, respectively, extracted from ref. 63.

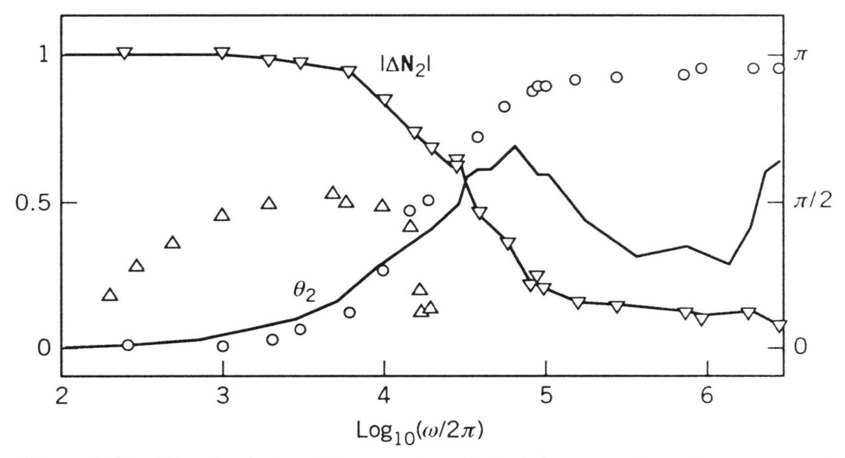

Figure 4.49. Plot of calculated θ_2 versus $\log_{10}(\omega/2\pi)$ for pentachlorodiphenyl on using the Kramers–Kronig method. (∇) and (o) are experimental points for $|\mathbf{\Delta N}_2|$ and θ_2, respectively, extracted from ref. 63. (Δ) denotes experimental points for θ_2 from ref. 59.

sponding maxima are relative to the main constituent of the liquid, that is, 70 and 30 kHz for Y_1 and Y_2, respectively. It should be noted that only one constituent was apparent from experiment, as well.

Tables IX and X list the four constituents of pentachlorodiphenyl. We find good agreement of their birefringence contributions for both

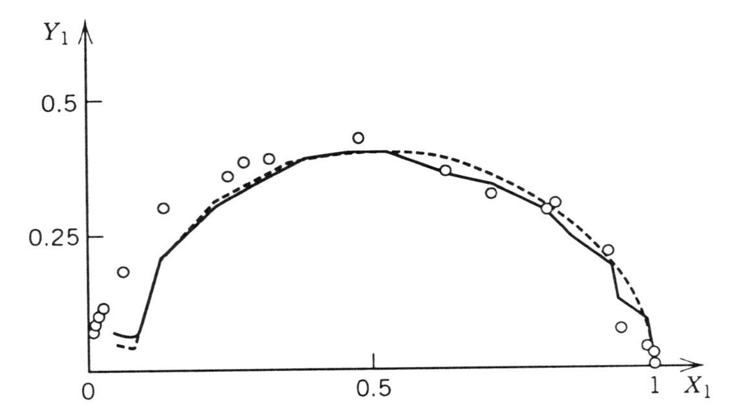

Figure 4.50. Cole–Cole diagram Y_1 versus X_1 for pentachlorodipheyny (full line). (o) are experimental points extracted from ref. 63. The broken line corresponds to our best-fit procedure.

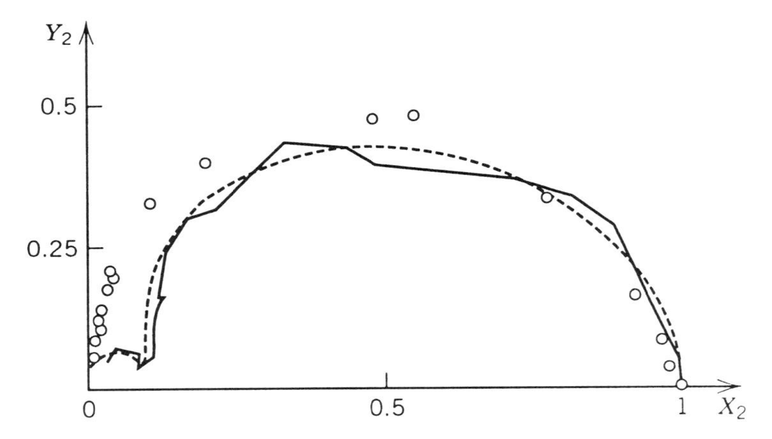

Figure 4.51. Cole–Cole diagram Y_2 versus X_2 for pentachlorodiphenyl (full line). (o) are experimental points extracted from ref. 63. The broken line corresponds to our best-fit procedure.

harmonic terms. Nevertheless, for increasing frequencies, more or less large deviations between these harmonics may be observed in $P_j^{(h)}$ and $D_j^{(h)}$ values and the previous interpretation may also be applied to that liquid.

For $h = 1, 2$, there are two kinds of aggregates with a large value of P indicating a disordered molecular association just as in linoleic acid. For $h = 3$, we get molecules of the pure liquid. Regarding $h = 4$, we

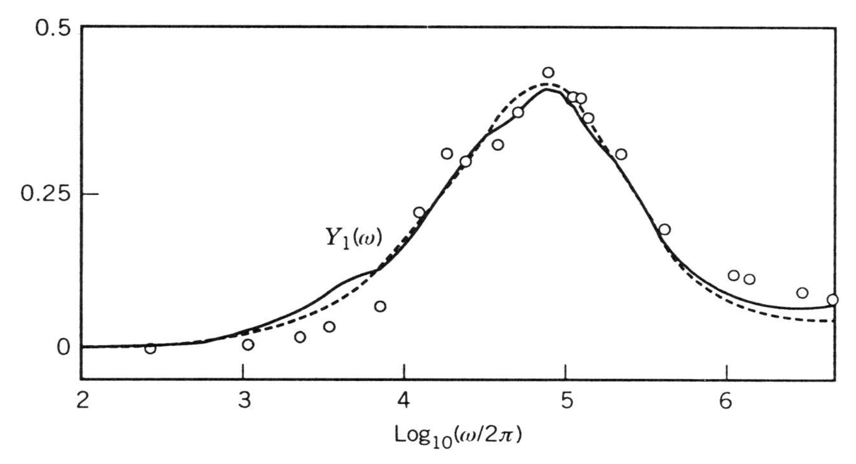

Figure 4.52. Plot of Y_1 versus $\log_{10}(\omega/2\pi)$ for pentachlorodiphenyl. (o) are experimental points extracted from ref. 63. The broken line corresponds to our best-fit procedure.

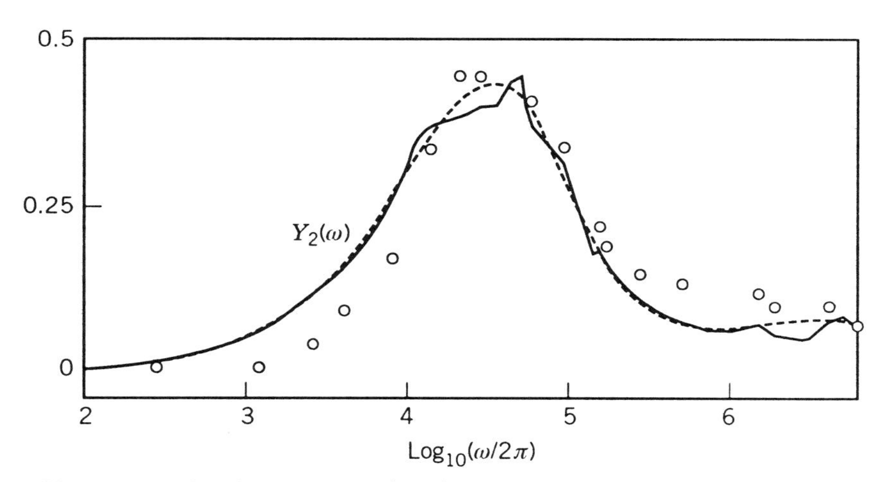

Figure 4.53. Plot of Y_2 versus $\log_{10}(\omega/2\pi)$ for pentachlorodiphenyl. (o) are experimental points extracted from ref. 63. The broken line corresponds to our best-fit procedure.

are in the highest frequency range and for that reason, we are faced with lighter molecules that probably belong to tetrachlorodiphenyl, a derivative of the same family as pentachlorodiphenyl. This fact is confirmed by one of us [49] by studying the Kerr effect response resulting from the application of a rectangular electric field pulse. In passing, it should be noted that the treatment used on Filippini's data would lead to one constituent at least, namely, a very big aggregate in

TABLE IX
Determination of Birefringence Parameters (first harmonic, $j = 1$) for Pentachlorodiphenyl

h	$C_1^{(h)}$	$P_1^{(h)}$	$D_1^{(h)}(s^{-1})$	$\tau_2^{(h)}(s)$
1	0.13	30	1.39×10^4	1.2×10^{-5}
2	0.26	30	3.88×10^4	4.3×10^{-6}
3	0.53	−0.30	2.56×10^5	6.5×10^{-7}
4	0.08	−0.20	8.33×10^6	2.0×10^{-8}

TABLE X
Determination of Birefringence Parameters (second harmonic, $j = 2$) for Pentachlorodiphenyl

h	$C_2^{(h)}$	$P_2^{(h)}$	$D_2^{(h)}(s^{-1})$	$\tau_2^{(h)}(s)$
1	0.12	30	1.39×10^4	1.2×10^{-5}
2	0.25	30	2.78×10^4	6.0×10^{-6}
3	0.53	0.35	1.28×10^5	1.3×10^{-6}
4	0.10	0.20	8.33×10^6	2.0×10^{-8}

the lowest frequencies, which is consistent with taking into account a lower temperature (0 instead of 10 °C).

Finally, the interest of Kramers–Kronig relations in determining the real and imaginary parts of the frequency-dependent complex Kerr function is emphasized and tested with a fair degree of success on linoleic acid and pentachlorodiphenyl. In particular, this powerful method gives more accurate results for θ_j than those obtained with experimental data and may be applied to any liquid. In this way, a main feature is evidenced for both liquids where three and four constituents are discovered, respectively, instead of only two and one by experiment. It is also found that the two harmonic components do not lead to identical values for P and D (or τ_2) parameters. This fact may be qualitatively explained by assuming a small fluctuation in the aggregating behavior during experimental measurements. However, as the response to an electrical stimulus resulting from an orientating continuous field superimposed on an alternating one is nonlinear, it is possible that the difference between the values of the parameters P and D, determined from both harmonic terms considered separately, may reflect that nonlinear response. The quality of our method is confirmed by a very careful analysis of spectra from a few experimental data, which makes it an efficient tool. Thus, the polydispersity of the medium is clearly established and characterized in a manner similar to that we obtained for dielectric relaxation. The Kramers–Kronig analysis, which we have applied to the stationary birefringence process, avoids the difficult measurement of the phase and also leads us to a better approach to the study of molecular mechanisms in a dielectric fluid.

The theoretical birefringence relations deduced from the rotational diffusion model in the meantime do not permit us to measure the moment of inertia of the aggregates, but their relative size may be estimated by the position of the frequency maxima in the dispersion curves $Y_j(\omega)(j = 1, 2)$.

From this fact, either by the study of the stationary regime or by the study of the transient regime or by the application of the Kramers–Kronig method, we have just shown the practical and theoretical interest of the electric birefringence, which allows one to get many microscopic parameters of a monodisperse or polydisperse fluid. In particular, the access to permanent and induced dipoles plays a usefully complementary role to dielectric relaxation measurements.

B. Nonlinear Kerr Effect Relaxation Including Inertial Effects

The preceding study of the birefringence where inertial effects are not taken into account shows that the variation of the phase angle with frequency associated with the Kramers–Kronig relations leads one to remarkable information concerning the fluid.

In dielectric relaxation, the theoretical relations including inertia are given by Eqs. (4.34), and measurements of phase angles at high frequencies allow one to reveal the effect of molecular inertia very quickly. In most cases, however, this last one is not very apparent in the Cole–Cole diagrams.

It is natural therefore to suppose that the high sensitivity of the phase will be very useful in Kerr effect relaxation when inertial effects are not ignored in the relations that describe this effect. We also know that inclusion of inertial effects enormously complicates the theoretical treatment of the birefringence, and that complete analytic expressions for an ellipsoid in a 3D medium taking account of both permanent and induced dipole moments are difficult to establish. However, some progress has been made on this subject. A new method is employed that consists of directly averaging the Langevin equation [28]. In what follows we shall present recent results obtained by Déjardin [27] for a plane rotator in 2D space (the disk model) including inertia to show how the evolution of phase with frequency can give significant results. This model has been chosen for the sake of mathematical simplification, and indeed, has limited physical applications. Nevertheless, it can provide us with a good approximation of inertial effects in electric birefringence. Finally, it is clear that the Kramers–Kronig method can be applied with much interest since we always use alternating fields.

In his theoretical treatment of Kerr effect relaxation including inertia, Déjardin generalized the work of Coffey and McGoldrick [26], who showed that the modified Smoluchowski equation is not valid for nonlinear responses, which is always so when the molecules only carry a permanent dipole. The only equation that may correctly describe the phenomenon is the Fokker–Planck–Kramers equation, which we have already presented in the historical summary of the Introduction.

Déjardin solves this equation by studying both polar and anisotropically polarizable molecules. The driving field is one that we previously considered, that is, a dc field superimposed on an alternating field, $E(t) = E_c + E_0 \cos \omega t$. The method starts by introducing the Fourier transform of the distribution function $W(\theta, d\theta/dt, t)$ in phase space, that is, $\Phi(\theta, u, t)$. By using this procedure, we obtain a set of matrix differential equations of the Brinkman type, that are easier to manipulate. This approach has the advantage that one does not need to use Weber functions. The expected value $\langle \cos 2\theta \rangle(t)$ characterizing electric birefringence in 2D space is calculated limiting the matrix to the order $n = 3$, to obtain tractable expressions for the amplitudes of the various components. This calculation is somewhat long and tedious by hand. Therefore, to avoid errors, we used a computer method for carrying out the various algebraic operations.

1. *Explicit Expressions for the Amplitudes of the Harmonic Components Resulting from the Simultaneous Action of Two Electric Fields*

The results so obtained for the stationary regime may be written in a form analogous to Eq. (4.103) and summarized as follows:

$$\langle \cos 2\theta \rangle(t) = \langle \cos 2\theta \rangle_{\mathrm{st}} + \sum_{j=1}^{2} [\Delta n_j(\omega) \cos(j\omega t) + \Delta k_j(\omega) \sin(j\omega t)] \tag{4.161}$$

where $j = 1$ or 2, just as before, corresponds to the first or the second harmonic. The term $\langle \cos 2\theta \rangle_{\mathrm{st}}$ is a time-independent component, but is dependent on the angular frequency.

Furthermore, the explicit expressions for the terms in the right-hand side of Eq. (4.161) are given by

$$\langle \cos 2\theta \rangle_{\rm st} = \frac{\lambda_0^2}{4\gamma\beta} \frac{-2\,\omega^2 a + 5b\beta\omega + 2\,\beta^2 a}{a^2+b^2} + \frac{\lambda_c^2}{2\gamma^2} + \frac{\beta_E}{2\gamma} \tag{4.162a}$$

representing the dc component

$$\Delta n_2(\omega) = 2\,\beta_0 \frac{\beta A + \omega B}{A^2+B^2} + 2\lambda_0^2 \frac{\mathscr{A}(-3\,\omega^2 + 2\,\beta^2) + 7\,\omega\beta\mathscr{B}}{\mathscr{A}^2 + \mathscr{B}^2} \tag{4.162b}$$

$$\Delta k_2(\omega) = 2\,\beta_0 \frac{\beta B - \omega A}{A^2+B^2} + 2\lambda_0^2 \frac{\mathscr{B}(-3\,\omega^2 + 2\,\beta^2) - 7\,\omega\beta\mathscr{A}}{\mathscr{A}^2 + \mathscr{B}^2} \tag{4.162c}$$

representing the second harmonic components, and

$$\Delta n_1(\omega) = 2\,\lambda_0\lambda_c \frac{\mathscr{A}'(-5\,\omega^2 + 4\,\beta^2) + 12\,\omega\beta\mathscr{B}'}{\mathscr{A}'^2 + \mathscr{B}'^2} + \left(\frac{\lambda_0\lambda_c}{\gamma} + 2\beta_{0c}\right) \frac{4\,\beta A' + 2\,\omega B'}{A'^2 + B'^2} \tag{4.162d}$$

$$\Delta k_1(\omega) = 2\,\lambda_0\lambda_c \frac{\mathscr{B}'(-5\,\omega^2 + 4\,\beta^2) - 12\,\omega\beta\mathscr{A}'}{\mathscr{A}'^2 + \mathscr{B}'^2} + \left(\frac{\lambda_0\lambda_c}{\gamma} + 2\beta_{0c}\right) \frac{4\,\beta B' - 2\,\omega A'}{A'^2 + B'^2} \tag{4.162e}$$

representing the nonlinear first harmonic components varying at the fundamental frequency. a, b, A, B, A', B', $\mathscr{A}$, $\mathscr{B}$, $\mathscr{A}'$, and $\mathscr{B}'$ are functions of ω representing the real and imaginary parts that arise from the calculation of the determinants of the 3×3 matrices. These are defined by

$$a = -\beta(3\,\omega^2 - 2\,\gamma) \qquad b = \omega(-\omega^2 + 2\,\beta^2 + 3\,\gamma) \tag{4.163a}$$

$$A = -4\,\beta(3\,\omega^2 - 2\,\gamma) \qquad B = 4\,\omega(-2\,\omega^2 + \beta^2 + 6\,\gamma) \tag{4.163b}$$

$$A' = -\beta(3\,\omega^2 - 8\,\gamma) \qquad B' = \omega(-\omega^2 + 2\,\beta^2 + 12\,\gamma) \tag{4.163c}$$

$$\mathscr{A} = -4[-2\,\omega^6 + \omega^4(14\,\beta^2 + 12\,\gamma) - \omega^2(27\,\beta^2\gamma + 2\,\beta^4 + 18\,\gamma^2) + 4\,\beta^2\gamma^2] \tag{4.163d}$$

$$\mathscr{B} = -4\,\beta\omega[-9\,\omega^4 + \omega^2(9\,\beta^2 + 33\,\gamma) - 6\,\gamma\beta^2 - 18\,\gamma^2]$$

$$\mathscr{A}' = -\omega^6 + \omega^4(13\,\beta^2 + 15\,\gamma) - \omega^2(60\,\beta^2\gamma + 4\,\beta^4 + 36\,\gamma^2) + 16\,\beta^2\gamma^2$$
$$\mathscr{B}' = -\beta\omega[-6\,\omega^4 + \omega^2(12\,\beta^2 + 55\,\gamma) - 20\,\gamma\beta^2 - 48\,\gamma^2] \tag{4.163e}$$

In Eqs. (4.162) and (4.163), we used reduced variables defined by

$$\gamma = \frac{kT}{I} \qquad \beta = \frac{1}{\tau_I} = \frac{\xi}{I} \qquad \lambda_0 = \frac{\mu E_0}{2\,I} \qquad \lambda_c = \frac{\mu E_c}{2\,I}$$

$$\frac{\Delta\alpha}{4\,I}\,E_0^2 = \beta_0\,, \qquad \frac{\Delta\alpha}{4\,I}\,E_c E_0 = \beta_{0c}$$

$$\frac{\Delta\alpha}{4\,I}\,\langle E^2(t)\rangle = \beta_E \qquad \langle E^2(t)\rangle = E_c^2 + \frac{E_0^2}{2}$$

where I is the moment of inertia of the molecule about a central axis perpendicular to itself, ξ is the rotational friction coefficient.

Equations (4.162) are not easy to apply to experiment. We have therefore sought to express them as functions of times τ_I (the friction time) and τ_1 (the Debye relaxation time), and the factor $P = (\Delta\alpha/\mu^2)kT$. We remark that we use τ_1 instead of τ_2 (the birefringence relaxation time) to use the same notations as Coffey and to compare our results with his results obtained in the case of a purely alternating field. By remarking that $\beta = 1/\tau_I$, and $\gamma = 1/(\tau_I\tau_1)$, we obtain

$$\langle \cos 2\theta \rangle_{\text{st}} = \frac{1}{8}\left(\frac{\mu}{kT}\right)^2 \langle E^2(t)\rangle \left\{ P + 1\Big/\left[1 + 2\left(\frac{E_c}{E_0}\right)^2\right] \times \left[\frac{2(2 - 3\,\omega^2\tau_I\tau_1) + \omega^2\tau_I^2(\omega^2\tau_I\tau_1 + 10\,\tau_1/\tau_I + 11)}{(2 - 3\,\omega^2\tau_I\tau_1)^2 + \omega^2\tau_I^2(-\omega^2\tau_I\tau_1 + 2\,\tau_1/\tau_I + 3)^2} + 2\left(\frac{E_c}{E_0}\right)^2\right]\right\} \tag{4.164a}$$

$$X_2(\omega) = \frac{\Delta n_2(\omega)}{\Delta n_2(0)} = \frac{2}{P+1}\left[\frac{(-3\,\omega^2\tau_I^2 + 2)\mathscr{A}_1 + 7\omega^2\tau_I\tau_1\mathscr{B}_1}{\mathscr{A}_1^2 + \omega^2\tau_1^2\mathscr{B}_1^2} + P\,\frac{(2 - 3\,\omega^2\tau_I\tau_1) + \omega^2\tau_I^2(-2\,\omega^2\tau_I\tau_1 + \tau_1/\tau_I + 6)}{(2 - 3\,\omega^2\tau_I\tau_1)^2 + \omega^2\tau_I^2(-2\,\omega^2\tau_I\tau_1 + \tau_1/\tau_I + 6)^2}\right] \tag{4.164b}$$

$$Y_2(\omega) = \frac{\Delta k_2(\omega)}{\Delta n_2(0)} = \frac{2\omega}{P+1}\left[\frac{(-3\,\omega^2\tau_I^2+2)\mathscr{B}_1\tau_1 - 7\tau_I\mathscr{A}_1}{\mathscr{A}_1^2+\omega^2\tau_1^2\mathscr{B}_1^2}\right.$$
$$\left. + P\tau_I\frac{(\omega^2\tau_I\tau_1+\tau_1/\tau_I+4)}{(2-3\,\omega^2\tau_I\tau_1)^2+\omega^2\tau_I^2(-2\,\omega^2\tau_I\tau_1+\tau_1/\tau_I+6)^2}\right] \tag{4.164c}$$

$$X_1(\omega) = \frac{\Delta n_1(\omega)}{\Delta n_1(0)} = \frac{2}{P+1}\left[\frac{(-5\,\omega^2\tau_I^2+4)\mathscr{A}_1' + 12\,\omega^2\tau_I\tau_1\mathscr{B}_1'}{\mathscr{A}_1'^2+\omega^2\tau_1^2\mathscr{B}_1'^2}\right.$$
$$\left. + \left(P+\frac{1}{2}\right)\frac{4(8-3\,\omega^2\tau_I\tau_1)+2\,\omega^2\tau_I^2(-\omega^2\tau_I\tau_1+2\,\tau_1/\tau_I+12)}{(8-3\,\omega^2\tau_I\tau_1)^2+\omega^2\tau_I^2(-\omega^2\tau_I\tau_1+2\tau_1/\tau_I+12)^2}\right] \tag{4.164d}$$

$$Y_1(\omega) = \frac{\Delta k_1(\omega)}{\Delta n_1(0)} = \frac{2\,\omega}{P+1}\left[\frac{(-5\,\omega^2\tau_I^2+4)\mathscr{B}_1'\tau_1 - 12\,\tau_I\mathscr{A}_1'}{\mathscr{A}_1'^2+\omega^2\tau_1^2\mathscr{B}_1'^2}\right.$$
$$\left. + \left(P+\frac{1}{2}\right)\tau_I\frac{2(\omega^2\tau_I\tau_1+4\,\tau_1/\tau_I+16)}{(8-3\,\omega^2\tau_I\tau_1)^2+\omega^2\tau_I^2(-\omega^2\tau_I\tau_1+2\tau_1/\tau_I+12)^2}\right] \tag{4.164e}$$

where $\Delta n_j(0)$ is the value of $\Delta n_j(\omega)$ when $\omega \to 0$, namely,

$$\Delta n_2(0) = \frac{1}{4}\,\lambda_0^2\tau_I^2\tau_1^2(P+1) \tag{4.165a}$$

$$\Delta n_1(0) = \lambda_0\lambda_c\tau_I^2\tau_1^2(P+1) \tag{4.165b}$$

$$\mathscr{A}_1 = -2\,\omega^6\tau_I^4\tau_1^2 + 2\omega^4\tau_I^2\tau_1^2\left(7+6\frac{\tau_I}{\tau_1}\right)$$
$$-\,\omega^2\tau_I^2\left(2\frac{\tau_1^2}{\tau_I^2}+27\frac{\tau_1}{\tau_I}+18\right)+4 \tag{4.165c}$$

$$\mathscr{B}_1 = 9\,\omega^4\tau_I^3\tau_1 - 9\,\omega^2\tau_I\tau_1\left(1+\frac{11}{3}\frac{\tau_I}{\tau_1}\right)+6\left(1+3\frac{\tau_I}{\tau_1}\right) \tag{4.165d}$$

$$\mathscr{A}_1' = -\omega^6\tau_I^4\tau_1^2 + \omega^4\tau_I^2\tau_1^2\left(13+15\frac{\tau_I}{\tau_1}\right) - \omega^2\tau_I^2\left(4\frac{\tau_1^2}{\tau_I^2}+60\frac{\tau_1}{\tau_I}+36\right)+16 \tag{4.165e}$$

$$\mathscr{B}_1' = 6\omega^4\tau_I^3\tau_1 - \omega^2\tau_I\tau_1\left(12 + 55\frac{\tau_I}{\tau_1}\right) + 4\left(5 + 12\frac{\tau_I}{\tau_1}\right) \tag{4.165f}$$

$X_j(\omega)$ and $Y_j(\omega)$ are the real and imaginary parts of the complex birefringence, normalized to unity. We have verified that when $\tau_I \to 0$, the results of the Debye disk model are found again (solution of the Smoluchowski equation).

2. *Cole–Cole Plots*

It is necessary to remark that the truncation at $n = 3$ implies that only small inertial effects are considered. These effects are characterized by the dimensionless parameter $\tau_I/\tau_1 \ll 1$, which is analogous to the quality factor Q defined in electrically resonant systems, so that for small values of τ_I/τ_1, this corresponds to the study of a very strongly damped oscillator. When inertia is taken into account, the Cole–Cole diagrams $Y_j(X_j)$ are deformed, particularly in the high-frequency region, as well as for small values [27] of τ_I/τ_1. Figures 4.54–4.61 illustrate the modification of these diagrams for both harmonics when one passes from $\tau_I/\tau_1 = 0$ (zero inertia) to $\tau_I/\tau_1 = 10^{-2}$. These figures are given for P values greater

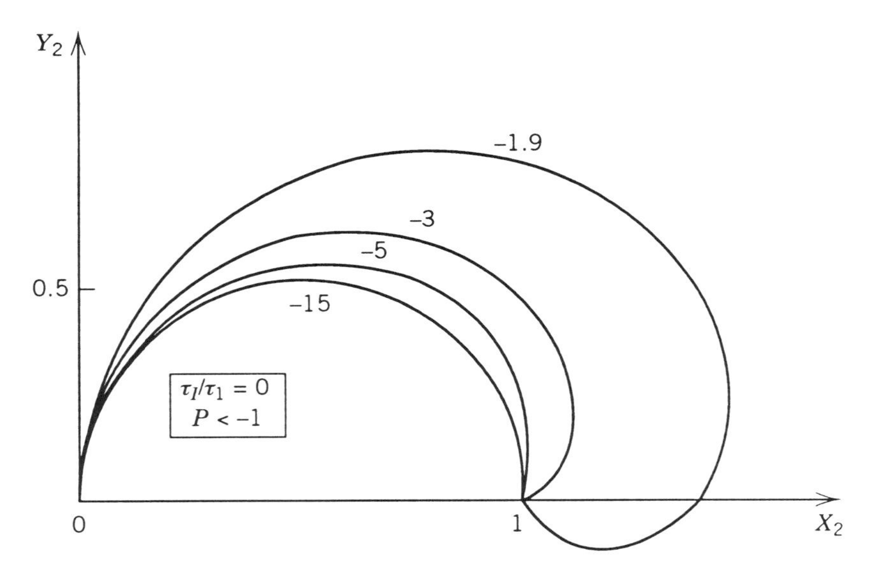

Figure 4.54. Plots of the imaginary part Y_2 versus the real part X_2 of the normalized birefringence function, for $P < -1$ and $\tau_I/\tau_1 = 0$. Numbers above or below plots represent various values of P.

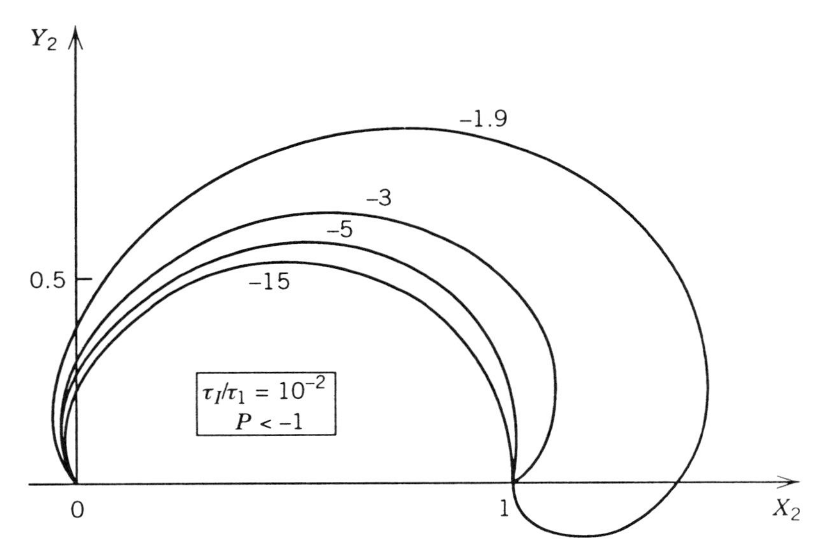

Figure 4.55. Plots of the imaginary part Y_2 versus the real part X_2 of the normalized birefringence function, for $P < -1$ and $\tau_I/\tau_1 = 10^{-2}$. Numbers above or below plots represent various values of P.

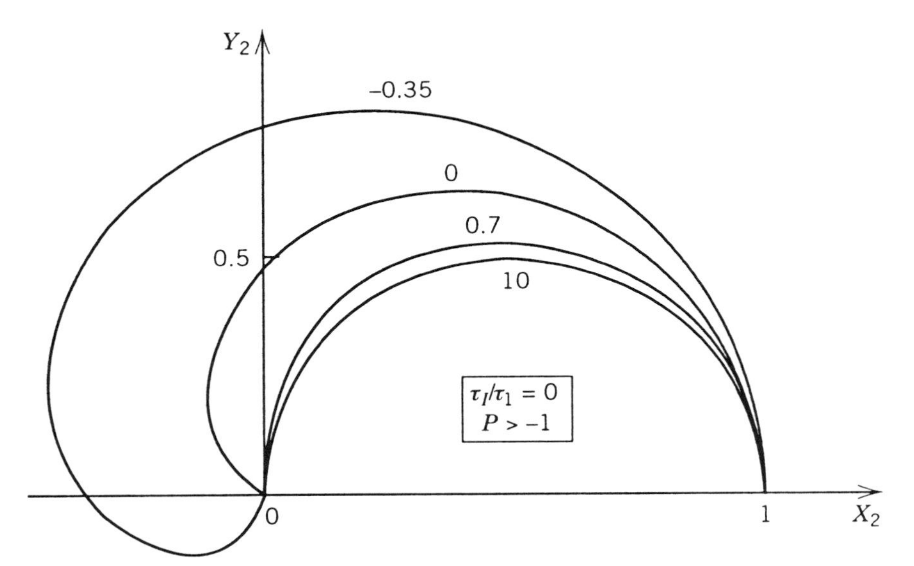

Figure 4.56. Plots of the imaginary part Y_2 versus the real part X_2 of the normalized birefringence function, for $P > -1$ and $\tau_I/\tau_1 = 0$. Numbers above or below plots represent various values of P.

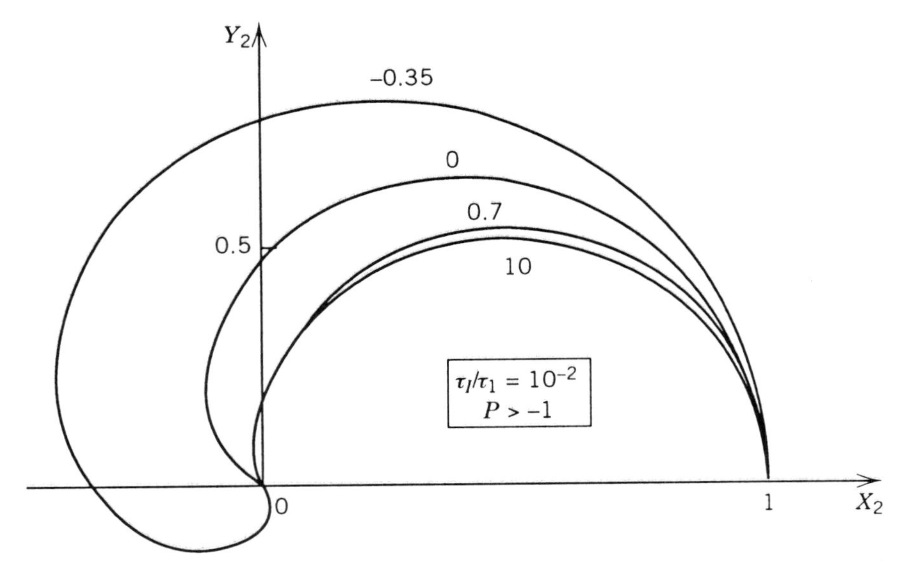

Figure 4.57. Plots of the imaginary part Y_2 versus the real part X_2 of the normalized birefringence function, for $P > -1$ and $\tau_I/\tau_1 = 10^{-2}$. Numbers above or below plots represent various values of P.

or less than -1 on account of the normalization factor of the harmonic components, which is proportional (here too) to $P + 1$. Déjardin chose a large spread of P values to obtain various curves but without seeking to reproduce those of Figs. 4.33, 4.34, 4.37, and 4.38 deduced from the solution of the Smoluchowski equation, because the 2D and 3D problems are not directly comparable with each other. We know, in effect, that the Debye relaxation time equal to ξ/kT for a disk becomes twice as small when one considers motion in 3D space.

Unfortunately, in searching the literature we have not found experimental measurements of the dynamic Kerr effect where inertial effects have not been ignored, that is, measurements that would eventually permit application of the dispersion relations [Eqs. (4.164)]. In addition to the apparatus problems that we already indicated, this is probably due to the difficulties in obtaining the birefringence modulus $|\Delta\mathbf{n}_j|$ that, for high ω values, is very small and indeed may be of the same order of magnitude as the experimental accuracy.

This is not true of the variation of the phase angle $\theta_j(\omega) = \tan^{-1}(Y_j/X_j)$, which we plotted in Figs. 4.62–4.71. This is much more important in the high-frequency region. In effect, the limiting values of these angles including inertia are in most cases twice as large as those with zero

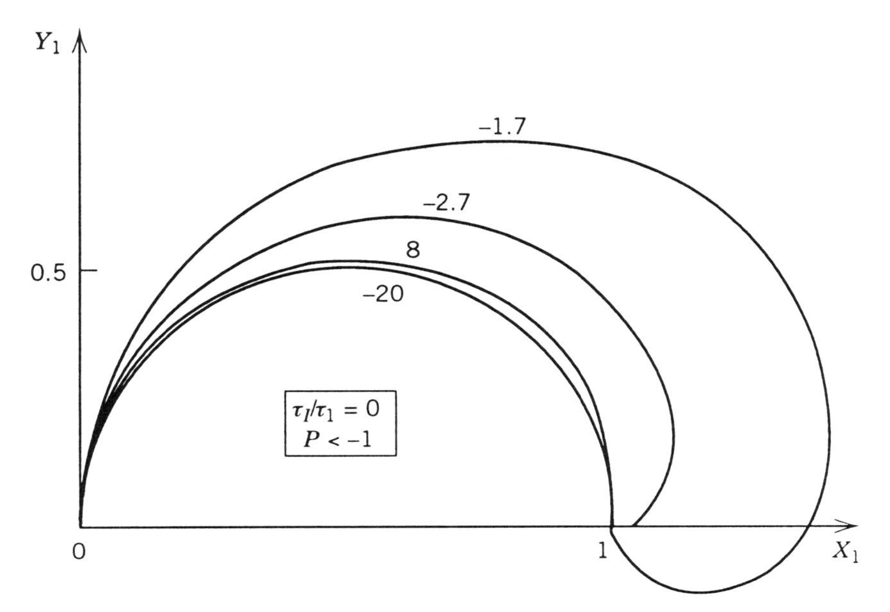

Figure 4.58. Plots of the imaginary part Y_1 versus the real part X_1 of the normalized birefringence function, for $P < -1$ and $\tau_I/\tau_1 = 0$. Numbers above or below plots represent various values of P.

inertia. It seems that such variations should be easily accessible and determined by experiment.

3. An Observable Consequence of Molecular Rotational Inertia from Phase Angles

With the object of practical use of the analytic expressions for the phase angle θ_j, we shall now study the plots of θ_j as a function of ω for several values of P and τ_I/τ_1 [74].

1. Case $j = 2$

We know that this case corresponds to the response of a purely alternating field. For high P values, either positive or negative, for which the induced dipole moment greatly exceeds the permanent one, the limiting value attained by $\theta_2(\omega)$ is $\pi/2$ for $\tau_I/\tau_1 = 0$ and π for $0 < \tau_I/\tau_1 \ll 1$. The curve of $P = 10$ shows this result (Fig. 4.62). For purely polar molecules ($P = 0$, Fig. 4.63), the limiting value goes over to π for $\tau_I/\tau_1 = 0$ to 2π for $0 < \tau_I/\tau_1 \ll 1$. For values of P close to the last case

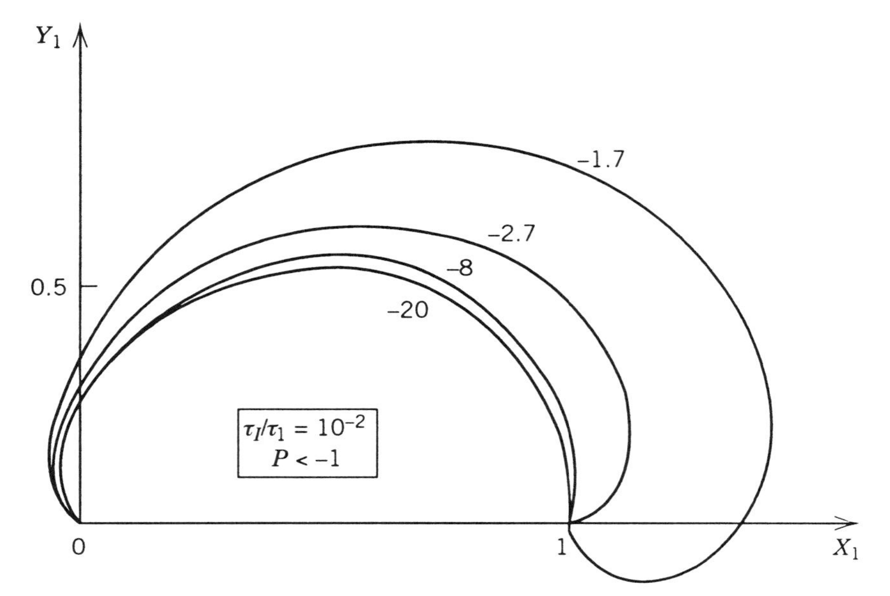

Figure 4.59. Plots of the imaginary part Y_1 versus the real part X_1 of the normalized birefringence function, for $P < -1$ and $\tau_I/\tau_1 = 10^{-2}$. Numbers above or below plots represent various values of P.

considered ($P = 0.1$, Fig. 4.64), the limiting values become equal to $\pi/2$ and π, respectively, but the curves exhibit a maximum somewhere between $\pi/2$ and π. Finally, around the value $P = -1$, we see in Figs. 4.65 and 4.66 ($P = -0.9$ and $P = -1.1$) that the responses are very different. For $P < -1$, $\theta_2(\omega)$ reaches $3\pi/2$ when inertia is discarded, and 2π with inertia. For $P > -1$, $\theta_2(\omega)$ is at high frequencies equal to $\pi/2$ and π, respectively. In the latter case, it passes through a negative minimum value that is situated at low frequencies.

2. Case $j = 1$

Figures 4.67–4.71 describe the evolution of $\theta_1(\omega)$ with $0 \leq \tau_I/\tau_1 \leq 0.012$, just as in the case where $j = 2$, for a particular value of P, and analogous conclusions may be drawn. However, the case of purely polar molecules now corresponds to $P = -0.5$ in place of $P = 0$, as shown in Eqs. (4.164d)

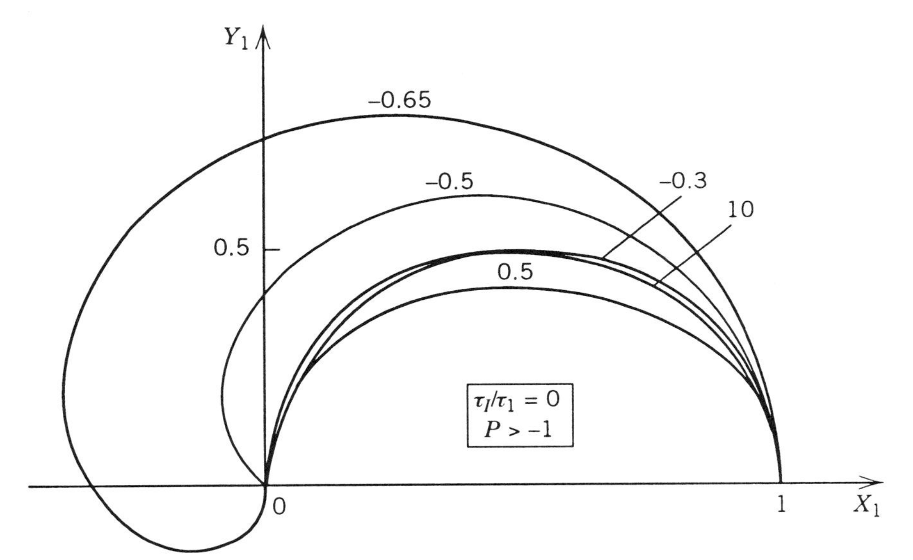

Figure 4.60. Plots of the imaginary part Y_1 versus the real part X_1 of the normalized birefringence function, for $P > -1$ and $\tau_I/\tau_1 = 0$. Numbers above or below plots represent various values of P.

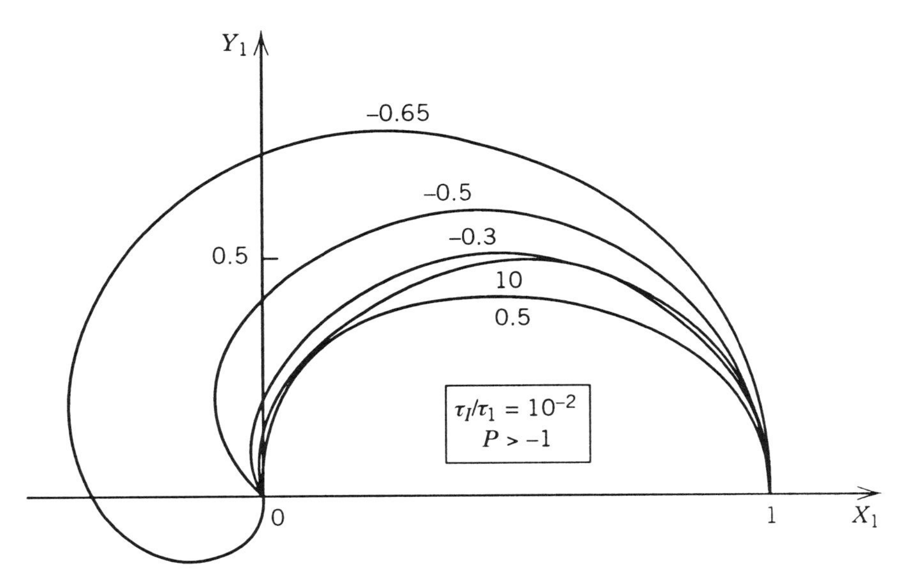

Figure 4.61. Plots of the imaginary part Y_1 versus the real part X_1 of the normalized birefringence function, for $P > -1$ and $\tau_I/\tau_1 = 10^{-2}$. Numbers above or below plots represent various values of P.

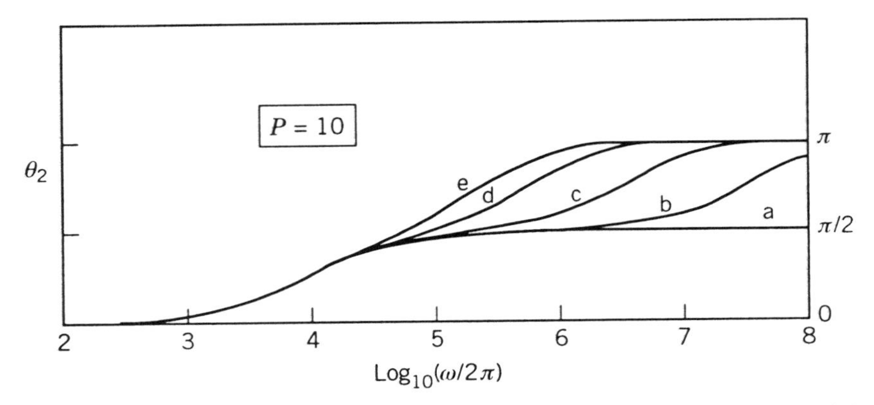

Figure 4.62. Plots of the phase angle θ_2 versus $\log_{10}(\omega/2\pi)$ for different values of the inertial parameter τ_I/τ_1 and $P = 10$. (a): $\tau_I/\tau_1 = 0$; (b): $\tau_I/\tau_1 = 8 \times 10^{-5}$; (c): $\tau_I/\tau_1 = 8 \times 10^{-4}$; (d): $\tau_I/\tau_1 = 0.005$; (e): $\tau_I/\tau_1 = 0.012$. (D has been chosen arbitrarily equal to $4 \times 10^4\ \text{s}^{-1}$).

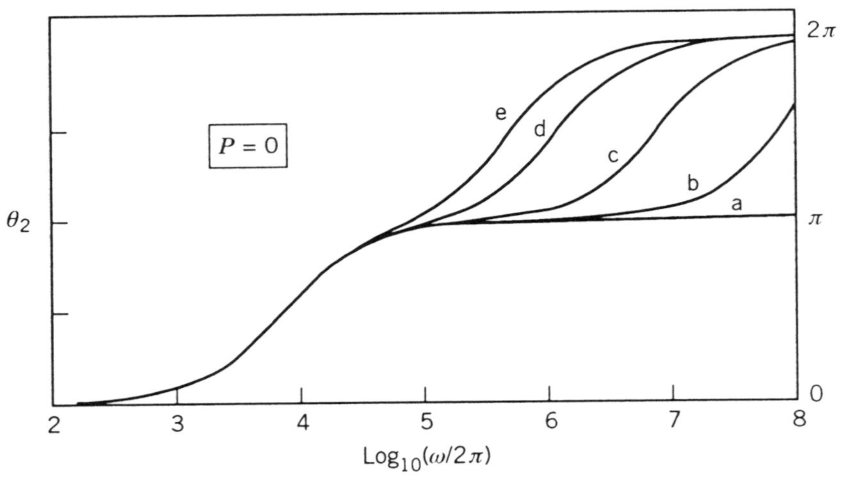

Figure 4.63. Plots of the phase angle θ_2 versus $\log_{10}(\omega/2\pi)$ for different values of the inertial parameter τ_I/τ_1 and $P = 0$. (a): $\tau_I/\tau_1 = 0$; (b): $\tau_I/\tau_1 = 8 \times 10^{-5}$; (c): $\tau_I/\tau_1 = 8 \times 10^{-4}$; (d): $\tau_I/\tau_1 = 0.005$; (e): $\tau_I/\tau_1 = 0.012$. (D has been chosen arbitrarily equal to $4 \times 10^4\ \text{s}^{-1}$).

and (4.164e). It is remarkable that these limiting P values are quite identical to those obtained in the 3D study with zero inertia. We note that the asymptotic value of $\theta_1(\omega)$ is attained less rapidly than that of $\theta_2(\omega)$. In other words, inertial effects manifest themselves more quickly in the high-frequency domain for the second harmonic component than they do

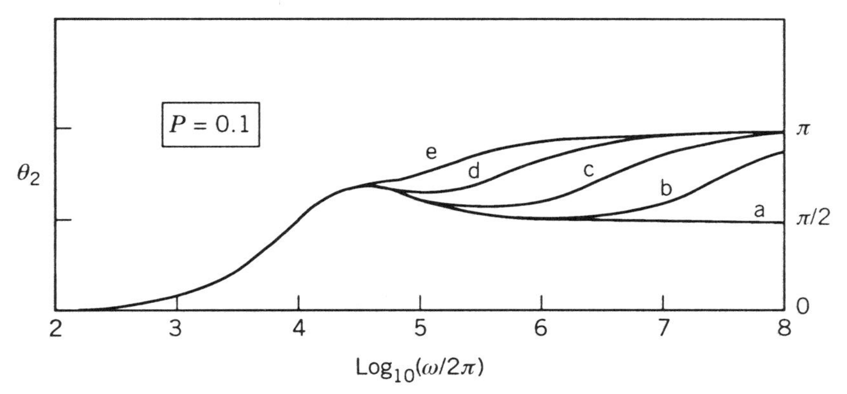

Figure 4.64. Plots of the phase angle θ_2 versus $\log_{10}(\omega/2\pi)$ for different values of the inertial parameter τ_I/τ_1 and $P = 0.1$. (a): $\tau_I/\tau_1 = 0$; (b): $\tau_I/\tau_1 = 8 \times 10^{-5}$; (c): $\tau_I/\tau_1 = 8 \times 10^{-4}$; (d): $\tau_I/\tau_1 = 0.005$; (e): $\tau_I/\tau_1 = 0.012$. (D has been chosen arbitrarily equal to $4 \times 10^4\ \mathrm{s}^{-1}$).

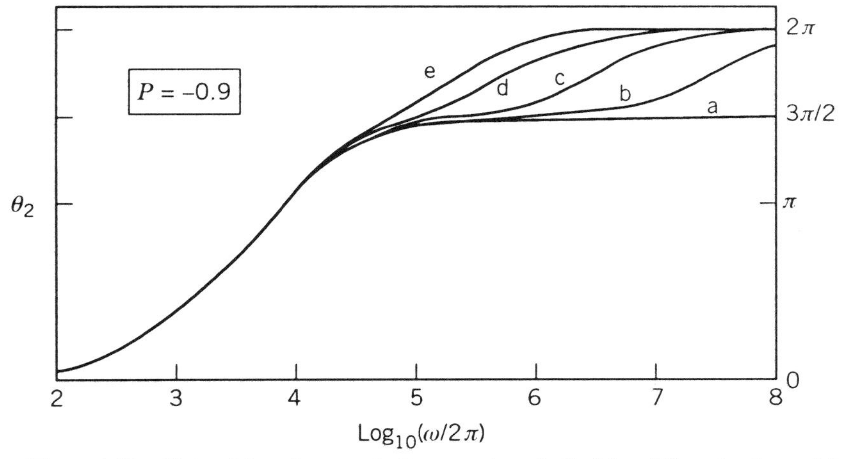

Figure 4.65. Plots of the phase angle θ_2 versus $\log_{10}(\omega/2\pi)$ for different values of the inertial parameter τ_I/τ_1 and $P = -0.9$. (a): $\tau_I/\tau_1 = 0$; (b): $\tau_I/\tau_1 = 8 \times 10^{-5}$; (c): $\tau_I/\tau_1 = 8 \times 10^{-4}$; (d): $\tau_I/\tau_1 = 0.005$; (e): $\tau_I/\tau_1 = 0.012$. (D has been chosen arbitrarily equal to $4 \times 10^4\ \mathrm{s}^{-1}$).

with the first one. This may be explained by the fact that the dc field has a much greater marked effect and causes a more reduced mobility of the molecules, with a consequent higher friction coefficient. Note that we have obtained a similar behavior in the phase angle φ' in dielectric relaxation including inertial effects, as shown in Fig. 4.5.

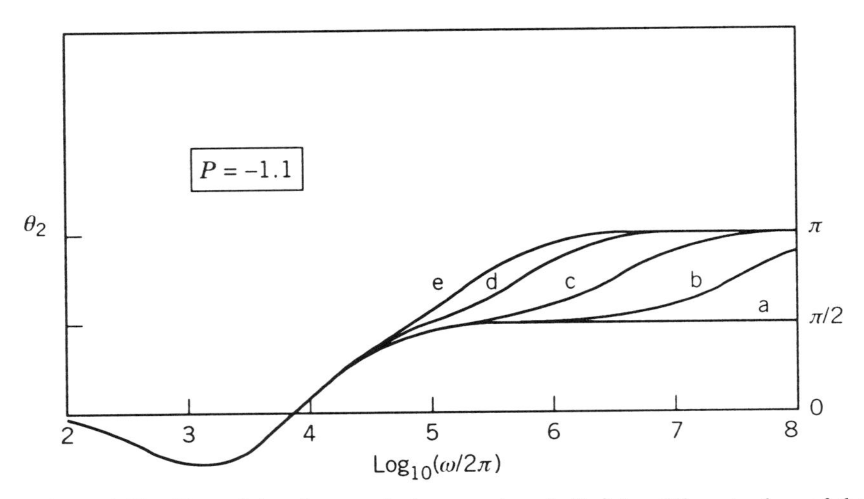

Figure 4.66. Plots of the phase angle θ_2 versus $\log_{10}(\omega/2\pi)$ for different values of the inertial parameter τ_I/τ_1 and $P = -1.1$. (a): $\tau_I/\tau_1 = 0$; (b): $\tau_I/\tau_1 = 8 \times 10^{-5}$; (c): $\tau_I/\tau_1 = 8 \times 10^{-4}$; (d): $\tau_I/\tau_1 = 0.005$; (e): $\tau_I/\tau_1 = 0.012$. (D has been chosen arbitrarily equal to $4 \times 10^4\ \text{s}^{-1}$).

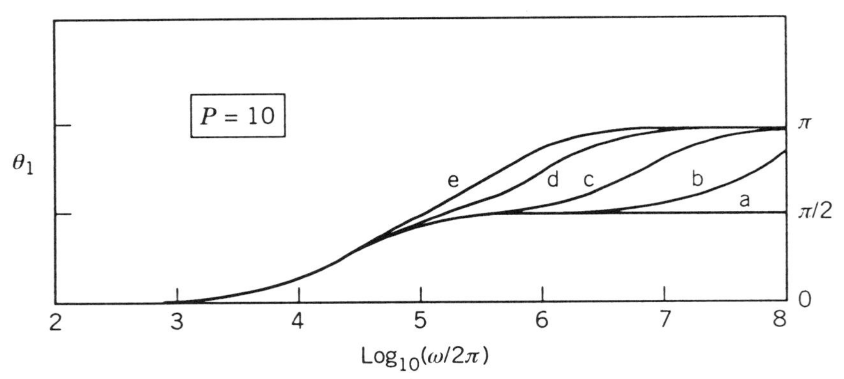

Figure 4.67. Plots of the phase angle θ_1 versus $\log_{10}(\omega/2\pi)$ for different values of the inertial parameter τ_I/τ_1 and $P = 10$. (a): $\tau_I/\tau_1 = 0$; (b): $\tau_I/\tau_1 = 8 \times 10^{-5}$; (c): $\tau_I/\tau_1 = 8 \times 10^{-4}$; (d): $\tau_I/\tau_1 = 0.005$; (e): $\tau_I/\tau_1 = 0.012$. (D has been chosen arbitrarily equal to $4 \times 10^4\ \text{s}^{-1}$).

To conclude, it appears from that the study of the dynamic birefringence by consideration either of the transient regime or of the stationary regime, leads us to important measurements of microscopic parameters such as the permanent dipole moment and the induced dipole one. Even if inertial effects are ignored, the Smoluchowski equation for which we

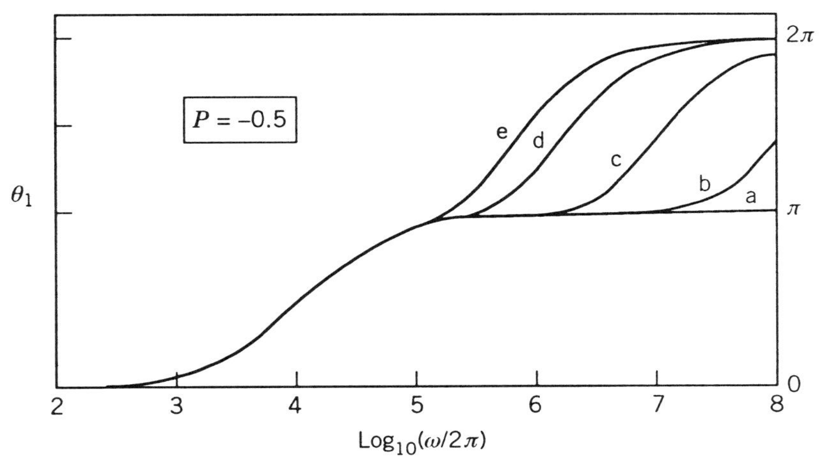

Figure 4.68. Plots of the phase angle θ_1 versus $\log_{10}(\omega/2\pi)$ for different values of the inertial parameter τ_I/τ_1 and $P = -0.5$. (a): $\tau_I/\tau_1 = 0$; (b): $\tau_I/\tau_1 = 8 \times 10^{-5}$; (c): $\tau_I/\tau_1 = 8 \times 10^{-4}$; (d): $\tau_I/\tau_1 = 0.005$; (e): $\tau_I/\tau_1 = 0.012$. (D has been chosen arbitrarily equal to $4 \times 10^4\ \text{s}^{-1}$).

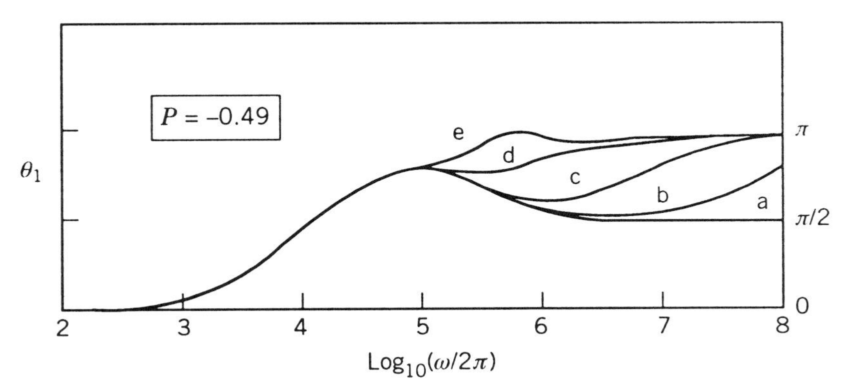

Figure 4.69. Plots of the phase angle θ_1 versus $\log_{10}(\omega/2\pi)$ for different values of the inertial parameter τ_I/τ_1 and $P = -0.49$. (a): $\tau_I/\tau_1 = 0$; (b): $\tau_I/\tau_1 = 8 \times 10^{-5}$; (c): $\tau_I/\tau_1 = 8 \times 10^{-4}$; (d): $\tau_I/\tau_1 = 0.005$; (e): $\tau_I/\tau_1 = 0.012$. (D has been chosen arbitrarily equal to $4 \times 10^4\ \text{s}^{-1}$).

have proposed an original method of solution, remains of interest and permits one to obtain these parameters by complementing information gained from dielectric relaxation. We have shown that the second-order response of the dynamic Kerr effect for a dc field coupled to an alternating field is rendered by two harmonic terms that express the nonlinearity of this response. We are interested in the two coupled fields

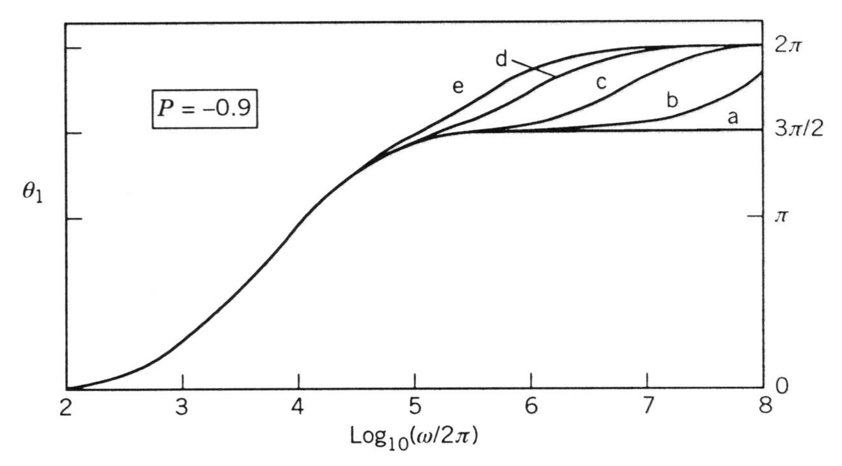

Figure 4.70. Plots of the phase angle θ_1 versus $\log_{10}(\omega/2\pi)$ for different values of the inertial parameter τ_I/τ_1 and $P = -0.9$. (a): $\tau_I/\tau_1 = 0$; (b): $\tau_I/\tau_1 = 8 \times 10^{-5}$; (c): $\tau_I/\tau_1 = 8 \times 10^{-4}$; (d): $\tau_I/\tau_1 = 0.005$; (e): $\tau_I/\tau_1 = 0.012$. (D has been chosen arbitrarily equal to $4 \times 10^4\ s^{-1}$).

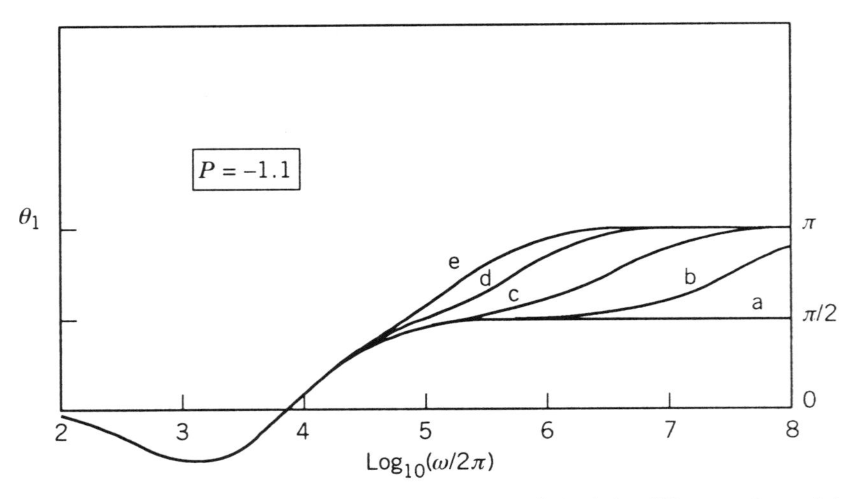

Figure 4.71. Plots of the phase angle θ_1 versus $\log_{10}(\omega/2\pi)$ for different values of the inertial parameter τ_I/τ_1 and $P = -1.1$. (a): $\tau_I/\tau_1 = 0$; (b): $\tau_I/\tau_1 = 8 \times 10^{-5}$; (c): $\tau_I/\tau_1 = 8 \times 10^{-4}$; (d): $\tau_I/\tau_1 = 0.005$; (e): $\tau_I/\tau_1 = 0.012$. (D has been chosen arbitrarily equal to $4 \times 10^4\ s^{-1}$).

for two reasons, which will be very beneficial for the measurements of these effects. The Kramers–Kronig relations, which we extended to treat these two harmonic terms and which we applied to linoleic acid and to pentachlorodiphenyl, have given good quality information, clearly better than that obtained by purely experimental consideration. So, they offer much more useful information than dielectric relaxation alone. In addition, we have shown that small inertial effects clearly manifest themselves in the curves representing the evolution of the phase angle with frequency, when they would be almost imperceptible in the Cole–Cole diagrams.

IV. CONCLUSIONS

In the study of dielectric relaxation, which we described in Section II.B, we interpreted some Cole–Cole diagrams with their center situated below the ε' axis (Fig. 4.7) as a result of the contribution of several oscillators that follow the generalized complex relation given in Eq. (4.34). This gives interesting results that may be exploited experimentally, as we have seen in its application to hemoglobin S.

In the simplest case, we consider two oscillators of the Debye type with negligible inertia and high $\gamma = \xi/(2IkT)^{0.5}$. To obtain maximum information, we attempted to find the polar equation of the relevant $\varepsilon''(\varepsilon')$ curve. To do this, we consider the normalized sums of the real ε' and imaginary ε'' parts of the complex dielectric constant $\boldsymbol{\varepsilon}$, such that

$$X' = \frac{\varepsilon_1' + \varepsilon_2' - (\varepsilon_{\infty 1} + \varepsilon_{\infty 2})}{S_1 + S_2} = \frac{S_1'}{1 + \omega^2\tau_1^2} + \frac{S_2'}{1 + \omega^2\tau_2^2} \tag{4.166a}$$

$$Y' = \frac{\varepsilon_1'' + \varepsilon_2''}{S_1 + S_2} = \frac{S_1'\omega\tau_1}{1 + \omega^2\tau_1^2} + \frac{S_2'\omega\tau_2}{1 + \omega^2\tau_2^2} \tag{4.166b}$$

where

$$S_1 = \varepsilon_{01} - \varepsilon_{\infty 1} \qquad S_2 = \varepsilon_{02} - \varepsilon_{\infty 2} \tag{4.167a}$$

$$S_1' = \frac{S_1}{S_1 + S_2} \qquad S_2' = \frac{S_2}{S_1 + S_2} \tag{4.167b}$$

In Eqs. (4.166) and (4.167), the subscripts 1 and 2 refer to the first and the second oscillator. The normalized curve is plotted in the reference frame $X'O'Y'$, as shown in Fig. 4.72.

By calculations similar to those that we have done for noninertial

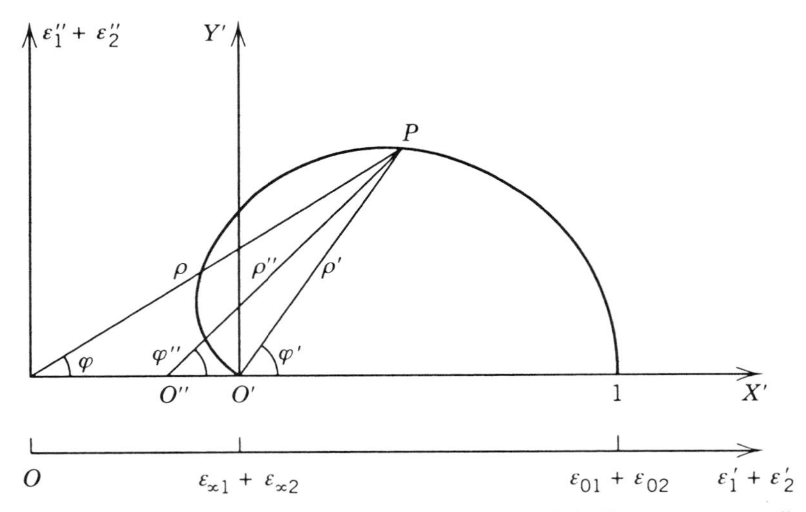

Figure 4.72. Definition of reference axes for the Cole–Cole diagram corresponding to the normalized sum Y' versus X' or to the non-normalized sum $(\varepsilon_1'' + \varepsilon_2'')$ versus $(\varepsilon_1' + \varepsilon_2')$ of dielectric constants of two oscillators.

birefringence, we obtain the following polar equation [75]

$$\rho'' = \left(\frac{1}{2} - C\right) \cos \varphi'' \pm \frac{1}{2}\sqrt{1 - \left(\frac{a-1}{a+1}\right)^2 \sin^2 \varphi''} \qquad (4.168)$$

where φ'' is the polar angle, and ρ'' is the radius from the origin O'', such that

$$\overline{O'O''} = C = \frac{S_1 a + S_2}{(a+1)(S_1 + S_2)} \qquad a = \frac{\tau_2}{\tau_1} \qquad (4.169)$$

Equation (4.168) characterizes a quasiconchoid of circle with an appearance analogous to that encountered all along in the study of the complex birefringence functions. Starting from this similarity between two well-distinguished physical phenomena, we sought to establish relations between birefringence and the dielectric constant.

Under the action of an electric field, we recall that an isotropic fluid medium becomes anisotropic and behaves like a homogeneous uniaxial medium where the optical axis is parallel to the direction of the electric field. Two dielectric constants are sufficient to describe it, namely, $\varepsilon_{\parallel}$ and $\varepsilon_{\perp}$. The first corresponds to vibrations of the electric field parallel to the optical axis and the second to vibrations perpendicular to this axis. These constants are generally complex so as to take account of the absorption, that is, energy dissipation arising from molecular friction in the medium during the rotational–vibrational movement. One can thus define extraordinary and ordinary complex refractive indexes $\mathbf{n}_e = \boldsymbol{\varepsilon}_{\|}^{0.5}$ and $\mathbf{n}_0 = \boldsymbol{\varepsilon}_{\perp}^{0.5}$. Consequently, the complex birefringence is exactly given by the difference $\mathbf{n}_e - \mathbf{n}_0$, denoted by $\Delta\mathbf{n}$.

We have often recalled that the orientational factor characteristic of Kerr effect relaxation is given by the expected value of the second Legendre polynomial while the orientational polarization that describes the dielectric relaxation is expressed by the expected value of the first Legendre polynomial. Rigorously speaking, it is not precisely the birefringence $|\Delta\mathbf{n}(t)|$, which is proportional to $\langle P_2(u)\rangle(t)$, as indicated by Eqs. (4.99) and (4.103), but really the modulus of the difference between the squares of the extraordinary and ordinary refractive indexes [52]. So, we can write

$$|\mathbf{n}_e^2(t) - \mathbf{n}_0^2(t)| = K\langle P_2(u)\rangle(t) \tag{4.170}$$

where K is a constant of proportionality. If $\mathbf{n}$ denotes the mean value of $\mathbf{n}_e$ and $\mathbf{n}_0$, $\mathbf{n} = (\mathbf{n}_e + \mathbf{n}_0)/2$, Eq. (4.170) becomes

$$|2\,\mathbf{n}\,\Delta\mathbf{n}(t)| = K\langle P_2(u)\rangle(t) \tag{4.171}$$

Moreover, one has

$$2\,\mathbf{n}\,\Delta\mathbf{n} = \mathbf{n}_e^2 - \mathbf{n}_0^2 = \boldsymbol{\varepsilon}_{\|} - \boldsymbol{\varepsilon}_{\perp} = \Delta\boldsymbol{\varepsilon} = \Delta\varepsilon' - i\,\Delta\varepsilon'' \tag{4.172}$$

Separating the real and imaginary parts, one obtains

$$\begin{aligned} 2\,n\,\Delta n - 2\,k\,\Delta k &= \Delta\varepsilon' \\ 2\,n\,\Delta k + 2\,k\,\Delta n &= \Delta\varepsilon'' \end{aligned} \tag{4.173}$$

where

$$n = \frac{n_e + n_0}{2} \qquad k = \frac{k_e + k_0}{2} \qquad \Delta n = n_e - n_0 \qquad \Delta k = k_e - k_0 \tag{4.174}$$

We see, following Eq. (4.172), that the expression of the real Δn and imaginary Δk parts of the birefringence is correctly expressed provided that the mean index n is a constant and the mean absorption k is negligible or zero. This is so on using a dc field or an impulse field, at very low frequencies, and in any region far from the absorption birefringence zone. Under these conditions, Eq. (4.173) reduces to

$$\begin{aligned} 2\,n\,\Delta n &= \Delta\varepsilon' \\ 2\,n\,\Delta k &= \Delta\varepsilon'' \end{aligned} \tag{4.175}$$

This result is simple and extremely interesting because it permits one to understand the above-mentioned procedure where we have shown that the sum of two oscillators may be represented in a Cole–Cole diagram by a quasiconchoid of a circle. Here, in birefringence, we no longer refer to the sum but to the difference between the permittivities, which is of no great influence on the final result. Equation (4.168) still holds and it suffices to replace S_2 by $-S_2$ in Eq. (4.169).

In this way, we can compare the relations coming from the difference of two oscillators with those established for Kerr effect relaxation. Furthermore, since the mean value of the second Legendre polynomial expresses only proportionality to the birefringence, we shall use normalized expressions. It is necessary to note carefully, from a physical point of view, that the comparison is possible in so far as the electric field, which allows one to make the measurement of the dielectric constant, plays the role of orientating the field for the birefringence, as well.

If the field is purely sinusoidal, the relations pertaining to the difference of the dielectric responses of two Debye type oscillators are

$$X'' = \frac{\varepsilon_1' - \varepsilon_2' - (\varepsilon_{\infty 1} - \varepsilon_{\infty 2})}{S_1 - S_2} = \frac{S_1''}{1 + \omega^2\tau_1^2} - \frac{S_2''}{1 + \omega^2\tau_2^2} \tag{4.176a}$$

$$Y'' = \frac{\varepsilon_1'' - \varepsilon_2''}{S_1 - S_2} = \frac{S_1''\omega\tau_1}{1 + \omega^2\tau_1^2} - \frac{S_2''\omega\tau_2}{1 + \omega^2\tau_2^2} \tag{4.176b}$$

where the subscripts 1 and 2 denote parallel and perpendicular vibrations, respectively, and

$$S_1'' = \frac{S_1}{S_1 - S_2} \qquad S_2'' = \frac{S_2}{S_1 - S_2} \tag{4.177}$$

The $2\,\omega$ birefringence response may be obtained from Eqs. (4.107a) and (4.107b), which we write in the form

$$X_{2\omega} = \frac{1}{P+1}\left[\frac{3}{1 + \omega^2\tau^2} + \frac{P - 2}{1 + \frac{4}{9}\,\omega^2\tau^2}\right] \tag{4.178a}$$

$$Y_{2\omega} = \frac{1}{P+1}\left[\frac{3\omega\tau}{1+\omega^2\tau^2} + \frac{(P-2)\frac{2}{3}\omega\tau}{1+\frac{4}{9}\omega^2\tau^2}\right] \qquad (4.178b)$$

where $\tau = 1/2D$ is the Debye relaxation time in 3D space. For the purpose of comparison, we deduce the following two equivalent relations

$$\tau_1 = \tau \qquad \tau_2 = \frac{2}{3}\tau \qquad \frac{\tau_1}{\tau_2} = \frac{\tau_\parallel}{\tau_\perp} = \frac{3}{2} \qquad (4.179a)$$

$$P = 2 - 3\frac{S_2}{S_1} = 2 - 3\frac{S_\perp}{S_\parallel} \qquad (4.179b)$$

Equations (4.176) are in agreement with many previous observations. We mention the longstanding ones of Raman and Sirkar [76] and of Kitchin and Müller [77] in 1928. They stated that the frequency range of the abnormal Kerr effect is the same as that of the anomalous dispersion of the dielectric constant. This seems to confirm that the anisotropy induced by the orientating electric field results from the formation of two groups of molecules vibrating with one perpendicular to the other. When the electric field of the laser beam allowing the measurement is parallel to the optical axis, the corresponding active group is more stable than the other active group vibrating in the perpendicular direction, because the time-constant $\tau_\parallel$ is greater than $\tau_\perp$. In passing, one must bear in mind that the orientating and investigating (laser) fields are two distinct entities.

Given a temperature, Eq. (4.179b) links the parameter P, which measures the ratio between induced and permanent dipole moments to the oscillator forces. It seems to us to be very important. It allows us, for example, to explain why the Cole–Cole diagram obtained for $P = 2$ in electric birefringence is exactly a semicircle, which we have simply stated before. This results from the fact that, in this case, the perpendicular oscillator is inactive, $S_\perp = 0$, or $\varepsilon_{0\perp} = \varepsilon_{\infty\perp}$. The parallel oscillator solely describes the birefringence, and its diagram is identical to that of Debye relaxation. In the same way, when P tends towards infinity, a semicircular plot is also obtained. This may be explained by the noncontribution of the parallel oscillator, $S_\parallel = 0$.

Then, we can also consider expressions for coupled fields (ac plus dc fields). We compare the parametric equations [Eqs. (4.176)]—for which we have not made any hypothesis about oscillator forces—with Eqs. (4.106)] related to the first harmonic component in ω which we rewrite in the normalized form

$$X_\omega = \frac{1}{P+1}\left[\frac{\frac{3}{4}}{1+\omega^2\tau^2} + \frac{P+\frac{1}{4}}{1+\frac{\omega^2\tau^2}{9}}\right] \quad (4.180a)$$

$$Y_\omega = \frac{1}{P+1}\left[\frac{\frac{3}{4}\omega\tau}{1+\omega^2\tau^2} + \frac{\left(P+\frac{1}{4}\right)\frac{\omega\tau}{3}}{1+\frac{\omega^2\tau^2}{9}}\right] \quad (4.180b)$$

The equivalence relations are now

$$\tau_1 = \tau \qquad \tau_2 = \frac{\tau}{3} \qquad \text{or} \qquad \frac{\tau_1}{\tau_2} = \frac{\tau_\parallel}{\tau_\perp} = 3 \quad (4.181a)$$

$$P = -\frac{1}{4} - \frac{3}{4}\frac{S_2}{S_1} = -\frac{1}{4} - \frac{3}{4}\frac{S_\perp}{S_\parallel} \quad (4.181b)$$

Equation (4.181a) shows a great stability for the population vibrating parallel to the field, for now we have $\tau_\parallel = 3\,\tau_\perp$. This is mainly due to the presence of the dc field. Equation (4.181b) allows one to understand, due to the inactivity of the perpendicular oscillator, the reason why we have a semicircle when $P = -\frac{1}{4}$ for the first harmonic of the birefringence. These encouraging results, however, give place to some other questions and prospects.

The expression for P is unique, and so Eqs. (4.179b) and (4.181b) show that the oscillator forces $S_\parallel$ and $S_\perp$ do not have the same form for two harmonic terms. This result is qualitatively evident since two types of populations vibrating parallel and perpendicular to the direction of the coupled fields are inevitably affected by the presence or absence of the dc field. It would be very interesting to know how the oscillator forces relate to the microscopic parameters. But, to our knowledge, no one has included the induced moment in the oscillator forces for either the alternating regime or for the coupled fields. We have at our disposal only the classical relation $S = N\mu^2/3kT$, which is not sufficient for our problem.

With the aid of these two relations we can at least indicate that the limiting case $P = -1$ corresponds to $S_\parallel = S_\perp$ for both harmonics. On the other hand, on considering negative and positive P values, it is understood that one of the two oscillator forces may become negative. This apparently surprising fact may be explained by the match existing between the permanent moment and the induced one. In effect, taking

$S_{\|} = N\mu^2/3kT$ in the case of a purely alternating field, we deduce that

$$S_{\perp} = \frac{N}{9}\left(\frac{2\mu^2}{kT} - \Delta\alpha\right) \tag{4.182}$$

an expression that may be negative. In this case, one might as well say that $S_{\perp}$ is positive, and that the birefringence is rendered by the sum of two oscillators. Such a hypothesis must be regarded with some caution and will not be applied to coupled fields. It is possible, nevertheless, in the entire domain where Eq. (4.175) remains valid, to include inertia. In effect, if one replaces Debye-type relations by oscillator relations that include inertia, one obtains a new formulation of the birefringence.

A final problem arises when the frequency range is enlarged, particularly including the absorption zone, which is as important as inertia is taken into account. In that region, the variation of the optical refractive indexes n_e and n_0 (or $n_{\|}$ and $n_{\perp}$) is significant and their mean values are not constant. It follows that the relations $2n\,\Delta n$ and $2n\,\Delta k$, that is, the expected value of the second Legendre polynomial, do not really reflect the birefringence. On account of this, only Δn and Δk must be considered. It is possible that this fact explains some disparity of the results that we found between both harmonics for P and D for linoleic acid and pentachlorodiphenyl, because the experimental study [63] of these liquids includes the anomalous dispersion region for the birefringence.

We know that

$$n = [\tfrac{1}{2}\,\varepsilon' + \tfrac{1}{2}(\varepsilon'^2 + \varepsilon''^2)^{1/2}]^{1/2} \tag{4.183a}$$

$$k = [-\tfrac{1}{2}\,\varepsilon' + \tfrac{1}{2}\,(\varepsilon'^2 + \varepsilon''^2)^{1/2}]^{1/2} \tag{4.183b}$$

Thus, in order to pursue the analogy between the birefringence and the dielectric constant, we rigorously write

$$\Delta n = n_{\|} - n_{\perp} \tag{4.184a}$$

$$\Delta k = k_{\|} - k_{\perp} \tag{4.184b}$$

where $n_{\|}$, $n_{\perp}$, $k_{\|}$, and $k_{\perp}$ may be calculated from Eqs. (4.183).

In conclusion we hope that this work may have successfully fulfilled the objectives that we set for ourselves. Essentially, we have seen that the phase angle calculated in accordance with inversion of the Kramers–Kronig relations leads us in all cases to fruitful information, which highlights inertial effects. It appears that the oscillator relations defining the dielectric constant are relations common to all states of matter. Here, the parameter γ, which reflects the duality between the friction coefficient and inertia for a given temperature, allows one to locate the state or the arrangement of matter, and equally the state in which the usual Debye relation is sufficient. Finally, we think that the attempt to find the correspondence between the dielectric constant and birefringence, which we have sketched out, should be continued because it contains many encouraging results.

APPENDIX A
MAXWELL'S EQUATIONS AND CLASSICAL THEORY OF OSCILLATORS IN CONDENSED MEDIA

Let m be the mass and q the charge of a particle in a condensed dielectric medium capable of moving under the action of the electric field $E(t) = E_0 \cos \omega t$. The nature of this particle depends on the wavefrequency of the electromagnetic radiation. In the ultraviolet and X-ray domain, the relevant particle is the electron. It is an ion or an atomic nucleus in the visible and IR region.

One knows that the vectorial equation that governs its movement is of the form

$$m d_t^2 \mathbf{s} + f d_t \mathbf{s} + K \mathbf{s} = q \mathbf{E} \tag{A.1}$$

where d_t and d_t^2 stand for d/dt and d^2/dt^2, respectively.

Due to the force $q\mathbf{E}$ arising from the electric field, the charge center has a displacement $\mathbf{s}$. It also experiences a drag force proportional to the speed $(-f d_t \mathbf{s})$ together with an elastic force that tends to return it to the equilibrium position proportional to the elongation $(-K\mathbf{s})$.

In a dielectric volume, we consider that all the same charge centers have identical evolution under the action of the field. If N denotes the number per unit volume, their movement corresponds to polarization current of density

$$\mathbf{i}_p = N q d_t \mathbf{s} \tag{A.2}$$

It is necessary to add the displacement current density to this current. This is

$$\mathbf{i}_p = \varepsilon_d d_t \mathbf{E} \tag{A.3}$$

The constant ε_d is real. It represents a contribution to the dielectric constant ε defined by the Maxwell's relation, so that the total current is

$$\mathbf{i}_t = \varepsilon d_t \mathbf{E} = Nqd_t \mathbf{s} + \varepsilon_d d_t \mathbf{E} \tag{A.4}$$

To determine ε it is necessary to find the speed of the charge centers, which is possible by starting from the solution of Eq. (A.1). We do not consider the transient regime but solely the steady-state or forced regime. Since $\mathbf{E} = \mathbf{E}_0\ \mathcal{Re}[\exp(i\omega t)]$ and all the equations are linear, solution using imaginary quantities is easy. The solution sought is periodic and is given by the real part of the exponential function

$$\mathbf{s} = \mathbf{s}_0\ [\exp(i\omega t)] \tag{A.5}$$

where $\mathbf{s}_0$ is a time independent vector. The velocity and acceleration may be expressed as

$$d_t \mathbf{s} = i\omega \mathbf{s} \qquad d_t^2 \mathbf{s} = -\omega^2 \mathbf{s} \tag{A.6}$$

By substituting Eq. (A.6) into Eq. (A.1), one easily finds that

$$(-m\omega^2 + if\omega + K)\mathbf{s} = q\mathbf{E} \tag{A.7}$$

$$d_t \mathbf{s} = \frac{q}{(-m\omega^2 + if\omega + K)} d_t \mathbf{E} \tag{A.8}$$

The dielectric constant of the medium is deduced from Maxwell's relation [Eq. (A.4)]

$$\boldsymbol{\varepsilon} = \varepsilon_d + \frac{Nq^2}{-m\omega^2 + K + if\omega} \tag{A.9}$$

This is a complex quantity with real part ε' and imaginary part ε''. It describes the absorption of energy by the medium that is characteristic of a damped oscillator. If $\omega \to 0$, one sees that $\varepsilon_\infty = \varepsilon_d$.

On setting

$$S = \frac{Nq^2}{K} \qquad \omega_0^2 = \frac{K}{m} \qquad \gamma = \omega_0 \frac{f}{K} \tag{A.10}$$

one has

$$\boldsymbol{\varepsilon} = \varepsilon_\infty + \frac{S\omega_0^2}{\omega_0^2 - \omega^2 + i\gamma\omega_0\omega} \tag{A.11}$$

If many charge centers interact with the electromagnetic wave, we may extend the relation to n oscillators, each of which is characterized by the respective parameters S_i, ω_i, and γ_i $(i = 1, 2, \ldots, n)$

$$\boldsymbol{\varepsilon} = \varepsilon_\infty + \sum_{i=1}^{n} \frac{S_i \omega_i^2}{\omega_i^2 - \omega^2 + i\gamma_i \omega_i \omega} \tag{A.12}$$

APPENDIX B
CARTESIAN AND POLAR EQUATIONS OF COLE–COLE DIAGRAMS IN DIELECTRIC RELAXATION INCLUDING INERTIAL EFFECTS

We set as variables

$$X = \varepsilon' - \varepsilon_\infty \qquad Y = \varepsilon'' \tag{B.1}$$

Taking account of Eqs. (4.34)

$$\varepsilon' = \varepsilon_\infty + \frac{S\omega_0^2(\omega_0^2 - \omega^2)}{(\omega_0^2 - \omega^2) + \gamma^2 \omega_0^2 \omega^2}$$

$$\varepsilon'' = \frac{S\omega_0^3 \gamma \omega}{(\omega_0^2 - \omega^2)^2 + \gamma^2 \omega_0^2 \omega^2}$$

we obtain

$$\frac{X}{Y} = \frac{\omega_0^2 - \omega^2}{\gamma \omega_0 \omega} \tag{B.2}$$

or

$$Y\omega^2 + X\gamma\omega_0\omega - Y\omega_0^2 = 0 \tag{B.3}$$

The discriminant $\Delta = (X\gamma\omega_0)^2 + 4Y^2\omega_0^2$ is always positive. So, the sole positive root that is physically possible is

$$\omega = \frac{-X\gamma\omega_0 + \sqrt{\Delta}}{2Y} \tag{B.4}$$

Hence,

$$(\omega_0^2 - \omega^2) = \frac{-(X\gamma\omega_0)^2 + X\gamma\omega_0\sqrt{\Delta}}{2Y^2} \tag{B.5}$$

Furthermore,

$$X\left((\omega_0^2 - \omega^2) + \frac{(\gamma\omega_0\omega)^2}{(\omega_0^2 - \omega^2)}\right) = S\omega_0^2 \tag{B.6}$$

so that

$$[-(X\gamma\omega_0)^2 + X\gamma\omega_0\sqrt{\Delta}]^2 + \gamma^2 Y^2\omega_0^2[2(X\gamma\omega_0)^2 + 4Y^2\omega_0^2 - 2X\gamma\omega_0\sqrt{\Delta}]$$
$$= \frac{2SY^2\omega_0^2}{X}[-(X\gamma\omega_0)^2 + X\gamma\omega_0\sqrt{\Delta}] \tag{B.7}$$

and finally we have the Cartesian equation

$$4A^2(-A + \sqrt{\Delta})^2 + \gamma^2 B^2(2A^2 + B^2 - 2A\sqrt{\Delta}) = 2S\gamma\omega_0 B^2(-A + \sqrt{\Delta}) \tag{B.8}$$

where

$$A = X\gamma\omega_0, \qquad B = 2Y\omega_0 \qquad \text{and} \qquad \Delta = A^2 + B^2 \tag{B.9}$$

Introducing the parameters r and θ, which are $r = \sqrt{\Delta}$,

$$A = r\cos\theta \qquad \text{and} \qquad B = r\sin\theta$$

one obtains

$$r = \frac{S\gamma\omega_0(-2\cos\theta\sin^2\theta + 2\sin^2\theta)}{8\cos^4\theta + \gamma^2\sin^4\theta + 2\cos^2\theta\sin^2\theta(2+\gamma^2) - 8\cos^3\theta - 2\gamma^2\cos\theta\sin^2\theta} \tag{B.10}$$

We further note that

$$\tan\varphi' = \frac{Y}{X} = \frac{r\sin\theta}{2\omega_0}\frac{\gamma\omega_0}{r\cos\theta} = \frac{\gamma}{2}\tan\theta \tag{B.11}$$

and

$$\rho' = \sqrt{X^2 + Y^2} = \frac{r\cos\theta}{2\gamma\omega_0}\sqrt{4 + \gamma^2\tan^2\theta} = \frac{r\cos\theta}{\gamma\omega_0}\frac{1}{\cos\varphi'} \tag{B.12}$$

Taking account of Eqs. (B.10) and (B.11), one has

$$\cos\theta = \frac{\gamma}{\sqrt{\gamma^2 + 4\tan^2\varphi'}} \qquad \sin\theta = \frac{2\tan\varphi'}{\sqrt{\gamma^2 + 4\tan^2\varphi'}} \tag{B.13}$$

and after some calculations, one finds that

$$\rho' = S\,\frac{\cos\varphi'\tan^2\varphi'(\sqrt{\gamma^2 + 4\tan^2\varphi'} - \gamma)}{\gamma^2\left(\gamma - \sqrt{\gamma^2 + 4\tan^2\varphi'} + \dfrac{2}{\gamma}\tan^2\varphi'\right)} \tag{B.14}$$

then

$$\rho'^2 - S\rho'\cos\varphi' - S^2\frac{\sin^2\varphi'}{\gamma^2} = 0 \tag{B.15}$$

or, only retaining the positive root

$$\rho' = S\,\frac{\cos\varphi' + \sqrt{\cos^2\varphi' + \dfrac{4}{\gamma^2}\sin^2\varphi'}}{2} \tag{B.16}$$

For $\gamma \to 0$ (negligible inertia), one finds that $\rho' = S\cos\varphi'$, and one easily recovers the Debye circle.

Equation (B.15) does not describe a classical curve. One can, however, compare it to the Descartes oval

$$(n^2 - 1)\rho^2 + 4(a - cn^2\cos\varphi)\rho + 4(c^2n^2 - a^2) = 0 \tag{B.17}$$

in which a, c, and n are positive or zero constants. The similarity between the two equations is more pronounced for γ values greater than 5.

APPENDIX C
DIELECTRIC CONSTANT AND KRAMERS–KRONIG RELATIONS

The dielectric constant $\boldsymbol{\varepsilon}$ is expressed by the relation that exists between the electric displacement $\mathbf{D}$ in a dielectric medium and the applied electric field $\mathbf{E}$

$$\mathbf{D} = \boldsymbol{\varepsilon}\mathbf{E} \tag{C.1}$$

If the medium is homogeneous and isotropic, $\boldsymbol{\varepsilon}$ is a scalar that depends on the thermodynamic state. If it becomes anisotropic under the action of a field or a sufficiently strong stimulus, $\boldsymbol{\varepsilon}$ becomes a tensor that describes a spatial dispersion. In a particular reference frame associated with the direction of the wavevector $\mathbf{k}$, $\boldsymbol{\varepsilon}$ is a symmetric tensor of second order

$$\boldsymbol{\varepsilon} = \begin{bmatrix} \varepsilon_{11} & 0 & 0 \\ 0 & \varepsilon_{22} & 0 \\ 0 & 0 & 0 \end{bmatrix} \tag{C.2}$$

This tensor completely describes the macroscopic properties of the medium, relative to the propagation along the wavevector $\mathbf{k}$ of a homogeneous monochromatic transverse wave of wavelength much greater than the microscopic dimensions.

The most general expression of the scalar dielectric constant (ε_{11} and ε_{22}), which we have simply called ε, as a function of the angular frequency, is derived from the generalized theory of the susceptibility. This links the properties of nonthermodynamic fluctuations with other values characterizing behavior of a medium subjected to certain external forces. It may be written

$$\boldsymbol{\varepsilon}(\omega) - \varepsilon_{\infty} = \int_{-\infty}^{+\infty} f(\tau) \exp(i\omega\tau) d\tau \tag{C.3}$$

τ has the dimension of time, ε_{∞} is the limiting value of $\varepsilon(\omega)$ as $\omega \rightarrow 0$, $f(\tau)$ is a function of time and of the properties of the medium that is finite for all τ. For dielectrics, this function tends to zero as $\tau \rightarrow \infty$, and the time in which $f(\tau)$ is sensibly different from zero is of the order of the relaxation time characterizing process. On account of the principle of causality—effect cannot precede cause—$f(\tau) = 0$ for $\tau < 0$, and Eq. (C.3) becomes

$$\boldsymbol{\varepsilon}(\omega) - \varepsilon_{\infty} = \int_{0}^{+\infty} f(\tau) \exp(i\omega\tau) d\tau \tag{C.4}$$

From this last expression, one may consider ω as a complex variable $\Omega = \omega - i\bar{\omega}$, and one deduces the following properties for $\boldsymbol{\varepsilon}(\Omega)$:

- $\boldsymbol{\varepsilon}(\Omega)$ is analytic in the upper half-plane Ω and on the real axis.
- $\boldsymbol{\varepsilon}(-\Omega) = \boldsymbol{\varepsilon}^*(\Omega)$, and for real Ω, $\boldsymbol{\varepsilon}(-\omega) = \boldsymbol{\varepsilon}^*(\omega)$, where the asterisk denotes the complex conjugate. In the latter case, one can separate the real and imaginary parts of $\boldsymbol{\varepsilon}$, and one finds that

$$\varepsilon'(-\omega) = \varepsilon'(\omega) \qquad \varepsilon''(-\omega) = \varepsilon''(\omega)$$

$\varepsilon'(\omega)$ is thus an even function and $\varepsilon''(\omega)$ an odd function of the frequency.

- In the upper half-plane, the function $\boldsymbol{\varepsilon}(\Omega)$ reduces only to the real part on the imaginary axis along which it decreases in monotonic fashion. In particular, $\boldsymbol{\varepsilon}(\Omega)$ has no zeros on the upper half-plane.

One can deduce a fourth property of the thermodynamic law of increasing entropy applicable to energy losses: to all media and all frequencies, ε'' is positive. The sign of ε', on the other hand, is not compelled to any such restriction.

Starting from these four properties, the Kramers–Kronig relations may be easily deduced. Let us consider, in effect, a point ω_c on the positive part of the real axis. In the case of dielectrics, the origin is not a pole of $\boldsymbol{\varepsilon}(\omega)$. The function

$$[\boldsymbol{\varepsilon}(\Omega) - \varepsilon_\infty](\Omega - \omega_c)^{-1}$$

is analytic in the domain shown in Fig. C1. Following Cauchy's theorem, the integral over the closed contour (C), namely, $(1 \rightarrow 2 \rightarrow 3 \rightarrow 4)$, is zero. If $\Omega \rightarrow \infty$, $[\boldsymbol{\varepsilon}(\Omega) - \varepsilon_\infty] \rightarrow 0$. So, if $R \rightarrow \infty$, the integral over the semicircle 4 is zero. It follows that

$$\lim_{R\rightarrow\infty} \int_C \frac{\boldsymbol{\varepsilon} - \varepsilon_\infty}{\Omega - \omega_c}\, d\Omega = \lim_{R\rightarrow\infty} \left[\int_1 \frac{\boldsymbol{\varepsilon} - \varepsilon_\infty}{\omega - \omega_c}\, d\omega + \int_3 \frac{\boldsymbol{\varepsilon} - \varepsilon_\infty}{\omega - \omega_c}\, d\omega \right] + \int_2 \frac{\boldsymbol{\varepsilon} - \varepsilon_\infty}{\Omega - \omega_c}\, d\omega = 0 \tag{C.5}$$

If the small radius r tends to zero, the term in brackets is the principal part of the integral in the interval $(-\infty, +\infty)$, and the last integration gives

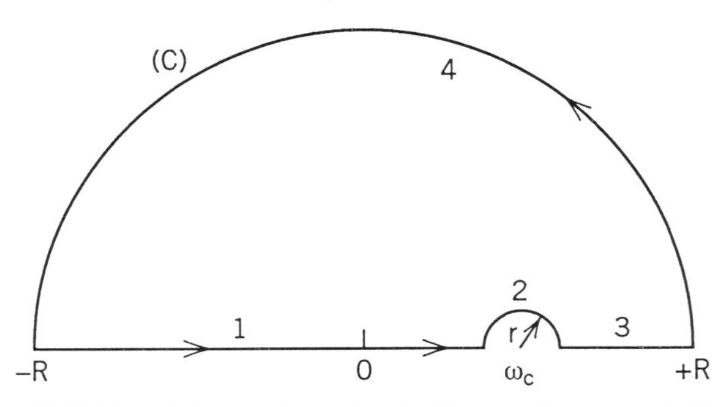

Figure C1. Definition of integration paths in the complex upper half plane for the dielectric function $\boldsymbol{\varepsilon}(\omega)$.

result

$$-i\pi[\boldsymbol{\varepsilon}(\omega_c) - \varepsilon_\infty]$$

Hence,

$$\varepsilon'(\omega_c) - i\varepsilon''(\omega_c) - \varepsilon_\infty = -\frac{1}{i\pi}\mathcal{P}\int_{-\infty}^{+\infty} \frac{\varepsilon'(\omega) - i\varepsilon''(\omega) - \varepsilon_\infty}{\omega - \omega_c}\, d\omega \tag{C.6}$$

and separating the real and imaginary parts

$$\varepsilon'(\omega_c) - \varepsilon_\infty = \frac{1}{\pi}\mathcal{P}\int_{-\infty}^{+\infty} \frac{\varepsilon''(\omega)}{\omega - \omega_c}\, d\omega \tag{C.7a}$$

$$\varepsilon''(\omega_c) = -\frac{1}{\pi}\mathcal{P}\int_{-\infty}^{+\infty} \frac{\varepsilon'(\omega) - \varepsilon_\infty}{\omega - \omega_c}\, d\omega \tag{C.7b}$$

These relations were obtained by Kramers and Kronig in 1927 [78].

Since $\varepsilon'(\omega)$ is an even function and $\varepsilon''(\omega)$ is an odd function of the frequency, one obtains

$$\varepsilon'(\omega_c) - \varepsilon_\infty = \frac{2}{\pi}\mathcal{P}\int_{0}^{+\infty} \frac{\omega\varepsilon''(\omega)}{\omega^2 - \omega_c^2}\, d\omega \tag{C.8a}$$

$$\varepsilon''(\omega_c) = -\frac{2\omega_c}{\pi}\mathcal{P}\int_{0}^{+\infty} \frac{\varepsilon'(\omega) - \varepsilon_\infty}{\omega^2 - \omega_c^2}\, d\omega \tag{C.8b}$$

The function $\ln \boldsymbol{\varepsilon}(\omega)$ obeys the same conditions for inversion as $\boldsymbol{\varepsilon}(\omega)$. Since $\ln \boldsymbol{\varepsilon}(\omega) = \rho(\omega) - i\varphi(\omega)$, one may deduce that

$$\varphi(\omega_c) = -\frac{2\omega_c}{\pi}\mathcal{P}\int_{0}^{+\infty} \frac{\ln \rho(\omega)}{\omega^2 - \omega_c^2}\, d\omega \tag{C.9}$$

and integrating by parts

$$\varphi(\omega_c) = -\frac{1}{\pi}\mathcal{P}\int_{0}^{+\infty} \frac{d \ln \rho(\omega)}{d\omega} \ln\left(\frac{\omega + \omega_c}{\omega - \omega_c}\right) d\omega \tag{C.10}$$

The calculation of this integral is in practice confined to the known interval (ω_1, ω_2). One can construct it by considering m points on the curve representing $\ln \rho(\omega)$, that is, on breaking that curve into $(m-1)$ rectilinear segments. Denoting the slope of one of these segments by p_i, one has

$$\varphi_2(\omega_c) = -\frac{1}{\pi}\left[\sum_{i=1}^{m-1} p_i \int_{\omega_i}^{\omega_{i+1}} \ln\left(\frac{\omega + \omega_c}{\omega - \omega_c}\right) d\omega\right] \tag{C.11}$$

This form suggests the introduction of the function $f(x_i)$ defined by the relation

$$f(x_i) = \frac{1}{\pi\omega_i}\int_0^{\omega_i} \ln\frac{\omega + \omega_c}{\omega - \omega_c}\, d\omega \tag{C.12}$$

where $x_i = \omega_c/\omega_i$. This function may be easily calculated in the following manner:

$$f(x_i) = \frac{1}{\pi}\left[(x_i + 1)\ln(x_i + 1) + (x_i - 1)\ln|x_i - 1| - 2x_i \ln x_i\right] \tag{C.13}$$

The expression for the phase angle then takes the form

$$\varphi_2(\omega_c) = \sum_{i=1}^{m-1} p_i[\omega_{i-1} f(x_{i-1}) - \omega_i f(x_i)] \tag{C.14}$$

This is useful for computer calculations. It is necessary to remark that in this calculation, the product $(x_i - 1)\ln|x_i - 1|$ tends to zero as x_i tends to 1, that is, ω_i tends to ω_c. The result is naturally more accurate the more dense one can make the number m of points.

APPENDIX D
CALCULATION OF THE HARMONIC COMPONENTS OF THE BIREFRINGENCE IN THE THEORY OF NONLINEAR RESPONSE

Equation (4.96) gives an expression for the distribution function $(f_{\gamma 2} + f_{\beta 2})(u, t)$ of the rotational Brownian motion of a symmetric molecule. We can write it in the form

$$(f_{\gamma 2} + f_{\beta 2})(u, t) = E_0^2 \mathbf{g}[I_1 + I_2] + E_c^2 \mathbf{g}[J_1 + J_2] + E_0 E_c \mathbf{g}[K_1 + K_2] \tag{D.1}$$

where

$$I_1 = \int_0^t \int_0^{t_1} \mathbf{A}_\gamma(t - t_1)\mathbf{f}_\gamma(t_1 - t_2)\cos\omega t_1 \cos\omega t_2\, dt_1\, dt_2 \tag{D.1a}$$

$$I_2 = \int_0^t \mathbf{f}_{\boldsymbol{\beta}}(t - t_1) \cos^2 \omega t_1 dt_1 \tag{D.1b}$$

$$J_1 = \int_0^t \int_0^{t_1} \mathbf{A}_{\boldsymbol{\gamma}}(t - t_1)\mathbf{f}_{\boldsymbol{\gamma}}(t_1 - t_2) dt_1 dt_2 \tag{D.1c}$$

$$J_2 = \int_0^t \mathbf{f}_{\boldsymbol{\beta}}(t - t_1) dt_1 \tag{D.1d}$$

$$K_1 = \int_0^t \int_0^{t_1} \mathbf{A}_{\boldsymbol{\gamma}}(t - t_1)\mathbf{f}_{\boldsymbol{\gamma}}(t_1 - t_2)[\cos \omega t_1 + \cos \omega t_2]\, dt_1 dt_2 \tag{D.1e}$$

$$K_2 = \int_0^t \mathbf{f}_{\boldsymbol{\beta}}(t - t_1) \cos \omega t_1 dt_1 \tag{D.1f}$$

We shall consider six integrals successively. The first one, I_1 may be written as follows:

$$I_1 = \int_0^t \mathbf{A}_{\boldsymbol{\gamma}}(t - t_1) \cos \omega t_1 f_1(x, t_1) dt_1 \tag{D.2}$$

where

$$f_1(x, t_1) = \int_0^{t_1} \mathbf{f}_{\boldsymbol{\gamma}}(t_1 - t_2) \cos \omega t_2 dt_2$$

Then,

$$I_1 = \int_0^t \mathbf{A}_{\boldsymbol{\gamma}}(t_1) \cos \omega(t - t_1) f_1(x, t - t_1) dt_1 \tag{D.3}$$

where

$$f_1(x, t - t_1) = \int_0^{t - t_1} \mathbf{f}_{\boldsymbol{\gamma}}(t - t_1 - t_2) \cos \omega t_2 dt_2$$

and

$$I_1 = \int_0^t \int_0^{t - t_1} \mathbf{A}_{\boldsymbol{\gamma}}(t_1) \cos \omega(t - t_1)\mathbf{f}_{\boldsymbol{\gamma}}(t - t_1 - t_2) \cos \omega t_2 dt_1 dt_2 \tag{D.4a}$$

$$I_1 = \int_0^t \int_0^{t - t_1} \mathbf{A}_{\boldsymbol{\gamma}}(t_1) \cos \omega(t - t_1)\, \mathbf{f}_{\boldsymbol{\gamma}}(t_2) \cos \omega(t - t_1 - t_2) dt_1 dt_2 \tag{D.4b}$$

Now, we use the trigonometric relation

$$\cos \omega(t - t_1) \cos \omega(t - t_1 - t_2) = \tfrac{1}{2}[\cos \omega t_2 + \cos \omega(2t_1 + t_2) \cos 2\omega t] + \tfrac{1}{2} \sin \omega(2t_1 + t_2) \sin 2\omega t \tag{D.5}$$

so that, for infinite time, we obtain

$$\begin{aligned} I_1 = & \frac{1}{2} \int_0^\infty \mathbf{A}_\gamma(t) dt \int_0^\infty \mathbf{f}_\gamma(t) \cos \omega t dt \\ & + \frac{1}{2}\left[\int_0^\infty \int_0^\infty \mathbf{A}_\gamma(t) \mathbf{f}_\gamma(t) \cos[\omega(2t_1 + t_2)] dt_1 dt_2\right] \cos 2\omega t \\ & + \frac{1}{2}\left[\int_0^\infty \int_0^\infty \mathbf{A}_\gamma(t) \mathbf{f}_\gamma(t) \sin[\omega(2t_1 + t_2)] dt_1 dt_2\right] \sin 2\omega t \end{aligned} \tag{D.6}$$

The integral I_2 may be written

$$I_2 = \int_0^t \mathbf{f}_\beta(t - t_1) \cos^2 \omega t_1 dt_1 = \int_0^t \mathbf{f}_\beta(t_1) \cos^2 \omega(t - t_1) dt_1 \tag{D.7a}$$

$$I_2 = \frac{1}{2} \int_0^t \mathbf{f}_\beta(t_1) dt_1 + \frac{1}{2} \int_0^t \mathbf{f}_\beta(t_1) \cos 2\, \omega(t - t_1) dt_1 \tag{D.7b}$$

The first integral of Eq. (D.7a) corresponds to the stationary part and the second to $2\,\omega$ part. On developing $\cos 2\,\omega(t - t_1)$ and making the time tend to infinity, we have

$$\begin{aligned} I_2 = \frac{1}{2}\Bigg\{ & \int_0^\infty \mathbf{f}_\beta(t) dt + \left[\int_0^\infty \mathbf{f}_\beta(t_1) \cos 2\, \omega t_1 dt_1\right] \cos 2\omega t \\ & + \left[\int_0^\infty \mathbf{f}_\beta(t_1) \sin 2\, \omega t_1 dt_1\right] \sin 2\omega t \Bigg\} \end{aligned} \tag{D.8}$$

For J_1 and J_2, one easily finds when $t \to \infty$

$$J_1 = \int_0^\infty \mathbf{A}_\gamma(t) dt \int_0^\infty \mathbf{f}_\gamma(t) dt \tag{D.9}$$

$$J_2 = \int_0^\infty \mathbf{f}_\beta(t) dt \tag{D.10}$$

The integral K_1 may be broken as follows:

$$K_1 = K_1' + K_1'' \tag{D.11}$$

where

$$K_1' = \int_0^t \int_0^{t_1} \mathbf{A}_\gamma(t - t_1)\mathbf{f}_\gamma(t_1 - t_2) \cos \omega t_1 dt_1 dt_2 \tag{D.11a}$$

$$K_1'' = \int_0^t \int_0^{t_1} \mathbf{A}_\gamma(t - t_1)\mathbf{f}_\gamma(t_1 - t_2) \cos \omega t_2 dt_1 dt_2 \tag{D.11b}$$

Furthermore,

$$K_1' = \int_0^t \mathbf{A}_\gamma(t - t_1) \cos \omega t_1 f_1(x, t_1) dt_1 \tag{D.12a}$$

where

$$f_1(x, t_1) = \int_0^{t_1} \mathbf{f}_\gamma(t_1 - t_2) dt_2$$

and

$$K_1'' = \int_0^t \int_0^{t - t_1} \mathbf{A}_\gamma(t - t_1)\mathbf{f}_\gamma(t_2) \cos \omega(t - t_1 - t_2) dt_1 dt_2 \tag{D.12b}$$

At infinity, one obtains

$$\begin{aligned} K_1' = &\left[\int_0^\infty \mathbf{A}_\gamma(t_1) \cos \omega t_1 dt_1\right] \cos \omega t \int_0^\infty \mathbf{f}_\gamma(t) dt \\ &+ \left[\int_0^\infty \mathbf{A}_\gamma(t_1) \sin \omega t_1 dt_1\right] \sin \omega t \int_0^\infty \mathbf{f}_\gamma(t) dt \end{aligned} \tag{D.13a}$$

$$\begin{aligned} K_1'' = &\left[\int_0^\infty \int_0^\infty \mathbf{A}_\gamma(t_1)\mathbf{f}_\gamma(t_2) \cos \omega(t_1 + t_2) dt_1 dt_2\right] \cos \omega t \\ &+ \left[\int_0^\infty \int_0^\infty \mathbf{A}_\gamma(t_1)\mathbf{f}_\gamma(t_2) \sin \omega(t_1 + t_2) dt_1 dt_2\right] \sin \omega t \end{aligned} \tag{D.13b}$$

Finally, we calculate the integral K_2

$$K_2 = \int_0^t \mathbf{f}_\beta(t_1) \cos \omega(t - t_1) dt_1 \tag{D.14a}$$

When $t \to \infty$, Eq. (D.14a) becomes

$$K_2 = \left[\int_0^\infty \mathbf{f}_\beta(t_1) \cos(\omega t_1) dt_1 \right] \cos \omega t + \left[\int_0^\infty \mathbf{f}_\beta(t_1) \sin(\omega t_1) dt_1 \right] \sin \omega t \tag{D.14b}$$

The expressions for all these integrals show that the birefringence up to second approximation is correctly given by the superposition of a stationary component, a component in ωt (first harmonic term) and a component in $2\omega t$ (second harmonic term). The three components are gathered in Eq. (4.98).

ACKNOWLEDGMENT

We wish to express our gratitude to Professor W.T. Coffey for many helpful discussions and careful correction of the manuscript.

REFERENCES

1. R. Brown, *Philos. Mag. NS*, **4**, 181 (1828).
2. E. Nelson, *Dynamical Theories of Brownian Motion*, Princeton University Press, Princeton, NJ, 1967.
3. W. T. Coffey, *Advances in Chemical Physics*, Vol. **63**, *Development and Application of the Theory of the Brownian Motion in Dynamical Processes in Condensed Matter*, Wiley-Interscience, New York, 1985.
4. A. Einstein, *Investigations on the Theory of the Brownian Movement*, R. H. Fürth, Ed., Dover, New York, 1956.
5. H. Risken, *The Fokker–Planck Equation*, Springer-Verlag, Berlin, 1984.
6. N. Wax, *Selected Papers on Noise and Stochastic Processes*, Dover, New York, 1954.
7. P. Debye, *Polar Molecules*, Chemical Catalog, 1929, Reprinted by Dover, New York.
8. J. R. McConnell, *Rotational Brownian Motion and Dielectric Theory*, Academic, London, 1980.
9. B. K. P. Scaife, *Complex Permittivity*, English University Press, London, 1971.
10. O. Klein, *Ark. Mat. Astr. Fys.*, **5**, 16 (1921).
11. E. P. Gross, *J. Chem. Phys.*, **23**, 1415 (1955).
12. M. W. Evans, G. J. Evans, W. T. Coffey, and P. Grigolini, *Molecular Dynamics and the Theory of Broad Band Spectroscopy*, Wiley-Interscience, New York, 1982.
13. R. A. Sack, *Proc. Phys. Soc.*, **B70**, 402, 414 (1957).
14. M. Y. Rocard, *J. Phys. Radium*, **4**, 247 (1933).
15. W. T. Coffey, M. W. Evans, and P. Grigolini, *Molecular Diffusion and Spectra*, Wiley-Interscience, New York, 1984.
16. H. C. Brinkman, *Physica*, **22**, 29 (1956).
17. R. A. Sack, *Physica*, 22, 917 (1956).

18. W. Alexiewicz, *Acta Phys. Polon.*, **A72**, 753 (1987).
19. W. T. Coffey and S. G. McGoldrick, *Chem. Phys.*, **120**, 1 (1988).
20. P. C. Hemmer, *Physica*, **27**, 79 (1961).
21. U. M. Titulaer, *Physica*, **A91**, 321 (1978).
22. J. L. Skinner and P. G. Wolynes, *Physica*, **A96**, 561 (1979).
23. G. J. Wilemski, *Stat. Phys.*, **14**, 153 (1976).
24. R. L. Stratonovitch, *Topics in the Theory of Random Noise*, Vol. **1**, Gordon & Breach, New York, 1963.
25. W. T. Coffey, S. G. McGoldrick, and K. P. Quinn, *Chem. Phys.*, **125**, 99 (1988).
26. W. T. Coffey, S. G. McGoldrick, and P. J. Cregg, *Chem. Phys.*, **125**, 119 (1988).
27. J. L. Déjardin, *J. Chem. Phys.*, **95**, 576 (1991).
28. W. T. Coffey, *J. Chem. Phys.*, **95**, 2026 (1991).
29. W. T. Coffey, J. L. Déjardin, Y. P. Kalmykov, and K. P. Quinn, *Chem. Phys.*, **164**, 357 (1992).
30. W. T. Coffey, *Mol. Phys.*, **39**, 227 (1980).
31. U. M. S. Murthy and J. K. Vij, *Liq. Cryst.*, **4**, 529 (1989).
32. C. J. F. Böttcher and P. Bordewijk, *Theory of Electric Polarization*, Vol. II, Elsevier, Amsterdam, New York, 1978.
33. A. Morita, S. Walker, and J. H. Calderwood, *J. Phys.*, **D9**, 2485 (1976).
34. W. T. Coffey, P. M. Corcoran, and M. W. Evans, *Proc. R. Soc. London*, **A410**, 61 (1987).
35. J. L. Déjardin, *Liq. Cryst.*, **3**, 1191 (1988).
36. J. L. Déjardin, *Liq. Cryst.*, **9**, 599 (1991).
37. F. A. Ferrone, J. Hofrichter, and W. A. Eaton, *J. Mol. Biol.*, **183**, 591 (1985). See also, J. Dean and A. Schechter, *N. Engl. J. Med.*, **299**, 752 (1978), for a detailed review on sickle-cell anemia.
38. Z. Delalic, S. Takashima, K. Adachi, and T. Asakura, *J. Mol. Biol.*, **168**, 659 (1983).
39. E. Hiedermann and R. D. Spence, *Z. Phys.*, **133**, 109 (1952).
40. C. Kittel, *Introduction to Solid State Physics*, Wiley, New York, 1976.
41. H. W. Bode, *Network Analysis and Feedback Amplifier Design*, Van Nostrand, New York, 1945.
42. G. Debiais, *Phys. Stat. Sol.*, **B107**, 729 (1981).
43. J. Ruiz-Pérez, Ph.D. Thesis, Toulouse (France), 1968.
44. G. Debiais, *Phys. Stat. Sol.*, **A83**, 269 (1984).
45. W. T. Coffey and B. V. Paranjape, *Proc. R. Ir. Acad.*, **78**, 17 (1978).
46. A. Morita, *Phys. Rev.*, **A34**, 1499 (1986).
47. J. L. Déjardin, *J. Chem. Phys.*, **98**, 3191 (1993).
48. G. Debiais and J. L. Déjardin, *Physica*, **A158**, 589 (1989).
49. J. L. Déjardin, Ph.D. Thesis, Montpellier (France), 1977.
50. H. Benoit, *Ann. Phys.*, **6**, 561 (1951).
51. R. H. Cole, *J. Chem. Phys.*, **76**, 4700 (1982).
52. H. Watanabe and A. Morita, *Advances in Chemical Physics.*, Vol. **56**, *Kerr Effect Relaxation in High Electric Fields*, Wiley-Interscience, New York, 1984.
53. J. L. Déjardin, G. Debiais, and A. Ouadjou, *J. Chem. Phys.*, **98**, 8149 (1993).
54. B. K. P. Scaife, *Specialist periodical reports*, *London*, *TheChemical Society*, **1**, 1 (1972).

55. J. Kerr, Phil. Mag., **50**, 337 (1875).
56. J. Kerr, *Philos. Mag.*, **9**, 157 (1875).
57. G. B. Thurston and D. I. Bowling, *J. Colloid Int. Sci.*, **30**, 34 (1969).
58. W. A. Wegener, *J. Chem. Phys.*, **26**, 368 (1969).
59. J. C. Filippini, Ph.D. Thesis, Grenoble, France, 1972.
60. S. Brunet, Ph.D. Thesis, Montpellier, France, 1971.
61. R. Marrony, Ph.D. Thesis, Montpellier, France, 1972.
62. C. Delseny, Ph.D. Thesis, Perpignan, France, 1980.
63. S. Bénet, Ph.D. Thesis, Perpignan, France, 1987.
64. W. T. Lotshaw, D. McMorrow, C. Kalpouzos, and G. A. Kenney-Wallace, *Chem. Phys. Lett.*, **136**, 323 (1987).
65. R. S. Wilkinson and G. B. Thurston, *Biopolymers*, **15**, 1555 (1976).
66. A. Morita and H. Watanabe, *Phys. Rev.*, **A35**, 2690 (1987).
67. W. Alexiewicz, *Mol. Phys.* **59**, 637 (1986).
68. J. L. Déjardin and G. Debiais, *Phys. Rev.* **A40**, 1560 (1989).
69. A. Morita and H. Watanabe, *J. Chem. Phys.* **75**, 1320 (1981).
70. A. K. Jonscher, *Dielectric Relaxation in Solids*, Chelsea Dielectric Press, London, 1983.
71. J. L. Déjardin and G. Debiais, *Physica*, **A164**, 182 (1990).
72. G. Debiais, Ph.D. Thesis, Perpignan, France, 1992.
73. J. L. Déjardin and G. Debiais, *Dynamic Behavior of Macromolecules, Colloids, Liquid Crystals and Biological Systems by Optical and Electrooptical Methods* (Electropto 88), H. Watanabe, Ed., Hirokawa Publishing Co, Tokyo, Japan, 1988.
74. J. L. Déjardin and G. Debiais, *J. Chemn. Phys.*, **95**, 2787 (1991).
75. G. Debiais and J. L. Déjardin, *J. Mol. Liq.*, **49**, 37 (1991).
76. C. V. Raman and S. C. Sirkar, *Nature* (London), **121**, 794 (1928).
77. D. W. Kitchin and H. Müller, *Phys. Rev.* **32**, 979 (1928).
78. H. A. Kramers, Atti del Congresso Internazionale dei Fisici, Como, **2**, 545 (1927). See also R. de L. Kronig, *J. Opt. Soc. Am.*, **12**, 547 (1926).

PREDICTING RARE EVENTS IN MOLECULAR DYNAMICS

JAMES B. ANDERSON

Department of Chemistry, The Pennsylvania State University, University Park, PA 16802

CONTENTS

A combination of transition-state theory and molecular dynamics yields the rare-event theory widely used for the simulation of simple and complex chemical reactions and related events that occur infrequently. For all but the simplest systems it is the *only* method capable of

Advances in Chemical Physics, Volume XCI, Edited by I. Prigogine and Stuart A. Rice.
ISBN 0-471-12002-2

simulating rare events without a prohibitive computational effort. It provides results that are rigorously correct for classical systems with reactants in thermal equilibrium. It makes possible the simulation of systems ranging in size from a few atoms to tens of thousands of atoms undergoing such processes as simple and complex reactions in the gas phase, exchange reactions in solution, diffusion and desorption at surfaces, diffusion of vacancies in solids, rearrangements of protein molecules in solution, enzyme-catalyzed reactions, epitaxial growth of crystals, membrane transport, and binary star formation. For many of these systems the development of the rare-event approach is at least as important in calculations of molecular dynamics as the development of fast computers. The framework of rare-event theory also provides the basis for semiclassical and approximate quantum treatments of these processes.

I. INTRODUCTION

Protein rearrangements are rare events, enzyme-catalyzed reactions are rare events, and even simple reactions in the gas phase are not all that frequent.

Indeed, most of the phenomena of interest in chemistry are events that occur infrequently. A typical molecule may undergo many millions of collisions, proceed through many millions of random fluctuations, and follow many millions of pathways before finding a route favorable to reaction. Since the computational simulation of these rare events in molecular dynamics in such a haphazard fashion is prohibitively expensive, the successful simulation of rare events requires guidance in finding favorable pathways. The combination of transition state theory and molecular dynamics in the "rare-event theory" proposed by Keck [1] provides the required guidance and makes such simulations possible. In reporting the first results from rare-event calculations Keck noted that the approach "opens an extremely interesting field for investigation."

The problem of computing the molecular dynamics of rare events is illustrated by the case of the chemical reaction $H + HF \rightarrow H_2 + F$. In conventional calculations [2] the starting points for trajectories are selected from distributions with the H and HF separated from each other at the edge of the interaction zone along a surface S_1, as shown in Fig. 5.1. Trajectories with sufficient total energy to cross the 33.0-kcal mol^{-1} barrier to reaction are followed until they reach product configurations by crossing surface S_2 or return to reactants by recrossing S_1. In this particular case the energy requirements for reaction are specific and only trajectories beginning with vibrationally excited HF are successful in

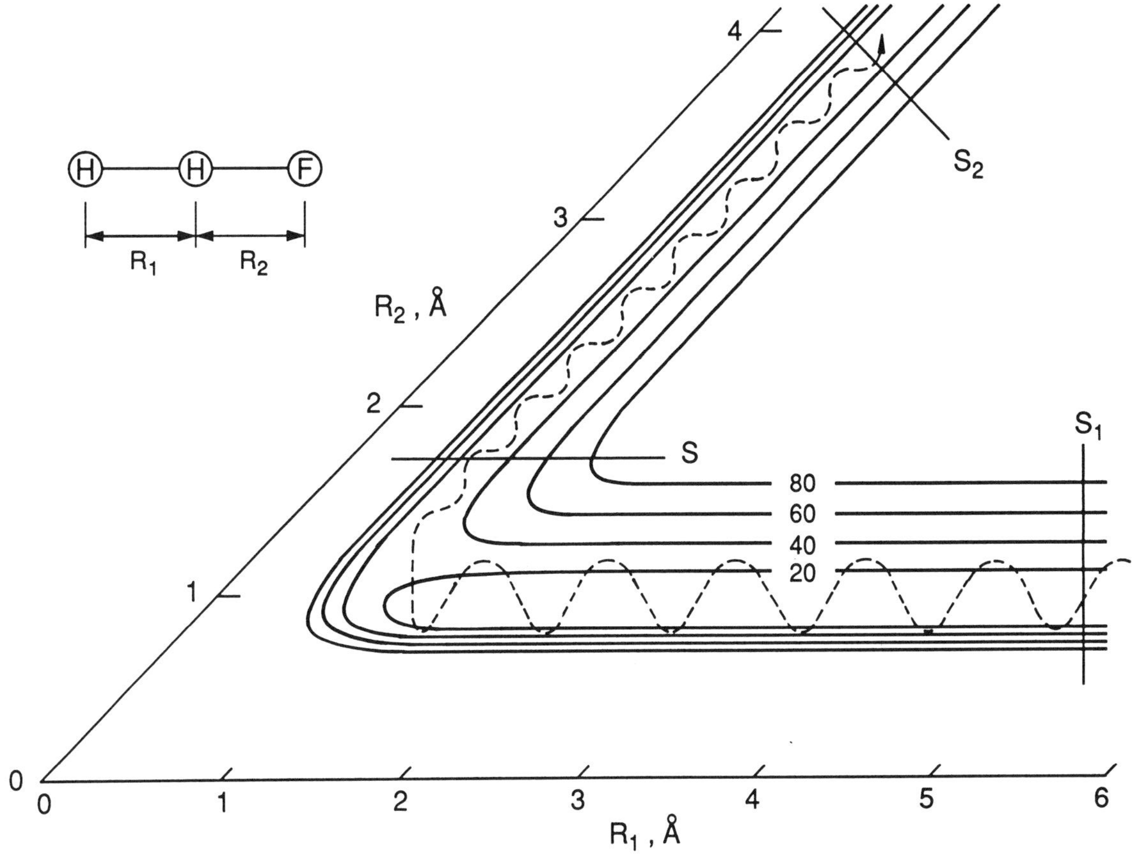

Figure 5.1. Potential energy surface for collinear H–H–F. A "successful" classical trajectory corresponding to the reaction $H + HF \rightarrow H_2 + F$ is shown by the dashed line. Potential energy contour lines are 20–80 kcal mol^{-1} above the minimum for H + HF. (From [2].)

reaching the products $H_2 + F$. Most of the trajectories return to reactants without crossing the barrier to reaction and a large number of trajectories must be computed to obtain a single reactive trajectory representative of those that occur in a system of thermally equilibrated reactants.

In the case of the reaction $HI + HI \rightarrow H_2 + I_2$ (or 2 I) at 700 K the reaction probability for sufficiently energetic reactants is about 1 in 10^6 and the determination of 1000 reactive trajectories starting from reactants would require the calculation of 10^9 trajectories [3, 4]. At $3 per trajectory such calculations were expensive in 1970 and even at a fraction of a cent per trajectory they remain expensive today. The solution to the problem is provided by the rare-event approach in which trajectories are selected from the equilibrium distribution of those crossing a surface S in the vicinity of the barrier and the probability of selecting a reactive trajectory is increased to about 1 in 2.

Computing the molecular dynamics of rare events is similar to the problem of finding the strings crossing the table shown in Fig. 5.2. The strings are laid on the table such that they enter at either end and exit at either end. Most of them enter and exit at the same end but a few cross the table and exit at the opposite end. One may find those rare strings crossing the table by tracing the path of each string entering at the left until it returns to the left end or exits at the right end. One may also find the same rare strings by forward and backward tracing of the path of each string crossing an arbitrary dividing line separating left and right. If the

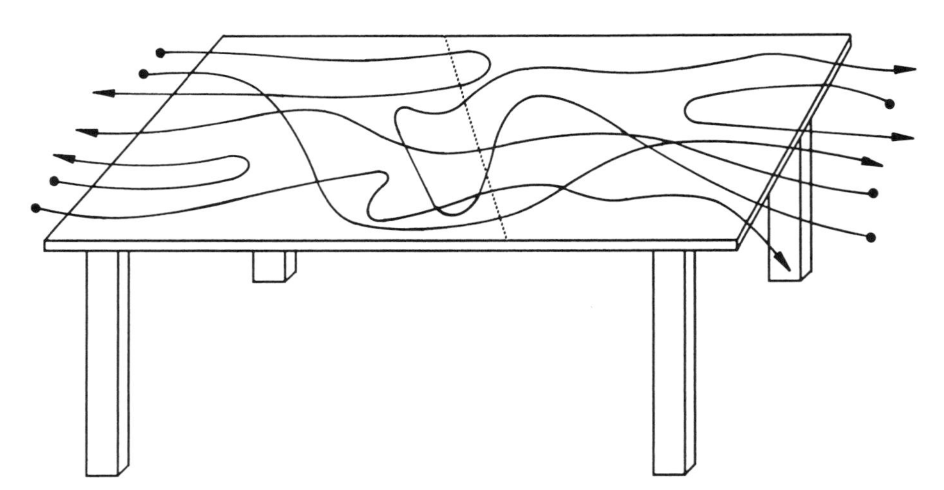

Figure 5.2. An illustration of the rare-event approach to finding the trajectories (strings) successfully crossing a table. (From [28].)

line is carefully placed, so as to minimize the number of strings crossing it, the number of paths that must be traced may be substantially reduced and those rare strings successfully crossing the table may be found more easily.

The classical transition state theory proposed by Marcelin [5] and later developed by Wigner [6], Horiuti [7], and Eyring et al. [8] provides the means for sampling trajectories. For systems to which classical mechanics applies the theory is rigorously valid and is a direct consequence of statistical mechanics [9]. Marcelin used the methods of Gibbs [10] to derive an expression for the rate at which species cross a surface in phase space separating reactants from products.

Marcelin's rate expression may be restated simply, following Hirschfelder and Wigner [9], in terms of the rate constant k,

$$k = \frac{1}{P_r}\left(\frac{P_t}{\delta}\right)\bar{v}\xi \tag{5.1}$$

where P_r is the probability of the system being in the initial reactant state, P_t is the probability of its being in the transition state, and $\bar{v}$ is the average velocity with which configuration points cross the surface. The transition state is a strip of width δ in configuration space along the dividing surface and (P_t/δ) is a probability density for points at the surface. Finally, ξ expresses the probability that a system that crosses the surface in the forward direction at complete equilibrium actually originated in the initial reactant state and will proceed directly to the final product state [11, 12]. As Hirschfelder and Wigner [9] noted, the transmission coefficient ξ is the only quantity in Eq. (5.1) that cannot be evaluated by well-known methods of statistical mechanics.

In 1962 Keck [1] proposed that the transmission coefficient could be evaluated by sampling configurations crossing a dividing surface and following their trajectories forward and backward in time with classical trajectory calculations. He demonstrated the efficiency of such calculations in investigating the mechanisms and rates of dissociation and recombination of H_2, O_2, and I_2 in collisions with Ar. These were the first demonstrations of the remarkable success that could be obtained from the marriage of transition state theory and molecular dynamics. Subsequent calculations by Anderson [12] and Jaffe and co-workers [3, 4] for simple and complex exchange reactions gave additional evidence.

The combined phase-space/trajectory or rare-event method is rigorously correct for classical systems in which the reactants are present in thermal equilibrium. In its various extensions the method is also rigorously correct for certain nonequilibrium systems. The rare-event method produces a sample of reactive trajectories representative of those occurring in the system. It provides all the details of reaction: the rate constant, the reaction pathways, collision details, energy requirements for reaction, and energy distributions among the products. Transition state theory by itself does not provide any of this information but merely gives an upper limit to the rate constant.

Classical mechanics has been found to be a very good approximation to quantum mechanics for many chemical reactions and a surprisingly good approximation even for reactions such as $H + H_2 \rightarrow H_2 + H$, which involve light hydrogen atoms. For reactions of heavier atoms neither exact potential energy surfaces nor exact quantum scattering calculations are available at present. For these the errors introduced by the use of classical mechanics cannot be directly evaluated, but it is likely that the errors in the potential energy surfaces lead to uncertainties greater than those due to the use of classical mechanics. Thus, one can argue that the errors of classical mechanics are not so serious in themselves and not as serious as other errors. And, of course, approximate quantum corrections to classical mechanics, such as tunneling corrections, can be applied to classical results. The rare-event method is a framework giving a basis for semiclassical and approximate quantum mechanical treatments of reaction processes.

Along with the assumption of classical mechanics is the assumption of a potential energy surface, that is, a Born–Oppenheimer potential energy surface for the motions of the nuclei. The motions of electrons are not considered except in special cases such as the electron-transfer reactions discussed below. In addition, the motions of the nuclei are assumed to occur on a single potential energy surface, although the possibility of switching from one surface to another could be included as it has been in conventional classical trajectory calculations.

In the following sections we describe in detail the theoretical basis for the rare-event approach and the calculation method. We discuss applications ranging from bimolecular reactions involving three atoms in the gas phase, to isomerization of small molecules in solution, to diffusion and reaction at surfaces, to enzyme-catalyzed reactions, to protein-folding reactions in solution.

II. PHASE-SPACE REPRESENTATION OF REACTION DYNAMICS

In 1915 Marcelin [5] observed that the state of a molecular system made up of n atoms—a molecule, several molecules, as complex as desired—could be characterized by $3n$ position coordinates $q_1, q_2, \ldots, q_{3n}$ and $3n$ momentum coordinates $p_1, p_2, \ldots, p_{3n}$ and that chemical reactions could be represented by the motions of points in the $6n$-dimensional phase space of those coordinates. An alternative statement is "The underlying order of any sort of complex system where there is confusion and unpredictability is essentially characterized by the movement of the system in phase space [13]." Marcelin divided the space into two parts with a $(6n-1)$ dimensional surface S and, following the statistical mechanical methods of Gibbs [10] as in hydrodynamics, obtained an expression for the number of points crossing the surface in a small time interval. Marcelin suggested that reaction would occur for a molecular system at an exceptional state attained upon reaching a critical surface S at the center of the system.

A modern version of the expression Marcelin wrote is

$$\frac{dN}{dt} = e^{-A/kT} \int_S e^{-E/kT} v_n \, dq_2 \cdots dq_{3n-1} \tag{5.2}$$

where dN/dt is the flow of points across the surface, $e^{-A/kT}$ is a normalizing factor, E is the sum of kinetic and potential energies for a point, and v_n is the velocity normal to the surface.

The development of transition state theory in its various forms proceeding from Marcelin's work has been described by Laidler and King [14]. We will not review the development here except as required to describe the rare-event approach. Horiuti [7] made several important advances in classical transition state theory. He specified the coordinates $q_1, q_2, \ldots$ as the orthogonal curvilinear coordinates of the nuclei of the chemical system and $p_1, p_2, \ldots$ as the conjugate momenta. He wrote the normalization constant in terms of the classical partition function for reactants. He also clarified the question of recrossings of the dividing surface and proposed the "variational" transition state theory in which an upper bound to the reaction rate is obtained by varying the location of the dividing surface to obtain a minimum value.

By following Horiuti's theory [7] the equilibrium rate constant k_e, (that for the flow of points in one direction across a dividing surface with reactants and products present at complete equilibrium) may be written in a form proposed by Keck [15] and modified by Jaffe et al. [16],

$$k_e(T) = Q^{-1} \int_{S, v_n > 0} e^{-E/kT} v_n \beta \prod_1^{3n} dp_j \prod_1^{3n-1} dq_j \tag{5.3}$$

where

$$Q^{-1} = \frac{1}{V} \int e^{-E/kT} \prod_1^{3n} dp_j \prod_1^{3n} dq_j \tag{5.4}$$

in which E is the total energy in center-of-mass coordinates, q_j and p_j are the conjugate position and momentum coordinates of the nuclei, v_{n} is the velocity in phase space normal to the dividing surface, β relates the differential surface area dS to the position and momentum coordinates,

$$dS = \beta \prod_{1}^{3n} dp_j \prod_{1}^{3n-1} dq_j \tag{5.5}$$

n is the number of nuclei, and Q is the classical partition function for reactants. The rate constant k_e is for $v_{\mathrm{n}} > 0$ and corresponds to reaction in the forward direction. Since it includes all crossings of the surface in the forward direction, it varies strongly with the location of the surface. Horiuti [7] pointed out that the dividing surface giving the lowest upper limit for the equilibrium rate constant k_e is not necessarily a surface passing through the saddle point on the potential energy surface.

The integrals of Eqs. (5.3) and (5.4) do not, in general, yield to analytic integration since the potential energy part of E is typically a complicated function of the coordinates q_j. However, unless the number of atoms is extremely large the integrals can be evaluated by any one of several types of Monte Carlo integration.

The familiar formulas of transition state theory developed by Wigner [6] and by Eyring [8] for approximating quantum systems may be obtained by replacing, where possible, the classical expressions of Eq. (5.3) with their corresponding quantum expressions to give

$$k_e = Q^{-1} \frac{kT}{h} Q^{\ddagger} e^{-E_0/kT} \tag{5.6}$$

where Q is the partition function for reactants, $Q^{\ddagger}$ is that for the transition state including all degrees of freedom except for the reaction coordinate, and E_0 is the potential energy reference for the transition state relative to that for reactants. In this expression for the equilibrium rate constant k_e the transmission coefficient is omitted.

The general expression of Eq. (5.1) includes both the classical and quantum transition state formulas as special cases.

The rate constant k (that for reactants in thermal equilibrium and products absent) is obtained by multiplying the equilibrium rate constant by a coefficient to account for a nonequilibrium distribution at the dividing surface and to account for those systems that fail to reach products after crossing the dividing surface. The transmission coefficient κ defined by Eyring [8] accounts only for the failure to reach products after crossing the dividing surface.

The transmission coefficient κ has often been carelessly defined. To avoid confusion we have chosen to use the term "conversion coefficient" and the symbol for the coefficient ξ defined by Wigner [6] to include the effects of failure of systems to originate from reactants and/or to reach products. This "conversion coefficient" ξ is defined as the ratio of the number of successful trajectories crossing a dividing surface in the forward direction to the total number of crossings of that surface in the forward direction at equilibrium. A successful trajectory crossing the surface in the forward direction more than once is counted as a single successful trajectory. This usage is entirely consistent with that of Horiuti [7] and Keck [1] and with our earlier usage [3].

When the conversion coefficient is defined as above it properly accounts for the failures of some trajectories to reach products and the failures of others to originate from reactants. The conventional rate constant k is given by the product of k_e and ξ,

$$k = \xi k_e \tag{5.7}$$

as in Eq. (5.1). Like k_e the coefficient ξ is strongly dependent on the location of the dividing surface. Since their product is a constant, they are inversely proportional to each other.

III. DISTRIBUTIONS IN THE TRANSITION REGION

Following the formulation of transition state theory in the early 1930s chemists argued for years over the question of equilibrium between reactants and activated complexes or transition states. When reactants and products are present in equilibrium with each other there is no question that there is an equilibrium distribution in the transition region, near the dividing surface, and throughout the accessible region of phase space. But, when products are absent nonequilibrium distributions may occur. Fortunately, the distributions for systems with reactants in thermal equilibrium and products absent have a very simple relationship to those in complete equilibrium: "When products are absent the distribution in the transition region is identical to an equilibrium distribution except that states lying on trajectories originating from products are missing [12]."

This simple fact may be seen with reference to the strings lying on the table in Fig. 5.2. Let the full set of strings represent the distribution of trajectories for complete equilibrium of reactants and products. Removal of products and trajectories originating from products leaves the trajectories originating from reactants unchanged. The unchanged trajectories are those originating from reactants at complete equilibrium.

Several authors [17] argued against the principle and others [18] argued that the principle is not proved. The basic principle is clearly stated by Horiuti [7] as a consequence of statistical mechanics and it is implicit in the papers of Wigner [6, 9]. Keck [19] showed the principle to be a consequence of Liouville's theorem. The principle follows more directly from the arguments we have given above [12] and is regarded by many as obvious [20]. Obvious or not, the principle is correct. An additional proof has been offered within the framework of flux–flux correlation functions by Yamamoto [21], and this has been restated in modern terms by Chandler [22].

IV. SAMPLING SUCCESSFUL TRAJECTORIES

As Keck [15] pointed out the local flow of trajectories across a dividing surface may be sampled and tested to obtain a "distribution of trajectories which reflects their a priori contribution to the reaction rate." These trajectories may be followed forward and backward in time to determine the conversion coefficient and to obtain a distribution of reactive or "successful" trajectories.

The sampling of crossing points can be carried out in conjunction with a Monte Carlo integration to determine the integral of Eq. (5.3). The trajectories are selected with probabilities proportional to the integrand of $v_n \exp(-E/kT)$ and thus depend on the flux rather than the density at the surface.

A procedure for determining the conversion coefficient is illustrated in Fig. 5.3. Several types of trajectories are shown with starting points on the dividing surface indicated by circles. A trajectory that (a) leads directly from a starting point on the surface to a product configuration without recrossing the surface and (b) originates from a reactant configuration (regardless of recrossings) is classified as a "successful" reactive trajectory. Trajectory 1 of Fig. 5.3 has a total energy insufficient to reach a product configuration and no calculation is required. Trajectories 3, 5, and 7 fail requirement (a) as determined by calculation. Trajectories 2, 4, and 6 meet requirement (a), but only 2 and 6 meet both requirements (a) and (b). The fraction of starting points that yield successful trajectories is 2 in 7 and the corresponding conversion coefficient is 2/7.

A trajectory calculation in the forward direction is terminated upon recrossing the dividing surface or upon reaching products. A trajectory calculation in the backward direction is terminated upon reaching either reactants or products. A product configuration is reached when products are formed and separate without any possibility of returning to each other. A reactant configuration is reached when reactants are similarly formed.

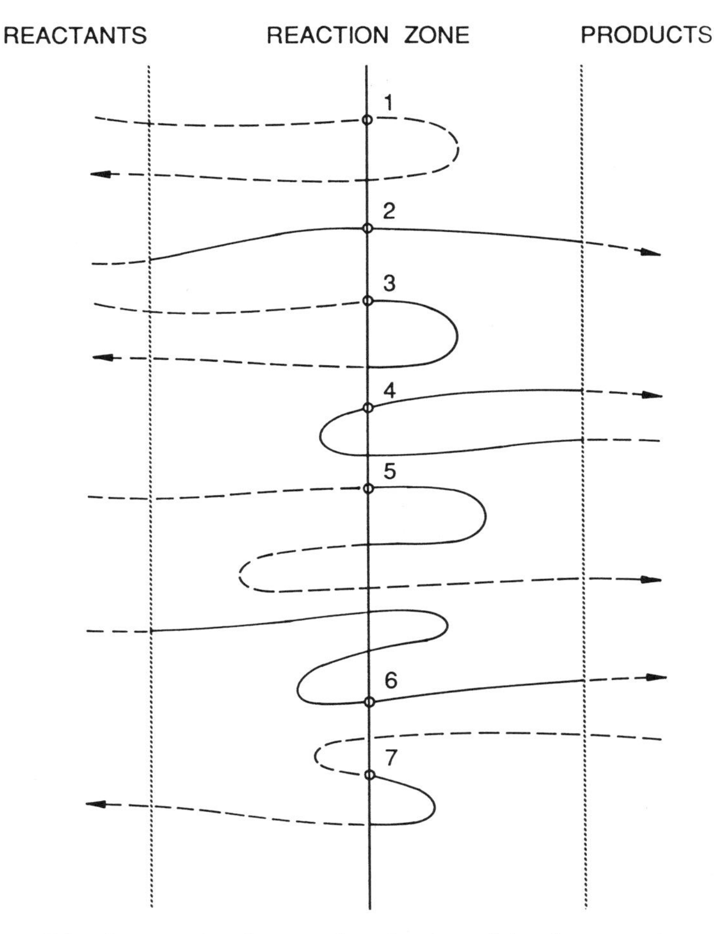

Figure 5.3. An example of a procedure for determining the conversion coefficient. Starting points for trajectories are circled. Calculated portions of the trajectories are indicated by solid lines; uncalculated by dotted lines. (From [16].)

The result is the same regardless of the location of the dividing surface provided it separates products from reactants. That is, the same set of successful trajectories is found from the complete set in Fig. 5.3 whether the dividing surface is at the edge of the reactant zone S_1, at the edge of the product zone S_2, or at an arbitrary location S across the interaction zone. For cases in which products or reactants do not separate, the criteria for reaching their states must be based on energy or momentum considerations.

A number of alternate schemes are available and the choice among them is governed by the characteristics of the event to be examined. Keck [1], for example, sampled trajectories crossing in the forward direction and followed them forward and backward regardless of recrossings. In this scheme the weight of a reactive trajectory making a single forward crossing is taken as unity and that of a reactive trajectory making n forward crossings is $1/n$. The conversion coefficient is equal to the sum of the weights divided by the number of forward crossings originally sampled.

Rosenberg et al. [23] and others used a scheme in which trajectories sampled crossing the surface in both directions are followed only forward in time. The conversion coefficient is given by the ratio of the net number (forward sampled less backward sampled) reaching products regardless of recrossings to the number sampled forward. To illustrate this scheme one must include in Fig. 5.3 the equally probable reverse of each trajectory shown. In the limit of a large number of trajectories a trajectory such as trajectory 4, which successfully reaches products, is canceled by its reverse. Trajectory 2 is not canceled by its reverse, which fails to reach products. The overall result is the finding of the same value for the conversion coefficient as found in other schemes, but a greater calculation effort is required and a sample of reactive trajectories is not produced directly.

The "local" conversion coefficient, given by the fraction of trajectories passing through a point on the dividing surface that are successful, may vary from 0 to 1 depending on its position on the surface. Identifying the regions of successful crossings is useful in understanding the reaction dynamics. For two-dimensional (2D) systems (i.e., having variables q_1, q_2, p_1, p_2) of fixed energy Pollak and Pechukas [24] (see also De Vogelaere and Boudart [25])] were able to show that successful and unsuccessful regions of the dividing surface are separated by boundaries that correspond to periodic orbital trajectories. A formula for three-dimensional systems has been derived by Toller et al. [26] and by Gillilan [27]. In principle, one may determine the successful regions of the dividing surface by finding the periodic orbits. In practice, it is far simpler to sample and test trajectories one by one.

A microcanonical version of the rare-event method has been developed by Anderson [28] for classical systems. It is applicable to systems in which reactants are in thermal equilibrium but limits the treatment to reactants of fixed total energy. The microcanonical rate constant k_E at specified energy E is defined in the sense of a distribution function

$$k_E = \frac{dk}{dE} \tag{5.8}$$

such that the complete rate constant is given by

$$k = \int_0^\infty k_E \, dE \tag{5.9}$$

The conversion coefficient is defined for each k_E by

$$k_E = \xi_E \, k_{eE} \tag{5.10}$$

where k_{eE} is the equilibrium rate constant.

The microcanonical version has been found useful in determining very small cross-sections for rare events occurring in collisions at specific energies [16]. Cross-sections as low as 10^{-24} Å^2 were determined for the reaction of H with HF at low energies.

V. TIME CORRELATION FUNCTIONS

The rate constants of various processes given by rare-event theory may be expressed in terms of time correlation functions. For the reaction $A \rightarrow B$ the rate constant may be written [29] as

$$k = \frac{1}{Z_A} \langle \dot{q}_\pm \delta(q^* - q) X_{A \rightarrow B}(\Delta t) \rangle \tag{5.11}$$

where Z_A is the partition function per unit volume for reactants, q is a position coordinate, q^* is the position coordinate of the dividing surface, the brackets indicate an integral over all position and momentum coordinates, the Dirac $\delta(q - q^*)$ converts the spatial integral to a surface integral along the dividing surface, and $X_{A \rightarrow B}(\Delta t)$ is the characteristic function for the reaction, which is unity for coordinates corresponding to B at time Δt and to A at time $-\Delta t$, and which is zero otherwise. This is the conventional formula applicable to sampling trajectories crossing the dividing surface in both directions and following them forward and backward in time to determine their characteristic functions $X_{A \rightarrow B}(\Delta t)$ for large Δt, that is, whether they are reactive or not.

A similar expression for the rate constant in terms of a time correlation function was derived by Yamamoto [21],

$$k = \frac{1}{Z_A} \langle \dot{q}_\pm \delta(q^* - q) X_B(\Delta t) \rangle \tag{5.12}$$

where the characteristic function $X_B(\Delta t)$ is unity for coordinates corresponding to B at time Δt. This is the expression applicable to sampling trajectories crossing the dividing surface in both directions and following them forward in time to determine their characteristic functions $X_B(\Delta t)$, that is, whether they reach B or not.

For sampling the flux across the dividing surface in the forward direction and following trajectories forward and backward in time to find those originating at A regardless of recrossings and passing directly to B without recrossing the rate constant is given by

$$k = \frac{1}{Z_A} \langle \dot{q}_+ \delta(q^* - q) X_{A \to (d)B}(\Delta t) \rangle \tag{5.13}$$

The characteristic function $X_{A \to (d)B}(\Delta t)$ is unity if the specified criteria are met and are zero otherwise.

For sampling the flux across the dividing surface in the forward direction and following trajectories forward and backward in time to find those originating at A regardless of recrossings and passing to B regardless of recrossings, the rate constant is given by

$$k = \frac{1}{Z_A} \langle \dot{q}_+ \delta(q^* - q) X_{A \to B, 1/n}(\Delta t) \rangle \tag{5.14}$$

The characteristic function $X_{A \to B, 1/n}(\Delta t)$ is equal to $1/n$ for a reactive trajectory, where n is the number of forward crossings for the trajectory, and zero for an unreactive trajectory.

The equivalence of the several expressions may be seen by considering the rate constants obtained for a set of trajectories such as those of Fig. 5.3, with consideration of microscopic reversibility. The equivalence of Eqs. (5.11) and (5.12) has also been shown in an analysis by Chandler [22].

For each of the many schemes that have been or could be devised for finding a set of reactive trajectories or for determining a conversion coefficient there is a corresponding time correlation function. There is also a corresponding procedure for finding the set of strings successfully crossing the table of Fig. 5.2.

VI. RELAXATION AND MULTIPLE TIME SCALES

For bimolecular reactions in the gas phase the meaning of an equilibrium distribution of reactant states is easily understood: the reactants cross into the transition region (at S_1 of Fig. 5.1) from an equilibrium distribution. This condition might be ensured experimentally by adding an inert gas to a reactant mixture to facilitate collisional relaxation among the species.

The essential requirement is that relaxation among the states be rapid compared to the depletion of reactive states.

Furthermore, it is assumed that the reactants do not undergo additional collisions in their passage through the transition region, that is, there are no termolecular collisions. Thus, for a bimolecular reaction in the gas phase one can identify three time scales with different characteristic times: τ_{rxn}, the time constant for disappearance of reactants; τ_{relax}, the time constant for relaxation of reactant states; and τ_{pass}, the time constant for passage through the interaction region. Under normal conditions one has $\tau_{rxn} \gg \tau_{\text{relax}} \gg \tau_{\text{pass}}$. Unless these conditions are met the rate constant itself will not have a clear meaning.

For reactions in condensed phases an entire phase can be treated in the same fashion as a bimolecular reaction above; but, one would prefer to simplify the treatment and separate as much as possible consideration of the internal energy relaxation of reactants, the exchange of energy with other species such as solvent molecules, the exchange among the other species, the separation of products, and all the other events associated with the rare event of interest. Each of these has a characteristic time scale and may be separable. Questions of relaxation and multiple time scales have been investigated by Widom [30], Berne [31], and others. Northrup and Hynes [32] denoted reactants and/or products in thermal equilibrium as "stable states."

VII. APPLICATIONS TO MODEL SYSTEMS

The rare-event approach may be applied in the prediction of rare events for systems obeying the laws of classical mechanics and these include not only real systems made up of atoms but also model systems. Typical of these model systems are those for reactions in solution in which the effects of solvent molecules are replaced by friction as in the Kramers approach to chemical kinetics [33]. The rare-event approach may be used to obtain (a) exact simulations of a real system as well as (b) exact simulations of a model system. The results of (a) and (b) may be used to assess the accuracy of approximate methods for treating a model system. At each level the rare-event approach provides a standard with which a lower level may be compared.

VIII. MULTIPLE BARRIERS (BOTTLENECKS)

The folding of a protein molecule occurs in a sequence of reactions or rearrangements each requiring the crossing of a potential energy barrier or passage through a bottleneck of one kind or another in configuration space. Many other processes occur as similar sequences of rare events. The barriers or bottlenecks separate individual identifiable species or types of configurations. Under conditions, such as intimate energy exchange with a solvent, which lead to thermal equilibration of the intermediate species, sequential reactions like these may be treated by successive applications of the rare-event method. Bennett [34–36] led the way in this area with a calculation in 1975 of the diffusion of vacancies in a solid and a full discussion in 1977 of the possibilities for other systems.

The basis for treating multiple barriers is illustrated in Fig. 5.4 for the sequential reaction

$$\mathrm{X} \underset{k_2}{\overset{k_1}{\rightleftharpoons}} \mathrm{i} \underset{k_4}{\overset{k_3}{\rightleftharpoons}} \mathrm{j} \underset{k_6}{\overset{k_5}{\rightleftharpoons}} \mathrm{k} \underset{k_8}{\overset{k_7}{\rightleftharpoons}} \mathrm{Y}$$

The five configurations X, i, j, k, and Y are separated by potential energy barriers. If the barriers are high relative to the value of kT and the various translational, rotational, vibrational, and electronic states of the species equilibrate rapidly (as with a solvent), then the rare-event approach may be used to calculate individual rate constants $k_1, k_2, \ldots, k_8$. From these the rates of individual reactions and of the overall reaction may be determined. In these applications the thermal equilibrium of each species is assumed, but chemical equilibrium among the species is not assumed.

For complex systems the rare-event method has reduced the calculation of trajectories to a problem that is small in comparison to the problems of finding reaction pathways and bottlenecks, calculating the equilibrium constants, and sampling the configurations crossing the

Figure 5.4. Bottlenecks in the pathway for the reaction X→Y. The reaction takes place across multiple barriers separating intermediate states. (Adapted from [36].)

dividing surface. Bennett [34–36] exhibited remarkable foresight in his early discussions of these problems. A variety of techniques has been developed for overcoming them: umbrella sampling methods [37], activated and branching-activated trajectory methods [38], and constrained reaction coordinate dynamics [39].

IX. ADDING QUANTUM EFFECTS

The rare-event approach provides a framework that is a basis for semiclassical and approximate quantum treatments of molecular dynamics. In the ideal situation one would be able to apply simple quantum corrections to classical rate expressions. It seems unlikely that such corrections could be both simple and accurate, but there is evidence for optimism in this area.

Truhlar, Isaacson, Skodje, and Garrett [40–42] have taken steps in this direction with a unified dynamical (UD) model, which combines recrossing corrections, quasiclassical ensembles, and microcanonical ensembles. These authors have compared results for the UD model with those of accurate full-quantum calculations. For a number of simple reactions, collinear and three-dimensional, good agreement was found. Alternate approaches to incorporating quantum effects have been reported by Tromp and Miller [43], by Smith [44] and Frost and Smith [45], and by Hwang and Warshel [46].

X. EXAMPLE: DISSOCIATION – RECOMBINATION

The first application of the rare-event approach was made by Keck [1] in calculating the rates of dissociation and recombination reactions of the type $X_2 + M \rightleftharpoons X + X + M$, specifically the dissociation and recombination of H_2, O_2, and I_2 in collisions with Ar. The potential energy surfaces were taken as the sums of the interaction potentials for individual pairs of atoms. The dividing surface was located at the top of the rotational barrier to reaction and trajectories were followed forward and backward until the argon atom was well separated from X_2 and/or both X atoms. Keck found, as he expected, that the fraction of trajectories giving reaction was "tremendously increased" over that for starting with separated reactants. Of 2400 trajectories sampled approximately one half were reactive.

Similar rare-event calculations were carried out by Kung and Anderson [47] for the dissociation–recombination of H_2 with H atoms as the third body. In this case an *ab initio* potential energy surface known to be reasonably accurate was available. The calculations were made for a temperature of 4000 K so that results could be compared with a number of experimental measurements at that temperature. The calculated rate constant was found to be at the center of the range of experimental values.

XI. EXAMPLE: THE REACTION $F + H_2 \rightarrow HF + H$

The first application of the rare-event approach to a chemical exchange reaction was in trajectory calculations of the molecular dynamics of the reaction of F with H_2 [16]. The reaction $F + H_2 \rightarrow HF + H$ is of special theoretical interest because it is one of the simplest examples of an exothermic chemical reaction.

Classical trajectory calculations [48] starting from reactants with a semiempirical potential energy surface have reproduced all the major features of a wealth of experimental measurements for this reaction. Use of the rare-event approach produced similar results with the only differences attributable to the use of quantized reactant states in starting from reactants. A comparison of fully classical microcanonical calculations starting from the dividing surface with similar calculations starting from reactants showed the results to be identical within statistical error. The calculation effort required in using the rare-event method to obtain 1000 trajectories representative of those for thermal reaction was about 1/50 the effort required in starting from reactants.

The reaction exhibits several interesting features of molecular dynamics that are clearly illustrated in classical trajectory calculations. Reaction occurs through a collinear (or nearly collinear) configuration F–H–H across a barrier of about 1 kcal mol^{-1} (perhaps higher) in the F—H–H entrance valley and proceeds with an energy release of about 32 kcal mol^{-1} in dropping to the F–H——H exit valley. Reaction is favored by a high translational energy in the approach of F and H_2 and energy is released primarily to vibration of the product molecule HF. Among all chemical reactions the reaction of F with H_2 is one of the fastest, one of the most exothermic, and one of the most specific in energy disposal. The reverse reaction is correspondingly specific in its energy requirements [2].

The rare-event calculations [16] for the reaction of F with H_2 were carried out with a semiempirical London–Eyring–Polanyi–Sato potential energy surface that is a reasonably good representation of the true surface. The surface is shown for collinear configurations by the contour plot of Fig. 5.5 for which the distances are expressed in terms of R_{HH}, the separation of H from H, and R_{F,H_2}, the separation of F from the center of mass of H_2. To simplify the analysis the dividing surface S was selected

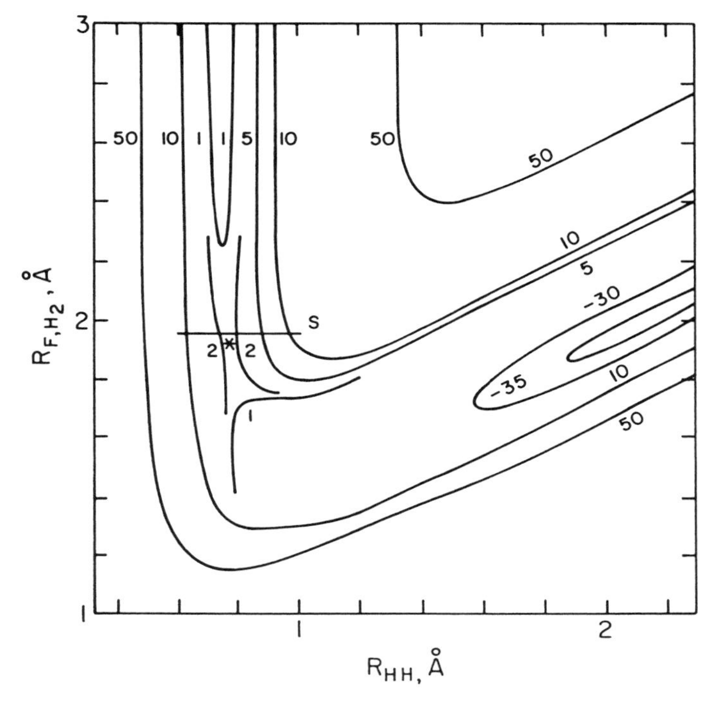

Figure 5.5. Potential energy surface for F–H–H in a collinear configuration. Equal potential curves are labeled in kilocalories per mole relative to the minimum for F separated from H_2. The dividing surface is indicated by S and the saddle point by an asterisk. (From [16].)

normal to the $R_{\mathrm{F,H_2}}$ coordinate spanning the entrance valley slightly upstream of the saddle point. The left and right edges were located in regions of high potential energy sufficient to eliminate any significant flow around the edges.

The system is most easily described in terms of particles A (one H atom), B (the other H atom), and C (the F atom) in spherical polar coordinates for A relative to B and C relative to the center of mass of AB. These coordinates are $R_{\mathrm{A,B}}$, $\theta_{\mathrm{A,B}}$, $\phi_{\mathrm{A,B}}$, $R_{\mathrm{C,AB}}$, $\theta_{\mathrm{C,AB}}$, and $\phi_{\mathrm{C,AB}}$. The q and p parameters of Eqs. (5.2)–(5.5) correspond to these positions and their conjugate momenta. The dividing surface is shown for these coordinates in Fig. 5.6. With the dividing surface fixed at constant $R_{\mathrm{C,AB}} = R'_{\mathrm{C,AB}}$ the integrations indicated in Eqs. (5.3) and (5.4) can be carried out over the orientation angles $\phi_{\mathrm{A,B}}$, $\theta_{\mathrm{C,AB}}$, and $\phi_{\mathrm{C,AB}}$ and over all the momenta to give an expression for the equilibrium rate constant:

$$k_e(T) = Q^{-1}(2\pi\mu_{\mathrm{A,B}}kT)^{3/2}(2\pi\mu_{\mathrm{C,AB}}kT)^{3/2}4\pi R'^2_{\mathrm{C,AB}}kT\, I_1 \quad (5.15)$$

where

$$I_1 = 2\pi \int_{-1}^{+1} \int_{R_{\mathrm{A,Bmin}}}^{R_{\mathrm{A,Bmax}}} e^{-V/kT} R^2_{\mathrm{A,B}}\, dR_{\mathrm{A,B}}\, d\cos\theta_{\mathrm{A,B}} \quad (5.16)$$

and

$$Q^{-1} = (2\pi\mu_{\mathrm{A,B}}kT)^{3/2}(2\pi\mu_{\mathrm{C,AB}}kT)^{3/2}\, I_2 \quad (5.17)$$

where

$$I_2 = 4\pi \int_0^{R''_{\mathrm{A,B}}} e^{-V/kT}\, R^2_{\mathrm{A,B}}\, dR_{\mathrm{A,B}} \quad (5.18)$$

The upper limit $R''_{\mathrm{A,B}}$ in the integration for I_2 is specified as the separation distance beyond which AB may be considered dissociated.

These equations may be combined to give a simpler expression for the equilibrium rate constant

$$k_e(T) = (8kT/\pi\mu_{\mathrm{C,AB}})^{1/2}\pi R'^2_{\mathrm{C,AB}}(I_1/I_2) \quad (5.19)$$

Numerical integration to determine I_1 and I_2 and substitution of these values into Eq. (5.19) gave a value of $3.29 \times 10^{12}\ \mathrm{cm^3\ mol^{-1}\ s^{-1}}$ for $k_e(T)$ at 300 K.

The trajectories crossing the dividing surface in the direction of F approaching H_2 were sampled from the equilibrium flux. In effect, points on the 11-dimensional dividing surface of Eq. (5.3) were selected with probabilities proportional to the integrand of Eq. (5.3). The spatial orientation could be specified arbitrarily as $\phi_{\mathrm{A,B}} = 0$, $\theta_{\mathrm{C,AB}} = 0$, and $\phi_{\mathrm{C,AB}} = 0$. The differential surface area for the remaining positions was then given by $R^2_{\mathrm{A,B}} dR_{\mathrm{A,B}} d\cos\theta_{\mathrm{A,B}}$ and crossing points on the surface shown in Fig. 5.6 were selected with probabilities proportional to $e^{-V/kT}R^2_{\mathrm{A,B}}$. The momentum $p_{z_{\mathrm{C,AB}}}$, normal to that surface, was selected from an equilibrium distribution weighted by the velocity across the surface,

$$f(p) = -p\exp(-p^2/2\mu kT) \qquad -\infty < p < 0 \quad (5.20)$$

The other momenta $p_{x_{\mathrm{C,AB}}}$, $p_{y_{\mathrm{C,AB}}}$, $p_{x_{\mathrm{A,B}}}$, $p_{y_{\mathrm{A,B}}}$, and $p_{z_{\mathrm{A,B}}}$ were selected

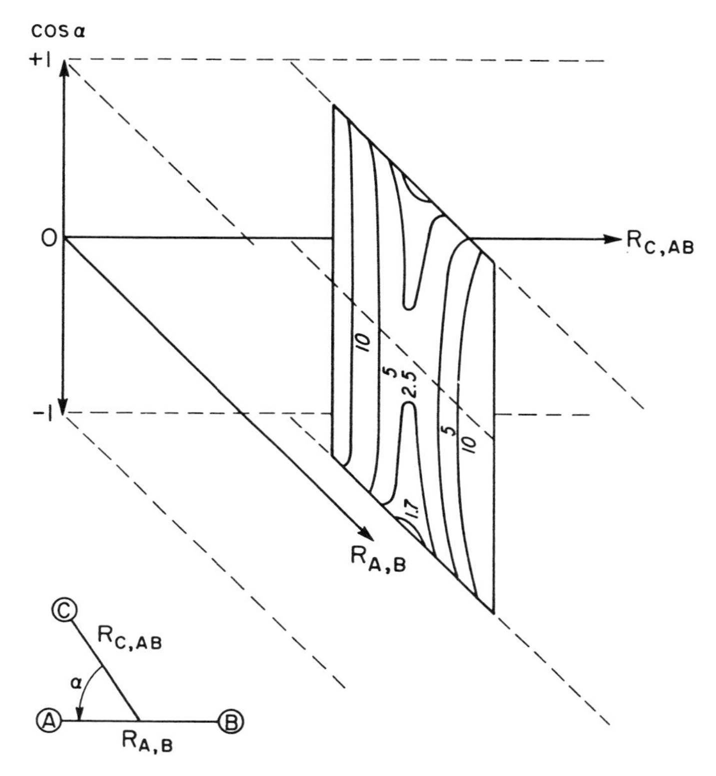

Figure 5.6. Location of dividing surface S for reaction of F with H_2, labeled as reaction of C with AB. Equal potential curves at the dividing surface are indicated in kilocarories per mole. (From [16].)

from an unweighted equilibrium distribution,

$$f(p) = \exp(-p^2/2\mu kT) \qquad -\infty < p < +\infty \tag{5.21}$$

The reaction zone was defined as the region with $R_{A,B}$ less than 6 Å in the direction of products and both $R_{C,A}$ and $R_{C,B}$ less than 6 Å in the direction of reactants. Trajectories sampled crossing the dividing surface were followed first in the direction of products HF + H. Trajectories found to recross the surface were terminated. Those that passed directly to products were followed backward from their starting points to determine if they originated from reactants. Of the 5117 trajectories sampled in crossing the dividing surface, all had sufficient energy to cross the barrier, and 1001 were found to be "successful" trajectories. None had multiple crossings. All unsuccessful trajectories failed by recrossing the surface in the direction back toward reactants. The conversion coefficient ξ was thus found to be 1001/5177 or 0.193 and the thermal rate constant, given by the product ξk_e, was 6.37×10^{11} cm^3 mol^{-1} s^{-1}. The sampling error for the set of 1001 trajectories was about 3%.

The set of 1001 reactive trajectories was examined to determine a number of different characteristics of the forward and reverse reactions. The energy distribution for reactants, plotted in Fig. 5.7, is peaked at the lowest values for both vibrational and rotational energies of H_2 and at about 4 kcal mol^{-1} for the initial translational energy for the approach of F and H_2. The energy distribution for reaction products, plotted in Fig. 5.8, is peaked at about 30 kcal/mol^{-1} for the vibrational energy of HF, at zero for the rotational energy of HF, and at about 10 kcal mol^{-1} for the translational energy of separation of the products HF and H.

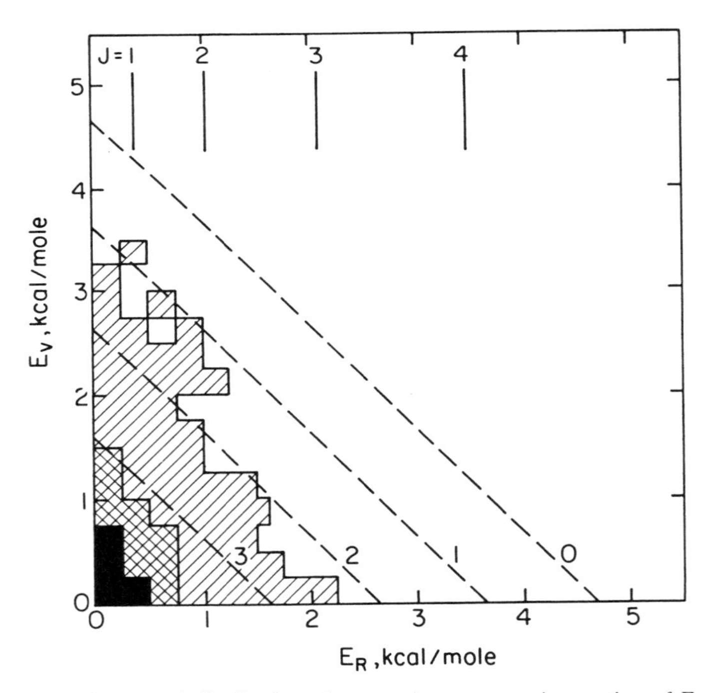

Figure 5.7. Calculated distribution of energy for reactants in reaction of F with H_2. Higher probabilities are indicated by darker shading. Approximate translational energies are indicated by the dashed diagonal lines. The rotational quantum levels for H_2 are indicated. (From [16].)

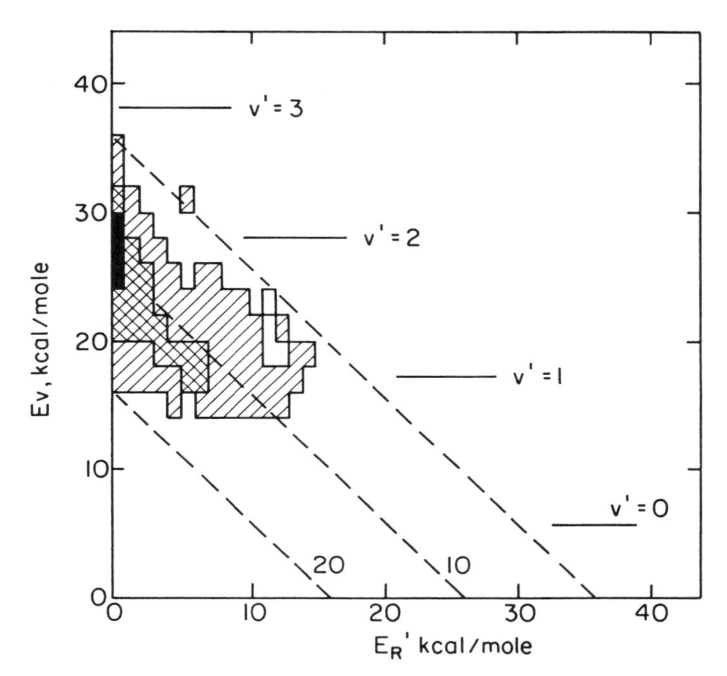

Figure 5.8. Calculated distribution of energy among products HF + H from reaction of F with H_2. Higher probabilities are indicated by darker shading. Approximate translational energies are indicated by the dashed diagonal lines. The vibrational quantum levels for HF are indicated. (From [16].)

Reaction cross-sections for the forward and reverse reactions were extracted from the results. These included, for example, the weighted average cross section $S_r(T, E_T)$ for the forward reaction as a function of translational energy for a thermal distribution of other parameters. Most interesting is the cross-section $S_r'(T, E_V', E_T')$ for the reverse reaction as a function of vibrational energy and translational energy for a thermal distribution of other parameters. In this case cross-sections as low as 10^{-24} Å^2 could be determined for the extremely rare reaction of H with HF at low vibrational energies.

Since the rate of reaction of F with H_2 is high, a reactive trajectory is not an especially rare event and the computational advantage of the rare-event approach might not be expected to be very large, but it was found to be a factor of about 50. The computational effort required to obtain 1001 reactive trajectories representative of reaction in a thermal system by sampling trajectories crossing a dividing surface in the vicinity of the saddle point was about 1/50 that for sampling trajectories starting from separated reactants. For the much slower reverse reaction of HF with H the computational advantage was found to be very much larger.

XII. EXAMPLE: THE REACTION $H_2 + I_2$ (OR 2I) $\rightarrow$ HI + HI

The advantages of the rare-event method are illustrated dramatically for the hydrogen–iodine reactions, overall $H_2 + I_2 \rightarrow HI + HI$, first investigated by Bodenstein 100 years ago. Two different groups carried out trajectory calculations for this reaction using the same semiempirical potential energy surface. Raff et al. [49] used the standard approach of starting trajectories with separated reactants. Jaffe and co-workers [3, 4] used the rare-event approach of starting trajectories at the dividing surface. The experiences of the two groups allow the two methods to be compared.

The potential energy surface used favored reaction along two paths: one through a collinear configuration I–H–H–I and the other through a trapezoidal configuration. Raff et al. [49] investigated many thousands of collisions of H_2 with bound and unbound I_2 but were unable to find any reactive trajectories except when statistically improbable collinear or very nearly collinear trajectories were specified. They predicted the overall rate of such processes to be five orders of magnitude lower than rates measured experimentally and concluded that reaction takes place by a mechanism involving an intermediate species H_2I. Jaffe and co-workers [3, 4] started 1349 trajectories at a dividing surface and found 693 reactive trajectories with approximately equal statistical weights. They observed the overall reaction to occur by both bimolecular and termolecular processes, $H_2 + I_2 \rightarrow HI + HI$ and $H_2 + I + I \rightarrow HI + HI$.

The advantages of the rare-event approach are clear. It found the correct mechanism and generated hundreds of successful trajectories. The conventional approach failed to find the mechanism and generated no useful trajectories.

The calculation effort to determine a sample of reactive trajectories was found to be a factor of 10^6 lower for the rare-event approach.

A modification of the potential energy surface to favor reaction along the collinear path produced a surface in better agreement with more advanced quantum calculations [50, 51]. On this surface later rare-event calculations [51] showed reaction by mechanisms involving highly vibrationally excited iodine molecules I_2(hi v) and unbound I atom pairs: $H_2 + I_2(\text{hi v}) \rightarrow HI + HI$ and $H_2 + I + I \rightarrow HI + HI$. Reaction was observed to occur by insertion of H_2 between the atoms of I_2(hi v), stretched to a separation of about 6 Å, or between the atoms of an unbound pair I + I separated by about 6 Å, followed by separation into HI + HI. A minor change in the potential energy surface might favor either the bimolecular or the termolecular mechanism and firm experimental evidence is absent, but it appears most likely that reaction occurs by both mechanisms as indicated by the rare-event calculations [50, 51].

XIII. EXAMPLE: VACANCY DIFFUSION IN SOLIDS

The first application of rare-event theory to problems of condensed phases was made by Bennett [34, 35] in 1975. He treated a problem of diffusion of point defects in solids as a problem of multiple stable states with the point defects assumed in thermal equilibrium in potential wells separated by barriers.

The system studied was a periodic crystal of 255 atoms arranged in a $4 \times 4 \times 4$ face-centered-cubic (fcc) cell with a single vacancy. The potential energy was given by pairwise-additive Lennard–Jones functions with conditions approximately those of argon at 80 K. The calculations were designed for investigating the mass dependence of the rate of diffusion of vacancies near the melting point of the crystal.

The dividing surface was described by a zero value of the reaction coordinate η, which could be simply expressed as a linear function of atomic coordinates r_i,

$$\eta = \text{const} + \sum_{i=1}^{N} a_i \cdot r_i \tag{5.22}$$

The coefficients a_i were determined by locating the exact N-body saddle point for the 255-atom system and the direction of steepest descent in $3N$-configuration space from that point.

The vacancy diffusion rate or spontaneous jump frequency Γ_j was given by the product of the equilibrium crossing frequency Γ' and the conversion coefficient ξ,

$$\Gamma_j = \Gamma' \xi \tag{5.23}$$

The variation of Γ' with atomic mass was known analytically so that only the variation of the conversion coefficient with mass was needed from the molecular dynamics calculations.

Bennett used a differential sampling method to investigate the small difference in ξ for masses m_1 and m_2 equal to 1.00 and 1.05 in units of the host atom mass. Starting conditions for trajectories with the different masses were selected with identical atom positions and with corresponding velocities (i.e., scaled to obtain identical kinetic energies). Points on the dividing surface were obtained as positions and velocities sampled from molecular dynamics calculations for $m_1 = 1.00$ constrained to the dividing surface with $\eta = 0$. The required velocity in the direction of the reaction coordinate was sampled from a flux-weighted Maxwellian distribution. Trajectories were calculated in the forward and backward directions for mass m_1 and, with the scaled initial velocity, for mass m_2.

The mean value of the conversion coefficient $\xi_{1.00}$ was determined to be 0.80 ± 0.06 and the mean difference $\xi_{1.05} - \xi_{1.00}$ was $+0.0018 \pm 0.0009$. The lower value of the conversion coefficient for the lower mass yields an isotope effect about 10% lower than given by uncorrected transition state theory and is "in agreement with most experimental data on fcc materials." Thus, Bennett's calculations indicate that previous treatments based on transition state theory alone without molecular dynamics calculations were in error due to the neglect of the effects of unsuccessful trajectories.

XIV. EXAMPLE: DIFFUSION ON SURFACES

The treatment of diffusion on surfaces is conceptually the same as that of diffusion in solids. At high temperatures diffusion for most systems is sufficiently rapid that conventional molecular dynamics methods are adequate. At lower temperatures hops between adjacent sites become rare events and conventional molecular dynamics calculations are impractical. Aside from classical behavior the fundamental assumptions required in the rare-event approach are that reactants (molecule and first site) are in thermal equilibrium and that products (molecule and second site) are immediately equilibrated thermally on formation. This is equivalent to the assumptions of thermal equilibrium at the reactant surface S_1 and no return from the product surface S_2 in the case of a gas-phase reaction. Correlated hops invalidate the assumptions.

The first applications were made by Voter and Doll [52] in 1985 with simulations of the diffusion of single rhodium atoms on a rhodium (100) surface. The calculations were performed for a system of 33 free atoms (1 adatom and 32 free surface atoms) and 96 fixed atoms (3 layers of 32 fixed atoms each), all interacting with pairwise-additive Lennard–Jones potentials. The agreement with experiment was excellent. Voter and Doll attributed the "surprisingly small" discrepancy between calculated rates of diffusion and those measured by field-ion microscopy to defects in the approximate potential energy function used in the calculations.

Voter and Doll [52] overcame the problem of multiple correlated hops by developing a method for treating multistate systems in a fashion similar to that for two-state systems. Their method accounts for thermal equilibration of an atom at sites not adjacent to the starting site. It could be applied to a number of other systems having similar characteristics.

The rare-event approach was used by Voter [53] to treat the diffusion of clusters of atoms on a surface to obtain classically exact overlayer dynamics. Using clusters of up to 75 rhodium atoms on a rhodium (100) surface, Voter found the dominant mechanism for the diffusion of larger clusters to be the diffusion of single atoms along the edges of the clusters.

XV. EXAMPLE: PROTEIN DYNAMICS—ROTATION OF THE TYROSINE 35 RING IN BPTI

The typical molecular dynamics simulation of a protein requires time steps of about 10^{-15} s and can be carried out for as many as 10^6 steps in a week or two on a modern supercomputer. The time span of 10^{-9} s is sufficient for the observation of many of the frequent (common) processes occurring in proteins, but many other processes of interest in proteins occur too infrequently to be observed in such a short span of time. These rare events may require simulations as much as 10^6 times longer unless the rare-event approach is used. As in the case of the hydrogen–iodine gas-phase reactions the rare-event approach provides the required gain in efficiency and makes the simulation of rare events occurring in protein molecules practical.

Reactions catalyzed by enzymes are in some cases limited in rate by the conformational changes required in one or more groups of the enzyme. For reaction to occur these groups must lie in favorable positions and a configuration in which all groups are located in favorable positions occurs only rarely.

The rare-event approach was first applied in simulations of protein dynamics in 1979 by McCammon and Karplus [54] in a study of the rate of rotation of a tyrosine ring in bovine pancreatic trypsin inhibitor (BPTI), a small protein of 58 amino acids. This model system is one of the first to be studied both experimentally and theoretically, and for that reason it has come to be called the "hydrogen atom" of protein dynamics [55]. The function of BPTI is to bind to the active site of the digestion enzyme trypsin and prevent its acting as a catalyst in the cleavage of protein molecules.

The transition studied in BPTI was the rotation of the phenol ring of tyrosine (Tyr-35)—a flat hexagonal ring of six carbon atoms located in a side chain of the amino acid tyrosine—buried in the interior of the globular protein structure. The ring is attached to the rest of the molecule by a single bond and rotation about that bond is hindered by steric interactions with nearby polypeptide chains. The structure of BPTI is well known from X-ray diffraction measurements of its crystals. The rotation of the ring is not particularly important to the function of the inhibitor, but the rate of rotation and its temperature dependence have been determined with measurements of nuclear magnetic resonance (NMR) line shapes, so that the energy and entropy of activation are known. At 300 K the measured rate constant is in the range of $1\text{–}10\,\mathrm{s}^{-1}$ and the activation energy is about $16\,\mathrm{kcal\,mol}^{-1}$. Rotation of the tyrosine ring is thus an activated process and it is representative of many transitions occurring in protein molecules.

The BPTI globule was treated as the system of the 454 heavy (nonhydrogen) atoms of BPTI together with the four heavy atoms of four internal water molecules. The hydrogen atoms were included indirectly by increases in the masses of the heavy atoms. Solvent molecules were not included. The (3×458)-dimensional potential energy surface was that of an empirical energy function giving the potential energy as a sum of terms involving bond lengths, bond angles, and dihedral angles for each bond, terms for hydrogen bonds, and terms for van der Waals and electrostatic interactions.

The rotation of the tyrosine ring is shown schematically in Fig. 5.9. The ring and a part of a nearby polypeptide chain are shown at two different times in the course of one of the successful ring-flipping trajectories calculated by McCammon and Karplus [54]. Movement of the nearby chain is necessary to create space for the rotation.

Locating the barrier to a transition in such a system is not the simple task of examining the 2D potential energy plot for the collinear reaction of F with H_2. It requires a major effort—computational and otherwise. McCammon and Karplus used a combination of molecular dynamics and Monte Carlo procedures to investigate the potential energy barrier to reaction. The globule was first allowed to relax to an equilibrated structure. The Tyr-35 ring was then rigidly rotated to a dihedral angle χ_{35}^{2} of 180° near the top of the barrier given by the simple dipeptide potential. With the ring held in place at or near the fixed dihedral angle the remainder of the globule was allowed to relax in a Metropolis Monte Carlo calculation, modified for more frequent sampling of moves for atoms near the ring, to produce samples from a Boltzmann distribution of constrained configurations. Trajectory calculations starting from these configurations were then carried out with initial ring rotation in forward and backward directions to determine whether a barrier was crossed. Barrier crossings were not observed for trajectories started at $\chi_{35}^{2} = 180°$, but additional calculations at other angles revealed successful barrier crossings at $\chi_{35}^{2} = 140°$, and that angle was chosen, in effect, as the

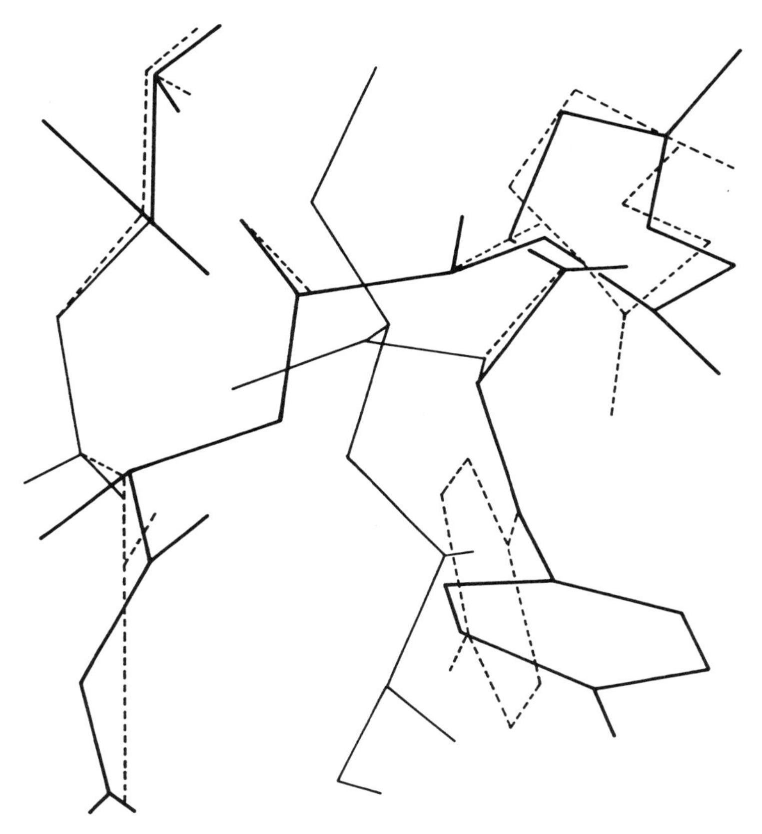

Figure 5.9. Rotation of tyrosine ring 35 in bovine pancreatic trypsin inhibitor. The ring and nearby structures are shown at two different times (solid and dashed lines) in a molecular dynamics simulation. (Adapted from [56].)

dividing surface. Monte Carlo calculations, with the system constrained to that angle, were then carried out to produce a sampling of configurations to be used as starting points (Set A) for more complete trajectory calculations. A second set (Set B) of configurations, with an angle $\chi_{35}^{2} = 200°$ corresponding to a second dividing surface, was determined in the same fashion using a different equilibrated structure prior to ring rotation.

Trajectories started at the dividing surfaces were followed in molecular dynamics calculations forward and backward in time. The initial velocities of the atoms of the ring were chosen as those for the collective rotation of the ring (i.e., corresponding to torsional motion of four of the six ring atoms with the appropriate moment of inertia, selected from a flux-weighted equilibrium distribution). The initial velocities of all other atoms and of the in-plane motions of the four rotating ring atoms were selected from Maxwellian distributions.

All 10 trajectories were successful in producing a ring rotation from one stable orientation to the other. In the barrier crossing the change in angle with time was monotonic for all trajectories, but the rate of change was varied and in several trajectories the motion was nearly stopped one or more times.

Examination of the details of the successful trajectories gave a complete picture of the combination of backbone, local polypeptide chain, and ring movements taking place in the transition. McCammon and Karplus [54] observed that the stresses due to nonbonded repulsion were relaxed by local adjustments, which were rapid compared to the rotation of the ring. They also observed that the torques on the rings were primarily impulsive in nature and that frictional effects were weak. The results of these early rare-event calculations for this "hydrogen atom" of protein dynamics provided valuable insights into the nature of activated transitions.

In succeeding calculations Northrup et al. [57] used umbrella sampling techniques to assist in locating the barrier. Their barrier height was about 10 kcal mol^{-1} and its crest was somewhat shifted from that of the earlier study. For 200 trajectories started forward and backward from the top of the barrier about 80% recrossed the dividing surface. The computed conversion coefficient was 0.22 and the rate constant had the value of about 10^5 s^{-1} (a value higher than the experimental value by a factor of 10^4–10^5). The high value for the rate is quite likely the result of the elimination of hydrogen atoms—and the additional steric hindrance associated with them—from the potential energy function used, but it might also be the result of eliminating the solvent or limiting the number of atoms allowed to move.

More recent calculations by Ghosh and McCammon [58] with an improved potential energy surface and the inclusion of solvent molecules in the system have given better agreement with experimental measurements of the rate of rotation. Hydrogen atoms were not included explicitly in these calculations, but 306 protein atoms and 54 water molecules were allowed to move. The barrier height was 12–13 kcal/ mol^{-1}. The conversion coefficient was found to be 0.25 and the rate constant was reduced to a value of 540 s^{-1}. Frictional effects due to solvent damping were found to be small. Ghosh and McCammon predicted that the remaining discrepancy between calculated and measured rate constants would be eliminated in future calculations with a more accurate model for the hydrogen atoms of the tyrosine ring.

XVI. EXAMPLE: DESORPTION FROM SURFACES

The first application of rare-event theory to processes occurring at surfaces was a computation of the thermal rate of desorption of single xenon atoms from a platinum surface carried out by Grimmelmann et al. [59]. The surface was modeled by 14 platinum atoms, with four central atoms free and the 10 others fixed, together with a continuum potential. The interaction potential for the xenon atom was the sum of pairwise interactions with the platinum atoms and the continuum term. Grimmelmann et al. [59] explored a variety of extensions of the rare-event approach to improve the efficiency of the calculations and were able to compute desorption rates efficiently even at temperatures for which the residence time on the surface was 10^{14} vibrational periods.

XVII. EXAMPLE: SIMPLE REACTIONS IN SOLUTION

By using the rare-event method one may simulate reactions in solution and include the motions of the reacting molecules, the motions of the solvent molecules, and all the interactions among them. Bennett [36] noted that a system of 1000 atoms or fewer is generally adequate for simulating most of the properties of matter. Modern molecular dynamics calculations can treat systems with as many as 100,000 atoms.

The simplest reactions in solution are recombination reactions such as $I + I \rightarrow I_2$, which are diffusion limited and have been examined experimentally in great detail. Much less is known about the details of more complex reactions in solution such as protein-folding reactions and enzyme-catalyzed reactions.

Grote and Hynes [60] used the rare-event approach in calculations to obtain rigorous classical results for several types of reaction models in the gas phase and in solution. They tested the classic Kramers model for a particle crossing a potential barrier under the influence of random thermal forces and frictional or damping forces to simulate the effects of a solvent. For several realistic models of time-dependent friction Grote and Hynes found the Kramers constant-friction model to be inadequate and they proposed a more effective model. The Grote–Hynes model incorporating the effects of time-dependent solvent friction on reaction rate has been found to give good agreement with full rare-event simulations.

At about the same time Montgomery et al. [61] used the rare-event method to investigate the isomerization of simple chain molecules in solution with solvent effects represented by a stochastic collision model. They also found behavior significantly different from that predicted by the Kramers model.

The first rare-event calculations of reactions in solution using complete molecular dynamics simulations were reported by Rosenberg et al. [23]. They simulated the isomerization of butane in liquid CCl_4 using a simple potential function for the butane molecule together with pairwise Lennard–Jones potentials for 125 spheres representing the CCl_4 molecules, all placed in a cubic box with periodic boundaries. A total of 7200 trajectories were started forward and backward from the top of the barrier separating trans and gauche configurations of the butane molecule. From these the conversion coefficient was found to be 0.276.

XVIII. EXAMPLE: AN S_N2 REACTION IN WATER

One of the important classes of reactions in organic chemistry is that of bimolecular nucleophilic S_N2 substitution: $X^- + RY \rightarrow RX + Y^-$. These reactions have most often been investigated in solution, but they have also been studied in the gas phase. Early experimental work indicated that solvent effects are extremely important for this class of reactions.

Bergsma et al. [62] used the rare-event approach to simulate the reaction $Cl^- + CH_3Cl \rightarrow ClCH_3 + Cl^-$ in water to investigate the role of polar solvent molecules and their effect on reaction trajectories and the rate constant. For the first time they were able to observe several important characteristics of reactions in polar solvents at the molecular level.

The potential energy surface used by Bergsma et al. [62] was not considered to be a good representation of the true surface for the reaction Cl^- with CH_3Cl in H_2O, but it was patterned after that reaction and it is a plausible surface for that reaction. In the gas phase the potential energy surface has the form of a double well, but in water the wells are largely removed by the greater stabilization of the reactants and products that have higher charge densities than intermediate configurations. The free energy profiles for reaction in the gas phase and in a polar solvent are illustrated schematically in Fig. 5.10.

The reaction system was simplified by the treatment of the methyl group as a single atom and by the use of an LEPS potential energy function for the reacting "atoms" Cl, CH_3, and Cl. The solvent potential was a semiempirical water intermolecular potential combined with an intramolecular water potential reproducing observed properties of the water molecule. This potential had previously given good representations of many properties of liquid water in molecular dynamics calculations.

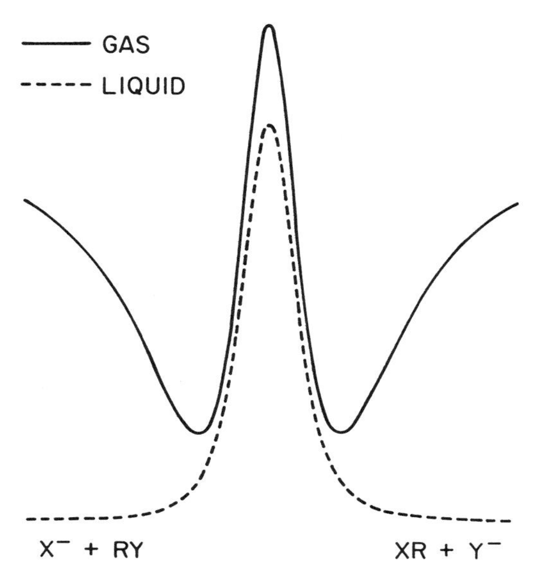

Figure 5.10. Illustration of a free energy profile along the reaction path for a bimolecular nucleophilic S_N2 substitution, $X^- + RY \rightarrow RX + Y^-$, in the gas phase and in a polar solvent. (Adapted from [62].)

The solvent–solute interaction potential was based on the interaction of charges on the reaction system with charges on the water molecules in combination with Lennard–Jones 6–12 interactions. The potential given by these functions was further adjusted with the use of switching functions to damp the interactions at large distances to avoid discontinuities and problems at the boundaries.

Trajectories were sampled from the thermal equilibrium distribution at a dividing surface defined in terms of the symmetric stretch and bend coordinates of the $[Cl–CH_3–Cl]^-$ complex passing through the saddle point of the gas phase LEPS potential energy surface. The asymmetric stretch corresponded to the reaction coordinate. The trajectories were calculated both forward and backward in time from the initial positions on the dividing surface.

The calculations were carried out for reaction in both the gas phase and in solution in liquid water. In most of the calculations for the solution the system contained 64 water molecules, but in a few cases the system was expanded to 263 water molecules in order to check on possible differences in the dynamics due to size effects. The number of trajectories followed in the gas phase was 128 and the number in solution was 400.

Most of the gas-phase trajectories in the forward direction were trapped in the potential well on the product side of the barrier for the duration of the 0.5 ps allowed reaction time. Those not trapped on the product side recrossed the barrier and were trapped on the reactant side. Of the 128 trajectories investigated 107 were successful, that is, having an initial configuration in the reactant well and a final state in the product well. Because of trapping the allowed time was not sufficient for investigation of the complete reaction in the gas phase from the approach of reactants to the separation of products.

In the liquid phase the absence of wells eliminated the possibility of significant trapping. Although recrossing of the dividing surface was common, those trajectories that recrossed did so within about 0.02 ps and most of these recrossed only once although a few recrossed several times. The fraction recrossing was 0.63 and with adjustment for multiple crossings the conversion coefficient was 0.55 for the 400 trajectories sampled. As pointed out by the authors, this is a significant dynamical correction to the rate constant given by transition state theory.

By examining the details of individual trajectories and comparing those for trajectories with and without recrossing of the barrier Bergsma et al. [62] were able to correlate their behavior with initial solvent configurations and with the strength of the solvent–solute interactions for trajectories sampled at the dividing surface. The rate of change of the atomic charge distribution along the reaction coordinate was found to have a major effect on the dynamics. In the region of the barrier the reaction was found to be characterized by nonadiabatic solvation, in which solvent configurations are frozen and fail to maintain a rapid adiabatic equilibrium with the solute. These and other factors controlling the dynamics were revealed by the trajectory calculations.

In later rare-event calculations of the same reaction in solution Gertner et al. [63] used the results to test the validity of several approximate methods for treating such reactions. In particular, they found the Grote–Hynes theory (based on the generalized Langevin equation) to give excellent agreement with the rare-event results and the Kramers theory (equivalent to the assumption of adiabatic equilibrium for the solvent) to give very poor agreement. Additional details of the energy-transfer processes associated with the reaction have also been examined. Gertner et al. [64] used the rare-event approach to investigate how the reactants obtain sufficient energy from the solvent to cross the barrier. They observed a vibrational excitation of the initial ion–dipole reactant complex and a gradual increase in kinetic and potential energies of reactants, followed by a fast shift of kinetic energy into potential energy of reactants at the top of the barrier. A substantial reorganization of the solvent was observed to be an essential part of the barrier climbing process. The characteristics of reaction for this strongly coupled ionic reactant/polar solvent system were found to be very different from those found in rare-event calculations for weakly coupled nonionic reactant–rare gas solvent systems [65].

XIX. EXAMPLE: LIGAND ESCAPE FROM THE HEME POCKET OF MYOGLOBIN

Another example of the application of the rare-event approach in protein dynamics is a calculation by Kottalam and Case [66] of the dioxygen ligand escape from the heme pocket of sperm whale myoglobin. Ligands such as O_2 are simple, the structures of the heme proteins are known, the reaction intermediates have been identified, and the rates of gaseous ligand binding to the heme proteins are known. To escape, an O_2 molecule released from binding to an iron atom in the heme pocket must find its way from the heme pocket through the interior of the protein to the solvent outside the protein. In a rigid protein with the structure given by X-ray crystallography the paths for ligand migration would be blocked by potential energy barriers of perhaps 100 kcal mol^{-1}. Migration is made possible by the movements of protein atoms to open "gates" for the passage of ligands.

Kottalam and Case used an AMBER united atom potential energy function with porphyrin and oxygen ligand parameters given by fitting several observed properties. They used umbrella sampling techniques to investigate the paths for escape and computed rate constants for escape using both classical transition state theory and the full classical dynamics of the rare-event approach. To reduce the computational effort the direct simulation was limited to residues within 12 Å of the initial oxygen position and the remainder of the system was approximated with Langevin dynamics. A contour map of potential energies encountered by an escaping molecule is shown in Fig. 5.11. Of course, a single plot such as that of Fig. 5.11 is merely a snapshot for a fixed protein configuration. A series is needed to show the relaxation of the protein structure to allow the molecule to pass.

One likely pathway for escape of an oxygen is indicated in Fig. 5.11. The bottleneck for this path lies in the region between the distal histadine and the valine residues. Kottalam and Case [66] determined potentials of mean force along this pathway for temperatures of 180–330 K and found the barrier to increase from 5.2 kcal mol^{-1} at 180 K to 7.2 kcal mol^{-1} at 330 K. Trajectory calculations, started forward and backward from near the top of the barrier, showed most of the crossing were "straight

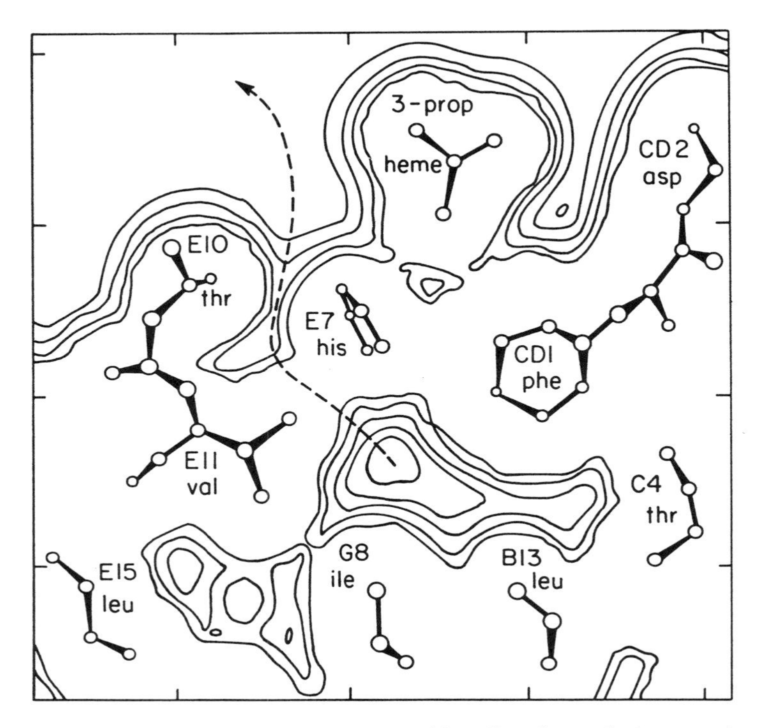

Figure 5.11. Potential energy surface encountered by a ligand near the heme pocket of myoglobin in its X-ray conformation. A likely path for its escape is indicated by the dashed line. (Adapted from [66].)

through" and did not recross the barrier. The conversion coefficients were found to be in the range of 0.8–0.9 for 30 trajectories at 180 and 300 K.

The computed rate constants were in good agreement with experimental rate constants at room temperature and their variation with temperature was consistent with experimental observations. The results clearly demonstrated the feasibility of rate constant calculations for ligand-binding processes.

XX. THE FULL RANGE OF APPLICATIONS

The full range of applications of the rare-event approach thus far is listed in Table I. The smallest system treated is the electron–atom system $X + e^- \rightarrow X^* + e^-$ [68]. The largest in terms of mass is a reaction of

TABLE I
Applications of Rare-Event Theory

Authors (date) [Ref.]	
Keck (1962) [1]	
dissociation and recombination reactions	$H_2 + Ar \leftrightarrow H + H + Ar$ also O_2, I_2
Woznick (1965) [67]	
dissociation and recombination	$X_2 + M \leftrightarrow X + X + M$
Keck (1967) [15]	
dissociation and recombination	$H_2 + Ar \leftrightarrow H + H + Ar$
also O_2, I_2 and He, Xe	
Mansbach and Keck (1969) [68]	
excitation and ionization of atoms by thermal electrons	$X + e^- \leftrightarrow X^* + e^-$
$X + e^- \leftrightarrow X^+ + e^- + e^-$	
Mansbach (1970) [69]	
binary star formation due to stellar three-body collisions	$A + B + C \leftrightarrow AB + C$
Shui et al. (1971) [70]	
dissociation and recombination	$X_2 + M \leftrightarrow X + X + M$
Shui (1972) [71]	
dissociation and recombination	$N_2 + Ar \leftrightarrow N + N + Ar$
Shui et al. (1972) [72]	
dissociation and recombination	$HCl + Ar \leftrightarrow H + Cl + Ar$
Anderson (1973) [12]	
intersecting channels, smooth curves, exchange reaction	$H + I_2 \leftrightarrow HI + I$
Jaffe et al. (1973) [16]	
exchange reaction	$F + H_2 \leftrightarrow HF + H$
Henry et al. (1973) [3]	
four-body exchange reaction	$H_2 + I_2$ (or 2I) $\leftrightarrow HI + HI$
Shui (1973) [73]	
dissociation and recombination	$H_2 + H \leftrightarrow H + H + H$

TABLE I (*Continued*)

Authors (date) [Ref.]	
Kung and Anderson (1974) [47]	
dissociation and recombination	$H_2 + H \leftrightarrow H + H + H$
Anderson (1974) [51]	
four-body exchange reaction	$H_2 + I_2$ (or 2I) $\leftrightarrow HI + HI$
Porter et al. (1975) [17]	
attempt to demonstrate errors in rare-event approach	$H + I_2 \leftrightarrow HI + I$
$H + Br_2 \leftrightarrow HBr + Br$	
Anderson (1975) [28]	
test of validity of rare-event approach	$A + BB \leftrightarrow AB + B$
$F + H_2 \leftrightarrow HF + H$	
Bennett (1975) [34]	
diffusion of vacancies in solid argon, 225 atoms	$Ar + (\ ) \leftrightarrow (\ ) + Ar$
Bennett (1975) [35]	
diffusion of vacancies in solids	solid argon
Jaffe et al. (1976) [4]	
(originally submitted to *J. Chem. Phys.* 1972)	
four-body exchange reaction	$H_2 + I_2$ (or 2I) $\leftrightarrow HI + HI$
Pollack and Pechukas (1978) [24]	
investigation of periodic orbital dividing surface for collinear reactions	$H + H_2 \leftrightarrow H_2 + H$
Snider (1979) [74]	
dissociation and recombination	$I_2 + M \leftrightarrow I + I + M$
$Br_2 + M \leftrightarrow Br + Br + M$	
Jaffe (1979) [75]	
exchange reaction	$ClO + O \leftrightarrow Cl + O_2$
McCammon and Karplus (1979) [54]	
activated processes in globular proteins	ring rotation in bovine pancreatic trypsin inhibitor

Chesnavich et al. (1979) [76]	
ion–molecule reactions	ion–dipole capture
Garrett and Truhlar (1979) [77]	
collinear exchange reactions	$X + H_2 \leftrightarrow XH + H$, X = H, F, Cl, Br, I
Montgomery et al. (1979) [61]	
isomerization reaction in a liquid	*n*-butane(*trans*) ↔ butane(*gauche*)
Chesnavich et al. (1980) [78]	
ion-molecule reactions	ion–dipole capture
McCammon and Karplus (1980) [79]	
activated processes in globular proteins	ring rotation in bovine pancreatic trypsin inhibitor
Garrett and Truhlar (1980) [80]	
exchange reactions	$H + H_2 \leftrightarrow H_2 + H$, others
Northrup and McCammon (1980) [38]	
barrier crossing	one-dimensional barrier problem
Grote and Hynes (1980) [60]	
gas-phase exchange reactions	
barrier crossing reactions in solution (Langevin model)	
Rosenberg et al. (1980) [23]	
isomerization reactions in liquids	*n*-butane(*trans*) ↔ *n*-butane(*gauche*)
Montgomery et al. (1980) [81]	
isomerization reactions in liquids	*n*-butane(*trans*) ↔ *n*-butane(*gauche*)
also *n*-pentane, *n*-decane	
Smith (1981) [44]	
collinear exchange reaction	$A + BC \leftrightarrow AB + C$
Grimmelmann et al. (1981) [59]	
desorption from a surface	Xe desorption from platinum
Adams and Doll (1981) [82]	
dissociative adsorption	model potential
Adams and Doll (1981) [83]	
dissociative adsorption	model potential

Table I (*Continued*)

Authors (date) [Ref.]	
Truhlar, Isaacson, Skodje, and Garrett (1982) [40] exchange reactions	$H + H_2 \leftrightarrow H_2 + H$, many others
Warshel (1982) [84] charge transfer reactions in solution	electron and proton transfers
Northrup et al. (1982) [57] rearrangement of globular proteins	isomerization of bovine pancreatic trypsin inhibitor
Su and Chesnavich (1982) [85] ion–molecule reactions	ion/polar–molecule capture collisions
Pollak (1982) [86] finding quasiperiodic orbits	$H + H_2 \leftrightarrow H_2 + H$
Swamy and Hase (1982) [87] ion-molecule reactions	$Li^+ + H_2O \leftrightarrow Li^+(H_2O)$, also Na^+, K^+
Warshel (1984) [88] enzymatic reactions	proton transfer in lysozyme
Warshel (1984) [89] enzymatic reactions	proton transfer in lysozyme
Connick and Alder (1983) [90] exchange of solvent molecules	water in first coordinate sphere of metal ion
Adams and Doll (1984) [91] desorption from surfaces	Ar and Ne from solid xenon surface
Tully (1985) [92] adsorption, desorption, trapping on a surface	Xe on a platinum surface
Straub et al. (1985) [93] activated barrier crossing	one-dimensional system with friction
Straub and Berne (1985) [94] barrier crossing	symmetric double well potential

Russell (1985) [95] dissociation and recombination involving ions	$He^+ + He + He \leftrightarrow He_2^+ + He$
Davis (1985) [96] intramolecular energy transfer	$OCS' \leftrightarrow OCS''$
Voter and Doll (1985) [52] diffusion on surfaces	Rh on a rhodium surface
Russell (1986) [97] ion–molecule dissociation and recombination	$He^+ + He + He \leftrightarrow He_2^+ + He$
Bergsma et al. (1986) [65] exchange reaction in solution	$A + BC \leftrightarrow AB + C$ in rare gas solvent
Warshel and Hwang (1986) [98] electron transfer reactions in solution	model $D + A \leftrightarrow D^+ + A^-$
Warshel, Russell, and Sussman (1986) [99] enzymatic reactions	catalytic reactions of trypsin
Bergsma et al. (1986) [100] exchange reaction in solution	$A + BC \leftrightarrow AB + C$ in rare gas solvent
Bhowmik and Su (1986) [101] ion–molecule interactions	ion/quadrupolar-molecule capture
Voter (1986) [53] diffusion of clusters on surfaces	Rh clusters on a rhodium surface
Lim and Tully (1986) [102] laser-induced desorption	desorption from a solid surface
Vertenstein and Ronis (1986) [103] membrane transport	diffusion of Na^+, Li^+, Rb^+ in a model membrane system
Hwang and Warshel (1987) [104] reaction in solution	electron transfer for two benzene molecules
Bergsma et al. (1987) [62] reaction in solution	$Cl^- + CH_3Cl \leftrightarrow ClCH_3 + Cl^-$

Table I (*Continued*)

Authors (date) [Ref.]	
Davis (1987) [105] exchange reaction	$H + H_2 \leftrightarrow H_2 + H$ (phase structure)
Frost and Smith (1987) [45] exchange reactions	$A + BC \leftrightarrow AB + C$ (collinear, PODS)
Snider (1987) [106] dissociation and recombination	$A_2 + M \leftrightarrow A + A + M$
Straub et al. (1987) [107] barrier crossing	two-dimensions with friction
Edberg (1987) [108] isomerization reactions in liquids	*n*-butane(*trans*) $\leftrightarrow$ *n*-butane(*gauche*)
Truhlar and Garrett (1987) [42] exchange reactions	$F + H_2 \leftrightarrow HF + H$, $O + H_2 \leftrightarrow OH + H$
Ghosh and McCammon (1987) [58] protein rearrangements in solution	isomerization of bovine pancreatic trypsin inhibitor
Kuharski et al., (1988) [109] isomerization reactions in liquids	cyclohexane(chair) $\leftrightarrow$ cyclohexane(boat)
Hwang et al. (1988) [110] charge-transfer reactions in solution	$X - Y \leftrightarrow X^- + Y^+$ in solution
Hwang et al. (1988) [111] S_N2 reactions in solution	$X^- + CH_3Y \rightarrow CH_3X + Y^-$
Harris et al. (1988) [112] associative desorption of atoms	H_2 desorption from copper
Kottalam and Case (1988) [66] enzyme-substrate interactions	ligand escape from the heme pocket of myoglobin
Warshel, Sussman, and Hwang (1987) [104] binding in genetically modified enzymes	anions with trypsin and subtilisin

Chandler and Kuharski (1988) [114] isomerization, electron transfer	cyclohexane(chair) $\leftrightarrow$ cyclohexane(boat) Fe^{2+}/Fe^{3+} electron transfer
Gertner et al. (1989) [63] exchange reaction in solution	$Cl^- + CH_3Cl \leftrightarrow ClCH_3 + Cl^-$
Ciccotti et al. (1989) [115] ion pair reactions in solution	$A^+ + B^- \leftrightarrow AB$
Cohen and Voter (1989) [116] diffusion on surfaces	self-diffusion on Lennard–Jones surface
Voter et al. (1989) [117] diffusion on surfaces	model atom on fcc(100) surface
Voter (1989) [118] diffusion on surfaces	general
Russell and Shyu (1989) [119] dissociation and recombination of ions and molecules	$He^+ + He + He \leftrightarrow He_2^+ + He$
Zichi et al. (1989) [120] electron–transfer reactions in solution	model of electron transfer in a polar solvent
Dumont (1989) [121] isomerization reactions	chaotic Siamese stadium billiard
L'Heureux and Kapral (1989) [122] model systems	noise-induced transitions in a bistable system
Warshel, Chu, and Parson (1989) [123] biological electron transfer	transfer steps in photosynthetic reaction centers
Agrawal et al. (1989) [124] diffusion on surfaces	Si diffusion on silicon
Li and Wilson (1990) [125] reactions in solution	$A + BC \leftrightarrow AB + C$ in argon solvent
Ciccotti et al. (1990) [126] ion pair interconversion in solution	$R^+\|X^- \leftrightarrow R^+\|\|X^-$

Table I (*Continued*)

Authors (date) [Ref.]	
Zhang et al. (1990) [127] diffusion on surfaces	H diffusion on nickel
Benjamin et al. (1990) [128] exchange reactions in solution	$Cl + Cl_2 \leftrightarrow Cl_2 + Cl$
Skodje and Davis (1990) [129] dynamics of short-lived species	model for unimolecular decay
Zhang and Metiu (1990) [130] diffusion on surfaces	atom diffusion on copper surface
Haug et al. (1990) [131] diffusion on surfaces	diffusion of H on copper
Brown and Clarke (1990) [132] isomerization in liquids	*n*-butane(*trans*) ↔ *n*-butane(*gauche*)
Glinsky and O'Neill (1991) [133] isomerization in liquids	*n*-butane(*trans*) ↔ *n*-butane(*gauche*)
Tucker et al. (1991) [134] barrier crossing model	barrier crossing for Langevin dynamics
Lazaridis et al. (1991) [135] conformational transitions in biological macromolecules	transitions of alanine dipeptide
Keirstead et al. (1991) [136] reactions in solution	$t\text{-BuCl} \leftrightarrow t\text{-Bu}^+ + Cl^-$ in water
Charutz and Levine (1991) [137] reactions in solution	$Cl + Cl_2 \leftrightarrow Cl_2 + Cl$ in liquid argon
Tully (1991) [138] reactions at surfaces	trapping of Ar on platinum
Benjamin et al. (1991) [139] reactions in solution	$A + BC \leftrightarrow AB + C$

Nagaoka et al. (1991) [140] reactions in solution	proton transfers in formamidine–water system
Gertner et al. (1991) [64] exchange reaction in solution	$Cl^- + CH_3Cl \leftrightarrow ClCH_3 + Cl^-$ in water
Rabani et al. (1991) [141] ion–molecule reactions	double minimum potential model
Jansen (1991) [142] desorption from surfaces	Xe desorption from palladium
Benjamin (1991) [143] reactions at the liquid–vapor interface	model isomerization, diatom in double well
Jansen (1992) [144] desorption from surfaces	Xe desorption from platinum and palladium
Marks and Thompson (1992) [145] nonadiabatic unimolecular reactions	N_2O predissociation
Jansen (1992) [146] desorption from surfaces	coverage-dependent Xe desorption from platinum
Cho et al. (1992) [147] nucleophilic substitution reactions	$Cl^- + CH_3Cl \leftrightarrow ClCH_3 + Cl^-$
Straus and Voth (1992) [148] activated rate processes in condensed matter	reactions in model systems
Nagaoka et al. (1992) [149] reactions in solution	proton transfer in formamidine–water systems
Gillian and Wilson (1992) [150] reactions in gas phase versus in solution	$A + BC \leftrightarrow AB + C$
Straus et al. (1993) [151] classical activated rate processes	model reaction systems
Ben-Nun and Levine (1993) [152] barrier descent dynamics	gas phase and liquid phase dynamics, coherent spectroscopy

Table I (*Continued*)

Authors (date) [Ref.]	
Schenter et al. (1993) [153]	
quantum analog of classical reaction	crossing Eckart barrier
Hwang and Warshel (1993) [46]	
proton transfer in solution	model system in water
Lazaridis and Paulaitis (1994) [154]	
coformational transitions of biomolecules	active site of tosyl-*a*-chymotrypsin
Ben-Nun and Levine (1994) [155]	
ion–molecule reactions in solution	ion capture model
Reese, Tucker, and Schenter (1995) [156, 157]	
unimolecular reactions in solvents	model with generalized Lengevin dynamics

binary star formation A + B + C → AB + C [69] and the largest in terms of complexity is ligand escape from the heme pocket of myoglobin [66]. The colinear reaction A + BB → AB + B [28] or the Siamese stadium billiard [121] might be regarded as the most basic. The laser-induced desorption of atoms from surfaces [102] or the transitions of the alanine dipeptide [135] might be regarded as the most applied.

In many of the applications listed in Table I a rare-event calculation has been used as a standard against which other methods are compared. In others the rare-event method has provided a fully classical case as a limiting case for testing semiclassical treatments. For many of these systems the development of the rare-event approach is at least as important in calculations of molecular dynamics as the development of fast computers. Since 1960 the speeds of advanced scientific computers have increased by a factor of about 10^5. The computational advantage provided by the rare-event method is larger than 10^5 for some systems. Of course, the advantages of faster computers and better computational methods are not competitive, they are cooperative. With still faster computers the range of applications will continue to increase.

ACKNOWLEDGMENTS

Support of this work by the National Science Foundation (Grant No. CHE-8714613) and the Office of Naval Research (Grant N00014-92-J-1340) is gratefully acknowledged. The author is indebted to the Humboldt Foundation and to the University of Kaiserslautern for the opportunity to carry out this work.

REFERENCES

1. J. C. Keck, *Discuss. Faraday Soc.*, **33**, 173 (1962).
2. J. B. Anderson, *J. Chem. Phys.* **52**, 3849 (1970).
3. J. M. Henry, J. B. Anderson, and R. L. Jaffe, *Chem. Phys. Lett.*, **20** 138 (1973).
4. R. L. Jaffe, J. M. Henry, and J. B. Anderson, *J. Am. Chem. Soc.*, **98**, 1140 (1976).
5. R. Marcelin, *Ann. Phys.* **3**, 120 (1915). See also K. J. Laidler, *J. Chem. Educ.* **62**, 1012 (1985).
6. E. Wigner, *Z. Phys. Chem.* **B19**, 203 (1932).
7. J. Horiuti, *Bull. Chem. Soc. Jpn.*, **13**, 210 (1938).
8. H. Eyring, *J. Chem. Phys.*, **3**, 107 (1935); M. G. Evans and M. Polyanyi, *Trans. Faraday Soc.*, **31**, 875 (1935).
9. J. O. Hirschfelder and E. Wigner, *J. Chem. Phys.*, **7**, 616 (1939).
10. J. W. Gibbs, *The Collected Works of J. Willard Gibbs*, Yale Univ. Press, New Haven, 1948.
11. Equivalent definitions of ξ may be used.
12. J. B. Anderson, *J. Chem. Phys.* **58**, 4684 (1973).

13. M. Crichton, *Jurassic Park*, Ballantine Books, New York, 1991, p. 75.
14. K. J. Laidler and M. C. King, *J. Phys. Chem.*, **87**, 2657 (1983).
15. J. C. Keck, *Adv. Chem. Phys.*, **13**, 85 (1967).
16. R. L. Jaffe, J. M. Henry, and J. B. Anderson, *J. Chem. Phys.*, **59**, 1128 (1973).
17. R. N. Porter, D. L. Thompson, L. M. Raff, and J. M. White, *J. Chem. Phys.*, **62**, 2429 (1975).
18. Similar opposition was encountered by Eyring. See J. O. Hirschfelder, *Adv. Chem. Phys.*, **34**, xv (1983).
19. J. C. Keck, *J. Chem. Phys.*, **32**, 1035 (1960).
20. I. Mayer, *J. Chem. Phys.*, **60**, 2564 (1974); J. B. Anderson, *J. Chem. Phys.*, **60**, 2566 (1974).
21. T. Yamamoto, *J. Chem. Phys.*, **33**, 281 (1960).
22. D. Chandler, *J. Chem. Phys.*, **68**, 2959 (1978).
23. R. O. Rosenberg, B. J. Berne, and D. Chandler, *Chem. Phys. Lett.*, **75**, 162 (1980).
24. E. Pollak and P. Pechukas, *J. Chem. Phys.*, **69**, 1218 (1978).
25. R. De Vogelaere and M. Boudart, *J. Chem. Phys.*, **23**, 1236 (1955).
26. M. Toller, G. Jacucci, G. DeLorenzi, and C. Flynn, *Phys. Rev.*, **B 32**, 2082 (1985).
27. R. E. Gillilan, *J. Chem. Phys.*, **93**, 5300 (1990).
28. J. B. Anderson, *J. Chem. Phys.*, **62**, 2446 (1975).
29. W. H. Miller, *Acc. Chem. Res.*, **9**, 306 (1976). See also P. Pechukas, in *Dynamics of Molecular Collisions*, W. H. Miller, Ed., Plenum Press, New York, 1976.
30. B. Widom, *J. Chem. Phys.*, **55**, 44 (1971).
31. B. J. Berne, in *Multiple Time Scales*, J. V. Brackbill and B. I. Cohen, Eds., Academic, New York, 1985.
32. S. H. Northrup and J. T. Hynes, *J. Chem. Phys.*, **73**, 2700 (1980).
33. P. Hänggi, P. Talkner, and M. Borkovec, *Rev. Mod. Phys.*, **62**, 251 (1990).
34. C. H. Bennett, *Thin Solid Films*, **25**, 65 (1975).
35. C. H. Bennett, in *Diffusion in Solids*, J. J. Burton and A. S. Nowick, Eds., Academic, New York, 1975, p. 73.
36. C. H. Bennett, in *Algorithms for Chemical Computation*, R. E. Christofferson, Ed., ACS Symposium Series, Vol. 46, American Chemical Society, 1977, p. 63.
37. G. M. Torrie and J. P. Valleau, *J. Comput. Phys.*, **23**, 187 (1977).
38. S. H. Northrup and J. A. McCammon, *J. Chem. Phys.*, **72**, 4569 (1980).
39. E. A. Carter, G. Ciccotti, J. T. Hynes, and R. Kapral, *Chem. Phys. Lett.*, **156**, 472 (1989).
40. D. G. Truhlar, A. D. Isaacson, R. T. Skodje, and B. C. Garrett, *J. Phys. Chem.* **86**, 2252 (1982).
41. D. G. Truhlar, A. D. Isaacson, and B. C. Garrett, in Theory of Chemical Reaction Dynamics, edited by M. Baer, (CRC Press, Boca Raton, Florida, 1985), p. 85.
42. D. G. Truhlar and B. C. Garrett, *Faraday Discuss. Chem. Soc.* **84**, 464 (1987).
43. J. W. Tromp and W. H. Miller, *Faraday Discuss. Chem. Soc.* **84**, 441 (1987).
44. I. W. M. Smith, *J. Chem. Soc., Faraday Trans. II* **77**, 747 (1981).
45. R. J. Frost and I. W. M. Smith, *Chem. Phys.* **111**, 389 (1987).
46. J.-K. Hwang and A. Warshel, *J. Phys. Chem.*, **97**, 10053 (1993).
47. R. T. V. Kung and J. B. Anderson, *J. Chem. Phys.*, **60**, 3731 (1974).

48. R. L. Jaffe and J. B. Anderson, *J. Chem. Phys.*, **54**, 2224 (1971).
49. L. M. Raff, L. Stivers, R. N. Porter, D. L. Thompson, and L. B. Sims, *J. Chem. Phys.*, **52**, 3449 (1970).
50. J. B. Anderson, *J. Chem. Phys.*, **100**, 4253 (1994).
51. J. B. Anderson, *J. Chem. Phys.*, **61**, 3390 (1974).
52. A. F. Voter and J. D. Doll, *J. Chem. Phys.*, **82**, 80 (1985).
53. A. F. Voter, *Phys. Rev.*, **B 34**, 6819 (1986).
54. J. A. McCammon and M. Karplus, *Proc. Natl. Acad. Sci. USA*, **76**, 3585 (1979).
55. M. Karplus, *Phys. Today*, 68, October 1987.
56. M. Karplus and J. A. McCammon, Sci. Am., **254**(4), 42 (1986).
57. H. Northrup, M. R. Pear, C.-Y. Lee, J. A. McCammon, and M. Karplus, *Proc. Natl. Acad. Sci. USA*, **79**, 4035 (1982).
58. I. Ghosh and J. A. McCammon, Biophys. J. **51**, 637 (1987); I. Ghosh and J. A. McCammon, *J. Phys. Chem.*, **91**, 4878 (1987).
59. E. K. Grimmelmann, J. C. Tully, and E. Helfand, *J. Chem. Phys.*, **74**, 5300 (1981).
60. R. F. Grote and J. T. Hynes, *J. Chem. Phys.*, **73**, 2715 (1980).
61. J. A. Montgomery, D. Chandler, and B. J. Berne, *J. Chem. Phys.*, **70**, 4056 (1979).
62. J. P. Bergsma, B. J. Gertner, K. R. Wilson, and J. T. Hynes, *J. Chem. Phys.*, **86**, 1356 (1987).
63. B. J. Gertner, K. R. Wilson, and J. T. Hynes, *J. Chem. Phys.*, **90**, 3537 (1989).
64. B. J. Gertner, R. M. Whitnell, K. R. Wilson, and J. T. Hynes, *J. Am. Chem. Soc.*, **113**, 74 (1991).
65. J. P. Bergsma, P. M. Edelsten, B. J. Gertner, K. R. Huber, J. R. Reimers, K. R. Wilson, S. M. Wu, and J. T. Hynes, *Chem. Phys. Lett.*, **123**, 394 (1986).
66. J. Kottalam and D. A. Case, *J. Am. Chem. Soc.*, **110**, 7690 (1988).
67. B. Woznick, AVCO-Everett Research Laboratory, Everett, MA, Report No. RR223 (1965).
68. P. Mansbach and J. C. Keck, *Phys. Rev.*, **181**, 275 (1969).
69. P. Mansbach, *Astrophys. J.*, **160**, 135 (1970).
70. V. H. Shui, J. P. Appleton, and J. C. Keck, Proceedings, Thirteenth International Symposium on Combustion, Combustion Institute, Pittsburgh, 1971, p. 21.
71. V. H. Shui, *J. Chem. Phys.*, **57**, 1704 (1972).
72. V. H. Shui, J. P. Appleton, and J. C. Keck, *J. Chem. Phys.*, **56**, 4266 (1972).
73. V. H. Shui, *J. Chem. Phys.*, **58**, 4868 (1973).
74. N. Snider, *Can. J. Chem.*, **57**, 1167 (1979).
75. R. L. Jaffe, *Chem. Phys.*, **40**, 185 (1979).
76. W. J. Chesnavich, T. Su, and M. T. Bowers, in NATO ASI Series B-Physics, Vol. 40, P. Ausloos, Ed., Plenum, New York, 1979.
77. B. C. Garrett and D. G. Truhlar, *J. Phys. Chem.* **83**, 1052 (1979).
78. W. J. Chesnavich, T. Su, and M. T. Bowers, *J. Chem. Phys.*, **72**, 2641 (1980).
79. J. A. McCammon and M. Karplus, *Biopolymers*, **19**, 1375 (1980).
80. B. C. Garrett and D. G. Truhlar, *J. Phys. Chem.* **84**, 805 (1980).
81. J. A. Montgomery, S. L. Holmgren, and D. Chandler, *J. Chem. Phys.*, **73**, 3688 (1980).
82. J. E. Adams and J. D. Doll, *Surf. Sci.*, **103**, 472 (1981).

83. J. E. Adams and J. D. Doll, *Surf. Sci.*, **111**, 492 (1981).
84. A. Warshel, J. Phys. Chem. **86**, 2218 (1982).
85. T. Su and W. J. Chesnavich, *J. Chem. Phys.*, **76**, 5183 (1982).
86. E. Pollak, *Chem. Phys. Lett.*, **91**, 27 (1982).
87. K. N. Swamy and W. L. Hase, *J. Chem. Phys.*, **77**, 3011 (1982).
88. A. Warshel, Proc. Nat. Acad. Sci. USA **81**, 444 (1984).
89. A. Warshel, Pontif. Acad. Sci. Script. Var. **55**, 59 (1984).
90. R. E. Connick and B. J. Alder, *J. Phys. Chem.*, **87**, 2764 (1983).
91. J. E. Adams and J. D. Doll, *J. Chem. Phys.*, **80**, 1681 (1984).
92. J. C. Tully, *Faraday Discuss. Chem. Soc.*, **80**, 291 (1985).
93. J. E. Straub, D. A. Hsu, and B. J. Berne, *J. Phys. Chem.*, **89**, 5188 (1985).
94. J. E. Straub and B. J. Berne, *J. Chem. Phys.*, **83**, 1138 (1985).
95. J. E. Russell, *J. Chem. Phys.*, **83**, 3363 (1985).
96. M. J. Davis, *J. Chem. Phys.*, **83**, 1016 (1985).
97. J. E. Russell, *J. Chem. Phys.*, **84**, 4394 (1986).
98. A. Warshel and J.-K. Hwang, *J. Chem. Phys.*, **84**, 4938 (1986).
99. A. Warshel, S. Russell, and F. Sussman, *Israel J. Chem.* **27**, 217 (1986).
100. J. P. Bergsma, J. R. Reimers, K. R. Wilson, and J. T. Hynes, *J. Chem. Phys.*, **85**, 5625 (1986).
101. P. K. Bhowmik and T. Su, *J. Chem. Phys.*, **84**, 1432 (1986).
102. C. Lim and J. C. Tully, *J. Chem. Phys.*, **85**, 7423 (1986).
103. M. Vertenstein and D. Ronis, *J. Chem. Phys.*, **85**, 1628 (1986).
104. J.-K. Hwang and A. Warshel, *J. Am. Chem. Soc.* **109**, 715 (1987).
105. M. J. Davis, *J. Chem. Phys.*, **86**, 3978 (1987).
106. N. Snider, *Chem. Phys.*, **113**, 349 (1987).
107. J. E. Straub, M. Borkovec, and B. J. Berne, *J. Chem. Phys.*, **86**, 4296 (1987).
108. R. Edberg, D. J. Evans, and G. P. Morriss, *J. Chem. Phys.*, **87**, 5700 (1987).
109. R. A. Kuharski, D. Chandler, J. A. Montgomery, F. Rabii, and S. J. Singer, *J. Phys. Chem.*, **92**, 3261 (1988).
110. J.-K. Hwang, S. Creighton, G. King, D. Whitney, and A. Warshel, *J. Chem. Phys.*, **89**, 859 (1988).
111. J.-K. Hwang, G. King, S. Creighton, and A. Warshel, *J. Am. Chem. Soc.*, **110**, 5297 (1988).
112. J. Harris, S. Holloway, T. S. Rahman, and K. Yang, *J. Chem. Phys.*, **89**, 4427 (1988).
113. A. Warshel, F. Sussman, and J.-K. Hwang, *J. Mol. Biol.* **201**, 139 (1988).
114. D. Chandler and R. A. Kuharski, *Faraday Discuss. Chem. Soc.*, **85**, 329 (1988).
115. G. Ciccotti, M. Ferrario, J. T. Hynes, and R. Kapral, *Chem. Phys.*, **129**, 241 (1989).
116. J. M. Cohen and A. F. Voter, *J. Chem. Phys.*, **91**, 5082 (1989).
117. A. F. Voter, J. D. Doll, and J. M. Cohen, *J. Chem. Phys.*, **90**, 2045 (1989).
118. A. F. Voter, *Phys. Rev. Lett.*, **63**, 167 (1989).
119. J. E. Russell and J. S. Shyu, *J. Chem. Phys.*, **91**, 1015 (1989).
120. D. A. Zichi, G. Ciccotti, J. T. Hynes, and M. Ferrario, *J. Phys. Chem.*, **93**, 6261 (1989).

121. R. S. Dumont, *J. Chem. Phys.*, **91**, 4679 (1989).
122. I. L'Heureux and R. Kapral, *J. Chem. Phys.*, **90**, 2453 (1989).
123. A. Warshel, Z. T. Chu, and W. W. Parson, *Science* **246**, 112 (1989).
124. P. M. Agrawal, D. L. Thompson, and L. M. Raff, *J. Chem. Phys.*, **91**, 6463 (1989).
125. Y. S. Li and K. R. Wilson, *J. Chem. Phys.*, **93**, 8821 (1990).
126. G. Ciccotti, M. Ferrario, J. T. Hynes, and R. Kapral, *J. Chem. Phys.*, **93**, 7137 (1990).
127. Z. Zhang, K. Haug, and H. Metiu, *J. Chem. Phys.*, **93**, 3614 (1990).
128. I. Benjamin, B. J. Gertner, N. J. Tang, and K. R. Wilson, *J. Am. Chem. Soc.*, **112**, 524 (1990).
129. R. T. Skodje and M. J. Davis, *Chem. Phys. Lett.*, **175**, 92 (1990).
130. Z. Zhang and H. Metiu, *J. Chem. Phys.*, **93**, 2087 (1990).
131. K. Haug, G. Wahnström, and H. Metiu, *J. Chem. Phys.*, **92**, 2083 (1990).
132. D. Brown and J. H. R. Clarke, *J. Chem. Phys.*, **92**, 3062 (1990); **93**, 4117 (1990).
133. M. E. Glinsky and T. M. O'Neil, *Phys. Fluids*, **B3**, 1279 (1991).
134. S. C. Tucker, M. E. Thompson, B. J. Berne, and E. Pollak, *J. Chem. Phys.*, **95**, 5809 (1991).
135. T. Lazaridis, D. J. Tobias, C. L. Brooks, and M. E. Paulaitis, *J. Chem. Phys.*, **95**, 7612 (1991).
136. W. P. Kierstead, K. R. Wilson, and J. T. Hynes, *J. Chem. Phys.*, **95**, 5256 (1991).
137. D. M. Charutz and R. D. Levine, *Chem. Phys.*, **152**, 31 (1991).
138. J. C. Tully, *Catalysis Lett.*, **9**, 205 (1991).
139. I. Benjamin, L. L. Lee, Y. S. Li, A. Liu, and K. R. Wilson, *Chem. Phys.*, **152**, 1 (1991).
140. M. Nagaoka, Y. Okuno, and T. Yamabe, *J. Am. Chem. Soc.*, **113**, 769 (1991).
141. E. Rabani, D. M. Charutz, and R. D. Levine, *J. Phys. Chem.*, **95**, 10551 (1991).
142. A. P. J. Jansen, *J. Chem. Phys.*, **94**, 8444 (1991).
143. I. Benjamin, *J. Chem. Phys.*, **94**, 662 (1991).
144. A. P. J. Jansen, *Surf. Sci.*, **272**, 193 (1992).
145. A. J. Marks and D. L. Thompson, *J. Chem. Phys.*, **96**, 1911 (1991).
146. A. P. J. Jansen, *J. Chem. Phys.*, **97**, 5205 (1992).
147. Y. J. Cho, S. R. Vande Linde, L. Zhu, and W. L. Hase, *J. Chem. Phys.*, **96**, 8275 (1992).
148. J. B. Straus and G. A. Voth, *J. Chem. Phys.*, **96**, 5460 (1992).
149. M. Nagaoka, Y. Okuno, and T. Yamabe, *J. Chem. Phys.*, **97**, 8143 (1992).
150. R. E. Gillilan and K. R. Wilson, *J. Chem. Phys.*, **97**, 1757 (1992).
151. J. B. Straus, J. M. G. Llorente, and G. A. Voth, *J. Chem. Phys.*, **98**, 4082 (1993).
152. M. Ben-Nun and R. D. Levine, *Chem. Phys. Lett.*, **203**, 450 (1993).
153. G. K. Schenter, M. Messina, and B. C. Garrett, *J. Chem. Phys.*, **99**, 1674 (1993).
154. T. Lazaridis and M. E. Paulaitis, *J. Am. Chem. Soc.*, **116**, 1549 (1994).
155. M. Ben-Nun and R. D. Levine, *J. Chem. Phys.*, **100**, 3594 (1994).
156. S. C. Tucker, *J. Chem. Phys.* **101**, 2006 (1994).
157. S. K. Reese, S. C. Tucker, and G. K. Schenter, *J. Chem. Phys.* **102**, 104 (1995).

THEORETICAL CONCEPTS IN MOLECULAR PHOTODISSOCIATION DYNAMICS

NIELS ENGHOLM HENRIKSEN

Chemistry Department B, Technical University of Denmark, Lyngby, Denmark

CONTENTS

Advances in Chemical Physics, Volume XCI, Edited by I. Prigogine and Stuart A. Rice.
ISBN 0-471-12002-2

I. INTRODUCTION

Photofragmentation involves chemical bond breaking due to absorption of radiation. The primary electronic transition in the process takes place, typically, in the ultraviolet or visible region of the spectrum. Photofragmentation is an important chemical reaction per se, for example, in the chemistry of the atmosphere.

A fascinating aspect, from a theoretical point of view, is the fact that photofragmentation encompasses essential elements of molecular spectroscopy as well as molecular reaction dynamics. The nature and scope of the relevant theoretical concepts is, accordingly, of quite broad interest. Photofragmentation is the focus of several of the state-of-the-art experiments in chemical physics including the observation of the real-time evolution of chemical bond breaking.

The modern theoretical description of the process begins with the work of Franck [1] in 1925. He proposed a mechanism for the direct dissociation of molecules by light absorption. In the words of Condon [2]: "If now a light quantum is absorbed which is of sufficient energy to bring the molecule into an electronic excited state, it is natural to suppose that in thus changing the electronic energy of the molecule no other specific action is exerted by the light on the molecule. The absorption of light merely substitutes a new law of nuclear interaction, say $V_2(r)$ for the old one $V_1(r)$. But since this new one has a different equilibrium position the atoms, at the instant after the absorption, will be away from equilibrium and so start to vibrate." He assumed in other words that the transition is to that point on the curve of the final state where the nuclei are still in the original positions. Franck first applied the method to dissociation of I_2. Experiments show, however, that while the most favored transition is to the energy level predicted, the actual absorption consists of a fairly broad continuous band.

A few years later (1928) Condon [3] could, using quantum mechanics, account for the discrepancy. Transition probabilities were calculated as overlaps between stationary nuclear wave functions and the continuous band spectrum (as well as discrete transitions to bound states) could, in principle, be explained. Thus, the classical picture was replaced by a description where the uncertainty in the nuclear positions is taken into account. The quantum mechanical description of electronic transitions in molecules came soon after Franck and Condon's papers (1935) referred to as the Franck–Condon principle [4] in honor of the scientists who first considered this problem.

The dynamical aspect of the electronic excitation present in the Franck–Condon principle, that is, the proposition that at the completion of an electronic transition in a molecule, the nuclei start to move away from their original positions, was appealing. This aspect was, however, lost in the quantum mechanical formulation by Condon.

The simple intuitive picture was reintroduced recently. Thus, a dynamical time-dependent formulation for nuclear dynamics induced by electronic excitation was introduced just 15 years ago [5]. This description gives the exact quantum mechanical counterpart of the Franck–Condon principle. The formulation gives, furthermore, theoretical expressions for observables—the total absorption probability as well as various final product distributions in photofragmentation—which can be formulated in such a way that the simple picture associated with the Franck–Condon principle shows up in the equations.

After the publication of the paper by Kulander and Heller [5] many papers used and expanded on the foundation given in that paper (see, e.g., [6–20] as well as references cited therein). The time-dependent approach has turned out to be very valuable. This is partly due to the intuitively appealing description that is not provided by standard formulations in terms of stationary states and partly due to the ease of making contact with (semi-)classical descriptions and pictures. In addition, the time-dependent approach addresses, in a natural way, issues that have become relevant due to the latest developments of laser technology, like: The effect of shape and duration of light pulses on dissociation dynamics and the option of doing time or energy resolved observations.

The purpose of this chapter is to give a detailed account of the *basic* theoretical concepts associated with the time-dependent formulation for nuclear dynamics induced by electronic excitation. The emphasis is on photodissociation dynamics. The account that follows is obviously strongly influenced by many of the papers written over the last 15 years and, in particular, by the pioneering work of Heller. A coherent account on this important new development in chemical physics, along the lines presented in the following, seems, however, to be timely.

For a comprehensive description of the many facets of photofragmentation dynamics that are not discussed in the present account, the reader should consult a recent book on the subject [21].

PART 1
PHOTOFRAGMENTATION DYNAMICS AND PRODUCT ANALYSIS

The aim of Part 1 is to establish a way of thinking about the subject in a form that allows us to exploit the intuitive picture associated with the Franck–Condon principle.

The photofragmentation of a triatomic molecule contains all the essential features of photofragmentation. For a triatomic molecule ABC, the general situation is schematically represented by

$$\mathrm{ABC}(n) + \text{radiation} \rightarrow \begin{cases} \mathrm{ABC}(m) \\ \mathrm{A} + \mathrm{BC}(m) \\ \mathrm{B} + \mathrm{AC}(m) \\ \mathrm{C} + \mathrm{AB}(m) \\ \mathrm{A} + \mathrm{B} + \mathrm{C} \end{cases} \tag{1.1}$$

where n and m denote the initial and final quantum numbers, respectively. We want to know exactly what happens in the transition from reactants to products in Eq. (1.1).

The results discussed in the following sections include

1. **Photofragmentation Dynamics**
   Dynamics of photoinduced bond breaking, that is the interplay between excitation and dissociation dynamics. We consider, for example, the effect of changing the characteristics of the radiation from a (conventional) continuous wave light source to a (laser) light pulse where the duration and shape can be controlled. Results like these are discussed in Section II.
2. **Product Analysis**
   We derive various expressions related to the outcome of the reaction—ranging from the total probability of the reaction to the ultimate details, that is, the probability of transforming the triatomic molecule ABC(n), in a given rotational, vibrational, and electronic state (denoted by n), into a given set of fragments on the right-hand side, again with complete specification of rotational, vibrational, and electronic states (denoted by m) as well as magnitude and direction of the relative momentum of the fragments. We will mainly focus on time-independent observables and show that even without doing time-resolved measurements, the observations can be related to the dynamics of bond breaking. Results like these are discussed in Sections III and IV.
3. **Control Schemes**
   We consider how the various parameters of laser light (i.e., frequency, phase, and pulse duration) can affect the outcome of the dissociation process. For example, we discuss a control scheme that allows us to do selective bond breaking thereby suggesting how one can experimentally steer which product is going to be formed. Results like these are discussed in Section IV.

The fundamental link to the Franck–Condon principle in all results to be presented is

$$|\psi_{\mathrm{FC}}(t)\rangle = \exp(-i\hat{H}_2 t/\hbar)|\phi(0)\rangle \tag{1.2}$$

This "Franck–Condon" wave packet evolves in the excited electronic state (denoted by "2"). At time $t=0$ it is $|\phi(0)\rangle$, which is a product between the initial nuclear state associated with the electronic ground state and the electronic transition dipole moment.

II. QUANTUM DYNAMICS OF MOLECULAR PHOTOFRAGMENTATION

We consider a molecule that interacts with a radiation field. An adequate description can be given by considering the molecule to be described by quantum mechanics and a classical description of the radiation field (see Appendix A). We let $|\Psi(t)\rangle$ denote the state vector of our molecule at time t. In the Schrödinger picture, the time evolution is given by [22],

$$i\hbar \frac{\partial|\Psi(t)\rangle}{\partial t} = (\hat{H}_{\mathrm{M}} + \hat{H}_{\mathrm{I}}(t))|\Psi(t)\rangle \tag{2.1}$$

where $\hat{H}_{\mathrm{M}}$ is the molecular Hamiltonian given by

$$\hat{H}_{\mathrm{M}} = \hat{T}_{\mathrm{n}} + \hat{T}_{\mathrm{e}} + V \tag{2.2}$$

$\hat{T}_{\mathrm{n}}$, $\hat{T}_{\mathrm{e}}$, and V are the kinetic energy of the nuclei, the kinetic energy of the electrons, and Coulomb interaction between all the electrons and nuclei, respectively, and $\hat{H}_{\mathrm{I}}(t)$ is the Hamiltonian for the interaction with the radiation field, which can be expanded in the form [23]

$$\hat{H}_{\mathrm{I}}(t) = \hat{H}_{\mathrm{ED}}(t) + \hat{H}_{\mathrm{EQ}}(t) + \hat{H}_{\mathrm{MD}}(t) + \cdots \tag{2.3}$$

where $\hat{H}_{\mathrm{ED}}(t)$, $\hat{H}_{\mathrm{EQ}}(t)$, and $\hat{H}_{\mathrm{MD}}(t)$ are the electric-dipole interaction, the electric-quadrupole interaction, and the magnetic-dipole interaction, respectively. The electric-dipole term is the dominating term in Eq. (2.3). Other contributions to $\hat{H}_{\mathrm{I}}(t)$ are much smaller and, normally, need only to be considered for the weak transitions, which are electric-dipole forbidden.

Now, we consider the explicit representation of Eq. (2.1) and the solution to first order in the radiation–matter interaction for a *prototype molecule* with

1. Three stationary electronic states.
2. Nonadiabatic coupling between electronic state 2 and 3.
3. Radiative coupling between electronic state and 1 and 2.

The state vector takes the form,

$$\langle \mathbf{q}, \mathbf{R}|\Psi(t)\rangle = \chi_1(\mathbf{R}, t)\psi_1(\mathbf{q}; \mathbf{R}) + \chi_2(\mathbf{R}, t)\psi_2(\mathbf{q}; \mathbf{R}) + \chi_3(\mathbf{R}, t)\psi_3(\mathbf{q}; \mathbf{R}) \tag{2.4}$$

where $\mathbf{q}$ and $\mathbf{R}$ denote the electronic and nuclear coordinates, respectively. The wave function ψ_i $(i = 1, 2, 3)$ is an electronic eigenstate

$$(\hat{T}_e + V(\mathbf{q}; \mathbf{R}))\psi_i(\mathbf{q}; \mathbf{R}) = E_i(\mathbf{R})\psi_i(\mathbf{q}; \mathbf{R}) \tag{2.5}$$

The equation is solved for fixed values of $\mathbf{R}$, which plays the role of a parameter. This is indicated by ";" in the equation. The physical motivation for this adiabatic form is the fast motion of the electrons as compared to the slow motion of the nuclei. For bound state problems, fast motion is connected with large spacing between energy states. Thus, the energy spacing between the lowest electronic states is normally large and we need to consider only a few electronic states.

We substitute this form of the state vector into Eq. (2.1) and use the orthonormality of the electronic eigenstates. The equation of motion for the nuclear degrees of freedom in a space of electronic states, can be written in a matrix form, thus we obtain the following equation for the *nuclear motion* in the presence of a radiation field,

$$\boxed{i\hbar\frac{\partial}{\partial t}\begin{bmatrix}|\chi_1(t)\rangle\\|\chi_2(t)\rangle\\|\chi_3(t)\rangle\end{bmatrix} = \left(\begin{bmatrix}\hat{H}_1 & 0 & 0\\ 0 & \hat{H}_2 & \hat{C}_{23}\\ 0 & \hat{C}_{32} & \hat{H}_3\end{bmatrix} + \begin{bmatrix}\hat{R}_{11}(t) & \hat{R}_{12}(t) & 0\\ \hat{R}_{21}(t) & \hat{R}_{22}(t) & 0\\ 0 & 0 & 0\end{bmatrix}\right)\begin{bmatrix}|\chi_1(t)\rangle\\|\chi_2(t)\rangle\\|\chi_3(t)\rangle\end{bmatrix}} \tag{2.6}$$

where

$$\begin{aligned}\hat{H}_i &= \hat{T}_n + E_i(\mathbf{R}) + \langle\psi_i|\hat{T}_n|\psi_i\rangle\\ \hat{C}_{ij} &= \langle\psi_i|\hat{T}_n|\psi_j\rangle - \sum_s \frac{\hbar^2}{M_s}\langle\psi_i|\nabla_s|\psi_j\rangle\cdot\nabla_s\\ \hat{R}_{ij}(t) &= \langle\psi_i|\hat{H}_{ED}(t)|\psi_j\rangle\end{aligned} \tag{2.7}$$

and integration in the matrix elements is over all electronic coordinates **q**. Accordingly, the matrix elements are functions of the nuclear coordinates **R**. The nuclear state vector is written as a three-component vector where each component can be given a simple physical interpretation. Thus, due to the orthonormality of the electronic eigenstates, the probability of finding the molecule in electronic state "i" at time t given the nuclear position **R** is simply given by $|\chi_i(\mathbf{R}, t)|^2$.

We assume that electronic state 1 is a bound state of the molecule and that the molecule at time t_0 is in this state. We consider the interaction with the radiation field as a first-order perturbation. The first-order correction to the time evolution can be formulated in the following way [22]: The molecule evolves from t_0 to t' in response to the (free) molecular time evolution operator, it interacts with the perturbation at time t' and responds subsequently to the (free) molecular time evolution operator till time t. In connection with the evaluation of this expression, we note that,

$$\exp\left(-i\begin{bmatrix}\hat{H}_1 & 0 & 0\\ 0 & \hat{H}_2 & \hat{C}_{23}\\ 0 & \hat{C}_{32} & \hat{H}_3\end{bmatrix}(t'-t_0)/\hbar\right)\begin{bmatrix}|\chi_1(t_0)\rangle\\ 0\\ 0\end{bmatrix} = \begin{bmatrix}\exp(-i\hat{H}_1(t'-t_0)/\hbar)|\chi_1(t_0)\rangle\\ 0\\ 0\end{bmatrix} \tag{2.8}$$

since for the free molecule, the nuclear motion in electronic state 1 is decoupled from the other electronic states. Multiplication with the interaction matrix gives

$$\begin{bmatrix}\hat{R}_{11}(t') & \hat{R}_{12}(t') & 0\\ \hat{R}_{21}(t') & \hat{R}_{22}(t') & 0\\ 0 & 0 & 0\end{bmatrix}\begin{bmatrix}\exp(-i\hat{H}_1(t'-t_0)/\hbar)|\chi_1(t_0)\rangle\\ 0\\ 0\end{bmatrix} = \begin{bmatrix}\hat{R}_{11}(t')\exp(-i\hat{H}_1(t'-t_0)/\hbar)|\chi_1(t_0)\rangle\\ \hat{R}_{21}(t')\exp(-i\hat{H}_1(t'-t_0)/\hbar)|\chi_1(t_0)\rangle\\ 0\end{bmatrix} \tag{2.9}$$

Thus, the physical interpretation of this equation is clear: The interaction with the radiation field at time t' transfers amplitude from electronic state 1 to 2. We assume that unbound (dissociative) nuclear motion is possible in electronic state 2 and/or 3. The nuclear motion on surfaces 2 and 3, to first order in the interaction with the radiation field, is given by

$$\begin{bmatrix} |\chi_2^{(1)}(t)\rangle \\ |\chi_3^{(1)}(t)\rangle \end{bmatrix} = -\frac{i}{\hbar}\int_{t_0}^{t} dt' \exp\left(-i\begin{bmatrix} \hat{H}_2 & \hat{C}_{23} \\ \hat{C}_{32} & \hat{H}_3 \end{bmatrix}(t-t')/\hbar\right) \times \begin{bmatrix} \hat{R}_{21}(t')\exp(-i\hat{H}_1(t'-t_0)/\hbar)|\chi_1(t_0)\rangle \\ 0 \end{bmatrix} \tag{2.10}$$

We choose $t_0 = 0$ and consider next the situation where $|\chi_1\rangle$ is a vibrational–rotational eigenstate (this assumption is relaxed in Section IV.D). We let ε_a denote the associated energy of the eigenstate.

We need to specify the electric field associated with the radiation field. A distinction between two situations is appropriate. That of laser light and classical (chaotic) light, respectively. The electric field associated with a single-mode laser source is written in the form,

$$\mathbf{E}(z,t) = \mathbf{E}_0 \cos(kz - \omega_l t + \phi) = \mathbf{E}_0(e^{i(kz-\omega_l t+\phi)} + \text{cc})/2 \tag{2.11}$$

where cc denotes the complex conjugate term. The field is linearly polarized in the direction given by the amplitude vector $\mathbf{E}_0$ and, for simplicity, assumed to propagate only in the z direction. We assume that the wavelength $\lambda_l = 2\pi c/\omega_l = 2\pi/k$ is large compared to the dimension of the molecule, that is, the spatial variation of the field is neglected. Thus,

$$\mathbf{E}(t) = \mathbf{E}_0 a(t)(e^{-i\omega_l t + i\phi} + \text{cc})/2 \tag{2.12}$$

where an unspecified envelope function, $a(t)$, is included, that is, short-pulse, long-pulse laser fields, and so on, are all under investigation.

The electric field associated with a classical light source corresponds to a large number of pulses with incoherent phases. The result of an actual experiment will correspond to a time average over many such pulses. Each pulse can be represented as a superposition of the Fourier components in Eq. (2.11) reflecting the frequency spread in the source. The incoherence means that the phases of the pulses are arbitrary and independent and, hence, interference terms associated with the different pulses tend to average out (see also Section IV.D). Accordingly, the result is equivalent to the situation where a single pulse is directed towards the molecule.

We focus next on a laser light source. The first electric field term in Eq. (2.12) is the one associated with absorption of radiation (Appendix

A). Retaining only this term, we get

$$\hat{R}_{21}(t') = \langle \psi_2 | -\boldsymbol{\mu} \cdot \mathbf{E}(t') | \psi_1 \rangle$$
$$= -\boldsymbol{\mu}_{21} \cdot \mathbf{E}_0 a(t') e^{-i\omega_l t' + i\phi}/2 \tag{2.13}$$

where $\boldsymbol{\mu}_{21}$ is the electronic transition dipole moment vector. We obtain now,

$$\begin{bmatrix} |\chi_2^{(1)}(t)\rangle \\ |\chi_3^{(1)}(t)\rangle \end{bmatrix} = \frac{ie^{i\phi}}{2\hbar} \int_0^t dt'\, e^{-i(\hbar\omega_l + \varepsilon_a)t'/\hbar} a(t')$$
$$\times \exp\left(-i \begin{bmatrix} \hat{H}_2 & \hat{C}_{23} \\ \hat{C}_{32} & \hat{H}_3 \end{bmatrix} (t - t')/\hbar \right) |\phi(0)\rangle \tag{2.14}$$

where

$$|\phi(0)\rangle \equiv \begin{bmatrix} \mathbf{E}_0 \cdot \boldsymbol{\mu}_{21} |\chi_1(0)\rangle \\ 0 \end{bmatrix} \tag{2.15}$$

Two limiting forms are possible for the envelope function of the light pulse. One such limit is the *δ-function limit*, defined by

$$a(t') = \delta(t' - t_p) \tag{2.16}$$

where $0 \le t_p \le t$. Thus, a light pulse is suddenly switched on and off at time $t = t_p$. From Eq. (2.14), we get

$$\boxed{\begin{bmatrix} |\chi_2^{(1)}(t)\rangle \\ |\chi_3^{(1)}(t)\rangle \end{bmatrix} = \frac{ie^{i(\phi - \alpha)}}{2\hbar} \exp\left(-i \begin{bmatrix} \hat{H}_2 & \hat{C}_{23} \\ \hat{C}_{32} & \hat{H}_3 \end{bmatrix} (t - t_p)/\hbar \right) |\phi(0)\rangle} \tag{2.17}$$

where $\alpha = (\hbar\omega_l + \varepsilon_a)t_p/\hbar$ is an overall phase factor. The term "vertical excitation" (often used synonymously with a Franck–Condon transition) is nicely illustrated in Eq. (2.17), that is, vertical refers to the instantaneous raising of the energy of the initial nuclear wave function, reflected by the appearance of the time-evolution operator associated with the electronically excited states. Note that the transition probability is proportional to the square of $\mathbf{E}_0 \cdot \boldsymbol{\mu}_{21}$ when the coordinate dependence of the transition dipole moment is neglected (Condon approximation).

The nonadiabatic coupling produces amplitude in electronic state 3. To first order in the coupling we obtain

$$|\chi_3^{(1,1)}(t)\rangle = \frac{e^{i(\phi-\alpha)}}{2\hbar^2}\int_{t_p}^{t} dt''\,\exp(-i\hat{H}_3(t-t'')/\hbar)\hat{C}_{32} \times \exp(-i\hat{H}_2(t''-t_p)/\hbar)\mathbf{E}_0\cdot\boldsymbol{\mu}_{21}|\chi_1(0)\rangle \tag{2.18}$$

Accordingly, we have the following important result in the *δ-function limit* for the light pulse: In this limit interaction with the radiation field, at time $t = t_p$, suddenly transfers amplitude from the ground- to the excited-state surface. Equation (2.17) shows that a wave packet, given as the product of the electronic transition dipole moment times the initial vibrational–rotational eigenstate of the molecule, is vertically excited. Subsequent to its creation this wave packet evolves on surface 2 and due to the nonadiabatic coupling terms, the wave packet evolution will ultimately produce wave packet amplitude on surface 3 as well (see Fig. 6.1).

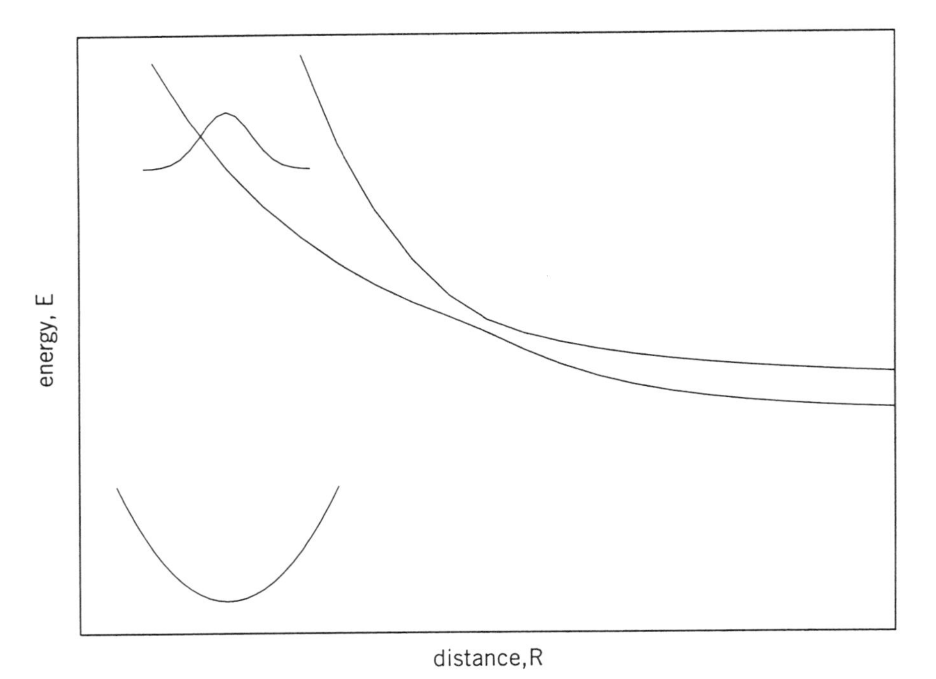

Figure 6.1. Schematic representation of Eq. (2.17), at $t = t_p$, for a molecule initially in the vibrational ground state.

Above we have treated the radiation-matter as well as the nonadiabatic coupling within first-order perturbation theory. The initial state for these two processes is a stationary and a nonstationary state, respectively. The nonadiabatic transition is, accordingly, not equivalent to the simple vertical transition induced by the radiation. The transition probability into state 3 can, under some simplifying assumptions, be related to the well-known Landau–Zener model [24].

The other limiting form for the light pulse is the *cw* (*continuous wave*) *limit*, defined by

$$a(t') = 1 \qquad t \rightarrow \infty \tag{2.19}$$

Equation (2.14) now takes the form

$$\begin{bmatrix} |\chi_2^{(1)}(t)\rangle \\ |\chi_3^{(1)}(t)\rangle \end{bmatrix} = \frac{ie^{i\phi}}{2\hbar} e^{-iE_l t/\hbar} \int_0^t du\, e^{iE_l u/\hbar} \times \exp\left(-i \begin{bmatrix} \hat{H}_2 & \hat{C}_{23} \\ \hat{C}_{32} & \hat{H}_3 \end{bmatrix} u/\hbar \right) |\phi(0)\rangle \tag{2.20}$$

where $t \rightarrow \infty$, $u = t - t'$, and $E_l = \hbar\omega_l + \varepsilon_a$.

Accordingly, we have the following important result in the *cw limit* for the light pulse: The interaction with the radiation field continuously transfers amplitude from the ground to the excited state. Equation (2.20) shows that wave packets, of the same form as in the δ-function limit, are vertically excited in the time interval from $u = 0$ to t. Subsequent to creation each wave packet evolves independently. The resulting nuclear state is obtained as a superposition of the wave packets.

Equations (2.17) and (2.20) are exact quantum mechanical counterparts to the classical picture that constitute the basis of the Franck–Condon principle. In Sections III and IV, where we calculate observables, we will show that these quantities can be expressed in terms of the wave packet describing the dynamics created in the *δ-function limit* [Eq. (2.17)].

Equation (2.20) is a half-Fourier transform of a time-dependent wave packet. We can characterize this state more when the non-adiabatic coupling terms can be neglected. We assume that $\hat{H}_2$ has a continuous spectrum and we get

$$|\chi_2^{(1)}(t)\rangle = \frac{ie^{i\phi}}{2\hbar} e^{-iE_l t/\hbar}|\mathscr{R}\rangle \tag{2.21}$$

where $t \to \infty$ and [25]

$$\begin{aligned} |\mathscr{R}\rangle &= \int_0^\infty du\, e^{iE_l u/\hbar} \exp(-i\hat{H}_2 u/\hbar)|\phi(0)\rangle \\ &= i\hbar \hat{G}(E_1^+)|\phi(0)\rangle \end{aligned} \tag{2.22}$$

the Green's operator, $\hat{G}(E_l^+)$, is given by [26]

$$\begin{aligned} \hat{G}(E_l^+) &= (E_l + i\varepsilon - \hat{H}_2)^{-1} \\ &= -\frac{i}{\hbar}\int_0^\infty du\, e^{iE_l u/\hbar} \exp(-i\hat{H}_2 u/\hbar) \end{aligned} \tag{2.23}$$

where $\varepsilon \to 0^+$. Thus, disregarding a phase factor the state $|\mathscr{R}\rangle$ is created in the *cw limit*. Using Eqs. (2.22) and (2.23)

$$(E_l - \hat{H}_2)|\mathscr{R}\rangle = i\hbar|\phi(0)\rangle \tag{2.24}$$

Assuming that $\langle \mathbf{R}|\phi\rangle$ is real, the real and imaginary part of this equation takes the form

$$\begin{aligned} \hat{H}_2\langle \mathbf{R}|\mathscr{R}\rangle_r &= E_l\langle \mathbf{R}|\mathscr{R}\rangle_r \\ \hat{H}_2\langle \mathbf{R}|\mathscr{R}\rangle_i &= E_l\langle \mathbf{R}|\mathscr{R}\rangle_i - \hbar\langle \mathbf{R}|\phi(0)\rangle \end{aligned} \tag{2.25}$$

The real part, $\langle \mathbf{R}|\mathscr{R}\rangle_r$, is accordingly an eigenstate of $\hat{H}_2$ at the energy E_l. This state is in addition the physically relevant part associated with observables. Thus, in the following sections, where we calculate observables, we will specifically show that in the *cw limit*, the total reaction probability, as well as the detailed final product distribution, can be expressed in terms of the real part of $\langle \mathbf{R}|\mathscr{R}\rangle$.

We are now ready to extract information from the state vector. Before starting on this in Section III, a few words about measurements—and in particular the notion of time and energy resolution—is pertinent.

The molecular state created after interaction with the radiation field is, in general, nonstationary. Some measurements can, accordingly, depend on time. Such time-domain measurements occur for observables represented by an operator, $\hat{A}$, where $[\hat{H}, \hat{A}] \neq 0$. Other measurements will be time independent. These are the energy-domain measurements that occur for observables represented by an operator, $\hat{A}$, where $[\hat{H}, \hat{A}] = 0$.

A fundamental result in quantum mechanics shows that, in general, time as well as energy resolution is limited and complementary in the sense [27],

$$\frac{\Delta A}{|d\langle \psi(t)|\hat{A}|\psi(t)\rangle / dt|} \Delta E \geq \hbar/2 \tag{2.26}$$

where ΔA is the uncertainty for an observable A and ΔE is the associated uncertainty in energy. The first factor defines the time resolution as the time required for the expectation value to change by its uncertainty (assuming that the change is constant). This factor is often referred to as the "uncertainty" in time or, better, the lifetime of a state. A high time resolution is, accordingly, obtained for an observable where the ratio between the uncertainty and the time derivative is small.

Short-lived states have poorly defined energies. A lower bound on the energy uncertainty is obtained for the operator where the time resolution attains its maximum. Likewise states with a well-defined energy must have long lifetimes, we observe that a lower bound on the time resolution is given by $\hbar/(2\,\Delta E)$. Thus, a large (small) spread in energy gives a small (large) uncertainty in time.

In the following sections we will focus mostly on energy resolved measurements. An important result is, however, that the outcome of such observations, in a transparent way, can be shown to be an *indirect* probe of the molecular time evolution (dynamics). We will also get in touch with time-resolved measurements (Section VIII.C), which constitutes a *direct* probe of the molecular time evolution.

III. THE TOTAL REACTION PROBABILITY

The total absorption probability, at time t, is

$$\begin{aligned} P_{\text{tot}}(\omega_l) &= \langle \chi_2^{(1)}(t)|\chi_2^{(1)}(t)\rangle + \langle \chi_3^{(1)}(t)|\chi_3^{(1)}(t)\rangle \\ &= \int |\chi_2^{(1)}(\mathbf{R}, t)|^2\, d\mathbf{R} + \int |\chi_3^{(1)}(\mathbf{R}, t)|^2\, d\mathbf{R} \end{aligned} \tag{3.1}$$

Using Eq. (2.14), we get

$$P_{\text{tot}}(\omega_l) = \frac{1}{4\hbar^2} \int_0^t dt'' \int_0^t dt' \, e^{iE_l(t''-t')/\hbar} a(t')a(t'') \times \langle \phi(0) | \hat{U}_M^{\dagger}(t, t'') \hat{U}_M(t, t') | \phi(0) \rangle \tag{3.2}$$

where

$$\hat{U}_M(t, t') = \exp\left(-i \begin{bmatrix} \hat{H}_2 & \hat{C}_{23} \\ \hat{C}_{32} & \hat{H}_3 \end{bmatrix} (t - t')/\hbar \right) \tag{3.3}$$

and integration in the matrix element is over the nuclear coordinates, **R**. Now,

$$\begin{aligned} \hat{U}_M^{\dagger}(t, t'') \hat{U}_M(t, t') &= \hat{U}_M^{-1}(t, t'') \hat{U}_M(t, t') \\ &= \hat{U}_M(t'', t) \hat{U}_M(t, t') \\ &= \hat{U}_M(t'', t') \end{aligned} \tag{3.4}$$

We introduce new variables,

$$\begin{aligned} u &= t' - t'' \\ v &= t' + t'' \end{aligned} \tag{3.5}$$

and the volume element transforms as

$$du \, dv = 2 \, dt' \, dt'' \tag{3.6}$$

and we get,

$$\boxed{P_{\text{tot}}(\omega_l) = \frac{1}{8\hbar^2} \int_{-t}^{t} du \int_{|u|}^{2t-|u|} dv \, e^{-iE_l u/\hbar} a((v-u)/2) a((v+u)/2) \times \langle \phi(0) | \exp\left(i \begin{bmatrix} \hat{H}_2 & \hat{C}_{23} \\ \hat{C}_{32} & \hat{H}_3 \end{bmatrix} u/\hbar \right) | \phi(0) \rangle} \tag{3.7}$$

It is instructive to consider an envelope function of the form,

$$a(t') = (2\gamma/\pi)^{1/4} e^{-\gamma(t'-t_p)^2} \tag{3.8}$$

which for $\gamma \to 0$ approach the *cw limit*. Assuming that $t_p \gg 2\sqrt{1/\gamma}$, this function is normalized so as $\int_0^\infty |a(t')|^2\, dt' = 1$, that is, it corresponds to a pulse with a fixed energy. Equation (3.7) now takes the form,

$$P_{\text{tot}}(\omega_l) = \frac{1}{4\hbar^2} \int_{-\infty}^{\infty} du\, e^{iE_l u/\hbar} e^{-\gamma u^2/2} \langle \phi(0) | \hat{U}_M(u, 0) | \phi(0) \rangle \qquad (3.9)$$

where $t \sim \infty$ means $t >> \sqrt{1/\gamma} + t_p$. This result can also be written in the form,

$$P_{\text{tot}}(\omega_l) = \frac{1}{4\hbar^2} \int_{-\infty}^{\infty} \sqrt{1/(2\pi\gamma)} e^{-E_l'^2/2\hbar^2\gamma}\, \widehat{au}\ (E_l - E_l')\, dE_l' \qquad (3.10)$$

Thus, the total absorption is given as the convolution between the Fourier transforms of the light pulse and the nuclear autocorrelation function

$$\widehat{au}\ (E_l) = \int_{-\infty}^{\infty} du\, e^{iE_l u/\hbar} \langle \phi(0) | \hat{U}_M(u, 0) | \phi(0) \rangle \qquad (3.11)$$

In the *cw limit* for the light pulse, defined by Eq. (2.19), Eq. (3.7) takes the form

$$P_{\text{tot}}(\omega_l) = \frac{t}{4\hbar^2} \int_{-t}^{t} du\, e^{iE_l u/\hbar} \langle \phi(0) | \hat{U}_M(u, 0) | \phi(0) \rangle \qquad (3.12)$$

where $t \to \infty$. Note that for $\gamma \to 0$ we obtain the same result from Eq. (3.9) (for the same pulse area).

It is easy to show that the right-hand sides in Eqs. (3.12) and (3.9) are real, as they should be. Since,

$$\begin{aligned} \langle \phi(0) | \hat{U}_M(u, 0) | \phi(0) \rangle^* &= \langle \phi(0) | \hat{U}_M^\dagger(u, 0) | \phi(0) \rangle \\ &= \langle \phi(0) | \hat{U}_M(-u, 0) | \phi(0) \rangle \end{aligned} \qquad (3.13)$$

we get

$$\begin{aligned} &\int_{-t}^{t} e^{iE_l u/\hbar} f(u) \langle \phi(0) | \hat{U}_M(u, 0) | \phi(0) \rangle\, du \\ &\quad = 2 \int_0^t \text{Re}\{ e^{iE_l u/\hbar} f(u) \langle \phi(0) | \hat{U}_M(u, 0) | \phi(0) \rangle \}\, du \end{aligned} \qquad (3.14)$$

where $f(u)$ is an even real-valued function. This equation shows, in addition, that a forward propagation in time, from $u = 0$ to $u = t$, is all that is needed.

Equations (3.9) and (3.10) make clear what we mean in terms of physics by the cw limit. The parameter γ should be so small that the function $\exp(-\gamma u^2/2)$ can be considered as constant on the time scale where the dynamics inherent in the autocorrelation function is important—or formulated in energy space—the width of the light pulse should be much smaller than the width of the features in the spectrum. When this condition is fulfilled, the envelope function in the integral of Eq. (3.9) can be replaced by unity and the total reaction probability is calculated as a Fourier transform of a nuclear autocorrelation function. Fast decay of the autocorrelation function implies that the cw limit for the light pulse is reached even for a quite short pulse. This limit defines the situation where the highest possible resolution in energy is obtained. We calculate the total probability from

$$\boxed{\begin{aligned} P_{\text{tot}}(\omega_l) &= \frac{1}{4\hbar^2}\int_{-\infty}^{\infty} e^{iE_l u/\hbar}\langle\phi(0)|\hat{U}_M(u,0)|\phi(0)\rangle\, du \\ &= \frac{1}{2\hbar^2}\int_{0}^{\infty} \mathrm{Re}\{e^{iE_l u/\hbar}\langle\phi(0)|\hat{U}_M(u,0)|\phi(0)\rangle\}\, du \end{aligned}} \tag{3.15}$$

where

$$\hat{U}_M(u,0) = \exp\left(-i\begin{bmatrix}\hat{H}_2 & \hat{C}_{23}\\ \hat{C}_{32} & \hat{H}_3\end{bmatrix} u/\hbar\right) \tag{3.16}$$

We observe that the total absorption probability in the *cw limit* is expressed in terms of the real-time dynamics of the molecule as created in the δ function limit. The dynamics is mapped out in the Franck–Condon region due to the overlap with the initial state. Measurement in the energy (frequency) domain of the total absorption spectrum is, accordingly, a way to get information about dynamics in the Franck–Condon region.

The dynamics inherent in the autocorrelation function is the dynamics of the molecule in the excited electronic states. Thus, we have signatures of dynamics even in energy-resolved observations. For large u, the wave packet will, in general, split into a bound and a dissociative part

$$\hat{U}_M(u,0)|\phi(0)\rangle = \hat{U}_M(u,0)|\phi_d(0)\rangle + \hat{U}_M(u,0)|\phi_b(0)\rangle \tag{3.17}$$

For simple dissociative motion $\hat{U}_M(u, 0)|\phi_d(0)\rangle$ will move away from the Franck–Condon region, that is, the area vertically above the initial state, and never return. The autocorrelation function will decay to zero when the wave packet is out of the Franck–Condon region. Actually, nodes in the wave packet associated with development of momentum will ensure that the decay to zero happens before the wave packet is out of the Franck–Condon region. Typically, it takes much less than a vibrational period for the molecule. The exact details depend of course on the steepness of the potential, the reduced mass of the separating fragments, and so on. We return to an explicit discussion of these matters in Section VIII.A. For more complicated dissociative motion, parts of $\hat{U}_M(u, 0)|\phi_d(0)\rangle$ can revisit the Franck–Condon region a few times. The autocorrelation function will decay to zero within the order of a vibrational period for the molecule. If a bound part of wave packet is present then $\hat{U}_M(u, 0)|\phi_b(0)\rangle$ will revisit the Franck–Condon region repeatedly. Many recurrences in the autocorrelation function will, accordingly, show up. However, a spectrum with "experimental" resolution can still be obtained by considering only a limited number of these recurrences [8].

If $\langle \mathbf{R}|\phi(0)\rangle$ is real, we can form the half-Fourier transform of the time evolved $|\phi(0)\rangle$ before taking the overlap with the initial state, then—within the limit of no non-adiabatic coupling—we can also express Eq. (3.15) in the form [cf. Eq. (2.22)],

$$P_{\text{tot}}(\omega_l) = \frac{1}{2\hbar^2} \int d\mathbf{R} \langle \phi(0)|\mathbf{R}\rangle \langle \mathbf{R}|\mathcal{R}\rangle_{\text{r}} \tag{3.18}$$

that is, an overlap between the initial state and the real part of $\langle \mathbf{R}|\mathcal{R}\rangle$ (the stationary state created in the *cw limit*). Note, however, that in the last expression for the absorption probability *the explicit reference to molecular dynamics is lost.*

The time-dependent approach based on the calculation of the nuclear autocorrelation functions has seen several applications. It has, for example, been used for the interpretation of complicated spectral features in absorption spectra of dissociating molecules [28–32].

IV. FINAL PRODUCT DISTRIBUTIONS

So far, we have not explored the fact that the molecule might dissociate and that free fragments can show up when it is excited electronically. The formulas derived in the previous sections are, in fact, also valid in situations where the molecule does not dissociate at all. In Section III we calculated the total probability associated with the transition from the left- to the right-hand side in Eq. (1.1).

Now we want to calculate the probability associated with the channels that correspond to bond breaking and focus on the chemical composition of the fragments, as well as more detailed questions concerning the exact quantum state of the fragments.

A. Detailed Final Product Distribution

The state of the fragments in arrangement channel α, is given by

$$\hat{H}_i^{\alpha}|E, i, n_{\alpha}\rangle = E|E, i, n_{\alpha}\rangle \tag{4.1}$$

where $\hat{H}_i^{\alpha}$ is the nuclear Hamiltonian of arrangement channel α [26] and electronic state i (= 2 or 3),

$$\hat{H}_i^{\alpha} = \hat{T}_{\text{rel}}^{\alpha} + \hat{H}_{\text{int}}^{\alpha} \tag{4.2}$$

which implies that $|E, i, n_{\alpha}\rangle$ is a product between eigenstates associated with the free relative motion between the fragments and the internal vibrational and rotational fragment states. The parameter n_{α} is a collective symbol for the vibrational and rotational quantum numbers and the momentum vector associated with the free relative motion of the fragments. By using Eq. (2.14), we can get an expression for the probability of finding fragments in these states. The relevant probabilities are obtained as,

$$\begin{bmatrix} P(E, 2, n_{\alpha}) \\ P(E, 3, n_{\alpha}) \end{bmatrix} = \begin{bmatrix} lim_{t\to\infty}|\langle E, 2, n_{\alpha}|\chi_2^{(1)}(t)\rangle|^2 \\ lim_{t\to\infty}|\langle E, 3, n_{\alpha}|\chi_3^{(1)}(t)\rangle|^2 \end{bmatrix} \tag{4.3}$$

To simplify, let us neglect the nonadiabatic coupling terms and focus on the nuclear dynamics associated with electronic state 2. Thus,

$$|\langle E, 2, n_{\alpha}|\chi_2^{(1)}(t)\rangle|^2 = \frac{1}{4\hbar^2}\left|\int_0^t dt' e^{-iE_it'/\hbar} a(t') \times \langle E, 2, n_{\alpha}|\exp(-i\hat{H}_2(t-t')/\hbar)|\phi_d(0)\rangle\right|^2 \tag{4.4}$$

Dissociation implies [26],

$$\exp(-i\hat{H}_2 t/\hbar)|\phi_d(0)\rangle \xrightarrow{t\to\infty} \sum_{\alpha} \exp(-i\hat{H}_2^{\alpha} t/\hbar)|\psi_{\text{out}}^{\alpha}\rangle \tag{4.5}$$

Equation (4.5) expresses the fact that after dissociation ($t \rightarrow \infty$) freely moving fragments can show up in the different arrangement channels.

We assume that the light pulse interacts with the molecule in a finite, but possibly very long, time t_l. Thus, $a(t') = 0$ for $t' > t_l$. The product distribution is then calculated for $t \rightarrow \infty$, which in practice means that $t - t_l > t_{\text{diss}}$, that is, $t > t_{\text{diss}} + t_l$. Using Eqs. (4.1) and (4.5) and the orthogonality of the states in different channels, we get

$$
\begin{aligned}
P(E, 2, n_\alpha) &= \frac{1}{4\hbar^2} \Bigg| \int_0^{t_l} dt'\, e^{-iE_l t'/\hbar} a(t') \\
&\quad \times \lim_{t \rightarrow \infty} \langle E, 2, n_\alpha | \sum_{\alpha'} \exp(-i\hat{H}_2^{\alpha'}(t - t')/\hbar) | \psi_{\text{out}}^{\alpha'} \rangle \Bigg|^2 \\
&= \frac{|\varepsilon(\omega)|^2}{4\hbar^2} \lim_{t \rightarrow \infty} |\langle E, 2, n_\alpha | \exp(-i\hat{H}_2^{\alpha} t/\hbar) | \psi_{\text{out}}^{\alpha} \rangle|^2 \\
&= \frac{|\varepsilon(\omega)|^2}{4\hbar^2} |\langle E, 2, n_\alpha | \psi_{\text{out}}^{\alpha} \rangle|^2
\end{aligned}
\tag{4.6}
$$

where

$$
\varepsilon(\omega) = \int_0^{t_l} dt'\, e^{-i\omega t'} a(t') \tag{4.7}
$$

and $\omega = (E_l - E)/\hbar$.

Equation (4.5) implies

$$
|\phi_d(0)\rangle = \sum_\alpha \hat{\Omega}_-^\alpha |\psi_{\text{out}}^\alpha\rangle \tag{4.8}
$$

where

$$
\hat{\Omega}_-^\alpha = \lim_{t \rightarrow \infty} \exp(i\hat{H}_2 t/\hbar) \exp(-i\hat{H}_2^\alpha t/\hbar) \tag{4.9}
$$

is the channel Møller operator for channel α [26]. Thus, the scattering state that evolves into channel α is

$$
|\phi_d^\alpha(0)\rangle = \hat{\Omega}_-^\alpha |\psi_{\text{out}}^\alpha\rangle \tag{4.10}
$$

which implies

$$
|\psi_{\text{out}}^\alpha\rangle = \hat{\Omega}_-^{\alpha\dagger} |\phi_d^\alpha(0)\rangle \tag{4.11}
$$

since the Møller operators are isometric, that is, $\hat{\Omega}_-^{\alpha\dagger} \hat{\Omega}_-^\alpha = \hat{I}$.

Thus, the probability of finding a product in the state $|E, 2, n_\alpha\rangle$ [Eq. (4.6)] is

$$\boxed{\begin{aligned} P(E, 2, n_\alpha) &= \frac{|\varepsilon(\omega)|^2}{4\hbar^2} |\langle E, 2, n_\alpha | \hat{\Omega}_-^{\alpha\dagger} | \phi_d^\alpha(0)\rangle|^2 \\ &= \frac{|\varepsilon(\omega)|^2}{4\hbar^2} \lim_{t\to\infty} |\langle E, 2, n_\alpha | \exp(i\hat{H}_2^\alpha t/\hbar) \\ &\quad \times \exp(-i\hat{H}_2 t/\hbar) | \phi_d^\alpha(0)\rangle|^2 \\ &= \frac{|\varepsilon(\omega)|^2}{4\hbar^2} \lim_{t\to\infty} |\langle E, 2, n_\alpha | \exp(-i\hat{H}_2 t/\hbar) | \phi_d(0)\rangle|^2 \end{aligned}} \tag{4.12}$$

where we have dropped superscript α in the last line, using the orthogonality of states in different channels.

It is interesting to note that the probability in the first line is expressed as an inner product with the vector $\hat{\Omega}_-^{\alpha\dagger}|\phi_d^\alpha(0)\rangle$ which is an interaction picture description of the scattering state in channel α. This vector is accordingly constant when the dissociation is over. In the last line of Eq. (4.12) a description in terms of the Schrödinger picture for the time evolution of the scattering state is adopted. Here the asymptotic state evolves in time. The interaction picture description can be very convenient in numerical implementations of the formula [33, 34].

Equation (4.12) can be considered for different forms of the light pulse.

In the *δ-function limit*,

$$|\varepsilon(\omega)|^2 = |e^{-i\omega t_p}|^2 = 1 \tag{4.13}$$

Thus, product states with various energies can be formed. The energies are determined by the energetic width of the asymptotic form of the wave packet. In practice, this limit is obtained when the energetic width of the light pulse is much larger than the energetic width of the wave packet. In time space this implies, $t_l \ll t_{\text{diss}}$.

In the *cw limit*,

$$|\varepsilon(\omega)|^2 = \left| \int_0^{t_l} dt' \, e^{-i\omega t'} \right|^2 = 4 \sin^2\left(\frac{\omega t_l}{2}\right) \Big/ \omega^2 \tag{4.14}$$

and

$$\begin{aligned} |\varepsilon(\omega)|^2 / t_l &= 4 \sin^2(\omega t_l/2)/(\omega^2 t_l) \xrightarrow{t_l \to \infty} 2\pi\delta(\omega) \\ &= 2\pi\hbar\delta(E_l - E) \end{aligned} \tag{4.15}$$

Thus, for sufficiently long times, the accessible product states have an energy that is determined by the energy of the photon. In practice, this limit is obtained when $t_l \gg t_{\text{diss}}$.

The intermediate case given by the Gaussian pulse shape defined in Eq. (3.8) gives

$$|\varepsilon(\omega)|^2 = |(2\gamma/\pi)^{1/4} \int_0^{t_l} dt' \, e^{-i\omega t'} e^{-\gamma(t'-t_p)^2}|^2 \tag{4.16}$$

For some given γ and t_l large ($t_l \gg \sqrt{1/\gamma} + t_p$), we get

$$|\varepsilon(\omega)|^2 = \sqrt{2\pi/\gamma} \, e^{-\omega^2/2\gamma} \tag{4.17}$$

which for a small value of γ approach $2\pi\delta(\omega)$, that is, the *cw limit.*

From Eq. (4.12) we observe that the detailed final product distribution is expressed in terms of the real-time dynamics of the molecule as created in the δ-function limit. The dynamics is mapped out in the product region. Measurement of the final product distribution is accordingly a way to get information about the dynamics acquired from the Franck–Condon region all the way to the product region.

In addition, the results of this section show that the product distribution is affected by the form of the light pulse. A form of control of the outcome of the reaction can accordingly be performed in this way. We return to this issue in Section IV.D.

The most detailed information obtainable from photodissociation dynamics is the state-to-state probabilities calculated in the present section. However, often less detailed information suffices. Such information can, of course, be obtained by appropriate summation over the detailed state-to-state information but a more direct calculational approach can be available. This is illustrated in the following two sections.

B. Recoil Distribution

Here we consider the probability that the fragments recoil with a given outgoing momentum **p**, irrespective of the final vibrational–rotational state of the molecular fragment. The derivation that follows is similar to the one used in [35] in a different context. Using Eq. (4.6) and writing the eigenstates and associated energy of the channel Hamiltonian in the form $|\mathbf{p}\rangle|n\rangle$ and $E = \varepsilon_t + \varepsilon_n$, respectively (i.e., the channel index is dropped to simplify the notation), we obtain in the *cw limit*

$$\begin{aligned} P_{\mathbf{p}}(\omega_l) &= \frac{\pi}{2\hbar}\sum_n |\langle \mathbf{p}|\langle n|\psi_{\text{out}}\rangle|^2 \delta(E_l - [\varepsilon_t + \varepsilon_n]) \\ &= \frac{1}{4\hbar^2}\sum_n \int_{-\infty}^{\infty} d\tau\, e^{i([E_l-\varepsilon_t]-\varepsilon_n)\tau/\hbar} |\langle n|\theta_{\text{out}}(\mathbf{p})\rangle|^2 \\ &= \frac{1}{4\hbar^2}\int_{-\infty}^{\infty} d\tau\, e^{i(E_l-\varepsilon_t)\tau/\hbar} \langle \theta_{\text{out}}(\mathbf{p})|\exp(-i\hat{H}_{\text{int}}\tau/\hbar)|\theta_{\text{out}}(\mathbf{p})\rangle \end{aligned} \tag{4.18}$$

where

$$|\theta_{\text{out}}(\mathbf{p})\rangle = \langle \mathbf{p}|\psi_{\text{out}}\rangle \tag{4.19}$$

and we made use of

$$\exp(-i\hat{H}_{\text{int}}\tau/\hbar)|n\rangle = e^{-i\varepsilon_n\tau/\hbar}|n\rangle \tag{4.20}$$

and the completeness relation

$$\sum_n |n\rangle\langle n| = \hat{I} \tag{4.21}$$

It is convenient to rewrite Eq. (4.18),

$$\boxed{\begin{aligned} P_{\mathbf{p}}(\omega_l) &= \frac{1}{4\hbar^2}\int_{-\infty}^{\infty} d\tau\, e^{i(E_l-\varepsilon_t)\tau/\hbar} \langle \theta_{\text{out}}(\mathbf{p}, t_f)| \\ &\quad \times \exp(-i\hat{H}_{\text{int}}\tau/\hbar)|\theta_{\text{out}}(\mathbf{p}, t_f)\rangle \\ &= \frac{1}{2\hbar^2}\int_0^{\infty} d\tau\, \text{Re}\{e^{i(E_l-\varepsilon_t)\tau/\hbar} \langle \theta_{\text{out}}(\mathbf{p}, t_f)| \\ &\quad \times \theta_{\text{out}}(\mathbf{p}, t_f+\tau)\rangle\} \end{aligned}} \tag{4.22}$$

where Eq. (3.14) was used in the last line, $|\theta_{\text{out}}(\mathbf{p}, t_f)\rangle$ is associated with $|\Psi_{\text{out}}(t_f)\rangle = \exp(-i\hat{H}_2 t_f/\hbar)|\psi_{\text{out}}\rangle$, which is the time-dependent form of the asymptotic state vector, and t_f is the final (finite) propagation time.

Thus, the desired quantity can be calculated without referring to the internal state of the fragment. When the correlation function is Fourier transformed at various translational energies (for a fixed photon energy) several peaks can show up with an energy spacing signifying the vibrational and rotational energy levels in the fragment. Note that the correlation function will depend on the orientation of the momentum vector.

C. Branching between Chemically Distinct Products

In this section we calculate the probability of forming given chemical products irrespective of the particular quantum state of the product [36].

Assume that two different arrangement channels are open, for example,

$$\mathrm{ABC}(n) + \hbar\omega_l \rightarrow \begin{cases} \mathrm{A} + \mathrm{BC}(n_1) \\ \mathrm{C} + \mathrm{AB}(n_2) \end{cases} \tag{4.23}$$

For the wave packet dynamics inherent in Eq. (4.12), this situation implies according to Eq. (4.5)

$$\exp(-i\hat{H}_2 t/\hbar)|\phi_d(0)\rangle \xrightarrow{t\to\infty} \exp(-i\hat{H}_2^1 t/\hbar)|\psi^1_{\text{out}}\rangle + \exp(-i\hat{H}_2^2 t/\hbar)|\psi^2_{\text{out}}\rangle \tag{4.24}$$

where $|\psi^1_{\text{out}}\rangle$ and $|\psi^2_{\text{out}}\rangle$ denotes the asymptotic form of the state vector in channel 1 and 2, respectively. Now, according to Eq. (4.6), the probability of finding the system in, say, arrangement channel 1 in the eigenstate $|E, 2, n_1\rangle$ is

$$P(E, 2, n_1) = \frac{|\varepsilon(\omega)|^2}{4\hbar^2} |\langle E, 2, n_1|\psi^1_{\text{out}}\rangle|^2 \tag{4.25}$$

where $\omega = (E_l - E)\hbar$, $E_l = \hbar\omega_l + \varepsilon_a$.

The probability of finding the system in arrangement channel 1 irrespective of the particular quantum state n_1 and integrated over the total energy E gives the absorption probability into arrangement channel 1, as a function of the photon frequency. We consider the *cw limit* for the electromagnetic radiation, and follow a well-known derivation [37] used for the total absorption spectrum,

$$
\begin{aligned}
P_1(\omega_l) &= \frac{\pi}{2\hbar} \int \sum_{n_1} |\langle E, 2, n_1 | \psi^1_{\text{out}} \rangle|^2 \delta(E - E_l)\, dE \\
&= \frac{1}{4\hbar^2} \int \sum_{n_1} \int_{-\infty}^{\infty} d\tau\, e^{i(E_l - E)\tau/\hbar} |\langle E, 2, n_1 | \psi^1_{\text{out}} \rangle|^2\, dE \\
&= \frac{1}{4\hbar^2} \int_{-\infty}^{\infty} e^{iE_l\tau/\hbar} \langle \psi^1_{\text{out}} | \exp(-i\hat{H}^1_2 \tau/\hbar) | \psi^1_{\text{out}} \rangle\, d\tau
\end{aligned} \tag{4.26}
$$

where

$$
e^{-i\hat{H}^1_2\tau/\hbar} |E, 2, n_1\rangle = e^{-iE\tau/\hbar} |E, 2, n_1\rangle \tag{4.27}
$$

and the completeness of the eigenstates of $\hat{H}^1_2$ was used, that is,

$$
\int \sum_{n_1} |E, 2, n_1\rangle \langle n_1, 2, E|\, dE = \hat{I} \tag{4.28}
$$

Again, in actual calculations what is known is not $|\psi^1_{\text{out}}\rangle$ but [cf. Eq. (4.24)],

$$
|\psi^1_{\text{out}}(t_f)\rangle = \exp(-i\hat{H}^1_2 t_f/\hbar) |\psi^1_{\text{out}}\rangle \tag{4.29}
$$

where t_f is the final propagation time. Therefore we rewrite to the form

$$
\boxed{
\begin{aligned}
P_1(\omega_l) &= \frac{1}{4\hbar^2} \int_{-\infty}^{\infty} d\tau e^{iE_l\tau/\hbar} \langle \psi^1_{\text{out}} | \exp(-i\hat{H}^1_2 \tau/\hbar) | \psi^1_{\text{out}} \rangle \\
&= \frac{1}{4\hbar^2} \int_{-\infty}^{\infty} d\tau e^{iE_l\tau/\hbar} \langle \psi^1_{\text{out}}(t_f) | \psi^1_{\text{out}}(t_f + \tau) \rangle \\
&= \frac{1}{2\hbar^2} \int_0^{\infty} d\tau\, \text{Re}\{ e^{iE_l\tau/\hbar} \langle \psi^1_{\text{out}}(t_f) | \psi^1_{\text{out}}(t_f + \tau) \rangle \}
\end{aligned}
} \tag{4.30}
$$

where Eq. (3.14) was used in the last line.

A completely equivalent formula can, of course, be derived for products in arrangement channel 2,

$$P_2(\omega_l) = \frac{1}{2\hbar^2} \int_0^\infty d\tau \, \mathrm{Re}\{e^{iE_l\tau/\hbar} \langle \psi^2_{\mathrm{out}}(t_f) | \psi^2_{\mathrm{out}}(t_f + \tau) \rangle \} \tag{4.31}$$

The ratio $P_1(\omega_l)/P_2(\omega_l)$ gives the branching ratio between the two chemically distinct products as a function of the photon frequency ω_l. The two wave packets $|\psi^1_{\mathrm{out}}(t_f)\rangle$ and $|\psi^2_{\mathrm{out}}(t_f)\rangle$ are according to Eq. (4.24) given as the part of the initial state evolving into channels 1 and 2, respectively. These wave packets are easily identified in practice [16]. The correlation functions for these wave packets will typically decay to zero very fast due to the translational motion of the fragments. The interpretation of the form of Eqs. (4.30) and (4.31) is simple. The Fourier transform of the autocorrelation function gives the energy spectrum of the state. For each of the states, $|\psi^1_{\mathrm{out}}(t_f)\rangle$ and $|\psi^2_{\mathrm{out}}(t_f)\rangle$, Eqs. (4.30) and (4.31), respectively, give the probability of finding the component of the state at energy $E_l = \hbar\omega_l + \varepsilon_a$. Thus, the branching ratio, at energy $E_l = \hbar\omega_l + \varepsilon_a$, is the ratio between the probabilities of finding the component of the states $|\psi^1_{\mathrm{out}}(t_f)\rangle$ and $|\psi^2_{\mathrm{out}}(t_f)\rangle$ at this energy.

D. Control Schemes

The final product distribution can be affected in various ways.

Passive control explores the possibilities embodied in Eq. (4.12). First, we consider a cw pulse. Equation (4.12) shows that control can be carried out via the frequency of the light that selects the accessible energy of the product states, or via the choice of the initial quantum state of the parent molecule. The theoretical and experimental results for HOD is a nice example of how the branching ratio between H + OD and D + OH is affected by these parameters [16]. However, we now explore nothing but the inherent dynamics of the photofragmentation process. In addition, Eq. (4.12) shows that the product distribution is affected by the form of the light pulse. The light pulse selects the accessible energies of the products. It is easily seen that the product distribution obtained with a δ-function pulse can be reproduced by a cw pulse when the frequency is scanned. Thus, changing the form of the light pulse gives a rather trivial effect and the distribution that is obtained with some particular pulse shape can always be reproduced by a consecutive set of dissociation processes induced by cw pulses with appropriate intensity and frequency. Thus, in principle, nothing new, with regard to product control, is obtained by applying non-cw pulses. In addition, it is clear that at a given energy, the form of the light pulse will not affect the relative probabilities for populating different degenerate channels. This branching ratio is determined by the inherent dynamics of the molecule as embodied in the Franck–Condon wave packet, Eq. (1.2). *Active* control schemes are designed to circumvent the dynamics of the Franck–Condon wave packet.

First, we consider a scheme [38] that is reminiscent of the Tannor–Rice pump–dump scheme [39]. It is based on thinking in the time-domain—more specifically on the spatial positioning of wave packets. It is a pump–pump scheme as illustrated in Fig. 6.2. Equation (2.10) is the starting point,

$$|\chi_2^{(1)}(t)\rangle = -\frac{i}{\hbar}\int_{t_0}^{t} dt'\, e^{-i\hat{H}_2(t-t')/\hbar}\hat{R}_{21}(t')\, e^{-i\hat{H}_1(t'-t_0)/\hbar}|\chi_1(t_0)\rangle \quad (4.32)$$

where

$$R_{21}(t') = -\boldsymbol{\mu}_{21}\cdot\mathbf{E}_0 a(t')e^{-i\omega_l t' + i\delta_l} \quad (4.33)$$

In the ultrashort pulse limit, $a(t) = \delta(t - t_p)$,

$$\boxed{|\chi_2^{(1)}(t)\rangle = \frac{ie^{i\alpha}}{\hbar} e^{-i\hat{H}_2(t-t_p)/\hbar}|\phi(t_p)\rangle} \quad (4.34)$$

where

$$|\phi(t_p)\rangle = \boldsymbol{\mu}_{21}\cdot\mathbf{E}_0|\chi_1(t_p)\rangle$$
$$|\chi_1(t_p)\rangle = e^{-i\hat{H}_1(t_p-t_0)/\hbar}|\chi_1(t_0)\rangle \quad (4.35)$$

and $\alpha = \delta_l - \omega_l t_p$. Thus, the state $|\chi_1(t_p)\rangle$ is transferred to the excited state at $t = t_p$, where it subsequently starts to evolve in time.

If now $|\chi_1(t_0)\rangle$ is a nonstationary state then $|\chi_1(t_p)\rangle$ will be a moving wave packet in the electronic ground state and the state created by the ultrashort laser excitation will depend on the time of excitation t_p. Equation (4.34) clearly constitutes a generalized form of the Franck–Condon wave packet (see Fig. 6.2).

A suitable form of the $|\chi_1(t_0)\rangle$ wave packet can be created by an intense infrared laser pulse. To enhance the insight, we compile below the results from a well-known simple analytically solvable model that addresses this phenomenon. The ingredients in this one-dimensional

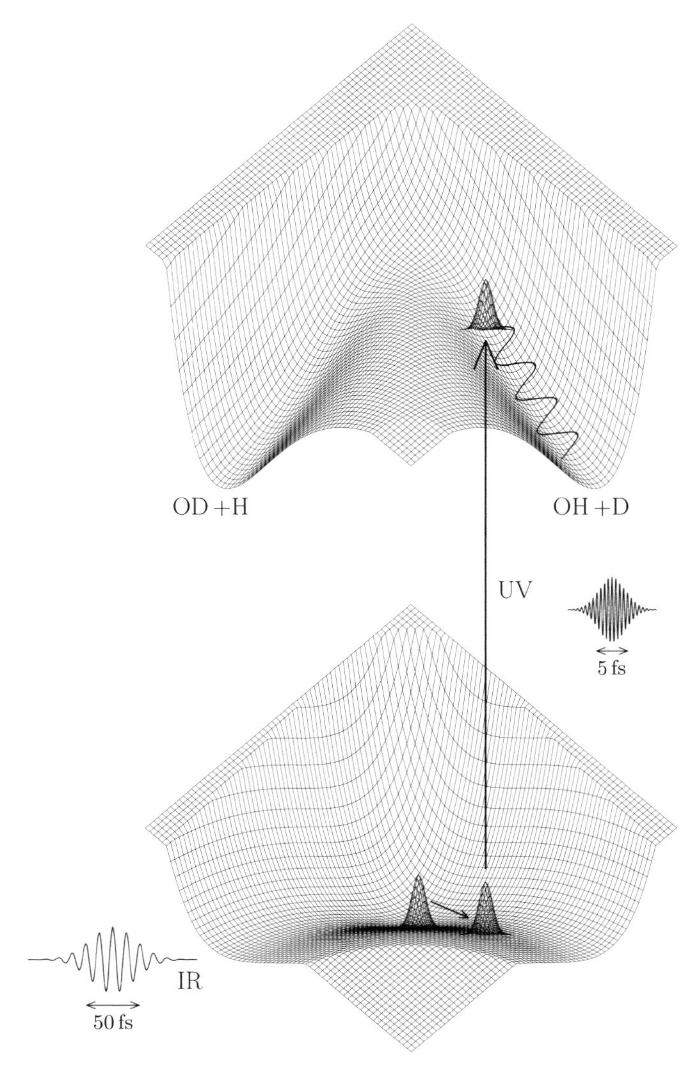

Figure 6.2. Schematic representation of a control scheme based on two laser pulses, an infrared (IR) and an ultraviolet (UV) pulse. The figure shows the scheme applied to the HOD molecule.

model are:

$$\hat{H}_1 = -\frac{\hbar^2}{2m}\frac{\partial^2}{\partial q^2} + \frac{1}{2}m\omega_k^2 q^2$$

$$|\chi_1(0)\rangle = |n\rangle$$

$$\hat{R}_{11}(t) = -\mu_{11}(x)E(t) \tag{4.36}$$

$$\mu_{11}(x) = \mu_{11}(x_e) + \mu'_{11}(x_e)(x - x_e) + \cdots$$

where $q = x - x_e$ and $\mu'_{11}(x_e) = \partial\mu_{11}/\partial x|_{x=x_e}$. Thus, a one-dimensional harmonic oscillator that is coupled to a radiation field via a dipole term where the dipole moment of the oscillator is linearized around the equilibrium position x_e. The total Hamiltonian is, accordingly, that of a linearly forced harmonic oscillator. The initial state is assumed to be one of the eigenstates of the unforced oscillator, $|n\rangle$. This problem can be solved for $|\chi_1(t)\rangle$. For $n = 0$, that is, when the oscillator starts out in the vibrational ground state, it can easily be solved by the Gaussian wave packet method to be discussed in Section VI and in general, for example, as in [40]. We obtain, for example,

$$\langle q\rangle = 2b\{\mathrm{Re}(\alpha^+)\sin\omega_k t - \mathrm{Im}(\alpha^+)\cos\omega_k t\}$$
$$(\Delta q)^2 = 2b^2(n + \tfrac{1}{2}) \tag{4.37}$$

where

$$\alpha^+ = -\frac{b}{\hbar}\int_0^{t_0} F(t')e^{i\omega_k t'}\,dt'$$

$$F(t) = -\mu'_{11}(x_e)E(t) \tag{4.38}$$

$$b = \sqrt{\frac{\hbar}{2m\omega_k}}$$

Thus, an oscillating wave packet where the uncertainty in position is time independent, that is, there is no spreading.

Now for $E(t) = E_1\cos(\omega t + \delta)$, we get

$$\langle q\rangle = \frac{E_1\mu'_{11}(x_e)}{2m\omega_k}\left\{\frac{\cos(\omega_k t - \delta - (\omega + \omega_k)t_0) - \cos(\omega_k t - \delta)}{\omega_k + \omega} + \frac{\cos(\omega_k t + \delta - (\omega_k - \omega)t_0) - \cos(\omega_k t + \delta)}{\omega_k - \omega}\right\} \tag{4.39}$$

At resonance, $\omega = \omega_k$,

$$\langle q \rangle \to \frac{E_1 \mu'_{11}(x_e)}{2m\omega_k} \left\{ \frac{\cos(\omega_k t - \delta - 2\omega_k t_0) - \cos(\omega_k t - \delta)}{2\omega_k} + t_0 \sin(\omega_k t + \delta) \right\} \tag{4.40}$$

The amplitude of the second term grows linearly with t_0. A simple linearly forced oscillation is obtained when $t = t_0$ and $\delta = 0$ (where the first term vanishes). The harmonic oscillator model gives a fair approximation to real anharmonic potentials when the interaction time t_0 is sufficiently short.

The present scheme was used in [38] where it was shown that complete control over the fragmentation pathway in HOD is obtainable, that is, it is possible to control which of the bonds is going to be broken in the photodissociation process. The vibrational local mode structure in the electronic ground state of HOD implies that bond selective vibrations (of the type described in the forced oscillator model) can be created with relative ease. These local mode states can then be transferred to the excited electronic state with a short laser pulse.

Finally, before closing this section, we briefly consider an alternative scheme termed coherent control [41]. The essential idea here is to control dynamics via quantum interference. From the discussion in the previous sections, it is clear that the phase of a single laser pulse does not affect the observables. The coherent control scheme suggested by Brumer and Shapiro [41] is based on two laser pulses for the electronic excitation step—where the phase of each laser pulse can be controlled—and, in addition, in this scheme the initial state of the molecule is prepared as a superposition of two eigenstates $|a\rangle$ and $|b\rangle$. Thus, in the notation of the previous sections, we have

$$\begin{aligned} e^{-i\hat{H}_1 t'/\hbar}|\chi_1(0)\rangle &= c_a|a\rangle e^{-i\varepsilon_a t'/\hbar} + c_b|b\rangle e^{-i\varepsilon_b t'/\hbar} \\ \hat{R}_{21}(t') &= -\boldsymbol{\mu}_{21} \cdot \mathbf{E}_0(a(t')e^{-i\omega_a t' + i\delta_a} + a(t')e^{-i\omega_b t' + i\delta_b}) \end{aligned} \tag{4.41}$$

where the frequencies are chosen such that $\hbar\omega_a + \varepsilon_a = \hbar\omega_b + \varepsilon_b$. We consider, in the *cw limit*, the probability that the fragments show up in the state $|E, 2, n_\alpha\rangle$. If we follow the arguments that were used to obtain Eq. (4.6), we get

$$P(E,2,n_\alpha) = 2\pi/\hbar\{|c_a|^2|\langle E,2,n_\alpha|\phi^a_{\text{out}}\rangle|^2 + |c_b|^2\langle E,2,n_\alpha|\phi^b_{\text{out}}\rangle|^2 + 2\,\text{Re}[c_a^* c_b e^{i(\delta_b-\delta_a)}\langle\phi^a_{\text{out}}|E,2,n_\alpha\rangle \times \langle E,2,n_\alpha|\phi^b_{\text{out}}\rangle]\}\delta(E-E_l) \tag{4.42}$$

Thus, the probability consists of three terms where the last term is an interference term. The magnitude of this term depends on the phase difference between the two pulses and, hence, this term is under experimental control.

We can, as in Section IV.C, derive a compact expression for the probability of forming a given chemical product irrespective of the particular quantum state. Thus, following the arguments that were used to obtain Eq. (4.30), we get the following expression for the probability that the fragments show up in channel α,

$$\begin{aligned} P_\alpha(\omega_l) = {} & \frac{|c_a|^2}{\hbar^2}\int_{-\infty}^{\infty} e^{iE_l\tau/\hbar}\langle\phi^{a,\alpha}_{\text{out}}|e^{-i\hat{H}^\alpha_2\tau/\hbar}|\phi^{a,\alpha}_{\text{out}}\rangle\,d\tau \\ & + \frac{|c_b|^2}{\hbar^2}\int_{-\infty}^{\infty} e^{iE_l\tau/\hbar}\langle\phi^{b,\alpha}_{\text{out}}|e^{-i\hat{H}^\alpha_2\tau/\hbar}|\phi^{b,\alpha}_{\text{out}}\rangle\,d\tau \\ & + \frac{2}{\hbar^2}\,\text{Re}\left\{c_a^* c_b\, e^{i(\delta_b-\delta_a)}\int_{-\infty}^{\infty} e^{iE_l\tau/\hbar}\langle\phi^{a,\alpha}_{\text{out}}|e^{-i\hat{H}^\alpha_2\tau/\hbar}|\phi^{b,\alpha}_{\text{out}}\rangle\,d\tau\right\} \end{aligned} \tag{4.43}$$

This equation contains two autocorrelation functions and a cross correlation function that is associated with the phase difference. This formula has been implemented for HOD [42]. The results showed that some control over the bond-breaking selectivity could be obtained by varying the phase difference.

The special properties of laser light are essential for the active control schemes discussed in the present section. The phase coherence of laser light is as essential to the coherent control scheme as the ultrashort laser pulses in the pump–pump scheme. In fact, an experiment with a classical light source, with a frequency spread that covers the frequencies ω_a and ω_b, would be reminiscent of the situation where a large number of pulses with *random phases* were directed towards the molecule. When a time average over many such pulses [of the form given in Eq. (4.41)] is performed, the term that depends on the phase difference will tend to average out due to the random oscillations in the complex exponential.

V. TIME-INDEPENDENT APPROACH, STATIONARY SCATTERING STATES

It is instructive to look into some of the results of the previous sections in terms of stationary states (see, e.g., [43–45]). To keep the discussion as simple as possible, we neglect the nonadiabatic coupling terms in this section. The dynamics is, accordingly, confined to the electronic state 2.

Equation (2.14) takes the form,

$$|\chi_2^{(1)}(t)\rangle = \frac{ie^{i\phi}}{2\hbar}\int_0^t dt'\, e^{-i(\hbar\omega_l+\varepsilon_a)t'/\hbar} a(t') \exp(-i\hat{H}_2(t-t')/\hbar)|\phi(0)\rangle \tag{5.1}$$

The time-evolution operator can be written in the form,

$$\exp(-i\hat{H}_2(t-t')/\hbar) = \sum_{\mathbf{n}} \int dE |E,\mathbf{n}-\rangle e^{-iE(t-t')/\hbar}\langle E,\mathbf{n}-| \tag{5.2}$$

where

$$\hat{H}_2|E,\mathbf{n}-\rangle = E|E,\mathbf{n}-\rangle \tag{5.3}$$

and continuum states are energy normalized, that is, $\langle E,\mathbf{n}-|E',\mathbf{n}-\rangle = \delta(E-E')$. The stationary states can be discrete and continuous—the integration over energy in Eq. (5.2) is understood to be taken as a summation for the discrete part of the spectrum. The parameter $\mathbf{n}$ is a set of quantum numbers that labels the stationary states at energy E. The minus sign indicates that we have chosen one particular state in this subspace of degenerate states. The precise specification of these states is not needed at the moment—we return to a closer discussion of this issue later in this section. We substitute Eq. (5.2) into Eq. (5.1),

$$|\chi_2^{(1)}(t)\rangle = \frac{ie^{i\phi}}{2\hbar}\sum_{\mathbf{n}}\int \langle E,\mathbf{n}-|\phi\rangle|E,\mathbf{n}-\rangle e^{-iEt/\hbar} \times \int_0^t dt'\, e^{-i(\hbar\omega_l+\varepsilon_a-E)t'/\hbar} a(t')\, dE \tag{5.4}$$

This equation represents the nuclear wave packet expressed in terms of stationary states. The amplitude associated with the stationary state at energy E is given by the Franck–Condon factor $\langle E, \mathbf{n}-|\phi\rangle$. The overall energetic width of the state is given by the integral over t', which in turn depends on the envelope function of the light pulse.

Let us again analyze the results for the two limiting forms for this function. The *δ-function limit*, defined by Eq. (2.16)

$$|\chi_2^{(2)}(t)\rangle = \frac{ie^{i(\phi-\alpha)}}{2\hbar} \sum_{\mathbf{n}} \int dE \langle E, \mathbf{n}-|\phi\rangle |E, \mathbf{n}-\rangle e^{-iE(t-t_p)/\hbar} \tag{5.5}$$

where $\alpha = (\hbar\omega_l + \varepsilon_a)t_p/\hbar$ is an overall phase factor. Thus, the state $|\phi\rangle$ is resolved on the stationary states of $\hat{H}_2$.

The *cw limit*, defined by Eq. (2.19)

$$|\chi_2^{(1)}(t)\rangle = \frac{ie^{i\phi}}{2\hbar} \sum_{\mathbf{n}} \int dE \langle E, \mathbf{n}-|\phi\rangle |E, \mathbf{n}-\rangle e^{-iEt/\hbar} \times \int_0^t dt'\, e^{-i(E_l-E)t'/\hbar} \tag{5.6}$$

or using Eq. (2.21)

$$|\mathcal{R}\rangle = \frac{2\hbar}{ie^{i\phi}} e^{iE_lt/\hbar} |\chi_2^{(1)}(t)\rangle = \sum_{\mathbf{n}} \int dE \langle E, \mathbf{n}-|\phi\rangle |E, \mathbf{n}-\rangle \int_0^t dt'\, e^{-i(E-E_l)t'/\hbar} \tag{5.7}$$

where $t \to \infty$. Now [22],

$$\int_0^\infty dt'\, e^{-i(E-E_l)t'/\hbar} = \pi\hbar\delta(E - E_l) + i\hbar\mathcal{P} \int 1/(E_l - E) \tag{5.8}$$

where $\mathcal{P}$ denote a principal value integral. Thus, we have the following expression for the state vector

$$\int_0^\infty du\, e^{iE_lu/\hbar} \exp(-i\hat{H}_2u/\hbar)|\phi\rangle = \pi\hbar \sum_{\mathbf{n}} \langle E_l, \mathbf{n} - |\phi\rangle |E_l, \mathbf{n}-\rangle + i\hbar\mathcal{P} \int \sum_{\mathbf{n}} \langle E, \mathbf{n} - |\phi\rangle |E, \mathbf{n}-\rangle/(E_l - E)\, dE \tag{5.9}$$

Now assuming that $\langle \mathbf{R}|\phi\rangle$ is real, it is easy to show

$$
\begin{aligned}
&\mathrm{Re}\left\{\left\langle \mathbf{R}\right| \int_0^\infty du\, e^{iE_l u/\hbar} \exp(-i\hat{H}_2 u/\hbar)\left|\phi\right\rangle\right\} \\
&= \frac{1}{2}\left\langle \mathbf{R}\right| \int_{-\infty}^\infty du\, e^{iE_l u/\hbar} \exp(-i\hat{H}_2 u/\hbar)\left|\phi\right\rangle \\
&= \pi\hbar \sum_{\mathbf{n}} \langle E_l, \mathbf{n}-|\phi\rangle \langle \mathbf{R}|E_l, \mathbf{n}-\rangle \\
&\equiv \mathrm{Re}\{\langle \mathbf{R}|\mathscr{R}\rangle\}
\end{aligned}
\tag{5.10}
$$

where Eq. (5.2) was used in the last line. We have then,

$$
\langle E_l, \mathbf{n}-|\phi\rangle = \int d\mathbf{R} \langle E_l, \mathbf{n}-|\mathbf{R}\rangle\, \mathrm{Re}\{\langle \mathbf{R}|\mathscr{R}\rangle\}/(\pi\hbar) \tag{5.11}
$$

The real part of $\langle \mathbf{R}|\mathscr{R}\rangle$ is accordingly the projection of the initial state, $|\phi\rangle$, on the eigenstates of $\hat{H}_2$ at energy E_l.

Let us now give expressions for the observables of Sections III and IV in terms of stationary scattering states. This allows us to make contact with the traditional expressions that were prevailing until the time-dependent formulation showed up.

First, we consider the total absorption probability. Using Eq. (5.2)

$$
\int_{-\infty}^\infty \exp\{i(E_l - \hat{H}_2)u/\hbar\}\, du = 2\pi\hbar \sum_{\mathbf{n}} |E_l, \mathbf{n}-\rangle\langle -\mathbf{n}, E_l| \tag{5.12}
$$

Then the total absorption probability, Eq. (3.15), takes the form

$$
\boxed{P_{\mathrm{tot}}(\omega_l) = \frac{\pi}{2\hbar} \sum_{\mathbf{n}} |\langle E_l, \mathbf{n}-|\phi(0)\rangle|^2} \tag{5.13}
$$

The stationary states used in this expression can be any set eigenstates of the full Hamiltonian [Eq. (5.3)].

To express the final product distributions in terms of stationary states, we need a few more details about an appropriate choice of stationary scattering states [26]. We begin by expanding the asymptotic state $|\psi^\alpha_{\mathrm{out}}\rangle$ on the eigenstates of the channel Hamiltonian, Eq. (4.1),

$$
|\psi^\alpha_{\mathrm{out}}\rangle = \sum_{n_\alpha} \int dE\, \langle E, 2, n_\alpha|\psi^\alpha_{\mathrm{out}}\rangle |E, 2, n_\alpha\rangle \tag{5.14}
$$

The corresponding scattering state at $t = 0$ is according to Eq. (4.10)

$$|\phi_d^\alpha(0)\rangle = \sum_{n_\alpha} \int dE\, \langle E, 2, n_\alpha | \psi_{\text{out}}^\alpha \rangle \hat{\Omega}_-^\alpha | E, 2, n_\alpha \rangle \tag{5.15}$$

Thus, the following definition is relevant

$$|E, 2, n_\alpha - \rangle \equiv \hat{\Omega}_-^\alpha | E, 2, n_\alpha \rangle \tag{5.16}$$

These states are the natural vectors for expanding the actual state at $t = 0$ since the expansion coefficients for the asymptotic and the full scattering state are the same.

Using $\hat{H}_2 \hat{\Omega}_-^\alpha = \hat{\Omega}_-^\alpha \hat{H}_2^\alpha$ [26], it follows that $|E, 2, n_\alpha - \rangle$ is an eigenstate of the full Hamiltonian $\hat{H}_2$. Since $|E, 2, n_\alpha - \rangle$ is a stationary state we can, accordingly, only in a loose sense speak of this state as the full scattering state that would evolve into the final state $|E, 2, n_\alpha\rangle$. In addition to satisfying the time-independent Schrödinger equation the state defined in Eq. (5.16) also satisfies another important equation that expresses $|E, 2, n_\alpha - \rangle$ in terms of $|E, 2, n_\alpha\rangle$. From Eq. (4.10) and a standard trick [26] where the Møller operator is rewritten as the integral of its derivative

$$\begin{aligned}
|\phi_d^\alpha(0)\rangle &= \hat{\Omega}_-^\alpha |\psi_{\text{out}}^\alpha\rangle \\
&= |\psi_{\text{out}}^\alpha\rangle + i \int_0^\infty dt\, e^{i\hat{H}_2 t/\hbar} V^\alpha e^{-i\hat{H}_2^\alpha t/\hbar} |\psi_{\text{out}}^\alpha\rangle \\
&= |\psi_{\text{out}}^\alpha\rangle + \lim_{\varepsilon \to 0^+} \sum_{n_\alpha} \int dE\, \hat{G}(E - i\varepsilon) V^\alpha |E, 2, n_\alpha\rangle \\
&\quad \times \langle E, 2, n_\alpha | \psi_{\text{out}}^\alpha \rangle
\end{aligned} \tag{5.17}$$

where $V^\alpha = \hat{H}_2 - \hat{H}_2^\alpha$ and $\hat{G}(z) = (z - \hat{H}_2)^{-1}$ is the Green's operator. Now Eqs. (5.14) and (5.15) imply that we can extract the following equation from Eq. (5.17):

$$\begin{aligned}
|E, 2, n_\alpha - \rangle &= |E, 2, n_\alpha\rangle + \hat{G}(E - i0) V^\alpha |E, 2, n_\alpha\rangle \\
&= |E, 2, n_\alpha\rangle + \hat{G}^\alpha(E - i0) V^\alpha |E, 2, n_\alpha - \rangle
\end{aligned} \tag{5.18}$$

This is the Lippmann–Schwinger equation. The operator $\hat{G}^\alpha(z) = (z - \hat{H}_2^\alpha)^{-1}$ and the last line in Eq. (5.18) is established by a well-known operator identity [26].

The detailed final product distribution Eq. (4.12) takes a simple form when it is expressed in terms of the scattering states introduced above. Thus,

$$\begin{aligned} P(E, 2, n_\alpha) &= \frac{|\varepsilon(\omega)|^2}{4\hbar^2} |\langle E, 2, n_\alpha | \hat{\Omega}_-^{\alpha\dagger} | \phi(0) \rangle|^2 \\ &= \frac{|\varepsilon(\omega)|^2}{4\hbar^2} |\langle \phi(0) | \hat{\Omega}_-^{\alpha} | E, 2, n_\alpha \rangle|^2 \\ &= \frac{|\varepsilon(\omega)|^2}{4\hbar^2} |\langle \phi(0) | E, 2, n_\alpha - \rangle|^2 \end{aligned} \tag{5.19}$$

Equation (5.19) has been the starting point for many computational approaches to the final product distribution (see, e.g., [43, 46]).

Alternatively, in the *cw limit* we can, via Eq. (5.11), express this result in terms of the real part of $\langle \mathbf{R} | \mathcal{R} \rangle$ which was the real part of the state vector created in this limit,

$$\begin{aligned} P(E, 2, n_\alpha) &= \frac{1}{2\pi\hbar^3} |\int d\mathbf{R} \langle E, 2, n_\alpha - |\mathbf{R}\rangle \mathrm{Re}\{\langle \mathbf{R} | \mathcal{R} \rangle\}|^2 \delta(E - E_l) \\ &= \frac{1}{2\pi\hbar^3} |\int d\mathbf{R} \langle E, 2, n_\alpha | \mathbf{R} \rangle \ \mathrm{Re}\{\langle \mathbf{R} | \mathcal{R} \rangle\}|^2 \delta(E - E_l) \end{aligned} \tag{5.20}$$

where Eqs. (5.16) and (2.25) were used in the last line.

Thus, the total reaction probability, Eq. (3.18), as well as the detailed final product distribution, Eq. (5.20), can be expressed in the *cw limit* in terms of the state vector created in this limit.

Before ending this section we note that the introduction of stationary scattering states implies that *the explicit reference to molecular dynamics is lost* in the expressions for observables [Eqs. (5.13) and (5.19)]. We return therefore to the main line of presentation where the formulation displays the dynamics in an explicit way.

PART 2
NUCLEAR MOTION: SEMICLASSICAL DESCRIPTION

The various relations derived in Part 1 are extremely useful because they give us a way of thinking about the subject. Such insight would not have been provided if we had simply done a frontal attack on Eq. (2.6). The next step to be taken in order to obtain insight in photofragmentation dynamics is to understand the quantum mechanical time evolution of nuclear motion—since this concept is involved in all the expressions derived in Part 1.

Again the frontal attack approach is possible, that is, a numerical solution of the time-dependent Schrödinger equation [47]. Such an approach is, however, only possible for molecules with a small number of nuclear degrees of freedom (i.e., diatomic or triatomic molecules). But more important, such an approach does not give much insight into why the nuclear motion takes a particular path. Instead, to provide a way of thinking about nuclear motion, we turn to semiclassical descriptions.

Our intuition about dynamics is strongly connected to classical mechanics. Therefore it is worthwhile to explore the possibility of describing the dynamics in terms of semiclassical mechanics. Here we use the term semiclassical mechanics in a pragmatic way. It denotes any approximation to quantum mechanics where classical mechanical quantities can be clearly identified.

The methods to be considered in Sections VI and VII are based on local harmonic approximations. The validity is therefore, in general, restricted to short-time dynamics where spreading of wave packets is moderate. These methods are, accordingly, ideal for fast processes such as photodissociation where bond breaking can take place in less than a typical vibrational period.

VI. GAUSSIAN WAVE PACKET DYNAMICS

We summarize below in Eqs. (6.1)–(6.9) the essential equations that constitute Gaussian wave packet dynamics [48]. The purpose is twofold: First to provide one appealing way of thinking about nuclear motion and, second, to present an important tool for the applications to be discussed in Sections VIII and IX.

We expand the initial state on a set of functions (a one-dimensional system is considered in this section, to simplify the notation)

$$\phi(q,0) = \sum_i c_i G_i(q,0) \tag{6.1}$$

then the linearity of the time-evolution operator implies that each function in the expansion evolves independently

$$\phi(q,t) = \sum_i c_i G_i(q,t) \tag{6.2}$$

The functional form of G_i is, in general, not preserved during the time evolution. Now we choose Gaussians of the form,

$$G_i(q,0) = \exp\{(i/\hbar)[(q - q_0^i)A_0^i(q - q_0^i) + p_0^i(q - q_0^i) + s_0^i]\} \quad (6.3)$$

as the set of functions at $t = 0$. The parameter A_0^i is a complex number, q_0^i and p_0^i are real numbers, and s_0^i is a complex phase factor. The imaginary part of s_0^i is by the normalization condition given as

$$\mathrm{Im}(s_0^i) = \frac{\hbar}{4}\ln\left[\frac{\pi\hbar}{2\,\mathrm{Im}(A_0^i)}\right] \quad (6.4)$$

We consider the time evolution of one of the Gaussians (index i is dropped, for simplicity, in the following) and assume that the Hamiltonian can be written as

$$\hat{H}_{\mathrm{LHA}} = -\frac{\hbar^2}{2m}\frac{\partial^2}{\partial q^2} + V_t(q) \quad (6.5)$$

where $V_t(q)$ is a (time-dependent) local harmonic approximation (LHA) to the true potential, $V(q)$, around $q = q_t$, that is,

$$V_t(q) = V(q_t) + V'(q_t)(q - q_t) + \tfrac{1}{2}(q - q_t)V''(q_t)(q - q_t) \quad (6.6)$$

where $V'(q_t) = \partial V/\partial q|_{q=q_t}$ and $V''(q_t) = \partial^2 V/\partial q^2|_{q=q_t}$. Then the time evolution of the Gaussian is given by

$$i\hbar\frac{\partial G(q,t)}{\partial t} = \hat{H}_{\mathrm{LHA}}G(q,t) \quad (6.7)$$

with the solution

$$G(q,t) = \exp\{(i/\hbar)[(q - q_t)A_t(q - q_t) + p_t(q - q_t) + s_t]\} \quad (6.8)$$

provided the parameters evolve in time according to

$$\begin{aligned}
\dot{q}_t &= p_t/m \\
\dot{p}_t &= -V'(q_t) \\
\dot{A}_t &= -2\,A_t^2/m - V''(q_t)/2 \\
\dot{s}_t &= i\hbar A_t/m + p_t^2/(2m) - V(q_t)
\end{aligned} \quad (6.9)$$

Thus, the Gaussian form is preserved when a harmonic approximation around the instantaneous q_t position parameter of the Gaussian is invoked. The first two equations in (6.9) are identical to Hamilton's equation of motion known from classical mechanics [49] and according to the last equation the wave packet acquires a phase given by the classical action along this trajectory. This phase gives rise to interference effects among the packets in Eq. (6.2).

The present scheme provides an exact solution to the dynamics in potentials of the form, $V(q) = k_0 + k_1 q + k_2 q^2$ (where k_0, k_1, and k_2 can be time dependent), since Eq. (6.6) offers an exact representation of such potentials. For anharmonic potentials, the validity of the local harmonic approximation depends on how localized the wave packet is in position space.

Equation (6.9) can be solved analytically for several potentials of relevance to photofragmentation dynamics. These applications are discussed in Sections VIII and IX. Note that the only assumption introduced above concerns the form of the Hamiltonian $\hat{H}_{\mathrm{LHA}}$. The kinetic energy operator in $\hat{H}_{\mathrm{LHA}}$ arises naturally when q is a Cartesian position coordinate. However, q need not be Cartesian since the same form of kinetic energy operator can show up for some radial and angular degrees of freedom problems.

The parameters of the Gaussian wave packet carry a simple physical meaning. The expectation value and the associated uncertainty of the position is

$$\begin{aligned} \langle q \rangle &= \langle G(q,t)|q|G(q,t)\rangle = q_t \\ (\Delta q)_t^2 &= \langle q^2 \rangle - \langle q \rangle^2 \qquad = \hbar/(4\,\mathrm{Im}(A_t)) \end{aligned} \tag{6.10}$$

The expectation value and associated uncertainty of the operator $\hat{p} = -i\hbar\,\partial/\partial q$ (which represent the momentum when q is a Cartesian coordinate) is

$$\begin{aligned} \langle \hat{p} \rangle &= -i\hbar \langle G(q,t)|\partial/\partial q|G(q,t)\rangle = p_t \\ (\Delta p)_t^2 &= \langle \hat{p}^2 \rangle - \langle \hat{p} \rangle^2 \\ &= \hbar([\mathrm{Re}(A_t)]^2 + [\mathrm{Im}(A_t)]^2)/\mathrm{Im}(A_t) \end{aligned} \tag{6.11}$$

The Gaussian wave packet approximation (i.e., the LHA) has been used in a variety of applications including the important paper by Kulander and Heller [5], which laid the foundation for the time-dependent formulation of photofragmentation dynamics. Other applications can, for example, be found in [6, 12, 17, 33, 24, 35, 37, 48]. Here we have only considered one-dimensional systems, and it should be stressed that a complete description of the dissociation dynamics of a triatomic molecule can be developed in terms of Gaussian wave packet dynamics [12]. Thus, the total absorption probability as well as detailed final product distributions out of vibrational–rotational eigenstates of the parent molecule has been calculated within the framework of this approach. All the calculations were, accordingly, based on just one approximation; the local harmonic approximation to the potential.

VII. WIGNER PHASE SPACE REPRESENTATION

Another convenient way to make contact with classical mechanics is via the Wigner phase space representation of quantum mechanics [50–54]. This is an exact representation of quantum mechanics that has the appealing feature of containing many elements, which appear to be very close to a classical description. Thus, operators are represented by functions on phase space and quantum mechanical states are represented by the so-called Wigner function [50]. This is a function defined on phase space that makes the proper transition from the classical description of the state, given by a single point in phase space, to the quantum mechanical description where uncertainty relations, and so on, have to be observed. We summarize below in Eqs. (7.1)–(7.5) some of the essential relations, that is, the phase space equivalent of expectation values and the phase space counterpart of time evolution.

The phase space equivalent of the expectation value of an operator is

$$\langle \psi(t)|\hat{A}|\psi(t)\rangle = \int\int a(\mathbf{p},\mathbf{q})\Gamma(\mathbf{p},\mathbf{q},t)\,d\mathbf{p}\,d\mathbf{q} \tag{7.1}$$

where $a(\mathbf{p},\mathbf{q})$ is the classical function corresponding to the operator $\hat{A}$. The function $a(\mathbf{p},\mathbf{q})$ is defined by

$$a(\mathbf{p},\mathbf{q}) = (2)^N \int d\boldsymbol{\eta}\langle \mathbf{q}-\boldsymbol{\eta}|\hat{A}|\mathbf{q}+\boldsymbol{\eta}\rangle \exp(2i\mathbf{p}\cdot\boldsymbol{\eta}/\hbar) \tag{7.2}$$

and $\Gamma(\mathbf{p},\mathbf{q},t)$ is the Wigner function, defined by

$$\begin{aligned}\Gamma(\mathbf{p},\mathbf{q},t) &= \left(\frac{1}{\pi\hbar}\right)^N \int d\boldsymbol{\eta}\langle \mathbf{q}-\boldsymbol{\eta}|\psi(t)\rangle\langle\psi(t)|\mathbf{q}+\boldsymbol{\eta}\rangle \exp(2i\mathbf{p}\cdot\boldsymbol{\eta}/\hbar)\\ &= \left(\frac{1}{\pi\hbar}\right)^N \int d\boldsymbol{\eta}\psi(\mathbf{q}+\boldsymbol{\eta},t)^*\psi(\mathbf{q}-\boldsymbol{\eta},t)\exp(2i\mathbf{p}\cdot\boldsymbol{\eta}/\hbar)\end{aligned} \tag{7.3}$$

where $\psi(\mathbf{q}, t) = \langle \mathbf{q} | \psi(t) \rangle$ and N is the number of degrees of freedom. A comparison of Eqs. (7.2) and (7.3) shows that the Wigner function is the phase space function associated with the projection operator, $(2\pi\hbar)^{-N} |\psi(t)\rangle \langle \psi(t)|$. It is easily seen that the Wigner function is always real-valued.

The time evolution of the Wigner function is given by

$$\frac{\partial \Gamma(\mathbf{p}, \mathbf{q}, t)}{\partial t} = \{H, \Gamma(\mathbf{p}, \mathbf{q}, t)\}_{\text{Moyal}} \tag{7.4}$$

where H is the *classical* Hamiltonian corresponding to $\hat{H}$ and the Moyal bracket is given by

$$\{H, \Gamma(\mathbf{p}, \mathbf{q}, t)\}_{\text{Moyal}} = \frac{2}{\hbar} \sin\left[\frac{\hbar}{2}\left(\frac{\partial}{\partial \mathbf{q}_H} \cdot \frac{\partial}{\partial \mathbf{p}_\Gamma} - \frac{\partial}{\partial \mathbf{p}_H} \cdot \frac{\partial}{\partial \mathbf{q}_\Gamma}\right)\right] H\Gamma(\mathbf{p}, \mathbf{q}, t) \tag{7.5}$$

where the subscripts indicate that the operator acts only on H or Γ respectively. The Moyal bracket is a complicated operator containing derivatives in $\mathbf{p}$ and $\mathbf{q}$ of infinite order. However, the power of Eq. (7.4) lies in its potential for making contact with classical mechanics. Thus, when H is at most quadratic in $\mathbf{q}$ the Moyal bracket reduces to the Poisson bracket.

It is instructive first to consider the Gaussian wave packet method in the phase space representation. The Wigner function corresponding to the Gaussian wave packet $G(q, t)$ is easily obtained via Eq. (7.3),

$$\begin{aligned}\Gamma(q, p, t) = &\frac{1}{\pi\hbar} \exp[-2(\Delta p)_t^2 (q - q_t)^2/\hbar^2 - 2(\Delta q)_t^2 (p - p_t)^2/\hbar^2] \\ &\times \exp[2(\text{Re}(A_t)/\text{Im}(A_t))(q - q_t)(p - p_t)/\hbar]\end{aligned} \tag{7.6}$$

The following relations hold,

$$\begin{aligned}\int \Gamma(q, p, t)\, dp &= (2\pi(\Delta q)_t^2)^{-1/2} \exp\left[-\frac{(q - q_t)^2}{2(\Delta q)_t^2}\right] = |G(q, t)|^2 \\ \int \Gamma(q, p, t)\, dq &= (2\pi(\Delta p)_t^2)^{-1/2} \exp\left[-\frac{(p - p_t)^2}{2(\Delta p)_t^2}\right] = |G(p, t)|^2\end{aligned} \tag{7.7}$$

Equation (7.7) is a special case of a relation between the Wigner function and the corresponding wave functions in position and momentum space that holds in general (can be shown directly from the definition of the Wigner function).

When $\text{Re}(A_t) = 0$, say for $t = 0$, Eq. (7.6) simplifies to the form,

$$\Gamma(q, p, 0) = \frac{1}{\pi\hbar} \exp\left[-\frac{(q - q_0)^2}{2(\Delta q)_0^2} - \frac{(p - p_0)^2}{2(\Delta p)_0^2} \right] \tag{7.8}$$

where we used that $(\Delta q)_0^2 (\Delta p)_0^2 = \hbar^2/4$. A contour plot of this function gives ellipses where the axes are parallel to the q and p axes, respectively. Equation (7.6) shows that, as time evolves, the center of the distribution rides on a classical trajectory, the contour ellipses start, however, to rotate and deform. This behavior is understood when it is noted that the time evolution of the Wigner function obeys a classical Liouville equation. The Gaussian wave packet method is based on a LHA around the instantaneous center of the packet. This approximation implies that the Wigner function associated with a Gaussian wave packet, evolves in time according to

$$\frac{\partial \Gamma(p, q, t)}{\partial t} = \frac{\partial H_{\text{LHA}}}{\partial q} \frac{\partial \Gamma(p, q, t)}{\partial p} - \frac{p}{m} \frac{\partial \Gamma(p, q, t)}{\partial q} \tag{7.9}$$

where

$$\frac{\partial H_{\text{LHA}}}{\partial q} = \left. \frac{\partial V}{\partial q} \right|_{q = q_t} + \left. \frac{\partial^2 V}{\partial q^2} \right|_{q = q_t} (q - q_t) \tag{7.10}$$

Thus, the time evolution can be interpreted as the classical evolution of a swarm of trajectories with weights given by Eq. (7.8) [51, 52].

Since the Wigner function is real-valued, one might get the impression that phases, and hence interference effects, are lost in the phase space representation. This is not the case. The Wigner function corresponding to, for example, a sum of Gaussians, Eq. (6.2), is

$$\Gamma(q, p, t) = \sum_i c_i^2 \Gamma_{ii}(q, p, t) + \sum_{i>j} c_i c_j (\Gamma_{ij}(q, p, t) + \Gamma_{ji}(q, p, t)) \tag{7.11}$$

where

$$\Gamma_{ij}(q, p, t) = \frac{1}{\pi\hbar} \int d\eta\, G_i(q + \eta, t)^* G_j(q - \eta, t) \exp(2ip\eta/\hbar) \tag{7.12}$$

The first term in Eq. (7.11) is independent of the phase of the wave packet, whereas the terms for $i \neq j$ are complex valued and these terms depend on the phase difference between wave packet i and j. Thus these terms carry the Wigner equivalent of interference.

An alternative to the Gaussian wave packet approximation is obtained if we at each phase space point neglect all anharmonic terms in the potential (i.e., a local harmonic approximation). Then the Moyal bracket reduces to the Poisson bracket and the time evolution of the Wigner function is given by a Liouville equation,

$$\frac{\partial \Gamma(p, q, t)}{\partial t} = \frac{\partial V}{\partial q} \frac{\partial \Gamma(p, q, t)}{\partial p} - \frac{p}{m} \frac{\partial \Gamma(p, q, t)}{\partial q} \tag{7.13}$$

This approximation is termed the Wigner method [52] and the equation is solved by running a swarm of trajectories.

The Gaussian wave packet approximation and the Wigner method are equivalent (and exact) for potentials free of anharmonic terms. When anharmonic terms are present, the two approximations are, obviously, nonequivalent [cf. Eqs. (7.9) and (7.13)]: they back off the harmonic limit differently. This aspect has, for example, been studied for the dynamics of Morse oscillators in [55] and [56]. Thus, it turns out that the accuracy of the wigner method is slightly better in this case.

The Wigner method has been used in several applications in photodissociation dynamics [57–59]. Although the first applications were promising, the contribution in [55, 59] makes clear that the validity of the Wigner method is limited to short times and situations where anharmonicities are unimportant. Introduction of the interaction picture for time evolution [60] can, however, extend the applicability considerably.

The expressions for the total absorption probability and the final product distributions takes, in the phase space representation, a form suggestive of classical mechanics for the dynamics as well as additional approximations for the initial and final states. These expressions are considered in the following sections.

A. Total Absorption Probability

The absolute square of an autocorrelation function can be written in the form

$$\begin{aligned} |\langle \phi(0)|\phi(u)\rangle|^2 &= \langle \phi(0)|\phi(u)\rangle \langle \phi(u)|\phi(0)\rangle \\ &= (2\pi\hbar)^N \int\int \Gamma(\mathbf{p}, \mathbf{q}, u)\Gamma(\mathbf{p}, \mathbf{q}, 0)\, d\mathbf{p}\, d\mathbf{q} \end{aligned} \tag{7.14}$$

However, to implement Eq. (3.15) directly we need the autocorrelation function itself. In the absolute value of the correlation function the phase is lost and it cannot be neglected in the subsequent Fourier transformation. Nevertheless, Eq. (7.14) clearly shows that position and momentum are on equal footing when the decay of the autocorrelation function is considered.

We can obtain an exact phase space equivalent of Eq. (3.15) in an alternative way. Thus, using Eqs. (5.13) and (7.1) to get the phase space equivalent of a projection operator, we find [61]

$$\begin{aligned} P_{\text{tot}}(\omega_l) &= \frac{\pi}{2\hbar} \sum_{\mathbf{n}}^{\text{open}} \langle \phi(0) | E_l, \mathbf{n}- \rangle \langle E_l, \mathbf{n} - | \phi(0) \rangle \\ &= \frac{\pi}{2\hbar} (2\pi\hbar)^N \int \int \sum_{\mathbf{n}}^{\text{open}} \Gamma_{\mathbf{n}}^{E_l}(\mathbf{p}, \mathbf{q}) \Gamma(\mathbf{p}, \mathbf{q}, 0)\, d\mathbf{p}\, d\mathbf{q} \end{aligned} \tag{7.15}$$

where $\Gamma(\mathbf{p}, \mathbf{q}, 0)$ is the Wigner function associated with $|\phi(0)\rangle$ and $\Gamma_{\mathbf{n}}^{E_l}(\mathbf{p}, \mathbf{q})$ is the Wigner function associated with the state $|E_l, \mathbf{n}-\rangle$.

We can consider approximations to the sum over Wigner functions or, equivalently, approximations to the phase space equivalent of the operator $\int_{-\infty}^{\infty} \exp\{i(E_l - \hat{H}_2)u/\hbar\}\, du$. The approximations derived in [61] are especially well suited to fast dissociation processes

$$(2\pi\hbar)^N \sum_{\mathbf{n}}^{\text{open}} \Gamma_{\mathbf{n}}^{E_l}(\mathbf{p}, \mathbf{q}) = \delta(H_2(\mathbf{p}, \mathbf{q}) - E_l) + \cdots \tag{7.16}$$

Thus, Eq. (7.15) takes the form

$$P_{\text{tot}}(\omega_l) = \frac{\pi}{2\hbar} \int \int \delta(H_2(\mathbf{p}, \mathbf{q}) - E_l) \Gamma(\mathbf{p}, \mathbf{q}, 0)\, d\mathbf{p}\, d\mathbf{q} \tag{7.17}$$

The delta function selects those points in phase space that have a classical energy equal to $E_l = \hbar\omega_l + \varepsilon_a$. Accordingly, the probability becomes the sum (integral) of weights [given by $\Gamma(\mathbf{p}, \mathbf{q}, 0)$] for these phase space points.

Additional approximations in Eq. (7.17) leads to the classical reflection approximation [61, 62]. Since the expectation value of the momentum for the initial state $\Gamma(p, q, 0)$ is zero, it can be a fair approximation to neglect the momentum contribution in H_2 when the phase space overlap is evaluated. Thus (for a one-dimensional system),

$$\begin{aligned} P_{\text{tot}}(\omega_l) &= \frac{\pi}{2\hbar} \int \int \delta(E_2(q) - E_l)\Gamma(p, q, 0)\, dp\, dq \\ &= \frac{\pi}{2\hbar} \int \delta(E_2(q) - E_l)|\phi(q)|^2\, dq \\ &= \frac{\pi/2\hbar}{|E_2'(q_{E_l})|} |\phi(q_{E_l})|^2 \end{aligned} \tag{7.18}$$

where q_{E_l} is the classical turning point, $E_l = E_2(q)$ (we have assumed that only a single turning point exists), and $E_2'(q_{E_l})$ is the derivative of the excited-state potential evaluated at the turning point. In this approximation the initial state is represented exactly by its wave function, whereas the final (continuum) state is approximated by a delta function at the classical turning point.

B. Branching between Chemically Distinct Products

To write down the phase space equivalent of Eq. (4.30), we proceed in the same way as for the total absorption [36]. We get a simple exact phase space expression for the partial probability $P_1(\omega_l)$, if we make explicit reference to the eigenstates of $\hat{H}_2^1$. Thus, we use

$$\exp(-i\hat{H}_2^1\tau/\hbar) = \int \sum_{n_1} |E, 2, n_1\rangle\langle n_1, 2, E| e^{-iE\tau/\hbar}\, dE \tag{7.19}$$

and Eq. (4.30) takes the form

$$\begin{aligned} P_1(\omega_l) &= \frac{1}{4\hbar^2} \langle \psi_{\text{out}}^1(t_f)| \int_{-\infty}^{\infty} \exp\{i(E_l - \hat{H}_2^1)\tau/\hbar\}\, d\tau |\psi_{\text{out}}^1(t_f)\rangle \\ &= \frac{\pi}{2\hbar} \sum_{n_1}^{\text{open}} \langle \psi_{\text{out}}^1(t_f)|E_l, 2, n_1\rangle\langle n_1, 2, E_l|\psi_{\text{out}}^1(t_f)\rangle \\ &= \frac{\pi}{2\hbar}(2\pi\hbar)^N \int \int \sum_{n_1}^{\text{open}} \Gamma_{n_1}^{E_l}(\mathbf{p}, \mathbf{q})\Gamma_{\text{out}}^1(\mathbf{p}, \mathbf{q}, t_f)\, d\mathbf{p}\, d\mathbf{q} \end{aligned} \tag{7.20}$$

where $\Gamma_{\text{out}}^1(t_f)$ is the Wigner function associated with $|\psi_{\text{out}}^1(t_f)\rangle$ and $\Gamma_{n_1}^{E_l}$ is the Wigner function associated with the state $|E_l, 2, n_1\rangle$. The summation runs over all open (degenerate) states at energy E_l.

We can now make contact with classical mechanics. First, classical mechanics for the time evolution that is the Wigner method gives

$$\Gamma^1_{\text{out}}(\mathbf{p}, \mathbf{q}, t_f) = \{\exp(\hat{L}t_f)\Gamma(\mathbf{p}, \mathbf{q}, 0)\}_1 \tag{7.21}$$

where $\Gamma(\mathbf{p}, \mathbf{q}, 0)$ denotes the Wigner function associated with $|\phi(0)\rangle$, the operator $\hat{L} = \{H_2,\}$ is the Poisson bracket and $\{\ \}_1$ indicates that we look at the trajectories that show up in channel 1. Thus, Eq. (7.20) with the approximation of Eq. (7.21), tell us that in order to find the probability of having a product in channel 1 at the energy $E_l = \hbar\omega_l + \varepsilon_a$, we have to run a swarm of trajectories with weights chosen according to the initial Wigner function $\Gamma(\mathbf{p}, \mathbf{q}, 0)$. Trajectories that end up in channel 1 are subsequently weighted by the function $\Sigma_{n_1} \Gamma^{E_l}_{n_1}$, and the result is obtained as a sum (integral) over all such weights associated with trajectories in the swarm. This summation will contain trajectories "off the energy shell," that is, trajectories with energies different from E_l.

Second, we can consider approximations to the sum over final state Wigner functions or, equivalently, approximations to the phase space equivalent of the operator $\int_{-\infty}^{\infty} \exp\{i(E_l - \hat{H}^1_2)\tau/\hbar\}\, d\tau$. We use the same analysis that gave Eq. (7.16), this is completely applicable in the present situation when the free translational motion of the fragments is fast

$$(2\pi\hbar)^N \sum_{n_1}^{\text{open}} \Gamma^{E_l}_{n_1}(\mathbf{p}, \mathbf{q}) = \delta(H^1_2(\mathbf{p}, \mathbf{q}) - E_l) + \cdots \tag{7.22}$$

Note that there is no reference to the quantum number n_1 on the right-hand side of this equation. Thus, using the approximations in Eqs. (7.21) and (7.22), Eq. (7.20) takes the form

$$P_1(\omega_l) = \frac{\pi}{2\hbar} \int\int \delta(H^1_2(\mathbf{p}, \mathbf{q}) - E_l) \times \{\exp(\hat{L}t_f)\Gamma(\mathbf{p}, \mathbf{q}, 0)\}_1\, d\mathbf{p}\, d\mathbf{q} \tag{7.23}$$

The delta function selects those trajectories in channel 1 that have the specified energy (E_l). Accordingly, the probability becomes the sum (integral) of weights [given by $\Gamma(\mathbf{p}, \mathbf{q}, 0)$] for these trajectories. Given that energy is conserved for a classical trajectory, this probability can also be calculated as the sum of weights for trajectories, which at $t = 0$ have the energy E_l and which sooner or later will move into channel 1. Thus, with the approximations introduced above the equation for the probability of having a fragment in channel 1 at energy E_l, takes a simple and intuitive form. The derivations leading to Eq. (7.23) have shown the nature of approximations involved in this equation. Therefore, it is possible to estimate the validity of this expression and correction terms can be included systematically, for example, by considering higher order terms in Eq. (7.22) as discussed in [61].

C. Detailed Final Product Distribution

Equation (4.12) takes the form,

$$\begin{aligned} P(E, 2, n_\alpha) &= \frac{|\varepsilon(\omega)|^2}{4\hbar^2} \lim_{t\to\infty} |\langle E, 2, n_\alpha| \exp(-i\hat{H}_2 t/\hbar)|\phi(0)\rangle|^2 \\ &= \frac{|\varepsilon(\omega)|^2}{4\hbar^2} \langle \psi_{\text{out}}(t_f)|E, 2, n_\alpha\rangle\langle E, 2, n_\alpha|\psi_{\text{out}}(t_f)\rangle \\ &= \frac{|\varepsilon(\omega)|^2}{4\hbar^2} (2\pi\hbar)^N \int\int \Gamma^E_{n_\alpha}(\mathbf{p}, \mathbf{q})\Gamma_{\text{out}}(\mathbf{p}, \mathbf{q}, t_f)\, d\mathbf{p}\, d\mathbf{q} \end{aligned} \tag{7.24}$$

where $\Gamma_{\text{out}}(\mathbf{p}, \mathbf{q}, t_f)$ is the Wigner function associated with the state,

$$|\psi_{\text{out}}(t_f)\rangle = \exp(-i\hat{H}_2 t_f/\hbar)|\phi(0)\rangle \tag{7.25}$$

and t_f is the final propagation time.

We can now introduce approximations. The Wigner method, that is, classical mechanics for the time evolution of phase space points imply

$$\Gamma_{\text{out}}(\mathbf{p}, \mathbf{q}, t_f) = \exp(\hat{L}t_f)\Gamma(\mathbf{p}, \mathbf{q}, 0) \tag{7.26}$$

Finally, the phase space formulation gives a clear justification of the vertical transition idea when it is applied simultaneously for positions as well as momenta. This version of the Franck–Condon principle is sometimes formulated in work that is based on a classical mechanical foundation [63, 64]. It should be clear from the discussion above that this proposition is present in the phase space equations before any classical approximations are introduced, accordingly, it is a consequence of the basic approximations—the adiabatic separation of electronic and nuclear motion and the use of first-order perturbation theory for the light-matter interaction.

PART 3
APPLICATIONS, SIMPLE MODELS

The third and final step to be taken to provide a way of thinking about photofragmentation dynamics, is to consider the application of the results of Parts 1 and 2 to small molecules. The aim is to establish basic insight into the form of the various observables considered in Part 1, without turning to specific molecules and extensive numerical work. We consider diatomic molecules in Section VIII and apply the Gaussian wave packet approximation of Part 2—and in Section IX triatomic molecules are considered where we, in addition, will invoke the time-dependent self-consistent-field description based on a Hartree product approximation.

It must be remembered that these dynamical approximations have a limited domain of validity and, therefore, quantitative agreement with experimental results can only be expected for certain simple systems. Nevertheless, a reference frame is established to which numerically exact results can be contrasted.

Since an exact foundation for the theory was presented in Part 1, we always have the option to eliminate, one by one, the various approximations introduced in Parts 2 and 3 and eventually end up with an exact implementation of the general results of Part 1.

VIII. THE DIATOMIC MOLECULE

The diatomic molecule is in some ways a very simple system. Nevertheless, very useful insight to important aspects of photofragmentation can be obtained by considering this system.

The nuclear Hamiltonian for the internal motion takes, in polar coordinates, the well-known form,

$$\hat{H} = -\frac{\hbar^2}{2\mu}\left(\frac{1}{r^2}\frac{\partial}{\partial r}r^2\frac{\partial}{\partial r} - \frac{\hat{L}^2}{\hbar^2 r^2}\right) + E(r) \tag{8.1}$$

where μ is the reduced mass of the two nuclei, r is the internuclear distance, $\hat{L}$ is the nuclear angular momentum, and $E(r)$ is the electronic energy. We consider the dynamics in a single excited electronic state, subscript 2 is dropped in this section to simplify the notation, that is, $\hat{H} \equiv \hat{H}_2$ and $E(r) \equiv E_2(r)$.

The rotational invariance implies

$$[\hat{H}, \hat{L}] = 0 \tag{8.2}$$

and

$$\hat{H}\left(\frac{u_l(r,t)}{r}Y_{lm}(\theta,\phi)\right) = \frac{Y_{lm}(\theta,\phi)}{r}\hat{H}(l)u_l(r,t) \tag{8.3}$$

where Y_{lm} are the spherical harmonics eigenstates associated with the angular momentum and $\hat{H}(l)$ is equivalent to a one-dimensional Hamiltonian given by

$$\hat{H}(l) = -\frac{\hbar^2}{2\mu}\frac{\partial^2}{\partial r^2} + V_l(r) \tag{8.4}$$

with the effective potential

$$V_l(r) = E(r) + \frac{\hbar^2 l(l+1)}{2\mu r^2} \tag{8.5}$$

where $l = 0, 1, 2, \ldots$. The time evolution of a state that is in a stationary rotational state, takes the form

$$e^{-i\hat{H}t/\hbar}\left(\frac{u_l(r,0)}{r} Y_{lm}(\theta,\phi)\right) = \frac{Y_{lm}(\theta,\phi)}{r} e^{-i\hat{H}(l)t/\hbar} u_l(r,0) \tag{8.6}$$

The initial state to be propagated is

$$\begin{aligned}\langle \mathbf{r}|\phi(0)\rangle &= \boldsymbol{\mu}_{21}(r)\cdot\mathbf{E}_0\psi_{l_i}(r)Y_{l_i m_i}(\theta,\phi)\\ &= E_0\frac{u_{l_i}(r,0)}{r}\cos\theta_r Y_{l_i m_i}(\theta,\phi)\end{aligned} \tag{8.7}$$

where $u_{l_i}(r,0) = \mu_{21}(r) r\psi_{l_i}(r)$. The parameters $l_i m_i$ are the quantum numbers for the angular momentum at $t = 0$. The parameter θ_r is the angle between the transition dipole vector, $\boldsymbol{\mu}_{21}$, and the direction of the electric field. The transition dipole vector is for a diatomic molecule either parallel (for electronic states of identical symmetry) or perpendicular (for electronic states of different symmetry) to the molecular axis. We consider here the parallel case. For a plane polarized field in the z direction, we then have $\theta_r = \theta$. We note [22],

$$\cos\theta\ Y_{l_i m_i}(\theta,\phi) = \sqrt{c(l_i m_i)}Y_{l_i+1,m_i}(\theta,\phi) + \sqrt{d(l_i m_i)}Y_{l_i-1,m_i}(\theta,\phi) \tag{8.8}$$

where

$$c(l_i m_i) = \frac{(l_i + m_i + 1)(l_i - m_i + 1)}{(2l_i + 1)(2l_i + 3)}$$

$$d(l_i m_i) = \frac{(l_i + m_i)(l_i - m_i)}{(2l_i + 1)(2l_i - 1)} \tag{8.9}$$

A. Total Absorption Probability

The total absorption probability in the *cw limit*, Eq. (3.15), takes the form,

$$\boxed{\begin{aligned} P_{\text{tot}}(\omega_l) = (E_0^2/4\hbar^2)c(l_i m_i) \int_{-\infty}^{\infty} e^{iE_l t/\hbar} \langle u_{l_i} | e^{-i\hat{H}(l_i+1)t/\hbar} | u_{l_i} \rangle \, dt \\ + (E_0^2/4\hbar^2)d(l_i m_i) \int_{-\infty}^{\infty} e^{iE_l t/\hbar} \langle u_{l_i} | e^{-i\hat{H}(l_i-1)t/\hbar} | u_{l_i} \rangle \, dt \end{aligned}} \tag{8.10}$$

where

$$\langle u_l | e^{-i\hat{H}(l')t/\hbar} | u_l \rangle = \int_0^{\infty} dr \, u_l(r, 0) e^{-i\hat{H}(l')t/\hbar} u_l(r, 0) \tag{8.11}$$

Thus, the problem reduces to the evaluation of this one-dimensional autocorrelation function. The total absorption probability consists of two terms, corresponding to the two final values of the angular momentum $l_i + 1$ and $l_i - 1$ (for $l_i = 0$ the second term is absent). The probability P_{tot} is a function of the laser frequency as well as the quantum numbers that specify the initial vibrational and rotational state of the diatomic molecule.

If the magnetic quantum numbers are initially unspecified, that is, if no special alignment is prepared, all m values are equally probable. The relevant observable is then obtained after a summation over these quantum numbers, which can be carried out analytically, using

$$\sum_{m_i=-l_i}^{l_i} c(l_i m_i) = \frac{(l_i + 1)}{3}$$

$$\sum_{m_i=-l_i}^{l_i} d(l_i m_i) = \frac{l_i}{3} \tag{8.12}$$

The evaluation of the autocorrelation function and its Fourier transform is considered now—with the aim of obtaining physical insight. Two approximate ways of obtaining this information in an analytical form are considered, for a molecule in the vibrational ground state with the assumption of a constant electronic transition dipole moment:

1. Short-Time Approximations to the Dynamics

For a molecule in the vibrational ground state, u_l is approximately a Gaussian wave packet with $p_0 = 0$, that is,

$$u_l(r) = \exp\{(i/\hbar)[A_0(r - r_0)^2 + s_0]\} \tag{8.13}$$

where $A_0 = i\,\mathrm{Im}(A_0) = i\mu\omega/2$. In the right-hand side of the equations of motion, Eq. (6.9), we put $t = 0$ and obtain (neglecting the spreading of the wave packet) [65]

$$\begin{aligned}
\frac{dr_t}{dt} &= 0 \\
\frac{dp_t}{dt} &= -\partial V_l(r)/\partial r|_{r=r_0} \\
\frac{dA_t}{dt} &= 0 \\
\frac{ds_t}{dt} &= i\hbar A_0/\mu + p_0^2/(2\mu) - V_l(r_0)
\end{aligned} \tag{8.14}$$

where $V_l(r)$ is the effective potential defined in Eq. (8.5). The change in a short time step τ starting at $t = 0$ is

$$\begin{aligned}
r_\tau &= r_0 \\
p_\tau &= -V_l'(r_0)\tau \\
A_\tau &= A_0 \\
s_\tau &= s_0 - (V_l(r_0) + \hbar\,\mathrm{Im}(A_0)/\mu)\tau
\end{aligned} \tag{8.15}$$

where $V_l'(r_0) = \partial V_l(r)/\partial r|_{r=r_0}$. Thus, the early dynamics of a wave packet with $p_0 = 0$ is governed by the time evolution of the momentum. The correlation function takes the form,

$$\langle u_l | e^{-i\hat{H}(l')\tau/\hbar} | u_l \rangle = \exp\left[-\frac{(V'_{l'}(r_0))^2}{8(\Delta p)_0^2}\tau^2 - i(V_{l'}(r_0) + (\Delta p)_0^2/\mu)\tau/\hbar \right] \tag{8.16}$$

The modulus of this complex function is

$$|\langle u_l | e^{-i\hat{H}(l')\tau/\hbar} | u_l \rangle| = \exp\left[-\frac{(V'_{l'}(r_0))^2}{8(\Delta p)_0^2}\tau^2 \right] \tag{8.17}$$

that is, a Gaussian decay. Assuming that the correlation function has effectively decayed to zero in the short time τ ($V'_{l'}(r_0)$ large and/or $(\Delta p)_0$ small), the Fourier transform, and hence Eq. (8.10), is easily obtained. The result for $l_i = 0$

$$\boxed{P_{\text{tot}}(\omega_l) = E_0^2 \mu_{21}^2/(12\hbar^2)\frac{\sqrt{2\pi}\hbar}{\sigma}\exp(-(\hbar\omega_l - V_1(r_0))^2/2\sigma^2)} \tag{8.18}$$

where

$$\sigma^2 = (V'_1(r_0)(\Delta r)_0)^2 \tag{8.19}$$

and the position uncertainty of the initial vibrational state is $(\Delta r)_0 = \sqrt{\hbar/(2\mu\omega)}$. Thus, the absorption probability is a Gaussian distribution in the laser frequency and the position and width of this distribution depends in a simple way on the value and slope of the excited-state potential as well as the width of the initial vibrational state. This result is closely related to the reflection approximation Eq. (7.18). Thus, with the additional assumption of a linear excited-state potential, we essentially obtain the same result as in the short-time approximation above (except for a minor shift in the spectrum towards lower energy corresponding to the zero-point energy $\hbar\omega/2$).

The dependence on the rotational state should be clear from the above results: The slope of the excited-state potential will tend to become more repulsive in situations with high rotational energy and the faster dissociation will result in a broader absorption spectrum. However, since the centrifugal potential typically is very small compared to the repulsive potential $E(r)$, due to the electronic energy, the dependence on the initial rotational state l_i will often be negligible.

2. Linearization of the Excited-State Potential

A second and potentially more accurate approximation is obtained by assuming that the effective excited-state potential can be linearized over the width of the initial Gaussian wave packet when the auto-correlation function is evaluated. Thus, denoting the slope by β,

$$V_l(r) = \beta_l(r - r_0) + V_l(r_0) \tag{8.20}$$

The time evolution of the Gaussian wave packet parameters Eq. (6.9),

$$\begin{aligned} r_t &= r_0 + (p_0 t - \beta_l t^2/2)/\mu \\ p_t &= p_0 - \beta_l t \\ A_t &= A_0/(1 + (2A_0/\mu)t) \\ s_t &= s_0 + (i\hbar/2)\ln(1 + (2A_0/\mu)t) - (V_l(r_0) - p_0^2/(2\mu))t \\ &\quad - \beta_l p_0 t^2/\mu + \beta_l^2/(3\mu)t^3 \end{aligned} \tag{8.21}$$

The correlation function takes the form ($p_0 = 0$)

$$\langle u_l | e^{-i\hat{H}(l')t/\hbar} | u_l \rangle = A(t)\exp\left[-\frac{\beta_{l'}^2}{8(\Delta p)_0^2} t^2 - i(V_{l'}(r_0)t - \beta_{l'}^2/(24\mu)t^3)/\hbar \right] \tag{8.22}$$

where

$$A(t) = [1 + i(\mathrm{Im}(A_0)/\mu)t]^{-1/2} \tag{8.23}$$

The modulus of this complex function is

$$|\langle u_l | e^{-i\hat{H}(l')t/\hbar} | u_l \rangle| = \frac{1}{[1 + (\mathrm{Im}(A_0)/\mu)^2 t^2]^{1/4}} \exp\left[-\frac{\beta_{l'}^2}{8(\Delta p)_0^2} t^2 \right] \tag{8.24}$$

We again have a Gaussian decay, which happens to be identical to the one obtained in the short-time limit, however, multiplied by another decaying function that is due to the spreading of the wave packet. The two decay mechanisms are independent in a linear potential, therefore a product form is obtained. The Fourier transform of the correlation function can be obtained, using

$$\int_{-\infty}^{\infty} \exp(ia_1 u - a_2 u^2 + ia_3 u^3)\, du = 2\pi |a| \exp\left(\frac{a_1 a_2}{3a_3} + \frac{2a_2^3}{27a_3^2}\right) \times \mathrm{Ai}\left\{a\left(a_1 + \frac{a_2^2}{3a_3}\right)\right\} \tag{8.25}$$

where $a = (3a_3)^{-1/3}$ and the Airy function, $\mathrm{Ai}(x)$, is defined by

$$\mathrm{Ai}(x) = \frac{1}{2\pi}\int_{-\infty}^{\infty} \exp(ixt + it^3/3)\, dt \tag{8.26}$$

If we neglect the decay due to the prefactor in the correlation function [i.e., $A(t) = 1$], the total absorption probability, Eq. (8.10), takes the form (for $l_i = 0$),

$$\boxed{\begin{aligned} P_{\mathrm{tot}}(\omega_l) &= \pi\varepsilon E_0^2 \mu_{21}^2/(6\hbar^2) \exp[-\zeta(E_l - V_1(r_0) - 2\eta/3)] \\ &\quad \times \mathrm{Ai}[-\varepsilon(E_l - V_1(r_0) - \eta)/\hbar] \end{aligned}} \tag{8.27}$$

where

$$\begin{aligned} \varepsilon &= \left(\frac{8\mu\hbar}{\beta_1^2}\right)^{1/3} \\ \zeta &= \frac{2}{\hbar\omega} \\ \eta &= \frac{\zeta^2\hbar^3}{\varepsilon^3} \end{aligned} \tag{8.28}$$

and $E_l = \hbar\omega_l + \varepsilon_a$. The simple symmetric Gaussian absorption band is accordingly replaced by a more complicated asymmetric band shape when the short-time approximation is abandoned. Closer inspection shows, in addition, that the maximum is shifted slightly towards higher energies.

Equation (8.27) is reminiscent of a well-known exact formula [66] for the total absorption probability derived via Eq. (5.13). Thus, when the spreading is included one obtains an expression that differs from Eq. (8.27) in a simple way—the square of the Airy functions is obtained and the argument of this function is divided by a factor of $4^{1/3}$.

B. Final Product Distribution

Dissociation of a diatomic molecule implies that a pair of recoiling atoms show up. The final product distribution is simply specified by the magnitude and orientation of the terminal relative momentum.

The asymptotic plane wave can be written in the form [26]

$$e^{ik\hat{\mathbf{k}}\cdot\mathbf{r}} = \frac{4\pi}{k} \sum_{l=0}^{\infty} \sum_{m=-l}^{l} i^l \frac{\hat{j}_l(kr)}{r} Y^*_{lm}(\hat{\mathbf{k}}) Y_{lm}(\theta, \phi) \tag{8.29}$$

where $\hat{j}_l(kr)$ is a Riccati–Bessel function, k and $\hat{\mathbf{k}} = (\Theta, \Phi)$ specify the magnitude and direction of the final relative momentum of the recoiling atoms, $\mathbf{p} = k\hat{\mathbf{k}}\hbar$. Then using Eqs. (8.7) and (8.8), the final product distribution, Eq. (4.12), takes the form [67]

$$\begin{aligned} P_{\mathbf{p}} = \mathscr{C}(\omega) \lim_{t\to\infty} \Bigg| & \sqrt{c(l_i m_i)} Y_{l_i+1,m_i}(\hat{\mathbf{k}}) \int_0^\infty dr\, \hat{j}_{l_i+1}(kr) \\ & \times e^{-i\hat{H}(l_i+1)t/\hbar} u_{l_i}(r, 0) \\ & + \sqrt{d(l_i m_i)} Y_{l_i-1,m_i}(\hat{\mathbf{k}})(-1) \int_0^\infty dr\, \hat{j}_{l_i-1}(kr) \\ & \times e^{-i\hat{H}(l_i-1)t/\hbar} u_{l_i}(r, 0) \Bigg|^2 \frac{(4\pi)^2}{k^2} \end{aligned} \tag{8.30}$$

where

$$\mathscr{C}(\omega) = E_0^2 \mathscr{N}^2 \frac{|\varepsilon(\omega)|^2}{4\hbar^2} \tag{8.31}$$

contains the information about the laser pulse and a normalization constant $\mathscr{N}$, for the plane wave corresponding to energy normalization.

For $l_i = 0$, only the first term is in play. We use the asymptotic form of the Ricatti–Bessel function,

$$\hat{j}_l(kr) \xrightarrow{r\to\infty} \sin(kr - l\pi/2)$$
$$= \frac{1}{2i}\left(e^{i(kr - l\pi/2)} - e^{-i(kr - l\pi/2)}\right) \tag{8.32}$$

The overlap with the first term is negligible since the wave packet has a (large) positive outgoing momentum, that is,

$$P_{\mathbf{p}} = \frac{\pi}{k^2}\mathscr{C}(\omega)\cos^2\Theta \lim_{t\to\infty}\left|\int_{-\infty}^{\infty} dr\, e^{-ikr} e^{-i\hat{H}(1)t/\hbar} u_0(r, 0)\right|^2 \tag{8.33}$$

Thus, the distribution is a product of a radial and an angular part. The radial part is simply the Fourier transform of the asymptotic wave packet.

For $l_i > 0$, introduction of the approximation, $\hat{H}(l_i) \sim \hat{H}(l_i - 1) \sim \hat{H}(l_i + 1)$, implies

$$P_{\mathbf{p}} = \frac{4\pi^2}{k^2}\mathscr{C}(\omega)\cos^2\Theta|Y_{l_i m_i}(\hat{\mathbf{k}})|^2$$
$$\times \lim_{t\to\infty}\left|\int_{-\infty}^{\infty} dr\, e^{-ikr} e^{-i\hat{H}(l_i)t/\hbar} u_{l_i}(r, 0)\right|^2 \tag{8.34}$$

If the magnetic quantum numbers are unspecified initially, that is, if no special alignment is prepared, all m values are equally probable. Since,

$$\sum_{m_i=-l_i}^{l_i} |Y_{l_i m_i}(\hat{\mathbf{k}})|^2 = \frac{2l_i + 1}{4\pi} \tag{8.35}$$

we observe that summation over these quantum numbers in the expression for $P_{\mathbf{p}}$ implies that the angular distribution again is the simple cosine distribution.

Now we consider the evaluation of the radial part of the distribution—with the aim of obtaining physical insight. We assume again that the molecule initially is in the vibrational ground state, approximated by the Gaussian wave packet in Eq. (8.13), and that the electronic transition dipole moment is constant.

The approximations introduced in Section VIII.A for the dynamics are obviously totally invalid in the present situation. Instead, we use the local quadratic approximation of the Gaussian wave packet method [i.e., Eq.

(6.6)]. The Fourier transform of a Gaussian, takes the form

$$\int_{-\infty}^{\infty} dr\, e^{-ikr} G_t(r) = \left(\frac{\pi\hbar}{-iA_t}\right)^{1/2} \exp[i\{-(p_t - \hbar k)^2/(4A_t) - \hbar k r_t + s_t\}/\hbar] \tag{8.36}$$

and evaluation of the absolute square gives the result

$$\boxed{P_p = \mu_{21}^2 \lim_{t\to\infty} \frac{\hbar\sqrt{2\pi}}{(\Delta p)_t} \exp\left[-\frac{1}{2(\Delta p)_t^2}(p - p_t)^2\right]} \tag{8.37}$$

The momentum distribution of the recoiling atoms is, accordingly, a Gaussian peaked around the final value of the momentum p_t. The width of the distribution is determined by $(\Delta p)_t$. The asymptotic values of these parameters are constants when the wave packet has entered the force free region.

The parameters evolve in time according to the equations of motion for a Gaussian wave packet. If we, for example, have an exponential interaction of the form,

$$V(r) = V_0 \exp(-ar) \tag{8.38}$$

the equations of motion for the position and momentum, Eq. (6.9),

$$\begin{aligned} \frac{dr_t}{dt} &= p_t/\mu \\ \frac{dp_t}{dt} &= -\partial V(r)/\partial r|_{r=r_t} \end{aligned} \tag{8.39}$$

can be solved analytically. The two equations are equivalent to the autonomous differential equation

$$\frac{d^2 r_t}{dt^2} = \frac{a}{\mu} V_0\, e^{-ar_t} \tag{8.40}$$

which with the initial condition $p_0 = 0$ implies

$$\begin{aligned} r_t &= r_0 - (2/a)\ln[\operatorname{sech}(ct)] \\ p_t &= \sqrt{2\mu V(r_0)}\tanh(ct) \end{aligned} \tag{8.41}$$

where $c = a\sqrt{V(r_0)/2\mu}$. The final value for the expectation value of the momentum is

$$p_\infty = \sqrt{2\mu V(r_0)} \tag{8.42}$$

Thus we obtain, as expected, a result for the expectation values that is equivalent to the classical prediction. When p_∞ is compared to the exact quantum mechanical result, it is often seen that the value obtained from the Gaussian wave packet approximation tends to be somewhat smaller than the exact value (see, e.g., [33]). This is related to the fact that in the Gaussian wave packet approximation the exponential potential is approximated by the less repulsive local harmonic potentials.

C. Dissociation Time

Now we consider some aspects of the notion of a dissociation time associated with the breaking of a chemical bond.

The probability that the bond length is larger than some value a is

$$P(t) = \int_a^\infty \langle u_l(t)|r\rangle\langle r|u_l(t)\rangle\, dr \tag{8.43}$$

If $|u_l(t)\rangle$ is a Gaussian wave packet,

$$\langle r|u(t)\rangle = \exp\{(i/\hbar)[A_t(r-r_t)^2 + p_t(r-r_t) + s_t]\} \tag{8.44}$$

where we have dropped index l, we obtain [cf. Eq. (7.7)]

$$P(t) = (2\pi(\Delta r)_t^2)^{-1/2} \int_a^\infty \exp\left[-\frac{(r-r_t)^2}{2(\Delta r)_t^2}\right] dr \tag{8.45}$$

which can be expressed in the form

$$\boxed{P(t) = \frac{1}{2}\left(1 + \operatorname{erf}\left(\frac{r_t - a}{\sqrt{2}(\Delta r)_t}\right)\right)} \tag{8.46}$$

where the error function is defined by

$$\operatorname{erf}(x) = \frac{2}{\sqrt{\pi}} \int_0^x e^{-u^2}\, du \tag{8.47}$$

and note that $\operatorname{erf}(-x) = -\operatorname{erf}(x)$. We observe that at the time $t = t_d$ given by $r_t = a$, i.e., the time where the center of the wave packet is at the position a, $P(t) = 1/2$.

The time derivative of $P(t)$ at $t = t_d$ is related to the time resolution. Thus,

$$\left.\frac{dP(t)}{dt}\right|_{t=t_d} = (2\pi)^{-1/2} \left.\frac{dr_t/dt}{(\Delta r)_t}\right|_{t=t_d} \tag{8.48}$$

In accordance with the definition of time resolution introduced in Section II, the ratio between the expectation value of momentum and uncertainty in position should be large in order to obtain a well-defined dissociation time. This gives the intrinsic resolution of dissociation time. Note that in a completely classical description $P(t)$ would be a Heaviside unit function that jumps from 0 to 1 at time t_d (see Fig. 6.3).

The dissociative motion can be probed in real time via Zewail's femtosecond "time-of-flight" (TOF) measurement and a dissociation time can be inferred [19, 20]. Two short pulses are used in a pump–probe experiment. The first (pump) pulse generates approximately a Franck–

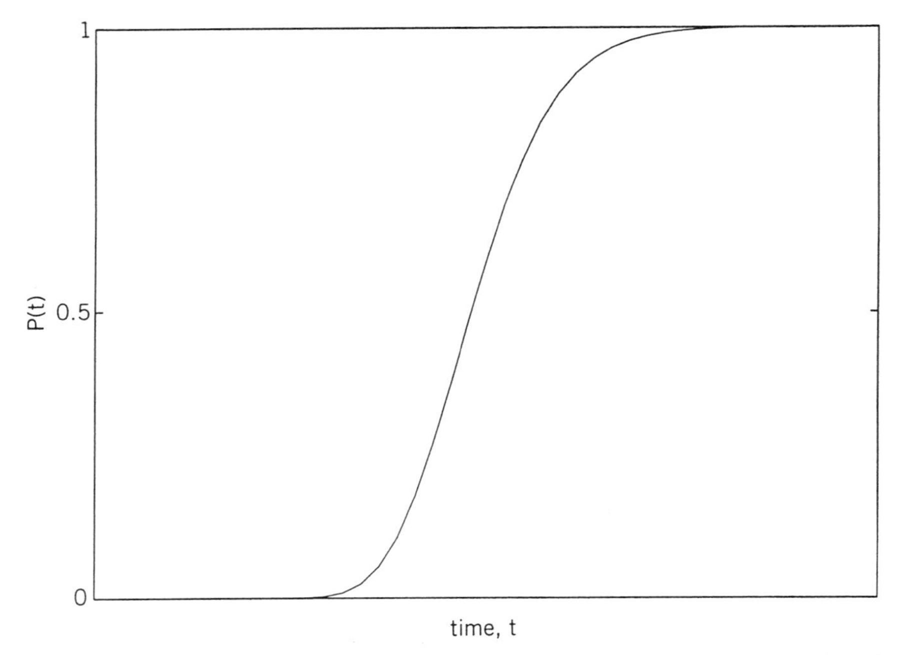

Figure 6.3. The probability that a bond exceeds some given length, as a function of time, according to Eq. (8.46).

Condon wave packet of the form given by Eq. (2.17) and the second (probe) pulse catches the moving wave packet in the excited electronic state, approximately as described by Eq. (4.34). The $P(t)$ function above is reminiscent of the experimentally obtained signal.

The dissociation time inferred from the width of the total absorption spectrum (Section VIII.A) is related to the decay of the autocorrelation function and is, obviously, shorter than the true dissociation time.

IX. THE TRIATOMIC MOLECULE

The dynamics of the triatomic molecule is much more complicated than in the diatomic molecule. To get useful insight we must identify the degrees of freedom of primary importance for a given measurement.

The identification of important degrees of freedom is closely connected to the choice of appropriate coordinates. We use internal (center-of-mass) coordinates without kinetic energy coupling. Assume that the molecule starts in the vibrational ground state. Normal mode coordinates typically give a convenient description of such a state. Thus, we can write (for details concerning Eqs. (9.1) and (9.3), see [12])

$$G_0(\mathbf{Q}) = \exp\{(i/\hbar)[\mathbf{Q} \cdot \mathbf{A} \cdot \mathbf{Q} + s_0]\} \tag{9.1}$$

where $\mathbf{A}$ is a diagonal matrix that contains the normal mode frequencies. The normal mode coordinates are, however, not the natural choice of coordinates when the dissociation process is to be described. In the simplest situation where one arrangement (chemical) channel is open, that is,

$$\mathrm{ABC} \rightarrow \mathrm{A} + \mathrm{BC} \tag{9.2}$$

Jacobi coordinates give a natural description. The six coordinates are associated with two vectors, one connecting atom A with the center-of-mass of BC and the other connecting the atoms in the diatomic fragment BC. Since the transformation from normal mode to Jacobi coordinates is linear the Gaussian form of the initial state is preserved when Jacobi coordinates are introduced. Thus,

$$G_0(\mathbf{R}) = \exp\{(i/\hbar)[(\mathbf{R} - \mathbf{R}_0) \cdot \mathbf{A}_0 \cdot (\mathbf{R} - \mathbf{R}_0) + s_0]\} \tag{9.3}$$

where $\mathbf{A}_0$ is a matrix which, in general, contains off-diagonal elements. Thus, the wave function contains position correlation.

We already know that the overall rotational state of the molecule is of minor importance in photodissociation dynamics (unless we consider the spatially resolved final product distribution). We write the Hamiltonian in a form that allows us to exploit this fact. Polar coordinates for the two Jacobi vectors are introduced and,

$$\hat{H} = \hat{T} + E(r, R, \phi) \tag{9.4}$$

where

$$\hat{T} = \frac{\hat{p}_R^2}{2\mu} + \frac{\hat{p}_r^2}{2m} + \frac{(\hat{L}_{\text{tot}} - \hat{L}_r)^2}{2\mu R^2} + \frac{\hat{L}_r^2}{2mr^2} \tag{9.5}$$

subscript 2 is dropped as in Section VIII to simplify the notation, that is, $\hat{H} \equiv \hat{H}_2$ and for the electronic energy $E \equiv E_2$. The parameter ϕ is the bond angle, R and r denotes the length of the two Jacobi vectors, that is, R is the distance from A to the center-of-mass of BC, and r is the internuclear distance in BC. The associated reduced masses are

$$\begin{aligned} 1/\mu &= 1/m_A + 1/(m_B + m_C) \\ 1/m &= 1/m_B + 1/m_C \end{aligned} \tag{9.6}$$

and the angular momenta associated with the two vectors are $\hat{L}_R$ and $\hat{L}_r$, respectively.

$$\hat{L}_{\text{tot}} = \hat{L}_R + \hat{L}_r \tag{9.7}$$

is the total angular momentum. We assume that the molecule initially is in the rotational ground state (and neglect the momentum of the photon). Thus, the appropriate kinetic energy operator is

$$\hat{T}_{J=0} = \frac{\hat{p}_R^2}{2\mu} + \frac{\hat{p}_r^2}{2m} + \left(\frac{1}{2\mu R^2} + \frac{1}{2mr^2}\right)\hat{L}_r^2 \tag{9.8}$$

The radial kinetic energy operators reduce to a one-dimensional form, exactly as in Section VIII, when the state $\Phi(r, R, \phi, t)/rR$ is propagated, since

$$\hat{T}_{J=0}(\Phi/rR) = \frac{1}{rR}\left(-\frac{\hbar^2}{2\mu}\frac{\partial^2}{\partial R^2} - \frac{\hbar^2}{2m}\frac{\partial^2}{\partial r^2} + \left(\frac{1}{2\mu R^2} + \frac{1}{2mr^2}\right)\hat{L}_r^2\right)\Phi \tag{9.9}$$

The angular momentum operator can be written in the form

$$\hat{L}_r^2 = -\hbar^2 \left\{ \frac{1}{\sin\theta} \frac{\partial}{\partial\theta} \left(\sin\theta \frac{\partial}{\partial\theta} \right) + \frac{1}{\sin^2\theta} \frac{\partial^2}{\partial\phi^2} \right\} \tag{9.10}$$

and the time evolution of the expectation value

$$\begin{aligned} \frac{d}{dt} \langle \hat{L}_r^2 \rangle &= \frac{i}{\hbar} \langle [\hat{H}, \hat{L}_r^2] \rangle \\ &= i\hbar \left\langle \frac{1}{\sin^2\theta} \left(\frac{\partial^2 E}{\partial\phi^2} + 2 \frac{\partial E}{\partial\phi} \frac{\partial}{\partial\phi} \right) \right\rangle \end{aligned} \tag{9.11}$$

Thus, the excitation of angular motion is related to $\partial E/\partial\phi$, that is, the anisotropy of the potential.

A. Total Absorption Probability

We focus on the radial degrees of freedom and assume that $\partial E/\partial\phi$ is small so that the bending motion can be neglected. We adopt an independent mode approximation and write the Hamiltonian in the form

$$\hat{H}_{\text{abs}} = \hat{H}_1(R) + \hat{H}_2(r) \tag{9.12}$$

where

$$\begin{aligned} \hat{H}_1(R) &= -\frac{\hbar^2}{2\mu} \frac{\partial^2}{\partial R^2} + V_1(R) \\ \hat{H}_2(r) &= -\frac{\hbar^2}{2m} \frac{\partial^2}{\partial r^2} + V_2(r) \end{aligned} \tag{9.13}$$

$\Phi(r, R, 0) = \phi_1(R, 0)\phi_2(r, 0)$, where ϕ_1 and ϕ_2 takes the Gaussian form given in Eq. (8.13), and the nuclear autocorrelation function can, accordingly, be factorized

$$\langle \Phi(0) | \hat{U}(t, 0) | \Phi(0) \rangle = \prod_{i=1}^{2} \langle \phi_i(0) | \hat{U}_i(t, 0) | \phi_i(0) \rangle \tag{9.14}$$

In a short-time approximation to the dynamics we obtain (as in Section VIII.A) correlation functions of Gaussian form. The Fourier transform is easily obtained, either by first multiplying the two correlation functions and observing that a new Gaussian function is obtained or by evaluating the convolution of the individual Fourier transforms. The total absorption probability in the *cw limit* accordingly takes the form

$$P_{\text{tot}}(\omega_l) = E_0^2 \mu_{21}^2/(4\hbar^2) \sqrt{\frac{2\pi\hbar^2}{\sigma_R^2 + \sigma_r^2}} \times \exp(-(\hbar\omega_l - V_1(R_0) - V_2(r_0))^2/2(\sigma_R^2 + \sigma_r^2)) \tag{9.15}$$

where

$$\begin{aligned} \sigma_R^2 &= (V_1'(R_0)(\Delta R)_0)^2 \\ \sigma_r^2 &= (V_2'(r_0)(\Delta r)_0)^2 \end{aligned} \tag{9.16}$$

Thus, we again obtain a Gaussian distribution. The width is simply the sum of the widths associated with the two degrees of freedom.

When the dynamics inherent in the correlation function is not solely describable by short-time dynamics more structured absorption bands can occur. If, for example, the dissociation is sufficiently slow there can be time for recurrences in the correlation function due to quasibound vibrational motion. Such a situation occurs, for example, in the photodissociation of water in the first absorption band [29]. In this situation the symmetric and asymmetric stretch normal mode coordinates are the appropriate independent modes. The short-time dynamics is dominated by motion along the symmetric stretch coordinate but for somewhat longer times spreading along the asymmetric stretch is important and essentially damps out recurrences that can occur along the symmetric stretch.

B. Vibrational Excitation

We focus on the radial degrees of freedom and assume that $\partial E/\partial\phi$ is small so that bending, and eventually rotation, can be neglected. The relevant Hamiltonian is, according to Eq. (9.9),

$$\hat{H}_{\text{vib}} = \hat{H}_1(R) + \hat{H}_2(r) + V_{12}(r, R) \tag{9.17}$$

where

$$\begin{aligned} \hat{H}_1(R) &= -\frac{\hbar^2}{2\mu}\frac{\partial^2}{\partial R^2} + V_1(R) \\ \hat{H}_2(r) &= -\frac{\hbar^2}{2m}\frac{\partial^2}{\partial r^2} + V_2(r) \end{aligned} \tag{9.18}$$

The asymptotic states ($R \rightarrow \infty$) are

$$\langle r, R|E, n\rangle = \mathcal{N} e^{ik_n R} \langle r|n\rangle \tag{9.19}$$

where the vibrational eigenstates are determined by

$$\hat{H}_2|n\rangle = \varepsilon_n|n\rangle \tag{9.20}$$

and the plane wave is an eigenstate of $\hat{H}_1(R \rightarrow \infty)$ with an energy normalization constant $\mathcal{N} = \sqrt{\mu/(2\pi\hbar^2 k_n)}$. The total translational and vibrational energy E is related to k_n by

$$k_n = \sqrt{2\mu(E - \varepsilon_n)/\hbar^2} \tag{9.21}$$

We assume that the wave function can be factorized initially (i.e., no correlation)

$$\Phi(r, R, 0) = \phi_1(R, 0)\phi_2(r, 0) \tag{9.22}$$

The TDSCF-approximation is introduced [68]

$$\hat{U}(t, 0)|\Phi(0)\rangle = \phi_1(R, t)\phi_2(r, t)e^{iF(t)/\hbar} \tag{9.23}$$

which implies

$$\begin{aligned} i\hbar \frac{\partial \phi_1(R, t)}{\partial t} &= (\hat{H}_1 + \langle \phi_2(r, t)|V_{12}(r, R)|\phi_2(r, t)\rangle)\phi_1(R, t) \\ i\hbar \frac{\partial \phi_2(r, t)}{\partial t} &= (\hat{H}_2 + \langle \phi_1(R, t)|V_{12}(r, R)|\phi_1(R, t)\rangle)\phi_2(r, t) \end{aligned} \tag{9.24}$$

with a phase factor

$$F(t) = \int_0^t du \langle \phi_1(R, u)|\langle \phi_2(r, u)|V_{12}(r, R)|\phi_2(r, u)\rangle|\phi_1(R, u)\rangle \tag{9.25}$$

which is real valued.

The final product distribution, Eq. (4.12), takes the form

$$\boxed{P(E, n) = \mathscr{C}(\omega)P(k_n)P(n)} \tag{9.26}$$

where

$$P(k_n) = \left| \int dR\, e^{-ik_nR} \phi_1(R, \sim\infty) \right|^2$$
$$P(n) = \left| \int dr \langle n | r \rangle \phi_2(r, \sim\infty) \right|^2 \tag{9.27}$$

and $\mathscr{C}(\omega)$ is defined in Eq. (8.31).

Thus, we obtain a product form where the first factor is a Fourier transform of the asymptotic wave packet, giving the associated momentum distribution, and the second factor is the vibrational probability distribution. The form of the momentum distribution can depend on $\phi_2(r, t)$ but is independent of the final vibrational state n. If the total probability distribution is plotted as a function of the total energy E, we obtain a series of momentum distributions each weighted by the vibrational distribution and shifted by the vibrational energy spacing of the diatomic fragment (since a given vibrational state is closed unless $E > \varepsilon_n$). This is reminiscent of the final product distribution obtained for simple systems (see, e.g., [58]). Various combinations of translational and vibrational energies can show up at a fixed total energy. A high degeneracy is expected when the vibrational energy spacing is small (see Fig. 6.4).

Insight concerning the form of $P(k_n)$ can be obtained if we use the local quadratic approximation in the equations of motion for $\phi_1(R, t)$. If the molecule starts out in the vibrational ground state $\phi_1(R, 0)$ is approximately a Gaussian, and we obtain

$$\boxed{P(k_n) = \lim_{t \to \infty} \frac{\hbar\sqrt{2\pi}}{(\Delta P)_t} \exp\left[-\frac{\hbar^2}{2(\Delta P)_t^2} (k_n - P_t/\hbar)^2 \right]} \tag{9.28}$$

where P_t is the Gaussian wave packet parameter associated with the expectation value of the momentum. If the effect of the time-dependent potential is small a fair approximation to the value of P_t can be obtained from the model based on an exponential potential.

Insight concerning the form of the vibrational probability distribution $P(n)$ is obtained by considering a description that is equivalent to the linearly forced harmonic oscillator model. The Hamiltonian $\hat{H}_2$ is approx-

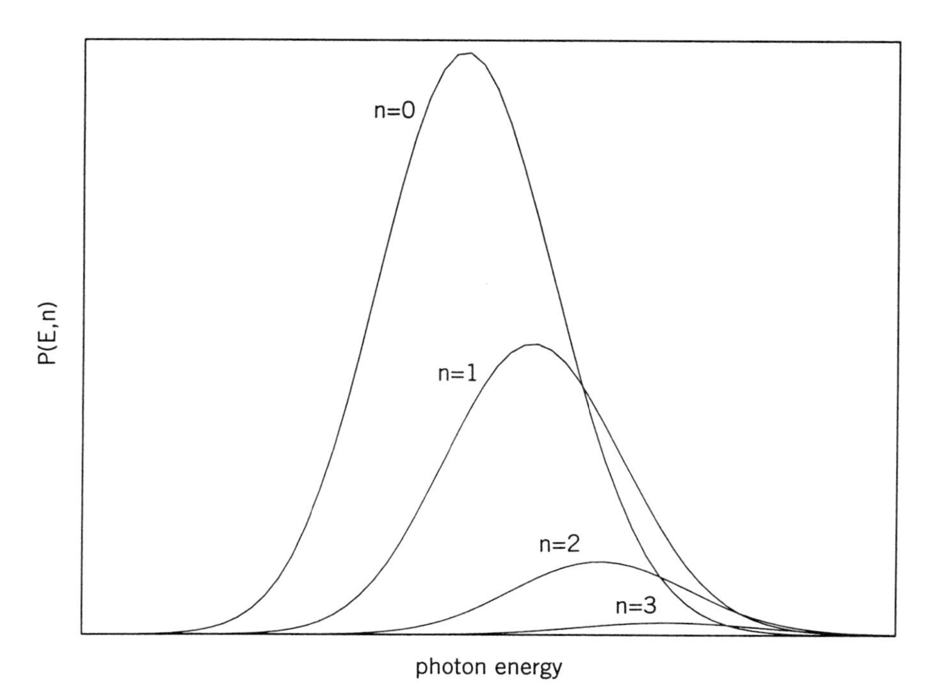

Figure 6.4. The final product distribution according to Eq. (9.26) in the cw limit. The figure shows the probability of finding a fragment in a given vibrational state n, as a function of the photon energy. The vibrational distribution is calculated from Eq. (9.37) for $\gamma = 0.5$.

imated by a harmonic oscillator

$$\hat{H}_2(r) = -\frac{\hbar^2}{2m}\frac{\partial^2}{\partial r^2} + \frac{1}{2}m\omega^2(r - r_e)^2 \tag{9.29}$$

and the interaction potential is expanded to first order in the oscillator displacement

$$V_{12}(r, R) = V_{12}(r_e, R) + V'_{12}(r_e, R)(r - r_e) \tag{9.30}$$

where $V'_{12}(r_e, R) = \partial V_{12}(r, R)/\partial r|_{r=r_e}$. The Hamiltonian, which determines the time evolution of ϕ_2, accordingly takes the form

$$\hat{H}_2(t) = \hat{H}_2(r) + F(t)(r - r_e) + f(t) \tag{9.31}$$

where

$$F(t) = \langle \phi_1(R, t)|V'_{12}(r_e, R)|\phi_1(R, t)\rangle \tag{9.32}$$

and $f(t)$ is a phase factor that can be dropped. Note that $F(0) \neq 0$ and $F(\infty) = 0$ since the interaction potential disappears at large separation. This Hamiltonian is exactly a quadratic form. Thus, a Gaussian wave packet continues to be a Gaussian during propagation. The center of the wave packet follows the equations of motion for the corresponding classically forced harmonic oscillator and the width is behaving exactly as in the corresponding unforced oscillator.

First, we consider the situation where $\phi_2(r, 0)$ is a Gaussian with a width that is identical to that of the (free) unforced harmonic oscillator ground state ($A_0 = im\omega/2$). Thus,

$$\phi_2(r, 0) = \left(\frac{m\omega}{\pi\hbar}\right)^{1/4} \exp\left[-\frac{m\omega}{2\hbar}(r - r_0)^2\right] \tag{9.33}$$

This assumption implies that the width is constant during the time evolution of the state, that is, the wave packet is a coherent state. When the equations of motion, Eq. (6.9), are solved, we obtain

$$\phi_2(r, t) = \left(\frac{m\omega}{\pi\hbar}\right)^{1/4} \exp\left[-\frac{m\omega}{2\hbar}(r - r_t)^2 - ip_t r/\hbar + i\gamma(t)\right] \tag{9.34}$$

where

$$\begin{aligned} r_t &= d\cos\omega t + 2b\,\mathrm{Im}(\alpha^- e^{i\omega t}) \\ p_t &= dm\omega\sin\omega t - 2b\,\mathrm{Re}(\alpha^- e^{i\omega t}) \end{aligned} \tag{9.35}$$

$d = r_0 - r_e$ is the initial displacement, $b = \sqrt{\hbar/2m\omega}$, $\gamma(t)$ is a real-valued function, and

$$\alpha^{\pm} = -\frac{b}{\hbar}\int_0^t F(t')e^{\pm i\omega t'}\,dt' \tag{9.36}$$

The vibrational probability distribution can be evaluated analytically, and takes the form

$$\boxed{P(n) = \gamma^n e^{-\gamma}/n!} \tag{9.37}$$

where

$$\gamma = \rho + (m\omega/2\hbar)d^2 + 2d\sqrt{m\omega/2\hbar}\,\mathrm{Re}(i\alpha^+) \tag{9.38}$$

and $\rho = \alpha^+\alpha^-$. Thus, the probability follows a Poisson distribution. The parameter associated with this distribution is a combination of the initial displacement d and a parameter ρ, which gives the strength of the excitation during dissociation.

Second, general solutions, including non-Gaussian initial states, can be handled, in part, analytically [40]. We write the vibrational probability distribution in the form

$$P(n) = \left| \sum_m \langle n|\hat{U}_2(\sim\infty, 0)|m\rangle\langle m|\phi_2(r, 0)\rangle \right|^2 \tag{9.39}$$

The time-evolution operator can be written in the form

$$\hat{U}_2(t, 0) = e^{-i\hat{H}_2 t/\hbar} e^{i(\alpha^+\hat{a}^\dagger + \alpha^-\hat{a} + \beta)} \tag{9.40}$$

where

$$\beta = \left(\frac{b}{\hbar}\right)^2 \int_0^t dt'\, F(t') \int_0^{t'} F(t'') \sin\omega(t' - t'')\, dt'' \tag{9.41}$$

and $\hat{a}$ and $\hat{a}^\dagger$ are the annihilation and creation operators for the harmonic oscillator. The matrix elements can be evaluated [40]

$$\langle n|\hat{U}_2(t, 0)|m\rangle = \begin{cases} e^{i\beta + i\varepsilon_n t/\hbar} e^{-\rho/2}(-i\alpha^+)^{n-m}\sqrt{\dfrac{m!}{n!}}\, L_n^{n-m}(\rho) & n \geq m \\[2ex] e^{i\beta + i\varepsilon_n t/\hbar} e^{-\rho/2}(-i\alpha^-)^{m-n}\sqrt{\dfrac{n!}{m!}}\, L_m^{m-n}(\rho) & n \leq m \end{cases} \tag{9.42}$$

where $L_i^j(\rho)$ are associated Laguerre polynomials, defined such that $L_i^i(\rho) = (-1)^i$. As a special limit we can consider the situation where $\langle m|\phi_2(r, 0)\rangle = \delta_{m,0}$, that is, $\phi_2(r, 0)$ is identical to the vibrational ground state of the diatomic fragment and the vibrational excitation is produced purely by the final-state interaction,

$$\begin{aligned} P(n) &= |\langle n|\hat{U}_2(\sim\infty, 0)|0\rangle|^2 \\ &= \rho^n e^{-\rho}/n! \end{aligned} \tag{9.43}$$

that is, a Poisson distribution that falls exactly within the description of

Eq. (9.37). This limit is obviously based on several assumptions that might not be valid in practice.

Various adaptations of the linearly forced oscillator model, related to the above description, have been suggested in the literature [66, 69]. It is obviously based on several assumptions. The underlying TDSCF approximation can, for example, be a poor approximation [70].

C. Rotational Excitation

Now we adopt a rigid-rotator description for the diatomic fragment. We assume, furthermore, that the motion takes place in a plane, that is, it is restricted to two dimensions. The Hamiltonian, Eq. (9.8), then takes the form

$$\hat{H}_{\text{rot}} = \hat{H}_1(R) + \hat{H}_2(\phi) + V_{12}(R, \phi) \tag{9.44}$$

where

$$\begin{aligned} \hat{H}_1(R) &= -\frac{\hbar^2}{2\mu}\frac{1}{R}\frac{\partial}{\partial R}R\frac{\partial}{\partial R} + V_1(R) \\ \hat{H}_2(\phi) &= -\frac{\hbar^2}{2I}\frac{\partial^2}{\partial \phi^2} \end{aligned} \tag{9.45}$$

$I = mr_e^2$ is the moment of inertia for the diatomic molecule and the centrifugal energy associated with the atomic motion is assumed to be small compared to $V_1(R)$, and consequently dropped. The radial kinetic energy operator simplifies when the state $\Phi(R, \phi, t)/\sqrt{R}$ is considered, since

$$\begin{aligned} \hat{H}_1(R)(\Phi/\sqrt{R}) &= 1/\sqrt{R}\left(-\frac{\hbar^2}{2\mu}\left(\frac{\partial^2}{\partial R^2} + \frac{1}{4R^2}\right) + V_1(R)\right)\Phi \\ &= 1\sqrt{R}\hat{H}_R\Phi \end{aligned} \tag{9.46}$$

The asymptotic ($R \to \infty$) fragment states are

$$\langle R, \phi | E, m \rangle = \mathcal{N} e^{ik_m R} \langle \phi | m \rangle \tag{9.47}$$

where the rotational eigenstates are determined by

$$\hat{H}_2 |m\rangle = \varepsilon_m |m\rangle \tag{9.48}$$

and the plane wave is an eigenstate of the operator $\hat{H}_R(R \to \infty)$ as defined in Eq. (9.46). Equation (9.48) implies

$$\begin{aligned} \langle \phi | m \rangle &= (2\pi)^{-1/2} e^{im\phi} \\ \varepsilon_m &= m^2\hbar^2/(2I) \end{aligned} \tag{9.49}$$

where $m = 0, \pm 1, \pm 2, \ldots$. The total translational and rotational energy E is related to k_m by

$$k_m = \sqrt{2\mu(E - \varepsilon_m)/\hbar^2} \tag{9.50}$$

We assume that the wave function can be factorized initially (i.e., no correlation)

$$\Phi(R, \phi, 0) = \phi_1(R, 0)\phi_2(\phi, 0) \tag{9.51}$$

The TDSCF-approximation is introduced

$$\hat{U}(t, 0)|\Phi(0)\rangle = \phi_1(R, t)\phi_2(\phi, t)e^{iF(t)/\hbar} \tag{9.52}$$

which implies

$$\begin{aligned} i\hbar \frac{\partial \phi_1(R, t)}{\partial t} &= (\hat{H}_R + \langle \phi_2(\phi, t)|V_{12}(R, \phi)|\phi_2(\phi, t)\rangle)\phi_1(R, t) \\ i\hbar \frac{\partial \phi_2(\phi, t)}{\partial t} &= (\hat{H}_2 + \langle \phi_1(R, t)|V_{12}(R, \phi)|\phi_1(R, t)\rangle)\phi_2(\phi, t) \end{aligned} \tag{9.53}$$

with a phase factor

$$F(t) = \int_0^t du \langle \phi_1(R, u)| \langle \phi_2(\phi, u)|V_{12}(R, \phi)|\phi_2(\phi, u)\rangle |\phi_1(R, u)\rangle \tag{9.54}$$

which is real valued.

The final product distribution, Eq. (4.12), takes the form,

$$\boxed{P(E, m) = \mathscr{C}(\omega)P(k_m)P(m)} \tag{9.55}$$

where

$$P(k_m) = \left| \int dR\, e^{-ik_mR} \phi_1(R, \sim\infty) \right|^2$$
$$P(m) = \left| \int d\phi\, e^{-il\phi/\hbar} \phi_2(\phi, \sim\infty) \right|^2 \tag{9.56}$$

where $l = m\hbar$ and $\mathscr{C}(\omega)$ is defined in Eq. (8.31).

Thus, we obtain a product form where the first factor gives the momentum distribution associated with translation, and the second factor is the rotational probability distribution.

Insight concerning the form of the rotational probability distribution $P(m)$ is considered next. We assume that the molecule starts in the bending ground state. The proper boundary condition is such that the wave function is invariant to an angle displacement of $2n\pi$ [71], which implies that

$$\phi_2(\phi, 0) = \sum_{n=-\infty}^{\infty} \psi_{\phi_0 + 2n\pi}(\phi) \tag{9.57}$$

where

$$\psi_{\phi_0}(\phi) = \exp\{(i/\hbar)[\alpha_0(\phi - \phi_0)^2 + \gamma_0]\} \tag{9.58}$$

ϕ_0 is the equilibrium bending angle of the electronic ground state (which might differ from the corresponding angle following vertical exitation). Each packet in Eq. (9.57) evolves independently due to the linearity of the time-evolution operator. The interaction potential V_{12} is now expanded to second order around the parameter ϕ_t. Thus, we can write

$$\langle \phi_1(R, t)|V_{12}(R, \phi)|\phi_1(R, t)\rangle = f(t) + F_1(t)(\phi - \phi_t) + \tfrac{1}{2}F_2(t)(\phi - \phi_t)^2 \tag{9.59}$$

where $F_1(t)$ and $F_2(t)$ are the expectation values of the first and second derivative of the V_{12} potential with respect to the angle ϕ, at $\phi = \phi_t$ (note that $F_1(\infty) = F_2(\infty) = 0$). This approximation implies that the Gaussian form is preserved at all times, and

$$\phi_2(\phi, t) = \sum_{n=-\infty}^{\infty} \psi_{\phi_t + 2n\pi, l_t}(\phi) \tag{9.60}$$

where

$$\psi_{\phi_t, l_t}(\phi) = \exp\{(i/\hbar)[\alpha_t(\phi - \phi_t)^2 + l_t(\phi - \phi_t) + \gamma_t]\} \tag{9.61}$$

The parameters evolve in time according to the equations of motion for a Gaussian wave packet, for example,

$$\frac{d\phi_t}{dt} = l_t / I$$

$$\frac{dl_t}{dt} = -\partial V(\phi, t)/\partial\phi|_{\phi=\phi_t} \tag{9.62}$$

where $V(\phi, t)$ is the potential given in Eq. (9.59).

The Fourier transform of Eq. (9.60) takes the form

$$\int_{-\infty}^{\infty} d\phi \, e^{-il\phi/\hbar} \phi_2(\phi, t) = \sum_{n=-\infty}^{\infty} e^{-i2n\pi l/\hbar} \left(\frac{\pi\hbar}{-i\alpha_t}\right)^{1/2}$$
$$\times \exp\{(i/\hbar)[-(l - l_t)^2/(4\alpha_t) - \phi_t l + \gamma_t]\} \tag{9.63}$$

The summation can be nonzero only when $l/\hbar$ is an integer (the Poisson summation formula), and evaluation of the absolute square gives

$$\boxed{P(m) = \lim_{t\to\infty} \frac{\hbar\sqrt{2\pi}}{(\Delta l)_t} \exp\left[-\frac{1}{2(\Delta l)_t^2}(l - l_t)^2\right]} \tag{9.64}$$

where $l = m\hbar$ is the discrete angular momentum. Thus, we obtain a Gaussian distribution centered around the final value of the angular momentum l_t (see Fig. 6.5). The diatomic fragment can, in the present model, rotate clockwise or anticlockwise depending on the initial orientation of the parent molecule. The Gaussian distribution is often, at least qualitatively, in agreement with experimental rotational state distributions [72].

ACKNOWLEDGMENTS

The major part of the work on which this chapter is based was carried out at Chemistry Laboratory III, H. C. Ørsted Institute, University of Copenhagen and Chemistry Department B, Technical University of Denmark. During 1987–1988, NEH spent one year in Eric

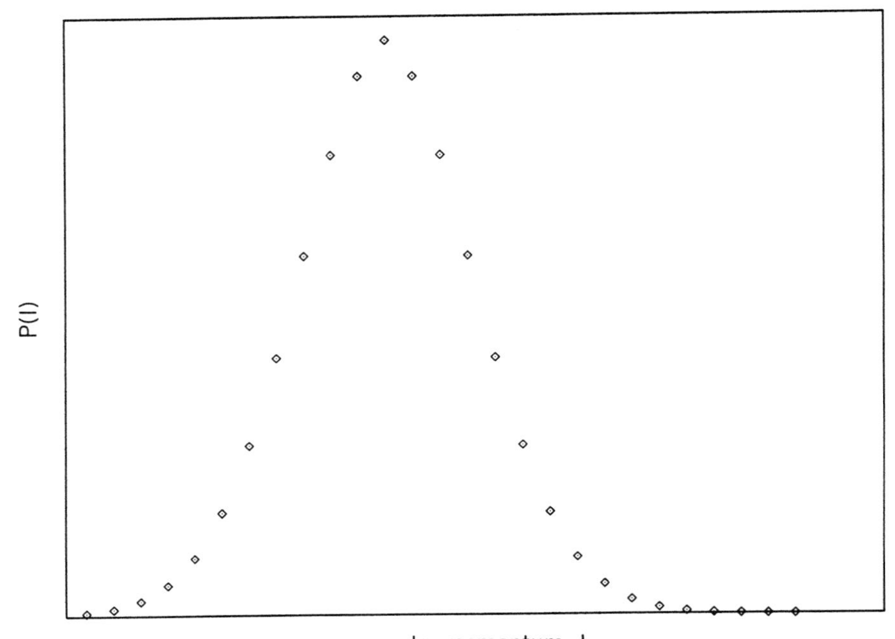

Figure 6.5. The probability distribution for the angular momentum of a diatomic fragment, produced in photodissociation of a molecule starting in the vibrational ground state, according to Eq. (9.64).

J. Heller's group at University of Washington, Seattle. This stay and Eric Heller's pioneering work in chemical physics has been of invaluable significance to me. I would like to express my gratitude to Jens Peder Dahl, Gert D. Billing, and Flemming Y. Hansen as well as my other colleagues and collaborators. Finally, I would like to thank the Danish Natural Science Research Council for financial support.

APPENDIX A
ON THE SEMICLASSICAL DESCRIPTION OF RADIATION-MATTER INTERACTION

We consider some aspects of the foundation of the so-called semiclassical description of radiation-matter interaction. The term semiclassical refers here to a quantum mechanical description of the matter part and a classical description of the radiation.

The total Hamiltonian, in the electric-dipole approximation is [23]

$$\hat{H} = \hat{H}_M + \hat{H}_R + \hat{H}_{ED} \tag{A.1}$$

where $\hat{H}_M$ takes the form given in Eq. (2.2) and, for a single-mode field (wave vector $\mathbf{k}$) having polarization specified by the unit vector $\boldsymbol{\varepsilon}$

$$\begin{aligned} \hat{H}_R &= \hbar\omega(\hat{a}^\dagger \hat{a} + \tfrac{1}{2}) \\ \hat{H}_{ED} &= \boldsymbol{\mu} \cdot \hat{\mathbf{E}} \end{aligned} \tag{A.2}$$

where the electric field operator is

$$\hat{\mathbf{E}} = i\sqrt{\hbar\omega/(2\varepsilon_0 V)}\boldsymbol{\varepsilon}\{\hat{a}e^{i\mathbf{kr}} - \hat{a}^\dagger e^{-i\mathbf{kr}}\} \tag{A.3}$$

V is the cavity volume and ε_0 the permitivity of free space. The parameter $\omega = c|\mathbf{k}|$, and with the understanding that a single-mode field is considered, subscript $\mathbf{k}$ is dropped throughout this Appendix. The operators $\hat{a}$ and $\hat{a}^\dagger$ are the photon annihilation and creation operators, respectively. Their action on a photon state with n photons is,

$$\begin{aligned} \hat{a}|n\rangle &= \sqrt{n}|n-1\rangle \\ \hat{a}^\dagger|n\rangle &= \sqrt{n+1}|n+1\rangle \end{aligned} \tag{A.4}$$

where the occupation number is an eigenvalue of the number operator $\hat{n} = \hat{a}^\dagger \hat{a}$,

$$\hat{n}|n\rangle = n|n\rangle \tag{A.5}$$

Now we introduce two approximations.

First, the TDSCF approximation

$$\Psi(t) = \Psi_M(t)\Phi_R(t) \tag{A.6}$$

implies

$$\begin{aligned} i\hbar\frac{\partial\Psi_M(t)}{\partial t} &= (\hat{H}_M + \langle\Phi_R(t)|\hat{H}_{ED}|\Phi_R(t)\rangle)\Psi_M(t) \\ i\hbar\frac{\partial\Phi_R(t)}{\partial t} &= (\hat{H}_R + \langle\Psi_M(t)|\hat{H}_{ED}|\Phi_M(t)\rangle)\Phi_R(t) \end{aligned} \tag{A.7}$$

where

$$\begin{aligned} \langle\Phi_R(t)|\hat{H}_{ED}|\Phi_R(t)\rangle &= \boldsymbol{\mu} \cdot \langle\Phi_R(t)|\hat{\mathbf{E}}|\Phi_R(t)\rangle \\ \langle\Psi_M(t)|\hat{H}_{ED}|\Psi_M(t)\rangle &= \langle\Psi_M(t)|\boldsymbol{\mu}|\Psi_M(t)\rangle \cdot \hat{\mathbf{E}} \end{aligned} \tag{A.8}$$

Second, the approximation that the field is unperturbed by the interaction. Thus, the occupation number is so large that the radiation field can be regarded as an inexhaustible source and sink of photons. This approximation implies that

$$\Phi_R(t) = \exp(-i\hat{H}_R t/\hbar)\Phi_R(0) \tag{A.9}$$

describes the time evolution of the field. Now,

$$\langle \Phi_R(t)|\hat{\mathbf{E}}|\Phi_R(t)\rangle = \langle \Phi_R(0)|\hat{\mathbf{E}}(t)|\Phi_R(0)\rangle \tag{A.10}$$

where

$$\hat{\mathbf{E}}(t) = i\sqrt{\hbar\omega/(2\varepsilon_0 V)}\boldsymbol{\varepsilon}\{\hat{a}e^{-i\omega t}e^{i\mathbf{kr}} - \hat{a}^\dagger e^{i\omega t}e^{-i\mathbf{kr}}\} \tag{A.11}$$

We now essentially have the description assumed in the semiclassical approach to radiation–matter interaction: A single Schrödinger equation describing the time evolution of the matter part where the Hamiltonian includes a dipole interaction term with an oscillating electric field.

We need, however, to see that the oscillating electric field can take the form used in the classical description of fields. It is well known that this happens in the situation where $\Phi_R(0) = |\alpha\rangle$ is a coherent state [23], that is,

$$|\alpha\rangle = \exp(-|\alpha|^2/2)\sum_{n=0} \alpha^n/\sqrt{n!}|n\rangle \tag{A.12}$$

where α is a complex number. To evaluate the expectation value of the electric field according to Eq. (A.10), we note that photon absorption on such a state implies

$$\begin{aligned}\hat{a}|\alpha\rangle &= \exp(-|\alpha|^2/2)\sum_{n=0} \alpha^n/\sqrt{n!}n^{1/2}|n-1\rangle \\ &= \alpha|\alpha\rangle\end{aligned} \tag{A.13}$$

Thus,

$$\begin{aligned}\langle\alpha|\hat{a}|\alpha\rangle &= \alpha \\ \langle\alpha|\hat{a}^\dagger|\alpha\rangle &= \alpha^*\end{aligned} \tag{A.14}$$

since, $\langle\alpha|\hat{a}|\alpha\rangle^* = \langle\alpha|\hat{a}^\dagger|\alpha\rangle$. The expectation value of the electric field now takes the form

$$\begin{aligned}\langle \alpha|\hat{\mathbf{E}}(t)|\alpha\rangle &= i\sqrt{\hbar\omega/(2\varepsilon_0 V)}\boldsymbol{\varepsilon}\{\alpha \exp(-i\omega t + i\mathbf{kr}) - \alpha^* \exp(i\omega t - i\mathbf{kr})\}\\ &= -2\sqrt{\hbar\omega/(2\varepsilon_0 V)}\boldsymbol{\varepsilon}|\alpha| \sin(\mathbf{kr} - \omega t + \theta)\end{aligned} \tag{A.15}$$

The amplitude of this wave is related to the expectation value of the photon occupation number since [23]

$$|\alpha| = \sqrt{\langle \alpha|\hat{n}|\alpha\rangle} \tag{A.16}$$

The uncertainty associated with the electric field is

$$\Delta E = \sqrt{\hbar\omega/2\varepsilon_0 V} \tag{A.17}$$

Thus, when the amplitude of the electric field is sufficiently large—corresponding to a large occupation number—this uncertainty is insignificant and the field is equivalent to a classical wave with a well-defined amplitude and phase.

REFERENCES

1. J. Franck, *Trans. Faraday Soc.*, **21**, 536 (1925).
2. E. Condon, *Phys. Rev.*, **28**, 1182 (1926).
3. E. U. Condon, *Phys. Rev.*, **32**, 858 (1928).
4. L. Pauling and E. B. Wilson, *Introduction to Quantum Mechanics*, McGraw-Hill, New York, 1935.
5. K. C. Kulander and E. J. Heller, *J. Chem. Phys.*, **69**, 2439 (1978).
6. Soo-Y. Lee and E. J. Heller, *J. Chem. Phys.*, **71**, 4777 (1979).
7. Soo-Y. Lee and E. J. Heller, *J. Chem. Phys.*, **76**, 3035 (1982).
8. E. J. Heller, *Acc. Chem. Res.*, **14**, 368 (1981).
9. M. V. Rama Krishna and R. D. Coalson, *Chem. Phys.*, **120**, 327 (1988).
10. R. D. Coalson, *J. Chem. Phys.*, **86**, 6823 (1987).
11. N. E. Henriksen, *Theor. Chim. Acta*, **82**, 249 (1992).
12. N. E. Henriksen and E. J. Heller, *J. Chem. Phys.*, **91**, 4700 (1989).
13. Xue-P. Jiang, R. Heather, and H. Metiu, *J. Chem. Phys.*, **90**, 2555 (1989).
14. V. Engel, R. Schinke, S. Hennig, and H. Metiu, *J. Chem. Phys.*, **92**, 1 (1990).
15. V. Engel and H. Metiu, *J. Chem. Phys.*, **92**, 2317 (1990).
16. J. Zhang, D. G. Imre, and J. H. Frederick, *J. Phys. Chem.*, **93**, 1840 (1989).
17. E. J. Heller, E. B. Stechel, and M. J. Davis, *J. Chem. Phys.*, **73**, 4720 (1980).
18. S. O. Williams and D. G. Imre, *J. Phys. Chem.*, **92**, 3374 (1988).
19. M. Dantos, M. J. Rosker, and A. H. Zewail, *J. Chem. Phys.*, **89**, 6128 (1988).
20. A. H. Zewail, *J. Phys. Chem.*, **97**, 12427 (1993).

21. R. Schinke, *Photodissociation Dynamics*, Cambridge University Press, Cambridge, 1993.
22. E. Merzbacher, *Quantum Mechanics*, Wiley, New York, 1970.
23. R. Loudon, *The Quantum Theory of Light*, Oxford University Press, Oxford, 1983.
24. N. E. Henriksen, *Chem. Phys. Lett.*, **197**, 620 (1992).
25. E. J. Heller, R. L. Sundberg, and D. Tannor, *J. Phys. Chem.*, **86**, 1822 (1982).
26. J. R. Taylor, *Scattering Theory*, Krieger, Malabar, 1983.
27. A. Messiah, *Quantum Mechanics, Vol. 1*, North-Holland, Amsterdam, 1961.
28. E. J. Heller, *J. Chem. Phys.*, **68**, 3891 (1978).
29. N. E. Henriksen, J. Zhang, and D. G. Imre, *J. Chem. Phys.*, **89**, 5607 (1988).
30. M. Dirke and R. Schinke, *Chem. Phys. Lett.*, **196**, 51 (1992).
31. R. Schinke and V. Engel, *J. Chem. Phys.*, **93**, 3252 (1990).
32. S. C. Farantos, *Chem. Phys.*, **159**, 329 (1992).
33. N. E. Henriksen and E. J. Heller, *Chem. Phys. Lett.*, **148**, 567 (1988).
34. S. Das and D. J. Tannor, *J. Chem. Phys.*, **92**, 3403 (1990).
35. G. Drolshagen and E. J. Heller, *J. Chem. Phys.*, **82**, 226 (1985).
36. N. E. Henriksen, *Chem. Phys. Lett.*, **169**, 229 (1990).
37. D. J. Tannor and E. J. Heller, *J. Chem. Phys.*, **77**, 202 (1982).
38. B. Amstrup and N. E. Henriksen, *J. Chem. Phys.*, **97**, 8285 (1992).
39. D. J. Tannor, R. Kosloff, and S. A. Rice, *J. Chem. Phys.*, **85**, 5805 (1986).
40. P. Pechukas and J. C. Light, *J. Chem. Phys.*, **44**, 3897 (1966).
41. M. Shapiro and P. Brumer, *J. Chem. Phys.*, **97**, 6259 (1992).
42. N. E. Henriksen and B. Amstrup, *Chem. Phys. Lett.*, **213**, 65 (1993).
43. M. Shapiro and R. Bersohn, *Annu. Rev. Phys. Chem.*, **33**, 409 (1982).
44. P. Brumer and M. Shapiro, *Adv. Chem. Phys.*, **60**, 371 (1985).
45. N. E. Henriksen, *Comm. At. Mol. Phys.*, **21**, 153 (1988).
46. M. Shapiro, *J. Chem. Phys.*, **56**, 2582 (1972).
47. R. Kosloff, *J. Phys. Chem.*, **92**, 2087 (1988).
48. E. J. Heller, *J. Chem. Phys.*, **62**, 1544 (1975).
49. H. Goldstein, *Classical Mechanics*, Addison-Wesley, New York, 1980.
50. E. P. Wigner, *Phys. Rev.*, **40**, 749 (1932).
51. J. E. Moyal, *Proc. Cambridge Philos. Soc.*, **45**, 99 (1949).
52. E. J. Heller, *J. Chem. Phys.*, **65**, 1289 (1976).
53. E. J. Heller, *J. Chem. Phys.*, **67**, 3339 (1977).
54. J. P. Dahl, in *Energy Storage and Redistribution in Molecules*, Hinze, J. (Ed.), Plenum Press, New York, 1983, p. 557.
55. N. E. Henriksen, G. D. Billing, and F. Y. Hansen, *Chem. Phys. Lett.*, **149**, 397 (1988).
56. F. Y. Hansen, N. E. Henriksen, and G. D. Billing, *J. Chem. Phys.*, **90**, 3060 (1989).
57. R. C. Brown and E. J. Heller, *J. Chem. Phys.*, **75**, 186 (1981).
58. N. E. Henriksen, *Chem. Phys. Lett.*, **121**, 139 (1985).
59. N. E. Henriksen, V. Engel, and R. Schinke, *J. Chem. Phys.*, **86**, 6862 (1987).

60. K. B. Møller, J. P. Dahl, and N. E. Henriksen, *J. Phys. Chem.*, **98**, 3272 (1994).
61. E. J. Heller, *J. Chem. Phys.*, **68**, 2066 (1978).
62. G. Herzberg, *Spectra of Diatomic Molecules*, Van Nostrand-Reinhold, New York, 1950.
63. S. Goursaud, M. Sizun, and F. Fiquet-Fayard, *J. Chem. Phys.*, **65**, 5453 (1976).
64. V. Engel and R. Schinke, *J. Chem. Phys.*, **88**, 6831 (1988).
65. E. J. Heller, in *Potential Energy Surfaces and Dyanmics Calculations*, D. G. Truhlar (Ed.). Plenum Press, New York, 1981, p. 103.
66. Y. B. Band and K. F. Freed, *J. Chem. Phys.*, **63**, 3382 (1975).
67. V. Engel and H. Metiu, *J. Chem. Phys.*, **89**, 1986 (1988).
68. A. D. McLachlan, *Mol. Phys.*, **8**, 39 (1964).
69. G. D. Billing and G. Jolicard, *J. Phys. Chem.*, **88**, 1820 (1984).
70. N. E. Henriksen, G. D. Billing, and F. Y. Hansen, *Chem. Phys. Lett.*, **199**, 176 (1992).
71. J. R. Reimers and E. J. Heller, *J. Chem. Phys.*, **83**,511 (1985).
72. R. Schinke, *Annu. Rev. Phys. Chem.*, **39**, 39 (1988).

CHAOS IN DISSIPATIVE SYSTEMS: UNDERSTANDING ATMOSPHERIC PHYSICS

C. NICOLIS

Institut Royal Météorologique de Belgique, 1180 Brussels, Belgium

G. NICOLIS

Center for Nonlinear Phenomena and Complex Systems, Université Libre de Bruxelles, 1050 Brussels, Belgium

CONTENTS

1. ATMOSPHERIC VARIABILITY: COMPLICATED OR COMPLEX?

From the standpoint of an observer caught in the middle of a blizzard, a flood, or a long drought, the atmosphere may appear as something erratic

Advances in Chemical Physics, Volume XCI, Edited by I. Prigogine and Stuart A. Rice.
ISBN 0-471-12002-2

or even malevolent. Yet the same laws that govern physical and chemical laboratory scale systems also apply to the massive system comprised by the atmosphere, the oceans, the land and ice masses, and the biosphere. The question therefore naturally arises, to what extent can the qualitative aspects and the quantitative behavior of weather and climate be satisfactorily comprehended on the basis of these elementary laws.

The state of the atmosphere is usually expressed in terms of the spatial and temporal distributions of wind components, temperature, pressure, density, mixing ratios of the various phases of water and concentrations of other substances such as ozone, carbon dioxide, or solid particles in the form of dust. The laws governing these observables are the basic laws of motion of fluid mechanics as applied to a rotating system, the laws of chemical kinetics and thermodynamics, as well as some more specific laws involving phase changes of matter and such processes as absorption, emission, and scattering of radiation by atmospheric constituents. These laws are commonly expressed as a system of partial differential equations of first order in time that one may write in the generic form [1]

$$\frac{\partial \mathbf{x}}{\partial t} = -\mathbf{v} \cdot \nabla \mathbf{x} + \mathbf{F}(\mathbf{x}, \lambda) \tag{1.1}$$

Here $\mathbf{x}$ is a shorthand vector notation for the entire set of relevant variables $\mathbf{x} = (x_1, \ldots, x_n)$; $\mathbf{v}$ is the three-dimensional wind vector; $\mathbf{F}$ stands for the effect of local physical and chemical processes responsible for the change of $\mathbf{x}$; and λ denotes a set of parameters interfering with the behavior, such as chemical rate constants, emission, or absorption coefficients.

Equations (1.1) exhibit a universal *nonlinearity*, expressing the fact that in the atmosphere a property $\mathbf{x}$ will be advected by the bulk velocity $\mathbf{v}$—itself a property of the system in hand. Additional nonlinearities may be present in the function $\mathbf{F}$ arising, for instance, from cooperative effects of chemical origin or from feedbacks associated with the interaction of radiation with matter (surface-albedo feedback, etc).

The atmosphere tends to be organized into identifiable structures, each having a typical size, shape, and life history [2]. A partial list of these structures, arranged in order of decreasing scale, would include circumpolar westerly wind belts, migratory extratropical cyclones, thunderstorms, or cumulus clouds. Still, despite this overall order there is a pronounced variability reflected by the lack of perfect or nearly perfect periodicity. Figure 7.1 depicts the daily record temperature (1982–1990) in Brussels.

One is struck by the extreme variability of this record. In addition to an overall regularity associated with the seasonal variation of solar flux

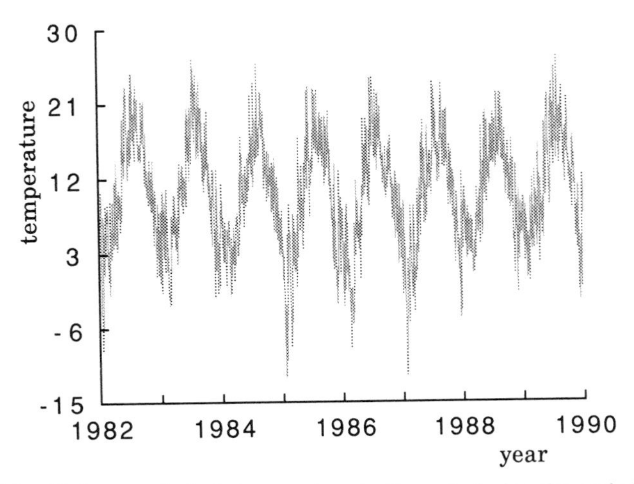

Figure 7.1. Daily record of air temperature in Brussels during the period 1982–1990.

arriving at the earth's surface as a result of the motion of earth in space, one observes irregular fluctuations. A clearcut signature of such fluctuations, common to all meteorological time series, is the existence of a broad band continuous part in the power spectrum (Fig. 7.2). This is very different from the line spectrum of periodic and quasiperiodic phenomena

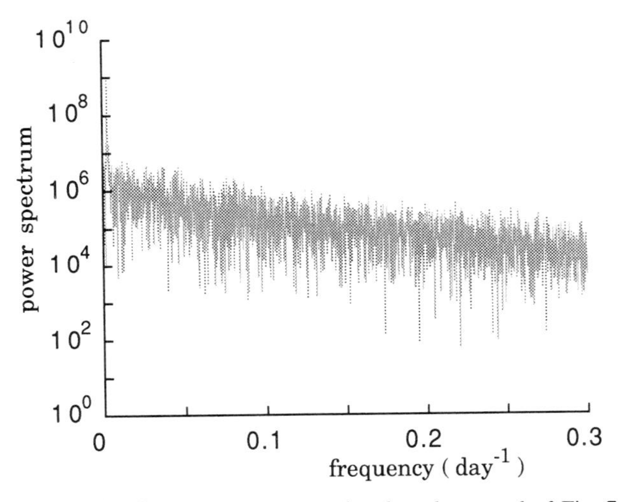

Figure 7.2. Power spectrum associated to the record of Fig. 7.1.

and shows that fluctuations are not limited to the immediate vicinity of a single scale but are actually manifested on a wide interval of time scales.

A straightforward consequence of the aperiodicity of the atmospheric dynamics is the difficulty to make predictions—a well-known fact in the meteorologist's everyday experience: contrary to periodic and quasi-periodic phenomena for which long-term predictions are possible, as amply demonstrated by the impressive success of astronomy, prediction of meteorological fields is very limited in time. Remarkable progress has been accomplished during the last decades, notably following the availability of high-performance computers, allowing one to perform reliable predictions for the next several days' interval. Yet the experience accumulated shows that the skill decreases with time, and that beyond 10 days or so prediction becomes essentially random, owing to the growth and accumulation of errors and uncertainties [3].

The question now naturally arises, is this lack of ability to perform reliable long-term predictions the result of the limited amount of data at our disposal and of imperfections in the currently available models? Or, rather, are there some fundamental mechanisms imposing irreducible limitations on the very possibility of making quantitative predictions beyond a certain well-defined lapse of time?

A first answer that comes to mind is to attribute these difficulties to the extreme complication of the atmosphere that encompasses phenomena on very different scales, from turbulence and thermal convection to the scale of global circulation. Thanks to the availability of powerful computers this problem has found a partial solution in the sense that it has been possible to include a wide range of scales in models of numerical weather and climate prediction, while at the same time sophisticated parameterization methods have been applied to account for the average effect of processes (such as turbulence) developing on scales smaller than the model grid. On the other hand, these unresolved scales also induce perturbations at the level of larger scales of essentially random character. Given enough time the accumulation of these random noises would end up modifying the large scale evolution in an appreciable manner. Then, in this view the difficulty to perform predictions would be a temporary drawback bound to disappear as the parameters become better known thanks to improved experimental techniques, and the computers become sufficiently powerful to resolve all the relevant scales.

If the dynamics described by Eqs. (1.1) were simple, the above mentioned difficulties could still be controlled in the sense that predictions whose accuracy is comparable to that of the coarseness of initial data could be issued. Now, as first realized by Thompson [4] and Lorenz [5, 6] in the atmosphere a small error in the initial data on a model of numerical forecasting will be amplified during the evolution generated by the model, and it is this error growth that will eventually set limits on the possibilities of satisfactory predictions, even with a well-defined, well-tuned perfectly deterministic model. We are thus led to inquire whether the finite precision of data and the concomitant neglect of subgrid phenomena are nothing but triggers revealing an instability that has nothing to do with practical limitations but is, rather, inherent in the laws governing the underlying system. Physical sciences provide us with a prototype capable of generating exactly this sort of complexity: deterministic chaos [7], which subtly combines a global, large-scale order and a local, small-scale variability arising from the ubiquitous property of sensitivity to the initial conditions. Our goal in this chapter is to examine the relation between nonlinear dynamics and chaos theories on the one side and the science of atmospheric phenomena on the other and, in particular, to sort out the new perspectives opened in this field by its cross-fertilization with one of the youngest and most rapidly growing branches of science.

In Section II we provide concrete evidence for nonlinear dynamics and chaos in the atmosphere coming both from mathematical modeling and from data analysis. This presentation will also offer the opportunity to summarize some of the concepts and techniques of dynamical systems theory briefly. As a representative case study a simplified atmospheric circulation model proposed by Lorenz [6] is taken up and analyzed in detail from the standpoint of nonlinear dynamics in Section III. Section IV is devoted to predictability, whose paramount importance stems from its relation to both fundamental theory and practice of operational forecasting. We successively review, the dynamics of error growth and the role of scales on predictability, and explore the possibilities of a statistical approach to the forecasting of complex systems. In Section V the passage from atmospheric dynamics, usually associated with short scale "weather" variability, to global climate dynamics is outlined. Evidence that nonlinear dynamics and chaos are ubiquitous in the context of global climate is presented as well. Some ideas aiming to connect these two scales of variability are also developed. The main conclusions and the perspectives opened by nonlinear dynamics and chaos theories in the fields of atmospheric and climate dynamics are discussed in Section VI.

II. NONLINEAR DYNAMICS AND CHAOS IN THE ATMOSPHERE

There exists a whole hierarchy of atmospheric models, each tailored to respond to specific needs, such as desired resolution, running time limitation, and space or time scales of the phenomenon to be described.

The majority of these models can be traced back, in one way or another, to a set of equations known as primitive equations [2]. These include

- Mass balance

$$\frac{d\rho}{dt} = -\rho \operatorname{div} \mathbf{v} \tag{2.1}$$

  where ρ is the mass density of the carrier fluid, $\mathbf{v}$ the velocity field relative to the earth and d/dt denotes the hydrodynamic derivative

$$\frac{d}{dt} = \frac{\partial}{\partial t} + \mathbf{v} \cdot \nabla$$

- Moisture balance

$$\frac{d\rho_{\mathrm{W}}}{dt} = -\rho_{\mathrm{w}} \operatorname{div} \mathbf{v} + S_{\mathrm{W}} \tag{2.2}$$

  where ρ_{W} is the density of water vapor and S_{W} represents the sources and sinks of water vapor.
- Momentum balance

$$\rho \frac{d\mathbf{v}}{dt} = -\nabla p - 2\mathbf{\Omega} \times \mathbf{v} + \boldsymbol{\rho}\mathbf{g} + \mathbf{F} \tag{2.3}$$

  where p is the hydrostatic pressure, Ω is the angular velocity of rotation of the earth, $\mathbf{g}$ is the acceleration of gravity, and $\mathbf{F}$ accounts for dissipative effects (viscosity, friction in the boundary layer, etc).
- Energy balance

$$c\frac{dT}{dt} + p\frac{d}{dt}\frac{1}{\rho} = Q \tag{2.4}$$

  where T is the temperature, c is the specific heat, and Q accounts for the heat sources and sinks arising from radiation, conduction, phase changes, or work of frictional forces.

These equations are to be completed by a number of relations specifying how p, S_{W}, $\mathbf{F}$, or Q are related to the state variables ρ, ρ_{W}, $\mathbf{v}$, and T. An example of such *diagnostic relations* is the equation of state

$p = p(\rho, T)$, which in the atmosphere reduces to the gas law

$$p = \rho RT \tag{2.5}$$

In addition, other variables may be introduced for particular applications, such as the concentrations of ozone and other trace constituents.

Equations (2.1)–(2.4) are usually mapped to a spherical geometry and simplified somewhat by assuming the vertical scale of the motion to be small compared with the horizontal one. The equations are further reduced to a set of *ordinary* differential equations by one of the following techniques [8]:

- Finite difference schemes, where the variables are represented on a discrete grid and their spatial derivatives are approximated by differences in the values of the variables at different grid points.
- Spectral schemes, where the various fields are represented in terms of truncated expansions in a function basis appropriate to the geometry of the problem under consideration (e.g., spherical harmonics).

To summarize, starting from the full dynamical system [Eqs. (2.1)–(2.4)] described by an infinite number of degrees of freedom one arrives at a representation in terms of a finite, though possibly very large (10^6 or so in operational weather forecasting) number of variables: the values of the meteorological fields at the grid points or the expansion coefficients in their representation in the appropriate function basis. A dynamical system of this type can be embedded in *phase space*, the space spanned by the full set of the relevant variables, in the sense that there is a 1:1 correspondence between the states reached in the course of its evolution in time and the points of this space. The first step in the process of forecasting consists in identifying the phase space point that most adequately represents the initial condition available from observation (initialization). The next step is to calculate numerically, by additional discretization in time, the trajectory of the dynamical system in phase space (model integration). To reach a high spatial resolution one includes the maximum number of degrees of freedom compatible with the computing power available. This brings, however, the complication that the essential traits may be masked by a multitude of secondary fluctuations. To sort them out special data analysis techniques are needed (e.g., principal component analysis). Furthermore the cost of a given integration considerably limits the number of trajectories that one may compute. This precludes reliable statistical analysis or a systematic exploration of the behavior in terms of the parameters.

A. Error Growth and Chaotic Dynamics in Numerical Models of Weather Forecasting

In 1982 Lorenz analyzed the operational forecasting model of the European Center for Medium Range Weather Forecasts (ECMWF), whose main purpose is to produce weather forecasts in the range extending from a few days to a few weeks. The operational routine involved daily preparation of a 10-day forecast of the global atmospheric state, using the present day's state as the initial condition. Since the equations are solved by stepwise integration, forecasts for intermediate ranges are automatically produced, and 1, 2, . . . , 10-day forecasts are routinely achieved.

Lorenz capitalized on the fact that the model produces rather good 1-day forecasts, so that the state predicted for a given day, 1 day in advance, may be regarded as equal to the state subsequently observed on the given day, plus a relatively small error. By comparing the 1- and 2-day forecasts for the following day, the 2- and 3-day forecasts for the day after, and so on, he determined how rapidly the error grows. Figure 7.3 summarizes the principal results, averaged over 100 consecutive days [9]. The upper curve measures the model's performance and indicates how rapidly the differences between two states, one governed by the model and one by the real atmospheric evolution, will amplify. The lower curve indicates how rapidly the difference between two states, both governed by the model, will amplify.

The lower curve clearly indicates sensitive dependence on initial conditions—the principal signature of deterministic chaos [7]. Extrapolation of the curve to very small differences suggested an error *doubling time* of about 2.5 days. Since the predictability analysis of Lorenz [9] the forecasting ECMWF model has been changed substantially, as the spatial resolutions have been increased and the physical parameterization schemes improved significantly. Yet, although there has been a considerable increase in the accuracy of forecasts for a few days ahead, it seems that the error doubling time is even shorter than the one estimated by Lorenz [3]. Detailed forecasting of weather states at sufficiently long range is therefore impractical, owing to the complexity inherent in the dynamics of the atmosphere.

B. Low-Order Models

The difficulties accompanying the integration of large numerical models of the atmosphere have prompted several investigators to explore another approach, in which one tries to reduce the number of degrees of freedom systematically while at the same time preserving the structure of the

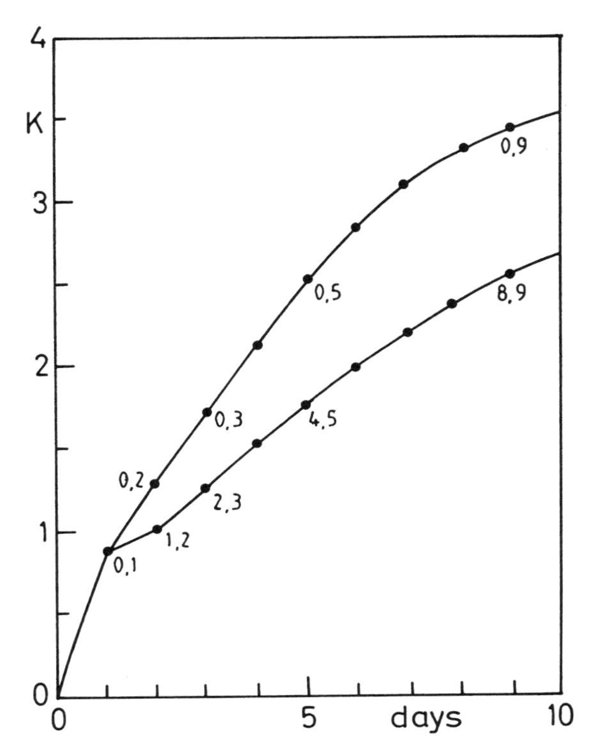

Figure 7.3. Root-mean-square difference between i-day and j-day forecasts of the 500-hPa temperature for the same day, made by the ECMWF operational model, averaged over 100 days. Numbers i, j appear beside selected difference values, which are plotted against values of j.

underlying equations. The merit of such simplified models is to concentrate on the most important qualitative phenomena; to allow for systematic exploration of the behavior in terms of the parameters; and to enable statistical analysis by generating, within reasonable time limits, a large number of phase space trajectories.

One of the most famous low-order models, which actually has been at the origin of modern ideas on deterministic chaos, is a simplified model of thermal convection developed by Lorenz [5]. Its starting point is in Eqs. (2.1), (2.3), and (2.4) as applied to a shallow horizontal fluid layer heated from below. This allows for additional approximations, such as $\mathbf{\Omega} = 0$; incompressibility, $\operatorname{div} \mathbf{v} = 0$; replacing ρ by a constant reference value ρ_0 in the left-hand side of Eqs. (2.3)–(2.4) and by a linearized equation of state $\rho = \rho_0(1 - \alpha(\mathrm{T} - \mathrm{T}_0))$ in the right-hand side of Eq. (2.3) where α is the coefficient of thermal expansion; replacing $\mathbf{F}$ in the same equation by the shear viscosity term $\eta \nabla^2 \mathbf{v}$; dropping the second term in the left-hand side of Eq. (2.4), and replacing Q in the right-hand side by Fourier's law $\kappa \nabla^2 T$, where κ is the thermal diffusivity. The resulting equations predict a transition from the state of rest to an organized flow when the *Rayleigh number* R, a suitable dimensionless combination of the parameters, crosses a threshold value R_c.

Under the above described *Boussinesq approximation*, expanding the velocity and temperature fields in Fourier series, keeping one Fourier mode for the vertical component of the velocity, and two Fourier modes for the temperature variation one arrives at the equations

$$\begin{aligned}
\frac{dx}{dt} &= \sigma(y - x) \\
\frac{dy}{dt} &= rx - y - xz \\
\frac{dz}{dt} &= xy - bz
\end{aligned} \tag{2.6}$$

The variable x measures the rate of convective (vertical) turnover, y the horizontal temperature variation, and z the vertical temperature variation. The parameters σ and r are proportional, respectively, to the Prandtl number (depending entirely on the intrinsic properties of the fluid) and to the Rayleigh number. Finally, the parameter b accounts for the geometry of the convective pattern. In addition to reproducing the thermal convection instability for $r \geq 1$, Eqs. (2.6) reveal a surprisingly rich variety of behaviors and have actually been used extensively to illustrate the onset of deterministic chaos in systems involving few variables [10].

From the standpoint of atmospheric dynamics thermal convection is a rather localized phenomenon (mesoscale variability). Low-order models can, however, be successfully designed for large scale phenomena related to global circulation [11]. Most of these models find their origin in a limiting form of Eqs. (2.1)–(2.4) in which:

- The horizontal wind components u and v are represented by a stream function ψ,

$$u = \frac{\partial \psi}{\partial y} \qquad v = -\frac{\partial \psi}{\partial x} \tag{2.7}$$

- The stream function and the temperature field are averaged vertically.

- The reduced fields, which now depend on the horizontal coordinates x and y only, are expanded in orthogonal functions.
- The resulting hierarchy of equations is truncated to the first few orders.

When thermal effects are discarded and information on topography (mountains and valleys) is included one arrives at the Charney–De Vore model [12, 13]

$$\begin{aligned}
\frac{dv_r}{dt} &= kv_i(u_0 - \beta/2k^2) - Cv_r \\
\frac{dv_i}{dt} &= -kv_r(u_0 - \beta/2k^2) + f_0 \frac{\tilde{h}}{2H} u_0 - Cv_i \qquad (2.8) \\
\frac{du_0}{dt} &= -f_0 \frac{\tilde{h}}{H} v_i - C(u_0 - u_0^*)
\end{aligned}$$

Here v_r, v_i are, respectively, real and imaginary parts of the first Fourier mode of the gradient of the stream function $k\psi$; u_0 is the velocity of the zonal mean flow that does not depend on y; f_0 is a mid-latitude value of the Coriolis parameter; β is a constant arising from the assumption that the Coriolis parameter depends linearly on y; C is a damping parameter; H is a scale height ($\sim 10^4$ m); $\tilde{h}$ is the first Fourier mode of the orographic profile, u_0^* stands for the forcing of the mean flow, and k is the wavenumber of the flow.

Equations (2.8) predict multiple steady-state solutions in the range of intermediate wavelengths $L(\sim 3 \times 10^6$ m). The great interest of this discovery is to provide a qualitative explanation of the phenomenon of persistence flow regimes in mid-latitudes, also referred as "blocking," which constitute one of the principal elements of low-frequency atmospheric variability (time scale of weeks or so). Their main characteristics are a preferred geographic localization (in contrast to the familiar "zonal" flows), a certain persistence, a tendency to recur, a variable lifetime, and a rapid transition toward the zonal regime. As these phenomena are poorly accounted for by large numerical models, here one has a clearcut example of the usefulness of the qualitative view afforded by simplified low-order models.

Another interesting family of low-order models is generated when thermal effects are incorporated by identifying vertical variations of the wind with horizontal temperature gradients. This is usually done in the framework of the *geostrophic relation*, in which momentum balance [Eq. (2.3)] is approximated by the balance between the Coriolis force and the pressure gradient. The equations assume the general form [8]

$$\frac{dx_i}{dt} = \sum_{jk} a_{ijk} x_j x_k + \sum_k b_{ij} x_j + C_i \tag{2.9}$$

A highly truncated version of Eqs. (2.9) leads to [6]

$$\begin{aligned} \frac{dx}{dt} &= -y^2 - z^2 - ax + aF \\ \frac{dy}{dt} &= xy - bxz - y + G \\ \frac{dz}{dt} &= bxy + xz - z \end{aligned} \tag{2.10}$$

in which x denotes the strength of the globally averaged westerly current, y and z the phases of waves carried along by the current and transporting heat poleward, and aF, G thermal forcings.

A detailed analysis of Eqs. (2.10) is carried out in Section III. It suffices to mention now that, depending on the parameter values, these equations predict the coexistence of two stable periodic solutions, of multiple steady states, of quasiperiodic behavior, or of deterministic chaos. Furthermore, in the presence of a seasonal variation of the forcing the model can simulate strong interannual variability marked by an irregular succession of "active" summers (strong oscillations of the westerly current) and "inactive" ones (oscillations remain weak) [6]. Again these phenomena, all of which are familiar from our own experience, are very poorly accounted for by the large numerical models of the atmosphere.

C. Dynamical Reconstruction from Time Series Data

A model necessarily provides only a schematic view of reality. In view of the uncertainties inherent in the modeling process it is, therefore, desirable to dispose of methodologies providing information on the character of a phenomenon on the sole basis of experimental data. Once available these methodologies can of course be applied, if appropriate, to the analysis of the output of a mathematical model.

The methods currently available to address this question are based on the fact that, as time elapses, the phase space trajectories of a physical system involving many degrees of freedom tend to an invariant manifold of phase space, to which one refers as the *attractor*. This reflects the dissipative character of the dynamics of such systems, a ubiquitous property that is also in direct relation with the second law of thermodynamics.

An invariant manifold embedded in a space may be conveniently characterized by its *dimension*. It is well known that the manifolds associated with chaotic attractors are *fractals*. They arise from a process of repeated fragmentations of phase space volumes concomitant to the expansion and folding of nearby phase space trajectories. In mathematical fractals the process of fragmentation may be carried out in a way to produce self-similar objects of noninteger dimension. Physical fractals are, on the other hand, multiscale objects in which global self-similarity is not verified. Such *multifractals* are characterized by an infinity of dimensions D_q, $q = 0, 1, 2, \ldots$, accounting, essentially, for local inhomogeneities [14, 15]. One of the most popular dimensions is the correlation dimension D_2, which provides information on the number of points within a certain hypersphere of radius r around a reference point $\mathbf{X}$, as the latter moves along the attractor.

A finer characterization of an attractor aims at the determination of dynamical properties. The most important of them, in view of the very definition of chaos, is the rate of divergence of two initially nearby trajectories in the course of time. The long time average of this divergence is the largest positive *Lyapunov exponent*, σ of the system [7]. In general, in a multivariate system there exist several directions in phase space along which an initial error is amplified, in other words, there exist several positive Lyapunov exponents. Furthermore, as pointed out above physical attractors are highly inhomogeneous. This entails a large dispersion of the local divergence rates over their averages, reflected by important fluctuations of the Lyapunov exponents as discussed further in Section IV.

When addressing the problem of dynamical reconstruction one disposes, as a rule, of the time series pertaining to the locally monitored value of a small number of observables, or even to a single observable. In atmospheric dynamics such observables may be the value of temperature, pressure, geopotential height ϕ/g_0 ($\phi = \int_0^z g dz$; z being a given altitude and g_0 the standard value of the acceleration of gravity g at the earth's surface) and so on. The question therefore arises, how to unfold this "one-dimensional" series into a multidimensional phase space. This is achieved by the method of delays [16], based on the realization that a time series of the form

$$X_0(t_1), X_0(t_1 + \Delta t), X_0(t_1 + 2\Delta t), \ldots$$

and the series generated by successive shifts

$$X_0(t_1 + \tau), X_0(t_1 + 2\tau), \ldots$$
$$X_0(t_1 + 2\tau), \ldots\ldots$$
$$\ldots.$$

where τ is an integer multiple of Δt, define in general linearly independent state vectors in the phase space spanned by them. By performing this unfolding for successively higher values of the lags $n\tau$ and by computing for each n the values of D_2, σ, and so on, one may check whether the results reach a saturation value beyond some n_{sat}. If so, the conclusion will be that the system at hand exhibits deterministic chaos if D_2 happens to be a noninteger higher than 2 and σ happens to be positive.

There exist several algorithms for carrying out the above program systematically [17, 18]. They all have their limitations, notably in connection with the finite number of data usually available when monitoring a real-world complex system like the atmosphere. Furthermore, one of their principal initial motivations—how to distinguish between deterministic chaos and random noise—is partially lost when one realizes that *correlated noise* may produce results that are hardly distinguishable from those expected in a deterministic system. Nevertheless, when applied in conjunction with theoretical ideas such algorithms do provide valuable information as they give an idea of the minimum number of key variables involved in the dynamics, and hence clues on a pragmatic modeling in which irrelevant modes are filtered out. Globally speaking, the emerging conclusion is that some aspects of atmospheric variability may indeed be governed by low-dimensional chaotic attractors. We particularly mention the analyses of geopotential time series at 500-h Pa and of surface pressure giving attractor dimensions of about 6[19,20], or of wind speeds giving dimensions of about 7.3 [21]. The output of a large numerical model of atmospheric circulation over 20 years has also given a dimension close to 6 for mid-latitudes and close to 11 for the tropics [22]. The fundamental question, how such drastic dimension reductions may arise in a system governed by millions of variables will be addressed in Section V. All in all, the status of low-dimensional chaos in atmospheric dynamics is an unfinished task that will certainly attract further attention in the future.

III. A CASE STUDY: LORENZ'S THREE-VARIABLE ATMOSPHERIC MODEL

To illustrate the type of complexity generated by low-order atmospheric models, in this section we give a detailed analysis of the set of Eqs. (2.10) from the standpoint of nonlinear dynamics [23]. These equations will also be considered further in Section IV, which is devoted to predictability.

An efficient method for revealing the behavior of a dynamical system is to determine the fixed points first (steady-state solutions); carry out next linear stability analysis around them thus determining the points in parameter space where local bifurcations can be expected; explore the possibilities of global bifurcations by identifying, through control of several parameters, high-codimension bifurcation points; and finally complete the picture obtained by this analytic investigation by computer simulations [24].

Let us first deal with the fixed points (x_0, y_0, z_0) of Eqs. (2.10)

$$\begin{aligned} 0 &= -y_0^2 - z_0^2 - ax_0 + aF \\ 0 &= x_0 y_0 - bx_0 z_0 - y_0 + G \\ 0 &= bx_0 y_0 + x_0 z_0 - z_0 \end{aligned} \tag{3.1}$$

From the second and third equation we find

$$\begin{aligned} y_0 &= \frac{G(1 - x_0)}{1 - 2x_0 + (1 + b^2)x_0^2} \\ z_0 &= \frac{bGx_0}{1 - 2x_0 + (1 + b^2)x_0^2} \end{aligned} \tag{3.2}$$

Substituting into the first equation (3.1) one finds

$$(1 + b^2)x_0^3 - [2 + (1 + b^2)F]x_0^2 + (1 + 2F)x_0 + \left(\frac{G^2}{a} - F\right) = 0 \tag{3.3a}$$

Introducing the new parameters

$$B = \frac{1}{1 + b^2} \qquad G' = \frac{G^2}{a} - \frac{F}{1 + b^2}$$

and the linear transformation

$$x_0 = \bar{x} + \frac{2B + F}{3}$$

we may write Eq. (3.3a) in the canonical form of a cubic

$$\bar{x}^3 + p\bar{x} + q = 0 \tag{3.3b}$$

where

$$p = B(1 + 2F) - \frac{(2B + F)^2}{3}$$

$$q = \frac{B(1 + 2F)(2B + F)}{3} + G' - \frac{2(2B + F)^3}{27}$$

The discriminant of Eq. (3.3b) is

$$\Delta = \frac{q^2}{4} + \frac{p^3}{27}$$

The curves labeled SN in Fig. 7.4, determined by $\Delta = 0$, separate the regions in the parameter plane (F, G) where the system (2.10) possesses either one of three fixed points. The precise site of the cusp of the curve SN is determined by the simultaneous vanishing of p and q. This occurs when

$$G = \frac{2\sqrt{12}b\sqrt{ab}}{3(1 + b^2)} \qquad F = \frac{1 + \sqrt{3}b}{1 + b^2}$$

The stability of the fixed points is determined by the eigenvalues ω of the Jacobian matrix of Eqs. (2.10), which satisfies

$$\begin{vmatrix} -a - \omega & -2y_0 & -2z_0 \\ y_0 - bz_0 & x_0 - 1 - \omega & -bx_0 \\ by_0 + z_0 & bx_0 & x_0 - 1 - \omega \end{vmatrix} = 0$$

leading to the characteristic equation

$$\omega^3 + u\omega^2 + v\omega + w = 0 \tag{3.4}$$

where

$$\begin{aligned} u &= a - 2(x_0 - 1) \\ v &= (x_0 - 1)^2 - 2(x_0 - 1)a + b^2x_0^2 + 2(y_0^2 + z_0^2) \\ w &= a(x_0 - 1)^2 + ab^2x_0^2 + 2(y_0^2 + z_0^2)(1 - x_0 - b^2x_0) \end{aligned} \tag{3.5}$$

Equation (3.4) has one zero root along the curves SN and one pair of

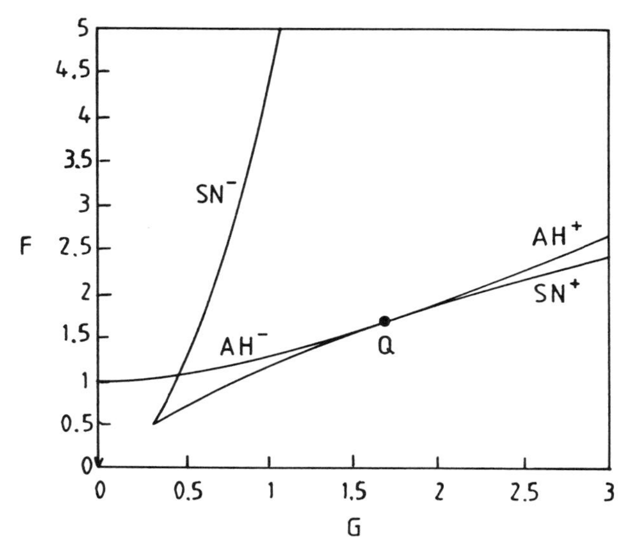

Figure 7.4. State diagram of model (2.10) in the (F, G) parameter plane. SN: locus of limit point bifurcation. AH: locus of Hopf bifurcation. Q: codimension 2 bifurcation point. The values of parameters a and b are set to $a = 0.25$, $b = 4$.

purely imaginary roots along curves AH^+ and AH^- of Fig. 7.4. The first set is the locus of limit point bifurcations with the exception of the cusp in which a pitchfork bifurcation takes place; the second set is the locus of Hopf bifurcations [24]. When $u = 0$ and $w = 0$ Eq. (3.4) has simultaneously one zero root and one pair of purely imaginary roots. By substituting the values of x_0, y_0, z_0 into Eqs. (3.2)–(3.3a) one identifies a point (F^*, G^*) in parameter space, denoted by Q in Fig. 7.4, which is a codimension two bifurcation point [24, 25]. It is well known from bifurcation theory that a system in the vicinity of such a point may give rise to very rich dynamical behavior that may subsequently extend far away in parameter space. In particular:

- *Quasiperiodic solutions* can be generated through a secondary bifurcation mechanism.
- Global bifurcations, such as *homoclinic bifurcations*, can take place [26], giving rise to closed infinite period orbits that are biasymptotic to a fixed point of the saddle-focus type, that is, they converge to it in the double limit $t \to t \pm \infty$.

Figure 7.5 depicts a *homoclinic orbit* arising in the system of Eqs.

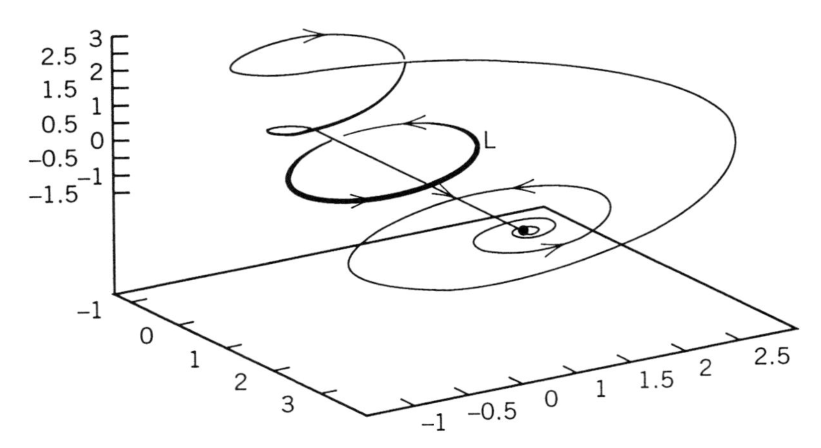

Figure 7.5. Homoclinic orbit to a saddle focus (above AH^- of Fig. 7.4) in the system of Eqs. (2.10).

(2.10) through the latter mechanism. Such orbits exist only for exceptional values of the control parameters. Nevertheless, they are of the utmost importance since for parameter values near those at which they bifurcate complex aperiodic behavior is bound to emerge. The mechanism underlying this emergence is as follows [26]. First, a stable fixed point behaving as focus (region between curves SN and AH^- in Fig. 7.4) loses its stability on crossing the curve AH^- and becomes a saddle-focus, giving rise to a stable limit cycle. For some range of parameter values this cycle may have complex linear stability eigenvalues (Floquet multipliers). This gives rise to the configuration depicted in Fig. 7.6: The unstable manifold W_u of the saddle-focus spirals onto the periodic orbit thereby forming a "whirlpool." The size of scrolls increases as the parameters vary further and eventually W_u touches the stable manifold W_s of the saddle-focus. In the vicinity of this situation a complex invariant set is generated, on which the behavior is chaotic. A two-dimensional projection of the corresponding phase space trajectory is given in Fig. 7.7. Actually, in this case the whirlpool is a repeller rather than an attractor. Attracting sets are, however, generated in other parts of parameter space. An example, to which we shall return in Section IV in connection with predictability, is given in Fig. 7.8. In the above described phenomenon the periodic orbit is stable. As the parameters cross a new boundary in the (F, G) space it loses its stability through a period doubling bifurcation signaling the

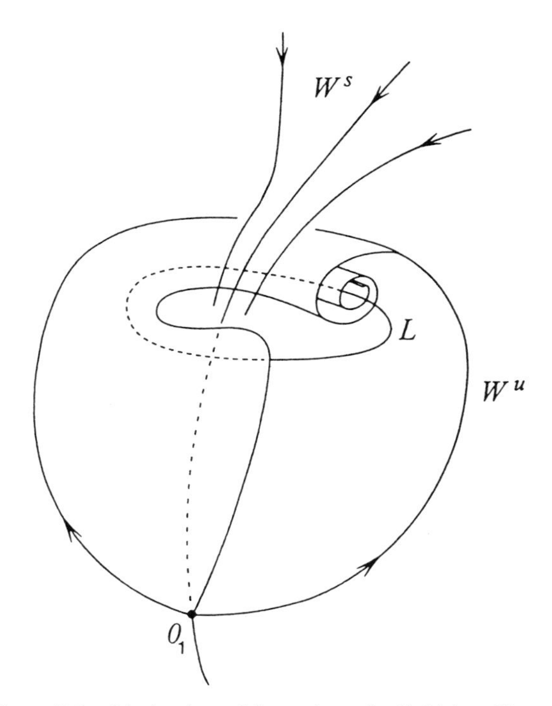

Figure 7.6. Mechanism of formation of a "whirlpool" repeller.

beginning of a Feigenbaum cascade [7], beyond which chaotic attractors or repellers of a different type may also emerge.

Chaotic attractors in three-variable (and of course higher order as well) systems are very rarely conventional attractors: Instead of being as usually expected the unique attracting set in some finite region of phase space, they typically contain an infinity of stable periodic orbits having very tortuous basins of attraction. Given enough time, a trajectory on such a *quasiattractor* [27] will pass close to these basins. This will transiently, result in a regular periodic-like behavior and may considerably complicate the applicability of the currently available algorithms of computation of dimensionalities and Lyapunov exponents [17, 18]. Figure 7.9 provides an example in which the dimension of the chaotic set has been computed numerically for $F = 6$ and 8 and for G in the range (0.2, 1.4). We see a clearcut signature of chaos in certain regions (e.g, for $F = 8$ and $G \approx 1$). On the other hand, for $F = 6$ the dimension drops

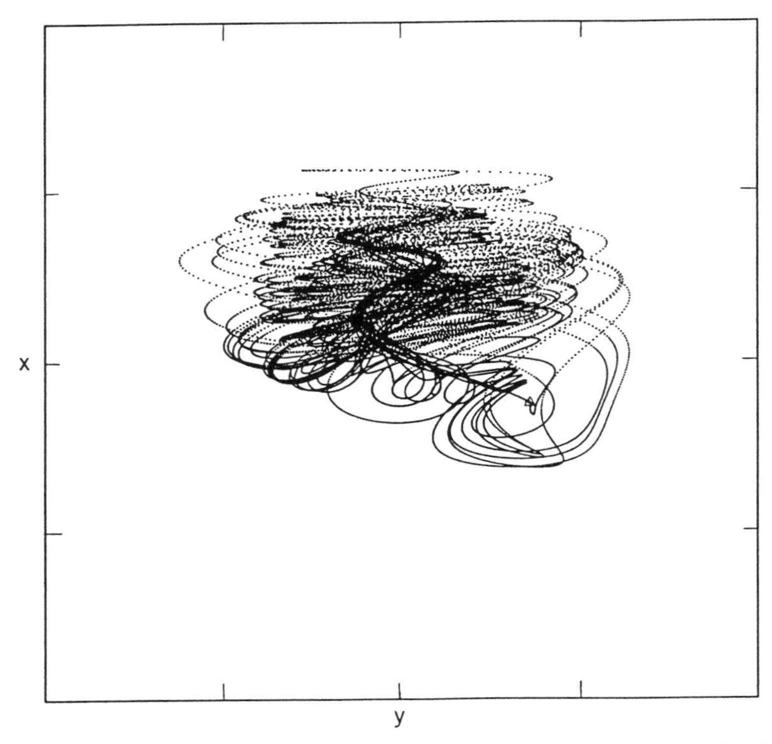

Figure 7.7. Two-dimensional projection of the chaotic repeller for $F = 5$, $G = 1.0778$, and a, b as in Fig. 7.4.

frequently to values between 1 and 2. This is likely to reflect the influence of the above mentioned stages of periodic-like behavior intermittently interrupting the chaotic evolution.

IV. PREDICTABILITY

We have already insisted on the fact that in chaotic systems sensitivity to initial conditions entails the growth of small initial errors—the latter always being present when dealing with the interaction between an observer and a complex system—thereby limiting the possibility of long-term predictions. This section is devoted to an analysis of the generic mechanisms governing error growth for both short and long times, and to their illustration on the atmospheric model of Section III.

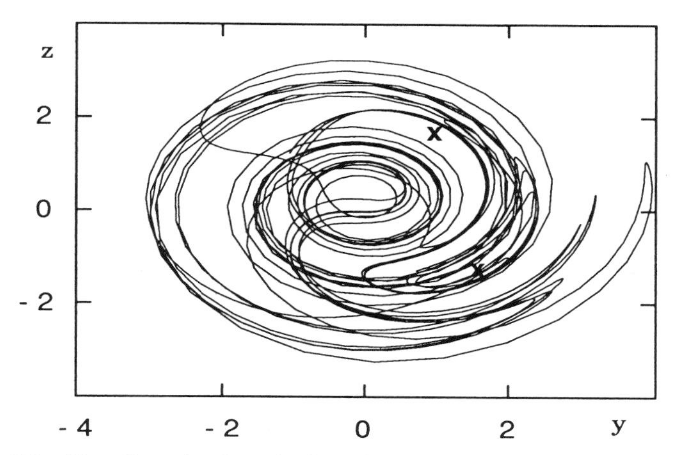

Figure 7.8. Two-dimensional projection of the chaotic attractor for model (2.10). Parameter values. Parameter values $a = 0.25$, $b = 6$, $F = 16$, and $G = 3$, crosses indicate two of the three fixed points of the system.

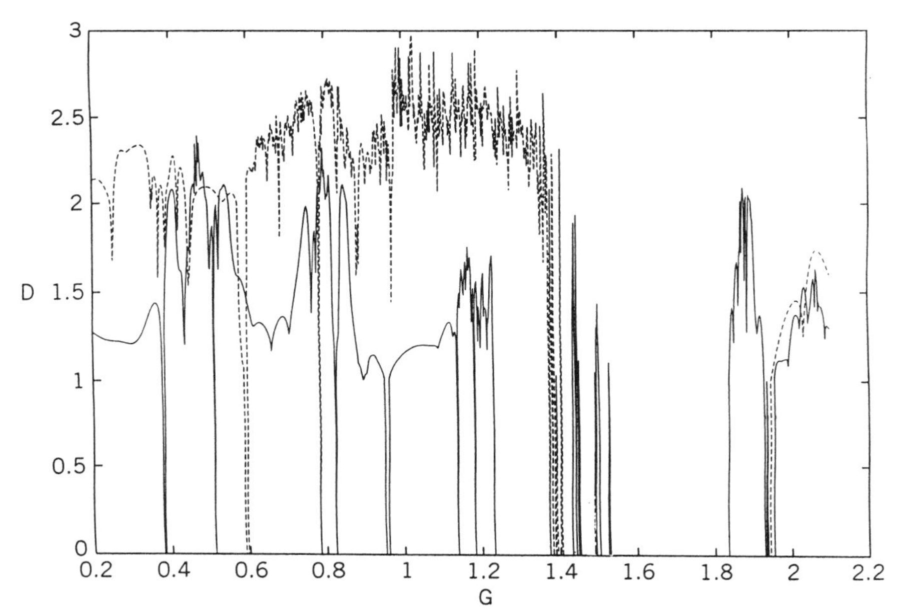

Figure 7.9. Parameter dependence of the dimension of the chaotic invariant manifolds of model (2.10) for $F = 6$ (full line) and $F = 8$ (broken line). Values of a and b as in Fig. 7.4.

A. General Formulation

Let $\mathbf{x}_0$ be an initial state of a dynamical system like the atmosphere, $\mathbf{y}_0 = \mathbf{x}_0 + \boldsymbol{\varepsilon}$ a state to which $\mathbf{x}_0$ is displaced as a result of an initial error $\boldsymbol{\varepsilon}$. By definition, the instantaneous error E_t is

$$E_t = |\mathbf{y}(\mathbf{t}; \mathbf{y}_0) - \mathbf{x}(\mathbf{t}; \mathbf{x}_0)| \tag{4.1}$$

where $\mathbf{y}(\mathbf{t}; \mathbf{y}_0)$, $\mathbf{x}(\mathbf{t}; \mathbf{x}_0)$ are obtained by integrating the evolution laws starting at $t = 0$ with the initial conditions $\mathbf{x}_0$ and $\mathbf{y}_0$, respectively, and $|\cdot|$ is the Euclidean norm in the phase space of the system.

Because of the complexity of the dynamics of typical atmospherically interesting systems, E_t as defined by Eq. (4.1) fluctuates considerably, over both time and the state space. To relate the error dynamics to the properties of the underlying system in an intrinsic manner and, more particularly, to the structure of its attractor, we adopt a probabilistic viewpoint: Specifically, the operation defined by Eq. (4.1) is repeated for initial conditions $\mathbf{x}_0$ running all over the attractor. The algebraic mean of the error over all these realizations, is then defined by

$$\begin{aligned}\langle E_t \rangle &= \int d\mathbf{x}_0 \rho_s(\mathbf{x}_0) |\mathbf{y}(\mathbf{t}; \mathbf{y}_0) - \mathbf{x}(\mathbf{t}; \mathbf{x}_0)| \\ &= \int d\mathbf{x}_0 \rho_s(\mathbf{x}_0) |\mathbf{f}^t(\mathbf{x}_0 + \boldsymbol{\varepsilon}) - \mathbf{f}^t(\mathbf{x}_0)| \end{aligned} \tag{4.2a}$$

where $\rho_s(\mathbf{x}_0)$ is the invariant probability distribution over the attractor and $\mathbf{f}^t$ is the formal solution of Eq. (1.1)

$$\mathbf{x}(t, \mathbf{x}_0) = \mathbf{f}^t(\mathbf{x}_0) \tag{4.2b}$$

with $\mathbf{f}^t$ hereafter referred to as the *resolvent operator*. Additional averaging over ε can also be performed if necessary.

In addition to the arithmetic mean featured in Eq. (4.2) other means can be considered. Of special relevance is the mean of the logarithm of the instantaneous error

$$\begin{aligned}\langle \ln E_t \rangle &= \int d\mathbf{x}_0 \rho_s(\mathbf{x}_0) \ln|\mathbf{y}(t; \mathbf{y}_0) - \mathbf{x}(t; \mathbf{x}_0)| \\ &= \int d\mathbf{x}_0 \rho_s(\mathbf{x}_0) \ln|\mathbf{f}^t(\mathbf{x}_0 + \boldsymbol{\varepsilon}) - \mathbf{f}^t(\mathbf{x}_0)| \end{aligned} \tag{4.3}$$

In chaos theory, expressions (4.1)–(4.3) are usually considered in the double limit of (in the indicated order) $\varepsilon \to 0$ and $t \to \infty$. This confines the dynamics to the linearized regime (tangent space) and allows one to identify the largest positive Lyapunov exponent σ_{max} as [7]

$$\sigma_{\text{max}} = \lim_{t\to\infty} \lim_{\varepsilon\to 0} \frac{1}{t} \ln \left| \frac{E_t}{\varepsilon} \right| \tag{4.4}$$

Actually, in most real world situations one is interested in the time behavior of small, but finite errors over a finite (and usually small) time interval. As stressed repeatedly, owing to sensitivity to the initial conditions, such errors must grow (in the mean) in time. On the other hand, as explosion to infinity is ruled out, $\langle E_t \rangle$ must eventually saturate to a finite value. Figure 7.10 qualitatively illustrates the balance between these two antagonistic trends. Three principal stages can be distinguished [28, 29]: a short-time regime during which errors remain small; an intermediate regime during which errors attain appreciable values at a time of the order of $t^* \sim (1/\sigma_{\text{max}}) \ln(1/\varepsilon)$ for which the $\langle E_t \rangle$ versus t curve presents an inflexion point; and a long-time regime where mean error reaches a saturation level representative of the global structure of the system's attractor. This behavior turns out to be quite general: It is shared by all chaotic systems that possess sufficiently strong ergodic properties, guaranteeing that given enough time the probability distribution on the attractor will tend to the invariant distribution ρ_s displayed in Eqs. (4.2) and (4.3).

In chaos theory, sensitivity to initial conditions and error growth are

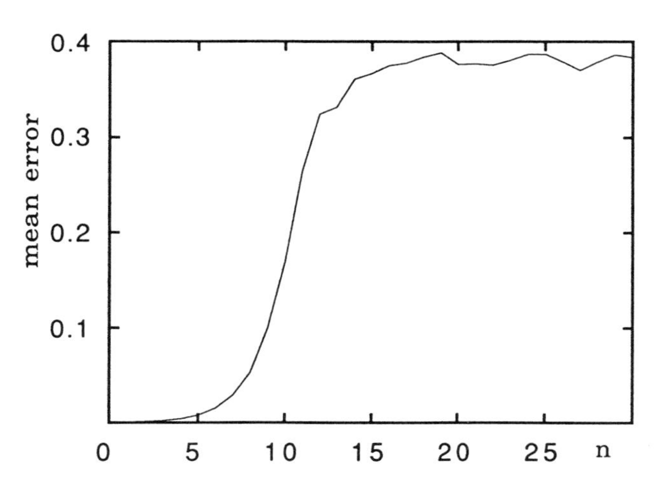

Figure 7.10. Schematic representation of the three stages of error growth in a prototype discrete time model giving rise to chaotic dynamics.

related to the Lyapunov exponents. Furthermore, it is traditionally stated that the very existence of a positive Lyapunov exponent as defined by Eq. (4.4) implies an *exponential growth* of a small initial error. Recent studies on large numerical models of the atmosphere revealed that the rate of growth of small initial errors is actually faster than the one predicted by the largest Lyapunov exponent, a behavior referred to as *superexponential* [30]. This conclusion has been corroborated further by low-order atmospheric models [31] such as the model of Eq. (2.10) and even by simple mathematical models of chaos [32]. The elucidation of this unexpected fact, which obviously further complicates the issue of prediction, is the object of Section III.B.

B. The Short-Time Behavior of Small Errors

To sort out the main ideas we first consider a one-dimensional recurrence

$$x_{n+1} = f(x_n) \tag{4.5}$$

which is the simplest type of dynamical system giving rise to deterministic chaos [7]. Equation (4.2a) becomes

$$\langle E_n \rangle = \int dx_0 \rho_s(x_0) |f^{(n)}(x_0 + \varepsilon) - f^{(n)}(x_0)| \tag{4.6}$$

Since errors remain small in the initial regime, an expansion of Eq. (4.6) in which only the lowest order term in ε is kept should be meaningful. At the level of Eq. (4.1) this leads to

$$E_n(\varepsilon, x_0) = \varepsilon \left| \frac{df^{(n)}}{dx_0} \right| = \varepsilon |f'(x_0)| \cdots |f'(x_{n-1})| \tag{4.7}$$

where $x_1, \ldots, x_{n-1}$ are the points visited by the "unperturbed" trajectory emanating from x_0. It is instructive to write this expression as

$$E_n(\varepsilon, x_0) = \varepsilon e^{n\sigma(n,x_0)} \tag{4.8a}$$

Substituting in Eq. (4.7) one is led to the identification

$$\sigma(n, x_0) = \frac{1}{n} \sum_{i=0}^{n-1} \ln|f'(x_i)| \tag{4.8b}$$

As mentioned earlier, for any finite ε, however small, the error is bound to leave the tangent space after a lapse of time of the order of $n^* \approx \ln 1/\varepsilon$. But in the idealized case, where the limit $\varepsilon \to 0$ is taken, n^* is pushed to infinity. The parameter $\sigma(n, x_0)$ is then reduced to the Lyapunov exponent, Eq. (4.4), considered in traditional chaos theory,

$$\lim_{n \to \infty} \sigma(n, x_0) = \langle \sigma(n, x_0) \rangle = \sigma \tag{4.9}$$

where in writing the second equality the assumption of ergodicity has been made.

We now turn to the general case of finite ε and n. From Eqs. (4.6)–(4.8) we have

$$\begin{aligned} \langle E_n \rangle &= \varepsilon \int dx_0 \rho_s(x_0) |f'(x_0)| \cdots |f'(x_{n-1})| \\ &= \varepsilon \int dx_0 \rho_s(x_0) e^{n\sigma(n,x_0)} \end{aligned} \tag{4.10}$$

This equation shows that in its most general version error growth amounts to studying the average over the attractor of an exponential function $\langle \exp n\sigma(n, x_0) \rangle$, where $\sigma(n, x_0)$ is given by Eq. (4.8b). To recover the picture of a Lyapunov exponent-driven exponential amplification of the error as usually stipulated in chaos theory one needs to identify Eq. (4.10) with the exponential of the long-time average of $\sigma(n, x_0)$, Eq. (4.9)

$$\varepsilon e^{n \langle \sigma(n,x_0) \rangle} = \varepsilon e^{n\sigma} \tag{4.11}$$

Now in a typical (multifractal) attractor this is not legitimate, since $\sigma(n, x_0)$ is x_0 dependent. This stems from the *fluctuations* of the local Lyapunov exponent

$$\sigma(x_0) = \ln |f'(x_0)| \tag{4.12}$$

It follows that for short-to-intermediate times significant deviations from a Lyapunov exponent-driven exponential error amplification are to be expected. We stress that this in no way implies a discrepancy with traditional theory: Both the theory and the concept of Lyapunov exponent are perfectly valid, but are unable to account for error dynamics in the relevant regime of small finite initial ε and finite time n.

In addition to the variability of local Lyapunov exponents, a second mechanism behind the superexponential behavior of error growth has been proposed in the literature. It has to do with the *nonorthogonality of the eigenvectors* of the linearized evolution operator of the system [29, 33], as a result of which certain linear combinations of the eigenvectors can grow faster than perturbations along a particular eigendirection.

The transient phase is dependent on the orientation of the perturbations in phase space and it is possible to identify those initial perturbations that will have the largest amplification over a given integration period.

The first nontrivial instance in which the configuration of the eigenvectors of the linearized operator in phase space becomes relevant is a two-variable system of constant coefficients.

$$\frac{d}{dt}\binom{x}{y} = \begin{pmatrix} a_{11} & a_{12} \\ a_{21} & a_{22} \end{pmatrix}\binom{x}{y} = \mathbf{A}\binom{x}{y} \tag{4.13}$$

Its general solution is of the form

$$\begin{pmatrix} x(t) \\ y(t) \end{pmatrix} = \mathbf{R}\begin{pmatrix} x_0 \\ y_0 \end{pmatrix} \tag{4.14a}$$

where the resolvent matrix is given by

$$\mathbf{R} = \begin{pmatrix} \frac{1}{D}(u_1 v_2 e^{\omega_1 t} - u_2 v_1 e^{\omega_2 t}) & \frac{u_1 u_2}{D}(e^{\omega_2 t} - e^{\omega_1 t}) \\ \frac{1}{D} v_1 v_2 (e^{\omega_1 t} - e^{\omega_2 t}) & \frac{1}{D}(u_1 v_2 e^{\omega_2 t} - u_2 v_1 e^{\omega_1 t}) \end{pmatrix} \tag{4.14b}$$

Here $\binom{u_1}{v_1}$ and $\binom{u_2}{v_2}$ are the eigenvectors of $\mathbf{A}$ corresponding to the eigenvalues ω_1, ω_2, and

$$D = u_1 v_2 - u_2 v_1 \tag{4.14c}$$

Applying Eq. (4.14b) to the initial conditions $\binom{x_0+\varepsilon_1}{y_0+\varepsilon_2}$ and $\binom{x_0}{y_0}$ and subtracting we obtain the error dynamics in the form

$$\begin{pmatrix} E_1(t) \\ E_2(t) \end{pmatrix} = R\begin{pmatrix} \varepsilon_1 \\ \varepsilon_2 \end{pmatrix} \tag{4.15}$$

or, using the Euclidean norm,

$$E_t^2 = (R_{11}^2 + R_{21}^2)\varepsilon_1^2 + (R_{12}^2 + R_{22}^2)\varepsilon_2^2 + 2(R_{11}R_{12} + R_{21}R_{22})\varepsilon_1\varepsilon_2 \tag{4.16}$$

where R_{ij} are the matrix elements of $\mathbf{R}$. Averaging next over the distribution of ε_1 and ε_2, assumed to be statistically independent random variables of mean zero and variances Δ^2, we arrive after the use of some algebra at

$$\langle E_t^2 \rangle = \frac{\Delta^2}{D^2} \{ (K^2 + D^2)(e^{2\omega_1 t} + e^{2\omega_2 t}) - 2K^2 e^{(\omega_1 + \omega_2)t} \} \quad (4.17a)$$

where D is defined in Eq. (4.14c) and K is the scalar product of the eigenvectors

$$K = u_1 u_2 + v_1 v_2 \quad (4.17b)$$

To assess the possibility of superexponential behavior we need to compare Eq. (4.17a) with $\langle E_0^2 \rangle e^{2\omega_1 t}$. The mean initial error follows directly from Eq. (4.17a), $\langle E_0^2 \rangle = 2\,\Delta^2$. We therefore inquire on the validity of the inequality

$$\frac{\Delta^2}{D^2} \{ (K^2 + D^2)(e^{2\omega_1 t} + e^{2\omega_2 t}) - 2K^2 e^{(\omega_1 + \omega_2)t} \} \rangle 2\,\Delta^2 e^{2\omega_1 t}$$

A detailed study leads to the following conclusions:

- When the eigenvalues ω_1, ω_2 are real nonorthogonality of the eigenvectors [$K \neq 0$ in Eq. (4.17b)] does not entail universal behavior of the error. Superexponential behavior is possible provided $(K/D)^2 > 1$. However, it is not manifested right from the beginning but only beyond a time t_c given by

$$t_c = \frac{1}{|\omega_2 - \omega_1|} \ln \frac{(K/D)^2 + 1}{(K/D)^2 - 1} \quad (4.18)$$

- When the eigenvalues ω_1, ω_2 are complex conjugate, the behavior is universally superexponential.

Now we illustrate the above ideas on the low-order atmospheric model of Eq. (2.10), and the parameters of Fig. 7.8. The mean Lyapunov exponents, computed using the available algorithms, are

$$\langle \sigma_1 \rangle = \sigma_{\max} = 0.56 \qquad \langle \sigma_2 \rangle \sim 0, \langle \sigma_3 \rangle = -1.41 \quad (4.19)$$

One can also evaluate the variability of the local Lyapunov exponents along the corresponding eigendirections, by monitoring the instantaneous values of the error at each integration step [34]. In this way one finds the following values of the variances:

$$\langle \delta\sigma_1^2 \rangle = 7.3, \langle \delta\sigma_2^2 \rangle = 6.8, \langle \delta\sigma_3^2 \rangle = 3.2 \quad (4.20)$$

showing the high inhomogeneity of the attractor. This is further confirmed by Fig. 7.11 depicting the histogram of the local exponents.

We now turn to error growth, following the general setting of this section. To capture the fine structure of short-time error evolution we register the values of every 0.01 time unit obtained from the integration of the model equations, Eq. (2.10), and compare it with a perturbed run in which small errors uniformly distributed around zero have been initially introduced in each of the equations. The full line of Fig. 7.12a summarizes the results of the simulation averaged over 100,000 realizations, for the arithmetic mean taken for convenience to be the quadratic mean $\langle E_t^2 \rangle$ rather than $\langle E_t \rangle$, whereas in the dotted line the effective amplification rate

$$\sigma_{\text{eff}} = \frac{1}{2t} \ln \left\langle \frac{E_t^2}{\varepsilon^2} \right\rangle \tag{4.21}$$

is plotted against time. [Notice that σ_{eff} in Eq. (4.21) is independent of time if $\langle E_t^2 \rangle$ is purely exponential.] For short times we observe a clearly superexponential behavior since σ_{eff} attains a maximum of about four times $\langle \sigma_1 \rangle$ at $t \sim 0.4$ units. This behavior is insensitive to the choice of the magnitude of the initial errors provided that the latter are sufficiently small (10^{-3} and smaller). For larger times the rate decreases steadily, but even for very small initial errors (at least up to 10^{-8}) it turns out that there is no time interval in which σ_{eff} attains a value close to $\langle \sigma_1 \rangle$ since in the mean time ($t \sim 12$ time units for $\langle \varepsilon \rangle \sim 10^{-5}$, Fig. 7.12a) the error has already increased sufficiently and approaches its saturation level. We also notice from Fig. 7.12b, which magnifies the short time behavior of error, the existence of slight oscillations.

Figure 7.13 summarizes the results of the time behavior of the logarithmic mean $\langle \ln E_t^2 \rangle$ (full line) together with its effective amplification rate (dotted line). Again the behavior is clearly superexponential although its deviation from a purely exponential function is less pronounced than in the case of the arithmetic mean of Fig. 7.12a.

We come now to the origin of the superexponential behavior in Figs. 7.12 and 7.13. Numerical evaluation of the instantaneous eigenvectors of the linearized matrix shows that they are, typically, nonorthogonal. Furthermore two of them correspond to a pair of complex conjugate eigenvalues. On the other hand the probability density of the local Lyapunov exponents, shown in Fig. 7.11, reveals pronounced variance, skewness, and kurtosis. On the grounds of the above observations we expect that in the present model both sources of superexponential

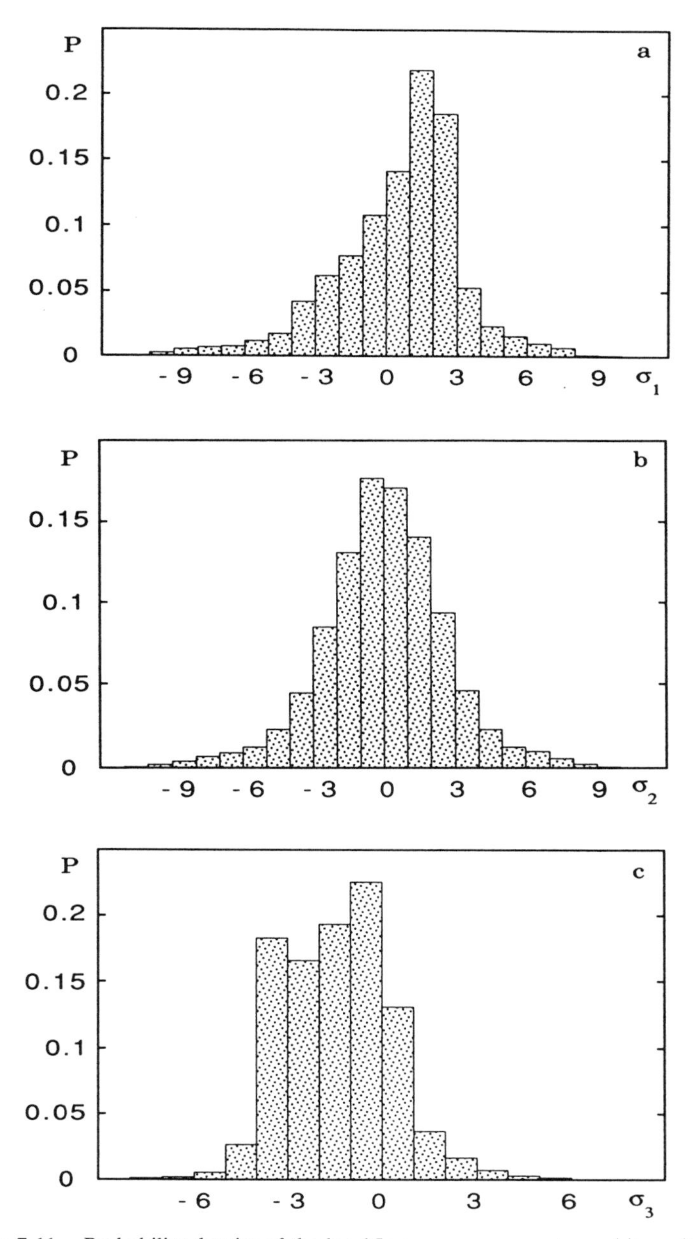

Figure 7.11. Probability density of the local Lyapunov exponents σ_1, (a); σ_2 (b), and σ_3 (c) for system (2.10) as obtained from 10,000 values sampled every 0.05 time units and the parameter values of Fig. 7.8.

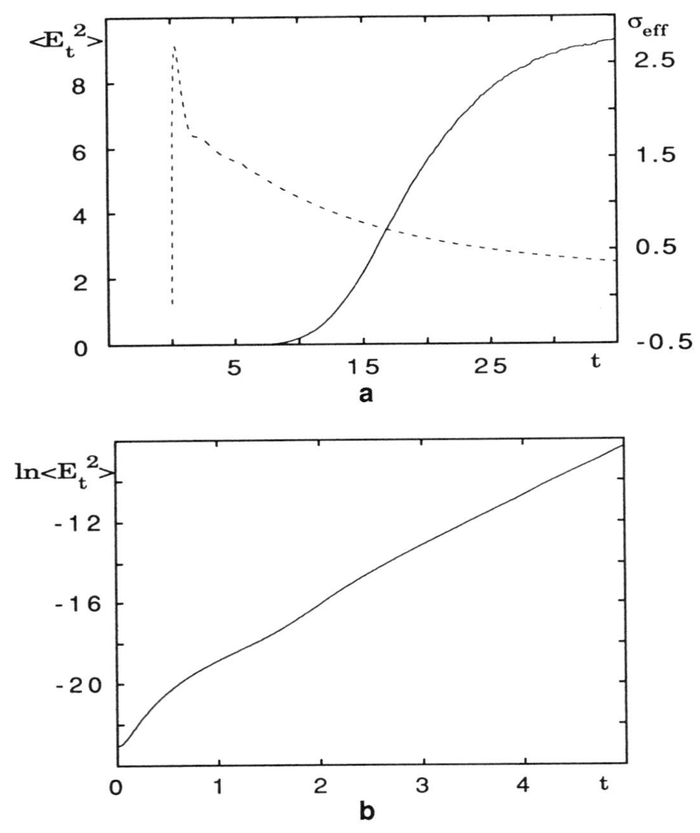

Figure 7.12. (a) Time dependence of the arithmetic mean of error, $\langle E_t^2 \rangle$, (full line) and $\sigma_{\rm eff}$ (dotted line) for model (2.10) averaged over 100,000 realizations: (b) short-time evolution of $\ln \langle E_t^2 \rangle$ for an initial error $\varepsilon \sim 10^{-5}$. Parameter values as in Fig. 7.8.

behavior—nonorthogonality of the eigenvectors and variability of Lyapunov exponents—are present. This suggests that the problem of prediction of a chaotic dynamical system in general, and of the atmosphere in particular, is more involved than usually thought: given a mean initial error, the knowledge of the leading Lyapunov exponent is in general not sufficient for estimating the validity of prediction n time steps ahead. In actual fact the skill of the model will depend on the entire hierarchy of moments of the probability distribution of the Lyapunov exponents as well as the orientation of the associated Lyapunov vectors in phase space.

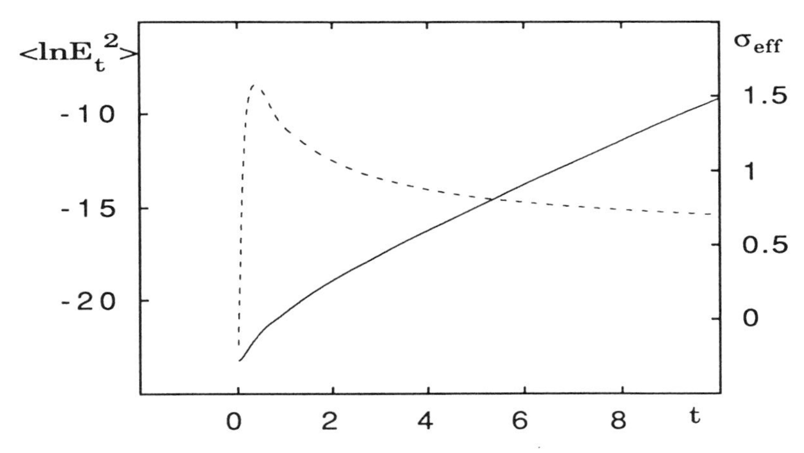

Figure 7.13. As in Fig. 7.12a but for the geometric mean of error, $\langle \ln E_t^2 \rangle$.

C. Spatial Scales and Predictability

The atmosphere is a spatially extended system giving rise to complex behavior in both space and time. At one extreme of the spectrum of complexity lies fully developed turbulence, which is known to prevail in certain atmospheric flows, notably in the boundary layer. We do not deal with this phenomenon in the present subsection. Rather, we address the problem of error growth and predictability in the presence of *spatio-temporal chaos* arising from the coexistence of several space scales in strong interaction, a property that no longer allows truncations retaining only the first few modes as in Eqs. (2.6), (2.8), or (2.10). More specifically we review the predictability properties of a simple prototype model of thermal convection which, as mentioned in Section I, is one of the processes participating in mesoscale ($\leq 10^3$ km) atmospheric variability.

The model is a contracted version of the Boussinesq equations referred to in Section II.B, which is valid for a two-dimensional roll pattern close to the instability threshold. It capitalizes on a result of bifurcation theory, according to which the dynamics near such a criticality reduces to a *normal form* equation displaying a key variable—the *order parameter* [24, 25]. All the system observables are related to this order parameter by equations similar to the "equation of state" familiar from physical chemistry, which are usually referred to in meteorology as "diagnostic relations." Without going into details we now write the final result leading to the normal form equation, which is called the *extended*

Kuramoto–Shivashinski equation [35] in the literature of nonlinear physics,

$$\frac{\partial \psi}{\partial t} + \eta\psi + \frac{\partial^2 \psi}{\partial x^2} + \frac{\partial^4 \psi}{\partial x^4} + 2\psi \frac{\partial \psi}{\partial x} = 0 \tag{4.22a}$$

supplemented with rigid boundary conditions

$$\psi(0) = \psi(L) = 0 \qquad \frac{\partial \psi(0)}{\partial x} = \frac{\partial \psi(L)}{\partial x} = 0 \tag{4.22b}$$

L being the system size. Figure 7.14 illustrates the spatiotemporal behavior of the variable ψ for $L = 100$ and $\eta = 0.01$. Its erratic character confirms that Eq. (4.22) admits chaotic solutions. This is further illustrated by the computation of the Lyapunov exponents—or more appro-

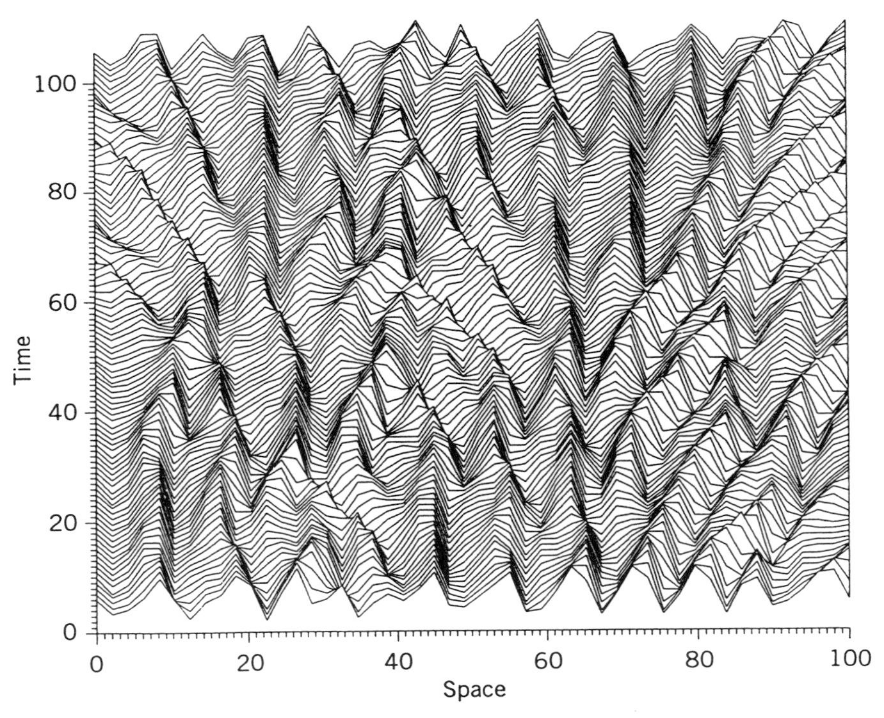

Figure 7.14. Spatiotemporal behavior of variable ψ of model (4.22) with $L = 100$ and $\eta = 0.01$.

priately of the *Lyapunov spectrum*, since there are now as many Lyapunov exponents as grid points in the integration of Eq. (4.22) by a finite difference method. The first 70 such exponents, 11 of which are positive, are depicted in Fig. 7.15. The largest is equal to $\sigma_{max} \approx 0.086$ time units^{-1}, the most negative one being about 1.8 time units^{-1}. This implies that stable, strongly dissipative modes evolve on a much faster scale than the leading unstable ones. It suggests that the dynamics could be reduced to a low-dimensional attractor, whose dimension can be estimated to be about 20.2 [36].

One of the basic issues in predictability studies pertains to the role of spatial scales in the growth of small errors. To address this question we selectively introduce an initial error in a particular spatial scale

$$\varepsilon_i = \xi \sin\left(2\pi k \frac{i-1}{N-1}\right) \qquad i = 1, \ldots, N \tag{4.23}$$

in which ξ is a uniform low-amplitude noise of zero mean, k is the wave number associated with the perturbation, and i labels the grid points.

In Fig. 7.16 the mean error evolution is represented for perturbations in wave numbers 4, 12, 20, 30, and 40 averaged over 2000 realizations for an initial error amplitude of 6×10^{-3}. We see that in the short time regime the repercussions of an error committed in the small scales on

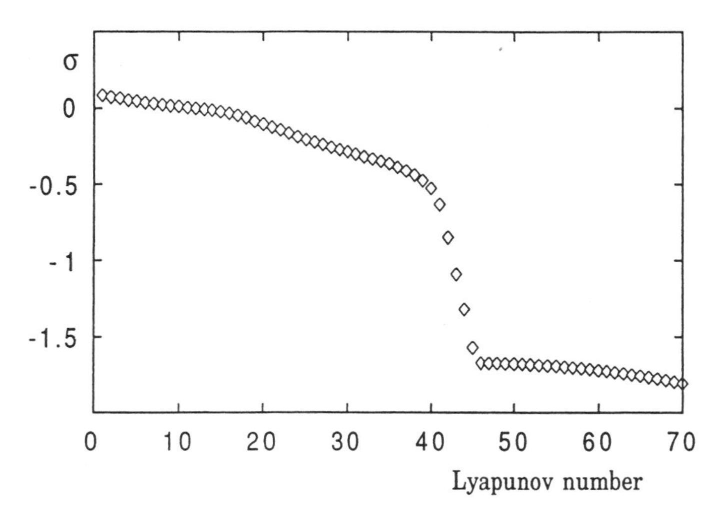

Figure 7.15. Lyapunov spectrum associated with Eq. (4.22) obtained after 10,000 time units. Parameter values as in Fig. 7.14.

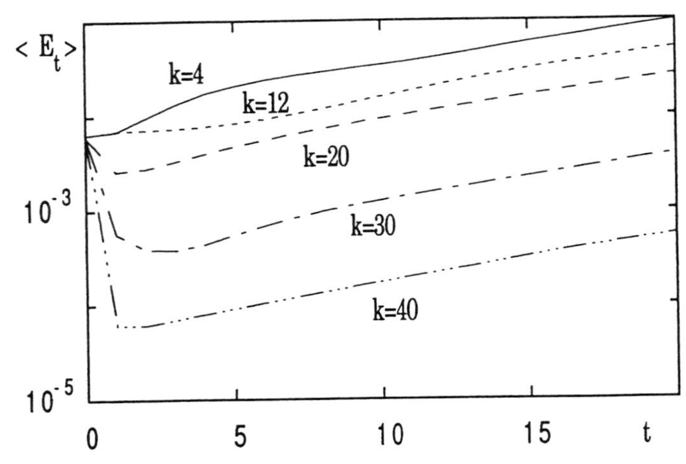

Figure 7.16. Short and intermediate time dependence of mean error $\langle E_t \rangle$ associated with model (4.22) averaged over 2000 realizations using a mean initial error $\langle \varepsilon \rangle = 6.10^{-3}$ introduced selectively in space.

predictability will be much less than if the error were introduced in the large scales.

To understand this unexpected behavior, which to some extent opposes the intuition that large-scale phenomena are more predictable than short-scale ones, it is necessary to realize that a spatially distributed system lives in a multidimensional phase space. An initial error spreads over a large number of linearly independent directions in this space, which we refer to as Lyapunov vectors. If the attractor is nonuniform, as is typically the case, these vectors are not fixed but rotate as the system runs over the attractor. To capture this variability in the Fourier space, we evaluate the spatial power spectrum of the local Lyapunov vectors. Figures 7.17a and b depict the results for five positive and six negative Lyapunov exponents [36]. We see that the spectra of the positive Lyapunov vectors are essentially confined to large scales ($k < 16$). In contrast, the negative vectors possess appreciable power at small scales: Those corresponding to the Lyapunov exponents 30 and 40 have a maximum at intermediate wavenumbers ($k \sim 20$) and those associated with 50, 60, and 70 are dominated by the small scales. These results suggest that an initial error concentrated on small scales will first be subjected, in the mean, to the influence of negative Lyapunov vectors: Such an error is thus bound to decrease until the positive Lyapunov vectors take over. In contrast, an initial error concentrated on large scales will be dominated, from the outset, by the positive Lyapunov vectors.

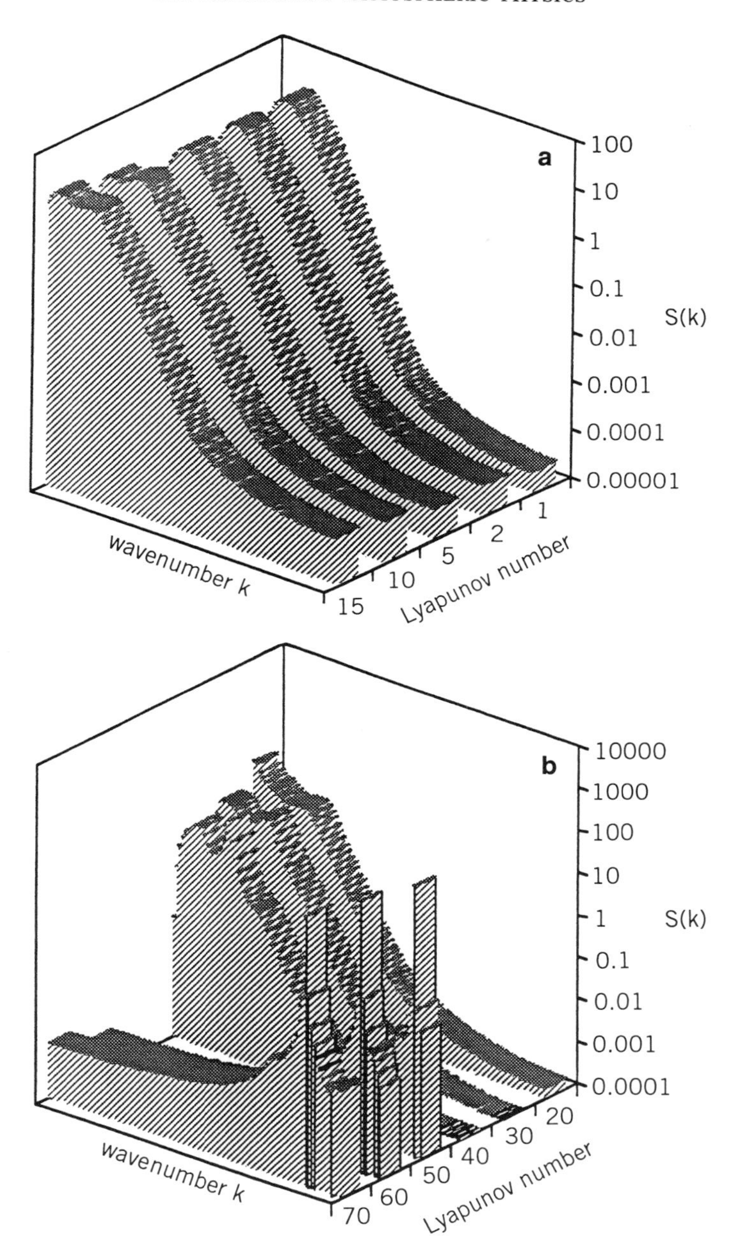

Figure 7.17. Spatial power spectrum averaged over 1000 time units of the local Lyuapunov vectors corresponding to Lyapunov numbers (a) 1, 2, 5, 10, and 12, and (b) 20, 30, 40, 50, 60, and 70.

D. Statistical Approach to Predictability

The analysis of the preceding sections has established that beyond a time of the order of the "Lyapunov time" $1/\sigma_{max} \ln 1/\varepsilon$ small initial errors ε reach a macroscopic value. Thereafter they behave in an essentially statistical manner, with a probability density induced entirely by the invariant density ρ_s on the attractor. Clearly, the idea of forecasting on the basis of the phase space trajectory loses its operational significance and one must resort to a new approach incorporating probabilistic elements.

Meteorologists were already aware of the practical necessity of such an approach well before the advent of chaos theory, notably in connection with the forecasting of events that could be classified in a limited number of possibilities: cyclonic or anticyclonic conditions, fraction of sunny, overcast, or rainy days over a certain period, and so on. More recently there has been an increasing interest in "ensemble forecasting," whereby statistical properties are deduced on the basis of several individual phase trajectories generated by a detailed numerical model and launched in parallel [37]. These interesting approaches are nevertheless confronted with a number of practical as well as conceptual difficulties: How do we achieve an honest statistics given the time limitations in running each of the individual trajectories? What is the nature of the process mediating the passage between the various discrete states of the set? How to use the ensemble method to predict the history of a particular reference trajectory? It is here that two recent developments of nonlinear dynamics—the statistical theory of chaos and nonlinear prediction—provide some valuable methods and ideas for meeting the challenge.

The statistical approach to chaos is based on the equation of evolution of the probability density $\rho(x, t)$ in the course of time. To simplify notation let us consider a dynamical system in the form of a one-dimensional recurrence [Eq. (4.5)]. The evolution of the density $\rho_n(x)$ is then described by the *Frobenius–Perron equation* [24, 38].

$$\rho_{n+1}(x) = \int dx_0 \delta[x - f(x_0)]\rho_n(x_0) = U\rho_n \tag{4.24}$$

Since we are interested in forecasting states that can be classified in phase space "clusters" we inquire whether the "fine grained" description based on Eq. (4.24) can be mapped, in a systematic way, to a "coarse-grained" one. One way to achieve this is to partition the phase space into a finite number of nonoverlapping cells $\{C_i\}$, $i = 1, \ldots, N$ representing the "measurable" states of our system (Fig. 7.18) and monitor the successive cell-to-cell transitions of the trajectory. To have a well-posed problem we

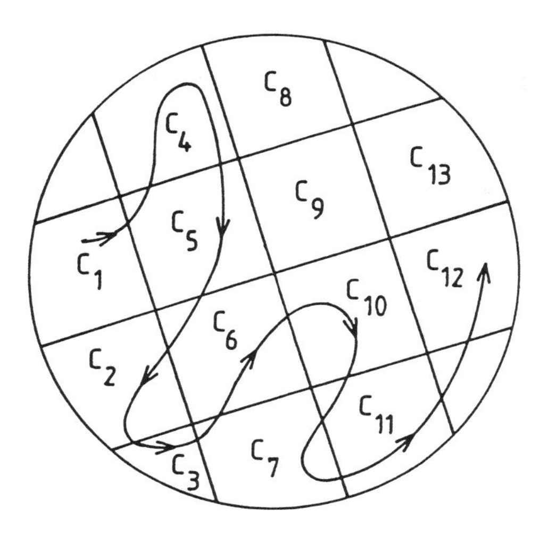

$C_1 \rightarrow C_5 \rightarrow C_4 \rightarrow C_5$
$\rightarrow C_6 \rightarrow C_2 \rightarrow C_3 \rightarrow C_7 \rightarrow \cdots$

Figure 7.18. Time evolution of a phase space trajectory viewed as a sequence of transitions between the cells of a "coarse-graining" partition.

need to make sure that the action of the dynamics does not alter the partition. This will be achieved if each element C_i is mapped by the transformation f on a union of elements (the partition is then referred to as the *Markov partition*).

To identify the nature of the process mediating the transitions between the cells we need to project Eq. (4.24) onto the partition. In general, the projector $\mathbf{E}$ will not commute with the evolution operator U: The state vector of the coarse-grained system $\mathbf{P}_n = (P_n(1) \cdots P_n(N))$

$$\mathbf{P}_n = \mathbf{E}\rho_n \tag{4.25}$$

will therefore not satisfy a closed equation. But when certain additional conditions depending on both the dynamics and the partition are met, closure can be achieved. Equation (4.24) then reduces to a *master equation* [39]

$$\mathbf{P}_{n+1} = \mathbf{W}\mathbf{P}_n \tag{4.26}$$

which is descriptive of a stochastic process. There is no contradiction whatsoever between this result and the deterministic origin of chaos: In the probabilistic view we look at our system through a "window" (phase space cell), whereas in the deterministic view it is understood that we are exactly running on a trajectory. This is, clearly, an unrealistic assumption in view of the finite precision of the monitoring process, of the roundoff errors, and (most importantly) of the sensitivity to initial conditions.

The simplest type of stochastic process satisfying Eq. (4.26) are Markovian processes, for which memory extends only one step backward in time. The converse is not true: a non-Markovian process can satisfy the master equation. On the other hand if the conditions ensuring the reduction of Eqs. (4.24)–(4.26) are not met the process will be manifestly non-Markovian, displaying long-term memory, and the problem of forecasting will be considerably complicated. Now, in the presence of discrete states, meteorologists are usually tempted to postulate transitions governed by Markov chains: A critical assessment of some ideas taken for granted is necessary here.

A master equation of the form of Eq. (4.26) is characterized by an irreversible approach to a unique invariant distribution $P_s(i)$. This property, which is reminiscent of the H theorem of statistical mechanics [40], shows that at the level of the probabilistic description the unpredictability of deterministic chaos is replaced by full predictability, in the sense that the statistical state of the system evolves in a regular manner: Whereas the deterministic description is nonlinear and unstable the probabilistic one is linear and stable. This suggests the interesting possibility of *statistical forecasting* of complex dynamical systems like the atmosphere.

As an example of the above described method we consider Lorenz's low-order model of thermal convection, Eq. (2.6). By plotting the successive maxima of the variable z versus their preceding values, it can be shown that the dynamics is mapped into the one-dimensional cusplike map depicted in Fig. 7.19, which can be represented analytically in a very satisfactory manner (in normalized variables) by [41]

$$f(z) = 1 - \xi |z|^{\theta} , \qquad -1 \leq z \leq 1 , \qquad 0 < \theta < 1 \tag{4.27}$$

The simplest Markov partition for this dynamical system is the two-cell partition provided by the end points of the interval and the fixed point $\bar{z}$. By using the explicit form of the cusp map, we can determine $\bar{z}$ from the relation

$$1 - \xi \bar{z}^{\theta} - z = 0$$

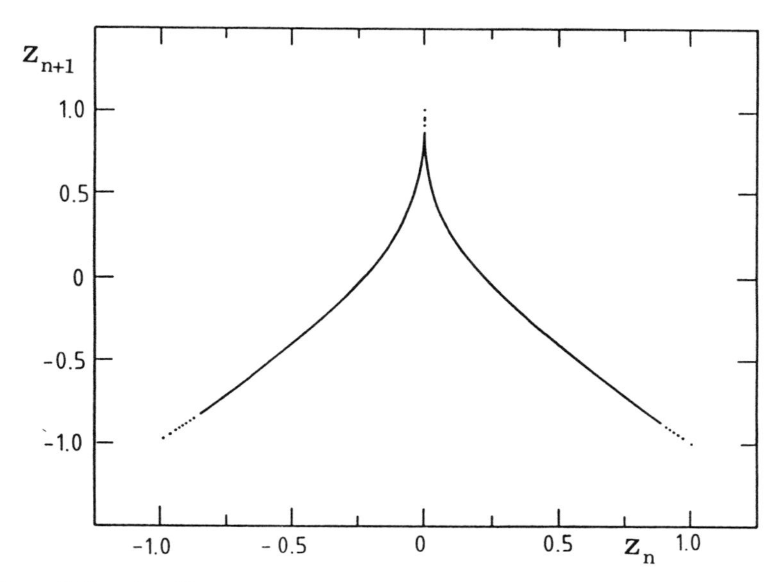

Figure 7.19. Cusplike return map constructed from the maxima of the variable z in Eqs. (2.6), in normalized units.

The solution of this equation belonging to the interval $(-1, 1)$ for $\xi = 2$, $\theta = 0.5$ is $\bar{z} \approx 0.17$. The question now arises as to whether the two-cell partition $\{C_i: -1 \leq z \leq 0.17, C_2: 0.17 \leq z \leq 1\}$ satisfies the Markovian conditions. The validity of these conditions cannot be proven analytically in this case. However, using a generalized χ^2 test one can numerically check that it is likely that they are indeed satisfied. This guarantees that the probability vector $\mathbf{P} = [P(1), P(2)]$ evolves according to Eq. (4.26). The explicit form of the transition probability matrix and of the invariant probability are found to be [42]

$$\mathbf{W} = \begin{pmatrix} 0.6 & 0.4 \\ 1 & 0 \end{pmatrix} \tag{4.28a}$$

$$P_\infty(1) = 0.7 \qquad P_\infty(2) = 0.3 \tag{4.28b}$$

Using this information and Eq. (4.26), one may now predict the probability that starting with a certain (probabilistic) initial condition, the maximum of the variable z will exceed or will be below the threshold value $\bar{z}$ at any particular moment. Similar predictions can be made for appropriate combinations of variables in other low-order atmospheric models.

We close this section with a short account of nonlinear prediction theory [43]. Let $X_1, \ldots, X_N$ be the values of a variable X at times $t_i, \ldots, t_N$ with $\Delta t = t_k - t_{k-1}$ fixed. We want to perform the "best" prediction of the value X_{N+1} on the basis of the knowledge of the time series $\{X_i, i = 1, \ldots, N\}$. To this end we unfold the series in phase space following the method of dynamical reconstruction outlined in Section II.C. In this space the state to be predicted is represented by a point surrounded by a certain number of close neighbors, representing earlier states: Their images after one or several time intervals Δt are therefore known. Introducing adequate (though at present to a great extent heuristic) time-independent weighting factors taking into account the distances between the predictee and its neighbors, one can in this way reconstruct the presumed position of this point after Δt. Numerical experiments conducted on both natural time series and time series generated by mathematical models show that in the presence of deterministic chaos the mean prediction error grows exponentially, whereas for random noise the law of growth is very different. It would certainly be interesting to apply this algorithm to meteorological time series of relevance in the problem of forecasting.

V. FROM SHORT-SCALE ATMOSPHERIC VARIABILITY TO GLOBAL CLIMATE DYNAMICS

Climate is the set of long-term statistical properties of the atmosphere. Everyday experience shows that the observables associated to this average description display new properties as compared to those of the fine-scaled variables. It is therefore legitimate to inquire whether the evolution laws of the climatic variables can be deduced from those of the meteorological fields, in much the same way that the hydrodynamic fields reflecting long-term statistical properties of molecular dynamics of large assemblies of interacting particles obey the well-known set of equations of fluid dynamics. This section aims to outline an approach for inferring these properties, which accounts, in addition, for dynamical aspects such as time correlations, power spectra, and predictability.

Let $\{X_i\}$ be a set of atmospheric short-scale variables evolving according to Eqs. (1.1) or (2.1)–(2.4). We define the averaged variables associated with climate as *running means* over a time interval ε [44] whose merit as compared to ordinary (infinite time) averaging is to keep track of the evolution that the averaged variables $\{\bar{X}_i\}$ might still undergo in time

$$\bar{X}_i(t) = \frac{1}{\varepsilon} \int_t^{t+\varepsilon} d\tau X_i(\tau) \tag{5.1}$$

A typical value of ε in climate theory is $\varepsilon \sim 10$ years. For simplicity we consider asymmetric averaging, to avoid the appearance of spurious negative time values in the integrand.

We notice that the averaging operation commutes with the time derivative. Indeed, by definition

$$\overline{\frac{dX_i}{dt}} = \frac{1}{\varepsilon}\int_t^{t+\varepsilon} d\tau \frac{dX_i}{d\tau} = \frac{1}{\varepsilon}[X_i(t+\varepsilon) - X_i(t)] \tag{5.2a}$$

On the other hand, by virtue of Eq. (5.1)

$$\frac{d\bar{X}_i}{dt} = \frac{1}{\varepsilon}\frac{d}{dt}\int_t^{t+\varepsilon} d\tau X_i(\tau) = \frac{1}{\varepsilon}[X_i(t+\varepsilon) - X_i(t)] \tag{5.2b}$$

which is identical to $\overline{dX/dt}$, Eq. (5.2a).

We now apply the averaging operator on both sides of Eq. (1.1)

$$\frac{d\bar{\mathbf{X}}}{dt} = \overline{-\mathbf{v}\cdot\nabla\mathbf{X}} + \overline{\mathbf{F}(\mathbf{X}\cdot\lambda)} \tag{5.3}$$

In view of Eqs. (5.2a and b) in the limit in which the right-hand side of Eq. (5.3) can be linearized for all practical purposes around a well-defined, time independent state, the properties of the averaged observables would follow straightforwardly from those of the initial variables. The situation is far more complex in the presence of nonlinearities, since now $\overline{\mathbf{v}\cdot\nabla\mathbf{X}} \neq \bar{\mathbf{v}}\cdot\nabla\bar{\mathbf{X}}$ and $\overline{\mathbf{F}(\mathbf{X})} \neq \mathbf{F}(\bar{\mathbf{X}})$, in other words to determine the evolution of $\bar{\mathbf{X}}$ one needs to know not only $\bar{\mathbf{X}}$ but also the entire hierarchy of moments $\overline{\delta\mathbf{X}^n} = \overline{(\mathbf{X}-\bar{\mathbf{X}})^n}$, $n > 1$. It follows that, in principle, there should be a strong feedback between local, short-scale properties, and global averages, unless the moments $\overline{\delta\mathbf{X}^n}$ could be neglected for $n \geq 2$. This is definitely not the case for a typical atmospheric variable where one witnesses large deviations around the mean, comparable to the value of the mean itself. More generally, this situation is rather typical for a nonlinear dynamical system as soon as the evolution equations admit multiple solutions or give rise to chaotic dynamics, the occurrence of which in atmospheric physics is by now well established, as we saw in the preceding sections. The lack of commutation of the averaging operator with the evolution law therefore constitutes a fundamental issue at the basis of the very relation between short-scale "weather" variability and global climate.

The traditional way to cope with this difficulty is to perform a truncation of the hierarchy to the first few orders or, more generally, to appeal to a *closure* relation linking higher order averages to lower order ones. Such closures essentially amount to introducing phenomenological, "diagnostic" type of relations between averaged momentum, energy, and fluxes and averaged observables, such as temperature and/or their spatial derivatives. Usually, to justify them one invokes the idea that the initial and averaged observables evolve on scales that are widely separated in both time and space. Such assumptions are difficult to justify from first principles, even in very simple cases. Nevertheless, in view of the success of this approach in capturing a good part of the relevant physics we shall first review some representative examples of low-order climate models obtained in this heuristic manner. Subsequently, we shall attempt a deeper look at the connection between atmospheric and climatic variability and sketch the rudiments of a systematic theory of averaging.

A. Low-Order Climate Models

We first survey some basic facts showing that, despite their statistical character, climate variables still display a very pronounced variability over a wide range of time scales. Most of the observations available come from the analysis of the isotopic composition of organic relics, giving direct information about paleotemperatures or global ice volume on the planet earth [45].

One of the first points that becomes obvious from such data is that the climatic conditions that prevailed in much of the last 200 or 300 million years were extremely different from those of the present day. During this period, with the exception of the Quaternary era (our era, which began about 2 million years ago) there was practically no ice on the continents and the sea level was about 80 m higher than at present. Climate was particularly mild, and the temperature differences between equatorial and polar regions were relatively weak.

During the Tertiary era, some 40 million years ago, a sharper contrast between equatorial and polar temperatures began to develop. At the beginning of the Quaternary era this difference was sufficiently important to allow for the formation and maintenance of continental ice. In the Northern hemisphere a series of *glaciations* took place in an intermittent fashion, sometimes pushing the glaciers as far as the middle latitudes. These climatic episodes present a preferred periodicity of about 100,000 years, though with considerable random looking variations, as shown in Fig. 7.20a, reflected by a well-defined pick in the associated power spectrum (Fig. 7.20b). Curiously, the scale of 100,000 years also happens to be the periodicity of change of the eccentricity of the earth's orbit around the sun because of the perturbing action of the other bodies of the solar system. This affects the mean annual energy received by the earth by a slight 0.1% [46]. How this small effect may be amplified by internal

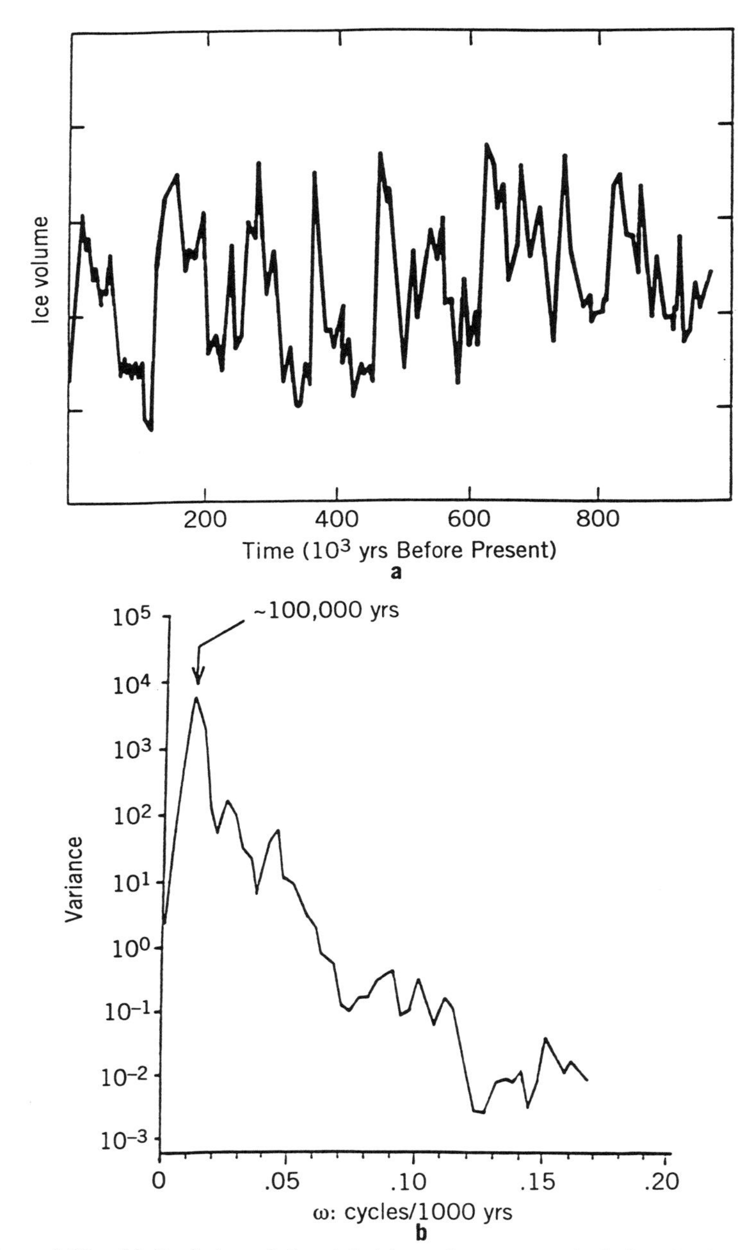

Figure 7.20. (a) Evolution of the global ice volume on earth during the last million years as obtained from the oxygen isotope record of deep sea core V28-238 [45]; (b) power spectrum associated to the record of (a).

dynamics to leave a clearcut signature in the glaciation record is a major paleoclimatology problem.

The last advance of continental ice in the Northern hemisphere attained its maximum some 18,000 years ago. There was at that time 70–80 million cubic kilometers of continental ice (compared to 30 million cubic kilometers of our times), and the sea level was some 120 m lower than today. Following this a rapid deglaciation took place, which essentially established (already 10,000 years ago) the same global situation as the present one. Climatic fluctuations on a shorter scale have then occurred such as the "climatic optimum" (some 6000 years ago), the "little ice age" (between AD 1550 and 1700), or the intermittent droughts in the Sahel region during the last centuries. Today, as is well known, there is increasing awareness of the possibility of human-induced climatic changes, notably in connection with the emission of greenhouse gases.

Climatic change essentially reflects the energy and mass balance of the planet. Energy balance [Eq. (2.4)] describes how the incoming solar radiation is processed by the system: What parts of that spectrum are absorbed and what are reemitted, how they are distributed over different latitudes and longitudes as well as over the continental and ocean surfaces. The most significant parts of mass balance [Eqs. (2.1) and (2.2)] as far as climate is concerned are the balance of continental and sea ice and of some key atmospheric constituents like water vapor and carbon dioxide.

Let us first argue in very global terms, considering the earth as a zero-dimensional object in space receiving solar radiation and emitting infrared radiation back to space. In such a view there is only one important state variable, the mean annual temperature T, which evolves in time according to the heat balance equation [47]

$$\frac{dT}{dt} = \frac{1}{C}\{Q[1 - a(T)] - \varepsilon_B \sigma T^4\} \tag{5.4}$$

Here C is the heat capacity of the system, Q is the solar constant, and a is the albedo, which expresses the part of the solar radiation emitted back to space. The temperature dependence of this quantity accounts for the surface-albedo feedback referred to briefly in Section II. The last term in Eq. (5.4) describes the emitted thermal radiation as a modification of the well-known Stefan–Boltzmann law of black-body radiation. The parameter σ is the Stefan constant and ε_B is an emissivity factor accounting for the fact that the earth does not quite radiate as a black body.

A study of Eq. (5.4), which is summarized in Fig. 7.21, shows that the

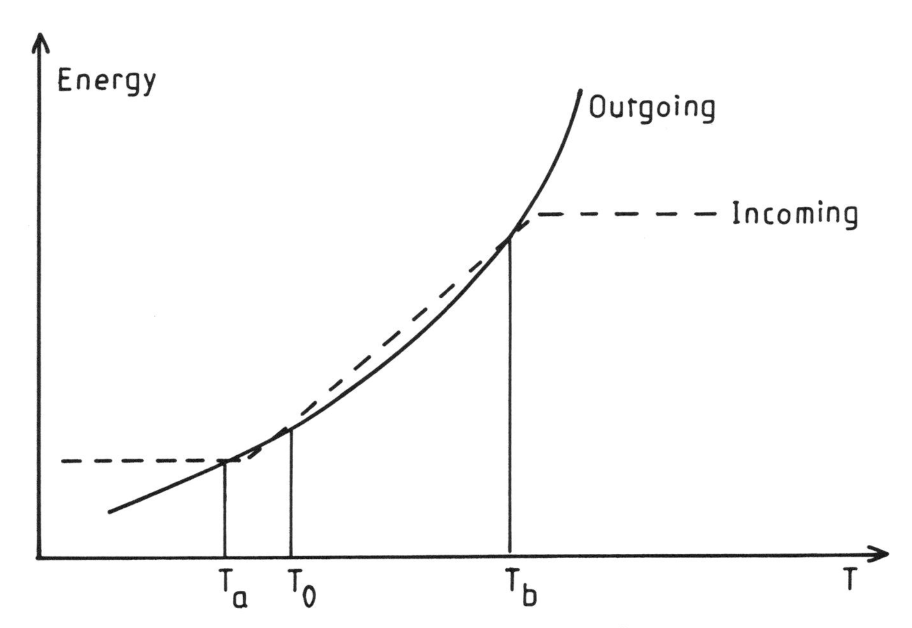

Figure 7.21. Incoming $[Q(1-a(T))]$ and outgoing $(\varepsilon_{\mathbf{B}}\sigma T^4)$ radiative energy curves as functions of T (global averaged surface temperature). For plausible parameter values their intersection gives rise to three steady states: T_a, T_0, and T_b.

system can admit up to three steady states. Two of them, T_b and T_a, are stable and correspond, respectively, to present-day climate and a cold climate reminiscent of a global glaciation. The third state, T_0, is unstable and separates the above two stable regimes.

The evolution of a system involving only one variable can always be represented as the evolution of a particle in a potential U which, in the present case, is defined by

$$U(T) = -\frac{1}{C}\int dT\{Q[1-a(T)] - \varepsilon_{\mathbf{B}}\sigma T^4\}$$

We call $U(T)$ the *kinetic potential*. Under the conditions of Fig. 7.21 this potential has two wells separated by a maximum.

Now the climatic system is continuously subjected to statistical *fluctuations*, the random deviations from deterministic behavior inherent in any physical system, which essentially account for the error committed in replacing Eq. (5.3) by an equation involving only averages. As we do not have sufficient information enabling us to write an evolution equation for the underlying probability distribution, we assimilate the effect of the fluctuations to a *random force*, $F(t)$. The energy balance equation, eq. (5.4), now becomes a stochastic differential equation of the form [48]

$$\frac{dT}{dt} = -\frac{\partial U(T)}{\partial T} + F(t) \tag{5.5}$$

The important new element introduced by this enlarged description is that different states become *connected* through the fluctuations. Stated differently, starting from some initial state the system will sooner or later reach any other state. This holds true as well if the stable states T_a and T_b are taken as initial states, whose deterministic stability is then transformed to some sort of *metastability*. One can show that the time scale of this phenomenon is determined by two factors: the *potential barrier*

$$\Delta U_{a,b} = U(T_0) - U(T_{a,b})$$

and the strength of the fluctuations as measured, typically, by the variance, q^2, of $F(t)$, assumed for simplicity to be a white noise process. The order of magnitude estimate is essentially given by the same arguments as those underlying Kramer's theory of chemical reactions in the liquid phase [49]: The mean transition time from state T_a or T_b via the unstable state T_0 is then

$$\tau_{a,b} \sim \exp\left(\frac{\Delta U_{a,b}}{q^2}\right) \tag{5.6}$$

It is clear that if fluctuations are reasonably small this time will be exceedingly large for any reasonable magnitude of the potential barrier. Typical values can be 10^4–10^5 years, that is to say, the range of the characteristic times of Quaternary glaciations. Still, we cannot claim that Eq. (5.6) provides an explanation of glaciation cycles, as the passage between the two climatic states T_a and T_b remains a random process, without a preferred periodicity. It is here that the influence of a weak periodic external forcing becomes crucial.

Let us repeat the above analysis by adding a small term of the form $Q\varepsilon \sin \omega t$, to the solar constant Q, whose period $\Delta = 2\pi/\omega$ represents the effect of variations affecting the mean annual amount of solar energy received by the earth, such as one of the orbital perturbations discussed earlier in Section V.A. As it turns out, if Δ is much smaller than the transition time $\tau_{a,b}$ the external forcing has practically no effect. But when Δ is comparable to $\tau_{a,b}$ the response of the system to the forcing is tremendously enhanced. Essentially, the presence of the forcing lowers the potential barrier and in this way facilitates the passage between climatic states. Moreover, because of the periodicity of the signal, the system is entrained and locked into a passage between two kinds of climate across the forcing-depressed barrier characterized by some average periodicity. We see here a qualitative mechanism of glaciations in which both internally generated processes and external factors play an important role. The simple model of *stochastic resonance* discussed above also allows us to identify the parameters that are likely to be the most influential in the system's behavior [50, 51].

Naturally, the schematic character of the model calls for the development of more sophisticated descriptions which, while still amenable to an analytical study, incorporate additional key variables in the description. A particularly interesting class of such models comprises those describing the interaction between average temperature and continental or sea ice [52]. For plausible values of the parameters they predict time-periodic solutions of the limit cycle type with periods comparable to the characteristic times of glaciation. However, by incorporating the effect of the fluctuations in the description, we find that any trace of coherence in the oscillating behavior will disappear after a sufficient lapse of time, owing to the weak stability properties of the phase of the oscillator [53]. The coupling with a periodic external forcing under certain conditions allows for the stabilization of the phase and thus, once again, for a regime of entrainment and phase locking between the climatic system and the external forcing. Furthermore, there exists a range of values of the amplitude and period of the forcing for which the response is chaotic [54], exhibiting a variability much like the one shown in Fig. 7.20.

Paleoclimatic time series similar to those of Fig. 7.20 have also been analyzed by the methods of dynamical reconstruction summarized in Section II.C. These investigations suggest the existence of low-dimensional chaotic attractors [55]. It is instructive to confront this conclusion with the results drawn from the models outlined briefly above. Two main points emerge. First, the instability of the motion on a chaotic attractor may well provide the natural mechanism for the amplification of the weak external signals arising from the earth's orbital variability. Second, as we saw in Section IV, when viewed through an appropriate window a chaotic attractor can generate a variety of stochastic processes. We therefore may regard the fluctuations introduced in Eq. (5.5) as a shorthand description of the variability of an underlying chaotic attractor.

B. Toward a Systematic Theory of Averaging

Now we come back to the question raised at the beginning of this section, namely, how to deduce the properties of averaged observables from those of the original, fine-scaled ones systematically [56]. We first comment on the range of the parameter ε over which averaging is carried out in Eq. (5.1). Let $\tau_1 = \tau_{\min}, \ldots, \tau_\kappa = \tau_{\max}$ be the spectrum of characteristic times appearing in the evolution equation (1.1). If $\varepsilon < \tau_{\min}$, $\bar{X}_i$ is essentially identified to X_i and averaging becomes trivial. In the opposite end of $\varepsilon > \tau_{\max}$, $\bar{X}_i$ becomes essentially the ergodic average of X_i in which any trace of the dynamics has been eliminated. Clearly, then, the relevant range over which ε is to be chosen must lie between $\tau_{\min}$ and $\tau_{\max}$ in such a way that fine scale properties are erased and attention is focused on evolution over a long time scale that is still shorter than $\tau_{\max}$. In the climatic context this would amount to considering averaging periods grossly corresponding to the distinct frequency bands of the variance spectrum of climatic change [44]. One might reasonably expect that the averaging process will effectively lead to an evolution of $\bar{X}_i$ over a characteristic scale longer than $\tau_{\min}$.

We shall now attempt to carry out the above program using Lorenz's low-order thermal convection model [Eqs. (2.6)] as a case study. As is well known for $\sigma = 10$, $r = 28$, and $b = 8/3$, Eqs. (2.6) give rise to the classical Lorenz attractor, the first clearcut demonstration of deterministic chaos in dissipative autonomous dynamical systems [5, 11]. Figure 7.22a–e summarizes the principal properties of the system in this regime: time series and phase space plot (Fig. 7.22a and b) invariant probability (Fig. 7.22c), time autocorrelation function (Fig. 7.22d), and power spectrum (Fig. 7.22e).

We turn next to the description in terms of the averaged observables. By applying the averaging operator, Eq. (5.1), on both sides of Eqs. (2.6) we obtain

$$\begin{aligned}
\frac{d\bar{x}}{dt} &= \sigma(\bar{y} - \bar{x}) \\
\frac{d\bar{y}}{dt} &= r\bar{x} - \bar{y} - \overline{xz} \\
\frac{d\bar{z}}{dt} &= \overline{xy} - b\bar{z}
\end{aligned} \tag{5.7}$$

These equations are no longer a closed set since they exhibit the quadratic averages $\overline{xz}$ and $\overline{xy}$. To obtain information on the latter one needs to multiply Eqs. (2.6) by x, y, z, take appropriate linear combinations, and again perform an averaging. This will link the quadratic averages to third-order ones. By repeating the process one will arrive at an *infinite hierarchy* for $\overline{x^{k_1}y^{k_2}z^{k_3}}$ which is, clearly, impracticable.

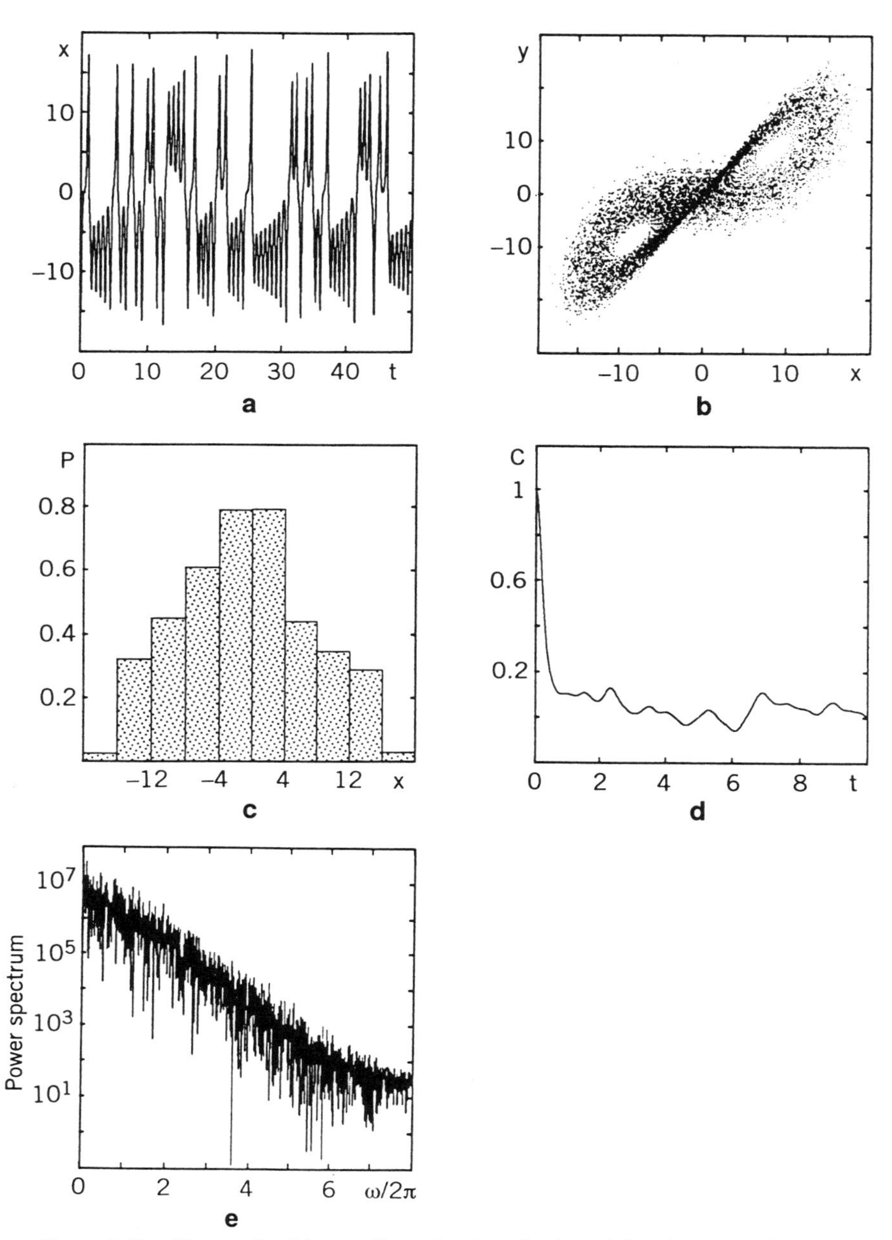

Figure 7.22. Time series (a); two-dimensional projection of the phase space in the x–y plane (b); asymptotic probability density (c); autocorrelation function (d); and power spectrum (e); of the variable x of model (2.6) with $\sigma = 10$, $b = 8/3$, and $r = 28$.

As mentioned earlier in this section, to cope with this difficulty one usually performs a truncation of the hierarchy to the first few orders. The idea that high-order moments, such as $\overline{\delta x^{k_1} \delta y^{k_2} \delta z^{k_3}}$ (with $\delta x = x - \bar{x}$, etc.), are either small or slowly varying in space and time is implicit in these operations. Such assumptions are difficult to justify from first principles, even in a system as simple as the Lorenz model. In this section, we take advantage of this simplicity to evaluate, numerically, the averaged observables from the original ones directly.

One of the first points to be settled when performing an averaging is to specify the value ε of the interval over which this operation is carried out, the choice of which depends, in turn, on the time scales present in the problem. In the case of the Lorenz model one is in the presence of two intertwined processes (see Fig. 7.22a and b): A small scale oscillation around the two unstable steady-states ($x_{\pm} = y_{\pm} = \pm\sqrt{b(r-1)}$, $z_0 = r - 1$); and a global scale motion associated to the intermittent transitions between these two states. The characteristic time scale of the first process can be estimated from the imaginary part Ω of the eigenvalue of the characteristic equation associated with Eq. (2.6) which, for the parameter values chosen, is

$$\tau_{\min} = \frac{2\pi}{\Omega} \approx 0.6 \tag{5.8}$$

To assess the time scale associated to the second process it is necessary to perform a transition time statistics. Figure 7.23 depicts the numerically computed distribution of residence times τ around one of the unstable states. We observe a highly structured pattern, reflecting the deterministic character of the dynamics. The mean residence time, which may legitimately be identified as the slowest relevant time scale of the system, turns out to be [57]

$$\tau_{\max} \approx 1.75 \tag{5.9}$$

We are therefore led to choose ε in the interval $0.6 < \varepsilon < 1.75$.

Figure 7.24a–e depicts the characteristics associated with the time series of the averaged observables of the model with $\varepsilon = 0.8$ time units [56]. We see that the attractor is now organized around two preferred regions of phase space (Figs. 7.24a and b). This is further confirmed by the shape of the invariant probability density (Fig. 7.24c), which shows a pronounced bimodality. In Fig. 7.24b and c a reduction of the attractor size is clearly coming out, in agreement with the idea that averaging erases the fine structure of the original dynamics. We may conclude that

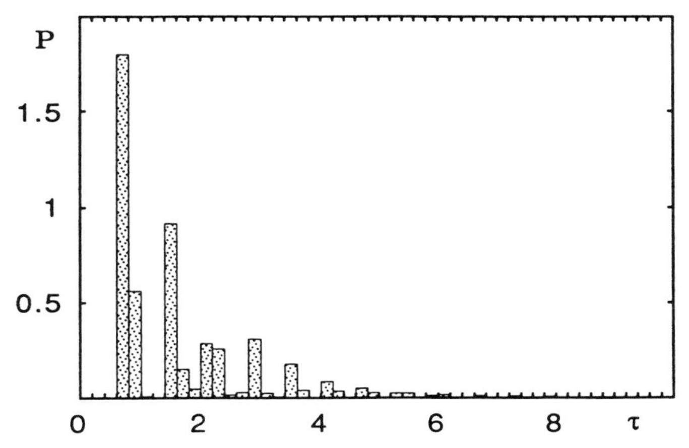

Figure 7.23. Probability density of residence times, τ around one of the nontrivial steady states of model (2.6) as obtained from 10,000 realizations and the parameter values of Fig. 7.22.

the "climate" of the Lorenz model, viewed through an appropriate window longer than the scale associated to local oscillations around $(x_{\pm}, y_{\pm}, z_0)$ (0.6 time units), exhibits transitions between two distinct states.

We turn now to the time-dependent properties. Figure 7.24d depicts the time autocorrelation function of $\bar{x}$. Comparison with the time autocorrelation function of x (Fig. 7.22d) reveals an initial enhancement of the correlations suggesting that $\bar{x}$ now varies over a longer characteristic time scale. The form of the power spectrum, shown in Fig. 7.24e, shows that this slowing down is accompanied by a greater regularity: the spectrum of $\bar{x}$ is organized around some well-defined peaks, whereas that of the initial variable x (Fig. 7.22e) is much closer to the spectrum of a noise process.

The reduction of attractor size, the initial enhancement of the correlations, and the more organized character of the power spectrum of the averaged observables of the Lorenz model suggest that these observables should be more predictable than the original ones. At this stage this conjecture is qualitative. We will now develop a more detailed analysis of the relation between predictability properties and time averaging.

The first question to be raised concerns the relation between the Lyapunov exponents of the original and averaged systems. Let us express

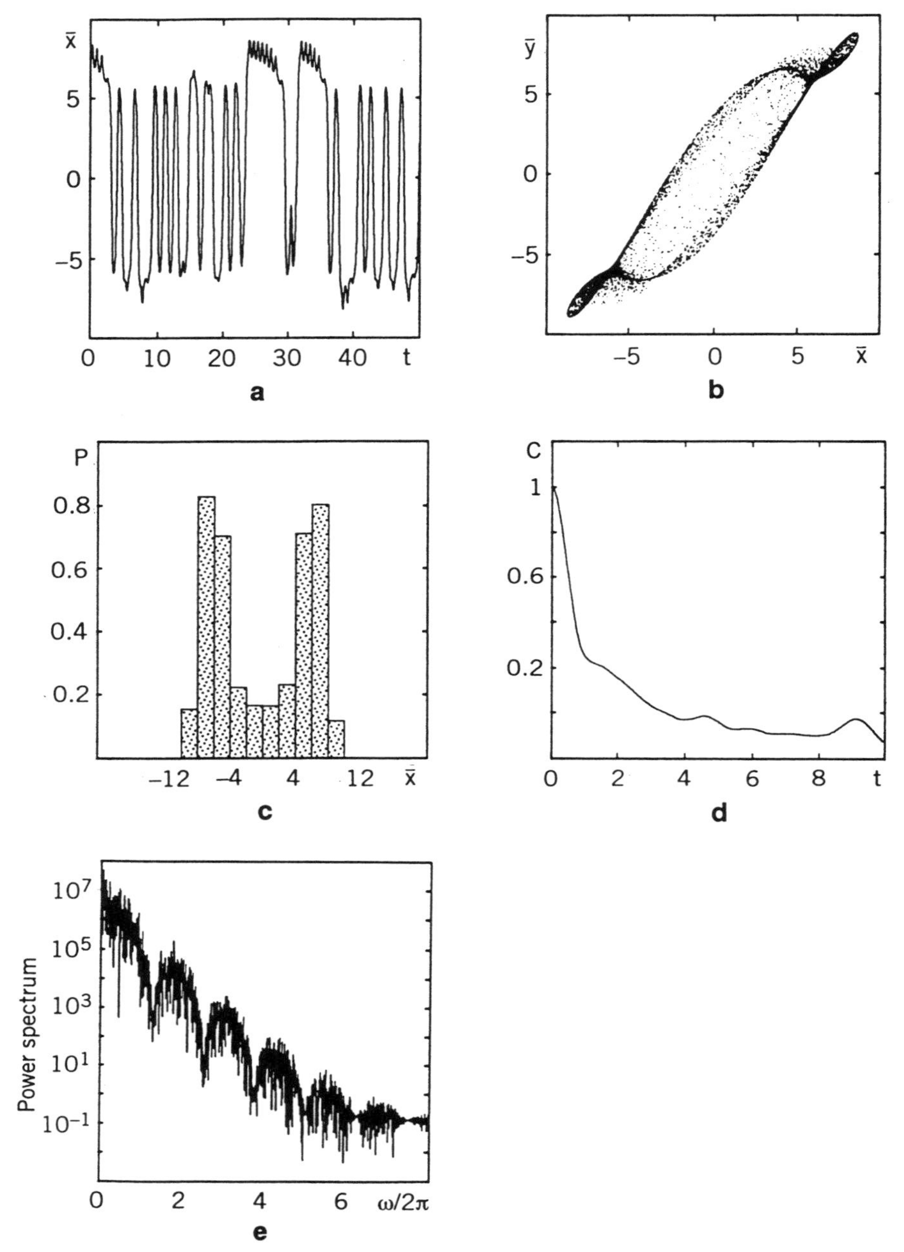

Figure 7.24. As in Fig. 7.22, but for the time averaged variables of model (2.6) with $\varepsilon = 0.8$ time units.

the averaged observables $\{\bar{X}_i\}$ [Eq. (5.1)] in a discretized form

$$\bar{X}_i(t_k) = \frac{1}{K} \sum_{j=0}^{K-1} X_i(t_{k+j}) \tag{5.10}$$

where it is understood that the time series data $\{\bar{X}_i(t_k)\}$ are sampled at discrete equally spaced time intervals t_k, $t_k - t_{k-1} = \Delta t$. Under the same conditions and provided Δt is sufficiently small the evolution equations of $\{X_i\}$ [Eqs. (1.1)] can be written in a finite difference form

$$X_i(t_k) = f_i(\{X_\ell(t_{k-1})\}) \tag{5.11}$$

Proceeding in the same way we may derive a set of equations generalizing Eq. (5.11) to all time shifts j entering into Eq. (5.10)

$$X_i(t_{k+j}) = f_i(\{X_\ell(t_{k+j-1})\} \qquad 0 \le j \le K-1 \tag{5.12}$$

The set of variables $\{X_i(t_{k+j})\}$ defines an extended, nK-dimensional phase space. For each i, the averaged observable $\bar{X}_i(t_k)$ is a particular linear combination of these new phase space variables. Since $\{\bar{X}_i(t_{k+j})\}$ are linearly independent one may construct for each i a set of $K-1$ linear combinations of them which, along with $\bar{X}_i(t_k)$, constitute yet another new set of phase space variables. By virtue of the Oseledec theorem the Lyapunov exponents of this new set are equal to those of $\{X_i(t_{k+j})\}$, which in turn are by construction equal to those of the original system $\{X_i(t_k)\}$, except for a K-fold degeneracy arising from the inflation of the phase space variables. In other words, the averaged observables are part of a dynamical system sharing the same Lyapunov exponents with the original one.

At first this result seems surprising, because of the comments on predictability previously made. The "paradox" is resolved if one realizes that what really counts in the issue of prediction of an observable is the rate of growth of an initially small error over a finite, short time interval. Let $\mathbf{X}(t)$ be a reference trajectory of the original system, and $\mathbf{X}(t) + \delta\mathbf{X}(t)$ the perturbed trajectory obtained from $\mathbf{X}(t)$ as a result of a small initial error $\varepsilon = |\delta\mathbf{X}(0)| \ll 1$. In Section IV.B we have seen that, owing to the variability of local divergence rates over a typical (multifractal) chaotic attractor, $|\delta\mathbf{X}(t)|$ increases in time for short t at a variable rate $\sigma_{\text{eff}}(t)$ which, typically, is larger than the maximum mean Lyapunov exponent σ,

$$|\delta\mathbf{X}(t)| = |\delta\mathbf{X}(0)| e^{\sigma_{\text{eff}}(t)t} \tag{5.13}$$

To see what happens at the level of the averaged observable $\bar{\mathbf{X}}(t)$ we first notice (switching back to the original description in terms of continuous time) that

$$\delta\bar{\mathbf{X}}(t) = \frac{1}{\varepsilon}\int_t^{t+\varepsilon} d\tau \delta\mathbf{X}(\tau) \tag{5.14}$$

or, taking the norm,

$$|\delta\bar{\mathbf{X}}(t)| \leq \frac{1}{\varepsilon}\int_t^{t+\varepsilon} d\tau |\delta\mathbf{X}(\tau)| \tag{5.15}$$

Using Eqs. (5.13)–(5.14) we may write this inequality in the form

$$e^{\bar{\sigma}_{\text{eff}}t} \leq \int_t^{t+\varepsilon} d\tau e^{\sigma_{\text{eff}}(\tau)\tau} \Big/ \int_0^{\varepsilon} d\tau e^{\sigma_{\text{eff}}(\tau)\tau} \tag{5.16}$$

To reach a more explicit expression we focus on the limit where the averaging time ε is small compared to the running time value t. By expanding the right-hand side of Eq. (5.16) in a Taylor series and retaining the leading term we obtain

$$\bar{\sigma}_{\text{eff}} \leq \sigma_{\text{eff}} \tag{5.17}$$

establishing that the effective amplification rate of an initial error in the averaged system is, typically, less than the one in the original system.

Figure 7.25 summarizes the results of the numerical experiments of error growth in the original (full lines) and averaged system for two different averaging periods. The curves have been obtained by considering the arithmetic mean over 20,000 realizations. In all cases, the error growth dynamics follows the general growthlike pattern found in Section IV.B [28–32]. However, the error growth curve of the averaged system is consistently smoother than, and remains below, the curve of the original system, entailing that both the initial amplification rate and the saturation level of the averaged observable are depressed. This is in full agreement with relation (5.17) and the results on the reduction of the attractor size. Notice, however, that the logarithmic mean of the error, depicted in Fig. 7.26, is essentially the same for the original and the averaged variable. This is again in full agreement with our earlier remarks on the effect of averaging on the mean maximum Lyapunov exponent.

One may wonder whether in the spirit of inequality (5.17), a relation linking the attractor dimension of the averaged and original systems may also be found. Numerical computation does not lead to a clearcut conclusion, owing to the small value ($d \approx 2.06$) of the attractor dimension of the original Lorenz model. On the other hand the following analytic

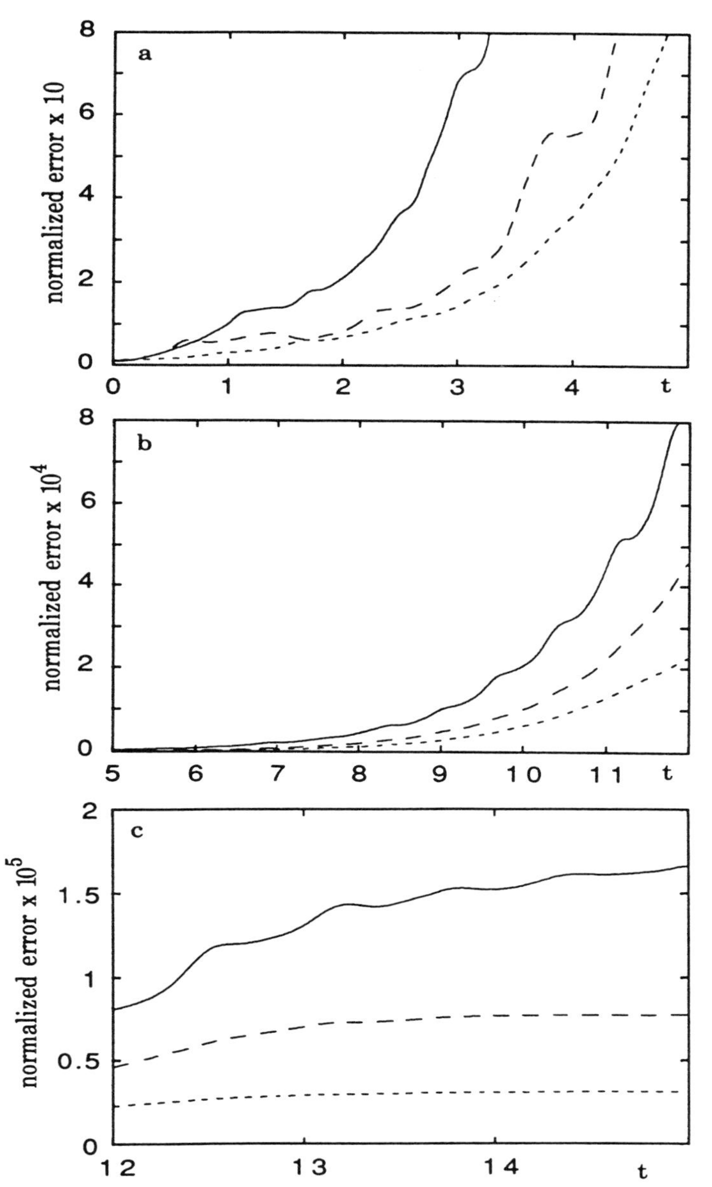

Figure 7.25. Short (a), intermediate (b), and long-time (c) evolution of the arithmetic mean of an initial error, normalized by its value at $t = 0$ for model (2.6). The full line represents the behavior of the error in the initial nonaveraged system, the broken and dotted lines depict the effect of an averaging over 0.8 and 2 time units, respectively. The corresponding initial errors are 9.5×10^{-5}, 1.3×10^{-4}, and 2.4×10^{-4}.

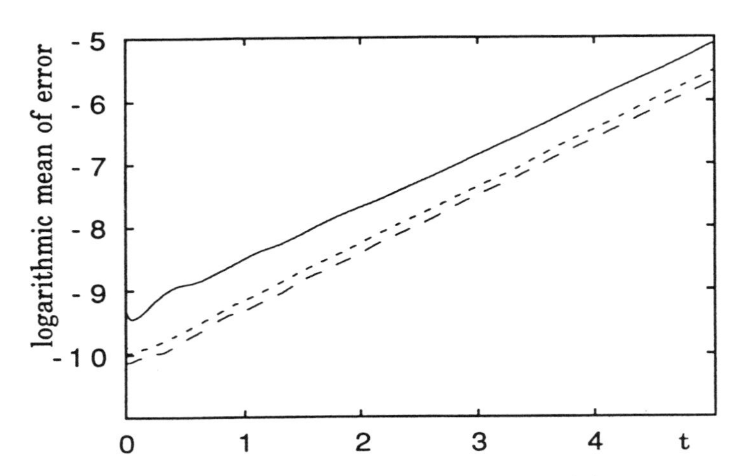

Figure 7.26. As in Fig. 7.25a but for the logarithmic mean of the error.

argument, strictly valid only in a multiperiodic system, can be advanced. Let $\{\omega_k\}$, $k = 1, \ldots, K$, be the frequencies (generally irrationally related) participating in the dynamics. The instantaneous value of x is of the form

$$x = \sum_{k=0}^{K} a_k \cos(\omega_k t + \phi_k) \tag{5.18}$$

where for simplicity we considered a uniform motion over the attractor, which in this case is a K-dimensional torus. The instantaneous value of the averaged variable $\bar{x}$ is then,

$$\bar{x} = \sum_{k=1}^{K} a_k \frac{1}{\varepsilon\omega_k} [\sin(\omega_k(t+\varepsilon) + \phi_k) - \sin(\omega_k t + \phi_k)] \tag{5.19}$$

Two cases may now be distinguished.

1. The averaging period ε is a rational multiple of one of the periods $T_\alpha = 2\pi/\omega_\alpha$, $\alpha \in (1, \ldots, K)$. In this "resonant" case one term in the brackets in Eq. (5.19) will be strictly vanishing. Effectively, the averaged system will therefore evolve on an attractor of reduced dimension—here a $K-1$ dimensional torus.
2. The averaging period ε does not satisfy the strict resonance condition formulated under (1), but is nevertheless larger than the first κ characteristic times $T_\alpha = 2\pi/\omega_\alpha$, $1 \le \alpha \le \kappa$. Under these conditions the corresponding contributions in Eq. (5.19) are "penal-

ized" by the factors $1/\varepsilon\omega_\alpha \ll 1$. They are thus effectively absent in the dynamics of the averaged system, which for all practical purposes sees its attractor dimensionality drop by κ units. Inasmuch as a chaotic attractor may be regarded as a fractalized torus [7], whose dimensionality is the torus dimensionality augmented by a fractional number, the argument is likely to carry through in systems undergoing chaotic dynamics. We suggest that this mechanism is at the basis of the dimensionality reduction of climatic and meteorological attractors, as discussed earlier in this chapter.

VI. DISCUSSION

This chapter has shown that the forecasting of atmospheric and climate dynamics is subjected to irreducible limitations arising from the ubiquitous property of sensitivity to the initial conditions. This realization, coming in a time in which the technological, societal, and political pressure to improve the period over which the future states of the atmosphere and climate can be reliably predicted is steadily increasing, calls for a serious reassessment of currently prevailing ideas and approaches to modeling and monitoring.

On the modeling side it must be realized that sophisticated, million variable numerical models run on the most powerful supercomputers acquire their full interest and significance only when a qualitative insight of the nature of the system at hand has been reached. Long-term climatic forecasts based on a static formulation in which the system adjusts to a "quasisteady state" level corresponding to the instantaneous value of the relevant parameters are, clearly, inadequate. They must be supplemented by studies on the system's dynamical response, capable of taking into account the entire repertoire of possible behaviors and formulated, whenever necessary, in probabilistic terms. Furthermore the very possibility of intrinsically generated complexity makes the distinction between "normal state" and "fluctuation," deduced from models, much subtler than usually thought, thereby placing the distinction between natural variability and externally induced systematic change on a completely new basis. Most importantly, it makes no sense to push a large numerical model, with all its inherent uncertainties arising from heavy parameterizations, beyond its own natural (and as yet largely unexplored) predictability limits.

Low-order models are likely to play an important role in this reassessment. As we saw throughout this chapter they can capture certain essential aspects of the complexity of atmospheric and climate dynamics while still remaining tractable and amenable to a comprehensive qualitative analysis. Their study should therefore be pursued in parallel to the development of detailed models. Furthermore, dynamical reconstruction techniques applied directly to data sets independent of any modeling provide one with *constraints* that must be satisfied by a detailed numerical model supposed to forecast the phenomenon at hand.

On the monitoring side, the idea that the underlying system is a complex, highly unstable dynamical system, is likely to have repercussions on the nature and number of data to be collected and in their sampling frequency. Furthermore, it is legitimate to inquire whether certain essential features of atmospheric dynamics, notably those associated with a fractal distribution of an observable in space and time, may be adequately reconstituted from the current global observational network. It is well known that this network has been extremely sparse in the past. The disparity between the past situation and the present-day one imposes severe constraints on our ability to reliably detect small-amplitude climatic changes, such as temperature increases, since the preindustrial period. Even today, despite impressive effort toward improving this state of affairs, the global observational network remains highly inhomogeneous and fractally distributed, a feature that may be responsible for missing events that are sparsely distributed in space, as, for instance, extreme events [58]. For a given (finite) number of stations it is therefore important to seek for an *optimal distribution*, if any, for which errors are minimized. Recent studies [59] suggest that the answer to this problem may be quite unexpected: An inhomogeneous distribution of sparsely distributed stations like those of the present network may produce, for certain types of phenomena, a record that is closer to reality than the homogeneous one. All in all it is our feeling that the new insights afforded by viewing atmospheric and climate dynamics from the standpoint of chaos theory will generate new ideas and new tools likely to revitalize and reorient the traditional approach to earth and environmental sciences in a healthy manner.

ACKNOWLEDGMENTS

This work is supported in part, by the Belgian Federal Office for Scientific, Technical and Cultural Affairs (O.S.T.C.) and by the Environment program of the European Commission.

REFERENCES

1. C. Nicolis and G. Nicolis, Eds., *Irreversible phenomena and dynamical systems analysis in geosciences*, Reidel, 1987.
2. J. R. Holton, *An introduction to dynamic meteorology*, Academic, New York, 1972. J. T. Houghton, *The physics of atmospheres*, Cambridge University Press, 1979.

3. A. J. Simmons, R. Mureau, and T. Petoliagis, Q. J. R. Meteorol. Soc., in press.
4. P. D. Thompson, *Tellus*, **9**, 275 (1957).
5. E. N. Lorenz, *J. Atm. Sci.*, **20**, 130 (1963).
6. E. N. Lorenz, *Tellus*, **36**, 98 (1984); E. N. Lorenz, Tellus **42**A, 378 (1990).
7. E. Ott, *Chaos in dynamical systems*, Cambridge University Press, 1993.
8. See, e.g., G. Fischer, Ed., *Numerical data and functional relationships in science and technology*, Vol. 4, Springer, Berlin, 1987; J. T. Houghton, Ed., *The global climate*, Cambridge University Press, 1984.
9. E. N. Lorenz, *Tellus*, **34**, 505 (1982); E. N. Lorenz, in *Irreversible phenomena and dynamical system analysis in geosciences*, C. Nicolis and G. Nicolis, Eds., Reidel, 159, 1987.
10. C. Sparrow, *The Lorenz equations*, Springer, Berlin, 1982.
11. E. N. Lorenz, *J. Meteorol. Soc. Jpn.*, **60**, 256 (1982).
12. J. G. Charney and J. G. De Vore, *J. Atm. Sci.*, **36**, 1205 (1979).
13. J. Egger, *J. Atm. Sci.*, **38**, 2606 (1981).
14. J. Feder, *Fractals*, Plenum Press, New York 1988.
15. H. Hentschel and I. Procaccia, *Physica*, **8D**, 435 (1983).
16. F. Takens, in *Lecture Notes in Mathematics*, Vol. **898**, Springer, Berlin, 1981 p. 363.
17. G. Mayer Kress, Ed., *Dimensions and entropies in chaotic systems*, Springer, Berlin, 1986.
18. N. Abraham, A. Albano, A. Passamante, and P. Rapp, Eds., *Measures of complexity and chaos*, Plenum Press, New York, 1989.
19. K. Fraedrich, *J. Atm. Sci.*, **43**, 419 (1986).
20. C. L. Keppenne and C. Nicolis, *J. Atm. Sci.*, **46**, 2356 (1989).
21. A. A. Tsonis and J. B. Elsner, *Nature (London)*, **333**, 545 (1988).
22. H. W. Barker and B. Van Zyl, *J. Climate*, **6**, 858 (1993).
23. A. Shil'nikov, C. Nicolis, and G. Nicolis, *Bifurcation and Chaos*, in press.
24. G. Nicolis, *Introduction to nonlinear science*, Cambridge University Press, 1995.
25. J. Guckenheimer and P. Holmes, *Nonlinear oscillations, dynamical systems and bifurcations of vector fields*, Springer-Verlag, (1983).
26. L. Shil'nikov, *Sov. Math. Dokl.*, **10**, 1368 (1969); I. Ovsyannikov and L. Shil'nikov, *math. Sbornik*, **2**, 415 (1992).
27. S. Gonchenko, D. Turaev, and L. Shil'nikov, *Sov. Math. Dokl.* **44**, 422 (1992); *Physica D*, **62**, 1 (1993).
28. C. Nicolis and G. Nicolis, *Phys. Rev.* **43A**, 5720 (1991); C. Nicolis, *Q.J.R. Meteorol. Soc.*, **118**, 553 (1992).
29. C. Nicolis, S. Vannitsem, and J.-F. Royer, *Q.J.R. Meteorol Soc.* **121**, 705 (1995).
30. *S. D. Schubert and M. Suarez*, *J. Atm. Sci.*, **46**, 353 (1989).
31. A. Trevisan, *J. Atm. Sci.*, **50**, 1016 (1993).
32. C. Nicolis and G. Nicolis, *Bifurcation chaos*, **3**, 1339 (1993).
33. J. F. Lacarra and O. Talagrand, *Tellus*, **40A**, 81 (1988).
34. H. D. I. Abarbanel, R. Brown, and M. B. Kennel, *J. Nonlinear Sci.*, **1**, 175 (1991):
35. P. Manneville, *Dissipative structures and weak turbulence*, Academic, San Diego, CA, 1990.

36. S. Vannitsem and C. Nicolis, *J. Geophys. Res.*, **99**, 10, 377 (1994).
37. R. Mureau, F. Molteni, and T. N. Palmer, *Q.J.R. Meterol. Soc.*, **119**, 299 (1993).
38. A. Lasota and M. Mackey, *Probabilistic properties of deterministic systems*, Cambridge University Press, 1985.
39. G. Nicolis and C. Nicolis, *Phys. Rev.*, **38A**, 427 (1988); McKernan and G. Nicolis, *Phys. Rev.*, **50E**, 988 (1994).
40. M. Kac, *Probability and related topics in physical sciences*, Interscience, New York, 1959.
41. P. C. Hemmer, *J. Phys. A: Math. Gen.*, **17**, 247 (1984).
42. C. Nicolis, *Tellus*, **42A**, 401 (1990).
43. G. Sugihara and R. M. May, *Nature* (*London*), **344**, 734 (1990).
44. B. Saltzman, *Adv. Geophys.*, **25**, 173 (1983).
45. H. H. Lamb, *Climate*: *present, past and future*, Vol. 2, Methuen, 1977; N. J. Shackleton and N. D. Opdyke, *Quat. Res.*, **3**, 39 (1973).
46. A. L. Berger, *Quat. Res.*, **9**, 139 (1978).
47. C. Crafoord and E. Källén, *J. Atm. Sci.*, **35**, 1123 (1978).
48. C. Nicolis and G. Nicolis, *Tellus*, **33**, 225 (1981).
49. C. Gardiner, *Handbook of stochastic methods*, Springer, Berlin, 1983.
50. C. Nicolis, *Tellus*, **34**, 1 (1982).
51. R. Benzi, G. Parisi, A. Sutera, and A. Vulpiani, *Tellus*, **34**, 10 (1982).
52. B. Saltzman, A. Sutera, and A. R. Hansen, *J. Atm. Sci.*, **39**, 2634 (1982).
53. C. Nicolis, *Tellus*, **36A**, 1 (1984).
54. C. Nicolis, *Tellus*, **36A**, 217 (1984).
55. C. Nicolis and G. Nicolis, *Nature* (*London*), **331**, 529 (1984); K. Fraedrich, *J. Atm. Sci.*, **44**, 722 (1987).
56. C. Nicolis and G. Nicolis, *J. Atm. Sci.*, in press.
57. C. Nicolis and G. Nicolis, *Phys. Rev.*, **34A**, 2384 (1986).
58. S. Lovejoy, D. Schertzer, and P. Ladoy, *Nature* (*London*), **319**, 43 (1986).
59. C. Nicolis, *J. Appl. Meteor.*, **32**, 1751 (1993).

AUTHOR INDEX

Numbers in parentheses are reference numbers and indicate that the author's work is referred to although his name is not mentioned in the text. Numbers in *italic* show the pages on which the complete references are listed.

SUBJECT INDEX